thomson.com

changing the way the world learns

To get extra value from this book for no additional cost, go to:

http://www.thomson.com/wadsworth.html

thomson.com is the World Wide Web site for Wadsworth/ITP and is your direct source to dozens of on-line resources. *thomson.com* helps you find out about supplements, experiment with demonstration software, search for a job, and send e-mail to many of our authors. You can even preview new publications and exciting new technologies.

thomson.com: *It's where you'll find us in the future.*

Social Forces and Aging
An Introduction to Social Gerontology

Eighth Edition

Robert C. Atchley
Miami University

Wadsworth Publishing Company
I(T)P® An International Thomson Publishing Company

Belmont, CA • Albany, NY • Bonn • Boston • Cincinnati • Detroit • Johannesburg • London • Madrid
Melbourne • Mexico City • New York • Paris • San Francisco • Singapore • Tokyo • Toronto • Washington

Sociology Editor: Eve Howard
Editorial Assistant: Deidre McGill
Marketing Manager: Mike Dew
Production: Ruth Cottrell
Print Buyer: Karen Hunt
Permissions Editor: Jeanne Bosschart

Copy Editor: Lura Harrison
Cover Design: Craig Hanson
Compositor: Conch Composition
Printer: Quebecor Printing/Fairfield
Cover Printer: Phoenix Color

Printed in the United States of America
    3 4 5 6 7 8 9 10

For more information, contact Wadsworth Publishing Company, 10 Davis Drive, Belmont, CA
94002, or electronically at http://www.thomson.com/wadsworth.html

International Thomson Publishing Europe
Berkshire House 168-173
High Holborn
London, WC1V 7AA, England

International Thomson Editores
Campos Eliseos 385, Piso 7
Col. Polanco
11560 México D. F. México

Thomas Nelson Australia
102 Dodds Street
South Melbourne 3205
Victoria, Australia

International Thomson Publishing Asia
221 Henderson Road
#05-10 Henderson Building
Singapore 0315

Nelson Canada
1120 Birchmount Road
Scarborough, Ontario
Canada M1K 5G4

International Thomson Publishing - Japan
Hirakawacho Kyowa Building, 3F
2-2-1 Hirakawacho
Chiyoda-ku, Tokyo 102, Japan

International Thomson Publishing GmbH
Königswinterer Strasse 418
53227 Bonn, Germany

International Thomson Publishing Southern Africa
Building 18, Constantia Park
240 Old Pretoria Road
Halfway House, 1685 South Africa

Library of Congress Cataloging-in-Publication Data

Atchley, Robert C.
    Social forces and aging : an introduction to social gerontology /
    Robert C. Atchley. — 8th ed.
        p.   cm.
    Includes bibliographical references and index.
    ISBN 0-534-50460-4
    1. Gerontology. 2. Aged—United States. I. Title.
HQ1061.A7798 1996
305.26—dc20   96-9699

Contents in Brief

Contents in Detail

Part 2

Basic Aspects of Individual Aging / 75

Chapter 4
Physical Aging / 76

Chapter 5
Psychological Aspects of Aging / 92

Chapter 6
Social Aspects of Aging / 133

Preface

This book provides a comprehensive introduction to social gerontology—the emerging social science dealing with human aging. The book's major strengths are:

- *Organization.* The chapters and parts of the book flow logically, from basic aspects of individual aging and adaptation, to aging in everyday life, to the effects of aging on needs and resources, to society's response to aging.

- *Flexibility.* The section organization allows various parts of the text to be recombined to suit the preferences of the instructor. For example, the sections dealing with family caregiving from Chapter 8 (Family, Friends, and Social Support) could be combined with the sections on long-term care from Chapter 14 (Health and Long-Term Care) and assigned as a block.

- *Coverage.* Major areas of theory, research, social policy, and practice are covered in significant depth. Although additional readings are always helpful, they are not necessary for readers to gain a full appreciation of the scope of social gerontology.

- *Conceptual clarity.* Sound definitions and frameworks are used to organize and analyze the literature in various areas. Terms are carefully defined, giving teachers and students a clear common vocabulary.

- *Analytical approach.* Original works are succinctly described and then analyzed and placed in context with other work.

- *Original research.* In many areas, gaps in knowledge have been filled with original research. For example, the chapters on demography and social inequality draw heavily on my own examination of census data. Chapter 9 (Employment and Retirement) relies to a large extent on my own theory and research in that area.

In other areas, the lack of theory led to the development of new theoretical approaches. For example, theories about the dynamics of self in later life, types of interpersonal bonds, the retirement process, stages of retirement, spheres of activity, the continuity theory of adaptation to aging, and the economic functions of retirement were developed to consolidate bits and pieces of research evidence and deal with gaps in existing theory. Existing theories are also modified and extended. Modernization theory, life course theory, theories about adaptation to role loss, the theory of societal disengagement, and age discrimination theory are examples of theories that I have modified and extended to improve their explanatory power and stimulate research.

- *Hot topics.* Frequent revision of the book allows timely topics such as the effect of the Republicans' Contract with America and its agenda on provision of services to older Americans, ethical conflicts and issues related to aging, and the debate about Social Security and Medicare to be included in an up-to-date fashion.

For this new edition, the order of the chapters has been extensively reorganized in response to feedback from readers.

The material on personal adaptation to aging now follows the three chapters on physical, psychological, and social aging—basic aspects of individual aging. The material on needs and resources and health and social services has been consolidated and reorganized into three new chapters:

- Income and Housing (Chapter 13)

- Health Care and Long-Term Care (Chapter 14)

- Community Social Services (Chapter 15)

This reorganization places individual and social policy material about health and long-term care into a single chapter. The same is true for community social services. In the earlier editions, social policy concerns were covered in separate chapters, which many professors found awkward.

To acknowledge an important increase in scholarship and interest, there is an entirely new chapter:

- Religion and Spirituality (Chapter 11)

There are new sections on:

- social factors in individual development (Chapter 6)

- adapting to chronic illness (Chapter 7)

- friendship patterns among the oldest-old (Chapter 8)

- those who reject retirement (Chapter 9)

- assisted living (Chapter 14)

- nursing home ombudsman programs (Chapter 15)

- elder abuse prevention and intervention (Chapter 15)

- aging as depicted in feature films (Chapter 16)

- ethical issues in aging (Chapter 16)

- dimensions of disadvantage Chapter 17)

- political trends (Chapter 19)

- social insurance (Chapter 19)

- structural lag theory (Chapter 20)

Heavily revised chapters include:

- The Scope of Social Gerontology (Chapter 1)

- The Demography of Aging (Chapter 2)

- Psychological Aspects of Aging (Chapter 5)

- Social Aspects of Aging (Chapter 6)

- Personal Adaptation to Aging (Chapter 7)

- Aging in Contemporary American Society and Culture (Chapter 16)

- Social Inequality (Chapter 17)

- Politics and Government (Chapter 19)

For this edition, the bibliography has been restricted to works specifically referenced in the text.

I am very grateful for the long-standing acceptance this book has received. To date, over 500,000 students have been introduced to social gerontology through previous editions of *The Social Forces in Later Life* (first through third editions) and *Social Forces and Aging* (fourth through seventh editions). These have included students from a wide variety of fields—sociology, psychology, home economics, nursing, social work, health, and adult development. The book has grown gradually larger, since its original publication in 1972, and the eighth edition bears little resemblance to the first.

I appreciate the many compliments and thank-yous I have received from both students and professors, and at the same time I accept the responsibility and challenge of working hard to see that *Social Forces and Aging* remains contemporary, and that it grows with the field of social gerontology. I am aided in this task by reviews commissioned by the publisher, but I also need your help. As you read, if you see errors or gaps in coverage, or if you think of topics that should be included or see an article that would be a particularly good research example, write to me. I cannot promise that I will agree with your point, but I want to know about it.

For the eighth edition, I was helped enormously in writing the new chapter on religion and spirituality by incisive and supportive reviews from Susan McFadden, Jeff Levin, and Gene Thomas. I also benefited from being part of the Omega Institute's Conscious Aging programs over the past three years. Finally, thanks to Rick Moody and Tom Cole for encouraging me to take on this difficult topic.

Reviewers for the eighth edition were Charlotte Dunham, Texas Tech University; Mary R. Holley, Montclair State University; Janet Hope, College of St. Benedict; David E. Redburn, Furman University; and Robbyn Wacker, University of N. Colorado.

Many colleagues have enriched, refined, and supported my intellectual and personal growth, including Bob Antonio, Bobby Applebaum, Sheila Atchley, Dick Campbell, Gloria Cavanaugh, Tom Cole, Steve Cutler, Ram Dass, Dave Ekerdt, Carroll Estes, Linda George, Lisa Groger, Joe Hendricks, Regula Herzog, Martha Holstein, Marty Jendrek, Bob Kastenbaum, Suzanne Kunkel, Powell Lawton, Jeff Levin, Chuck Longino, George Maddox, Pam Mayberry, Susan McFadden, Kathy McGrew, Hazel Markus, Vic Marshall, Shahla Mehdizadeh, Meredith Minkler, Rick Moody, Leslie Morgan, Carol Ryff, Zalman Schacter, Warner Schaie, Carol Segrave, Millie Seltzer, Diana Spore, Debbie Stanley, Harvey Sterns, Lillian Troll, Cathy Ventura-Merkel, Larry Weiss, and Susan Whitbourne.

I owe special thanks to Dave Lewis, whose enthusiastic devotion to social justice, humanism, and social science attracted me to sociology and demography. Fred Cottrell served as my mentor for thirteen years. Wise and unselfish, Fred taught me that knowledge is most valuable when cast in a form that can be used to better people's lives. Millie Seltzer was a fountainhead of creativity and good humor, and it was my extraordinary good fortune to have her for a work partner and best friend for nearly thirty years. She exemplified the power of unwavering support. Sheila Atchley continues to teach me the value of having an equal partner with whom to experience all of life. Melissa Atchley is my connection to many new ways of looking at the world. Chris Atchley exemplifies hope and positive energy. Chris and Tara Miller radiate the power of love and basic human kindness. And through Carl Adlon and Nisargadatta Maharaj I learned who is creating and using all this knowledge. I have been truly blessed by all of these people.

I owe a tremendous debt to the staff of the Scripps Gerontology Center. Cheryl Johnson helped with many of the tables and figures in this book. Lisa Haston fended off the phone callers so I could concentrate on getting the book completed. Thelma Carmack, who has been with this book longer than anyone besides me, did her usual wonderful job with the bibliography. Their dedication, competence, and good humor are contagious. Finally, I would like to thank the many generations of students at Miami University for continuing to challenge me to grow and improve my understanding.

Part 1

Setting the Stage

This book is an introduction to *social gerontology*, the discipline devoted to the study of the nonphysical aspects of human aging. As you will see, social gerontology covers a wide range of topics. Most of the book deals with up-to-date, factual information and current perspectives on aging. Part 1, consisting of three chapters, provides a context for this material.

Chapter 1 describes the *scope* of social gerontology: the general subjects included in social gerontology, how aging and the older population are defined, and some of the broad issues involved in the study of aging. Chapter 2 presents the *demography* of aging. It covers how aging becomes widespread in a population and how population aging is measured. It then considers the growth of the older population in the United States as well as its size, composition, geographic distribution, mobility, and life expectancy. Chapter 3 is a *historical overview* of aging in the United States, from colonial times to the present. Because aging today is often contrasted with aging in earlier times, it is important that such comparisons be based on fact rather than myths about "the good old days." A knowledge of history also helps us better understand how contemporary society works. Together, these three chapters provide the background needed for the study of present-day social gerontology.

1 The Scope of Social Gerontology

The process of **AGING*** has been around as long as life itself. Provided, of course, that illness or violence does not end life before its genetically programmed span, all living organisms pass through three broad stages from conception to death: maturation, maturity, and aging. And even though the *average* length of human life in most societies did not extend into "old age" until the twentieth century, there have been old people on earth for thousands of years.

Although biology forms the primary basis of aging, the *significance* of aging is largely social. Physical changes associated with aging, such as declining eyesight or graying hair, have little significance except as they relate to what is expected of people. For example, declining eyesight is only a problem if it (1) cannot be corrected and (2) interferes with a person's ability to function normally. And gray hair is significant only because it is used conceptually to assign people to a particular social category. Thus, we need to consider not only what aging does to us but what we do with aging.

Aging also influences how a society or group is itself viewed. We speak of an "aging society" when the average age of its members is increasing. "The graying of America" is not about individuals who are aging but about the United States as a whole.

Aging affects everyone because nearly everyone has the potential to grow old and all the groups in which we live have older members. But although aging has always been a part of human life, the systematic study of aging, especially its social aspects, is relatively recent. For example, the Gerontological Society of America—an organization of researchers, practitioners, and educators interested in aging—was not founded until 1945. The behavioral and social sciences section was not established until 1956, and social gerontology as a concept was not developed until the late 1950s (Tibbits 1960, 3). However, since 1960, research on aging has expanded so rapidly that in 1990 there were research and education on social aspects of aging at more than 1,300 colleges and universities in the United States alone.

What Is Gerontology?

Gerontology is the use of reason to understand aging. The term was first used to refer to the scientific study of aging, but contemporary usage also includes the study of aging using methods from various other disciplines, such as humanities, social policy, and human services. For us to understand and cope with aging, knowledge from a variety of sources is needed. Thus, gerontology includes the results of research on aging from all academic disciplines and fields of

---

*All boldfaced terms are defined in the Glossary.

professional practice. Biologists study the effect of age on the body's immunity to disease. Physicians search for effective ways to treat disease in older people whose immunity has been reduced. Psychologists study changes in bodily coordination with age. Occupational therapists search for ways to retrain older people whose coordination is impaired. Economists study income requirements of middle-aged and older people. Retirement counselors gather information about how to stretch retirement income. Sociologists study how aging affects social roles. Recreation workers develop ways to help older people get involved in new roles. The list goes on. Almost every area of study or practice that deals with people or their needs has a branch that is devoted to aging. And all these branches of all these fields come together under the label of *gerontology*.

There are four interrelated aspects to the study of aging: physical, psychological, social psychological, and social.

- The study of *physical aging* examines the causes and consequences of the body's declining capacity to renew itself; the physical effects of bodily aging; and the means for preventing, treating, or compensating for illness or disability caused by or related to physical aging.

- The study of *psychological aging* focuses on sensory processes, perception, coordination, mental capacity, human development, personality, and coping ability as they are affected by aging.

- *Social psychological aging* focuses on the interaction of the individual with his or her environment and includes such topics as attitudes, values, beliefs, social roles, self-image, and adjustment to aging.

- *Social aging* refers to the nature of the society in which individual aging occurs, the influence that society has on its aging individuals, and the impact aging individuals

have on their society. Social aging also includes interactions among society's various social institutions, such as the economy or health care, as they apply to the needs of an older population. *Society* as used here is not a single thing. The word refers to the shared ideas and common actions of the residents of a nation, and it includes messages in the mass media, common beliefs, typical ways that people solve problems, laws and regulations, administrative procedures, ideologies, and a host of other factors. Society does *not* act as a unit but as a loose structure of individuals, each with a slightly different view of beauty, truth, and goodness. Resist the trap of thinking of society as a single entity capable of single-minded action.

Social Gerontology

All four aspects of aging are so interrelated in everyday life that it is often difficult to distinguish one from another. Yet subdividing gerontology is useful for the systematic study of aging. **Social gerontology** is the subfield of gerontology that deals primarily with the *nonphysical* side of aging. Physical aging interests social gerontologists only as it influences the ways individuals and societies adapt to one another. Yet, because physical aging is at the root of all aspects of aging, social gerontologists need to understand as much as they can about it.

What Is Human Aging?

Aging is a broad concept that includes *physical changes* in our bodies over adult life, *psychological changes* in our minds and mental capacities, *social psychological changes* in what we think and believe, and *social changes* in how we are viewed, what we can expect, and what is expected of us.

It is common to describe the human life cycle in terms of three broad periods: (1) *maturation,* in which the person develops, (2) *full maturity,* in which the person exercises full powers, and (3) *aging,* in which the person gradually declines. However, this view, based on very general biological characteristics of life in animals and plants, is actually too simplistic.

Biological, Psychological, Social Psychological, and Social Aging

Biological aging is the result of numerous processes, most of which do not progress at the same rate. For example, the kidneys typically show diminished functioning much earlier than does the skin. In addition, different physical functions reach maturity at different ages. For instance, we usually reach sexual maturity biologically several years before we attain full height. To complicate matters, most physical functions vary quite a bit from person to person at all stages of life. As an illustration, diastolic blood pressure ranges from 45 to 105 even in men age 18 to 24 (80 is considered "normal").

When we examine *psychological aging* in adulthood, we find that certain functions diminish with age while others increase or remain relatively constant. For example, mathematical problem-solving ability generally declines with age, vocabulary usually increases, and habits tend to remain relatively constant. Variability is as great for psychological aging as it is for physical aging.

Social psychological aging results from the long-term interactions people have with the social world around them. For example, each person's concept of self is greatly influenced initially by feedback from others, but over time adults often become less dependent on others for evaluations of the self. Adults also tend to experience shifts in values as they age. Early on, they are heavily influenced by achievement values, but later they may become less interested in personal achievement and material concerns and more interested in the quality of relationships with family and friends.

Social aging, on the other hand, is largely an arbitrary process of establishing what is appropriate to or expected of people of various ages. Often age definitions are contained in legislation or administrative guidelines for programs and are usually not based on research information about what people of various ages are typically capable of. Thus, school begins at age 6, even though most children are quite capable of beginning sooner (and many do). Airline pilots must retire at 60 even if they can still fly a plane perfectly well. Ideas about when a person is old enough to marry vary from 13 to 16 for the rural working class to 25 to 30 in the upper middle class. Concepts of life stages and the life course represent socially defined ideals concerning how people are to progress through life. However, there are many versions of these ideals and many people do not follow these ideals for a variety of reasons. As a result, social aging adds yet another level of variability to an already complex set of dimensions.

The Two Faces of Aging

Aging is not one process, but many, and it has many possible outcomes, some positive and some negative. On the one hand, increasing age brings greater experience and expanded opportunities for wisdom or skill at a variety of activities ranging from politics to music. Wisdom and experience can give an older person the kind of long-range perspective that is invaluable in an adviser. Older people can also be keepers of tradition. They know about many unrecorded events that have taken place over the years in families, at workplaces, in communities, and in the nation. Aging can also bring a personal peace and mellowing. Later life can be a time of extraordinary freedom and opportunity once the responsibilities of employment and child rearing are set aside.

On the other hand, aging is a losing proposition for some people. They may lose physical or mental capacities, good looks, opportunity for employment and income, or positions in organizations to which they belong. They may outlive their spouse and friends.

Aging is neither predictably positive nor predictably negative. For some people it is mainly positive, for others it is mainly negative, and for still others it is somewhere in between.

How aging is viewed by society also reflects the two-sided nature of aging. Some realms, such as politics, stress the advantages of age. Other areas of life, such as employment, emphasize the disadvantages of aging. Still others, such as the family, incorporate both positive and negative aspects of aging. The double-edged nature of aging is also reflected in the current literature on the subject. Some researchers emphasize the negative when they study aging people, focusing on sickness, poverty, loss of social roles, isolation, and demoralization. Their theories seek to explain how people arrive at such an unhappy state. They tend to see aging as a problem. Other writers emphasize the positive, viewing most elders as being in good health and in frequent contact with family, and as having at least adequate incomes and a high degree of satisfaction with life. The theories they develop try to explain how aging can have such positive outcomes. For these researchers, the problems of aging apply to a minority of older people.

Because aging can have both positive and negative outcomes, *neither view is wrong.* Certainly, both kinds of outcomes exist, and understanding both is essential. However, it is also important to acknowledge that in the older population as a whole, *positive outcomes outnumber the negative by at least 2 to 1.* Even in advanced old age, there is usually a balance of positive and negative aspects of existence.

The positive–negative nature of aging is further reflected in the fact that aging is both a social problem *and* a great achievement. For a sizable minority of older Americans, the system does *not* work. These people have

difficulty securing an adequate income, are discriminated against because of their age at work and in social programs, lack adequate health care, and need better housing and transportation. That these problems recur regularly is a significant social problem.

Yet the majority of older Americans do not encounter such problems, so for them the system works. They are in good health, have modest but adequate retirement pensions, own their own homes, drive their own cars, and need little in the way of social services. Believing that all older people are needy or that they all are self-sufficient is a pitfall to be avoided. *Both types of people exist.* The fact that most older people do not need assistance makes it possible to do something for those who do. For example, in times of strong resistance to new taxes, most state governments are fortunate that less than 5 percent of the older population needs publicly funded long-term care.

Defining the Aging and the Older Populations

Aging begins long before it becomes obvious. But to use age as a social attribute, it is necessary to identify specific *indicators* of aging, such as chronological age, functional capacity, or life stage that can be used to classify people into various ages.

Chronological Age

Chronological age, for which the birth certificate is an unambiguous source, satisfies the need to set a point at which bureaucratic rules and policies can be applied and to separate people who are eligible for something from those who are ineligible. However, because the relationship between chronological age and the consequences of aging is not strong, all such chronological definitions tend to misclassify some proportion of the population.

For example, 65, the most common age for classifying people as aged or elderly, is currently the age at which people become eligible for full retirement benefits from **Social Security** as well as for health benefits from **Medicare.** Since 65 was historically the most common age at which people retired, when Social Security was developed 65 seemed the chronological age at which most people would need such benefits and assistance. Yet, thousands of people become unable to work because of poor health *before* 65, and thousands more continue employment *after* 65.

Thus, selection of age 65 to define eligibility for these programs included some people (such as the employed) who perhaps did not need the programs and unfairly excluded others (such as those under 65 and in ill health) who did need the programs' benefits. These are the costs of using chronological age definitions.

The problem is compounded by the fact that many other chronological ages are used to define "older." The U.S. Department of Labor defines a worker as "older" at the age of 40. At 60, people are eligible to participate in most senior centers. Under Social Security, widows are eligible for survivor benefits at 60. At 62, people become eligible to live in housing for elders. Age 65 is the minimum age at which retirees can draw maximum Social Security retirement pensions. At 70, restrictions on earnings by Social Security recipients no longer apply. These are only a few of the government's definitions. The range in age for *all* government as well as local community agencies is amazing, resulting in a confusing hodgepodge of definitions that reveal no consensus about when "old age" begins chronologically.

Similar confusion reigns in the research literature. For example, in the 1995 volume of the *Journal of Gerontology,* operational starting points for the age category *older adults* ranged from 50 to 70.

Apart from being used to determine eligibility for programs or research, chronological age categories are also used to describe individuals. For example, people between the ages of 13 and 19 are described as teenagers, and adults in their 20s are differentiated from those in their 30s. The rationale for using chronological age this way is that in an age-graded society, people of similar age are likely to be in similar situations and confronting similar problems. Thus, people in their teens are likely to be in school, and people in their 30s are likely to be dealing with issues of family and jobs.

Unfortunately, this approach is not very useful for the older population. The age span is too big, and experiences of older people, even those of the exact same age, are too variable. The older population ranges in age from 65 to 105, or more; people who are 65 usually have little in common with those who are 100 because of different living situations. To cope with this problem, gerontologists sometimes divide the older population into those under 75, those 75 to 84, and those 85 and older. These categories are referred to as the young-old, middle-old, and old-old or oldest-old, respectively. This differentiation is better than none, but the demarcation points are arbitrary and based more on how the U.S. Bureau of the Census reports data on the older population than on common experience among people in these age categories. The common practice of referring to people by age decades—50s, 60s, 70s, 80s, 90s, and centenarians—is probably as good a way of subdividing the older population as any, but data are seldom reported this way.

Gerontologists widely recognize that chronological definitions and categories have serious limitations. Nonetheless, the age of 65 remains dominant as the *legal* definition of when a person becomes "older"; it also dominates in research as the most-used demarcation point for *aged, older,* and *elderly.* However, there are many exceptions to this usage.

Functional Age

Definitions of **functional age** rely on observable individual attributes to assign people to age categories. Physical appearance, mobility, strength, coordination, and mental capacity are examples of such functional attributes. Commonly used general criteria for categorizing people as old include gray hair, wrinkled skin, and stooped posture. Adults who move stiffly, tentatively, and with poor coordination exhibit the physical frailty that we associate with old age. And people who are quite forgetful, sometimes confused, and hard of hearing have some of the psychological frailties associated with old age. Anyone who has all these attributes is undoubtedly "old" regardless of his or her chronological age. Since only a small percentage of people have more than a few of these attributes, classifying people into age categories based on functional attributes is an uncertain process. Functional age definitions also vary from one environment to another. For example, professional baseball players usually become functionally old as early as 30 or 35, whereas grandmothers can still be functionally capable at 90.

Because functional age definitions are so difficult to assess, they are seldom used in research, legislation, or social programs. Indeed, there is considerable doubt that effective summary measures of functional age could be developed, even within relatively narrow areas. Nevertheless, in everyday life such definitions give us a general feeling of where to place people along a continuum of age categories.

Life Stages

Very often we use a combination of physical and social attributes to categorize people into broad **life stages,** such as adolescence, young adulthood, adulthood, middle age, later maturity, and old age. These life stages are heuristic ideal-types.* Each type reflects an array of physical, psychological, and social attributes or circumstances that are commonly thought to characterize that life stage. The

chronological boundaries of many life stages are fuzzy. Let us first look at middle age.

Middle Age **Middle age** is the life stage during which most people first become aware that physical aging has noticeably changed them. During middle age, people often begin to seek less physically demanding activities. Recovery from exertion takes longer. Minor chronic illness becomes more prevalent. Vision and hearing begin to decline.

What else typically happens during middle age? Job careers often reach a plateau of routine performance. Middle-aged parents' responsibilities change as their young adult children leave their parents' households to establish their own. Married couples often grow closer. People sometimes make midlife job changes. Community involvement may increase. Stay-at-home mothers whose children are grown often rejoin the workforce. More and more people retire in middle age with no continuation of employment. Middle age is also a time when people experience death of those close to them, especially parents.

Middle age is a stage marked primarily by social transitions—at home, on the job, in the family. Significant physical transitions usually come later. Midlife can also be a time of reflection and the beginning of an inner process of evolving a personal life meaning. For most people, middle age is an exciting time because many of its transitions lead to a more satisfying, sometimes less hectic life.

Chronologically, middle age for most people usually begins at about 40, although this varies a great deal, depending on when people or those around them perceive that they are *symptomatically* middle-aged. Still, middle age is the stage at which people become part of the *aging* population (as opposed to the *older* population).

Later Maturity The declines in physical functioning and energy availability that begin in middle age continue in **later maturity** (usually considered as beginning sometime in

---

*Ideal-types are abstract composites of reality that are created by scholars to provide general descriptive pictures of complex patterns.

Women in three life stages.
Photograph by Marianne Gontarz

the 60s). Chronic illness becomes more common. Activity limitations become more prevalent, although most people continue to be relatively active. Mortality begins to take its toll among family and friends, which often brings home the fact of one's own mortality.

As with middle age, the major changes associated with later maturity are social. Retirement typically occurs during this stage. For most people this is a welcome and beneficial change, but one that sometimes brings a reduction in income. By contrast, people in later maturity increasingly find themselves involved in caring for their older parents.

Most people retain a fair measure of physical vigor in later maturity that, coupled with freedom from responsibilities, makes this life stage one of the most open and free for those prepared to take advantage of it.

Old Age   Chronologically, the onset of **old age** typically occurs in the late 70s (although

many people in their 80s or 90s show few signs of it). Old age is characterized by extreme physical frailty. Disabling chronic diseases are more common. Mental processes slow down; chronic organic brain disease becomes more prevalent.

Individuals in old age feel that death is near. Activity is greatly restricted. Social networks have become decimated by the deaths of friends and relatives and by the individual's own disabilities. But even in old age, most people have frequent contact with family and friends. Physical dependency and institutionalization are common. Old age is apt to be unpleasant, at least externally. Whereas middle age and later maturity are defined mainly by social factors such as the empty nest or retirement, old age is defined more by the physical or mental frailty that accompanies it. Most people die before they reach extreme disability.

These stages of middle and later life are based on sets of characteristics that seem to be related in many cases, although seldom will a particular individual show each and every characteristic typical of a given stage. Also, these categories are not based on chronological age. Different people show characteristics of old age at different chronological ages—one person could be symptomatically in old age at 55 while another might still be symptomatically in later maturity at 85.

In this book we use the terms **midlife** and *middle age* to refer to people in their 40s, 50s, and early 60s. **Aged, elderly, elder,** and **older people** will be used interchangeably for people age 65 and over. These chronological definitions misclassify some people, but they are necessary in order to summarize information and make comparisons. Just remember that we are using them for convenience; *their value in helping us relate to specific individuals is limited and varies depending on the context.*

Social Gerontology Is a Unique Field of Study

Many of the concepts, theoretical perspectives, and research issues in social gerontology are unique to gerontology and largely unknown in other disciplines. To be sure, social gerontology shares a great deal of its vocabulary, ideas, perspectives, and research techniques with other social sciences, but it also has many of its own. In addition, many of the tools borrowed from other social sciences must be modified, sometimes substantially, before they can be utilized effectively in the study of aging.

For example, the term **self,** as used in sociology, social psychology, and other social sciences, usually refers to ideas about oneself that are developed in interaction with others. This view is useful in explaining how young people develop a sense of self. However, in gerontology we much more often study people whose self-concept is already solid and not very sensitive to outside influence. Thus, to be of use to gerontologists, the concepts and theories about the self that come from general social science literature must be modified and extended. For gerontology, the issue is more often how individuals *maintain* or *defend* their self-concepts than how their self-concepts are initially established.

A great deal of what passes for fact about aging is actually the perpetuation of false stereotypes. For example, the literature on the family tends to treat as fact the belief that the transition from a rural, agrarian society to an urban, industrial one caused families to become more mobile and to lose their capacity for including or caring for their elderly members. However, the view of this issue in the gerontology literature is quite different. From more than three decades of careful historical and survey research it is clear that families in general do include their older members, have not abandoned them, are in frequent contact with them, and, when necessary, provide them with physical, financial, and emotional support. Likewise, historical research has shown a greater variety of family patterns, some quite antagonistic to older members of the family, than the "good old days" stereotypes reflect. Despite the remarkable consistency of these gerontological findings, they have been slow to make their way into the general social science and family literature.

A final difference between social gerontology and most other social sciences is that gerontology covers a much broader portion of the human lifespan. As a result, research that follows the same individuals or categories over an extended period of time (longitudinal research) is much more prevalent and important in gerontology.

In matters having to do with aging, social gerontology has unique perspectives and information that are not widely known in other social sciences. As a result of its more in-depth focus on aging, social gerontology's information base about aging tends to be much more accurate and detailed than material on aging found in general social science disciplines.

The remainder of this section provides an overview of concepts and theories, facts, and research issues specific to social gerontology.

Concepts and Theoretical Perspectives

Concepts are created by humans to organize thought, observation, and communication. Like all tools, concepts must be precise if they are to be effective. Of course, complete conceptual precision remains a very elusive goal. In social gerontology, you will encounter new concepts as well as familiar concepts with new meanings. You may also need to modify or reject some of your existing concepts. A major goal of this book is to provide a sound conceptual basis for thinking, observing, and communicating about aging.

Concepts are usually grouped into classifications and **theories.** This book includes numerous classifications, from types of physical changes that accompany aging to types of human services for older people. These classifications help put specific observations in an appropriate context and identify valid points of comparison.

In the social sciences, theory and research methods are often considered separate fields. In graduate education, theory and research methods are taught separately by faculty members with very different orientations to knowledge. This practice leads to theories that are difficult to apply and to research that is not well-informed by theory. This book takes a different approach. Few theories attempt to explain everything, and those that do are usually unsuccessful. Theories come to life in the process of applying them to an appropriate context. In this book, we discuss the major theories of social gerontology in the context of the substantive issues they are intended to explain. For example, Cumming and Henry (1961) touted their disengagement theory as a general theory of individual and societal adaptation to human aging. But researchers who wanted to test this theory found that the individual disengagement component dealt mainly with adaptation to loss of social roles and the societal disengagement component attempted to explain society's loss of interest in the contributions of its older members. Accordingly, we discuss each component of disengagement theory in the appropriate context. Individual disengagement is discussed as one of several theories that try to explain how people adapt to role loss in Chapter 7, Personal Adaptation to Aging. Societal disengagement is discussed in Chapter 16, Aging in Contemporary American Society and Culture, as one of several theories of how society responds to its aging population.

Theoretical perspectives help us understand not only *what* is happening but *how* and *why*. Table 1-1 shows some of the major concepts and explanatory perspectives you will encounter in social gerontology and how they relate to one another. It is obvious from the table that social gerontology is loaded with theories and perspectives, which runs contrary to the notion that gerontology is atheoretical. Do not be concerned if many of these concepts and theories are unfamiliar now; by the time you finish this book, they will seem like longtime acquaintances. *There is not a single chapter in this book that does not make use of concepts and theoretical perspectives,* many of which are unique to social gerontology.

Remember, concepts and theoretical perspectives are merely *tentative*—always in need of verification. Of course, there are some theories (such as the theory of gravity) that have been reconfirmed so many times they hardly seem tentative. But we must *always* be alert to change. (People were once certain that the earth was flat.)

Another point to remember is that perspectives are seldom mutually exclusive. For instance, we can say, "People cope with role loss by consolidating their efforts within their remaining roles," *and* we can say, "People cope with role loss by withdrawing." Since some people follow one pattern and some the other, the real question is not which perspective is true and which false, but what

Table 1-1 Concepts and perspectives in social gerontology.

Individual Aging		
Mental Aging (Ch. 5)	Social Aging (Ch. 6)	Personal Adaptation to Aging (Ch. 7)

Senses
  Thresholds
Perception
  Stimulus generalization
Motor capacity
  Central processes
Mental ability
  Decrement
  Increment
  Cohort differences
Personality
  Human development
  Stage theories
  Process theories
	Social roles
	Self-concept (Ch. 5)
	Self-esteem (Ch. 5)
	Life course
	Age norms
	Age grading
	Social support (Ch. 8)
	Exchange theory (Ch. 8, Ch. 16)
		Coping
		Compensation
		Continuity theory (Chs. 5, 6, 7)
		Coping with role loss
		  Activity theory (substitution)
		  Consolidation
		  Disengagement theory
		Social breakdown theory

Aging and Society			
Aging Society	Social Status of Elders (Ch. 16)	Elder's Response as Category (Ch. 19)	Care of Elders (Ch. 14)

Political economy
  (Ch. 16)
Moral economy (Ch. 9)
Retirement ans a social
  institution (Chs. 9, 18)
Critical gerontology
  (Ch. 12, Ch. 16)
	Age discrimination
	Societal disengagement
	Modernization theory
	  (Ch. 3)
	Age stratification theory
		Elders as subculture
		Elders as minority group
		Old-age voluntary associ-
		  ations
		Elders as interest group
			Medical care model
			Social care model
			Holistic care model
			Filial responsibility
			  (Ch. 8)
			Continuum of care

proportion of people who experience role loss fit into *each* perspective (and why do some people withdraw while others do not)?

Factual Information

Facts are relatively objective statements about what is true. To be factual, information about aging needs to be *representative*. Most of us get quite a bit of information from our everyday experience, information that is often correct as far as it goes. Unfortunately, we are seldom exposed to a complete cross section of people or situations. For example, suppose someone asks, "Do people have difficulty adjusting psychologically to retirement?" And suppose we have a group of four experts respond. The first expert is a psychiatrist (who, after all, should know a lot about psychological adjustment). She says, "Yes. I

see many people who are highly distressed about retirement." The second respondent is the director of a planned retirement community (who has experience with retired people). He says, "No. In my ten years at Covebrook Center, I have seen only two people who had a significant problem in adjusting to retirement." The third to respond is a director of a senior center, "Yes. Many of the people who come to my center are quite anxious. They feel cut adrift and unsure of what to do." The fourth person, another senior center director, says, "That's interesting. At my center the people are all retired from Xenon Corporation and they're all delighted with retirement."

Whom is one to believe? The difficulty here is that the experience of each expert is *selective*. The psychiatrist has little experience with well-adjusted people—they do not need her services. The retirement community staffer sees people who are healthy, wealthy, and committed to an active lifestyle. The two senior center directors' conflicting views reflect the fact that the clientele of senior centers differ in their social, ethnic, occupational, regional, and community backgrounds.

Information collected via scientific research has at least three major advantages over information gathered from everyday experience. First, social science research examines samples of people or situations whose representativeness can be evaluated, thus giving us a basis for weighing evidence from various sources. Second, the procedures used to collect scientific information and arrive at scientific conclusions are explicit, thus providing another tool for weighing evidence. Weighing evidence is a skill that requires experience and practice. The facts seldom speak for themselves; they need to be interpreted and put into context. Third, scientific culture provides an agreed-upon body of standards by which to judge research adequacy. For example, statistical tests of significance allow us to assess the probability that particular research results are merely chance occurrences rather than meaningful and repeatable facts.

Research Issues

With only minor modification, most of the research methods of the other social sciences are applicable to social gerontology. For example, the statistical techniques that students learn in psychology, sociology, geography, and so on are essentially the same as those social gerontologists use. But in social gerontology, the emphasis differs from that of the other social sciences. Let us look at examples from the logic of research, the types of research, sampling, and measurement.

Logic of Research   Much social science research involves making comparisons among groups or categories, or examining relation-

ships among variables measured at the same moment in time. If we want to know the effect of race on income, for example, we might examine the average incomes of various racial categories in a given year. But if we are to learn about aging, we cannot simply compare people of different ages at a single point in time; we also need to observe how the same people change over time. In addition, historical periods can influence the effect of being a particular age. In gerontology, we call these different concepts *age differences, age changes,* and *period effects*.

Age differences result from comparisons of people who are different ages at the same moment in time. For example, in the United States in 1990, the median income of men 55 to 64 was $24,804, while for men 65 and over it was only $14,183 (U.S. Bureau of the Census 1991b). From these data, one might conclude that aging causes income to go down, which may indeed be the case. But it was also true that the proportion of men having graduated from high school was 73.2 percent at age 55 to 64 and only 57 percent at age 65 and over (U.S. Bureau of the Census 1992b). Should we conclude that aging causes level of education to drop? Of course not. This illustrates a problem: *There are factors other than aging that can cause age differences.* The preceding age difference in education was not caused by aging but by two other factors: (1) The two age categories (called **cohorts**) received their educations at different historical times, and (2) ideas about how much education is appropriate and society's capacity to provide education both change over time. Even in the income example, aging is probably responsible for only part of the age difference; part is also probably the result of a higher average wage when the younger cohort began employment.

Age changes occur in the same individual over time. For example, when Bob was 19, he had 20/20 vision, but by the age of 40 it was 20/40, good enough to drive legally without glasses and to perform all the tasks required in his everyday life. Nevertheless, he had experienced an age change in visual acuity.

Period effects are always with us, and they interact with aging. For instance, one of the major transitions to adulthood is getting one's first full-time job. But the nature and quality of this transition are very much affected by the climate of the times in which it occurs. Those who first sought employment during the Great Depression generally had a tough time and may have had to settle for a much less desirable job than they were qualified for. Those who entered the labor market in the mid-1960s generally found good jobs readily available, provided they were not drafted for the Vietnam War. In the early 1990s, economic recession, along with large numbers of young adults from the baby boom, created an extremely tight job market.

The effect of a time period on an individual depends on how old the person is when experiencing that period. At the onset of the Great Depression, for example, the suicide rate increased among white males in America, and the amount of increase went up for each successive five-year cohort. From 1925 to 1930, the suicide rate of white males age 30 to 34 went up 38 percent; for those age 40 to 44, 49 percent; and for those age 50 to 54, 51 percent (Atchley 1980a).

Whether period effects, age changes, and age differences can be completely disentangled is questionable (Palmore 1978; Schaie 1987b). Nevertheless, social gerontologists must be aware of these different types of effects and design research that tries to control and separate them as much as possible.

Types of Research Social scientists use several types of research to gather information, including direct observation in the field, analysis of existing statistics or records, experiments, and social surveys. Social gerontologists use these same types of research, but not necessarily in the same proportions as in other social sciences. A comparatively large number of social gerontology topics are still in exploratory stages, which affects the types of research being done.

When research into a new field begins, investigators are much like explorers charting unknown territory; ideas about the new territory are vague and maps are crude. As the investigators gradually explore the territory more fully, they are able to identify major routes and important locations. For example, research in cognitive psychology has progressed to the point where tight experiments can be designed to answer detailed questions about such topics as how aging affects memory or problem solving. On the other hand, research on how aging affects marriage is still mainly descriptive, trying to identify the major facts, issues, and perspectives that will produce useful research. In social gerontology, which is a relatively new specialty within social science, much research is necessarily more descriptive. Thus, small-scale field studies and pilot surveys are more numerous than in other fields.

Ralph Waldo Emerson defined a scholar as a person in the process of hard thinking (Emerson 1837). In this context, research means gathering information and then reflecting on it. It also follows that the more methods the scholar uses to gather information, the more likely the results of reflection will be accurate. In my own studies of individual retirement, I have observed retired people in their homes and in their communities, interviewed them, examined survey data I collected from over 10,000 retired people, read the research results of dozens of others who have studied individual retirement, and looked at various kinds of demographic data. After more than thirty years of such work, I am continuing to learn about individual retirement. At the same time, I have become aware that comparative, cross-national, and historical research on retirement are still in their early development. The point is that *every* type of research is necessary in order to gain an understanding of the social world.

Unfortunately, most research fields experience fads, with particular types of research being favored. In the social sciences right now, large national sample surveys are in vogue, justified in part by the perception that they are more representative than small-scale surveys. Yet analysis for various regions of the country reveals that local response rates to these large surveys are often as low as 40 percent, certainly an inadequate representation. This in turn means that such surveys, although national in scope, are not as representative as they might appear to be. National surveys are certainly useful, but they also benefit from cross-checking with the results of other types of research. And there are important research questions that do not lend themselves to surveys.

In addition, the mystique of the large national sample by implication casts a shadow on the value of surveys conducted on more limited populations or on community-based studies. Yet for many purposes, community-based studies are superior to national surveys. For example, in any study of the effect of retirement on leisure activities, it would be important to hold constant the availability of recreational programs and facilities. This would be vastly easier to do in a study of a single community than in a national survey of people in hundreds of communities.

Sampling Because older people constitute a small percentage of the total population, sampling the older population is sometimes difficult, particularly if we are looking for something that occurs in only a minority of older people. For example, mental health professionals cite cases of depression associated with retirement. However, in studies that have followed representative samples of people through retirement, only a tiny fraction experienced depression associated with it. This means that if one wanted to study those who become depressed as a result of retirement, finding an adequate sample would be difficult indeed. There are no lists one could use, and going door-to-door would be too expensive. Unfortunately, gerontologists must often settle for the samples they

can get rather than the samples they would like to have. And while this is true some of the time in any field, it is true more often in social gerontology.

Measurement Measurement involves translating observations into meaningful categories or numbers. The adequacy of the procedures we use to measure social variables is judged by two important criteria: validity and reliability. The term **validity** refers to the degree of correspondence between what is supposed to be measured and what is actually measured. The closer the correspondence, the more valid the measure. Validity can be assessed by comparing the results of the measure in question with an already accepted measure, but often the researcher must settle for measures that only *appear* to most observers to be valid, without real proof. Since validity often cannot be established concretely, the investigator must be constantly on the alert for indications of validity.

The term **reliability** refers to the extent to which a given measure yields results that are stable over successive trials. For example, in the *test-retest method* of assessing reliability, a test is given to the same people more than once and the results compared. Note that a measure can be reliable without being valid, but it can never be valid without being reliable.*

Questions of validity and reliability are important for gerontological research because many of the measures we want to use were established for young research participants, not older ones. It is quite possible for a measure to be valid and reliable for college students but not for retired professors. The investigator who wants to use an already-established measure will ordinarily want to reestablish the validity and reliability of the measure when it is tried with older people. Gerontological researchers have paid too little attention to this problem. These and many

---

*For a detailed discussion of validity and reliability, see Babbie (1995, 122–129).

other methodological issues related to the study of aging are covered in more detail by Maddox and Campbell (1985) and Lawton and Herzog (1989).

Despite some methodological difficulties, social gerontology has made enormous strides over the past two decades. As you will see, we know quite a bit about aging. Much of what we know has been found repeatedly, so we can be relatively confident about it. But, as in any field, knowledge is constantly changing, which should keep us on our toes.

Some Important Gaps in Knowledge

Most aspects of later life have been researched far more than one might expect, but there are nevertheless two very important gaps in our present knowledge: the lack of cross-national data and the lack of data on older minority group members. Since aging becomes a visible social problem primarily in industrialized societies, there has been very little research on social gerontology in the nonindustrialized nations of the world. Cross-national, comparative reports are sparse (Cowgill 1986; Sokolovsky 1983; Palmore 1980b; Oriol 1982; Holmes and Holmes 1995; Keith et al. 1994). In addition, the data for industrialized nations are quite variable, with the United States by far the most widely researched. Thus, the body of knowledge we call *social gerontology* is heavily biased in terms of the U.S. situation.

Moreover, despite the fact that older Americans have been researched far more than any other older population in the world, less is known about significant subgroups of older Americans. Not only are there individual differences that produce heterogeneity in the older population, but there also are subgroup differences in culture and behavior that create diversity among older people. Older Americans cannot be understood apart from their earlier lives; if they are members of a minority group, their experience has probably been quite different from

that of most of their fellow older Americans. Thus, older people who also happen to be African American, poor, Appalachian, foreign-born, or members of an ethnic minority probably face an old age different from that of the majority of the older population. Although research on aging among minorities is much more prevalent now than two decades ago, much of what we have to say about older people in this book still may not necessarily apply to older members of minority groups.

Focus on the United States

The aging experience results from an interaction of physical, mental, social, and cultural factors. As a consequence, aging varies considerably across cultures, even for the economically developed nations. The range of variation across all cultures and historical periods is truly mind-boggling and cannot be addressed adequately here. This book focuses primarily on aging in the United States in the 1990s. Many of the major points probably apply to aging in other developed societies, but unless specific confirming data are presented, do not assume so. Cross-cultural comparisons are used primarily to highlight significant aspects of life in the United States, not to provide a survey of cross-cultural gerontology. As we will see, the United States is a complex, multicultural society that is difficult to comprehend, and presenting an overview of aging in American society is a significant challenge in itself.

Social Policy Issues

The term *social policy* refers to actions taken by government and the political processes through which collective, goal-directed decisions and allocations are made. Analysis of social policy involves studying the policy choices available, the strategies that could be used to implement these choices, and their actual or anticipated consequences (Hudson 1987a).

Social gerontology includes a wide range of social policy issues that affect aging and older people. For example, without Social Security there could be no widespread opportunity for retirement in the United States. Medicare and Medicaid policies have a profound effect on the availability of and access to health care by older people. The Older Americans Act created a national network of agencies providing social services to older people. Other federal policies affecting older people involve nutrition, community-based long-term care services, transportation, housing, recreation, and many other areas. State and local government policies are also important to the well-being of the older population. Policy issues are covered throughout this book but receive special attention in Chapters 3, 9, and 13 through 20.

Unlike most social sciences, in which social policy is either ignored or considered a separate and distinct subspecialty, social gerontology integrates social policy issues related to aging into the mainstream of its knowledge base. This is a significant step because it sensitizes all social gerontologists to important social policy issues and improves the likelihood that research results and practice issues will be brought into the policy arena, and that policy and practice issues will be brought into the research arena.

Professional Practice

Social gerontology also includes knowledge about how to work most effectively in professions that provide services to aging or older people. A *profession* is a vocation that requires training in the liberal arts and sciences as well as advanced study in a specialized field. Although research, teaching, and social policy involvement all satisfy this definition, in social gerontology we generally reserve the term *practice* to refer to those professions involved in the design, administration, and delivery of direct services. Practice concepts are emphasized in Chapters 8, 11, 12, 13, 14, and 15.

A unique feature of social gerontology is the high degree of communication among those interested in research, social policy, and professional practice as they relate to aging or the older population. As a result, research in social gerontology deals with policy and practice issues to a much greater extent than do the other social sciences. Thus, well-educated students of social gerontology must be familiar not only with the concepts, theories, and factual base generated by research but also with the major social policy issues and arenas, and the primary professions and practice environments in which gerontological knowledge is both needed and used.

Careers in the Field of Aging

Defined broadly, the field of aging consists of occupational positions that require knowledge about aging or older people. The size of this pool of jobs is enormous because almost all areas of employment in business, voluntary organizations, and government have a substantial number of jobs that include at least some responsibility in relation to aging or older people. Jobs devoted primarily to aging are the core of the field. Such jobs include teaching gerontology; doing gerontological research; developing and marketing products or services geared to aging or older adults; developing or administering government programs aimed at older people; influencing social policies on aging; providing direct services to older people; or informing the public about aging via books, television, or newspapers. In addition, a great many jobs include the concerns just listed as part of their duties.

In planning a career in the field of aging, there are two basic routes: specialization in gerontology or specialization in another profession with gerontological knowledge and skills attached. Right now, **gerontologists**— people who specialize in aging—work mainly in colleges and universities, agencies on aging at various levels of government, and organizations that serve older people. People in other professions who are often required to have expertise in gerontology include lawyers, physicians, dentists, nurses, social workers, physical therapists, speech therapists, audiologists, clinical psychologists, and psychiatrists. In the corporate world, people in management, marketing, product development, and sales often need gerontological knowledge to perform effectively.

In the past, credentials in the field of aging consisted mainly of on-the-job experience, but formal education in gerontology is increasingly required in various areas of practice, especially in education and health and social services (Peterson and Douglass 1990). People with an undergraduate background in gerontology often secure jobs in agencies that promote, organize, or provide services to older people. Many of the entry-level administrative and professional service jobs in organizations discussed in Chapters 13, 14, and 15 are for people with baccalaureate degrees. Graduate training in gerontology is essential for those who wish to teach or do research in gerontology in a university setting and is often a prerequisite for professional practice with older adults in fields such as medicine, nursing, social work, and long-term care administration. Many planning and administrative jobs in the field of aging are occupied by people with a master's degree in gerontology.

As an academic field of study, gerontology is still in its developmental stages. Over the past two decades, gerontology has grown substantially. Colleges and universities are increasingly offering gerontology as part of their liberal education programs, and many have undergraduate minors and majors in gerontology. About thirty schools offer master's programs in gerontology, and in 1995, three universities offered doctoral study in gerontology. As society ages, interest in gerontology can be expected to continue to grow.

Summary

Although aging has its roots in the nature of humans as biological organisms, the significance of aging is often defined socially. Aging not only affects what people are capable of doing, but it also influences what they are expected to do, allowed to do, or prohibited from doing. Large increases in the proportion of the population that is older also influence the issues that societies must resolve and lead us to speak of "aging societies."

Gerontology is a relatively new field devoted to the study of aging. It includes knowledge about aging from all academic disciplines and fields of professional practice as well as that developed within gerontology itself. Social gerontology deals with the nonphysical side of aging and includes the psychological, social psychological, and social aspects of aging.

Human aging is not one process but many. Even at the biological level there is great variation within a single individual as to when aging begins in various organs or systems and the rate at which it progresses. Psychologically, some functions decline with age, some remain relatively stable, and some improve. Social aging brings the aging individual both advantages and disadvantages. Because aging varies both within and between individuals and it can have both positive and negative results, the consequences of aging for a particular individual are not predictable. The two faces of aging—positive and negative—are reflected in society's treatment of older people, in the literature on aging, and in our theories about aging and older people.

To study aging, we must know who to study. We can define aging by chronological age, functional age, or life stage. Each type of definition has advantages and limitations. Because chronological definitions are the easiest to use, they dominate aging research and social policy about aging. Chronologically, middle age begins at about 40 in most cases.

The age of 65 is the most common demarcation age for when a person becomes aged, elderly, or older, although different age markers are used in numerous governmental laws and regulations as well as in a great deal of research into aging. We refer to members of the older population as elders or older people.

As a field of study, social gerontology shares concepts, theoretical perspectives, classifications, factual knowledge, and research methods with other social sciences, but it also has developed many that are uniquely its own. Theories are numerous in social gerontology, and most of them are relatively unknown outside gerontology. Social gerontology is also unique in that it integrates scientific knowledge about social policy and professional practice into the knowledge base created by basic social scientific research. The field of social gerontology thus includes people who are employed not only in universities or research institutes but in social policy and professional practice professions as well.

Science is the main method used to collect factual information in social gerontology because, compared with everyday observation, scientific information is usually more representative, easier to evaluate, and can rely on a set of rules and conventions for making decisions about whether or not something is true. However, social gerontology is not exclusively a social science because it incorporates knowledge from the humanities, social policy, and various areas of professional practice as well.

Research methods in social gerontology differ from those in other social sciences in the attention paid to differentiating age differences from age changes and period effects. Age differences derive from comparing individuals of different ages at a given point in time and can result from a number of factors other than aging. Age changes are changes in a given individual over time. Period effects result from variations in social conditions over time. Age and period effects interact so that the consequences of being a particular age depend somewhat on when in history one experiences that age.

Social gerontology, like any new discipline, needs an abundance of research to explore new territory, which means a larger proportion of small-scale pilot research than in more established disciplines. Nevertheless, social gerontology, like all disciplines, benefits from research of all types. Researchers in social gerontology are often frustrated by difficulties in sampling—particularly for relatively rare events, characteristics, or processes—in what is already a relatively small segment of the total population. As a result, less than ideal samples are more common in gerontology than in some other disciplines. Researchers in social gerontology also have to beware of measuring older people with instruments that were developed for use with other age groups. It is important to reestablish the validity and reliability of such measures for use with older people.

Our knowledge of social gerontology is based primarily on studies of aging in the United States. Studies of aging in industrial societies of Europe and Asia are comparatively sparse, and well-done studies of aging in Third World countries are rare. Studies of aging among members of minority groups within the United States have increased in recent years, but there are still many gaps compared to what we know about aging in the general population.

Social gerontology also includes information about social policies that affect older people and that have been enacted through governmental and political processes. In addition, social gerontology includes professional practice knowledge about how to work most effectively with aging or older people.

Careers in gerontology are becoming more prevalent. The core of the field consists of occupations that are focused on research and teaching about aging; developing, implementing, and influencing social policies aimed at aging or older people; creating and distributing products and services designed primarily for aging or older people; doing professional practice with aging or older adults; staffing organizations that directly serve older people; and informing the public about aging through mass media. This represents a significant field of occupational opportunity. Formal education in gerontology is an increasingly important credential for many occupations in the field of aging.

2 ▮ The Demography of Aging

One of the main forces behind the development of social gerontology is the aging of the population, the increasing proportion of older people in many countries of the world. This chapter discusses how the aging of a population is measured and what demographic forces produce it and the details about the composition and geographic distribution of the older population of the United States. The chapter ends with a discussion of how the older population is influenced by the demographic processes: fertility, mortality, and migration. Together, these topics allow us to understand demographic factors as an important *context* for both societal and individual adjustment to aging. The chapter also illustrates how population issues can influence society's institutions as they relate to the older population.

Measuring Age Structure

To understand the age structure of a society, we need some means of summarizing it. After all, in 1995 the U.S. population was made up of over 260 million individuals. One effective way to summarize the age structure of the United States is to classify the population into five-year age categories and then compute the percentage of the population in each category. Because the number of males relative to the number of females at any age can also be important, percentages are usually compiled by gender for each age category.

Table 2-1 shows the percentage distribution of the projected 1995 U.S. population, by age and sex. From it we can determine that older people then represented 12.8 percent of the population and that older women outnumbered older men by a considerable margin. Another 27.6 percent of the population was between the ages of 40 and 64. Thus, using common chronological definitions, over 40 percent of the U.S. population was middle-aged or older in 1995.

Population Pyramids

A better way to visualize the age and sex structure of a population is to convert the percentage distribution into a **population pyramid**. Figure 2-1 is a population pyramid constructed from the data in Table 2-1.* The figure shows better than the table the effect of the post–World War II "baby boom" and the subsequent drop in birthrates. We return to the age structure of the United States later.

Population pyramids are useful for making comparisons between age and sex categories within a single population and between populations, including populations of very different

---

*This type of chart is called a pyramid because at the time it was first devised, most population age and sex structures resembled a pyramid—there were a great many young children and a steady erosion of numbers occurred because deathrates were high at all ages.

Table 2-1 Population of the United States, by age and sex: 1995.

AGE (Years)	Total Number (in thousands)	Percent	Males Number (in thousands)	Percent	Females Number (in thousands)	Percent
Under 5	19,553	7.4[a]	10,014	3.8	9,539	3.6
5–9	19,225	7.3	9,848	3.7	9,376	3.6
10–14	18,895	7.2	9,678	3.7	9,217	3.5
15–19	18,024	6.9	9,238	3.5	8,786	3.3
20–24	17,885	6.8	9,084	3.5	8,802	3.3
25–29	18,994	7.2	9,522	3.6	9,471	3.6
30–34	21,850	8.3	10,888	4.1	10,962	4.2
35–39	22,267	8.5	11,080	4.2	11,187	4.3
40–44	20,233	7.7	10,004	3.8	10,229	3.9
45–49	17,440	6.6	8,560	3.3	8,880	3.4
50–54	13,642	5.2	6,630	2.5	7,012	2.7
55–59	11,089	4.2	5,321	2.0	5,767	2.2
60–64	10,064	3.8	4,739	1.8	5,324	2.0
65–69	9,948	3.8	4,521	1.7	5,427	2.1
70–74	8,852	3.4	3,850	1.5	5,002	1.9
75–79	6,693	2.5	2,710	1.0	3,983	1.5
80–84	4,461	1.7	1,590	0.6	2,871	1.1
85–89	2,320	0.9	700	0.3	1,620	0.6
90–94	1,002	0.4	250	0.1	752	0.3
95–99	263	0.1	53	0.02	210	0.08
100+	54	0.021	10	0.004	44	0.017
Total	262,754	99.9[b]	128,292	48.7	134,461	51.2
65+	33,594	12.8	13,685	5.2	19,908	7.6
85+	3,638	1.4	1,013	0.4	2,625	1.0

[a] Row total may not match male and female percentages due to rounding.
[b] Column total does not equal 100 percent due to rounding.
Source: Compiled by the author from projections performed by the U.S. Bureau of the Census (1992g).

sizes. For example, Figure 2-2 shows population pyramids for Afghanistan and Sweden in 1988. From these graphs we can see that Afghanistan had a very high percentage of children and a relatively low percentage of older people, while Sweden had a very high percentage of older people and a relatively low percentage of children. The reason for these differences is discussed later in the chapter; for now it is enough to note the value of population pyramids in making comparisons between societies' age–sex structures.

Sometimes we want to compare many different societies or the same society at many different points in time. To do so using population pyramids or percentage distributions is too cumbersome; we need a single summary measure of the degree of population aging. The three most commonly used measures are the percentage of the population that is age 65 and over, the aged

Figure 2-1 Population of the United States, by age and sex: 1995.

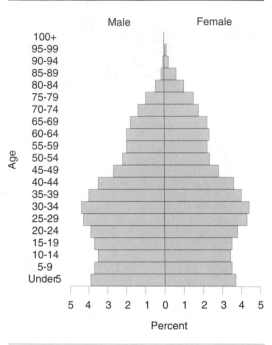

Source: Compiled by the author from projections done by the U.S. Bureau of the Census (1992g).

dependency ratio, and life expectancy at birth.

Proportion of Older People

The **percentage of older people in the population** is simple to compute and is based on data readily available for most countries of the world. It is also readily available for nearly all subdivisions of the United States, which allows us to compare the United States with other countries and to compare subdivisions of the United States. Table 2-2 shows the percentage age 65 and over in selected countries, and it indicates that the United States, along with most of Europe, has a relatively high proportion of older people in its population. Argentina and the two former Soviet Republics of Russia and Georgia are in an intermediate position, and the remainder of the countries have relatively low percent-

Figure 2-2 Populations of Sweden and Afghanistan by age and sex: 1988.

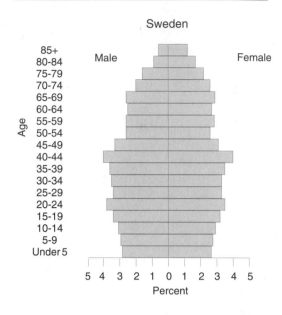

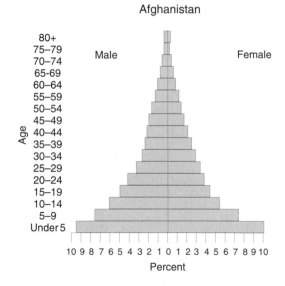

Source: Tabulated by the author from United Nations (1990).

ages of older people. (The dependency ratio columns of Table 2-2 are considered shortly.)

When we look at population data within the United States (see Table 2-6 on page 33), we see obvious variations among the states in

Table 2-2 Percentage of population 65 and over and dependency ratios: selected countries, 1992.

Country	Percentage Age 65 and over	Dependency Ratio		
		Aged	Youth	Total
Nigeria	2	.04	.85	.89
Kenya	2	.04	1.00	1.04
Nicaragua	4	.07	.96	1.03
El Salvador	4	.08	.85	.93
Brazil	5	.02	.58	.60
Taiwan	6	.09	.40	.49
Argentina	9	.15	.49	.64
Georgia[a]	9	.14	.38	.52
Russia[a]	10	.15	.34	.49
Australia	11	.16	.33	.49
Canada	11	.16	.31	.47
Japan	13	.19	.26	.45
United States	13	.20	.34	.54
Netherlands	13	.19	.26	.45
Germany	15	.22	.23	.45
United Kingdom	16	.25	.29	.54
Denmark	16	.24	.25	.49
Sweden	18	.28	.28	.56

[a] Countries that were part of the former U.S.S.R.
Source: Compiled by the author from Population Reference Bureau (1992).

the percentage of older people. For example, the elderly population ranges from 4.1 percent in Alaska, through Utah, next lowest at 8.7 percent, to 18.3 percent in Florida. We discuss this distribution in more detail in a later section of this chapter. For now it is enough to realize that the percentage of older people is an effective statistical tool for looking at population aging cross-nationally and within the United States.

Dependency Ratios

The **aged dependency ratio,** another common measure of population aging, is the ratio of the number of older people in the population to the population in the age categories most closely associated with employment (usually 15 to 64 in an international context). It is designed to give a very rough index of the size of the older population in comparison with the size of the population

that could be expected to pay taxes to support benefits for the older population. Obviously, the aged dependency ratio misclassifies a large number of older people who are employed and a large number of people 15 to 64 who are unemployed or not in the labor force. For this reason, the aged dependency ratio is a much-less-than-perfect indicator of what it was designed to show. Nevertheless, it is widely used, so you need to be familiar with it.

The aged dependency ratio is often used in conjunction with the **youth dependency ratio,** the ratio of the number of youths under 15 to the total number of employable age (15 to 64). Obviously, in the United States, age 15 is too young as a boundary for the economically active population. Here a better age bracket would be 18 to 64. The youth and aged dependency ratios combined form a **total dependency ratio,** or the total "dependent" population (youth and the aged) relative to the "working-age" population.

Table 2-2 shows the aged, youth, and total dependency ratios for selected countries in 1992. Notice that the youth dependency ratios fluctuate much more than the aged dependency ratios. Once again, the United States is among the countries with a high ratio of older people. Notice also that countries with high aged dependency ratios tend to have relatively low youth dependency ratios, which means, at least hypothetically, that those countries with high demand for services for older people should also have lower demand for services for young people.

Much has been written recently about the "burden of dependency" that an aging population brings. However, Table 2-2 makes it absolutely clear that the U.S. population could age considerably (so that it resembled Sweden in its age–sex structure) without the total dependency ratio being altered very much. Indeed, countries with the highest "dependency burden" are those with high youth dependency ratios. Thus, high birthrates, not aging populations, translate into high total dependency ratios. If we are concerned with the dampening effect of economic dependency on economic growth, for example, then we should be much more concerned with countries with high birthrates than countries with aging populations.

Table 2-2 also lets us compare the percentage of the population that is older with the aged dependency ratio as indicators of population aging. Notice that the table is in rank order by proportion of older people in the population. We see that while aged dependency ratios generally parallel the increases in percent aged, there are some exceptions, which are related to abnormal age distributions. The measure selected for use depends mainly on convenience because both are effective. I prefer the percentage of older people in the population because it is more readily available and less subject to misinterpretation. Although the aged

dependency ratio has an appealing logic, because of the misclassification problem mentioned earlier it does not accurately measure economic dependency, and this in turn raises the potential for misinterpretation.

Life Expectancy

Life expectancy, or the average number of years an age category is expected to live given the base-year mortality rates, can be computed for any age category, but life expectancy at birth is the one most commonly used. As an index of aging, life expectancy at birth shows the average length of life a cohort is expected to live given the mortality rates in the year of its birth. For example, in 1990, life expectancy at birth was 75.4 for the entire U.S. population—71.8 for men and 78.8 for women. For white males it was 72.7, and for white females, 79.4. For blacks, male life expectancy at birth was 64.5 compared with 73.6 for females (National Center for Health Statistics 1995).*

Because the mortality rates from which it is computed are among the most widely available population statistics, life expectancy has the advantage of allowing comparisons across a wide range of societies. The main disadvantage is that life expectancy at birth is very sensitive to infant mortality rates, which do not necessarily parallel mortality rates at other ages. As Table 2-3 shows, life expectancy at birth fluctuates much more from country to country than does life expectancy at the age of 65. As a result, life expectancy at birth is a highly imprecise indicator of population aging.

Perhaps the most significant limitation of **life table** statistics as indicators of population aging is that they do not take into account *fertility rates* in a population. For a population to grow older, the number of older people *must* increase at a faster rate than the

---

Table 2.3 shows data for the United States in 1988; the 1990 figures reflect slight increases for both sexes.

Table 2-3 Life expectancy at birth and at age 65, by sex: selected countries, latest available year as of 1990.

| | Life Expectancy | | | |
| | At Birth | | At Age 65 | |
Country	Male	Female	Male	Female
Malawi (1977)	38.2	41.2	10.6	11.4
Rwanda (1978)	45.1	47.7	10.9	11.8
Zambia (1980)	50.4	52.5	11.5	12.1
Nepal (1981)	50.9	48.1	11.0	11.5
Botswana (1981)	52.3	59.7	10.2	12.5
India (1980)	52.5	52.1	11.7	13.2
Guatemala (1980)	55.1	59.4	13.5	14.2
Colombia (1985)	63.4	69.2	14.0	15.8
Ecuador (1985)	63.4	67.6	14.3	15.8
Hungary (1987)	65.7	73.7	12.0	15.2
Ukraine (1986)	65.9	74.5	12.4	15.5
Venezuela (1985)	66.7	72.8	14.2	16.7
Poland (1987)	66.8	75.2	12.3	15.9
Chile (1990)	68.0	75.0	13.8	16.7
United Kingdom (1987)	71.2	77.5	13.3	17.2
Austria (1987)	71.5	78.1	14.2	17.4
U.S.A. (1988)	71.5	78.3	14.9	18.6
France (1987)	72.0	80.3	14.9	19.4
Australia (1986)	72.8	79.1	14.6	18.5
Canada (1986)	73.0	79.8	14.9	19.2
Sweden (1987)	74.1	80.2	15.0	18.9
Japan (1987)	75.6	81.4	16.1	19.7

Source: Compiled by the author from United Nations (1990).

number of children entering the population. If the birthrate is high, there can be a substantial reduction in mortality at older ages yet little change in the proportion of older people. Thus, the most useful measure of population aging must not only indicate the prevalence of older people in the population but also give some indication of the relationship between the size of the older population compared with other age categories. Both the proportion who are 65 and over and the aged dependency ratio satisfy this criterion; life expectancy does not.

How Populations Age

The demographic forces behind population aging are straightforward. As human societies gained the ability to control disease, death was postponed for more and more people. And the forces that allowed control of disease, particularly economic development, also promoted lowering of the birthrate by lowering the economic value and raising the economic cost of having children. The result was a larger number of survivors moving into the older ages and a smaller number of infants entering the population, which of course resulted in an aging population.

Control of disease began with the discovery that clean water and sewage control could reduce the spread of infectious diseases. Disease control was furthered by the development of national systems of transportation, which reduced the effects of local famine, and of sanitary food-processing and storage methods. Finally, the introduction of medical interventions such as antibiotics and immunization ended centuries of terror associated with diseases such as smallpox, typhoid, cholera, polio, plague, and a host of others. In most countries, deathrates went down gradually, followed some time later by a fall in birthrates.

When the birthrate is higher than the deathrate, the population increases. For example, assuming no migration, a population with a birthrate of 37 per 1,000 per year and a deathrate of 9 per 1,000 per year has an annual rate of natural increase of (37 – 9)/1,000, or 2.8 percent. If this percentage does not seem like much, remember that populations increase geometrically (2, 4, 8, 16), so that a population increasing at 2.8 percent per year would *double* its size in just 25 years! Many Third World countries today are growing this fast or faster. For example, in 1994 most of Central America was growing at 3 percent per year or more (Population Reference Bureau 1994).

The falling deathrates that began in Europe in the late 1600s produced extremely high population increases until birthrates also began falling. Today, in most countries of Europe, Japan, Canada, and the United States, population growth is low because deathrates and birthrates are both low.

Low deathrates mean that more people survive to enter old age; combined with low birthrates, this fact means that the number of children born into a population is not substantially different from the number of people dying. For example, in 1991 Germany had a birthrate of 11 and a deathrate of 11 (Population Reference Bureau 1992). The eventual result (assuming no migration), if birthrates and deathrates continue to be equal, would be a rectangular age structure such as Sweden's (see Figure 2-2), whose population is quite evenly distributed across the various age categories. By contrast, high birthrates and deathrates produce an age structure like Afghanistan's (Figure 2-2), in which a very large proportion of the population is under 15 and a very small proportion survives to 65. Thus, we say that Sweden has a relatively "old" population and Afghanistan a relatively "young" one.

Population aging can also occur *within* an older population, usually the result of lowered deathrates at the older ages, but possibly also the result of the entry of a relatively small birth cohort into the older population. For example, during the depression years of the 1930s, birthrates in the United States were lower than in the preceding decades. In 1995, when that cohort begins to enter the older population, the average age of the older population will go up even if deathrates and the proportion of older people in the population remain the same.

Growth of the Older Population

When the first U.S. Census was taken in 1790, about 50,000, or 2 percent, of the 2.5 million Americans were 65 or older. One hundred years later, the older population had grown to 2.4 million and composed just less than 4 percent of the population. As Figure 2-3 shows, from 1890 to 1920 the older population grew relatively slowly, but from 1920 on the rate of increase accelerated, so that by 1990 there were 31.1 million older Americans—12.5 percent of the population.

Figure 2-3 also includes projections of the older population to the year 2050; these projections were made under conservative as-

Figure 2-3 Older population (in millions): United States, 1890–2050 (actual and projected).

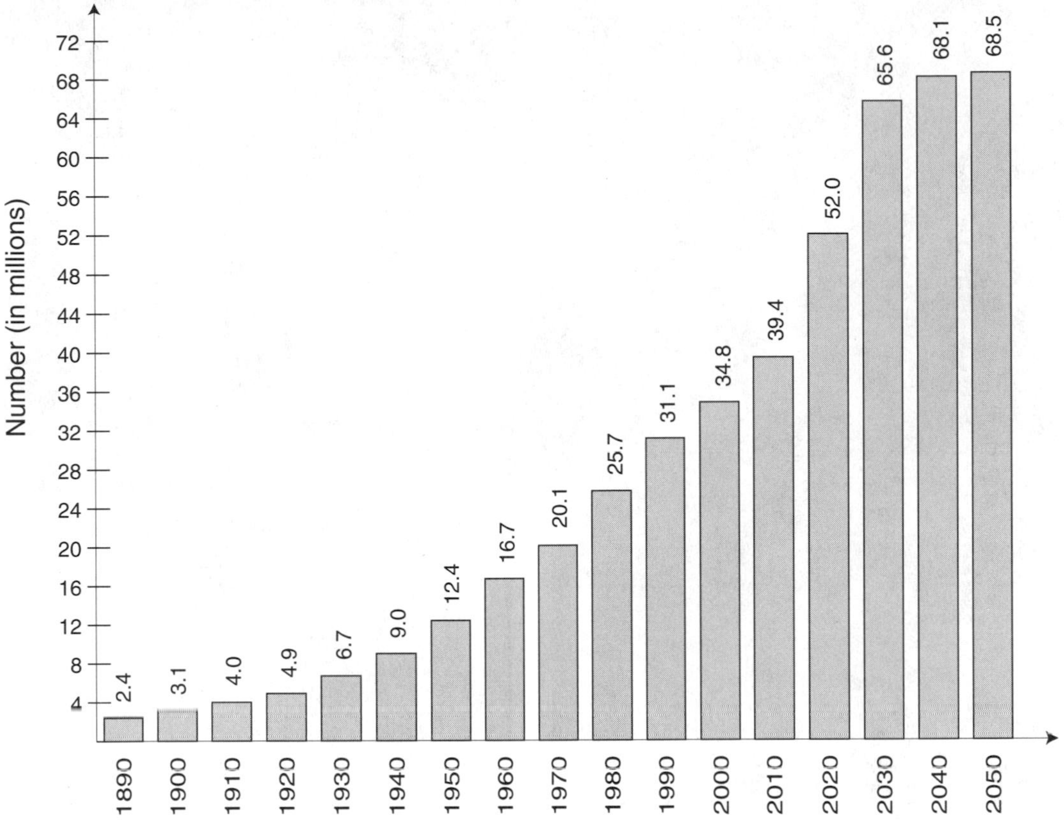

Source: U.S. Bureau of the Census (1989a, 1992g).

sumptions that birthrates would remain low and deathrates would decline modestly. The projections also assumed a small net out-migration of older people, mostly foreign-born elders moving back to their country of origin. Given that mortality rates have consistently fallen faster than projected, the projections in Figure 2-3 may underestimate the number of older people in the future. Of course, assumptions about the birthrate affect projections of the older population only slightly because nearly everyone who will be part of the older population in 2050 has already been born. The 68.5 million older people projected for 2050 represent 22.9 percent of the total population projected for that year.

Composition of the Older Population

Members of the public often think of the older population as a relatively homogeneous category, but nothing is further from the truth. The older population can be categorized and compared across a number of important dimensions, including age, gender, marital status, education, and income. Each of these dimensions shows a wide range of variation among older people. Differences in the composition of the older population are also discussed in more detail in later chapters, especially health (Chapter 4), income

The older population is very diverse.
Photograph by Harvey Stein/Black Star

(Chapter 13), living arrangements (Chapter 13), and race and ethnicity (Chapter 17). The older population is *very diverse* and becoming more so.

Age and Sex

The older population is itself growing older. And as it does, older women increasingly outnumber older men. If we divide the older population into the *young-old* (65 to 74), the *middle-old* (75 to 84), and the *old-old* (85 and over), the effects of aging within the older population can be seen more clearly. As Table 2-4 shows, in 1960 nearly two-thirds of the older population was under 75, but by the year 2000, the young-old will have dropped to 50.5 percent of the older population. At the same time, the old-old will have increased from 6 percent of the older population in 1960

to 14.7 percent in 2000. In absolute numbers, the entire older population will more than double over the forty-year period, while the number of old-old will nearly quintuple.

The **sex ratio** is the number of males per 100 females. Aging of the older population itself will cause the overall sex ratio in the older population to drop from 82 men per 100 women in 1960 to only 65 men per 100 women in 2000 (see U.S. Bureau of the Census 1992g, 1989a, 1977). In the old-old population, the drop in sex ratio is especially dramatic, from 67 men per 100 women in the 85-and-over age category in 1960 to only 38 men per 100 women in that age category in the year 2000. The reason behind this shift in the sex ratio is the well-known difference in mortality rates for men and women. As Figure 2-4 indicates, women

Table 2-4 Changing age composition of the older population: United States, 1960–2000.

Age Category	1960	1970	1980	1990	2000[a]
Young-old					
(65–74)N[b]	11.0	12.4	15.6	18.0	17.7
Percent[c]	66.2	62.3	60.8	57.9	50.5
Middle-old					
(75–84)N	4.6	6.1	7.8	10.1	12.3
Percent	27.7	30.7	30.3	32.5	34.9
Old-old					
(85+)N	1.6	1.4	2.3	3.0	4.9
Percent	6.0	7.0	8.8	9.6	14.7

[a] Projected
[b] Number (in millions)
[c] Percent of the older population
Source: Compiled by the author from U.S. Bureau of the Census (1992g, 1989a, 1977).

Figure 2-4 Probability of dying during selected age intervals: United States, 1988.

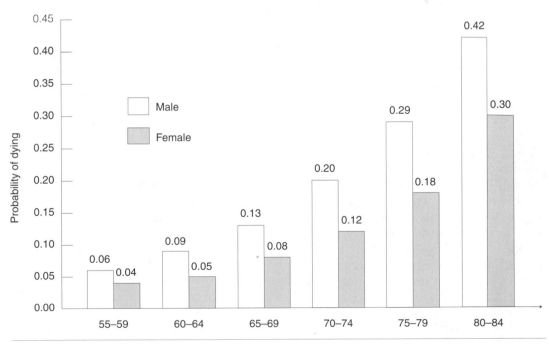

Source: Compiled by the author from National Center for Health Statistics (1991b).

have a substantially lower probability of dying at every age. The underlying reason for this difference is unknown, but our best guess is that about half is caused by gender differences in environment and social roles and about half by genetic sex differences (Palmore 1980b).

Marital Status

Three-fourths of older men are married, and more than half of older women are widowed. Even among those 75 and over, nearly 70 percent of men have spouses living with them. On the other hand, at age 75 and over, two-thirds of women are widows. Being widowed is more common among African-American older people than among whites (see Figure 2-5). These data have far-reaching implications for social support—advice, approval, and caregiving. Most older men have spouses to rely on, but most older women must find other sources of social support. However, as age increases, the proportion of widowers among men increases substantially, so by age 85 and over nearly half of men are widowed.

Education

The median number of years of school completed by people age 25 and over in 1991 was 12.8. The figure for the total older population of 12.2 was not substantially different. However, as Table 2-5 shows, there was considerable educational diversity within the older population connected to age, gender, racial, and ethnic differences in educational attainment. For example, age differences in median number of years of school completed are much greater for African American men than for white or Hispanic American men. (Read down the first column of Table 2-5.) Age differences in median education were smaller for women than for men in all racial and ethnic categories (compare column 1 with column 2 in Table 2-5).

Through a type of analysis called *cohort analysis,* we can look at how changes in educational systems over time have affected specific age cohorts differently. For example, the

cohort in Table 2-5 that was age 65 to 69 in 1991 had higher median years of school completed and a larger proportion completing high school compared to the cohorts that were 70 to 74 or 75 and over. This is the effect of the spread over time of compulsory public education. However, if we look at the proportion completing college, we see that more recent cohorts are better educated among men but not among women. Thus, this cohort effect was concentrated mainly among men. If we look at the proportions completing high school, we can see that more recent cohorts have made substantial gains among both African Americans and whites but not among Hispanic Americans. Thus, the spread of public education did not affect age cohorts uniformly; rather, these effects were mediated by race, ethnicity, and gender.

Education affects employability, access to information about political issues, knowledge of benefits and services available, and a host of other factors that influence quality of life. Although educational differences are smaller for the younger age cohorts within the older population compared with the older cohorts, the differences are still substantial, especially with respect to the proportion of high school and college graduates.

Income

Although most older people have an adequate income, a substantial minority does not. In addition, distribution of income is skewed very much toward the low end of the income range, especially compared with the middle-aged population. For example, Figure 2-6 shows the percentage distribution of money income in 1990 for individuals in two age categories: 45 to 54 and 65 and over. The differences are striking. Income was distributed reasonably evenly across the range of income categories in the middle-aged, 45 to 54 population, whereas 68.3 percent of the older population was concentrated in the four income categories under $20,000 per year.

We cover income in more detail in Chapter 13, "Income and Housing," but the

Figure 2-5 Widowhood of people 55 and over, by race and gender: United States, 1991.

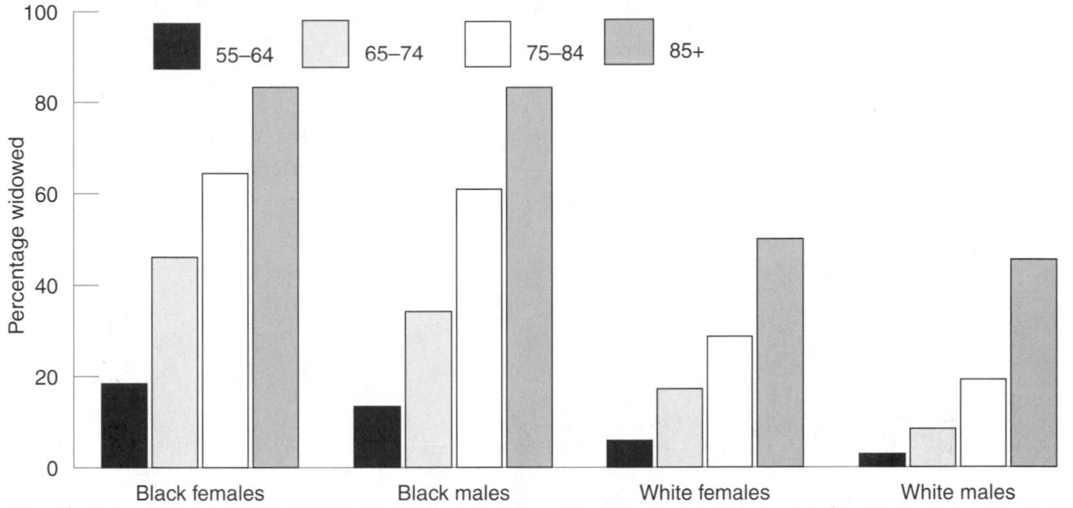

Source: U.S. Bureau of the Census (1992d).

Table 2-5 Educational attainment, by age and sex, for total, white, black, and Hispanic populations: United States, 1991.

Age	Median Years of School Completed		Percent High School Graduates		Percent College Graduates	
	Male	Female	Male	Female	Male	Female
Total						
65–69	12.4	12.4	63.8	67 1	19.0	9.4
70–74	12.3	12.3	60.0	61.7	15.4	10.7
75+	11.3	12.0	47.3	50.1	12.8	9.0
White						
65–69	12.5	12.4	66.3	70.1	19.9	9.5
70–74	12.4	12.3	63.0	64.7	16.4	10.8
75+	12.0	12.1	50.3	53.2	13.7	9.4
Black						
65–69	10.2	10.5	36.9	39.0	7.7	6.3
70–74	8.6	10.0	25.8	33.0	4.1	4.8
75+	6.1	8.2	13.6	19.9	3.3	4.6
Hispanic						
65–69	8.2	8.1	28.8	27.2	8.4	5.0
70–74	7.7	8.0	24.6	27.8	6.2	5.9
75+	7.4	7.2	26.9	22.8	5.8	5.5
Total population						
Age 25+	12.8	12.7	78.5	78.3	24.3	I8.8

Source: Compiled by the author from U.S. Bureau of the Census (1992b).

Figure 2-6 Distribution of household income, by selected ages of householder: United States, 1990.

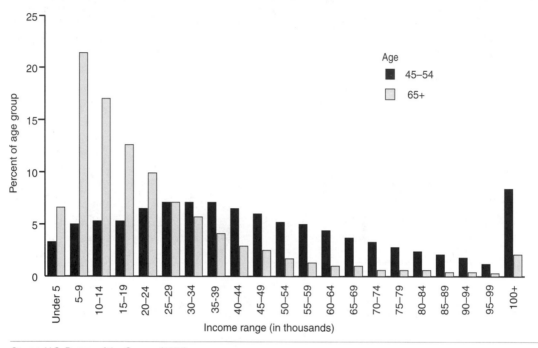

Source: U.S. Bureau of the Census (1991).

data in Figure 2-6 show clearly that the common image portrayed in the media that the older population consists mainly of people who are well-off economically is incorrect. Most older people have incomes that are just barely adequate, and older people are much more vulnerable to poverty than the middle-aged population.

Implications

The older population forms a very heterogeneous category. It varies widely on all the demographic characteristics we have examined so far: age, sex ratio, marital status, education, and income. This variability means that planning to meet the needs of this heterogeneous population is a difficult task. Heterogeneity stands in the way of legislators, administrators, and the public when they try to get a picture of "older Americans." On the other hand, this diversity of the older popu-

lation has advantages. Since many older people have adequate educational and financial resources, their need for benefits and services is minimal, which eases the financial and administrative job involved in assisting those in need. We return to this theme in Chapter 19, "Politics and Government."

Geographic Distribution of the Older Population

The geographic distribution of the older population in the United States is similar to that of the general population. The most populous states—California, New York, Texas, Pennsylvania, Illinois, Ohio, Florida, Michigan, and New Jersey—are also the states with the largest populations of older people (see Table 2-6). The concentration of older people in the population is another matter because it depends on the size of the

Table 2-6 Rank, number, and percentage of each state's total population that is age 65 and over: 1990.

Rank,[a] State		Population Age 65+	Percent Age 65+
United States		31,241,831	12.6
1	Florida	2,369,431	18.3
2	Pennsylvania	1,829,106	15.4
3	Iowa	426,106	15.3
4	Rhode Island	150,547	15.0
4	West Virginia	268,897	15.0
6	Arkansas	350,058	14.9
7	South Dakota	102,331	14.7
8	North Dakota	91,055	14.3
9	Nebraska	223,068	14.1
10	Missouri	717,681	14.0
11	Kansas	342,571	13.8
11	Oregon	391,324	13.8
13	Massachusetts	819,284	13.6
13	Connecticut	445,907	13.6
15	Oklahoma	424,213	13.5
16	New Jersey	1,032,025	13.4
17	Montana	106,497	13.3
17	Wisconsin	651,221	13.3
17	Maine	163,373	13.3
20	New York	2,363,722	13.1
20	Arizona	478,774	13.1
22	Ohio	1,406,961	13.0
23	Alabama	522,989	12.9
24	District of Columbia	77,847	12.8
25	Tennessee	618,818	12.7
25	Kentucky	466,845	12.7
27	Illinois	1,436,545	12.6
27	Indiana	696,196	12.6
29	Minnesota	546,934	12.5
29	Mississippi	321,284	12.5
31	North Carolina	804,341	12.1
31	Delaware	80,735	12.1
33	Idaho	121,265	12.0
34	Michigan	1,108,461	11.9
35	Washington	575,288	11.8
35	Vermont	66,163	11.8
37	South Carolina	396,935	11.4
38	Hawaii	125,005	11.3
38	New Hampshire	125,029	11.3
40	Louisiana	468,991	11.1
41	Maryland	517,482	10.8
41	New Mexico	163,062	10.8
43	Virginia	664,470	10.7
44	Nevada	127,631	10.6
45	California	3,135,552	10.5
46	Wyoming	47,195	10.4
47	Texas	1,716,576	10.1
48	Georgia	654,270	10.1
49	Colorado	329,443	10.0
50	Utah	149,958	8.7
51	Alaska	22,369	4.1

[a] Rank is by percentage age 65 and over.
Source: Taeuber, 1992.

older population compared with that of other age categories. As Table 2-6 shows, high concentrations of older people occur primarily in Florida, the Northeast, and the Midwest. Extremely low concentrations occur only in Utah and Alaska. As we will see shortly when we consider migration, variability in the proportion of older people in the various states has been caused mainly by two types of migration, in-migration of older people to the Sunbelt and out-migration of younger people from economically disadvantaged areas in the Midwest and Northeast.

Locally, older people are slightly less likely than the general population to live in metropolitan areas, and those who do are more likely to live in the suburbs. In 1980, for the first time more older people lived in suburbs than in central cities. Older people tend to live in older suburbs that were established before 1960 (U.S. Senate Special Committee on Aging 1986).

In nonmetropolitan areas, older people are more likely to live in small towns. Nearly 20 percent of older people live in counties with no town with a population of over 25,000 and no city with a population of 50,000 or more in any adjacent county. This means that a large minority of older Americans live in areas that are likely to lack extensive facilities and services. Another 20 percent of older people live in inner-city neighborhoods where fear of crime and urban decay present major problems.

Living Arrangements

Of the older population, 94 percent live in ordinary community households, and the other 6 percent mainly in nursing homes, homes for the aged, and other group quarters. Older people represented about 88 percent of nursing home residents in 1985 (National Center for Health Statistics 1989b). A very small percentage (much less than 1 percent) live in mental hospitals, tuberculosis

hospitals, prisons, and institutions for the developmentally handicapped.

Among older people who live in community settings, the vast majority either live alone or with a spouse (see Figure 2-7). The proportion living with relatives other than a spouse is higher in older age cohorts, increasing from about 23 percent at age 65 to 79, to 28 percent at 80 to 84, to 34 percent at 85 to 89, to 50 percent at 90 and older (Coward, Cutler, and Schmidt 1989). More than 80 percent of older adults who live with relatives other than their spouse live with adult offspring, and living with adult children does not exceed 20 percent until age 85 or older. Coresidence is covered in detail in Chapter 8, "Family, Friends, and Social Support."

As a result of the mortality differences discussed earlier, older women are much more likely than older men to be widowed and living alone. Among those 75 and older, nearly 80 percent of older men live in a family setting, compared with only 47 percent of older women (see Figure 2-7).

Some people assume that older people are forced to live alone because their families do not want them, but research on this issue has shown that older people prefer the independence of living alone as long as they are financially and physically able to do so. In addition, when older people need assistance, family members are by far the most likely to provide it. For example, Mehdizadeh and Atchley (1992) found that in Ohio, family provided about 67 percent of the care being given to disabled elders, with the other 33 percent coming from nursing homes and community agencies that provide home-based assistance.

In 1990, about 6 percent of the older population lived in nursing homes or homes for the aged—4 percent in nursing homes and 2 percent in homes for the aged and other group quarters such as board-and-care homes. As might be expected, the percentage of older people living in long-term care facilities and homes for the aged increases with

Figure 2-7 Living arrangements of elders: 1991.

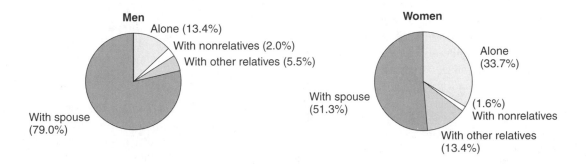

Age 65 to 74

Men
Alone (13.4%)
With nonrelatives (2.0%)
With other relatives (5.5%)
With spouse (79.0%)

Women
Alone (33.7%)
With spouse (51.3%)
(1.6%) With nonrelatives
With other relatives (13.4%)

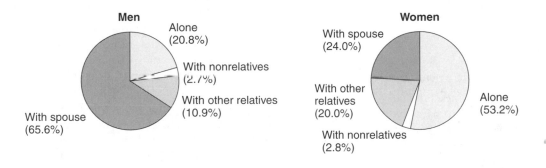

Age 75 and over

Men
Alone (20.8%)
With nonrelatives (2.7%)
With other relatives (10.9%)
With spouse (65.6%)

Women
With spouse (24.0%)
With other relatives (20.0%)
Alone (53.2%)
With nonrelatives (2.8%)

Source: U.S. Bureau of the Census (1992d).

age, from 2 percent at age 65 to 74, to 8 percent at 75 to 84, to 25 percent at 85 or older (Longino 1986).

Manton and Soldo (1985) projected that the older population living in long-term care facilities would increase from 1.2 million in 1985 to 2.2 million by the year 2000, mainly due to the aging of the older population. Since the 85-and-older age category is growing fastest within the older population and rates of institutionalization are highest for that category, even with greater availability of community care alternatives, there will be a dramatic increase in the need for space in long-term care facilities for the old-old.

Residences that provide long-term care come in more forms each year. The old dichotomy of nursing home versus community residence has long-since lost significance. Today, we have in-home care, adult day care, assisted living apartments, small group homes, nursing facilities, and continuing care retirement communities, all providing long-term care. This topic is covered in more detail in Chapter 13, "Income and Housing," and Chapter 14, "Health Care and Long-Term Care."

Population Processes and the Older Population

The processes that change any population include fertility, mortality, and migration. All these affect the older population too, but the effects are often different from the effects on the general population. Discussing all the possible interrelationships among these forces and population size, composition, and distribution is well beyond the scope of this book. Instead, we will look at only some of the more obvious and significant relationships. We first consider the influence of past fluctuations in birthrates on the age and sex composition of the older population. We then look at changes that are occurring in deathrates at the older ages and the implications of these changes. Finally, we consider migration and its influence on the geographic distribution and concentration of the older population.

Fluctuating Birthrates

Most Americans are familiar with the "baby boom"—the period of very high birthrates that occurred in the United States between 1947 and 1967. What most people do not realize is that birthrates have fluctuated periodically over the nation's entire history. Let us look at the relative size of several birth cohorts: a relatively small cohort born in the depression years of 1931 to 1940, a large 1956 to 1965 cohort born within the baby boom, and a relatively small "baby bust" cohort born between 1971 and 1980. figure 2-8 shows six population pyramids that follow these cohorts from 1990 to the year 2060.*

The first thing worth noting about Figure 2-8 is that the depression era cohort (people between 50 and 59 in 1990) is smaller than

---

*These population pyramids show actual population numbers instead of percentages. Although percentages are generally more useful, actual numbers are better for the kinds of comparisons we want to make here.

the cohorts either before or after it. This was probably advantageous to the members of the cohort as they grew up because they had fewer age peers to compete with for opportunities in education and employment. And when this cohort retires, probably between 1990 and 2005, the baby boom cohort will be in the labor force to help provide retirement income. The depression cohort will also have less competition for scarce services to older people than the cohorts just ahead of it. As pointed out earlier, when the depression cohort enters the older population, the average age of the older population will increase because of the relative smallness of the depression cohort.

In contrast, the baby boom cohort (people between 25 and 34 in 1990) will have experienced more competition for opportunities in education and employment than the cohorts either before or after it. However, because it is large, the pressure on the baby boom cohort to provide retirement income for the small depression era cohort will be lower than for most cohorts, other things being equal. Yet when the baby boom cohort retires, probably between 2015 and 2030, it would have had to rely on the relatively small baby bust cohort to help provide retirement income. That is exactly the reasoning Congress used in 1983 when the retirement age under Social Security was designated to be gradually raised to 67 beginning in the year 2000. This legislation also created the plan to amass a very large reserve in the Social Security trust fund that could be used to help fund the baby boom cohort's retirement and thus not overburden the baby bust cohort. Also, when the baby boom cohort enters its later years, the average age of the older population will drop.

The baby bust cohort (people between 10 and 19 in 1990) will also have less competition among age peers for opportunities, especially in comparison with the baby boom cohort. However, it faces stiff competition for

Figure 2-8 Projected population (in millions), by age and sex: United States, 1990–2060.

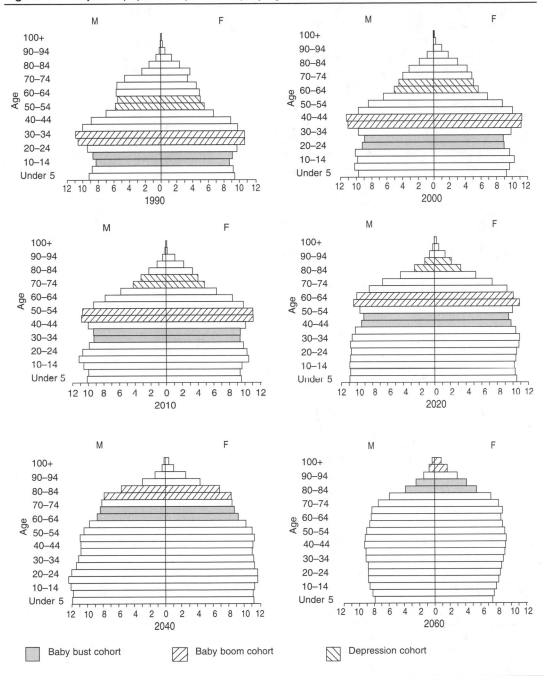

Source: U.S. Bureau of the Census (1992a, 1992g).

jobs from the large baby boom cohorts just ahead of it and it may be taxed more to provide for the baby boom cohorts' old age. And when the baby bust cohort itself reaches 65, the large baby boom cohort will be competing with them for facilities and services for older people. Obviously, both the size and age structure of the older population will be affected by both the depression and the baby boom cohorts. Thus, the effects of fluctuating fertility are an important force in the composition of the older population and these effects last a long time.

Changes in Mortality

Mortality rates in the United States are already quite low and still falling. For example, from 1970 to 1991, the age-adjusted deathrate fell from 7.1 to 5.1 per 1,000 population per year (U.S. Bureau of the Census 1994). Age-adjusted deathrates are computed using a standardized age distribution in order to measure the effect of changes in mortality holding changes in age distribution constant. Further declines in mortality are expected. For example, the projections in Table 2-7 show sizable and steady increases in the concentration of mortality in the older ages, especially 85 and older.

Fries (1980) estimated that by the year 2045, we will have reached the maximum possible survival curve, at which point 95 percent of all deaths will occur between the ages of 77 and 93, with the average age of death (for those who do not die from trauma such as accidents or violence) approaching 85. Fries's analysis is controversial, particularly his idea that the age of 85 is a significant limit to the average length of life. Rothenberg, Lentzner, and Parker (1991) looked at patterns of mortality in the United States from 1962 to 1984 and concluded that the general shape of the survival curve did not change; instead it was merely pushed upward by significant delays in mortality, which of course goes against the squaring off of the survival curve predicted by Fries. Similar

Table 2-7 Actual and projected percentage of all deaths occurring at 65 and over and at 85 and over: United States, 1960–2080.

YEAR	Age 65+	Age 85+
1960	59.0	10.8
1965	60.8	11.9
1970	61.5	12.9
1975	64.3	15.1
1980	67.2	17.9
1985	70.5	20.1
1990	73.1	22.1
1995	74.7	24.2
2000	74.8	26.8
2005	74.4	28.9
2010	73.9	30.5
2020	76.4	29.1
2030	81.8	28.4
2040	84.2	35.9
2050	84.9	43.5
2080	86.3	46.2

Source: U.S. Bureau of the Census (1989a).

findings were reported earlier by Manton (1986, 1988).

As more progress is made in preventing premature death from heart disease and related disorders of the circulatory system, more of the older population will live out their lives in better health, especially in the later years. On the other hand, we will also have larger numbers of people reaching the point of frailty associated with advanced old age. These old-old people will require more assistance even in their own homes compared with the relatively young older population of today. These concerns with not only the length of life but its quality led to the development of the concept **active life expectancy,** the number of years of functional well-being that people could expect given current age-specific rates of disability (Rogers, Rogers, and Belanger 1989). At this point it appears that declining mortality rates at the older ages apply primarily to relatively healthy people who are less likely to become disabled, thereby increasing active life expectancy. However, there will continue to be a significant minority of elders surviving in a highly disabled state. We discuss this phenomenon in more detail in Chapter 4, "Physical Aging and Health."

Migration

The age pattern of migration closely parallels early adulthood changes in life circumstances such as leaving parents' households, graduation from school, getting a job, and getting married. Figure 2-9 shows that a large majority of the residential moves occurring between March 1989 and March 1990 involved people age 20 to 34 and their young children. These people were resolving issues of young adulthood. The proportion moving declined sharply from ages 30 to 45 and mobility continued to decline more slowly after that. For men, there was an increase in overall mobility in the 50s, but these were mainly local

Figure 2-9 Total, intracounty, and interstate residential mobility, by sex and age: United States, 1989–1990.

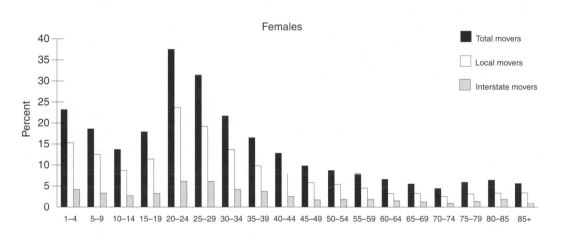

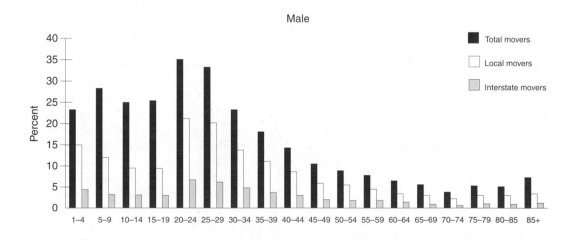

Source: Compiled by the author from data collected by the U.S. Bureau of the Census (1991c).

moves. Note that there is no reflection in these one-year mobility data of increased mobility connected to the usual retirement age of 65. However, there was an increase in both local (intracounty) and interstate residential mobility after age 75 for both men and women.*

Earlier data on residential mobility showed more migration of older people than the data shown in Figure 2-9. For example, Rogers (1988), reporting on mobility from 1975 to 1980, found much higher annual interstate migration at around age 65. Golant (1990) reported that residential mobility of elders declined considerably for the period from 1980 to 1985, compared with the 1975 to 1980 period. He offered several possible explanations for this decline. First, the mortgage interest situation in the early 1980s made residential mobility more difficult. In the late 1980s the collapse of real estate values in many parts of the country substantially eroded the return older people could get by selling their current housing, another factor that may have served to inhibit mobility. Second, the numerical size of the young-old cohorts (who were the most likely to be making interstate moves) declined in the Northeast and Midwest, which were the main areas sending interstate migrants. Third, Sunbelt areas that had attracted migrants earlier came to be seen as congested, overpriced, and vulnerable to crime. Finally, developers in the Midwest and Northeast began to offer retirement housing options that represented intervening opportunities for elders who wanted the convenience of living in a retirement community without having to sever long-standing community or regional ties.

On this latter point, for example, Cuba (1991) found that previous experience with

vacation living on Cape Cod led many Massachusetts older people to choose Cape Cod over other possible retirement destinations. He also suggested that kinship ties and experience with nearby retirement locations could combine to restrict the geographic distance over which the search for a retirement destination might take place. As more retirement communities were developed in the Midwest and Northeast, there may well have been a drop-off of interest in even considering more distant locations.

Clark and Davies (1990) reported that there was also a category of older people, particularly older African Americans, who were "trapped" in inner-city housing. For example, some elders who bought homes in inner-city areas of Detroit, Chicago, Milwaukee, and other areas of the Midwest, because of crime in the neighborhood, could not sell them at any price. For these people, mobility was not an option. Burkhauser et al. (1995) found that older homeowners in distressed neighborhoods were three times less likely to leave these neighborhoods compared with younger homeowners. Based on all of these factors, Golant (1990) suggested that the decline in interstate migration of elders might be a long-term trend.

To help explain patterns of elders' residential mobility in the United States, Litwak and Longino (1987) developed a model of migration based on various life course pressures and opportunities. In their model, the *first move* occurs primarily among married older people in their 60s who are relatively healthy and have adequate retirement incomes. Retirement frees these people from having to live near their places of employment. Reasons for moving at this stage are complex and may involve perceptions of negative features of the place of origin, especially crime (Serow 1987), as well as attractive amenities such as climate, facilities, and services at the place of destination (Meyer 1987). Maintaining friendship networks may also be involved if many of the couples' friends have

---

*This pattern holds mainly for whites. Pitcher, Stinner, and Toney (1985) found that there was very little migration among older blacks after age 50 and no age differences in migration rates among older blacks.

already moved. In this relatively young category of older people, kinship ties often can be maintained at a distance through periodic visits and telephone contacts.

Although Litwak and Longino did not mention them, older seasonal migrants also fit into this first move pattern. Sometimes called "snowbirds," older seasonal migrants often have two residences, living in their community of origin during the warmer months and moving to a residence in the Sunbelt during the cold months. Snowbirds are also likely to be married and healthy and to have above-average retirement incomes. In April 1980, the U.S. Bureau of the Census enumerated over 500,000 older people who listed themselves as nonpermanent residents (Hogan 1987). Two-thirds of these people were concentrated in Florida (250,000), California (25,000), Arizona (39,000), and Texas (25,000). About 20 percent of snowbirds are from Canada, and the remainder come from the northern half of the United States (Sullivan and Stevens 1982).

In a representative survey of older people in Minnesota, conducted in the summer of 1988, Hogan and Steinnes (1994) found that just over 9 percent headed south for a major part of the previous winter, mostly to Arizona, California, Florida, and Texas. They began leaving in November, but the largest proportions were absent from Minnesota during the months of January, February, and March. Compared with elders who stayed home in Minnesota, the seasonal migrants were more likely to have high incomes, college educations, and good health, and to be married and retired. Thus, they fit the pattern described for Litwak and Longino's first move.

According to Litwak and Longino (1987), a *second move* is likely to occur when older people develop chronic disabilities that create difficulty in carrying out everyday household tasks, such as shopping, cooking, cleaning, and taking care of finances. If there is a spouse to help with these tasks, a move may be delayed, but once widowed, older people coping with disabilities often move to be closer to kin to get the help they need. Since kin can be located anywhere, some proportion of second moves could be expected to be interstate moves.

A *third move*, to an institution, most often occurs in response to severe disability or lack of kin. Most disabling chronic conditions become progressively worse and, in a small proportion of older people, reach the point at which care at home is no longer feasible. Most third moves are local rather than long-distance, but about 10 percent of return migration to the Northeast from the Sunbelt is to nursing homes, probably nearer to family who can arrange placement, visit, and monitor the quality of care.

The research findings generally show that retirement migration is less common than might be supposed. Speare and Meyer (1988) found that only 10 percent of older migrants in the 1983 Annual Housing Survey cited retirement or amenities as the reason for their move. Henretta (1986) found that retirement migration occurs very close to retirement, which suggests that retirement migration is an acting out of long-standing plans.

Migration in later life is usually the result of a long-term process of consideration rather than an immediate response to an event such as retirement. Haas and Serow (1993) found that most older interstate migrants to North Carolina had thought about moving for some years. Factors that predisposed them to want to live somewhere else were primarily climate, urban problems such as crime, congestion and pollution, and cost of living factors. Criteria used to select their destination stressed scenic beauty, mild climate, recreational and cultural opportunities, and lower cost of living. Most had visited the area, read literature, and talked with friends in the area before committing themselves to the move. Most considered only two or three possible destinations for their later-life migration.

The prevalence of Litwak and Longino's second category of moves connected with increased moderate disability can be inferred from the increase with age in moves to be nearer kin and moves tied to widowhood. Speare and Meyer (1988) found that kinship migration steadily increased as a reason for mobility with each successive age cohort, with nearly 28 percent of migrants 75 or older citing kinship as a reason for migration. As expected, widowhood caused more moves in older cohorts, but the effect was small (see Table 2-8). However, Longino et al. (1991) looked at the relationship between functional disability and residential mobility and found that among elders disability doubled the probability of moving; this provides strong evidence for Litwak and Longino's second move.

Evidence on moves into institutions shows that many such moves occur after a lengthy period of home care. Litwak and Longino (1987) found that among older migrants from Florida to New York, 20 percent had moved in with an adult child and 12 percent had moved into a nursing home in New York. Among older migrants from Florida to Ohio and Pennsylvania, about 13 percent had moved into institutions and 11 to 12 percent had moved in with adult children.

In all, these data show that the types of moves Litwak and Longino (1987) hypothesized do in fact account for a substantial amount of residential mobility among older people. However, the evidence also suggests that there are other types of moves, such as moves to smaller housing units in the local area, not included in Litwak and Longino's model. For example, most respondents in Speare and Meyer's (1988) study gave reasons for noninstitutional migration that did not fit into Litwak and Longino's categories. More research on this topic is needed.

Migration of older people has produced major changes in the concentration of older people in the various states. For example, Florida's general population ranks fourth in size (up from seventh in 1980), but its older population is second in the nation, largely due to migration of older people to Florida. Rogers and Watkins (1987) reported that migration to Florida, California, and Arizona has had a remarkably dominant influence on the redistribution of America's elders.

From 1965 to 1970, the major migration streams involved older people who were moving from northern states into Florida and California. By the 1975 to 1980 period, Arizona was moving up on California as a destination for older interstate migrants, and Texas was receiving significant numbers of older migrants from the Plains states (Biggar 1984).

Apart from Florida, on the East Coast, from 1960 to 1980, North Carolina moved from twenty-seventh to the seventh most popular destination for older migrants (Biggar 1984). On the Gulf Coast, Texas ranked fourth among Sunbelt states in attracting older migrants. In the West,

Table 2-8   Percentage engaging in different types of migration, by age: United States, 1983.

Age	Types of moves					
	Amenity	Retirement	Kinship	Widowhood	Other	Total
45–54	1.4	1.1	7.3	0.2	89.9	100.0
55–64	4.5	5.9	15.1	1.4	73.1	100.0
65–74	4.1	5.5	23.4	2.3	64.6	100.0
75+	5.4	3.5	27.8	3.5	59.4	100.0

Source: Adapted from Speare and Meyer (1988).

Table 2-9 Net income gained or lost in 1979 due to migration of people age 60 and over, selected states.

Top Ten States in Net Income Gained	Net Gain (in millions)
Florida	$3,468
Arizona	692
Texas	388
North Carolina	170
Arkansas	168
Oregon	127
South Carolina	117
Nevada	116
Georgia	115
Washington	113

Top Ten States in Net Income Lost	Net Loss (in millions)
New York	$1,956
Illinois	838
Ohio	500
Michigan	471
Pennsylvania	387
New Jersey	351
Massachusetts	228
Indiana	161
Maryland	160
Connecticut	150

Source: Adapted by the author from Longino and Crown (1990).

California and Arizona were the states that received major streams of older migrants.

The older people who moved into Florida and Arizona tended to be relatively young, to be members of a couple, to live independently, and to have relatively high Social Security incomes. By contrast, older migrants to California and Texas were much more likely to be over 75, to live with their adult children or in an institution, and to have low Social Security incomes. This pattern suggests that older migrants to Florida and Arizona were predominantly young-old couples who resembled people on extended vacations (Longino 1981); they moved to enjoy a leisure lifestyle and they had the economic resources to do so. On the other hand, a large minority of migrants to California and Texas were older, poorer, single individuals who needed assistance by living either with their children or in an institution. Of the Sunbelt states, Texas and California have received large streams of migration among people of all ages; thus part of the migration of older adults to these states probably consisted of moves older people made to be nearer adult children who could provide caregiving or arrange for and monitor the provision of long-term care.

Biggar (1984) found that out-migration of older adults from Florida tended to be "return" migration to the states from which they originally moved. She suggested that these streams consisted mainly of people who migrated to Florida in their young-old years and were returning to their states of origin in response to the death of a spouse or to be nearer family to cope with health or financial problems. This is consistent with Litwak and Longino's (1987) second type of move.

The effects of these trends are not trivial. The northeastern states have sent some of their ablest and most economically independent older people to Florida, but those who are moving from Florida to the Northeast are older, poorer, and needier.

The economic impact of elder migration can be substantial. For example, Crown (1988) estimated that Florida gained more than $1 billion in income from the net migration exchange of older people from New York State alone. Happel et al. (1988) found that snowbirds spent nearly $500 million in the Phoenix area in the winter of 1986 to 1987. In a more ambitious study, Longino and Crown (1990) used data on consumer expenditures and migration to estimate the net gains and losses to states in 1979 as a result of the migration of people age 60 and older. Not surprisingly, they found that Florida topped the list of net gainers with an annual gain of nearly $3.5 billion. Arizona was a distant second with a net gain of $692 million (see Table 2-9). At the other extreme, New

York was the big net loser, with nearly $2 billion in net loss due to migration of elders. Illinois, Ohio, and Michigan were also heavy net losers.

These bits and pieces suggest that the overall economic effects on both states of origin (as losses) and states of destination (as gains) are enormous, and these effects may well be exacerbated in times of economic recession, when economic growth cannot offset as well the losses from migration of elders. Of course, if migration of elders is in fact entering a long-term decline, as discussed earlier, then the economic effects of migration would decline accordingly. This is another subject in need of research.

Summary

This chapter covers a lot of ground, including how to measure age structure, how populations age, and a detailed portrait of the older population in the United States. It also addresses how population aging is influenced by fertility, mortality, and migration.

Age structure can be measured in a number of ways: by simple percentage distributions of the population, classified by sex and age categories; by population pyramids; or by summary indicators such as the proportion of the population that is older. Indicators such as life expectancy at birth and the aged dependency ratio can also be used, but each has more liabilities than does the simple proportion of older people in the population.

Populations age as a result of low deathrates and low birthrates. Low deathrates mean that more people survive to enter the older population; low birthrates mean that the proportion of children born into the population is not very different from the proportion dying.

The older population of the United States has grown rapidly in this century and will grow even more rapidly over the next fifty years. By 2050, there will probably be more than 68 million older Americans, representing nearly 23 percent of the total population.

The older population is quite heterogeneous and includes a great range of ages. The male-to-female sex ratio drops substantially in each successively older age cohort, a result of much lower deathrates for women than for men.

Three-fourths of older men are married, while more than half of older women are widowed. At age 75 and over, two-thirds of women are widows. Education varies widely in the older population: about 9 percent have very little or no formal education, while nearly 9 percent are college graduates. The gap between the education of the older population and that of the general adult population is closing. In addition, many older people are self-educated. The great variation within the older population on nearly every population characteristic is a challenge to planners and policy makers.

The older population is evenly distributed throughout the United States, with Florida being the only state to have a very heavy concentration of older people (18.3 percent, see Table 2-6) and Alaska being the only state to have an unusually low concentration (4.1 percent). Older people are more likely than other age groups to live in small towns and rural areas. The percentage of older people living in suburbs is increasing.

Only about 6 percent of older people live in institutions, over half of these in nursing homes. Among those not in institutions, 80 percent of men and just under half of women live with a spouse. Women are much more likely than men to be widowed and living alone. Women age 75 or older are much more likely than men in the same age category to live with a relative other than a spouse. Older people generally live alone because they prefer to do so; adult offspring rarely desert or abandon their elderly parents.

Fluctuations in the birthrate in the past have caused ebbs and flows in the age structure of the population and in the relative size of the older population compared with other age categories. Deathrates will continue to drop during the next fifty years, resulting in more survivors *in* old age as well as more survivors *to* old age.

Migration influences the concentration of older people in various parts of the country, through both out-migration of the young from the Midwest and in-migration of older people to the Sunbelt. However, migration is not common among older people, and those residential moves they do make tend to be local moves within the same county. Older people's long-distance moves tend to be in the direction of the Sunbelt.

Both permanent and seasonal migrants to the Sunbelt tend to be retired couples who are both healthy and have above-average retirement incomes. Once disability begins to affect the lifestyle of an older household, moves to be nearer kin are likely. Moves to institutions usually occur after a period of in-home care by family.

3 The History of Aging in America

A common misconception about older people is that they had it better in "the good old days," but it is important to separate the rhetoric and ideology of aging from the realities. Indeed, ideas about aging and the realities of growing older have probably always had both positive and negative elements. Certainly, this has been the case in Western civilization. For example, Aristotle had the following to say about elderly men:

They have lived many years; they have often been taken in and often made mistakes; and life on the whole is a bad business. The result is that they are sure about nothing and under-do everything. They "think," but they never "know"; and because of their hesitation they always add a "possibly" or a "perhaps," putting everything this way and nothing positively. They are cynical; that is, they put the worse construction on everything. . . . They are small-minded, because they have been humbled by life: their desires are set upon nothing more exalted or unusual than what will help them keep alive.

(McKee 1982,11)

In contrast, Cicero said:

Those, therefore, who allege that old age is devoid of useful activity adduce nothing to the purpose, and are like those who would say that the pilot [navigator] does nothing in the sailing of the ship, because, while others are climbing the masts, or running about the gangways, or working the pumps, he sits quietly in the stern and simply holds the tiller [which steers the ship]. He may not be doing what the younger members of the crew are doing, but what he does is better and much more important. It is not by muscle, speed, or physical dexterity that great things are achieved, but by reflection, force of character, and judgment; in these qualities old age is not only not poorer, but is even richer.

(McKee 1982,26)

And Plato said:

[A]t the age of 50 those who have survived the tests and proved themselves altogether the best in every task and form of knowledge must be brought at last to the goal. We shall require them to turn upward the vision of their souls and fix their gaze on that which sheds light on all, and when they have thus beheld the good itself they shall use it as a pattern for the right ordering of the state and the citizens and themselves throughout the remainder of their lives.

(McKee 1982,53)

In truth, aging is not predictable. It is sometimes a positive force and sometimes a negative one, even in the same individual. For thousands of years, writers and orators have sometimes emphasized the positive qualities that aging can bring and sometimes the negative. So it is with culture. For example, Simmons (1945), in his extensive review of aging in seventy-one cultures, found that in some cultures the aged were revered and in others they were not. As we will see, Americans have always been ambivalent about aging. And because both positive and negative elements of aging are always present, in any era either could be emphasized. But what factors influenced the weight of emphasis? Was it the realities of aging, or was it our untested beliefs and ideologies, or both? This is a main issue we will address in this chapter.

Fortunately, the volume of historical material about aging in the United States has increased dramatically in recent years. Fischer (1978), Achenbaum (1978, 1983, 1986), C. Haber (1983), Graebner (1980), Gratton (1986), Quadagno (1988), Hushbeck (1989), Haber and Gratton (1992; 1994), Cole (1992), and Kertzer and Laslett (1995), in particular, have assembled a great deal of evidence about the history of aging in America, and this chapter relies heavily on these resources.

Modernization Theory

The interpretation of historical materials can be guided by theories that provide frameworks for organizing various fragments of history. In the case of aging, **modernization theory** has been frequently used by sociologists to organize ideas about how aging and treatment of older people as a social category may have changed during the past 200 years. Historians, on the other hand, have been less than enthusiastic about modernization theory (Haber and Gratton 1992).

The central thesis of modernization theory is that the processes that caused societies to evolve from rural and agrarian social and economic systems to urban and industrial ones also caused change in the positions that older people occupy in the society and the esteem afforded to them both individually and as members of a social category. The direction of change is usually *assumed* to be for the worse.

Modernity: Central Ideas

In the context of this discussion, modernization does not mean merely the evolution of contemporary or present-day society. It refers instead to a sociocultural shift which began in the latter part of the 1700s and which brought new ways of thinking about how best to pursue material progress in the form of more and better goods and a higher general level of living. The core of ideas that constitute the modern view include the development of science and technology as ways of knowing, and the technical specializations they gave rise to; rationalization in the form of standardization, consistency, and coordination in work organizations, especially through the hierarchical bureaucracy; the rapid spread of ideas through new media of communication; increased social mobility both up and down in new technology-driven systems of production and distribution; increased geographic mobility through new systems of transportation; an increase in the size and scale of organizations; and the development of new forms of organization such as corporations and the technocratic political state. *As an ideal,* the modern society dominated by techno-scientific rationality, universal education, mass organizations and communications, cosmopolitan urban populations, and political democracy was supposed to create abundance for all and thereby assure peace and prosperity.

The changes caused by the pursuit of modernization as an ideal happened neither suddenly nor uniformly. For example, advances in science and technology affected transportation a good fifty years before they had profound effects on medicine (Johnson 1991; Cole 1992). In addition, modernization was a transition that had a beginning and a middle, and perhaps will have an end. Thus, although the beginnings of modernization occurred in the period from 1780 to 1830, modernization was not widespread in its effects until the early 1900s, and it did not fully mature until the 1950s.

It is not altogether clear when people began to conceive of a postmodern alternative to modernism, but *postmodernism* certainly did not become widely discussed in the United

States until the 1980s. Postmodernism is an idea that rejects modernism's principles of universalism and rationalism as well as traditionalism's appeal for continuity with the past. In the 1990s many elements of modernism still dominate our political and economic life even though there are clear signs that a postmodern ideology of diversity has undercut the legitimacy of many of the ideas associated with modernism. The point is that ideal-types such as modernism rarely occur in their pure type. Instead, elements of modernism coexisted with elements of traditionalism early on, and now elements of modernism coexist with elements of a nihilistic postmodernism.

As we examine modernization theory in relation to aging, the central question is modernization's effect on how older people were regarded as a social category. Did age matter as much or more than gender, class, race, or ethnicity in determining life chances at any point during the transition called modernization or in a historical period that could be called the modern age? Can we identify aspects of modernization that were directly related to changes in how older people as a category were treated and the esteem and power afforded to them?

Aging and Modernization

Anthropologist Leo Simmons (1945) was probably one of the first researchers to address the issue of modernization's effect on older people. Based on a cross-cultural study of seventy-one societies, he concluded that in relatively stable agricultural societies, elders usually occupy positions of favor and power, mainly because of the concept of seniority rights. But when the rate of change increases, Simmons said, older people lose their advantaged status. He did not specify how or why this happened.

Using a sociological perspective, Cottrell (1960b) viewed modernization as a result of the growing use of fossil fuels and technology to increase human productivity. To Cottrell, the most significant aspect of the historical

shift from agrarian to high-energy industrial forms of production was its effect on the organization of society. Agrarian societies revolved around the village, which itself was a collection of families. According to Cottrell, the power of elderly men, and occasionally elderly women, in the agrarian system stemmed from their positions as heads of families, which in turn admitted them to the council of elders that ran the community. In addition, tradition was the main way that people decided issues in agrarian societies, which gave elders value as keepers of knowledge and tradition. Heads of families made decisions in all realms of life: economic, political, religious, and social. Cottrell himself grew up in the Utah frontier culture of the early 1900s, and his view of traditional patriarchal societies was most definitely conditioned by this experience.

With the growing use of wind, water, steam, and electric power, the relationship between human effort and the volume of goods produced weakened. Because production depended less on land and more on technology, decision-making power shifted from many landowners to relatively fewer owners, managers, and financiers, which eroded the power of family patriarchs. In addition, as the volume of scientific and technical knowledge grew beyond the capacity of any one human to know it all, the value of the elders as keepers of knowledge diminished (Cottrell 1960b).

Fischer (1978,108–112) advanced the notion that for the new egalitarian type of society to emerge, the traditional hierarchical type had to be undercut. In the process, because they were usually in control of traditional societies, older people *as a category* came under attack by those who wanted to change the system. Thus, it was not their capabilities or a lack of them that caused older people to lose their advantaged positions but the idea that they were symbols of an outdated social order.

All these are plausible explanations of why elders lost their hold on the privileged posi-

Figure 3-1 Pressures toward lower social status of older people.

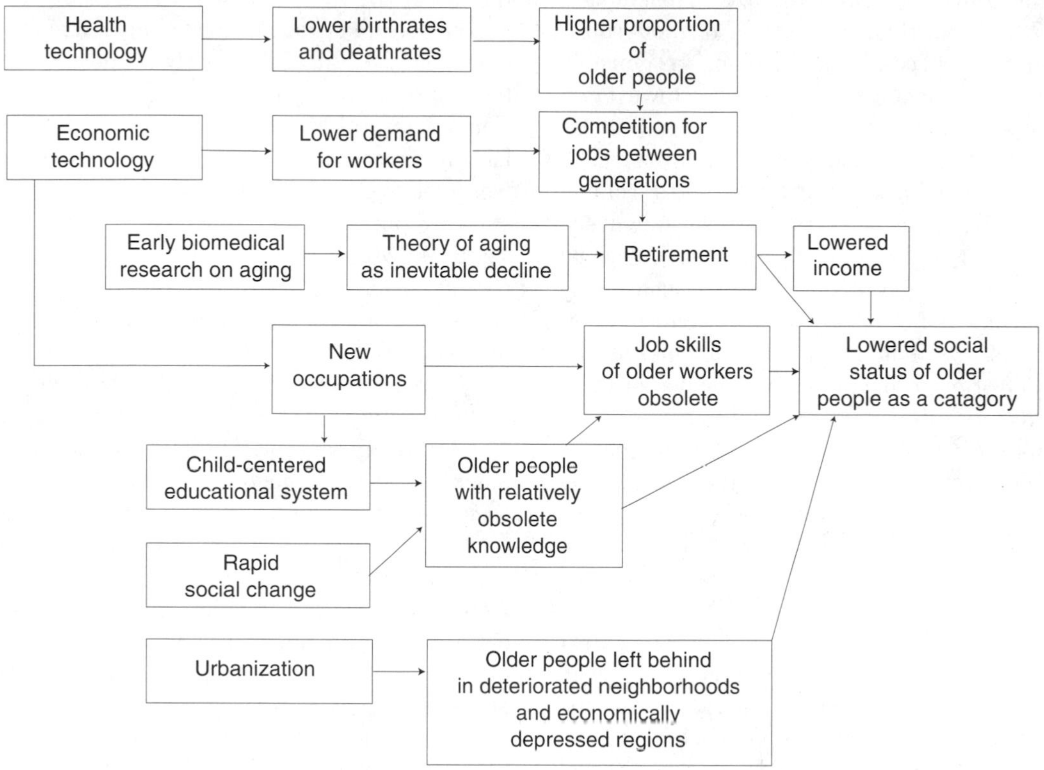

```
Health            →  Lower birthrates    →  Higher proportion
technology           and deathrates         of
                                            older people
                                               ↓
Economic          →  Lower demand        →  Competition for
technology           for workers            jobs between
                                            generations
                                               ↓
   Early biomedical   →  Theory of aging   →  Retirement  →  Lowered
   research on aging      as inevitable decline              income

        New              →  Job skills      →  Lowered social
        occupations          of older workers    status of older
                             obsolete            people as a catagory

   Child-centered      →  Older people
   educational system     with relatively
                          obsolete
   Rapid                  knowledge
   social change

   Urbanization        →  Older people left behind
                          in deteriorated neighborhoods
                          and economically
                          depressed regions
```

Source: Adapted and revised by the author from Cowgill (1974b,141).

tions in industrialized society. But older people did not become merely equal to everyone else; they became less valued than other age categories. Why? Sociologist Donald Cowgill (1972, 1974, 1986) developed a theory to explain why older people were devalued by the process of modernization (see Figure 3-1). He felt that several factors associated with modernization combined to reduce the desirability of elders as participants in society. First, as we saw in Chapter 2, demographic trends produced a higher proportion of older people in the population of the United States. This, coupled with a lower demand for workers because of the increased use of technology, heightened the competition for jobs between the old and the young. In addition, the growing number of new kinds of jobs reduced the value of experience and practiced skills, which

were older people's main ways of offsetting their relative lack of physical dexterity, according to Cowgill. Retirement lowered the value of elders because it was based on the assumption that they were no longer capable and because it dropped them into a less desirable income category. Rapid social change and child-centered education outside the family made obsolete much of the knowledge that had formerly been a foundation of esteem for elders. Finally, urbanization often left older people behind in rural areas or economically depressed areas of the country, causing them to be viewed as "backward." For these reasons, older people presumably lost a great deal of power and prestige in the process of modernization.

Some analysts contend that the decline in the social position of older people occurred *before* many of the effects Cowgill cited became obvious (Fischer 1978; Achenbaum 1978). If so, then the aspects of modernization Cowgill cited could not be responsible for lowering the status of elders. However, if we consider that modernization is a process that could occur in stages and if we consider that modernization might first and foremost be a revolution of *ideas,* then we might see both industrialization and the changing social position of older people as being caused by the same shift in ideas. The central point here is that modernization is not merely the result of more people or greater use of technology or of bureaucratic organization. Efforts to assist today's Third World nations to modernize have amply illustrated that new ways of *thinking* must also be present if new forms of economic and social organization are to take root.

Let us examine these ideas about modernization and the older population in the context of a brief review of the history of aging in America, beginning with the period before the American Revolution and proceeding to the present. We will not consider every period of American history, only significant happenings that influenced aging or older people.

Aging in Colonial America

Summarizing American life in any time period is risky because we are and always have been a diverse people. The colonial settlers came from a variety of backgrounds and made their livings in a variety of ways. In the 1600s, the dominant economic activity was trade—in furs and skins, salted fish, and wood for England's growing fleets of ships. Settlements were small, and work organizations, although numerous, were also small. Life was organized around the community. By the mid-1700s, farming had eclipsed logging and trapping as a major economic activity in the New World. In 1770, for example, tobacco, wheat and flour, dried fish, and rice were the leading export commodities (U.S. Bureau of the Census 1975,1183–1184).

The social heritage of the colonies was primarily British. In 1790, nearly 80 percent of the population of the new United States claimed one of the countries in the British Isles as their nation of origin. Older people were relatively rare in the New World, constituting only about 2 of every 100 people. Yet those who did survive to old age often occupied an advantaged position, partly because position was loosely tied to an age hierarchy.

The cornerstones of life in the New World were religion, politics, family, agriculture, and trade. In each of these arenas, capable older people had distinct advantages over their younger counterparts. Even incompetent elders were sometimes protected by the seniority principle.

Religious beliefs prevalent in the colonial era supported a positive view of the contributions of elders. For example, the Puritans and Pilgrims conceived of old age as a life stage contained within the social and spiritual journey through life. They felt that elders, through their struggle to understand and carry out the will of God, were more developed spiritually compared with the young, and they expected elders to set a moral and social example. Early American religion had high expectations of elders, including dignified behavior and appearance and continued social contribution (Cole 1992).

The sacred scriptures' many references to respect for elders were translated in the pulpit and in popular literature into an ethic of veneration of elders. This ethic was based in the belief in both Puritan and Calvinist Protestantism that God was in control of every detail of life. Someone favored with a long life must be morally superior. Thus, in a society dominated by concern with demonstrating one's moral worth, aging elevated people to a revered and exalted position.

However, this veneration was a mixed blessing for older people. The distance that the hierarchical moral order imposed between family generations sometimes stood in the way of friendship and affection between older parents and their offspring. In addition, many people viewed the demand that elders continually set a moral example for the young as a hindrance to relaxing and enjoying life.

However, it was not just religious exhortations that gave elders positions of power. Their esteem and integration into the social fabric was tied to their control of valued assets and continued skill at valued tasks. Older people who were poor were treated with disrespect, and those with physical frailties were treated with disdain (Haber and Gratton 1992).

In agriculture, power and control then—as now—were based on land ownership and position in the family. Older men in colonial America often determined when the management of family lands would be turned over to the younger generation; adult men could be denied economic independence by their fathers until well into adulthood.

Colonial elders also had a potential advantage in trade based on maintaining skills developed through experience in trading. Skillful trading required both wide knowledge of the relative worth of the objects being traded and wisdom in dealing with people. Experience was a very distinct advantage in this system. In addition, young people learned how to trade by watching their elders. The lore of the period is filled with stories illustrating the poor performance of youth against the wisdom of age in the marketplace.

Because of the veneration of elders, the importance of age in acquiring position, and the importance of experience in the market, it was probably natural that the colonists turned to the elders for leadership in times of crisis. For example, Fischer recounted the story of Deacon Josiah Haynes, who on his eightieth birthday

turned out with the rest of the Minutemen of his town of Sudbury and marched eight miles to Concord bridge with blood in his eye and a long stride that left his younger neighbors puffing and struggling behind him.

> (Fischer 1978,50)

Haynes took leadership of his company from a young man who was less eager to fight:

Twenty-one elite companies of British infantry fled down the Boston road with an infuriated eighty-year-old Congregational deacon close at their heels—so close, in fact, that Josiah Haynes was shot and killed while reloading his musket, and another "grey champion" entered New England's folklore.

> (50)

Fischer goes on to say that "the grey champion" was not just a cultural myth. In times of crisis, older men joined the ranks of Selectmen and Revolutionary Committeemen even before younger men left for the army. The grey champion was institutionalized as "Uncle Sam" and had become a popular symbol of the new United States of America by the 1830s (Achenbaum 1978, 25).

Traditional societies have a unique approach to training their members to make decisions. Riesman (1950) developed a classification scheme based on how people make decisions: those who emphasize tradition as a way of deciding he called *tradition-directed;* those who emphasize their own internal principles he called *inner-directed;* and those who base their decisions on what their contemporaries consider proper he labeled *other-directed.* Riesman contended that different economic and demographic eras require different human qualities and therefore create a differential emphasis among these various ways of coming to decisions. He also believed that the predominant mode of decision making in any period has strong effects on how parents rear their children.

In Riesman's framework, colonial America emphasized tradition in the midst of rapid and sometimes chaotic social change. In the traditional society,

Stages of a man's life (1848).
Photograph courtesy of the Library of Congress Archives

the parents train the child to succeed *them,* rather than to succeed by rising in the social system. Within any given social class, society is age-ranked, so that a person rises as a cork does in water: it is simply a matter of time, and little *in him* needs change.

> (Riesman 1950,39; italics in the original)

In such societies, the family is the major force shaping the individual's approach to life; imitation is the major process through which people learn to cope; and older family members tend to run the family. The authority of age was supported by both custom and law.

Obviously, then, aging had certain advantages in colonial America, but do not assume that these advantages were widely realized or automatic. Indeed, even though many people did live to become old, age was seldom used to elevate a person's social position suddenly. More often, age simply added a bit more prestige to people who already had earned the respect of their families and communities. In addition, the chronological point at which age brought status was not clear-cut since legislative definitions of old age did not exist and role changes that could be used to define old age were less common (C. Haber 1983).

In addition, the major metaphor used to describe life stages during the colonial era was two staircases: one rising from infancy to maturity and the other descending from maturity to old age and death. Each step of the staircase was a life stage with specific social responsibilities, and there were separate versions of the life course for women and men (Cole 1992). This life-stage imagery, which in the 1700s was circulated widely even among the poor via cheap reprints, clearly shows that the model of aging as decline was a part of the culture prior to modernization.

Stages of a woman's life (1848).
Photograph courtesy of the Library of Congress Archives

Although there were both positive and negative images of aging in the colonial period, historians generally conclude that the social position of elders was at least neutral and often advantageous. This emphasis on the positive aspects of aging has been linked to the traditional social structure, which emphasized a stable social hierarchy in which people rose to increasingly responsible positions even in the family by waiting their turn. That there were many exceptions to this pattern in reality (Haber and Gratton 1992) does not alter the conclusion that age-graded advancement was the ideal.

From the Revolution to the Civil War

At the time of the American Revolution, a great expansion of wealth in the colonies was under-

way, due primarily to the growth of markets and systems of exchange (Fischer 1978, 110). With this increase in wealth came an increase in the range of income inequality: More wealth was concentrated in the hands of a few (who were often older but not as predictably so).

During the years following the Revolution, the population of the new country expanded very rapidly, from just under 4 million in 1790 to nearly 13 million in 1830. Such a large increase in population has several effects on a society. As the size and density of a population increase, pressures are exerted for more specialization within society, roles tend to become more formalized, and wider variations tend to develop in what is considered proper behavior. Population expansion also tends to increase the number of levels of influence and authority, the number of prestige and influence hierarchies, and the potential for friction and conflict (Atchley 1971b).

Ideologically, the Revolution marked an acceleration of emphasis on individual achievement, religious secularism, equality, and the free market. These *ideological* shifts along with economic and demographic forces were setting the stage for the evolution of a new type of society, an evolution in which hierarchical relations of family were slowly replaced by social relationships based on achievement. These new relationships were held together not by loyalty to family or community but by contract. These changes took away the advantages of age, but they did not leave older people at a disadvantage—yet.

As guiding principles, the ideals of liberty and equality gained ground steadily in the new United States, replacing traditional hierarchical ideas of social position. As Fischer stated:

There was a radical expansion of the idea of equality—not equality of possessions but equality of condition, equality of legal status, equality of social obligation, equality of cultural manners, equality of political rights. . . .

American society became a world in which "Liberty" (now a singular noun and capitalized) was itself a prior condition for most men (women, slaves, and paupers excepted). The growth of that idea destroyed the authority of age by dissolving its communal base, for the powers of elders in early America had rested upon a collective consciousness—upon the submergence of individuality in the family, the town, the church.

(1978,109)

Between 1790 and 1830, the land area of the United States more than doubled. This increase, coupled with rapid population growth and increased wealth, encouraged a great deal of geographic mobility. Many adult offspring no longer remained in the communities in which they had grown up. In New England, the relatively isolated communities of the mid-1700s began to expand and overlap one another.

The legitimacy of the traditional hierarchy was further eroded by religious changes in the New World. By the mid-1800s, Methodist, Baptist, and Presbyterian doctrines—with their denial of predestination and belief in individual salvation as a possibility for all—injected egalitarian and individualistic thought into the religious sphere (Cole 1992). In these new ideologies, long life was not necessarily evidence of moral superiority, nor were elders necessarily superior guides for religious life. These new religious ideologies stood in stark opposition to the prevailing religious ideologies, which held that elders were morally superior because of their vast experience and their opportunity for a lifetime of self-improvement (Achenbaum 1978,35).

The evangelical religious ideas that began to gain hold in the mid-1700s assumed that salvation was there for the taking and did not require an uncertain lifelong struggle; therefore, aging people had no special knowledge or virtues that could serve as an example to the young. Faith replaced experience and striving as the cornerstone of spirituality. God, often visualized as an old man who moved in mysterious and harsh ways, was overshadowed by the young, ever-forgiving Christ (Cole 1992). Attention shifted from the old as a moral example to the young as candidates for conversion.

Aging was seen dualistically as *either* positive or negative, rather than both. Positive aging was seen as the result of virtue and self-control; negative aging was seen as stemming from sin and irresponsibility. This view presumed that the individual could control the outcome of her or his aging by behaving morally and responsibly and thereby avoiding negative outcomes. Therefore, dependence and disability in later life came to be defined as a product of failure and sin. Cole (1992) contended that these views provided the moral and philosophical foundation for negative ideas about aging that were later transformed into widespread prejudice and discrimination against aging and older people.

Authority in the traditional society was based on the aristocratic and religious hierarchical standards of the times. To establish a new republic in the new egalitarian society, some new form of legitimacy was required, one that established the "rightness" of authority among free and equal people. The ideology of *the social contract,* given shape by John Locke in the late 1600s, provided the needed legitimacy. Authority came not from on high, but from the consent of the governed through the social contract. As Klapp pointed out:

[T]he idea of the social contract has unique advantages as a legitimizing device. . . . It provides a natural and reasonable explanation for getting together to set up an authority over free [people].

(1973,136)

Because it is based on the theory that people are "endowed by their creator with certain inalienable rights," the social contract, although revocable, has ultimate authority. It is an ideal device to serve as a social bond among people who are not kin. Under the social contract, elders have neither more nor fewer rights than anyone else.

While the Declaration of Independence set forth the ideology of the social contract, Adam Smith's *The Wealth of Nations,* also published in 1776, set forth the free market as an ideal way of determining the social order. The concept of the free market includes

any competitive offering which leads to the fixing of a value by collective response, which allocates resources, and which commits people to action and organization as the result of their willingness to "buy."

(Klapp 1973,162)

Self-interest, not moral principles, could be depended on to provide the balance that would produce an effective social system. As Smith put it:

It is not from the benevolence of the butcher, the brewer, or the baker that we expect our dinner, but from their regard for their own interest. . . . I have never known much good done by those who affected to trade for the public good. It is an affectation, indeed, not very common among merchants, and very few words need to be employed in dissuading them from it. . . . The natural effort of every individual to better his own condition, when suffered to exert itself with freedom and security, is so powerful a principle, that it is alone and without any assistance, not only capable of carrying on the society to wealth and prosperity, but of surmounting a hundred impertinent obstructions with which the folly of human laws too often incumbers its operations.

(1776,151)

In other words, once free of meddling by government and church, individual selfish interest would lead us all to a better world because each person would look out for personal interests. This is individualism in its most basic economic form. Yet as long as the market was dominated by old-fashioned trade, elders could hold on to the advantages of experience. As we will see, only in the late nineteenth century did the ideal of the free market become coupled with faith in progress through technology and efficiency and medical lore stressing disadvantages of aging, thus putting older people at a disadvantage.

The growing trend toward individualism, the westward movement of large numbers of young adults, and the growing complexity of economic, political, religious, and social life in America also changed the emphasis on how people made decisions. In Riesman's (1950) terms, the emphasis shifted from tradition-directed to inner-directed decision making. In a pluralistic world,

the growing child soon becomes aware of competing sets of customs—competing paths of life—from which he is, in principle, free to choose. And while parentage and social origins are all but determinative for most people, the wider horizon of possibilities and of wants requires a character which can adhere to rather generalized and more abstractly defined goals. . . . A society in which many

people are internally driven—and are driven toward values, such as wealth and power, which are
by their nature limited—contains in itself a dynamic of change by the very competitive forces it
sets up. Even those who do not care to compete
for higher places must do so in order not to de-
scend in the social system, which has become a
more open and less age-graded and birth-graded
one.

(1950,40–41)

In this inner-directed mode, the parent is

not satisfied with mere behavioral conformity . . .
but demands conformity of a more subtle sort,
conformity as evidence of characterological fitness
and self-discipline.

(1950,42)

In such a system, ethics and work performance depend not as much on external pun-
ishments and pressures, but are

instilled as a drive in the individual, and tremendous energies are unleashed toward the alteration
of the material, social, and intellectual environment and toward the alteration of the self.

(1950,43)

Thus, parents retained essential character-
molding functions. They, more than anyone,
were responsible for creating internalized
standards and principles that would see
their children through an unpredictable
future.

As we pointed out earlier, little in the new
ideologies inherently put older people at a
disadvantage. However, because many held
responsible positions in the old order, which
had to be undercut to make way for the new
one, older people became the target of social
disapproval. Less positive traits of age (such
as physical infirmity), which earlier had been
glossed over, were used to discredit elderly
officials. In addition, the new religions were
beginning to neutralize the role of experience
in moral development. It was not that every
characteristic of older people had become
negative. Indeed, what is more likely is that a

more realistic balance was perceived between the advantages and liabilities of aging,
compared with the overly positive viewpoint
of the early 1700s.

The Beginnings of Industrialization

Although industrialization did not take hold
on a large scale in America until after the
Civil War, the advent of industrial production methods in the 1840s brought with it
signs of trouble for older workers (Hushbeck 1989). The rationalization of labor in
new, larger factories and the introduction of
mass production techniques both put into
play labor processes that were to contribute
to the spread of age discrimination after the
turn of the century. Rationalization of labor
depended on job standardization. People in
the factory were expected to adhere to a
formalized division of labor, use a uniform
work schedule, and work under close supervision in return for an hourly wage. Mass
production involved large batch production
of interchangeable parts, which eliminated
many of the skilled fitting positions that
were filled by older, more experienced
craftsmen. As a result of better science and
engineering, machines were created that
had the precision needed to displace the fit-
ters who had corrected the defects in products generated by earlier, cruder machines.
This theme of improved machines displacing
skilled labor was to be repeated in industry
after industry as industrialization accelerated. The upshot of rationalization was job
standardization, which reduced the capacity
of organizations to accommodate the changing capacities of workers as they aged or,
more often, were disabled by working under
unhealthy or dangerous working conditions
over an extended period of time.

Although these trends began in the ante-
bellum period, it was not until the emergence
of a true national economy after the Civil

War that a large proportion of older workers were affected by them.

Civil War to 1900

Numerous changes occurred toward the end of the nineteenth century, most importantly: the rise of the scientific method; the development of an ideology of progress through efficiency, technology, and bureaucracy; the acceleration of industrial development; the emergence of retirement as a social institution; and massive immigration from Europe. As we will see in this section, it was *ideological* change, not change in objective circumstances or increases in the proportion of older people in the population, that gave rise to both industrialization and age discrimination. Thus, modernization theory needs to be revised to take into account these ideological components.

Rise of Science

The first and perhaps most sweeping change of the nineteenth century was the rise of science as a method of knowing. In the scientific method, observation replaces tradition as the source of ideas about the world and how it works. What this new method proposed

was to produce a rigorously purified, value-free body of information for society, screened by the scientific method rather than by censors. Such reliable information would presumably allow any society to adapt most rapidly to whatever problems it had to face.

(Klapp 1973,47)

Science caught on quickly in a society bent on material progress as its highest priority. Unfortunately, science generated new information much faster than society could develop the decision rules or values needed to sort out this new information. As we will see shortly, the image of elders was one of the first casualties of science.

Scientists working in Europe advanced the notion that diseases in old age could be understood only by clinically investigating and localizing specific diseases. By the 1880s, American medicine had evolved to the point that research using this approach was practical, leading to a detailed documentation of the physical decrepitude associated with aging in most of the body's organs and systems. There was almost total preoccupation with establishing age differences in *average* levels of functioning; no consideration was given to variety *within* given age categories. And there was no information on the extent to which the effects of these changes could be alleviated. For example, declines in eyesight are practically insignificant as long as they can be offset by wearing corrective eyeglasses.

Researchers looking at the effects of aging on mental capacities came to similar negative conclusions. Achenbaum (1978,46) cited the work of George Beard in the 1860s and 1870s, which concluded that mental abilities were at their peak between the ages of 30 and 45, after which they declined. William James, the pioneering psychologist, contended that as people age they become set in their mental ways and unable to keep up with the changing times.

Out of these medical and psychological *assumptions* about aging came the scientific variant of the long-standing theory of aging as inevitable decline, which held that people go through a period of maturation, a period of maturity during which their powers are greatest, and a period of aging during which both physical and mental resources are diminished. As in most fledgling fields of study, the initial facts were sparse, and so the conclusions drawn from them were sometimes great fanciful leaps rather than carefully taken scientific steps. Not until the 1960s did alternatives to the theory of aging as inevitable decline, such as compensation theory and continuity theory, appear and begin to find wide support in the scientific literature. Thus, for nearly one hundred years the theory of aging as decline was the dominant scientific model of the physical and mental effects of aging.

Ideology of Material Progress

The same optimistic and pragmatic urges that gave rise to science also gave rise to a new economic ideology of material progress through efficiency, bureaucracy, and technology. Human effort was seen as the key to progress. Achievement in science and technology and in the organizing of effort was seen as leading to a better lot for humankind. Many of the evils of social inequality that appeared at the beginnings of industrialization, such as child labor, would naturally disappear through the increasing material well-being that would arise from the application of technology in a free market. Another key concept in the formula for progress was efficiency: Only by making the most efficient use of both people and machines could the material growth necessary for an affluent society be achieved.

Numerous practical offshoots of this ideology of progress appeared in the late nineteenth century. First was a dramatic rise in demand for achievement from the individual. Second was the growing domination of work by bureaucracy. Third was growth in technology. Each of these changes set the stage for massive age discrimination.

Individual Achievement   Individual achievement was an important aspect of the new organization of work. Although America has always been thought of as a land of rugged individualists, the reality has often been otherwise. As we saw earlier in the chapter, early America emphasized community effort to solve community problems. Fitting in and doing one's assigned job were not related to individual achievement or creativity but to conformity. Although the move toward achievement standards began in the 1700s, it accelerated dramatically after the Civil War. For example, deCharms and Moeller (1961)

sampled every third page of four representative children's reading texts for each twenty-year interval from 1800 to 1960 and calculated the frequency with which individual achievement imagery occurred in these texts (see Figure 3-2). They found that achievement imagery was uncommon prior to 1850 but multiplied sharply after 1850, reaching a peak in 1890 and declining thereafter. The rise in achievement imagery very clearly occurred *before* industrialization began to dominate the American economy (about 1890).

Bureaucracy   When the focus of social standards becomes individual achievement rather than group achievement, competition increases and every individual in a work organization is expected to produce to a certain level. But in organizations where everyone works together, such comparisons are often difficult to make. After the Civil War, there was a sharp increase in the size of work organizations and a rise in the use of bureaucratic principles. The main features of bureaucracy are a highly specialized division of labor, formal rules, impersonal procedures, hierarchical chain of command, and centralized authority. The job standardization that accompanied bureaucracy created ideal standards against which the achievement of each employee could be measured.

New Technologies of Production and Distribution   The development of the railroads in the late 1800s created a true national economy and made possible massive centralization of production in large factories located in urban centers. Economies of scale also promoted the increase in size of work organizations. Large organizations needed large amounts of capital, and corporations replaced the partnership or proprietorship as forms of company ownership; and with this change came an insulation of management from face-to-face contact with the workers. Absentee ownership and hired management

Figure 3-2 Mean frequency of achievement imagery in children's readers: United States, 1800–1960.

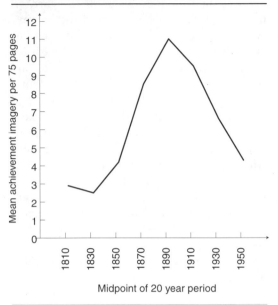

Midpoint of 20 year period

Source: Adapted from deCharms and Moeller (1961,139).

replaced the owner-manager of the nineteenth century. By 1927, the 200 largest corporations (exclusive of banking) had 49 percent of the total assets of all American corporations, leaving the remaining 300,000 corporations to divide the remaining 51 percent (Hushbeck 1989).

During this same time, the mechanization of farming and land policies that brought vast regions of the Midwest into agricultural production following the Civil War both contributed to a relative drop in prices for farm commodities. This began a movement toward larger farms and a declining proportion of the population engaged in farming that continues today. The decline in the viability of farming as an enterprise that could support the self-sufficiency of most of the middle class disproportionately affected older people because before 1900 most older people lived on farms and relied on agriculture for their livelihood. Younger people who were displaced from farm areas could harbor some hope of employment in the nation's new industries, whereas most displaced older farmers could not.

The growth of bureaucracy and the rationalization of jobs did not immediately affect older people in the urban workforce. For one thing, bureaucracy increased the levels of authority in organizations, giving older workers more managerial slots. For another, there was still a spirit of community among those who worked in these new bureaucratic organizations, just as there is today. Supervisors were very reluctant to discharge longtime employees, even when they were disabled and obviously unable to perform (Graebner 1980,90). The lack of provisions for either public or private retirement pensions meant that to force someone to retire was to force him or her into poverty. Finally, older people were still a very small percentage of the population, which meant that the number of infirm older workers who needed to be accommodated in growing organizations was relatively small.

Beginnings of Retirement

Although age discrimination was not common before 1900, the beginnings of a retirement institution began to emerge in the late nineteenth century. The most significant fact about the emergence of pensions and mandatory retirement policies is that they had little to do with aging. The forerunner of modern private retirement pensions was instituted by the Baltimore & Ohio Railroad in 1884. The plan stipulated that at the age of 65 retirement was compulsory and that workers with at least ten years of service were entitled to a pension. Pensions were paid entirely by the company, and the monthly amount was 1 percent of the worker's final wage, multiplied by the number of years of service (Fischer 1978,166). Between 1885 and 1900, eight other companies instituted pension programs. Pensions tended to follow the "railroad formula," but the minimum retirement age, length of service requirements, and mandatory age rules varied. Some plans required employee contributions, while others did not (Achenbaum 1978).

Private pensions were developed to discourage workers from changing jobs, not to provide old-age income security. There was a great deal of labor mobility in the early days of industrialization. Pensions, with their length-of-service requirements, were established to make changing from one job to another unprofitable. In the words of one official:

The pension operates as an incentive to hold men between the age of 40 and 50 when they have acquired the experience and skill which makes them especially valuable and prevents their being tempted away by slightly increased wages.

(Quoted in C. Haber 1978,83)

During the period from the Civil War to the turn of the century, industrial unions were trying to gain management acceptance of the idea of seniority as the basis for layoffs or cutbacks. Seniority rules in effect established a property right to a job, with the most senior (and usually the oldest) workers having the greatest right to job security. Compulsory retirement developed as a way to limit the protection seniority rules provided. Management opposition to seniority was based on two facts: senior workers were usually more costly, and seniority created problems in dealing with workers whose age-related infirmities made them no longer able to perform. Mandatory retirement solved both problems, limiting the protection given by seniority and relieving the employer from having to confront workers too infirm to continue. Labor unions were willing to trade mandatory retirement for seniority because at the time few workers lived to reach mandatory retirement age, especially compared with the huge numbers who would benefit from seniority (C. Haber 1978).

Immigration

A final factor operating to change the situation for older workers was massive migration into the United States from southern and eastern Europe. Of the 76 million residents of the United States in 1900, nearly 12 percent (9 million) had entered the country after 1880. Unlike immigrants from earlier eras, these people did not settle primarily in the rural frontiers; they settled in cities and secured jobs in the nation's growing manufacturing industries. And they often worked for wages much lower than native workers would accept. To employers looking for profits, the contrast between the wages of older workers and the wages of immigrant workers had obvious implications. For example, in the glass industry in Ohio, skilled workers from England, Belgium, and Germany formed the backbone of the labor force in the 1880s, but by 1900, most of these people had been replaced by machines and unskilled immigrants from eastern Europe (Hushbeck 1989). "The tendency of newer workers from abroad to displace older Americans . . . received official notice prior to the turn of the century, and the fate of the older worker served to bolster the arguments for restrictive immigration legislation" (Hushbeck 1989, 70).

1900 to 1929

From the turn of the century to the Great Depression, the pace of industrialization quickened substantially. The prosperity of the period was reflected in Herbert Hoover's campaign promise of "a chicken in every pot and two cars in every garage." The ideology of progress through technology and efficiency in the free market seemed to be working. From the point of view of social gerontology, the most important developments during this period were the continued high rate of immigration from Europe; the acceleration of the use of fossil fuels in industry; the introduction of increasingly more machine technology; the consolidation of the "aging as decline" thesis in medicine; the development of the idea that science and medicine could alter the rate and course of aging; the rise of "scientific management" as a means of managing work; the growth of age discrimination in the workforce; the increase in the number of older people in poverty; the discovery of aging as a social problem; and the development of early ways of coping with the problems of the older population.

Changing Ways of Doing Work

From 1900 to 1930, the use of fossil fuels in industrial production tripled, a rate far higher than at any time before or since (U.S. Bureau of the Census 1954; 1978b). These increased energy resources were used to run machines.

Using the printing industry as an example, Graebner (1980) traced the impact of technology on older workers. In 1885, printing was a craft industry in which type was set by hand in numerous small shops. But in 1884, Mergenthaler invented the Linotype, a machine that allowed direct setting of lines via a typewriter-like keyboard. By 1895, the Linotype had swept the printing industry. Between 1895 and 1915, **age discrimination** resulted in the displacement of many older workers in the printing industry. But why were older workers at such a disadvantage in an industry in which skill and experience still played a large part? Graebner attributed it to three primary factors: a speedup in the pace of work, the theory of aging as decline, and principles of "scientific management."

The speedup of work was the combined result of labor agitation for a shorter workday and management concern for getting a full, productive return on money invested in machinery. Unions pressed for a reduction in the workday from ten hours to nine by offering to work faster; the success of their plan placed many workers—older workers, in particular—at a disadvantage. Graebner reported that those affected

understood that not the machine but the demands placed on its operator by the shop owner were behind "the grind the old boy has to undergo today in order to hold his job." Although printers agreed that good eyesight and supple fingers were requisites of Linotype operation, . . .

(1980,23)

The "unnatural pace" of the work was held responsible for the difficulties older workers had in performing satisfactorily.

While labor unions accepted work speedups as a trade-off for shorter working hours, management's motives involved both efficient return on investment and new ideology about the ideal industrial worker. "Scientific management" was not really scientific; it was merely systematic and logical. Many of its tenets were developed by deduction from untested assumptions about human performance. Nonetheless, as a body of ideas about how to organize work, scientific management influenced decisions in the workplace. For our purposes, the most important principle had to do with what sort of worker industry was looking for. One key to successful management was "the rejection of ordinary workmen and the employment only of unusual ones" (Graebner 1980,32). Thus, the speedup in work was justified in part by the principle that only exceptionally fast operators should be operating the machinery in the first place.

The spread of mass production caused the replacement of skilled workers, who had largely directed their own work, by semiskilled workers, whose work was directed by management. As Gratton stated:

This transformation led to a general degradation of skill in blue-collar work in the early 20th century and diminished the opportunity of older workers to use experience and skill to retain their positions.

(1986,176)

As long as older workers were part of a face-to-face labor process, norms of respect for elders and the valuing of experience might counterbalance concerns about the effects of aging on performance. But as organizations increased in size and the nexus of policy making shifted from the relatively small town to the massive urban area, the hierarchical nature of bureaucracy and mass communication combined to alter the possibility that personal qualities of individual elders would enter into the dialogue at all. As C. Wright Mills put it:

In a mass [society], far fewer people express opinions than receive them; for the community . . . becomes an abstracted collectivity of individuals who receive impressions from the mass media. The communications that prevail are so organized that it is difficult or impossible for the individual to answer back immediately or with any effect. The realization of opinion in action is controlled by authorities.

(1963,355)

In the early twentieth century, the authorities were the medical profession and the "scientific managers."

Age Discrimination and Poverty

In selecting employees in an age of technology, the manager is always faced with the problem of screening applicants so that only those who stand a reasonable chance of filling the bill will be seriously considered. The theory of aging as inevitable decline did its damage by identifying older workers as a category that justifiably could be ruled out for hiring and singled out for firing or layoff (C. Haber 1983).

For example, in 1905 William Osler, physician-in-chief of Johns Hopkins University Hospital and about to become Regius Professor of Medicine at Oxford University, gave a widely publicized address in which he stated that "the effective, moving, vitalizing work of the world is done between the ages of twenty-five and forty" (Graebner 1980,4–5). He went on to say that after age 40 men were comparatively useless and after

age 60 men were completely useless and should be retired. Although these ideas aroused considerable opposition, they represented the general view of medicine at that time. And, as is currently the case, medical opinion was influential. For example, impressed by Osler's speech, steel industrialist Andrew Carnegie established the Carnegie Foundation for the Advancement of Teaching, which had as one of its cornerstone policies for progress in education the development of retirement pensions to allow older teachers to be moved into retirement (Cole 1992).

About this same time, maximum hiring ages were developed. The employees of new industrial enterprises were not a cross-section of those seeking jobs; they were primarily young men. Hushbeck (1989) reported that in the 1920s, 45 percent of firms with fewer than 25 workers had overt rules against hiring men older than 40 or 45 and 95 percent of firms with 1,000 or more workers had such rules. Age discrimination against women was even worse. In 1910, the U.S. Bureau of Labor reported that women's earnings peaked at age 24 and that after age 30 women not only had trouble finding jobs but also keeping them. Although the evidence on employment in the period before World War II is not of the quality one might desire, Hushbeck concluded that the evidence we do have points to declining labor force participation of workers after age 45 beginning in the period from 1900 to 1910.

Hushbeck found that not only were workers over 45 being excluded from the labor force, but those who managed to stay in it were concentrated in low-paid unskilled jobs and casual and seasonal jobs. This was mainly the result of a volatile economy and high turnover rates among the new mass production industries. Overproduction often led to sudden plant shutdowns and a rise in unemployment. Indeed, unemployment rates were higher among workers who worked for large-scale concerns. Once unemployed, older workers faced a disadvantage in trying to get another job. High turnover rates also disconnected older workers from any prospect of a more personalized labor process that might take their individual capabilities into account. Between 1910 and 1929, in large corporations, labor turnover rates of 100 percent or more were not unusual.

Older workers were not only pushed out of the labor force by restrictive hiring practices and discriminatory management philosophies but also by the harsh realities of industrial work during this period. The average industrial worker worked ten hours per day, six days per week, a demanding number by today's standard. Regulations and standards for occupational health and safety were nearly nonexistent prior to 1929. As a result, the cumulative effect of years of exposure to risks of accidents, harmful gases, foundry dust, coal dust, and the like decreased many workers' capacity to hold jobs.

However, even with these pressures, many older workers were successful in keeping their jobs. The exclusion of elders from the workforce was gradual rather than sudden. Nevertheless, labor force participation of elders declined from nearly 70 percent in 1890 to 55 percent in 1920, and by 1929 there were substantial numbers of older people without jobs and without pensions.*

The majority of residents of the "poorhouse" in most communities were old, and avoiding the poorhouse became a major security goal of Americans who grew up in the 1920s and saw older people relegated there. Many elders who were not on any form of welfare were desperately poor. Thus, it was not the physical or mental effects of individual aging that made aging a social problem but the economic effects of age discrimination in the absence of adequate retirement pensions. Even as late as 1920, fewer than 1 percent of elders were eligible for pensions (Achenbaum 1978,83).

Haber (1983) pointed out that, in addition to the "aging as decline" image propounded by the medical profession, social workers and others who worked with elders had "discovered" aging as a social problem. Based on their experience with elders in almshouses, welfare programs, and hospitals (a very limited, mostly urban, slice of the older population), social activists advanced the stereotype of elders as separated from their families by migration and rendered poor by age discrimination in employment (Haber and Gratton 1992). We have since learned that neither of these views represented the experience of most elders, although there was certainly a growing population of elders who were poor, and age discrimination was certainly a reality.

The approach used to solve the social problems posed by a growing population of elderly poor assigned the responsibility first to the family and then to the community (either a locality or a labor union). The community solution was the old-age home. Between 1875 and 1919, more than eight hundred benevolent homes for the aged were founded. These local institutions usually catered to elderly members of religious denominations or nationality groups (Achenbaum 1978,82). Also during this period, most U.S. counties built homes for the indigent, and more than two-thirds of the residents of these institutions were elders. But by 1929, it was clear that the older population in general, and the population of elderly poor in particular, was growing too fast to be handled by building more institutions. In addition, "old folks' homes" had become dreaded institutions, so expanding their availability had little general support.

The ideological climate of the period from the turn of the century to the depression fostered three issues that still pervade the politics of aging: age discrimination, old-age

---

*Historians disagree about the extent to which elders were excluded from the workforce. See Haber and Gratton (1992).

financial insecurity, and retirement. Age discrimination and financial insecurity in old age produced the most intense conflicts.

The Great Depression

The 1930s was a disastrous decade for America and the rest of the industrialized world. As bad as it was, the Great Depression set in motion social forces that eventually shifted the status of older people from desperate to more hopeful. The main elements of this shift were attempts to reform the image of aging, a changing definition of the causes of poverty, the collapse of many private pension schemes, the inability of benevolent homes and almshouses to cope with the rising numbers of elderly poor, political pressure from organizations of older people, the emergence of a clear federal role in public welfare, and the advent of Social Security.

Beginning in the 1920s, popular books with titles such as *Life Begins at Forty* began appearing; these books were intended to foster more positive identities and constructive lifestyles for aging people. In addition, scientists were developing alternatives to the theory of inevitable decline with age (Cole 1992). But despite these early attempts, the negative images of aging prevailed in the 1930s, no doubt in large part because so many older people were poor.

The Spread of Poverty

Prior to the depression, poverty seemed immoral to most Americans, who assumed that people were poor in their old age because they had failed to save money. The prevailing ideology also held that people could achieve anything if they only had willpower. If people were poor, it was because they chose to be poor. Therefore, those who were poor deserved to be poor. Part of the justification of the stark conditions of the almshouses was that they would motivate people to choose not to be poor. There was little recognition that our beliefs about the relative worth of individual attributes like age, race, gender, ethnic group, and social class were being used as discriminatory screening devices in the labor market and that these same biases severely limited the occupational success that individual achievement could bring.

The depression challenged the common conceptions of the causes of poverty and made people acknowledge the role of social forces. Too many people who had faithfully followed the prescription of industry, prudence, and thrift nonetheless lost their jobs, saw their pensions disappear, lost their life savings in failed banks, lost their homes to foreclosure, and so on. Too many "decent" people were on relief. It was finally obvious that something beyond individual failure was at work, and that "something" was society's institutions.

The depression greatly worsened the situation of many retired people. As a result of capital losses, thousands of business failures, collapsed banks, and a host of other economic disasters, most private pension plans were disrupted. Many plans curtailed benefits, and 100,000 workers lost pension coverage altogether. In 1930, the New York Old Age Security Commission reported that most private pension plans were practically insolvent (Achenbaum 1978,129).

The New Deal

The depression also called into question the long-standing assumption that a free market would automatically set society on the right course. After all, the free market had produced the depression. But what could replace the free market in guiding national development? The New Deal was essentially a government-guided national effort to lift the country out of the depression. In accepting it, the public accepted for the first time a role for the federal government as provider for the general well-being of the American public. Government leadership was to replace the invisible hand on the free marketplace.

Even before the Civil War, the federal government had taken an active role in promoting various kinds of economic activity.

Older men in a bread line during the Great Depression.
Photograph by AP/Wide World Photos

For example, government was central in providing financial subsidies and low-cost land that facilitated the development of a national rail network. Federally imposed import duties played a major role in the evolution of American industry by protecting American industries from low-cost products from Europe's growing industries. Government policies promoted massive immigration to expand the pool of labor for industry. What the government had *not* done was take an active role in coordinating the economy or protecting the public from risks associated with an industrial economy.

One of the major social problems of the depression was the steady growth in the number of older people in poverty. By 1940, a substantial proportion of the nation's 9 million older people was on some form of relief (Fischer 1978,174), and unemployment among older men approached 30 percent (Achenbaum 1978,128). This situation gave rise to several national organizations pressing for some sort of government intervention on behalf of elders, the most well known of which was the *Townsend Movement.* Francis Townsend was a California physician who proposed that everyone over the age of 60 be given an old-age pension of $200 a month, with the stipulation that they refrain from employment and spend the pension within 30 days (Achenbaum 1978,129). The plan was to be financed through a sales tax (Fischer 1978,180). By the mid 1930s, Townsend claimed over 3 million followers. Although Townsend's was the largest such organization, there were numerous others also agitating for some sort of old-age security (Quadagno 1988).

The Social Security Act In 1934, President Roosevelt's Committee on Economic Security began drafting the legislation that would become the *Social Security Act of 1935.* It was a broad-based social insurance plan designed to address two of the main evils of the day: old-age income insecurity and unemployment.* We are concerned here only with the old-age security aspects of the program.

---

*Accounts differ as to whether this was primarily humanitarian legislation or an example of shrewd political manipulation. (See Achenbaum 1983,47–48; Graebner 1980; and Quadagno 1988.)

Title I of the Social Security Act provided for a federal–state program of public assistance specifically for older people. Called *Old-Age Assistance,* this program was for the aged poor, regardless of employment history. It imposed national standards on states that were to receive federal funds. For example, a state could not require local communities to carry all of the state's share of the program; some funds had to come directly from the state. The state also had to administer its program centrally, with a guaranteed right to appeal for those denied assistance (Achenbaum 1978,134).

Title II of the Social Security Act set up a national social insurance system to provide pensions for retired workers, disabled workers, and (in 1939) survivors of workers. Individuals had to have held jobs covered by the program for a specified length of time to be eligible for retirement benefits at the age of 65. Initially, the Title II program did not cover farm laborers, domestic workers, government employees, or the self-employed. In 1939, the act was amended to provide spousal and other dependents' benefits in addition to the worker's retirement pension. Eligibility for full benefits required that, after retirement, the worker restrict earnings from employment to less than $15 per month.

The funding mechanisms chosen for Old-Age Assistance and retirement/disability/survivors benefits reflect their very different purposes. Old-Age Assistance was to be funded from general tax revenues, whereas retirement, disability, and survivor benefits were to be financed through a payroll tax, half from the employee and half from the employer. The primary difference between the two programs was the nature of the social contract. The Title I contract involved a direct federal–state partnership for providing assistance to those who were both old and poor. The Title II social contract was quite different: In return for workers' participating in the Social Security system, the U.S. government was obligated to provide them a

pension. The level of retirement or disability benefits was tied to their level of earnings over the working years.

Rather than establish an individual savings account for every participant in the system, tax contributions from current workers were to be used to pay benefits to those who had already retired. This format was selected for at least two reasons. First, there was stiff opposition from business leaders to putting such a large amount of savings capital under the control of the federal government. Second, several European systems had already tried the savings capital approach and had found it lacking: reserve funds were too susceptible to erosion from inflation. Thus, the solvency of the Social Security retirement/disability/survivor system does not depend on money in individual accounts; its solvency derives completely from the formula used to collect the needed revenues. As we will see later, the mistaken contention among many of today's journalists and politicians that Social Security is "going broke" has seriously eroded public confidence in the system. The capacity of the U.S. government to fulfill its Social Security obligations is not in doubt; the debate is over how high benefits should be and how they should be financed, not over whether the government has the resources to do it.

The Social Security Act was a small step in the direction of income security for the aged, but because it was founded on the premise that the aged would not be employed, it has inadvertently supported the ideology of age discrimination. Indeed, one of the major benefits that New Deal politicians saw coming from Social Security was a reduction in unemployment brought about by changing the status of many older people from unemployed to retired (Graebner 1980,186).

1942 to 1965

From 1942 to 1965 numerous changes affected the older population, the most important ones being the rapid aging of the population, steady improvements in retirement income, new attitudes toward consumption and leisure, lower retirement ages, and efforts to improve both the image and the circumstances of older people. Although the growth of the older population was overshadowed by the baby boom that followed World War II, the fact is that from 1940 to 1960 the older population nearly doubled. During the war, older people stayed on the job. The government developed screening programs that brought older people back into the labor force, and people who otherwise would be retired were not, thus slowing down the rate at which older people were leaving the labor force.

Growth of Programs for Elders

The war also strengthened the nation's willingness to put its faith in the federal government. The postwar period saw rapid growth in federal economic policy. The federal government assumed increasingly more of the primary responsibility for the well-being of the population. As a result, increases in coverage and adequacy of benefits under Social Security were enacted.

Court cases improved the coverage of private pensions. By 1950, corporations were allowed tax deductions for their contributions to private pension plans. In addition, for the first time unions were allowed by the courts to include pensions in collective bargaining with management. From 1945 to 1955, the number of private pension plans rose from 7,400 covering 5 million workers to 23,000 covering more than 15 million workers (Atchley 1982c,272). Such expansion did little for those already retired, but it laid the foundation for adequate retirement incomes for those who were to retire in the 1970s.

Inequities Amid Affluence

Achenbaum (1983,55) pointed out that "statistics barely capture the surge in productiv-

ity and prosperity that ensued as the economy readjusted to peacetime needs." From 1945 to 1960, the gross national product more than doubled, and except for three short recessions, the period from 1945 to 1970 was the "longest segment of uninterrupted prosperity this country has ever enjoyed." The economic prosperity of the period had many effects. It allowed the average American worker's standard of living to rise to unprecedented heights while also allowing business an ever-increasing volume of profits. The accelerated use of technology led to shorter workweeks and earlier retirement. High income and easier credit stimulated desires for increasingly more consumer goods. Even retirement was marketed as a commodity, and eventually many people came to think of it as prepaid leisure. The growing amount of time over which the individual had discretion and the general affluence of the period led to a new kind of individualism. For many people, survival was assured, so they turned to developing the skills and knowledge needed for a satisfying life.

Yet within this general aura of prosperity was an undertow of discontent (Achenbaum 1983,66–83). Galbraith (1958) decried the lavish spending on selfish consumption in the absence of America's willingness to use some of its vast wealth to improve public institutions for the common good. Whyte's *The Organization Man* (1956) and Mills' *White Collar* (1951) pointed to problems of over-conformity and alienation from work. Riesman (1950) wrote about other-direction, in which internal direction gives way to social chameleons capable of adopting any ethic their compatriots demand.

In 1961, the White House Conference on Aging drew attention to the majority of older Americans who were poor in the midst of plenty. In the 1972 edition of this book, I wrote that 60 percent of older people in 1965 had incomes below, at, or very near the poverty level and that 80 percent of them had no income whatever apart from Social

Security (Atchley 1972,139). These dismal figures for the older population in 1965 were even worse for older members of minorities, particularly African Americans. In addition, there were no means available to pay for medical care for needy older people. In short, neither the older population nor minority groups participated much in the surge of affluence in America from 1942 to 1965.

The civil rights movement of the early 1960s raised the nation's consciousness about massive injustice beneath the aura of the "affluent society." At the root of this movement were a doctrine of social justice through law and an assertion of individual rights. The older population was among several disadvantaged categories identified as being in special need.

1965 to 1980

Between 1955 and 1965, America had become aware of the massive injustice lying beneath the gloss of prosperity. From 1965 to 1975, policy makers tried to do something about it, chiefly through federal legislation providing elders a wide variety of supports and services and protecting them from discrimination.

Legislation on Behalf of Older People

In 1965, Congress passed the **Older Americans Act**, which established an Administration on Aging as part of the Department of Health, Education and Welfare (now the Department of Health and Human Services) and authorized grants for planning, coordination of services, and training. As an amendment to the Social Security Act, *Medicare* was also established in 1965 to provide financing for many types of health services for older Americans. The *Age Discrimination Employ-ment Act* of 1967 prohibited the use of age as a criterion for hiring, firing, discriminatory treatment on the job, and referral by employment agencies. It also

prohibited job advertisements reflecting age preferences, age discriminatory pressures by labor unions, and retaliation against workers who asserted their rights under the act.

In the 1970s, the legislative onslaught on problems of the older population continued. In 1972, the Older Americans Act was amended to provide for a national network of agencies serving older people. Old-Age Assistance was replaced by **Supplemental Security Income (SSI),** which established a more adequate floor of income for the elderly poor in *all* the states. In 1974, the *Employees Retirement Income Security Act* imposed federal standards on private pension plans and set up a national Pension Benefit Guarantee Corporation to ensure that private pension promises would be kept. In 1978, the Age Discrimination in Employment Act was amended to raise the age at which retirement could be made mandatory for most workers from 65 to 70.

The 1970s also saw rapid membership growth in organizations of older people. By the mid-1970s, the American Association of Retired Persons (AARP) claimed a membership of over 6 million; the National Council of Senior Citizens claimed 3 million (H. Pratt 1974). The combined membership of these organizations represented at least one-third of the entire older population. It would be a mistake to assume that political advocacy was the major goal motivating older people to join such organizations. Some joined to take part in group purchasing programs for commodities and services such as drugs, travel, and insurance. Others joined for purely social purposes. Still, the large number of members allowed these groups to do at least two things of historical consequence. First, magazines such as *Modern Maturity,* which emphasize a positive view of aging, were produced. Second, these organizations had the resources to employ skilled lobbyists to protect the legislative gains of 1965 to 1975. We return to this subject in Chapter 19, "Politics and Government."

Research on Aging

On the scientific front, evidence was being amassed that contradicted the theory of aging as inevitable decline, at least in terms of aging's effect on social functioning. In 1961, Cumming and Henry published *Growing Old: The Process of Disengagement,* in which they argued that society's abandonment of older people and the corresponding withdrawal of older people from social life was necessary, inevitable, and desired by individuals as well as society.* This book ran directly against the "keep active" school of successful aging, and the conflict between the two views sparked an unprecedented volume of research. The results of this research showed that the vast majority of older people were capable people who remained integrated in social networks and that disengagement was mainly a result of age discrimination.

The Older Americans Act also marked the beginning of systematic government support for research on aging and programs to train specialists in the field of aging. In 1974, the National Institute on Aging was added to the National Institutes of Health and given the charge of setting priorities and funding basic research on aging.

We will not go into more detail on these developments here; they are covered in the remainder of the book. Suffice it to say that the period from 1965 to 1980 was one of unprecedented growth in concern for the general well-being of the older population and the number of programs designed to promote it.

The 1980s

The beginning of the 1980s was marked by a full assault on the gains made on behalf of older Americans in the previous fifteen years. The Reagan administration went all out to convince the American public that older Americans had been lifted out of poverty and no longer needed government assistance; it chose to ignore that this assistance was exactly *why* the situation of elders had improved. The president promised that program cuts would be fairly shared by all Americans, but in fact he proposed severe cuts in assistance to the very neediest of older people—the poor, the disabled, the institutionalized—while providing substantial tax cuts for the well-to-do.

These changes in support levels were not made on the basis of age prejudice; indeed, older people fared better in the budget cutting than many other categories of Americans. The cuts were the result of a change in ideology about the appropriate role of the federal government in providing for the well-being of the population.

In stark contrast to the political and economic philosophy that had guided the role of the federal government since the New Deal, Ronald Reagan's administration stood for economic conservatism, especially a pullback of government from a coordinating role in the economy and from protecting the public interest through regulation of business and industry. The government's main active role was one of increasing the supply of flexible capital available to private business by sharply reducing taxes. Many felt that these policies answered the need to curb alarming levels of inflation.

Reagan's solution to the problems of old-age dependency was to minimize federal responsibility and call on families, communities, and private charities to assume the burden and at the same time to continue paying taxes to support programs that benefited mainly the economic elite, on the presumption that this would eventually benefit the entire population through increased employment. These expectations were unrealistic. Families and communities had already assumed the lion's share of the cost of caring for elders, and by 1986 it was clear that the "trickle down"

*See Chapters 7 and 16 for further discussions of disengagement theory.

theory of economic support for the poor did not work.

Reagan's solution to the problem of financing Social Security was to propose that older people take a cut rather than increasing taxes. This cut was accomplished by increasing the retirement age for full Social Security benefits from 65 to 67, beginning in the year 2000 (which lowers the total amount of benefits through the individual's years of retirement), and by postponing cost-of-living increases (which meant a loss in real income—purchasing power) for those already retired. Likewise, rising Medicare costs were dealt with in part by substantially increasing the older person's share of hospital charges. These conservative solutions shifted the burden to the individual, regardless of capacity to bear it.

As we have seen throughout this chapter, shifting ideologies have been as important as demographic or economic circumstances in shaping society's treatment of its elders. The 1980s saw an ideological shift throughout the developed world away from concern with social justice and entitlements of citizenship—the ideological bases for the legislative programs that prevented poverty for much of the older population, provided health-care financing, and protected elders somewhat from age discrimination—toward concern with individual financial achievement and societal economic growth (Dahrendorf 1988). As a result of this shift, programs for older people, such as Social Security and Medicare, which provide economic relief to all generations within the population, have come to be *viewed* as responsible for a wide array of social ills, such as unaffordability of housing for young families, poverty among children, escalating health-care costs, and the huge federal budget deficit. As we will see, none of these opinions stand up to the facts, but as we saw in this chapter, facts are often less important than assumptions in shaping our national consciousness.

The debate over Social Security and Medicare will not be settled by the time this book is published because it reflects some very deep divisions within American society about how best to deal with an aging population. These divisions are based on differing views of the realities of aging, the appropriate role of government, the capacity of individuals and families to fend for themselves without federal protective regulations and tax relief, and a host of other issues. Somehow we must find a way to resolve these conflicting views if we are to avoid severe problems as our older population increases in both proportion and age.

In the early 1990s, economic stagnation and a shifting economic structure accelerated the attack on benefits for elders and at the same time increased pressures to retire workers at ever earlier ages. Economic stagnation increases unemployment, which decreases government revenues. The rapid decline in manufacturing jobs in the 1980s caused large portions of the workforce to shift downward in occupational status and income (Reich 1991), which also reduced government revenue. The rapid military buildup and large cuts in individual and corporate income taxes which occurred in the early 1980s produced a staggering increase in the national debt, and interest on that debt coupled with declining revenues reduced discretionary government funds available to cope with the nation's many problems. In such a climate it is understandable that advocates and policy makers would look longingly at the large pool of government revenues set aside for social insurance programs. But Social Security revenues cannot be legally used for other purposes, and reducing Social Security payments in no way guarantees that general revenues could be increased accordingly.

Paradoxically, Social Security has increasingly suffered from its success. In difficult economic times, Social Security is solvent, efficient, free of corruption, and responsible for a modest level of income security for

America's elders. When other institutions in society are struggling, anxiety, fear, and envy can combine to produce antagonism toward those programs that are not in distress. Nevertheless, Social Security continues to have strong support from the general public, perhaps because most realize that practically every family in America benefits from it in one way or another.

In the economic upheavals of the 1980s and early 1990s, there was widespread "downsizing" of work organizations in America. Early retirement incentive programs were one of the major tools used by management to cope with the need to reduce the number of employees. As a result, the average age of first retirements declined, which of course increased the number of people retired under Social Security.

The nature of the postmodern condition is clearly illustrated by the paradox of using economic problems to justify an attack on programs that support people in retirement while at the same time relying on retirement, which depends on these very programs, as a major means of coping with economic problems.

Modernization Theory Revisited

We have looked generally at several variants of modernization theory and many details and trends of American history since colonial times. Now we can return to the question of the usefulness of modernization theory as an organizing framework for explaining how aging interacted with social change in the United States during the transition from a largely agrarian colonial society to the urban, industrial giant of the 1950s. Did modernization negatively change older people's access to power and esteem and opportunities to earn a living? Did modernization increase the probability that older people would be treated categorically rather than individually? Were different forces operating during the beginnings of modernization compared with its maturation?

We must recognize that *modernization* is a complex concept. It includes fundamental shifts in values, attitudes, beliefs, and knowledge as well as changes in practical notions about how best to organize ourselves to produce goods and services, and improvements in technical know-how. It seems clear from the record that before we could develop modern concepts of how to organize production and distribution, we first had to develop less rigid ways of thinking about people. Individualism, achievement, and interdependence replaced conformity, imitation, and economic self-sufficiency of the family as ways of thinking about how people fit into groups long before the advent of high-energy industrial technology. Thus, for example, individualistic religious philosophies and egalitarian political ideals played their most important role in modernization early by attacking the moral basis for the hierarchical, age-graded society. This occurred well before industrialization.

The notion that modernization brought about a disintegration of the family and therefore eroded the social position of elders is not well-supported. It is true that with industrialization and urbanization there was a reduction in the extent to which families were units of economic production. However, families continued to play an important role in the distribution of income, and the potential for elders' loss of status was offset to a large extent by the emergence of the family as a means of coping with the economic uncertainties of wage labor and shortages of urban housing. This is an important lesson. Families have consistently adapted to changing circumstances by altering the array and emphasis of functions performed by the family, and elders have remained remarkably integrated in American families while upheavals were going on in religion, science and technology, and the economy. Modernization theory is thus not very helpful in explaining what happened to older people in families.

In the economy, Cowgill's version of modernization theory applies mainly to industrialization in the United States during the period from 1900 to 1930. Throughout that era, the ideal of modernization as rationality, efficiency, technology-driven productivity, and abundance created by elite workers engaged in mass production in large, impersonal organizations was clearly dominating the American economy. There can also be little doubt that during this time medical beliefs and the assumptions made by "scientific management" advocates caused a rapid deterioration in the potential for productivity imputed to older workers and increased prejudice and discrimination against them. Proving that these ideas were common is relatively easy, but establishing the magnitude of their effect is not. Some industries were well on the way toward modernization in 1890, whereas others did not get under way until the 1920s. The pace of technological change was very uneven across different industries. Given the small proportion of elders in the population, the lengthy time over which modernization was spread, and the large number of separate industries affected, it is not surprising that aggregate employment data do not show wholesale exclusion of elders from the workforce. But the decline was steady even in the period before Social Security. The weight of the evidence seems to support Cowgill's version of how older workers fared during the industrialization taking place from 1900 to 1930. It applies less well to both earlier and later periods. This is also an important lesson. We all hope for our theories to explain everything, but they seldom do. Thus, it is important to specify the conditions under which our theory could be expected to apply. We also need to be sensitive to the need for different theories to explain different historical eras.

Modernization theory is of little use in helping us understand what happened to the life chances of elders after World War II. What we need is a theory that explains how older people came to be identified as a social category

having a high priority for public support. *Social justice theory* is a much more helpful perspective because it focuses attention away from what elders can contribute to the economy today, and toward what their contribution has been over a lifetime and what sort of support is their just due. The social policy rhetoric that supported the development of programs aiding elders certainly stressed the extent of disadvantage in the older population, but it also stressed the perspective that these same elders had built a powerful nation and were owed first consideration in return. We will return to this perspective again in Chapter 19, "Politics and Government."

Modernization cost older people status as a social category, and programs based on an ideology of social justice returned a good measure of it. But social ideology changes, and recent challenges to programs benefiting the older population illustrate their vulnerability in the political process. Like the past, what the future holds will no doubt reflect our ideologies. The conditions out of which those ideologies will evolve are examined in the remainder of this book.

Summary

The fact that human aging can produce both positive and negative results has been reflected in our ideas about aging since at least the time of the Greeks. Aging can bring wisdom or senile dementia, perspective or anxiety, skill or infirmity. This basic duality is reflected in the ideas about aging contained in most cultures. But as we saw in this chapter, social ideologies often have had more influence on how older people as a social category were viewed and treated than have the objective circumstances of older people.

Modernization theory has been used to account for changes in the treatment afforded older people. It begins with the notion that in preindustrial societies elders had certain advantages, which they lost in the process of industrialization. In colonial America, seniority rights gave elders first claim to important social positions, and the communal and hierarchical nature of preindustrial society also gave older people a moral as well as a social advantage that was embodied in the ethic of veneration of elders. In addition, elders controlled the family, which in turn led to control of the community. All these perspectives led to a view that at one time Americans who were in the "proper" social category (that is, free, white, male, and from the appropriate social class) could expect their social status to rise with age. For the poor, aging could be a miserable lot.

Modernization—the processes by which societies moved from rural, agrarian economic and social systems to urban, industrial ones—caused older people to lose their advantaged position as a social category, even though some elderly people might have succeeded in retaining power and privilege. Several factors have been advanced as possible causes of this loss of status, but the status of older people did not suddenly change, nor did it change as a result of a single factor.

Before the American Revolution, the image of elders was probably overly positive, for it tended to underplay the negative aspects of aging. The social position of elders was assured by various social institutions—property law, religious doctrine, trade, and the image of the heroic "grey champion."

It appears that over time the moral and social advantages older people enjoyed were neutralized by a host of factors: increased inequality and expansion of wealth and trade, population increase, egalitarian social ideology, evangelical Christian theology, free market ideology, and individualism. But not until the late nineteenth century did the theory of aging as inevitable decline in both physical and mental capacities solidify. Early scientific research detailed countless negative changes, without considering their effects in the context of the whole person or the possibility of compensation.

Industrialization did not spring forth suddenly. Many decades of agricultural and mercantile economic growth provided the wealth necessary for industrialization. Rapid population growth from both natural increase and immigration provided an ample labor force. But new ideas were mainly responsible for the growth of industrialization—ideas about how to organize work, the importance of individual achievement, faith in progress through efficiency in a free market, increased use of technology, and less reliance on human energy. These ideological shifts produced a competitive society, and the theory of aging as inevitable decline put older people at a disadvantage in the competition.

These shifts occurred in the 1890s, prior to very high levels of industrialization and prior to rapid growth in the older population, but they did not displace large numbers of older workers. Thus, even though elders had lost status as workers, not until 1900 to 1930, when mass production became more common and impersonal bureaucratic personnel procedures replaced the personal, communal approach to hiring and firing, does age discrimination appear to have become widespread.

Advocates of modernization theory usually assume that underlying economic and demographic factors were the main influences on the status of the elderly. But powerful ideological factors were also at work, often prior to the economic and demographic changes. Thus, any adequate formulation of modernization theory must include these ideological influences.

In the 1930s, social insurance advocates and elders organized for political action against age discrimination and the resulting enforced poverty, and thereby raised the country's consciousness and provided impetus for the Social Security Act of 1935. Thus, the explanatory power of modernization theory was essentially negated by 1940. After that, an ideology of social justice and the growth of the federal government's role in public programs became more important forces.

After World War II, the United States entered an era of enormous prosperity, an era in which the relatively free market of the pre–Great Depression era was replaced by a federally regulated market. In addition, social welfare increasingly became a federal responsibility. Accordingly, interest groups vied with one another to influence federal legislation.

The civil rights movement advanced an ideology of social justice through law, and the 1961 White House Conference on Aging focused attention on the dire circumstances of most older Americans. By the mid-1970s, prosperity, interest group politics, social justice ideology, and the large numbers of elderly poor had all combined to produce an impressive array of financial support, services, and programs for older people, those who work with elders, and those who study aging.

In the 1980s the shift toward a more restrictive role for the federal government and declining revenues from a sluggish economy combined to cause cutbacks in benefits, particularly from Social Security and Medicare. In the early 1990s, a well-organized political drive to curtail federal government programs in general challenged the future of all federal benefits for older Americans, including Social Security and Medicare.

The history of aging in America is not set in stone. The review in this chapter is based on current evidence, but researchers are still digging to uncover answers to important questions about what happened to elders in the past. This is important research, because unless we understand what actually happened and why, we cannot learn and benefit from our past experience as a nation.

Part 2

Basic Aspects of Individual Aging

Part 2—Chapters 4, 5, 6, and 7—examines personal aspects of aging; that is, the ways that physical, psychological, and social aging affect individuals, and how individuals adapt to aging. It begins with physical aging, its causes, general physical effects, and effects on health in particular. Part 2 then looks at psychological aging in terms of specific mental functions, human development, and mental illness. It then looks at the extent to which individual aging is influenced by social factors, such as the use of age as a social attribute, that determine an individual's life chances. Chapters 4 through 6 are concerned with models and typologies used to identify various aspects of individual aging and the data that result from their application in research. Part 2 concludes with a chapter on personal adaptation to aging, how individuals cope with the physical, psychological, and social changes associated with aging. It also looks at what can happen when people do not adapt to aging and identifies criteria associated with successful adaptation to aging.

4 | Physical Aging

Many people are uncomfortable about the subject of aging because aging ultimately brings people nearer death. Since what people notice about physical aging tends to be decremental, they may also fear that their own aging will entail a lengthy period of serious physical and mental decline.

Although physical aging is inevitable, physical disability and dependency are not. Except for a relatively brief terminal episode of illness, a large majority of people grow old and die without *ever* experiencing significant disability. But, you may ask, how can this be true, given that physical aging seems to have affected the appearance, alertness, and agility of older people we see in everyday life?

We tend to misperceive the general effects of physical aging because we tend to ignore healthy, vital older people. For instance, when you are at a department store, you may not notice the healthy older people who are there because they move about like everyone else, with posture and alertness like those of other adults. But you probably do notice those older people who are frail or disabled because they often move slowly and with difficulty, and they may be accompanied by someone who assists them. Because of the relatively high visibility of frail older people, we tend to overestimate their prevalence in the older population; and because of the relatively low visibility of able-bodied older people, we underestimate their prevalence. As we will see, even at 85 years of age or older, a majority of older people are still able-bodied.

In our exploration of physical aging, we first discuss why the body grows old and then look at general influences that aging has on bodily functioning. Finally, we examine the effects of aging on physical health.

Why We Grow Older

Ideas about physical aging are tied to the concept of the **life span**—the maximum length of life that is biologically possible for a given species. Among animal species there are wide variations in life span, which are thought to be a function of differences in the genetic makeup of each species. Human life span is thought to be about 120 years. Of course, since numerous environmental and social factors intervene, the average length of life—life expectancy—is usually much lower than the theoretical maximum, as we saw in Chapter 2.

Interestingly, one cause of aging is civilization. As Hayflick (1987) pointed out, aging rarely happens to animals in the wild because physical decrements quickly make animals vulnerable to disease and predators. Few wild animals other than primates live long enough after sexual maturity to experience being old; only with the protection of culture do animals, including humans, live long enough to grow old. In this sense, aging can be considered as having been produced by cultural intervention. Because human culture has provided people with protection for thousands of years, we tend to forget that our prehistoric fore-bears lived a much more precarious life, one that did not involve much aging.

The significance of physical aging lies in what happens to the human body during three broad periods of the life cycle. We pass first through a period of **maturation,** during which the body grows and develops to its peak level of physical functioning; next, a period of **maturity,** during which physical functioning remains at peak level; and finally, a period of **aging,** during which the body gradually loses its capacity for peak performance. *Each of the body's systems and organs is on a slightly different schedule for maturation, the duration of maturity, and the onset and rate of aging.* Summary measures of physical aging are not feasible even when functional measures (instead of chronological ages) are used because functions vary so much within individuals.

Because aging is so variable and is not one process but a group of them, biologists have had difficulty separating aging from other phenomena, such as disease, that can affect the body. Strehler (1977) pinpointed several useful criteria in differentiating aging from other biological processes. First, to be a part of the aging process, a phenomenon must be universal; that is, it must eventually occur in all people. This means that the fact that older people are more likely to have a condition does not make that condition a part of aging. For example, older people are more likely than the young to get lung cancer, but since not all older people get it, lung cancer is *not* a part of aging. On the other hand, as people age, they all experience a decline in the ability to fight off disease, which is caused by a decrease in the effectiveness of the body's immune system. Thus, a decline in bodily immunity *is* a part of aging. Strehler's second criterion of aging says that aging comes gradually from within the body. This rules out environmental factors such as cosmic radiation or situational factors such as accidents, which affect the body, but only from outside. This criterion is obviously linked to the genetic basis of aging. Third, according to Strehler, aging must have a *negative* effect on the body's functioning; otherwise we would call it maturation (growth) or maturity (no change). Thus, physical aging is something that happens to all people, comes on gradually from within the organism rather than from the external environment, and has a negative effect on physical functioning. To be a part of aging, any physical change must meet *all* three criteria.

Rowe and Kahn (1987) challenged the traditional view of aging by arguing that there are three kinds of aging: optimal, usual, and pathological. *Optimal* aging is pure aging, in which both the genetic expression of development and positive environmental influences combine to produce aging as a gentle slope with little disease or disability and minimal impact on functional status. *Usual* aging involves positive or neutral genes interacting with negative environmental factors to produce chronic disease and functional limitations that are noticeable and limiting in minor ways. *Pathological* aging combines a negative genetic predisposition with a negative environment to produce pronounced susceptibility to chronic disease and disability. This categorization is not easily translated into research operations, but it suggests a new positive pole to the continuum of aging that we are beginning to observe in healthy older people of advanced age (Machemer 1992; Garfein and Herzog 1995).

Biologists studying what aging is and why it occurs have looked at every level of physical functioning: genes, cells, molecules, tissues and organs, and physiological systems such as the immune and endocrine systems. Let us now briefly look at some of these areas.*

---

*The biology and physiology of aging are extremely complex topics. For a more complete treatment, see recent reviews in Finch and Schneider (1985), Whitbourne (1985), Schneider and Rowe (1990), I. Ross (1995), and Spirduso (1995).

Genetic Functions

Genetically, aging can be seen as either a pre-programmed result of inherited genes or a result of gene changes that occur with age. It has been known for a long time that, regardless of social or cultural influences, various animal species differ considerably in life span. Thus, longevity appears to be *preprogrammed into the genetic structure* of the species. Although the rate of aging varies considerably from individual to individual, those with similar heredity show similarities in aging. Within the human species, variations in genetic heredity influence both life expectancy and the probability that aging will be optimal as opposed to usual or pathological. Both optimal and pathological aging runs in families.

The idea that aging may result from *changes in genetic functioning* has centered on errors that occur as the body synthesizes needed proteins or enzymes according to the genetic code. Presumably, this synthesis results in the body's manufacturing proteins and enzymes that are not up to genetic specifications and so cannot function properly. As the organism ages, these errors accumulate and eventually result in aging and death of the cell.

Reff (1985) concluded that the weight of the evidence does not support the "error" theory. Accuracy of synthesis of proteins and enzymes does not appear to decline with age. Most of the altered proteins and enzymes that have been identified in aging animals are probably the result of chemical reactions that occur *after* genetically programmed protein synthesis. Because the half-life of these proteins increases with age, older animals are more likely to have altered proteins, but these alterations are probably not caused by errors in genetic transcription.

Another line of genetic research involves changes in cells' capacity to repair damage to their genetic structures. For example, cells from species with long life spans are more likely to repair radiation damage than are cells from species with short life spans (Sinex 1977). However, aging does not appear to be caused by a decline in genetic repair capability (Tice and Setlow 1985). The extent to which unrepaired genetic proteins accumulate in aging organisms and the potential negative effects this might have are unclear.

Still another aspect of genetic research concerns the mapping of the human genome. Because many chronic diseases such as diabetes or hypertension are heavily influenced by genetic predisposition, recent work has focused on identifying the specific genetic markers for specific diseases. These genetic markers make early diagnosis, prevention, and treatment possible, which in turn may allow individuals with a genetic predisposition toward pathological aging to avoid this negative outcome (Butler 1995).

Cross-Links and Free Radicals

Another line of research deals with the effects over time of adverse reactions within cells and molecules. Over time, certain molecules in the body develop links, either between parts of a given molecule or between molecules. These *cross-links*, which are very stable and which accumulate over time, result in physical and chemical changes that affect how the molecules function.

Much of the research evidence on the relationship between cross-links and aging comes from the study of collagen. Collagen is a gelatinous substance present in connective tissue. Through a chemical process called *glycosylation,* sugars in the body cross-link with collagen. Cross-linked cells accumulate in collagen with age, leading to a loss of elasticity in various tissues, including cartilage, blood vessels, and skin. This accumulation appears to occur because the body is not completely efficient in recognizing and eliminating cross-linked cells (Lee and Cerami 1990).

Free radicals are unstable oxygen molecules that are produced when cells metabolize oxygen. By joining themselves to various proteins, free radicals can create biologically abnormal molecules. In addition, their instability can lead to chain reactions of abnormal molecule formation. The free radical theory of aging (Harman 1956) held that free radicals could over time lead to an accumulation of ineffective abnormal molecules that could impair the functional capacity of the organism. Fortunately, it was later discovered that the body produces enzymes that metabolize the molecules that are produced by free radicals, thus reducing their concentration and potential for accumulation. The action of this enzyme does not appear to decrease with age (Rothstein 1987).

Both cross-links and free radicals have the potential for accumulating damage to the body over time and could therefore be related to a reduction in functional capacity. But in both cases there are other natural processes at work in the body to offset or ameliorate the negative effects of cross-links and free radicals. This illustrates the importance of having an overview of biological functioning while we examine the effects of specific biological processes.

The Immune System

One of the body's primary defenses, the immune system is geared to recognize and destroy substances from both inside and outside the body that are not part of the "normal" self. Thus, immune cells produce antibodies that attack foreign substances, viruses, bacteria, and even cancer cells, to keep them from interfering with the body's functioning. The immune system matures in early childhood, remains on an extended plateau into early adulthood, and begins to decline in effectiveness thereafter. Part of the increased vulnerability to disease associated with aging results from this decline in the immune system. Indeed, among inbred strains of animals, long-lived strains retain competent immune systems longer than do short-lived strains (Hausman and Weksler 1985).

There are three ways that age changes in the immune system influence physical functioning: (1) If the immune system over time loses its capacity to recognize deviations in substances produced within the body, then mutated cells that formerly would have been destroyed by the immune system survive to the potential harm of the organism. This could explain why susceptibility to cancer increases with age. (2) Sometimes the immune system produces antibodies that destroy even normal cells, a process known as *autoimmune reaction*. The prevalence of autoimmune antibodies in the blood increases with age. Autoimmune reactions have also been linked with several age-related diseases, including rheumatoid arthritis and late-onset diabetes. (3) A failure of the immune system to defend against disease or too slow an immune response can increase the physical debilitation associated with both acute and chronic diseases. The decline in the body's immune system and autoimmune reactions are both promising areas of research that could lead to the identification of ways to influence the rate of immunological decline (R. Miller 1990).

Physiological Controls

Human beings are incredibly complex biological organisms. Our survival requires continual, simultaneous monitoring and control of thousands of systems, structures, and processes within our bodies. In fact, the aging we see in the total human being may result more from a breakdown in integrative mechanisms than from changes in individual cells, tissues, or organs (Shock 1977b,639). The endocrine glands and the nervous system are the two primary bodily mechanisms for integrative control over homeostasis—the maintenance of a reasonably uniform internal environment, including body temperature, acid-base balance, blood glucose, and blood oxygen.

Endocrine glands, such as the thyroid, pancreas, or pituitary, produce hormones that are transmitted through the bloodstream to organs throughout the body. With aging, there is not only a decline in organ response to these controlling hormones but also an increase in the amount of deviation from normal that is required to trigger the production of the hormones needed to restore homeostasis. There are some indications that these declines are the result of age changes in the brain centers for endocrine control and that endocrine declines can be prevented or even reversed to some extent by introducing chemicals that improve the brain's capacity to trigger the production of hormones (I. Ross 1995).

Endocrine systems show a wide variety of age changes, ranging from trivial effects to effects so profound that they resemble major diseases. For example, carbohydrate intolerance increases with age, and increased blood glucose may reach the point typical of diabetes and have similar physiological effects (Minaker, Meneilly, and Row 1985). Declines with age in endocrine responsiveness can also affect the body's reserve capacity, which in turn may increase susceptibility to disease as well as impair functional capacity.

Homeostasis also relies on the nervous system, which contains most of the mechanisms for sensing changes that require adjustment as well as the network along which signals to carry out an adjustment must pass. Cotman and Holets (1985) concluded that aging produces a variety of effects on the pathways across which nerve impulses must travel (dendric spines and synapses). In some parts of the central nervous system, dendrites and synapses decline in prevalence with age; in other parts they increase; and in still others there is no age change. Synaptic growth and restructuring probably occur well into old age.

Aging is associated with a noticeable decline in the efficiency of neuromuscular coordination (Spirduso 1995). However, there is less evidence about how other aspects of coordination by the nervous system are related to aging (Saxon and Etton 1994).

The body's control systems, endocrine and nervous, are obviously of prime importance to human survival. But whether they play a central role in physical aging remains to be seen. This is one of the most active and promising areas of current biological research.

In Sum

Physical aging involves a group of processes that reduce the viability of the body and increase its vulnerability to disease. The reduction in viability occurs at all levels—cells, molecules, tissues and organs, and control systems. The increase in vulnerability to disease results mainly from a decline in the functioning of the immune system.

The search for explanations of aging has produced numerous possibilities. Aging may result from:

1. A hereditary genetic program that sets limits on growth, aging, and longevity

2. Age-related changes in the functioning of the genetic program

3. An age-related buildup of cross-links or free radicals resulting in lowered molecular functioning

4. An age-related lowered efficiency of the immune system in identifying and destroying potentially harmful or mutated cells

5. An age-related increase in autoimmune reactions in which antibodies that destroy normal cells are produced

6. An age-related decrease in the efficiency of endocrine control of various vital functions

7. An age-related decline in the capacity of the nervous system to speedily and efficiently maintain bodily integration and prevent bodily deterioration

It will be some time before these various explanations are tested and sorted out. It is also likely that they all play some part in the process of aging and that there is no single key that will unlock the mystery of physical aging.

Physical Consequences of Aging*

Beliefs about physical changes that result from aging are used to develop and justify ideas about how elders ought to behave and how others ought to behave toward elders. Throughout this book, we continually try to establish whether various social patterns exist because of *actual* limitations imposed by age or because of *presumed* limitations that have no basis in fact. At this point, let us examine basic facts about the consequences of physical aging.

Physical aging affects the functioning of our bodies in many ways. It alters the amount of energy we can mobilize; our stature, mobility, and coordination; our physical appearance; and our susceptibility to illness. As we look in detail at these various kinds of changes, we need to ask three basic questions: (1) What usually happens? (2) Can decrement be prevented, treated, reversed, or compensated for? (3) What are the *social* consequences of the typical change?

Physical Energy

The amount of physical energy that a person has is a function of the capacity of the blood and organs and other bodily systems to deliver oxygen and nutrients throughout the body and to remove waste products, a process that requires complex coordination among the body's various systems. And because aging reduces both the body's capacity to coordinate its systems and the level of functioning of these systems, aging reduces the supply of physical energy that the body can mobilize. For example, Spirduso (1995)

reported that the ability to get oxygen into the blood peaks at about 20 years of age, declines sharply until about 45, and then remains steady in active people through 70. At age 70, the volume of oxygen that can be gotten into the blood by sedentary people is only about half as much as at age 25. However, this performance difference can be reduced through exercise training (Buskirk 1985).

The socially important issue here is the extent to which this decline in physical energy interferes with typical social functioning. Spirduso reported that at low to moderate levels of physical work, age does not affect the ability to perform work. The only effect associated with age was the somewhat longer time required to recover from work. Most of the work that adult Americans perform is in the low to moderate range—well within the physical capacities of older adults who have no disabling chronic condition (at least 60 percent of older people).

Sleep

Sleep is also affected by aging. Cross-sectional research has found dramatic age differences in both amount and quality of sleep (Woodruff 1985). For example, the daily average sleep time at age 25 is just over 7 hours, at 40 it is just under 7 hours, at 60 it is below 6 hours, and at 75 it is below 5 hours. In addition, age brings a marked increase in awakenings during the night. Older people often complain of disrupted sleep, and this problem is more common among older men than older women. However, these gender differences appear to be merely an exaggeration of gender differences that exist much earlier in the life span (Dement et al. 1985).

Snoring is also more common among men, and heavy snoring increases with age. Loud snoring is more than an inconvenience; it may seriously disrupt the sleep of others. Lugaresi et al. (1975) found that all the heavy snorers they studied showed blood pressure

---

*For an excellent review of the physiology of aging, see Whitbourne (1985).

Exercise enhances the ability to mobilize energy.
Photograph by Suzanne Murphy/Tony Stone Images, Inc.

abnormalities during sleep. They concluded that lifelong snoring may be an important risk factor in the development of heart disease.

Apnea is the temporary suspension of respiration. Sleep apnea also increases with age and has been found to be associated with excessive daytime sleepiness. Irregular heartbeat and high blood pressure are both more likely to occur during sleep apnea.

The cause of these age differences in sleep patterns is unknown. More people are seeking treatment for sleep disorders, which should improve our research base. However, research on sleep deprivation suggests that the quality of sleep after the age of 50 could be improved if people would spend less time in bed and rely less on medications (Woodruff 1985).

Stature, Mobility, and Coordination

Physical activity depends partly on the structure of the body and its ability to move effectively. As adults grow older, they actually get shorter, partly because bones that have become more porous develop curvature and partly because some older people carry themselves with a slight bend at the hips and knees. A height loss of 3 inches is not uncommon. Older women are especially likely to find themselves too short to reach conveniently in environments designed for the common "adult standard." For instance, standard kitchen cabinets, sinks, and counters tend to be too high and too deep for many older women.

The ability to move may also be influenced by aging. For example, we know that with age, arthritis increases and connective tissue in joints stiffens. However, we do not know how these changes influence the ability to flex and extend arms, legs, and fingers. This is a major area in need of research. Muscle strength also affects the capacity to move. There is little age change in strength until about age 50, and the loss through 70 appears minimal—about 15 percent. But by age 80, loss of strength may reach 40 percent (Whitbourne 1985,21). Again, exercise can partially prevent or reverse this loss (Buskirk 1985).

Physical coordination is a complex process that involves taking in sensory information, attaching meaning to it through perception, selecting appropriate action based on that perception, transmitting instructions to the various parts of the body that need to act, and initiating action. Coordination depends on several body organs and systems: sensory organs provide information; the nervous system transmits that information to the brain; various parts of the brain handle perception and selection, initiation, and monitoring of action; and various muscle groups perform action under the control of the nervous system and the brain. In most cases, these separate functions occur in such rapid succession that the interval between sensation and response action is quite short. This rapidity is especially true of practiced skills such as typing, playing a musical instrument, and operating familiar equipment such as automobiles.

Aging can influence coordination through its effects on any of the systems that support physical activity. In fact, however, sensory organs and muscle groups generally perform well into old age. Coordination and performance are much more likely to be influenced by age changes in the brain-based functions, especially by a slowdown in decision making related to performance (Spirduso and MacRae 1990).

Physical Appearance

Wrinkled skin, "age spots," gray hair, and midriff bulge are common examples of age changes in appearance. Although it is true that the skin becomes wrinkled with age and is more susceptible to dryness and loss of hair, it is still quite able to perform its protective function much longer than the theoretical life span of 120 years. The significance of aging skin is mainly social rather than functional.

Although people do not die of "old skin," two-thirds of people 70 or older have skin disorders that require medical treatment (Kligman, Grove, and Balin 1985). Multiple skin conditions are common among the very old and tend to be different from the skin conditions that affect the young. Severe skin problems such as impaired wound healing can hamper physical functioning.

Aging skin, gray hair, and a thicker waistline are significant primarily because they symbolize membership in a less desirable social category—older people. Oil of Olay, Clairol, and Buster's Magic Tummy Tightener enjoy brisk sales not because people want to be young, but because they want to avoid being classified as old. This, in turn, has nothing to do with any genuine physical effect of aging; it reflects the fact that older people are often looked down on and discriminated against.

Age changes in appearance are also feared for what they may do to physical attractiveness. Yet aging does not predictably affect physical attractiveness. Certainly, if we take a narrow, idealized view of physical beauty, then not very many people fit the bill at any age. But in terms of practical attractiveness—the ability to draw positive attention through one's physical appearance—there are plenty of unattractive 25-year-olds. And many people find that they become more attractive as they grow older.

Aging rarely transforms a silk purse into a sow's ear; more often it provides interesting facial lines, the image of experience and character. Nevertheless, people tend to fear what aging will do to their appearance, especially if they see attractiveness as one of their major assets. Harris (1994) surveyed adults of all ages in New Mexico and found a gender difference in terms of the importance attributed to various effects of aging on appearance. Balding was seen by both men and women as the most negative appearance factor associated with aging, but only when it happened to women. Balding was perceived as neutral for men. Neck wrinkles were also seen as negative for women but not for men by both genders. Men thought that changing body shape was negative for women, but women did not share this view. Interestingly, neither gray hair nor white hair were seen negatively by respondents of either gender. Nevertheless, 34 percent of the women respondents and 6 percent of the men colored their hair, and about 10 percent of both genders admitted that they lie about their age. Face wrinkles were also seen as a neutral feature, but 24 percent of women reported using wrinkle creams.

Ironically, the respondents also held negative opinions of both men and women who tried to alter the effects of aging on their appearance. These results suggest that insecurity about the influence of aging on appearance is widespread, but the proportion of aging people who take steps to alter their appearance is much smaller than most people think because of the negative evaluations attached to such actions. For example, Harris's respondents estimated that two-thirds of women color their hair, but the proportion of women who actually did so was only 34 percent. Most respondents of both genders who took steps to alter their appearance did so for vanity and to feel better about themselves, not to appear better to others. Women also cited job reasons as important factors motivating their attempts to look younger, but men did not.

Some people try to alter the effects of aging surgically. The main reasons cited for having cosmetic surgery in middle age or later are job security, marital security, and social attitudes about appearance (Wantz and Gay 1981,154). Goodman (1994) studied a small sample of women who had undergone cosmetic procedures and a matched sample who had not. She found that younger women tended to opt for surgical procedures that either augmented or reduced their breasts or that removed fat from stomach or thighs, whereas women over 50 mostly had face-lifts or facial chemical peels. Goodman found that women who had cosmetic surgery had higher self-esteem than women who had not had cosmetic surgery, even before their surgery. The popular press contains accounts of increased numbers of men opting for cosmetic surgery for job reasons, but so far there is no research literature supporting this contention.

Most people tend to be preoccupied with physical attractiveness and to ignore other attractive features. Yet the secret of many of the most attractive people is their enthusiasm, attention, and personality, characteristics that can improve with age. Still, some people experience aging as a loss of attractiveness, a change they can find difficult to deal with. Fortunately, most of us grow older with mates and friends whose actions reassure us that changes in physical appearance are not nearly as important as we have been led to think they are.

Other Physical Changes

The cessation of menstruation in women is called **menopause.** While it usually occurs in middle age (often between 45 and 55), young women frequently associate menopause with later life. Menopause is a transition, not a single event. It is caused by a gradual decline in the body's production of the "female" hormones estrogen and progesterone. As this hormonal decline occurs, women can experience a wide variety of physiological and emotional responses. Apparently there is no typical pattern. Irregular menstrual periods, "hot flashes," vaginal dryness, and mood changes can occur in many different combinations during menopause. To make matters more difficult, the time required to complete the transition varies from a few months to several years. Menopause is generally considered to have been completed when a woman has not had a menstrual period for one year.

Medical treatment for menopause generally consists of estrogen replacement therapy, which involves taking estrogen as a medication. Estrogen replacement therapy often reduces the unpleasant symptoms of menopause, but it remains controversial. While estrogen replacement alleviates symptoms of menopause and provides some protection against heart disease and osteoporosis, it may also cause unpleasant side effects such as weight gain or nausea and may increase the risk of breast and uterine cancer. Recent studies of the effects of estrogen replacement therapy of more than five years duration suggest that the increased long-term risks of breast cancer may outweigh the potential benefits of a lowered risk of heart disease.

Aging reduces the speed with which the nervous system can process information or send signals for action. This decline in the speed of operation of the nervous system is a widespread problem even in middle age. Most people begin to notice lagging reflexes and reaction time in their late 40s. Changes in the nervous system also influence most psychological processes, which are dealt with in Chapter 5.

Failure of the circulatory system is the most common cause of death for people over 40. Heart disease or interrupted blood flow to the brain or heart are prevalent among older people as a result of reduced cardiac output, reduced elasticity of the large arteries, and general deterioration of the blood vessels. At age 75, the probability of death from cardiovascular disease is 150 times higher than at 35.

The respiratory, digestive, excretory, reproductive, and temperature-control systems all decline with age. But while these are among the most common and major effects of aging, only rarely do these systems decline enough with age to produce disability (Whitbourne 1985).

Aging and Physical Health*

Health is a central factor in everyone's life. Most people, including older people, are fortunate to be able to take good health for granted. Good health enables people to participate fully in their customary lifestyle. Poor health, on the other hand, negatively affects life satisfaction, participation in most social roles, and the way others treat us. As age in-

*This section deals with health characteristics; health behavior is considered in Chapter 14, "Health Care and Long-Term Care."

creases, the proportion of people with health problems increases steadily, but even among people in their 80s, a majority rate their health as good and suffer no serious disabling effects of chronic illness. That is the good news. The bad news is that individually it is very difficult to predict who will experience physical disability as part of their aging. Although declining health and disability are more common and occur at earlier ages among people at the lower socioeconomic levels, these negative outcomes occur at all socioeconomic levels within the population. This unpredictability is probably responsible for much of our fear of aging.

Health is a general concept that is difficult to define. It refers not merely to the absence of disease or disability but also to positive states, such as mental, physical, and social well-being. Rather than consider health as something one does or does not have, it is more useful to look at health as a continuum. Figure 4-1 shows several points along the continuum from good health to poor health. At the positive end is complete social, physical, and mental well-being, where individuals not only have no illness or disability, but they also experience the vitality that comes from a well-functioning body and mind. The next point on the health continuum is absence of physical disease or impairment. Here individuals are okay in that they are not ill or impaired, but their health is a neutral rather than a positive force in their lives. Next is the onset of health conditions that cause illness or impairment. If such conditions become serious enough, individuals may find it difficult to maintain their customary lifestyle and may have to limit their activities. Beyond limitation of activities, people may find that their physical condition becomes impaired to the point that they need help from others. If the help needed is minor or only sporadic, then the level of disability might be termed mild. If the person needs help with several tasks of daily living, we might call the disability severe. Extreme

Figure 4-1 The human continuum.

Excellent health						Poor health
Social, physical, and mental well-being	Absence of disease or impairment	Presence of a health condition	Restriction of activity	Mild disability	Severe disability	Extreme frailty

need help from others

frailty is typified by severe disability coupled with a need for constant monitoring by others. These points along the health continuum are arbitrary to some extent, but they are a useful framework for comparing the health of older people with that of people in other age categories.

The two general ways of measuring health are subjective global assessments of health and measures based on information about specific health conditions. We look first at the age pattern in general self-assessments of health. In the 1990 National Health Interview Survey, 72 percent of older respondents rated their health as good to excellent, and 28 percent reported their health as fair to poor (National Center for Health Statistics 1991a). By contrast, the proportion rating their health as fair to poor was 16 percent for those age 45 to 64 and 6 percent for those age 25 to 44. Thus, the proportion rating their health as fair to poor was higher among older people but still a distinct minority.

Self-rated health among older people is strongly related to income. For example, 39 percent of older Americans with 1990 annual incomes below $10,000 rated their health as fair to poor, compared with 19 percent among those with incomes of $35,000 or more (National Center for Health Statistics 1991a). But note that even among the poor, most older people rate their health as good to excellent.

If we look at the prevalence of health conditions, it becomes clear that a large proportion of older people who rate their health as good to excellent have health conditions that make their health less than ideal. Only about 15 percent of the older population reports no

disease or impairment. For these people, aging is a gradual change; physical decrements can be accommodated within their customary lifestyle and good health is a reality. The proportion with no disease or impairment declines steadily with increasing age, but even among 90-year-olds there are people who have no health conditions that limit their activities.

Heredity, lifestyle, and chance all have a part in producing these positive outcomes. People from healthy families tend to be healthy themselves. Absence of risk factors such as smoking and substance abuse promote positive outcomes. And luck plays an important part in avoiding illness and injury. For example, people often cannot know in advance the health hazards of the occupations they select, so having a job that is not health-threatening is to some extent a matter of chance.

Health Conditions

The three basic varieties of health conditions are acute conditions, which are expected to be temporary; chronic conditions, which are long-term or permanent; and injury conditions, such as bone fractures, lacerations, or burns.

An **acute condition** is expected to be of limited duration and may be as mild as a bruised foot or as serious as pneumonia. Within the population, the incidence of acute conditions decreases with age. For example, in 1990, Americans under age 5 had an average of 3.7 acute conditions per year; the incidence decreased to 1.1 acute conditions per year for those age 65 and over. While the incidence of all types of acute conditions decreases with age, the decrease for influenza is not as pronounced as the decrease for "childhood" diseases such as mumps and measles and for the common cold. However, when older people experience acute conditions, the conditions disable them longer (see Table 4-1).

There is a substantial gender difference in the incidence and duration of acute conditions. Compared with men, women at all adult ages show a higher incidence of acute conditions and a longer duration of restricted activity. However, this gender difference narrows after age 45 (National Center for Health Statistics 1991a).

Chronic conditions include conditions that are long-term, leave residual disability, require special training for rehabilitation, or may be expected to require a long period of supervision, observation, or care. Chronic conditions include *diseases* such as asthma, high blood pressure (hypertension), diabetes, heart disease, and arthritis, and *impairments* such as deafness, paralysis, and permanent stiffness in joints. Chronic conditions are long-term but not necessarily disabling. For example, chronic sinusitis is prevalent in today's older population, yet it is seldom disabling.

Figure 4-2 shows the most prevalent chronic conditions among middle-aged and older people. Arthritis, hypertension, and hearing impairment are by far the most prevalent chronic conditions. The proportions of older people afflicted by them are higher at older ages. By age 75 or older, over half have arthritis, and well over one-third suffer from hypertension, one-third have hearing impairments, and about one-third have heart conditions (National Center for Health Statistics 1991a). Among the top ten chronic conditions, only sinusitis is not potentially disabling.

In the general population in 1990, the incidence of *injuries* declined with age, from 29 per 100 people at age 17 and under to 15 per 100 at age 65 and over (National Center for Health Statistics 1991a). More than half of

Table 4-1 Average number of acute conditions per person per year and average duration of restricted activity, by age: United States, 1990.

Age	Average Number of Acute Conditions	Average Number of Days of Restricted Activity per Acute Episode
0–4	3.7	2.6
5–7	2.4	2.8
18–24	1.7	4.5
25–44	1.5	4.4
45–64	1.1	5.3
65+	1.1	7.9

Source: Compiled by the author from National Center for Health Statistics (1991a) data.

Disability need not be an end to happiness.
Photograph by Paul Damien/Tony Stone Images, Inc.

the injuries experienced by older people occurred at home.

Older people are much more susceptible to injury from *falls* than younger people. Burnside (1981) found falls to be the leading cause of accidental death among people 75 and over. Hornbrook et al. (1994) reported that a quarter of community-dwelling elders

Figure 4-2 Prevalence of the top nine chronic conditions, by age: United States, 1990.

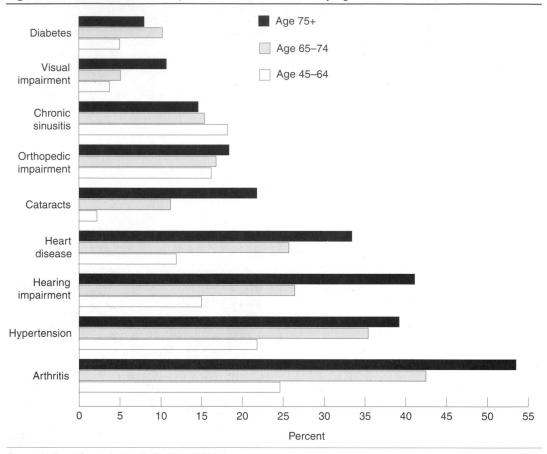

Source: National Center for Health Statistics (1991a).

age 65 to 74 experienced at least one fall during the past year. The incidence of falls increased with age, and over half of frail elders reported falling during the preceding year. Two-thirds of elders who experienced falls had a second fall within six months. Older women are especially likely to fall. The most common cause of falls is tripping or stumbling over something at home, no doubt related to poor coordination and failing eyesight. Falls also occur as a result of dizziness, which is probably related to chronic conditions such as high blood pressure and to side effects of the medications used to treat these conditions. Medications used to treat anxiety, depression, or insomnia also contribute to the problem. Efforts to prevent falls have so far had only minimal success.

Fortunately, most chronic conditions can be effectively managed through rehabilitation and health care. As a result, only about 45 percent of older people with chronic conditions find their activities limited by them, and severe limitation of activity is uncommon. As Figure 4-3 shows, severe limitation of activity occurs for only about 2 percent of people 65 to 74, and at age 85 and over, severe limitation of activity occurs for only about 12 percent of women and just under 8 percent of men.

Another way to look at the effects of health conditions is to examine the prevalence of need for help with various activities involved in daily living. Figure 4-4 shows the proportions of older people not living in institutions who needed help with selected ac-

Figure 4-3 Percentage of the older population having severe limitation of activity, by age and sex: United States, 1983.

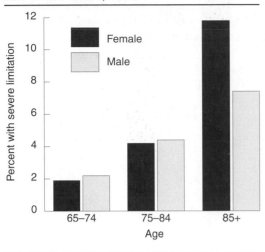

Source: U.S. Senate Special Committee on Aging (1986).

tivities. At 65 to 74, only about 3 to 5 percent of people needed help with any of the activities listed, but by age 85 and over, more than 25 percent needed help with shopping, more than 20 percent needed help bathing, and more than 15 percent needed help with light housework and preparing meals. In general, then, we can say that severe disability is uncommon among older people, but there is a significant proportion who need help, especially among the oldest-old. Still, those who need help are a small although significant minority.

Manton, Corder, and Stallard (1993) used the longitudinal National Long-Term Care Surveys of 1982, 1984, and 1989 to examine changes in the prevalence of functional limitations such as those shown in Figure 4-4. They found a substantial reduction in the

Figure 4-4 Percentage of older population needing help in selected activities, by age: United States, 1984.

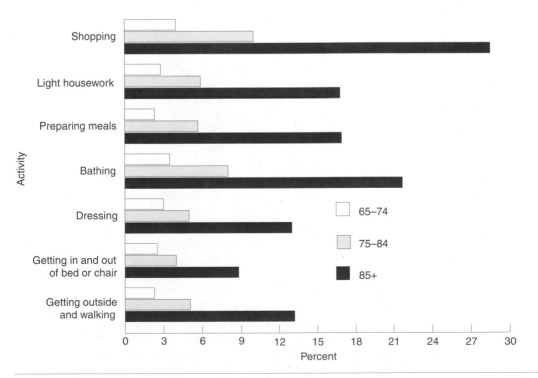

Source: National Center for Health Statistics (1989a).

prevalence of disability from 1984 to 1989, even though the panel had grown older. These findings indicate that improvement in health more than offset the effects of the aging of the study panel. If these declines in disability continue, then active life expectancy will increase even in the face of population aging.

Many health-care professionals, such as physicians and nurses, assume that a certain amount of limiting illness is normal for an aging person, an assumption that is *absolutely untrue*. Yes, limiting chronic conditions are common among older people, but many of these conditions are preventable, most are treatable, and all can be compensated for to some extent.

The two types of social consequences of health conditions are: (1) social limits imposed by other people's perceptions of the condition, and (2) functional limits imposed by the condition itself. Social limits are often much more severe than the physical limits. For instance, take the case of Dr. T. He suffered a heart attack at the age of 49 and was disabled for about two months. Still, it took him almost two years to convince his colleagues that he had recovered and that they could resume referring patients to him. The major functional limit caused by health conditions in later life is in the area of employment. Parnes and Nestel (1981) found that 51 percent of the 2,016 men who retired during their ten-year longitudinal study had retired because of poor health. Poor health also poses by far the greatest limitation on both participation in the community and leisure pursuits, although the proportion of older people affected in nonemployment areas is under 20 percent.

Compensation is a crucial concept in social gerontology because people attempt to intervene when aging produces negative effects in their lives. As a process, compensation involves taking action to offset or counterbalance undesired outcomes. Thus, aging people do not simply put up with chronic disease, they attempt to manage their chronic conditions so as to eliminate or minimize negative impacts on their lifestyles. As we will see, compensations occur in all areas of individual aging.

Summary

The basic causes of physical aging are still a mystery. The maximum life span of the species is probably genetically determined, and differences in individual longevity are definitely related to differences in heredity. However, these factors do not explain why, even when free of disease, we age as we approach the end of our allotted life span. Numerous factors may explain physical aging. Errors in the production of new cells may make these cells less effective and, in turn, create less viable tissues, organs, and physiological systems. The body's immune system becomes less effective at warding off disease, and it also produces increasing numbers of autoimmune antibodies that attack normal cells. The endocrine glands become less efficient at maintaining homeostasis. The nervous system's capacity to control and integrate various functions declines. These changes, often cumulative, can all interact with one another.

The consequences of all these physiological changes include reduced capacity to mobilize physical energy, shorter stature, and lessened mobility and coordination. However, in the absence of disease, these changes seldom reduce physical resources below the minimum needed for normal adult functioning.

Physical changes such as wrinkled skin, gray hair, and midriff bulge are functionally unimportant. But they are important symbolically because they are used to label people as belonging to that less desirable social category—older people. Changes in appearance can also be threatening to those who idealize physical attractiveness.

As a cohort grows older, the incidence of acute, short-term illness declines while chronic conditions become more prevalent. But although these changes accompany aging of the *cohort,* they do not necessarily accompany aging for the *individuals* in the cohort. The older a cohort is, the larger the percentage in the cohort whose activities are limited by chronic conditions. Nonetheless, the proportion of older people with no limitations on activity has actually gone up substantially in recent years, indicating that more people are enjoying a healthy aging. The major causes of activity limitation in later life are arthritis and heart disease, both moderately preventable and treatable now and likely to become more so in the future.

Thus, while physical aging certainly involves loss of capacity, the key question is how much capacity is left. Those aging people who avoid or successfully manage disabling chronic conditions are left with ample physical resources to continue leading enjoyable lives, and many disabled elders lead quite satisfying lives despite their disabilities.

5 Psychological Aspects of Aging

Psychology is a vast field of study that includes specific capacities such as vision, memory, thinking, problem solving, intelligence, reaction time, and the like. It also covers the more holistic study of the mind: the nature and content of consciousness, the role of the unconscious, and the relation of mind to behavior. Psychology is also concerned with the interchange between the person and her or his physical and social environments and with the person's adaptation to internal and external experiences of the world. In addition, psychology is concerned with mental illness. Aging has been studied in relation to all these concerns of psychology. In this chapter we review concepts and research findings in three very different areas: (1) the effects of aging on specific psychological functions, (2) aging and adult development, and (3) aging and mental health.

In recent years, the volume of research in psychology has exploded, especially concerning specific psychological functions, making a thorough review well beyond the scope of this book. However, we will look at the major findings from each area and refer those of you interested in more depth to recent, extensive reviews of the literature, several of which are particularly valuable: Birren and Schaie (1985, 1990); Birren, Sloane, and Cohen (1992); Salthouse (1991c); Alexander and Langer (1990); Whitbourne (1985); Schaie (1987); Craik and Salthouse (1992); and Kastenbaum (1993).

Aging and Specific Psychological Functions

Many of the basic psychological processes, such as sensation, perception, and motor performance, have obvious links to physical aging. As a result, they tend to follow the maturation/maturity/aging model presented in Chapter 4. However, when we begin to examine psychological processes in which experience and practice can play a part, such as intelligence, problem solving, and learning, then the picture is less clear-cut. The more closely a function is tied to physical capacities, particularly physical coordination, the more likely it is to decline with age. The more heavily a function depends on experience, the more likely it will increase and the less likely it will decline with age. In general, practiced skills tend to be maintained to a greater extent than unpracticed ones. In addition, although many laboratory studies show declines in various aspects of cognition, such as memory and learning, data from real-life performance of older adults show much less serious age-related changes. Finally, as with physical functions, the point of onset of psychological aging and the rate of change connected with it vary widely from person to person.

As we examine various psychological functions, keep in mind that we are interested mainly in the practical effects of age on the social functioning and psychological well-being of aging people. However, these effects are much more difficult to establish than might appear. For one thing, aging is only one of many factors that can cause changes in individual functioning. For example, physical illness can cause dramatic changes in behavioral functioning, but as we saw earlier, physical illness is not aging. In addition, aging-related changes are often very modest in magnitude. They may produce a certain amount of drama on a laboratory researcher's graph, but their practical significance may be much more limited. Finally, we cannot completely disentangle age, period, and cohort effects to isolate the "pure" effect of aging. (See Chapter 1 for a discussion of this issue.)

Sensory Processes

The senses are the means by which the human mind experiences the world outside and inside the body. To adapt to and interact with the environment, the individual depends on the senses to gather information about it. Sensory organs pick up information about changes in the internal or external environment and pass this information on to the brain, where all the input from the sensory organs is collected and organized. The subjective result is called *sensory experience.*

The minimum amount of stimulation a sensory organ must experience before sensory information is taken in is called a *threshold;* the higher the threshold, the more stimulus there must be for information to be detected. Thus, for example, some people have a low threshold for sound and require very little sound to hear, but others have a high sound threshold and require a considerable amount. In studying the operation of the sensory processes as a function of age, we are concerned not only with changes in threshold that come with advancing age but also with the possible failure of a particular sensory process.*

---

*For a more detailed discussion of aging and sensory processes, see Birren, Sloane, and Cohen (1992); Birren and Schaie (1985, 1990); and Whitbourne (1985).

Vision Vision is a particularly adaptable sense. It can receive experience over a wide range of color, intensity, distance, and field angle. The eye has an iris for controlling the amount of light that reaches the optic nerve via the retina and a lens for bending the light that enters the pupil (the opening in the center of the iris) to focus the light pattern on the retina.

To focus light from both near and distant objects, the eye lens must change shape. As age increases, the ability of the lens to change shape and thereby focus on very near objects decreases. Hence, many older people need glasses for reading, if for nothing else. The tendency toward farsightedness (presbyopia) increases about tenfold between the ages of 10 and 60, but not much more thereafter (Whitbourne 1985).

Age also influences the size of the pupil, which is important because the pupil controls the amount of light that enters the eye. Lens shape and pupil size are important because proper focusing of an image on the retina requires both suitable quality of light (controlled by the lens) and the proper amount (controlled by the iris). Since both lens and iris decline in function with increasing age, vision as a whole gets poorer. The average diameter of the pupil also narrows with age, greatly decreasing the amount of light entering the eye. The eye of the average 60-year-old admits only about one-third as much light as the eye of the average 20-year-old.

Visual acuity is relatively poor for young children but rapidly increases in quality, hitting its peak at about age 30. Thereafter, a marked decline begins. By 60, few people have normal vision, even with correction. The most common problem is loss of ability to focus on near objects, which hampers ability to read. However, there are many exceptions to this general trend because visual acuity is one of the most variable human senses.

But this is not all of the story. We are actually less interested in peak performance than in ability to perform to a socially defined minimum level of performance. For example, to get an unrestricted driver's license in Ohio, an adult must have distance visual acuity of 20/40 or better. The percentage of people with *uncorrected* distance vision of 20/40 or better increases slightly from age 20 to age 40 and then drops sharply until age 70, when it begins to level off. Aging reduces the percentage who are visually qualified to drive from over 90 percent at age 40 to just 30 percent at age 80. However, if we look at *corrected* vision, at age 60 more than 90 percent are still visually qualified for unrestricted driving, and at age 80, 70 percent are still qualified. In addition, people with 20/60 to 20/45 corrected distance vision are qualified by law for a "daytime only" driver's license, which adds another 15 percent—making a total of 85 percent qualified to drive at the age of 80. Thus, *compensation* increases visual qualification at age 80 from just 30 percent without compensation to 85 percent with it.

Visual adaptation to dark involves both the speed and adequacy of adaptation. Older people seem to adapt about as fast as the young, but their adaptation is not nearly as good. They also find it more difficult to distinguish between levels of brightness. In addition, the stimulus threshold of the optic nerve increases, meaning that more light is required to produce vision. Color vision also changes as the individual grows older: the lens gradually yellows and thereby filters out violet, blue, and green. The threshold for these colors increases significantly as people grow older, which is why older people usually see yellow, orange, and red more easily than darker colors.

Because of all these changes to the eye, older people need either glasses or large-print books for reading, and close work of all kinds becomes more difficult. General levels of illumination must be significantly higher for older people to get the same visual impact as young people. Older people need more visual contrast compared with young adults. Poor adaptation

to darkness means that older people have difficulty driving at night—which does not mean that older people should not drive at night, merely that they must exercise extra care. Finally, changes in color vision mean that for older people to get the same satisfaction from looking at colors as young people, their environment must present more yellow, orange, and red and less violet, blue, and green.

About 25 percent of people 75 and over have cataracts, a clouding of the lens that diffuses light, heightens sensitivity to glare, and impairs vision. Treatment of cataracts, which is successful in about 95 percent of cases, involves surgically removing the natural lens and replacing it with a contact lens.

Only a small percentage of older Americans are severely visually impaired or legally blind. However, visual impairment is strongly age-related. Among about 1.5 million Americans who are unable to read a newspaper with or without glasses, about 70 percent are age 65 or over. Nearly half of the people who are legally blind in the United States are 65 or older (Hooper 1994). Among those who are legally blind, 85 percent have low vision rather than total blindness. *Low vision* is severe vision loss that cannot be corrected by glasses, medication, or surgery. Lowman and Kirchener (1979) projected that by the year 2000, with the aging of the older population, over 350,000 older Americans would be legally blind and another 1.8 million would suffer from severe visual impairment. In general, the prevalence of visual impairment "increases exponentially with age" (Kline and Schieber 1985,295). Horowitz (1994) pointed out that visual impairment is a significant contributor to functional disability for both community-dwelling and institutionalized elders. *

Hearing   Hearing, the second major sense, has many aspects, but the most essential ones are the detecting of the pitch (frequency) of

---

*For more on vision, see Hooper (1994), Orr (1992), or Weber (1991).

sound, its intensity, and the time interval over which it occurs. As people grow older, their reactions to frequency and intensity change, but there is no evidence to indicate that ability to distinguish time intervals changes significantly. Why these changes occur is not clear (Olsho, Harkins, and Lenhardt 1985).

Presbycusis refers to age-related changes in hearing ability. Presbycusis is a syndrome that includes inability to hear high-frequency tones and impaired frequency discrimination. About 80 percent of people with impaired hearing are over 45, and more than half are over 65.

Hearing loss begins at about the age of 20. Very gradually, people lose their ability to hear certain frequencies and to discriminate among adjacent frequencies. This type of hearing loss is only slight for low-pitched sounds but tends to be considerable for high-pitched sounds. Thus, as age increases, higher sounds become relatively harder to hear. About 28 percent of the older population experiences severe hearing loss (Scheiber 1992). Most loss of hearing involves loss of sensory receptors in the ear and loss of nerve cells in the auditory pathway to the brain (Marsh 1980).

Older people are more susceptible to ear damage than the young. Thus, industrial noise produces greater hearing loss among older workers than among younger ones. Corso (1981) concluded that both noise and aging contribute to the development of presbycusis. Older people, even those with relatively slight hearing loss, have greater difficulty making the fine distinctions required to comprehend speech. After about the age of 55, there is also a consistent gender difference in hearing ability, which prior to that point is about equal for men and women. After 55, considerably more men show hearing loss.

In summary, as people grow older they have more difficulty hearing high-pitched sounds and low-intensity sounds. That may explain why organ music (rich in low tones) is popular among older people, why they need to play their radios and televisions louder, and why background noise is more distracting to them.

Impaired hearing is a hearing loss of sufficient magnitude to reduce the individual's capacity to interact successfully with the environment. Loss of hearing does not always impair comprehension. For example, an individual with good eyesight can compensate for hearing loss by reading lips. Schieber (1992) concluded that 30 to 40 percent of people at age 65 will have hearing loss sufficient to produce functional disability. Whether hearing loss causes reductions in social involvement is not clear. Hearing impairment does not appear to produce paranoia, although depression is slightly more likely among those with hearing loss.

Hearing loss can have a major impact on speech communication. Age-related changes in ability to discriminate among sounds make speech more difficult to comprehend, especially when people talk fast, there is background noise, or there is distortion or reverberation of sound. Changes in capacity to make vocal sounds add to speech problems, although these changes can often be offset by retraining.*

Loss of either vision or hearing can be partly compensated for by the other, but when both senses decline simultaneously, as is sometimes the case among older people, adaptation to the environment can be a serious problem. Not only is employability adversely affected, but those who deal with such doubly disabled people must learn to take these limitations into account in order to help them make maximum use of their capabilities and opportunities.

Taste, Smell, and Body Sensations The evidence indicates that all four *taste* qualities—sweet, salt, bitter, and sour—show an elevated threshold after the age of 50. Not

---

*For greater detail on hearing, see Schieber (1992) or Hooper (1994).

only do individual taste buds become less sensitive with aging, but the number of functional taste buds also drops slightly. People in later life are apt to require more highly seasoned food to receive the same taste satisfactions as when they were 20. They also seem to prefer tart tastes and to show less interest in sweets. Nevertheless, age changes in taste sensitivity tend to be small (Moore, Nielsen, and Mistretta 1982; Weiffenbach, Baum, and Burghauser 1982; Bartoshuk et al. 1986).

Research on the sense of *smell* indicates sharp declines after age 65 in capacity to smell and to assess both strength and type of odor. After age 80, more than 75 percent of elders show major impairment in their sense of smell (Scheiber 1992).

The senses of taste and smell are of great importance because they are major components of a person's capacity to enjoy. They are also important to the capacity to survive, for example, by detecting spoiled food, smoke, or natural gas.

The so-called general *body sensations* include touch, pain, muscle movement, and vibration. Sensitivity to touch appears to increase from birth to about the age of 45, after which the threshold rises sharply. Pain is important in alerting the body to emergency, something within or outside the body that threatens its well-being. Older people appear less sensitive to pain than do young people. However, it is difficult to measure pain because both the feeling and reporting of pain are conditioned by cultural and personality factors. Laboratory data are few, and they often show no age change in pain threshold (Hooper 1994).

As people grow older, they make more errors in estimating the direction of muscle action. The ability to detect vibrations appears to decline with age only for the legs and feet.

The Perceptual Processes

The senses provide the means for assembling and classifying information, but not for evaluating it. The process of evaluating information gathered by the senses and giving it meaning is called *perception*. Not all sensory input is perceived, for perception is a conscious process and some senses, such as the sense of balance, are mainly nonconscious. But for every sense in which there is consciousness of the stimulus received, there is an evaluative perception. In perception, not all the information collected may be given equal weight. In visual perception, for example, shape appears to be a more important characteristic than color in evaluating an object.

Closure is the ability to come to a decision concerning the evaluation of a stimulus. People become less capable of closure as they grow older. This tendency may partly account for the apparent indecisiveness of some older people. Moreover, people seem to experience a decline with age in the general speed with which they can organize and evaluate stimuli.

Most research shows an age decrement in perception. One explanation is that aging affects the speed with which the nervous system can process one stimulus and make way for the next. The nervous system "trace" of the initial stimulus interferes with perception of subsequent stimuli. This explanation is consistent with a wide variety of research findings that show that with increasing age, the time between successive stimuli must be increased for accurate perception to occur. Also, certain perceptual illusions are more easily produced in older people. For example, older subjects see a continuous motion picture at about 34 frames per second, while 20-year-olds still see discrete pictures at 39 frames per second (Botwinick 1978,158).

For many of the sensory and perceptual processes, it would appear, declines in function seldom seriously hamper behavior until after the age of 75. Functional disability prior to that time is much more likely to occur as a result of disease or injury than as a result of aging per se.

Psychomotor Performance

The term *psychomotor performance* refers to a complex chain of activities that begins with a sensory mechanism and ends with a reaction, usually muscular. A muscle that acts as a part of a stimulus-response chain is called an *effector.* Between the sense organs and the effectors lies a chain of brain mechanisms called *central processes.* Ideally, psychomotor performance involves sensory input, attaching meaning to that input through perception, incorporating the perceived information into the mind alongside other ideas (integration), making a decision concerning any action required by this new information, sending neural signals to the appropriate effector, and activating the effector (response). This process is much more complex than simple sensation or perception.

Psychomotor performance is limited by the capacities of the various parts of the sensorimotor system. In most cases, the sensory and effector mechanisms are quite capable of handling most tasks. Hence, the limits on performance are usually set by central processes dealing with perception, translation from perception into action, and detailed control of action. Decision making related to performance is particularly slowed by age (Salthouse 1985).

Performance capacity is a function not only of the available pathways in the brain but also of the strength of the sensory "signal" and its relative strength compared with other signals entering the brain at the same time. A decline in absolute or relative signal strength can be partially offset by taking more time to integrate incoming data. Such *compensation* may explain much of the slowness of performance associated with aging. Errors are likely to result if the older individual cannot take the extra time needed—too little information is being processed between the moment sensory input is received and the time when action must be taken.

The most serious sensory limitation on psychomotor performance is the general rise in sensory thresholds. Once a sensory response is triggered, however, the senses pose little problem. Research findings suggest that changes in performance would remain even if there were some way to eliminate sensory decline completely (Salthouse 1991c). There is only a negligible decline in the speed of nerve conduction as age increases.

The effectors usually are not a prime weakness in the psychomotor chain. In fact, any poor muscle performance in old age may be more a result of poor coordination than decline in muscular strength or endurance (Whitbourne 1985). Since coordination is under the control of the central processes, any inefficiency of effectors in old age could be considered a result rather than a cause of poor psychomotor performance.

The important limitations on psychomotor performance, then, appear to come from the central processes. But regardless of the cause of the limitations, there are definite changes in observed psychomotor performance as the individual ages. The most important changes from the point of view of social functioning are in reaction time, speed and accuracy of response, and ability to make complex responses.

Reaction Time Reaction time is usually defined as the period that elapses between the presentation of a stimulus and the beginning of the response to that stimulus; traditionally it has been considered a measure of the time used by the central processes.

Verbal reaction time tends to be similar for young and older adults, but motor reaction time increases beginning at about age 50 (Spirduso and MacRae 1990). Although very slight for simple tasks, this increase becomes greater as tasks grow more complex. The more choices involved in a task, the longer it takes a person to react. The slowing of reactions in older people may result mainly from a tendency toward care and accuracy that seems to characterize this group. Since older people tend to spend more than an average amount of time checking results, part of their slowness may be the difference between the time required for *accuracy* and that required for *certainty*.

The slowing of reaction time appears to involve chiefly the central processes. Neither the speed of input nor the speed of output is responsible, and, as mentioned, slowing results partly from a desire for certainty rather than a physical inability to act quickly. However, response time does not have a clear influence on higher processes such as cognition. In addition, exercise, increased motivation, and practice can partly counteract the effects of slowed reaction time (Spirduso and MacRae 1990). Finally, individual differences in reaction time are so great that even with age changes, some older people are quicker in response time than some young adults.

Speed and Accuracy   *Speed* of movement also tends to decrease with age. Again, the evidence points to the central processes as the source rather than to any loss of muscle ability. In fact, when older people try to hurry, their control capabilities are often so poor that their movements appear jerky. The central processes are further implicated because for simple movements the slowing with age is only slight, whereas for complex movements, in which the same muscles must be more controlled, the slowing with age is more marked.

These same observations apply to *accuracy*. Accuracy of movement declines as the individual grows older, unless more time is taken to compensate for the greater difficulty in controlling the response (Salthouse 1991b).

Complex Performance   A complex performance is a series of actions in response to a complex stimulus. For example, Salthouse (1984) examined the performance of professional transcript typists of various ages to find out if practice prevented deficiencies in the component processes involved in typing or if the older typists used some other means of compensation. He found that older typists had much slower reaction times compared with younger typists but that their times between striking keys were nearly identical, which meant that, although the older typists were slower on one of the components of typing, they somehow made up for this deficiency. In further investigation, Salthouse found that there was a significant increase with age in the size of the gap between the character being read and the character being typed, with typists in their 60s reading on average 4.8 characters ahead compared with only 2.4 characters for typists in their 20s. Thus, the older typists had reorganized their visual scanning to allow them more time to process each character, which *compensated* for their slower reaction times and produced typing speeds equivalent to those of the younger typists. This compensation had occurred gradually and nonconsciously.

In summary, it appears that the psychomotor performance of older people is limited more by the central processes than by any other factor. The central processes can handle only a limited amount in a given time. Any loss in capacity can be offset by taking longer, and as the typists demonstrated, strategies for taking longer may not affect overall performance speed. When taking longer is not possible, a much larger percentage of errors will occur among older people than among young adults. The more complex the integrating and controlling functions must be, the more aging slows down performance, particularly for unfamiliar tasks. To compensate, older people often shift their emphasis from speed to accuracy.

These various factors in psychomotor performance have important implications. First, factors other than changes in the central processes have relatively little influence on psychomotor response. Also, the central processes are very difficult to offset mechanically, the way glasses or hearing aids can offset sensory loss. Instead, compensation often takes the form of revised processing strategies. People can also maintain effective functioning by concentrating activities in domains where competence has remained high and avoiding areas of declining performance (Salthouse 1990b). Also, remember that performance varies considerably within age groups. Therefore, age is a poor predictor of individual performance. Overall, age changes in psychomotor performance have little effect on the capacity of older adults to function in typical adult activities such as employment, work around the home, and leisure (Salthouse 1982).

Mental Functioning

Sensation, perception, and psychomotor performance are all very important for the functioning of the individual, yet in human beings as in no other animal these processes primarily serve mental functioning. The term *mental functioning* refers to a large group of complex processes, subdivided for convenience into learning, memory, thinking, problem solving, creativity, intelligence, expertise, and wisdom.

Learning   *Learning* is the acquisition of information or skills and is usually measured by improvements in task performance. All studies of performance indicate a decline in learning with age. Clearly, however, a number of factors other than learning ability affect performance, including motivation, performance speed, ill health, and physiological states. In practice, it is extremely difficult to separate these components of performance in order to examine the "pure" influence of aging on changes in learning ability, although a number of studies have attempted to do so. What this means is that we are in a poor position to say just what effect aging has on learning ability.

Although learning performance tends to decline with age, the decline is not substantial until past the age of 70 (Arenberg and Robertson-Tchabo 1980). All age groups can learn. Given a bit more time, older people can usually learn anything that other people can learn. Extra time is required to learn information or skills as well as to demonstrate that learning has occurred. Tasks that are particularly conducive to good performance by older people involve manipulation of concrete objects or symbols, distinct and unambiguous responses, and low interference from prior learning.

Arenberg and Robertson-Tchabo pointed out that the literature on learning and aging often confuses age changes in learning *ability* with age changes in *competence*. Because everyday environments seldom require people to use their maximum learning capacity, even substantial age changes can occur without affecting social competence. In addition, we should not assume that age declines in learning performance can never be modified. For example, Sterns and his colleagues (1977) trained older drivers to perform better, and improvements were still there six months later.

Memory   Memory is intimately related to learning, since to remember is partial evidence of learning. Thus, a person who does not learn has nothing to remember, and conversely, a person who cannot remember shows no sign of having learned.

There are three *stages of memory:* encoding or registration, retention or storage, and recall or retrieval. *Encoding* is the active processing of information into memory and may include both simple perception and registration of input as well as the organization and integration of new information with existing knowledge (Salthouse 1991c). *Retention* refers to the ability to sustain the encoded information over time. *Recall* is retrieval of material that has been encoded and retained. A failure at any of these stages results in no *measurable* memory.

It is commonly believed that all aspects of memory decline with advancing age. However, some older people escape memory loss altogether, and some people retain excellent memory functioning well into old age.

Although laboratory studies show declines in memory capacity with age, the amount of information stored in memory continues to increase over time for all but very ill older people (Perlmutter et al. 1987). In addition, new information is linked to already existing conceptual structures in memory, which means that the new information is associated with familiar information and therefore easier to remember. This strategy can compensate for much of the decrement in raw memory performance.

Cutler and Grams (1988) studied the prevalence of memory problems in a national sample of adults 55 and over. They found that 25 percent said they experienced no memory problems at all, and only 15 percent reported frequent problems associated with forgetting things. The incidence of memory problems did increase with age, but even at 85 and over, only 23 percent reported frequent memory problems. Health problems and sensory impairments were the factors most strongly related to memory problems. People who reported deteriorating health over the past year reported more memory problems, and people who had trouble hearing had the highest incidence of memory problems. Cutler and Grams suggested that uncorrected deafness is a frightening and debilitating condition that interferes with registration of information and contributes greatly to everyday memory problems. Perlmutter et al. (1987) pointed out that there are numerous reversible conditions that produce memory problems, including depression, vitamin deficiency, alcohol use, and use of some tranquilizers, particularly diazepam. In all, these results suggest that memory problems can be managed at least to some extent by effective prevention and compensation.

Thinking Via learning and memory, humans have at their disposal a great many separate mental images. *Thinking, problem solving,* and *creativity* all apply to the development and manipulation of ideas and symbols. Differentiation, stimulus generalization, categorization, and concept formation are fundamental thought skills that allow us to order the chaos brought into the mind by learning and perception.

Differentiation occurs at two separate levels. The first is the level of sensation, perception, or learning. Age declines in these functions would serve as effective barriers to concept formation. A second level of differentiation involves a process psychologists call *stimulus generalization*. Stimulus generalization makes different stimuli functionally equivalent; and the more similar the stimuli, the more nearly equal the responses. For example, individual oranges are unique in weight, thickness of skin, number of seeds, number of bumps on their skins, and so on, yet we have the mental capacity to consider all oranges alike. This capacity is stimulus generalization, and it is an essential prerequisite for the process of *categorization*.

The capacity for stimulus generalization appears to decline with age. Compared with younger people, elders produce more specific differentiations. Once mental information has been differentiated, it must be categorized. *Categorization* allows information to be dealt with in general terms, which is much easier than trying to deal with everything in specifics. Thus, when we encounter a stop sign, we can deal with it as a member of a class of objects rather than trying to figure out why it is there, who put it there, and so on. The assumptions we make when we encounter a stop sign are the results of stimulus generalization and categorization.

Older people seem to be particularly poor at forming concepts. *Concept formation* often involves making logical inferences and generalizations. Older people have been found to resist forming a higher-order generalization and to refuse choosing one when given the opportunity. Most studies seem to agree that as age increases, ability at concept formation and its components declines.

Yet common sense tells us that older people form concepts, and that some do it exceedingly well. What does it mean, then, when we say that older people consistently show a decline in performance on tests of differentiation and categorization? Data from studies where education level was controlled suggest that both level and type of education may modify, but not eliminate, the relationship between age and measured thinking ability. Likewise, studies that have held IQ constant have shown a reduction in the association between age and concept formation, although not completely. Some studies have also shown that declines in memory function do not account for the age decline in ability to form concepts. Other studies suggest that those who retain the greatest degree of verbal facility in old age are also those who retain the greatest skill in concept formation (Schaie 1980).

In their entirety, these data suggest that concept formation is not completely independent of other skills such as learning and intelligence, but at the same time it is not completely dependent on them either. A substantial part of the decline with age in measured ability to form concepts appears to be genuine, and not caused by artifacts of the measurement process or by the influence of intervening variables (Salthouse 1991c).

Problem Solving Problem solving involves making choices and decisions through the application of logical and analytical skills. It entails making logical deductions about categories, their properties, and differences among them. Problem solving differs from learning in that learning is the *acquisition* of skills and perceptions, while problem solving is *using* these skills and perceptions to make choices.

Older people are at a disadvantage in solving problems if many items of information must be dealt with simultaneously. They have more difficulty giving meaning to stimuli presented and more trouble remembering this information later when it must be used to derive a solution. The number of errors made in solving problems rises steadily with age.

In general, problem solving shows the same trend of decline that has been observed in the other mental processes. Again, although slight decrements occur earlier, socially meaningful decrements seldom take place until the 70s.

Creativity Creativity is problem solving that is unique, innovative, original, and inventive. Humans take great pride in their ability to innovate, to engage and transform the environment rather than merely react to it. To understand the effect of aging on creativity, we must first define creativity. Two major approaches have been used. The first uses psychometric tests to assess an individual's capacity to generate innovative alternative responses to a stimulus problem. In both longitudinal and cross-sectional studies, creativity measured in this way tends to peak somewhere in the 30s and either remain at a plateau or decline thereafter (Simonton 1990a). However, Simonton found that the *validity* of psychometric measures of creativity is questionable, because scores on these tests do not correlate with actual creative production.

A second approach to studying creativity relies on evaluations of actual creative performance. Panels of judges look at the output of scholars, artists, writers, composers, architects, engineers, and so on, in terms of both quantity and quality of their productions. Analysis has focused on three aspects of creative production: the age curve in creative production over the course of a career, the relationship between productivity and creativity, and the relationship between age of first contributions, longevity, and rate of output (Simonton, 1990a).

The general age curve of creative production resembles an inverted J. There is a slow buildup to a peak, and then a decline thereafter. However, the age at which the peak occurs and the magnitude of decline differs significantly depending on the field of endeavor. For example, in pure mathematics and theoretical physics, fields that require intense concentration of mental energy, peak creativity tends to occur in the early 30s, and there tends to be a sharp decline thereafter. On the other hand, in novel writing, history, and philosophy, peak creativity occurs in the late 40s or early 50s, with little or no drop-off well into late life (Simonton 1990a).

Part of the differences in the age pattern of creativity across fields may relate more to how work is organized in the field than to age differences in creative potential. For example, among the electrical engineers who develop new computers, eighteen-hour workdays are not unusual during the lengthy time a new machine is being created and software written for it. People who work in this field are highly susceptible to burnout, a psychological syndrome characterized by failing self-confidence, fatigue and listlessness, impaired decision-making ability, and a crisis in work-related motivation. In fields where many people are burned out by age 35, it is unsurprising that creativity drops off steeply.

When Simonton looked at how quantity and quality of creative production were related, he concluded that, although total output showed an inverted J pattern, the ratio of creative production to total production did not show an age pattern. Indeed, the age at which a particular work was done was not useful at all in predicting the value of that particular contribution. Simonton hypothesized that each person has a relatively stable probability of producing a work that will be deemed creative; therefore, those with the greatest quantity of production will also have the largest number of creative productions.

Simonton also found that individual differences in lifetime output were enormous. About 10 percent of the most prolific elite in any field produced about 50 percent of all contributions. Beginning at an early age, having a long working life, and having high output

over the entire career are all ways that creative people can amass a substantial record of creative production. Simonton concluded that the age at which a person is likely to show a peak within the lifetime record is entirely a function of the form of creative expression and the peak age typical for that form. Again, this illustrates an interaction between the creative potentials of individuals and the social organization of particular fields of work.

Simonton (1990b) summarized his more than fifteen years of studying aging and creativity with the following optimistic picture. Creativity may decline with age, but creative people seldom become devoid of creativity in old age. Creativity is motivated by the prospect of self-actualization, which is a strong motive for overcoming or compensating for even the most disabling conditions of old age. People who are extremely creative in young adulthood tend to remain above average in creativity in later life. There are many domains in which creativity does not decline appreciably in later life. There can be a resurgence of creativity in the last years of life. For example, Simonton (1989) studied 1,919 musical compositions by 172 classical composers and found that as they approached their final years of life, composers tended to produce pieces that were briefer and less complex but that were rated high on aesthetic significance by musicologists.

A crucial issue in the study of creativity revolves around social supports of creative effort. Zuckerman and Merton (1972) pointed out that within the academic science disciplines, for example, the political cards are stacked in favor of older members, who are overrepresented in committee posts, on proposal review committees, among journal referees, and so on. On the other side, however, is the pressure placed on established scholars to move into supervisory and leadership positions, which often takes them away from their own individual creative work. Thus, social factors may mediate age differences in raw creativity.

Intelligence When we think of intelligence, we usually mean an ability to acquire and apply knowledge that is both *potential* and *actual*. In practice, however, we always deal with *measured* intelligence as defined by responses to items on an intelligence test. Yet no matter how extensive or well-prepared a test is, there is always a margin of error in its measurement of actual mental ability. In addition, intelligence is a complex of abilities, not merely a single factor. For example, Baltes and Labouvie (1973) found studies reporting age patterns on as many as twenty primary mental abilities, ranging from visuomotor flexibility to inductive reasoning. How aging relates to measured intelligence depends to some extent on the specific abilities emphasized by the specific test.

Woodruff-Pak (1989) found that studies of the relationship between aging and intelligence have evolved through four stages. In stage 1, investigators looked exclusively at cross-sectional age patterns in overall intelligence test scores and concluded from these studies that intelligence declined with age. In stage 2, investigators began to look at longitudinal data and to consider different dimensions of intelligence. These studies showed that intelligence was maintained by respondents in longitudinal studies to a much greater extent than the cross-sectional results had suggested. When age patterns were examined separately for different mental abilities, some declined with age, some increased, and some remained stable—there was no uniform age pattern. In stage 3, researchers tested various interventions to see if adult intelligence test scores could be improved. They found that practice and familiarity with IQ tests did improve IQ test scores, but no more so for older people than for the young (Salthouse 1991c). In stage 4, scholars are examining the possibility of "higher" mental abilities that may increase with age. For example, tolerance of ambiguity, integrative thinking, and wisdom are all abilities that might increase with age, at least under some circumstances.

Schaie looked at the proportion of his twenty-eight-year longitudinal study participants who improved or maintained their mental abilities over a series of seven-year periods, by age at the end of the interval. Among those who were age 81 at the end of the interval, more than 70 percent maintained or improved their scores on inductive reasoning, and more than 60 percent remained stable or improved their scores in verbal meaning, spatial orientation, number, and word fluency (see Figure 5-1). The proportions improving or remaining stable to age 60 were about 80 percent for all abilities. This picture does not support the notion that intelligence declines with age. For a majority, the pattern is one of stability or increment.

On the other hand, these data indicate that nearly one-fifth of the subjects did show a decline in intelligence scores before age 60, and the proportion showing a decline increased with age. The general pattern appears to be one of increment until the late 30s or early 40s, stability until the late 50s or early 60s, small decrements until the 70s, and sharper declines thereafter. Just as we saw both positive and negative outcomes with regard to the results of physical aging, we see a similar pattern with regard to intelligence. A majority of people maintain their intelligence into later life, but a significant minority experiences a significant decline. At this point, we do not understand why measured intelligence declines with age for some people and not for others.

Can we intervene to improve intellectual performance of elders? Many studies have attempted to manipulate performance, and the results generally show that elders' performance is improved both by training targeted to specific mental skills and by general practice taking tests (Woodruff-Pak 1989). However, older people do not appear to benefit more from training or practice than younger people do. If training is given to people of various ages, all age groups improve in performance and age differences in measured

Figure 5-1 Proportion with stable performance over time, by age and type of performance.

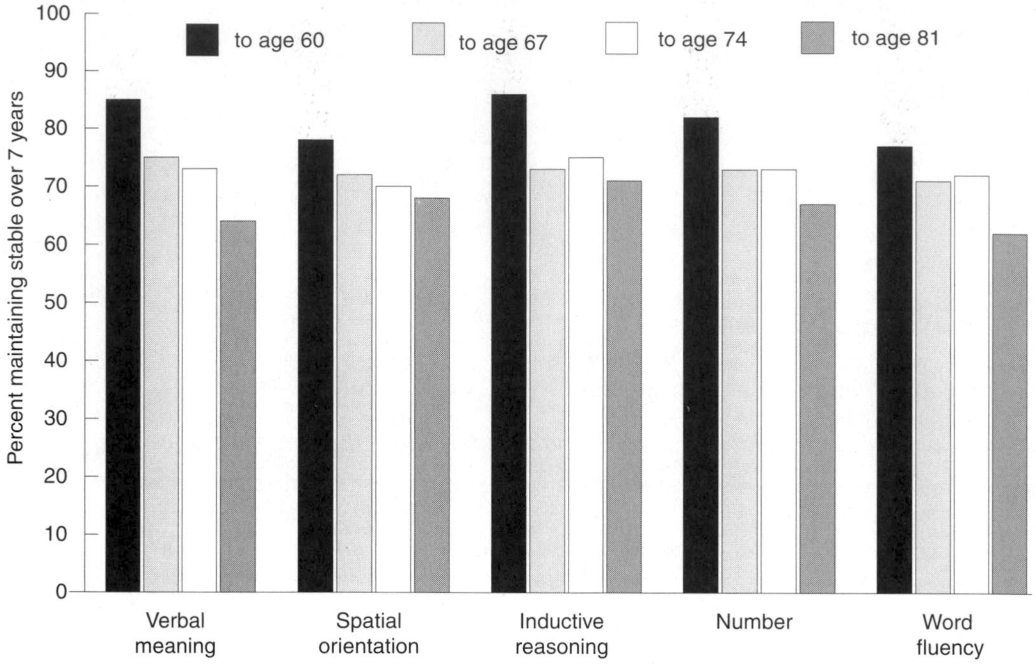

Source: Schaie (1990).

ability tend to remain, although at a higher level (Salthouse 1991c).

Most of the theories that have thus far informed the study of aging and intelligence were constructed from the outside in. That is, investigators collected information on a large number of specific mental operations and then used statistical techniques such as factor analysis of intercorrelations among the items to identify dimensions of mental ability. To begin a search for mental abilities that may grow in later life, researchers may need to ask if perhaps the detailed level of operation typical of most intelligence test items did not cause early researchers to miss important integrative dimensions. Mental qualities such as tolerance for ambiguity or wisdom or meditative consciousness may exist at a level of abstraction greater than that usually tapped by intelligence tests.

For example, Pascual-Leone (1990) contrasted "practical" intelligence with "meditative" intelligence. Whereas practical intelligence is a calculative process aimed at solving specific problems, meditative intelligence is a detachment that allows us to view any problem-solving process in a larger context. One important quality of meditative intelligence is that it allows the person to release the connection between thought and emotion and thereby to see problems from a less biased point of view. Thus, meditative intelligence may be a prerequisite for wisdom. But how could we devise a test for meditative intelligence? This is the challenge facing those who seek to study mental capacities, such as wisdom, that have traditionally been thought to grow in later life.

Expertise Expertise can be defined as extremely high levels of skill or knowledge. Usually, expertise is a function of ability plus experience. In addition, experience is often systematically provided by training and practice (Salthouse 1987). Expertise requires extensive factual and how-to knowledge. In addition, expert knowledge is better organized than that of novices. Because experience is often age-related, older people who are operating in areas of their expertise often equal or better the performance of younger people, despite decrements in raw abilities such as memory or reaction time. Occupational studies have shown that experienced older workers often perform as well as experienced younger workers. Salthouse (1987,152) concluded his review of the expertise literature as follows: "Research based on adults with little or no experience in the activities being measured may not be particularly meaningful for the purpose of predicting competence of experienced individuals." The point is that expertise is a complex of skills that tends to be maintained and added to by practice as long as illness does not interfere.

Wisdom Ancient texts of both the East and the West contain references to the wisdom of age. But only recently have psychologists begun to develop concepts and measures and to study the relation of age and wisdom. The research of Paul Baltes perhaps best illustrates this new line of inquiry. Baltes (1993,586) defined wisdom as "excellent judgment and advice about important and uncertain matters of life." In Baltes's model, wisdom is an expert knowledge system involving five elements: *factual* knowledge about the pragmatics of life; *procedural* knowledge about the pragmatics of life; knowledge needed to place situations in an appropriate *context*; knowledge that considers the situational *relativism* of life goals or values; and knowledge that considers the *uncertainties* of life.

Baltes hypothesized that wisdom was rare, that not all older people develop wisdom, but that aging creates conditions conducive to wisdom in two ways. First, development of the knowledge required for wisdom, and also practice in using these domains of knowledge as an ensemble, takes time and requires experience in diverse sets of circumstances. Second, motivation to practice wisdom requires development of a concern for others that is most often manifested in middle age or later.

Baltes felt that cognitive declines limited the number of elders who could become wise. Others may be limited by their narrow life experience. Achenbaum and Orwoll (1991) contended that development of wisdom stems from a transcendent point of view, first in terms of one's self, then in terms of interpersonal relationships, and finally in relation to the entire planet.

Baltes (1993) and his colleagues developed a series of dilemmas that they presented to samples of adults. Examples included, "Imagine a good friend of yours is calling you up to tell you that she can't go on anymore, that she has decided to commit suicide" or "A 15-year-old girl wants to get married right away. What should she consider and do?" The researchers first trained their respondents to "think aloud" about how they would respond to the dilemmas presented and then taped their responses to the various dilemmas. Using the criteria for wisdom cited above, raters assigned a "wisdom score" to each respondent's performance. They rated three groups of people whose training or life experience could reasonably be expected to create a capacity for wisdom: young clinical psychologists (age 30 or older), older clinical psychologists (mean age of 71), and a panel of distinguished citizens who were believed by people in the community to be wise (mean age of 67). They also tested a control group made up of well-educated professionals both old and young. They found no age differences in the wisdom performances of the young and older clinical psychologists. The older clinicians produced as many responses in the top 20 percent as the younger clinicians. The performance of older clinicians was especially good when they responded to dilemmas more typical of their own stage of life. People whose training prepared them to function in the five wisdom domains (clinical psychologists) performed better than the control group members; and the panel of people thought to be wise, who had learned their skills primarily from life experience, turned in the best performances of all. Both the wisdom nominees and the older clinical psychologists showed a high prevalence of wisdom, especially on the suicide task. All of the most highly rated performances were accomplished by people over 50.

In general, these research findings suggest that wisdom is a function of life experience and the capacity to learn from it. Not all older people become wise, but the wisest people tend to be middle-aged or older.

Practical Implications of Research into Mental Functioning

The foregoing research on aging and mental functioning has practical implications in many areas. Here we consider several areas: general social functioning, automobile driving, job performance, sports performance, and everyday performance. We also look at effects of diversity and social structure on mental functioning.

General Social Functioning Age-related decrements probably affect only a small proportion of the population under 75 because there is a continuity of lifestyle and habits; that is, most people continue to practice existing skills rather than learn new ones. In addition, for many, freedom from job demands means that learning, problem solving, and creative activity can be paced to suit the individual. There is no reason to believe that aging (as opposed to chronic disease or disability or generational differences) significantly limits social functioning.

Automobile Driving The driving performance of older people is often questioned. In recent years, media coverage has drawn attention to cases of older drivers who lost control of their vehicles with disastrous results. Such stories almost always include questions about whether older people should be al-

lowed to continue to drive, based on the presumption that aging makes people less safe as drivers. Interestingly, such questions are not raised about younger drivers who are involved in similar accidents.

Older people tend to drive more slowly and cautiously, which often irritates impatient drivers and leads to the charge that older drivers are a hazard. Using large national data files, Evans (1988) examined the age patterns in driver fatalities, pedestrian fatalities, and severe crash involvements. Compared with all other drivers, male drivers below the age of 35, and particularly men around the age of 20, have by far the highest likelihood of being killed in a car crash, killing a pedestrian, or being involved in a severe crash. Part of this pattern is due to the fact that young men drive more miles than young women do, but even when the data were controlled for miles driven, young women have a substantially lower incidence of driver fatalities. Men 20 years old had nearly five hundred driver fatalities per million drivers in 1983, which was more than three times higher than the incidence for men at age 65. After the age of 20, fatal and severe crash incidences decline sharply with age and reach their lowest levels between the ages of 55 and 65. After age 70, the incidence of driver fatalities per million drivers increases slightly for both men and women, but even at 85, the incidence is less than half that for younger drivers. Obviously, there are poor drivers of all ages, but the evidence clearly shows that older drivers as a category are actually less hazardous than younger drivers as a category.

Charness and Bosman (1992) concluded that the higher accident rates and incidence of fatalities among drivers age 20 and under were due mainly to inexperience and excessive risk taking. Unsafe speed and alcohol use are the leading factors implicated in accidents for young male drivers. For older drivers, right-of-way violations are the leading cause of accidents, which implies a breakdown in the cognitive-perceptual components of driving such as estimating the speed of oncoming cars or reacting too slowly to unexpected events.

They also found that visual acuity is not related consistently to accident records. Indeed, older people with poor driving records were found to actually have better vision than those with good driving records. However, those who had an impaired visual field, such as limited peripheral vision, were more than twice as likely as elders with normal visual fields to have been involved in accidents or cited for traffic violations. Older adults had much greater difficulty reading and responding to traffic signs, especially on dark streets at night. Older adults also had more difficulty shifting their focus between the road and instrument panel, especially at night. Sterns and his colleagues (1977) found that the visual mapping of older drivers could be improved considerably by specific driver training and that these gains in performance were maintained over time; therefore, it is possible to compensate for some of these age changes with training. Illuminating the road and road signs would make night driving much safer for older drivers. Thus, environmental compensations could help, too.

Some functions essential to safe driving do not decline with age. For example, older drivers were equal to younger drivers in reacting to sudden stops by a driver in front of them and in their response time to surprise obstacles in their path. Field studies testing the driving ability of younger and older drivers have not found significant differences (Charness and Bosman 1992).

As with other functions we have examined, most older drivers are quite capable of safely operating a motor vehicle. But some of them have visual, motor, or cognitive impairments that make them dangerous drivers. The challenge is to develop fair ways to screen drivers to identify those who are unsafe. We have just seen that simple visual exams are probably not sufficient. How so-

phisticated must the tests be? At what age should we begin to screen drivers? And to be fair, would we have to develop tests that take risk-taking drivers off the roads as well? Can you imagine it? "I'm sorry, Mr. Colt. Your driving personality profile shows that you are a danger to yourself and others. Please surrender your license." In a society such as ours in which the personal automobile is a prerequisite for leading an independent adult lifestyle in most parts of the country, it may be impossible to keep unsafe drivers from driving even if their licenses have been revoked. The data on repeat drunk driving offenders does not lead to optimism.

Some older people voluntarily stop driving, but we do not know how common this pattern is. In a small pilot study, Persson (1993) conducted focus groups with fifty-six retirement community residents who had voluntarily stopped driving. About one-fourth did so on the advice of a physician, but generally physicians are reluctant to assess medical competence to drive. Over one-third decided to stop driving because of increased nervousness behind the wheel and trouble seeing pedestrians and cars. About 16 percent quit on advice from family members, and 18 percent quit because of medical conditions that interfered with driving. A few people stopped driving because of a sudden change in functioning, such as that following a stroke. But most quit after a gradual process of adaptation. People first stopped driving at night or in heavy or fast traffic. Some couples adapted to increasing difficulties by driving in tandem, with one spouse acting as copilot. People gradually reduced the number of miles driven.

Job Performance   McEvoy and Cascio (1989) reviewed ninety-six studies covering more than 38,000 employees and concluded that when all types of performance measures are used, there was no correlation between age and job performance. However, when objective data (as opposed to subjective ratings by supervisors) were used, job performance increased slightly with age (Waldman and Aviolo 1986). Caution must be used in interpreting these data, however, because less effective older workers are more likely to be moved out of the labor force and their productivity or performance is not included in data based on workplace studies. What we can say is that those older workers who remain in the workforce seem to be able to perform as well as younger workers. The effects of experience on work performance occur early in the learning of most jobs, so experience is not a major advantage for most older workers (Park 1992).

We are thus left with a dilemma. The literature shows a positive correlation between cognitive ability and job performance. Cognitive performance declines with age for many functions. So why is there no decline in job performance with age? Park suggested some possibilities. First, cognitive deficits may predict poor job performance only for skills that require full mobilization of cognitive resources, as in learning in an entirely new job environment. However, the performance deficit in a new job may be only temporary. For example, older adults with no prior experience with computers have been found to take up to three times as long to learn to use standard computer software packages as similarly inexperienced young adults. Nevertheless, the older adults were eventually able to achieve proficiency equal to that of young people.

Older workers may also be more likely to use others in the environment to help them maintain their performance. For example, high-level professionals and executives may be quite astute at picking staff to help them with tasks that they themselves find difficult, while at the same time maintaining their essential judgment, interpersonal, and organizational skills at an expert level (Park 1992). Even routine production workers often have some flexibility to share work with others in ways that allow each of them to optimize their specific skills.

Sterns and McDaniel (1994) reported that older workers appear to benefit much less from job training than younger workers do. Older workers take longer to complete training tasks, especially if they involve use of computers. In addition, in post-tests assessing retention of training material, older workers retained significantly less than did younger workers. Reasons for the disparity between positive job performance and negative training performance of older workers are unclear. This is an area much in need of research.

One important caution is in order. Studies of aging and work have often defined "aging" and "older" at relatively young ages. The data on cognitive aging suggest that we would not expect to see widespread changes in job performance related to aging until at least the middle 70s. But most of the studies of this subject have defined "older workers" as young as 45. Park (1992) rightly pointed out that we need more studies of job performance of older workers well into their 70s before we can confidently address the issue of the effects of cognitive aging on job performance. Of course, as long as workers continue to retire in their early 60s, the issue of the effect of cognitive aging on work performance may not push to the forefront, but if we begin to think of retaining a significant proportion of workers beyond age 70, then studies of job performance at advanced ages would be more relevant.

A key to effective use of employees at any age is matching the strengths of the job prospect with the demands of the job. Tests consisting of small work samples are more effective in measuring job potential than general tests of aptitude or ability. The use of job-specific testing could result in better job performance at all ages (Sterns and Alexander 1987).

Sports Performance Sports generally involve peak performance. Performance of athletic record holders is established under rigorously controlled conditions, so age differences in athletic records for similar events offer important practical data on psychomotor combined with physical performance changes. Spirduso and MacRae (1990) reviewed the literature on the relation between age and peak athletic performance and found that track athletes generally adapt to slowing of motor reaction times by increasing the distances over which they compete. The data clearly show why. In the 1500-meter run, the fastest 40-year-olds could run at a rate only 88 percent as fast as the world record holder. But in the marathon, the fastest 40-year-olds could run at a rate that was virtually identical to that of world-class marathoners. Age effects were much more pronounced for field events, such as the shot put, that require explosive output of power. For example, the masters shot put record for 60-year-olds was only 49 percent as far as the world record; whereas the age-60 marathon record was 76 percent of the world record.

However, cross-sectional comparisons exaggerate the effects of age-based declines in athletic performance. Cross-sectional age differences are about twice as large as longitudinal changes in the same athletes over time. In addition, some older athletes show improved performance over time. For example, longitudinal research that followed runners over time found that times of runners in their 40s remained stable over a six-year period, whereas times for runners in their 50s and 60s improved (Spirduso and MacRae 1990).

In an interesting study, Molander and Bäckman (1994) studied age differences in highly skilled Swedish miniature golfers ranging in age from 15 to 73. When they compared performance under training conditions with performance in competition, the younger players improved their performance in competition whereas the older players' performance declined. During competition, younger players' motor performances and

level of concentration improved; but they declined among players over 50. They attributed these differences to higher levels of distractibility in the older players caused by overarousal in the competitive situation. An optimal level of arousal improved performance, but overarousal hampered it.

Everyday Performance Salthouse (1982, 202–203) was impressed with the *absence* of age effects in most normal activities, given the consistent laboratory study findings of decline. He gave several reasons to be cautious when drawing conclusions about everyday life from laboratory results. First, there is a bias against reporting no change because such results often seem less exciting or noteworthy. Second, the variability *within* age groups is usually much greater than variability *between* age groups. Studies have seldom found age differences as high as 50 percent, but differences of this magnitude *within* age groups are not uncommon. Third, most activities of daily living are probably minimally demanding, which means that adults could experience substantial decrements in peak performance and still function well socially. Such performance deficits then become obvious only in emergencies, when time pressure increases, tasks become more complex, and experience is of less help. Fourth, home and job activities of older people have been highly practiced for literally thousands of hours.

Diversity Thus far we have looked at general patterns of aging in relation to various mental functions. The extent to which these patterns differ systematically by gender, race, social class, or ethnicity is difficult to assess because most studies do not use these variables in the analyses of their results. Huyck (1990) reviewed the literature on gender differences in cognitive functioning and concluded that there is no evidence that men and women differ significantly in the age pattern of overall intelligence test scores. There were small gender differences in the functions that showed decline. Men tended to decline more than women in speed of processing, whereas women tended to decline in accuracy of response. Jackson, Antonucci, and Gibson (1990) found no studies of mental ability and aging that included African Americans or other racial or ethnic minorities. This represents a huge gap in our knowledge.

Effects of Social Structure   Schooler (1990) demonstrated that psychological performance is very much conditioned by the social roles and social situations people confront and the psychological skills transmitted, evoked, and rewarded in those situations. For example, access to jobs is strongly related to parents' education and occupation, and jobs that promote occupational self-direction increase both men's and women's intellectual flexibility. Substantive complexity of work also influences the intellectuality of both men's and women's leisure activities. Spatial thinking, statistical thinking, logical thinking, and expertise have all been shown to be positively related to environmental resources and reinforcement, which are correlated with social class lifestyles. Substantively complex home and work environments increase intellectual flexibility by encouraging the internalization by the individual of a self-direction orientation that tends to be applied across a wide variety of social situations. Self-direction orientations are correlated with higher performance motivation.

Different social worlds encourage different aspects of psychological and psychomotor performance. Some subcultures emphasize math skills, others verbal skills, others sports skills, and some a balance of all these skills. The point is that different social situations demand different skills.

Schooler suggested that some of the declines in intellectual flexibility attributed to aging may in fact stem from the fact that many elders are not faced with substantively complex demands in their everyday lives. Longitudinal research is needed to assess this issue.

Drives, Motives, and Emotions

Thus far we have considered how aging affects the various capabilities necessary for effective functioning. But individuals must also be willing to act. Energy for such action is mobilized by drives, motives, and emotions (Wigdor 1980). Physical aging appears to increase the level of stimulation required to arouse human drives, motives, and emotions. In addition, age-related social factors, such as opportunity for stimulation or boredom, can also affect arousal.

Drives   Drives are unlearned bodily states that are frequently experienced as tension or restlessness and make people want to act. When a person is hungry, for example, feelings of tension and restlessness do not have to be learned, they just appear. The primary drives that have been studied in relation to age changes include hunger, sex, and activity.

There has been little systematic study among humans of age changes in *hunger,* but data from animal studies indicate that older animals are less driven by hunger and can withstand greater food deprivation than younger animals. The literature on the food habits of older people contradicts this picture somewhat. Some older people appear to have less appetite than their younger counterparts, but many older people continue to enjoy eating and do not reduce their food intake appreciably with age.

This seeming enigma raises an interesting point. Past learning associated with the satiation or reduction of drives is frequently overlooked as a factor influencing responses to drives. Certain drives appear regularly, and human culture contains patterns for satisfying them. These cultural patterns become ingrained in the individual as habits, which have a way of acting like drives. A reduction in a physiological state (drive) need not necessarily lead to a change in behavior. Hence, eating habits may persist even though the physiological basis for eating has diminished. When food consumption remains constant or increases with age, it is probably a result more of culture than of the hunger drive.

Another facet of this same question concerns the fact that behavior can serve more than one function. In measuring the hunger drive, for example, we observe only behavior toward food. However, this behavior can result from habit, from a desire for taste sensation, or from a desire to socialize at mealtime, as well as from hunger. These dilemmas are present in the study of human drives, regardless of which one we study.

Regarding *sex,** age changes in sexual behavior are not merely a reflection of changes in sex drives. Whatever instinctive drives we have to seek sexual gratification are shaped by culture so fundamentally that drives can be the effect as much as the cause of sexual behavior. For example, cuddling and hugging one's partner can produce sexual arousal where before there had been none. In addition, our culture contains many images that contain sexual suggestions that can heighten sensitivity to and desire for sexual pleasure.

The evidence that we have on age patterns in sex drives comes mainly from Masters and Johnson (1966), who studied the physiology and psychology of sexual response in a small cross-sectional sample of adults, including older adults. Their research showed that people over 50 of both genders took longer to become aroused and experienced less pronounced and fewer physical spasms at orgasm compared with adults in the 20- to 40-year-old range. Older men also felt less need to ejaculate.

Masters and Johnson concluded that physical aspects of normal aging had a significant effect on sexuality of older women. In aging women, they found that steroid starvation and hormonal imbalance following menopause often led to changes in the female sex organs that made intercourse and orgasm painful. The uterus did not elevate as much as in younger women, creating less room for

the vagina to expand to accommodate the penis. Amount and quality of vaginal lubrication were more variable in the older women, and uterine contractions at orgasm were more apt to be painful. Masters and Johnson also found that estrogen replacement therapy could alleviate most if not all of these painful symptoms, but as we saw in Chapter 4, there are negative sides to this therapy as well. They concluded that most older women who continue to have sex regularly are able to retain much higher capacity for sexual expression than those who do not. Whitbourne pointed out that sexuality of older women is often limited by the scarcity of men at older ages, combined with "an aversion to masturbation and homosexual relationships among current generations of older adults" (1985,112).

Among older men, sexual interest declines much less than frequency of intercourse (Pfeiffer, 1977). Although men may retain much of their interest in sex, the performance of their sex organs becomes less predictable, particularly the amount of erection of the penis. Masters and Johnson (1966) found that older men took longer to achieve erection and experienced a lesser degree of erection compared with younger men. They concluded that a big part of the difficulties older men experienced were psychological. To be sure, ill health and medication can both influence the physical capacity for sexual arousal. But by the same token, fear of impotence can literally make that fear come true. It is common for older men to interpret delay in achieving an erection as evidence that they are becoming impotent as a result of aging. Whitbourne (1985) suggested that when the time required to achieve an erection increases by even two or three minutes, men accustomed to quicker performance are apt to panic. Masters and Johnson cited fear of failure as the main cause of inadequate sexual response in healthy older men. If the man can relax and be patient, then impotence is

---

*This topic is considered again in Chapter 8.

much less likely. As with women, one of the major factors associated with being able to maintain effective sexuality for older men was having sex regularly.

Thus, the sex drive appears to wane with age, especially in men. But interest in sex is much more than simply satisfying a biological drive. Sex meets deep emotional needs as well, and with age there is a shift away from emphasis on physical and emotional passion toward emotional togetherness and physical pleasuring.

Most animals appear to have a drive toward undirected, spontaneous *activity* (sometimes called *curiosity*). Animal studies have found that this drive first increases with age and then decreases. In rats, for example, spontaneous activity increases from birth to puberty and declines from then on. This drive has not been studied in humans, but at least part of the lethargy to be found in some older people probably results from a decline in the drive toward activity. Nevertheless, sedentary lifestyles and the decline in available energy are probably more important than any decline in drive in reducing the spontaneous activity of older people.

In summary, it appears that drives wane as age increases, but because a good many factors intervene between drives and overt behavior, drives probably play a relatively small independent part in explaining behavior.

Motives Motives are closely related to drives. Drives are highly generalized dispositions to action. Superimposed on these drives are motives—patterns of learned behavior that give specific direction to these general tendencies to act. Motives are thus specific, goal-directed, and learned. Since motives are learned, even relatively satiated human desires can be aroused, given a sufficient amount of stimulus; and motives can also disappear if there is little opportunity to satisfy or reinforce them. Goals that have been studied as the targets of motives include achievement, affiliation, and power (Kogan 1990).

There is some evidence that competitive *achievement* motivation is lower for older age categories. For example, Veroff, Reuman, and Feld (1984) studied age differences in a large sample of adults ranging in age from the early 20s to age 80 and over. They found a linear decline with age in achievement motivation, and the decline was more pronounced for women than for men. However, longitudinal research has shown increases in achievement motivation from young adulthood to early middle age (Stevens and Truss 1985). This suggests that achievement motivation may rise to a peak in middle age and that declines actually occur in later maturity or later. Maehr and Kleiber (1981) cautioned that in later life the meaning of achievement may change from competition and social comparisons to affiliation and intergenerational continuity, from winning as an individual to surviving as a group. More longitudinal research is needed to assess this possibility.

Veroff, Reuman, and Feld (1984) found that motivation for *affiliation* showed cross-sectional declines with age for women. The prospect of *power* increasingly motivated men until midlife and then declined. Veroff and his colleagues also found a strong situational component to motives. For example, the drop in achievement motivation for women occurred only among employed women, not among homemakers.

Emotions Emotions are strong feelings. They can be pleasant, as with joy, love, or pride; they may be unpleasant, as with sadness, fear, anger, frustration, or guilt. Emotions are such an important part of life that the ability to feel emotions and the capacity to cope with them are key indicators of mental health. Early cross-sectional studies of aging and emotional experience concluded that there was a decline with age in emotionality—the experiencing of emotions—and that unhappiness was more common among older people in comparison with the young (Dean 1962; Lakin and Eisdorfer 1962; Gurin,

Veroff, and Feld 1960; and Bradburn and Caplovitz 1965).

More recent studies have not found increased emotional negativity with age. Cameron (1975) reviewed five surveys, totaling 6,700 respondents, on which the item "How would you characterize your mood in the last half hour? (happy, neutral, sad)" was included. He found that a happy mood was much more common than an unhappy or neutral mood. He found no age differences.

One of the better studies to date on emotional experience and age was conducted by Malatesta, Zander, and Kalnok (1984). They surveyed 240 white, middle-class suburban residents in three age categories: 17 to 34, 35 to 56, and 57 to 88. Their questionnaire covered frequency and intensity of negative emotions such as fear, anger, disgust, and sadness and positive emotions such as joy and contentment. They also asked about age norms* for displaying emotions, the centrality of emotions in the person's life, and perceptions of change with age in emotional experiences. They also asked open-ended questions about what triggered various kinds of emotional experiences.

They found that there was no tendency for their older respondents (mean age 66) to have more negative responses. They also found more similarities than differences across their age categories. Gender differences were small. Most older respondents did not feel that their emotionality had changed with age. Emotional experiences were just as important to the older respondents as to the middle-aged ones, but less important than to the young adults. Sadness was caused mostly by physical problems for adults of all ages. For example, 55 percent of the young reported sadness connected with physical problems, compared with 66 percent of the middle-aged and 79 percent of the

elders. On the other hand, personal losses caused sadness for 45 percent of the young, 34 percent of the middle-aged, and only 21 percent of the elders. These findings suggest that health becomes a bigger source of sadness with age, but that the common assumption that personal losses are a more common source of distress in later life may not be accurate. They also found that the older age category was much more likely to agree that they should not express their emotions, but, at the same time, they did not report any greater incidence of inhibition of emotions compared with the younger age categories. This is an example in which older people do not abide by age norms.

Orientations: Time and Control

In addition to drives and motives, people have stable mental orientations toward themselves and their environments. Terms such as orientation, attitude, value, and belief are used to differentiate among these various types of mental predispositions. In this section, we look at two orientations that have been examined extensively in relation to aging: time orientation and beliefs about control.

Time   Aging is widely thought to influence the subjective experience of time. In particular, time seems to pass more quickly with advancing age, and the horizon of future time shortens in later life. Although most of us measure time in terms of clock time or calendar time, psychologically time is not absolute, but relative and subjective (Schroots and Birren 1990). We have all had the experience of subjective time passing much more quickly or more slowly than objective clock time. Perceptions that time is passing slowly tend to be associated with negative emotional experiences and understimulation, whereas perceptions that time is passing quickly tend to be linked to positive experiences and high levels of mental concentration. As people grow older there is a general increase in the proportion of positive experiences and a

---

*Age norms specify what is or is not allowed for people of various ages. They are discussed in detail in Chapter 6.

corresponding decline in negative ones. This can be found in the literature on emotions, conflict in marriages, life satisfaction, productive activity, and so on. On the other hand, we would expect that elders who experience disabling conditions that limit their capacity to be absorbed by satisfying experiences would experience time as passing more slowly. Thus far only very rudimentary studies have been done on the relation of aging to the speed of subjectively experienced time, and the results have been inconsistent. More research is needed in this area.

How is the present time linked to the past and to the future, and does this linkage change with age? As Hendricks (1982) pointed out, conceptually both the past and the future are mental constructions that are developed by people to serve a purpose. Markus and Nurius (1986), for example, noted that people use conceptions of their past, present, and future selves to make life choices. That is, people make choices that are consistent with past and present selves and that promote hoped-for future selves while avoiding feared or disvalued future selves.

Time orientation can be defined as the relative balance of past, present, and future in an individual's thinking (Kastenbaum 1983). Elders are commonly thought to dwell in the past more than in the present or future, whereas children are thought to live mainly in the present. However, very little research has been done to test this conventional wisdom. Thomae (1992) concluded that having a positive image of one's personal future was common among elders and was an important aspect of coping with a variety of changes associated with aging. Sill (1980) found that having the perception of a foreshortened future was related to an expectation of death in the near future, not to advanced chronological age. In my longitudinal study of aging and adaptation (Atchley 1982b), I found little evidence of preoccupation with the past among elders living in independent households.

Control Many of the subjective responses to changes associated with aging are thought to be mediated by the individual's generalized beliefs about who or what produces important events in their lives (Skinner and Connell 1986). The term *locus of control* (Rotter 1966) is an either-or construct in which the person has either an internal locus of control (the individual's enduring personal qualities influence events) or an external locus of control (events are produced by powerful others, chance, or fate). *Self-efficacy* is the perception that the individual is capable of behaving in such a way as to produce a desired outcome. The relationship between control and aging has received increasing attention since Rodin and Langer (1977) reported that institutionalized older people who were given some personal control over their environment showed a slowing of physical decline and lived longer.

How do perceptions of control change with age? Lachman (1986) examined age patterns of control beliefs in a series of longitudinal studies. She found no evidence of a decline in belief in personal efficacy. In fact, belief in personal efficacy decreased in early middle age, remained stable until the early 60s, and then began to *increase*. Lachman linked these findings to overload in early middle age from the demands of conflicting social roles related to jobs and family. On the other hand, Lachman found that when control beliefs were focused on specific dimensions of life, elders were less likely to feel in control. For example, most elders felt that the effect of aging on their intellectual abilities was a matter of chance and that the future course of their health was heavily dependent on the actions of health-care providers. As usual, the picture is not simple. Older people can believe themselves to be effective and at the same time acknowledge dimensions over which they have little control. Lachman's research illustrates the importance of treating control as a multidimensional construct.*

One important outcome of the increased attention being paid to the possible salutary effects of feelings of personal control has been the infusion of a philosophy of autonomy into programs for older people. For example, Regnier and Pynoos (1992) articulated twelve principles that should be used by professionals designing environments to be used by elders, and most of them were based on an assumption that promoting control, choice, and autonomy should be a primary value. For instance, being able to choose privacy or social interaction, having an environment that facilitates control in finding one's way about, having fixtures and physical features that facilitate manipulation by the disabled, compensations for sensory impairments, and personalization of the living environment could all be expected to support a sense of personal efficacy.

Adult Development: Personality, Self, and Life Structure

Adult development encompasses personality, self, and life structure across the adult life span. The **self** is your perception of yourself—what you think you are like, what you think you should or should not be like, and how you feel about the fit between the two. **Personality** is how you appear to others, especially your attitudes, values, beliefs, habits, and preferences. **Life structure** is the pattern or design of a person's life that results from the interaction between the person and the external world at a given time in the life span (Levinson 1990). Scholars have looked at personality, self, and life structure through the life span in terms of stages, processes, and the persistence of structures over time. The literature on adult development and

*For a more detailed discussion of control and aging, see Baltes and Baltes (1986).

aging is so filled with separate theories and concepts that even just a summary of all of them lies beyond the scope of this book. Instead, we look at prominent examples of two types of theories: stage theories and process theories. We then look at some research results concerning age changes in personality structure throughout adulthood.

Stage Theories

The idea that people pass through stages of development has been around a long time. Perhaps the most influential stage theory of adult development was formulated by Erik Erikson (1963). Erikson's theory is mainly concerned with how people develop an identity in childhood and adolescence, but it also considers development in adulthood and old age.

Identity refers to a set of characteristics that differentiates self from others and that persists over time. Identity can also be a goal through which people try to arrive at a conception of themselves as loving, competent, and good (Whitbourne 1986). To Erikson, identity is built on a foundation of trust, autonomy, initiative, and industry. These qualities, developed in childhood, allow individuals to form a view of themselves as capable, worthwhile, and safe. People who do *not* develop these qualities have difficulty trusting others, have doubts about their abilities, feel guilty about their poor performance, feel inferior to others, and have no confidence in their ability to face the changes of adulthood. Between these two extremes is a continuum along which individual identity can vary considerably.

According to Erikson, the main issue of human growth and development in early adulthood is learning to establish *intimacy*—close personal relationships. Developing intimacy involves learning to unite one's own identity with that of another person. People who do not develop intimacy remain isolated, still relating to others yet lacking a sense of union.

The developmental issue of middle adulthood is *generativity versus stagnation.*

Generativity—the ability to support others, particularly one's children and other members of younger generations—involves caring and concern for younger people and also an interest in making a contribution to the world one lives in. Stagnation—when an individual does not learn to contribute to others—is typified by a lack of interest in others, especially the young, a feeling of having contributed nothing, and the appearance of just going through the motions.

To Erikson, the developmental issue of late adulthood is *ego integrity versus despair.* Ego integrity involves being able to see one's life as having been meaningful and to accept oneself as having both positive and negative characteristics, and to be unthreatened by this acceptance. Integrity provides a basis for approaching the end of life with a feeling of having done the best one could under the circumstances. Despair is the result of rejecting one's life and oneself and includes the realization that there is not enough time left to alter this assessment. The despairing person is prone to depression and is afraid of dying.

Erikson's stages are both incremental and contingent. That is, life is a process of advancement through hierarchical stages, provided one adequately resolves the central life issue at each successive stage. To develop intimacy, one must first have developed a positive identity. To develop generativity, one must have the capacity for intimacy. And to develop integrity, one must feel the connectedness and contribution that come from intimacy and generativity. The progression of human growth in childhood is closely tied to chronological age by the expectations of home and school, but in adulthood individuals are freer to move at their own paces. Thus, although generativity may typify middle adulthood, many middle-aged people are still learning to deal with identity or intimacy; therefore they feel irritated by demands to exhibit generativity.

The early formulations of Erikson's theory had an either-or character to them. Either one developed the capacity for intimacy or one was doomed to a life of isolation. However, Erikson, Erikson, and Kivnick (1986) modified this aspect of the theory to acknowledge that the polar opposites typical of the various developmental stages usually coexist. This new formulation adds important dimensions that result from the resolution and balancing of the developmental dilemmas presented in each life stage. For example, in adolescence the young person does not just develop an identity, she or he must also learn to live with periods of confusion through *fidelity,* remaining committed to an identity ideal. *Love* provides a sense of continuity in relationships that allows intimacy and isolation to coexist in young adulthood; *caring* about both self and others balances generativity and self-absorption; and *wisdom* is a quality that allows the person to experience integrity even in the midst of a period of despair.

Erikson's theory is most useful for what it suggests rather than for any scientific validity. Indeed, many of the theory's concepts, such as integrity and self-acceptance, are so difficult to measure that it may never be possible to test the theory scientifically. Nevertheless, Erikson's framework can be an effective tool for relating to others. By listening to people, we can get a sense of where they stand on the developmental issues Erikson raised, which in turn may help us understand not only their behavior but also their priorities and their aspirations.

Another stage theory of adult development is that of Daniel Levinson (Levinson et al. 1978; Levinson 1990). According to Levinson, the adult life cycle is organized as a sequence of eras. For example, *middle adulthood* is an era lasting from about age 40 to age 65. In this era, physical capacity is sufficient for an energetic, personally satisfying, and socially valuable life. People become senior members of their social worlds, whatever form these worlds may take. They may become responsible for not only doing their own work but also nurturing the next generations of young adults (1990).

Within each era, the individual alternates between structure-building periods and transitional or structure-changing periods. In structure-building periods an adult elaborates and refines the relationships with people, groups, institutions, or environments that constitute the design or pattern of his or her life in that era. The primary task is to maintain and enhance life within the structure the person has devised out of the opportunities and constraints presented to her or him. This structure arises out of interaction between the person and his or her social world. Although there are many forms life structure can take, Levinson found that marriage-family and occupation were the most common centerpieces of adult life structure, at least through late adulthood. However, life structure is inherently unstable because both the person and the social world are constantly changing. After about five to seven years of living with a relatively stable life structure, the person begins a transitional period in which the existing life structure is reappraised, potential new directions are evaluated, and commitments are perhaps made to new choices that will be the basis for a revised life structure. Levinson theorized that transitional periods usually last about five years.

The developmental periods of early and middle adulthood were organized by Levinson (1990,44) as follows:

1. Early adult transition (17–22): the developmental bridge between preadulthood and early adulthood

2. Early life structure for early adulthood (22–28): the time for building and maintaining an initial mode of adult living

3. Age-30 transition (28–33): reappraising and modifying the early adult structure to create the basis for the next life structure

4. Culminating life structure for early adulthood (33–40): the structure that provides the basis for completing the era and realizing young-adult goals

5. Midlife transition (40–45): a great cross-era shift that terminates young adulthood and initiates middle adulthood

6. Entry life structure for middle adulthood (45–50): provides the initial basis for life in a new era

7. Age-50 transition (50–55): offers a mid-era opportunity to modify and perhaps improve the entry life structure for middle adulthood

8. Culminating life structure for middle adulthood (55–60): maintaining the framework with which middle adulthood is completed

9. Late adulthood transition (60–65): the boundary period between middle and late adulthood, which separates the two periods and begins the building of a life structure for late adulthood

In transitional periods, the person has three developmental tasks, according to Levinson: bringing the previously existing life structure to a close (termination), developing a concept of individuation appropriate to a new life structure, and initiating the new life structure. In transitions there are always times when new and old life structures coexist. In Levinson's theory, individuation in all adult transitions or life structure building periods occurs along four key dimensions: young-old, destruction-creation, masculine-feminine, and attachment-separateness. Each era involves an "appropriate" mix of young and old qualities, of destruction and creation, of masculine and feminine qualities, and of attachment to others and separateness. Thus, Levinson's theory does not have the cumulative and contingent quality that we found in Erikson's stage theory. In Erikson's theory, development proceeds incrementally, but in Levinson's theory, evolution occurs cyclically. Erikson's stages are tied to specific developmental tasks, whereas

Levinson's stages are the product of an underlying instability to any life structure. The cyclic decay and rebuilding of life structures must go on, in Levinson's view, regardless of how well one disposed of the developmental tasks of the previous stage.

As with Erikson's theory, Levinson's approach to the stages of adult development is useful mainly for the clues it gives about where to look for significant evolution in a human being. Its weaknesses are its lack of rationale for the five- to seven-year developmental cycles it assumes, unclear reasons why transitional periods overlap chronologically with structural periods at some points in the sequence and not at others, and its overuse of the cultural life course to link significant eras to chronological age. As we will see in the next chapter, many adults do not follow the sequential pattern set forth by Levinson. Nevertheless, Levinson gave us some important ideas to use in examining the dynamics of whatever patterns people actually do follow.

Process Theories

For many psychologists, adult development does not involve discrete stages but instead is the result of the continuous operation of various developmental processes. Riegel (1976) contended that adult development is a dialectical process that is itself caused by contradictory ideas or actions produced by the constant changes occurring in the person and the person's environment. These contradictions are not deficiencies to be corrected but invitations to a new level of integration. For example, you might notice that at times you are cautious and at other times adventurous. Accepting both parts of yourself as valid is a higher level of integration than putting them in opposition to one another in an effort to make yourself one or the other. "Developmental and historical tasks are never completed. At the very moment when completion seems to be achieved, new questions and doubts arise in the individual and in society" (1976,697). To Riegel, de-velopment is a process of resolving the doubts, questions, and contradictions that constantly flow through our consciousness. Riegel's theory is insightful in its focus on the everyday character of development. However, it says very little about the specific mechanisms for achieving the necessary resolution.

Whitbourne and Weinstock (1979) addressed more specifically the dynamics of continuous development. They argued that adult identity is an integration based on a person's knowledge about her or his physical and mental assets and liabilities, ideas (motives, goals, and attitudes), and social roles.* Identity serves to organize the interpretation of experiences, the assigning of subjective meaning to them. And identity can be modified by experience. The day-to-day contradictions that appear between the identity we bring to experience and the feedback we get from experience can be responded to in a number of ways. If the person is flexible about the content of his or her identity, then identity is a theory of the self that is constantly being tested, modified, and refined through experience (G. Kelly 1955). To the extent that one's observations of oneself are honest, one's theory of self gets better as time goes on. That is, the results our identity leads us to expect are the results we get. And the more times our theory works, the greater our confidence in it.

Sometimes people develop an identity but refuse to test it. They use their identity to decide how to act, but they do not let the results modify their ideas about themselves. Other people never quite develop a firm identity that could be tested; they do not quite know what to expect of themselves in various situations. The result is that they behave in an inconsistent and confused way.

Aging affects the stability of identity in several important ways. First, the longer one has an adult identity, the more times one's theory

---

Social role is defined in detail in Chapter 6.

An older violinist demonstrates continuity.
Photograph by Charles Gupton/Tony Stone Images, Inc.

of self can be tested across various situations. This experience usually results in a stable personal identity that stands up well to the demands of day-to-day living. Second, the reduction in social responsibilities, such as retirement, associated with later adulthood can reduce the potential for conflict among various aspects of identity. Third, aging for most people means continuing familiar activities in familiar environments. Most people have long since developed skill and accumulated accomplishments in these arenas; all they need is to maintain them.

The identity perspective also provides a basis for predicting when change might reach crisis proportions. When change in either the individual or the environment is so great that it cannot be integrated without a fundamental reorganization of one's theory of self, an identity crisis results. An identity crisis in these terms means reassessing the very foundations of one's identity. The changes that precipitate the crisis may occur in the individual or in the social situation. Consider these examples:

Mr. F. has learned that within six months to a year he will be totally blind.

Mrs. G.'s husband died six weeks ago. They had been married forty-seven years.

Mr. M. retired from teaching after forty-one years.

Mrs. B. has become increasingly frail. After sixteen years of living alone as a widow, she must move in with her divorced middle-aged daughter.

None of these changes need *automatically* trigger an identity crisis. Whether an identity crisis occurs depends on how central the dimension is to the individual and whether the change was anticipated. Profound physical changes usually have less ambiguous outcomes than social changes. Going blind would be a profound adjustment for most of us. But how much Mrs. G. is affected by her husband's death depends on how her relationship with him has fit into her identity. Whether they were inseparable companions whose selves were completely intertwined or whether theirs had been a marriage of convenience that had endured mainly by force of habit makes a lot of difference to whether Mr. G.'s death brings on an identity crisis in Mrs.G. Mr. M. may easily be able to give up teaching if he feels he has completed what he set out to do as a teacher and that it is time to move on. However, he may not be able to leave so easily if he is being forced to retire when he feels he still has something to contribute. How Mrs. B. resolves her dilemma depends on how she feels about being dependent—what it means for her identity—particularly in her relationship with her daughter. An identity crisis is not the product of events alone; it stems from the *interpretation* of events in the context of a particular identity.

Continuity theory (Atchley 1971a, 1989, 1995d) also assumes that adult development is continuous. Whereas most theories of adult development have been constructed by extending child development theories into adulthood, continuity theory grew out of studies of older adults. The older adults interviewed showed great resilience in the face of physical, mental, and social losses. The research revealed that most older people had, over the years, developed patterns of thinking and acting that continued to evolve and that were used continuously in adapting to changing needs and circumstances.

Continuity theory is concerned with the evolution of both life structure and personality and self. As adults develop, they become invested in the mental pictures that they use to organize their ideas about themselves and the external world. They *actively construct* these patterns continuously in order to adapt to change and to seek their goals.

Continuity theory assumes that adults have goals for *developmental direction*. They have ideals about themselves, their activities, their relationships, and their environments toward which they want to evolve. These goals for developmental direction are influenced by the person's location in the social structure—by their family ties, gender, social class, and so on—but they are also profoundly influenced by the individual's life experiences. Adults use life experiences to make decisions about selective investments: which aspects of themselves to develop to the fullest, which relationships to focus their attention on, which activities to engage in, which careers to pursue, which groups to join, what community to become a part of, and so on.

Developing, maintaining, and preserving *adaptive capacity* is the core of adult development, according to continuity theory. As adults reach middle age, they tend to have increasingly firm ideas about what their adaptive strengths and weaknesses are. They use these ideas to make choices that *they* see as taking advantage of their perceived strengths and avoiding their perceived weaknesses. As they continue to evolve, adults also have increasingly firm ideas about what gives them satisfaction in life, and they fashion and refine a life structure that delivers the maximum life satisfaction possible given their circumstances. The ideas middle-aged and older adults have about what it takes to adapt to life are thus presumed to be the result of a lifetime of learning, personal evolution, and selective investment. It should not be surprising, then, that in adapting to change, people tend to continue to use the internal and external patterns they have spent so much time and energy developing.

Continuity theory is about continuous evolution and individuation, not stability or lack of change. It is a feedback systems theory* which assumes that people are constantly learning about themselves and their world and using that information to increase, maintain, or retain their adaptive capacity. This process is never completed. Some people do not learn much from life, so continuity for them does not lead to "successful aging." Others learn an enormous amount and become increasingly satisfied both with their level of understanding of themselves and the world and with their ability to adapt to whatever life brings. For these people, continuity does bring successful aging.

Continuity theory is not deterministic. Unlike Erikson's theory, for example, continuity theory does not predict what the developmental issues will be for particular life stages or what the specific developmental results will be. Instead it points us to the *processes* people are likely to use in seeking their goals and in adapting to change whenever it comes. Continuity theory tells us where to look, not what we will find. It is a general model that can be used to understand why particular people have developed in the way that they have. It also predicts that they are likely to try to continue to develop in that direction and to resist pressures from others to change their direction.

To study continuity requires longitudinal data on the internal and external adaptive structures of individuals. Each of these structures is personalized to some extent, but each includes values, beliefs, self-image, lifestyles, and coping skills.

This review of theories of adult development covers only a very few of the many and varied approaches that have been used to address the issues of adult development and how it occurs. Each of these approaches has value, none of them is sufficient to explain all aspects of adult development, and there is no danger that the debates surrounding adult development will be resolved soon. Now let us turn to a more detailed examination of the research results concerning two important psychological dimensions: personality and self.

Age Changes in Personality

The term *personality* refers to what the individual typically says and does and how the individual characteristically interacts with her or his physical or social environment. Personality develops out of an interaction between a person's innate dispositions, capacities, and temperament on the one hand and the physical and social environment on the other. The individual actively participates in this process rather than simply being affected by internal or external forces beyond his or her control.*

Personality is difficult to study because it is unique for each individual and is made up of hundreds of patterns of thought, motivation, and response. Every theory of personality tends to focus on different aspects of personality, and every person using a personality theory tends to promote his or her focus as the most effective. This difficulty is further compounded by the fact that some theorists are primarily researchers while others are psychotherapists; these two specialties look at very different samples of people for very different reasons. Another problem in the study of personality is that the object of study—the inner workings of the mind—cannot be observed directly. For all these reasons, the scientific study of personality and aging has advanced very little toward developing a specific profile of age changes in personality (Kogan 1990; Thomae 1992).

Nevertheless, much of the research done on the relation between aging and personality has conceptualized and measured personality

---

*Feedback systems theory presumes that structural elements of any interactive system are modified mainly by information (feedback) that comes from observing the results of action (Buckley 1967).

---

*For alternative views of and assumptions about personality and how it is affected by aging, see Kogan (1990).

quantitatively as a complex of dimensions made up of specific measurable personality traits rather than in terms of stages of personality development or processes of development. Dimensions of personality that have been studied in relation to aging include neuroticism, extroversion, openness to experience, agreeableness, and conscientiousness (Kogan 1990). Two themes emerge from these studies: (1) There is continuity of personality* in that individual differences in typical thoughts, motives, and emotions tend to be maintained over time, and (2) there is also considerable evolution of personality over adulthood, so the longitudinal correlation between personality traits in early childhood and late adulthood is only modest (median correlation of .25 across personality components) (1990,334).

On the subject of *adaptation patterns,* the research findings suggest that personality may be adaptable early in adulthood but that it becomes more stable by late middle age. The research evidence is exceedingly strong that, after middle age, stability of personality greatly outweighs personality change (Kogan 1990; Thomae 1992). This pattern of stability in personality after middle age has been confirmed by many investigators (Schaie and Parham 1976; Neugarten 1977; Thomae 1975; McCrae and Costa 1982; Costa and McCrae 1988). Stability of personality in later adulthood is related to two factors. First, over time there is an "institutionalization" of personality, meaning that individuals expect themselves to respond in ways that are consistent with their past. Second, people build up around themselves familiar social networks that facilitate dealing with the social world in habitual ways (Atchley 1989).

The issue of continuity versus change stems mainly from differences in the level of study. Although inconsistent, the evidence points most clearly to a large amount of continuity in the *global* aspects of personality. At the same time, there is substantial support for the idea that aging adults perceive a great deal of change in the *details* subsumed under these global conceptions (Bengtson, Reedy, and Gordon 1985).

Those global aspects of personality that remain with the individual across various social situations tend to persist despite substantial changes in the details of everyday life. Once formed, the core or global personality tends to be resilient, and the components of personality can apparently contain a multitude of conflicting details without causing the individual to doubt the validity of the global concept (Kaufman 1986). Continuity is also supported by what happens to the self with advancing age.

Age Changes in the Self

The objective self—the *me* (as contrasted with the subjective self, the *I*)—consists of what we think and feel about ourselves. It can be viewed in terms of four components. *Self-concept* is what we think we are like; the *ideal self* is what we think we ought or ought not to be like; *self-evaluation* is a moral self-assessment of how well we live up to our ideal self; and *self-esteem* is whether we like or dislike ourselves, and how much. Both self-concept and ideal self are often tied (more closely early in life) to the social positions we occupy, the roles we play, and the norms associated with personal characteristics such as sex, race, and social class.

With regard to the dynamics of the self, how does aging differ from maturation? To answer this question we need to consider how the self works in young adulthood. In young adulthood, we play many roles, and it is often difficult to tell which is the role and which is our self, partly because there seldom are any written or verbal descriptions or rules that dictate what to do in a particular role. Instead, we tend to base our behavior on what we have seen others do; we try to imitate them. The

---

*That there is a continuity does *not* mean that there is not also change. The implication is simply that change occurs in the *context* of considerable continuity.

people we imitate are called *role models.* We also watch the reactions of others to our own performance; if we want to improve the impression others have of our performance, we may gradually change the way we play the role. We may attempt to change what others expect of us by introducing personal information about ourselves; at the same time, others may analogously attempt to alter what we expect of them. This taking of roles, finding role models, dropping roles, and transforming roles—all in a wide variety of settings (job, family, community)—is so common in early adulthood that it is often difficult to develop a firm sense of self. The self gets too mixed up with this complexity of roles.

Aging introduces a crucial variable into this confusing situation: *Our role relationships increase in duration.* Over time, adults often drop relationships that are not what they want or are unrealistic, given their capacities or interests. With time, we can also become more adept at differentiating self from role, mostly because we are able to view the self across a variety of roles. And the more we know about the self, the more potential we have for avoiding roles that do not suit us. These patterns are not ironclad laws, they are merely tendencies that probably fit the aggregate experience of a large number of people. They are ideas about what can happen when a person is open to learning about the self and using that knowledge to inform life choices. This process need not even be conscious or calculated—it may appear to "just happen."

By the time we reach middle age, we have a huge backlog of evidence about what we are like and should be like in a wide variety of roles and situations and how this self may or may not fluctuate from time to time. While we are trying to separate self from role, messages from others about our self and personality may greatly influence what we think we are or ought to be. But as time goes by, more and more people develop a firm idea of what they are like. Thus, although the self may be relatively open to outside influence during the process of self-discovery, it tends to be relatively closed to it during the process of self-maintenance.

Self-evaluation and self-esteem always go hand in hand. They are the moral and emotional reactions to self-assessment of the fit between self-concept (what you are) and the ideal self (what you ought or ought not to be). The noted psychologist William James (1890) expressed the relationship as follows:

$$\text{self-esteem} = \text{success/pretensions}$$

We can rewrite this as:

$$\text{self-evaluation or self-esteem} = \\ \text{self-concept/ideal self}$$

If this relationship is valid, then self-esteem could be raised by either increasing success (self-concept) or decreasing pretensions (ideal self).

Since the ideal self contains ideas about what one ought to be like as well as ideas about what one ought not to be like, success can be defined in terms of both achieving the "oughts" and avoiding the "ought-nots." If achievement motivation declines with age, then perhaps leading a moral life by avoiding the ought-nots assumes more importance in later life compared with early adulthood, when achieving positions and status is stressed by society. In other words, self-esteem in later life may depend more on moral qualities than on social achievement. Although much more research is needed in this area, the theory fits very well the age data on the self.

Many people have low self-esteem in early adulthood. They are unsure of themselves, what they are capable of, and what they ought to be capable of. And if one is unsure, it is difficult to bring any of the elements into focus. Over time, self-esteem tends to increase. Bengtson, Reedy, and Gordon (1985) found that a majority of research studies had observed higher self-esteem among older people than among the young. This difference may be accounted for in the following way. Once a person has a firm self-concept, then the person can bring the ideal self more in line with the self-concept. In other words, self-acceptance increases with age. A person may be freer to achieve this self-acceptance because of the reduction in role responsibilities (mainly child rearing and employment) that usually occurs around late middle age and because of the increased potential for personalizing role relationships over time. Systems *do* bend to fit people, but this usually takes time. And people can bend to fit systems, again, with time.

For those who experience such a resolution (the majority), the outcome is a solid concept of self that stands up well to the vicissitudes of later life. This continuity in self couples with the continuity of personality to serve as a basis for defense against the assaults of ageism.

Defending Personality and the Self

The gerontology literature is full of assertions that ageism leads to a declining self-image. But numerous studies have shown that most older people do not have a negative self-image and that self-esteem tends to increase with age (Bengtson, Reedy, and Gordon 1985). This apparent contradiction occurs because the first assertion is incomplete; it should be, *"In the absence of adequate defenses,* exposure to ageism erodes the basis for self-esteem."* Fortunately, most people are adequately braced to defend their ideas about themselves and their personalities.

The continuity of personality and self through time means that how we see ourselves involves not only an assessment of our current performance and characteristics but those from the past as well. And people differ a great deal in how much weight they give to the past when making their current assessments of self.

Because the past is available for use in defending the self, roles that a person no longer

plays can provide the self with important evidence of past competence and success that is often not apparent to an outsider. Greenwald (1980) went so far as to argue that people revise or fabricate their personal histories in ways that virtually guarantee a positive evaluation. First, the past is recalled as a drama in which the self was the leading player, supporting a perception that one had more control and influence than the other actors—even that one was the center of the other actors' actions.

Were it not for a second process, this tendency might lead to a negative self-evaluation for someone whose earlier roles had many negative results. But people tend to take credit for positive results and to blame outside factors for the negative, as a way of preventing poor results in the past from reaching the current self-image. If this hypothesis is true, then as time goes by one can take a great deal of credit and very little blame. This ability also has important positive implications for the success element in the self-esteem equation.

Greenwald also pointed out that the self is a very conservative historian, one who insists that new information fit and support the prior view of self. This theory does not imply that the self is impervious to contradictory new information but that, as with most ideas, the greater the amount of prior confirmation of self, the greater the amount and credibility of new information that will be needed to change one's existing self-image. For the vast majority of older people, the outcome matches what we would expect from Greenwald's theory: Most older people have positive self-images, quite resistant to disconfirmation (Markus and Herzog 1991).

Not only are there psychic tools for controlling information about our pasts, but there are also techniques by which we control information about the present. Over time, people tend to conclude that they know themselves better than anyone else does. This tendency can lead us to assign more weight to our own notions about ourselves and to discount what others say about us. We may also feel that stereotypes about some category we might be assigned to do not apply to us. The gerontology literature has consistently shown that older people can accept the stereotypes about aging and older people and at the same time consider these stereotypes as having no applicability to them personally.

Continuity of self is reinforced and defended when people remain in familiar environments that allow them to exercise well-practiced skills, providing in turn an experience of competence. Everyday tasks such as doing housework, driving a car, and shopping can become important symbols of self-sufficiency (Lawton, Moss, and Grimes 1985).

Relative appreciation can also help define a condition or situation as more positive than it might otherwise appear; for example, an older person may acknowledge that aging has reduced the amount of heavy work that he or she can do but at the same time recognize that *compared with others of the same age,* the decrement has been insignificant. This process is logically opposite to that of *relative deprivation,* in which people feel that they are less fortunate than others. Older people probably show a sizable bias toward relative appreciation over relative deprivation (George and Klipp 1991).

The preceding defense techniques are usually effective in insulating us from general stereotypes such as "college student," "professor," or "older person." However, when someone attacks us directly, then another type of defense is needed. We can find something with which to discredit our attacker and thereby discount what has been said. Or we can literally forget such comments. We can also avoid such attacks by interacting selectively; that is, by interacting as much as possible with people who are not so negative. Also, mildly disabled people sometimes avoid negative responses by venturing forth only on days when they are sure their physical capabilities will not occasion comment.

People often restrict their interactions to those whom they trust to support a positive moral and emotional self-assessment. In this respect, the freedom from required involvement with people outside one's own circle that often typifies later life is a boon. The more ageist and hostile the environment, the greater the advantage to voluntary involvement.

When Defenses Fail

The defense techniques just described are enough to protect about 80 percent of older people most of the time in most everyday situations. Yet the remaining 20 percent has low self-esteem. Why?

First, some older people never develop a solid identity or self-image. These people are defenseless and unsure of themselves their entire lives, and what we see in them currently may have nothing to do with aging. Other older people have the notion that they are *supposed* to be worthless—that a bad self-image is good for them. Ageism gives them more opportunity to accumulate negative credits and they soak up the insults like a sponge, even seeking them out sometimes. These are the masochists in our midst. But probably the biggest category of people with lifelong low self-esteem is made up of those whose expectations for themselves have far outdistanced their capacity for success and whose moral rigidity has prevented the necessary rapprochement. These are the lifelong perfectionists. For all these people, the problem is not a failure to defend self-esteem but a failure ever to develop it. There are also people who do not experience a loss of self-esteem until later life. The key factors responsible for this loss in later life are (1) decline in physical capacity, (2) an already vulnerable self-image, and (3) loss of control over one's physical environment.

A *gradual* decline in physical capacity can be incorporated gradually into the self-concept, and the ideal self can be modified gradually so as to take gradual declines into account. Thus, gradual changes can sometimes be experienced without loss of self-esteem. One crucial issue here is whether one must simply cut back on activity or cut it out altogether. In other words, physical changes that disrupt continuity of preferred activities have a much greater potential for affecting self-concept and self-esteem than do physical changes that allow continuity at a reduced level. For example, a man who sees himself as a capable handyman who can do his own home repairs can continue to hold this image even if he is forced to slow down or get a little help from time to time. But when he wants to repair a loose doorknob and is unable to hold the screwdriver tight enough to do the job, it becomes more difficult to retain the image of current capability. His past accomplishments would still be available as a source of self-esteem, but not his current accomplishments. If being a capable handyman today is important to this person's sense of self, then a loss of self-esteem could occur. When activity losses occur gradually, a person can anticipate them, and the results for the self may not be as serious as when the change is sudden and substantial. For example, people with progressive arthritis have much more time to adjust to the idea of cutting down activities than do those who suffer incapacitating strokes.

Some people reach later life with adequate yet vulnerable self-esteem. High vulnerability can arise in several ways, most of them related to the innumerable human characteristics that form the basis for the ideal self, such as physical ability, mental ability, appearance, roles, activities, groups one belongs to, personal qualities such as honesty, and relative prestige, honor, or wealth. The vulnerability of self to the changes that aging can bring depends largely on the vulnerability of the specific bases for self-image. For example, people who idealize physical appearance are more vulnerable to aging than those whose ideal selves are based mainly on interpersonal qualities such

as warmth toward others or trustworthiness. Generally, those whose self-images are based on structural aspects of life such as roles or social position are more vulnerable because aging predictably brings role loss and reduction in income.

As said earlier, most older Americans do not have low self-esteem. Thus, if the foregoing explanation is true, then most older people could be expected to value personal qualities more than positions and roles. Indeed, one study found that the top-ranking personal goals of older people were being dependable and reliable, having close family ties, and being self-reliant and self-sufficient (Atchley 1980b,74). The emphasis was on personal qualities and on relationships that one is least likely to lose. Goals such as being prominent in the community or having an important position were ranked at the very bottom. Another study found that adults across all ages tended to stress interpersonal qualities such as being honest, forgiving, and helpful over instrumental qualities such as being intelligent and having imagination (Rokeach 1973).

People who have a diversified basis for self-esteem are less vulnerable to aging than those whose selves depend on only two or three characteristics. For example, for some people their job is the primary basis for their ideal self, their definition of success, and their self-esteem. Such people are rare, but if they are forced to retire, either because of health or social policy, then the results could be severe. Again, as the research evidence shows, the self-esteem of most older Americans is based on a highly diversified pattern of activities, roles, personal qualities, and other characteristics.

Loss of control over one's home or community environment can pose serious problems for maintaining self-image. As long as people remain in a familiar environment, skills and knowledge developed in the past can often compensate for decrements in functioning. But people who move to a new environment can be confronted with changes in their ability to move about or gather and process new information, changes they might not have been aware of before they moved. And if the person has difficulty figuring out the new environment and coping with it, there can be a very unsettling lapse in the person's usual sense of competence. In addition, the person may have left behind a group of friends or family that supported his or her use of past history as a source of self-esteem, a problem that can be aggravated if the person also comes into contact at this time with ageist service providers who think that most older people are helpless or incompetent. Nevertheless, changing environments does not necessarily rob people of all their defenses. They can still attach more weight to their own experiences, discount what others say to them, selectively interact with others, and selectively participate in their new social environment. In addition, only relatively few older people who move to new environments know no one at the new locale. Having friends already there is one of the major factors in selecting a new environment (Longino 1981).

But entering an institution is another matter. When people enter nursing homes, they are usually at the mercy of the staff, not only for medical and personal care needs but for support for their sense of self as well. They usually find themselves among people who do not know their past histories and who cannot be counted on to support this source of self-esteem. They are in an unfamiliar environment, which might be difficult to learn to negotiate even if they were not sick or disabled. And in a nursing home it is very difficult to avoid or discount negative responses from others, for the resident is usually not in control of who will be seen, where, and when.

A large majority of adults continue to develop in adulthood but in ways that are tied to and supportive of their already-existing personalities and self-concepts. Most older adults have developed effective adaptive techniques for dealing with change and with

threats to the security of their life structure, self-concept, or personality. However, there is an important minority of people who experience mental disorder in later life, and we now turn our attention to this topic.

Mental Disorders*

Mental disorder is a noticeable and dysfunctional behavioral or psychological pattern that occurs in an individual and is associated with personal distress or impaired ability to function in everyday social roles. Age affects the incidence and seriousness of mental disorders. Of the nonorganic disorders such as schizophrenia, paranoia, anxiety, and depression, depression is the most likely to have its onset in later life, often in response to losses either in physical functioning or of social resources through death. Depression can also be a side effect of some medications used to treat diseases common to later life (LaRue, Dessonville, and Jarvik 1985). Parmalee, Katz, and Lawton (1989) reported that 12 percent of nursing home residents suffered from major depression and another 30 percent showed marked symptoms of depression. Gallagher et al. (1989) found that more than 20 percent of caregivers of elders with Alzheimer's disease showed symptoms of depression. Finally, depression is very prevalent among people with chronic physical pain (Luggen 1985; Parmalee, Katz, and Lawton 1991). Thus, much depression in later life is a response to negative circumstances. Fortunately, depression is the most treatable nonorganic mental disorder.

Organic mental disorders, also called **dementias,** are typified by mental confusion, loss of memory, incoherent speech, poor orientation to the environment—and sometimes poor motor coordination, agitated behavior,

*For more on mental disorders, see Birren, Sloane, and Cohen (1992).

depression, or delirium as well. About 55 percent of dementia in the older population is caused by **Alzheimer's disease**—a progressive and irreversible deterioration of brain tissue. The remainder is caused by a variety of factors, some of which can be reversible, such as strokes, brain tumors, nutritional deficiencies, alcoholism, and adverse reactions to drugs (Katzman 1982–1983). Estimates of the prevalence of dementia vary, but nearly all studies agree that the prevalence of most types of dementia increases with age, especially for Alzheimer's disease. Raskind and Peskind (1992) reported that 4 to 8 percent of the population 65 and over has symptoms of dementia significant enough to impair their ability to live independently, and at age 85 or over as high as 37 percent of elders have been found to exhibit such impairments.

The implications of this rise in the prevalence of chronic organic brain disorders are controversial. Wershow (1977) contended that most of these cases are beyond treatment and that the situation raises some ethical questions:

Must we be bound to keep alive as long as possible those poor souls who sit tied into their chairs, babbling incoherently, doubly incontinent, whose dining consists of having usually cold food stoked into them as rapidly as possible by overworked aides who must move on and stoke the next bed's prisoner? Is this living, much less living with dignity?

(1977,301)

In addition, Raskind and Peskind (1992) reported that people with dementia often engage in agitated behaviors such as screaming, pacing, and violent outbursts. Staff of nursing homes are often expected to cope with substantial physical and verbal abuse from residents suffering from dementia.

On the other side, Hellebrandt's (1978) observations provide a less stereotypic and more optimistic view of life for people with chronic organic brain disorder who have not reached the severe level of deterioration. She pointed out that most people suffering from early stages of dementia manage to live at home, cared for by family members and friends. Infections or other complicating diseases often end life before the individual reaches a "vegetative existence."

Hellebrandt, herself 76 at the time, also described life for sixteen patients with dementia who live in a locked ward, part of the convalescent facility of a large retirement village:

Oblivious to their plight, without apprehension or concern, they move about at will, accepting the locked door with matter-of-fact aplomb on all but the most exceptional occasions. . . . The idea of the locked door may distress a visitor [or the reader] but it is a meaningless concept to virtually all patients deteriorated enough to require segregation in this unit.

(1978,68)

The average age of these residents was 85. They were generally cooperative and in good general health. They were among the most contented residents of the nursing facility:

They move about freely, often in pairs, showing evidence of concern one for another even though they never address each other by name and cannot identify the person with whom they are walking. . . . They are clean, neat, and groomed appropriately, for the most part. On occasion they may wear two or three dresses at once. . . . The casual visitor would find the group deceptively normal.

All patients in the locked ward are completely disoriented as to place, time, day, date, year, season, or holidays. They have a poor memory for ongoing events. . . . None knows that their behavior is in any way aberrant. Neither do they realize that they have been institutionalized.

(1978,68)

This group was classifiable as borderline severe brain damage, according to the Kahn-Goldfarb Mental Status Questionnaire. Yet much of what we consider human still was present in these people—concern for others, friendliness, optimism.

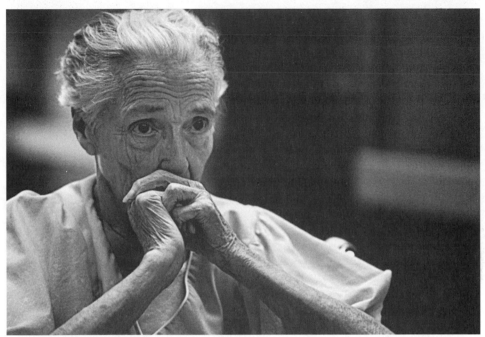

Dementia is like being lost.
Photograph by Gordon Baer/Black Star

On the other hand, symptoms of dementia include incoherent or nonresponsive speech, wandering, nonfunctional repetitive actions such as screaming or banging, removing clothing at inappropriate times and in inappropriate places, wandering into other people's rooms and poking about, and a host of other "difficult" behavior. "Normal" elders often refuse to associate with demented elders in meals programs, senior centers, nursing homes, retirement communities, and the like.

The experiences of all those who encounter demented elders in community settings probably have one thing in common: frustration. When we interact with people, we expect them to be able to pay attention and to take part of the responsibility for the interaction. When people are running programs for a large number of older people, their level of staffing seldom allows them to provide close monitoring for individual participants. Unfortunately, elders with dementia do not fit these expectations, and as a result, demented elders have few choices,

even if they could make them. No one wants them.

Frankfather (1977) found that mentally impaired elders living in the community tended to be shuffled from one place to another in a "loop" of agencies, making temporary stops at various institutions—usually nursing homes or hospitals, institutions that operate on a "medical model" and where the usual response to problems in "managing" the behavior of the demented person is to prescribe tranquilizers. Hospital discharge planners find it very frustrating to try to place these people in appropriate facilities. As one said, "Her problem is that nobody wants a screamer. The previous social worker tried three times to place her" (1977,150).

Caring for people with dementia is not easy. Student visitors are often upset by patients' inability to remember them from one visit to the next. Many people lack the patience to listen to incoherent ramblings. Incontinence disgusts most people. And it is difficult to regularly face this reality with the knowledge that for those of us who live long enough, there is a one in five chance of experiencing dementia ourselves.

Summary

Aging produces higher sensory thresholds and lowered sensory acuity. Yet there is a great deal of individual variability in the extent of the decline, and few people experience sensory limits on activities prior to age 75. The evidence with respect to age changes in perception is less clear, but various perceptual functions appear to decline with age. Age changes in the central processes also seem to limit psychomotor responses, reaction time, and complex performance. However, many of these changes can be at least partially prevented or compensated for.

The evidence on the relationship between aging and mental functioning is even more confusing. Cross-sectional studies tend to show a drop with age in measured intelligence, while longitudinal studies frequently report stability or increase over time. Intelligence is made up of several dimensions, some of which do not change with age, some of which improve, and some of which decline. In general, age changes in intelligence are minimal in comparison with cohort differences. The evidence on learning is difficult to interpret, but there is no doubt that most older people can learn effectively. Memory declines with age, with the prime problem appearing to be the retrieval function. Aging also seems to reduce the ability to categorize and make logical inferences, although these trends may be partly social in origin. The evidence with respect to creativity in older people also suggests that creative people often retain their creativity into very old age and that social factors are at least as relevant as psychological age changes.

Drives appear to wane as age increases, but because many factors intervene between drives and behavior, declining drives seem to produce less behavior change with age than might be expected. Motives guide and direct action toward specific goals. There is some indication that as people grow older they define achievement less in terms of competition and more in terms of cooperation. However, motives can be changed, even in older people.

The ability to experience emotions does not appear to change much with age, and emotional experience is important to people of all ages. Poor health is a source of negative emotions at all ages, but personal losses cause more problems for younger adults than for middle-aged or older adults. Older people feel that they are supposed to refrain from expressing their emotions, but few of them do so.

Many psychologists might consider the fundamental subject of this chapter to be psychobiology, particularly the early work on aging. In recent years, however, there has been a growing recognition of the important role that social factors play in basic psychological functioning. Throughout this chapter we repeatedly saw instances where social factors, particularly historical and generational factors, were more important than age changes in producing the cross-sectional relationships between age and various psychological functions. The psychology of aging truly is moving from assuming that unless proved otherwise, age differences are a result of psychobiological aging to assuming that unless proved otherwise, social factors are at least as important as physical aging. It has also been made clear that the range of individual differences *within* various age categories is usually much greater than the average differences *between* age categories.

From the adult development perspective, aging is either a movement through discrete stages or a continuous unfolding. Erikson's influential stage theory assumes that the individual personality is built by confronting a predictable series of life issues that establish capability, worth, and safety. The first life issue of adulthood is intimacy versus isolation, which is usually resolved in early adulthood. The second is generativity versus stagnation, which is often an issue of middle adulthood. The final issue, ego integrity versus despair, is usually an issue of later life. In later formulations, Erikson's theory was modified to acknowledge that the polar opposites coexist and are resolved through balancing factors such as fidelity to an ego ideal, love, caring, and wisdom. Erikson's theory is most useful in helping to identify how far a person has progressed in the developmental process. It is less useful as a means of predicting the central issues of a particular person's life at a particular time.

Levinson's stage theory of adult development presumes that the successive life stages require different life structures, which consist of patterns of thinking, relating, and behaving. Because the individual and the social world around her or him are constantly changing, life structure is inherently unstable, and the individual must undergo periodic transitional phases in which new life structures are created that are more suited to new life stages and changed circumstances. Levinson's theory contends that development alternates between periods of stability and change; thus, the theory is cyclic. The theory predicts the general phases a person will experience, but not the specific content of the life structure.

Process theories of adult development view aging as a matter of an individual's continuously taking in information about the self, interpreting that information in light of a theory of self, and refining the theory based on experience. The concern here is with individual identity and how it evolves over the life cycle. Aging can solidify one's theory of self by providing a large amount of experience about the self. Reduced job and parenting responsibilities that often accompany aging can lower the potential for conflict among the various aspects of one's identity. Thus, aging is more a matter of identity maintenance than of identity formation. Identity crises result from age changes in the individual or the environment that cannot be accommodated without a complete reorganization of the person's self-concept. No life change predictably causes identity crises; identity crises result from the individual's interpretation of life events.

Continuity theory holds that adults evolve through the use of information about the effectiveness of the life structure, self-concept, and personality. Individuals shape their own development and have goals for developmental direction. Developing, preserving, and maintaining adaptive capacity is the core of adult development, according to continuity theory. Development occurs when adults use life experiences to select which aspects of themselves to develop, which activities to engage in, which careers to pursue, which groups to join, and so on.

After young adulthood, there is more continuity than discontinuity in both self and personality, and each becomes more stable with age. This consolidation is a natural consequence of adult development, in which the duration and multiplicity of relationships and environments provide the information needed for the individual to develop a consistent view of her or his personality and self, including a customary strategy for coping with the demands and opportunities of life. Most people enter later life with a stable personality and positive self-esteem.

Coping with aging is usually a matter of defending a positive self-image. Using past successes, discounting messages that do not fit our existing self-concept, refusing to apply general beliefs about aging to ourselves, selectively interacting with others, and selectively perceiving what we are told are all ways to defend against ageism. But some people develop neither a stable personality nor a positive self-image. And there is little in the situation of older people that is conducive to doing so at that advanced stage.

A person who loses self-esteem in later life tends to do so because (1) physical changes have become so pronounced that the person is forced to accept what he or she sees as a less desirable self-image, (2) the person's self-esteem is precarious or vulnerable as a result of being either too dependent on social positions and roles or too narrow, or (3) the individual has lost control over her or his home or community environment to such an extent that she or he is essentially defenseless. This last possibility is especially likely when a person moves into a nursing home. Most people have the resources and defenses needed to retain or maintain self-esteem into old age, a fact made very clear by the multitude of studies showing that older people living in community settings have stable, well-defined personalities and high self-esteem, and that self-esteem increases with age.

Mental illness is best defined as a loss of mental functioning significant enough either to cause the individual concern or to impair the person's social functioning. Serious mental ill-ness can be divided into functional mental disorders (which have no apparent organic basis), reversible dementia, and irreversible dementia, such as Alzheimer's disease. Depression is the only functional disorder whose prevalence increases with age. Social losses, such as death of friends and relatives, are the most common source of temporary depression, and chronic pain is a leading cause of long-term depression among older people.

Reversible dementia can arise from medication interactions, alcoholism, heart disease, or diabetes. Irreversible dementia is caused by brain tissue loss or poor circulation. No mental disorder is a normal part of aging. However, the proportion with dementia increases sharply after age 75. Developing effective prevention and treatment measures for dementia is an important goal if our society is to maintain a high quality of life for older people.

6 ▌ Social Aspects of Aging

Our individual lives are structured by our social environments as well as by our physical and mental capacities. Social attributes such as gender and race affect individual life chances. Group memberships influence access to knowledge and opportunities. We interact with one another not only as individuals but in groups as role players. And the roles we play are sometimes organized sequentially into a life course tied to age or life stage. In addition, social processes such as socialization engage us with our social world. All these aspects of the social environment can be influenced by age.

In this chapter, we begin by examining some of the ways that social age can be defined; then we look at how aging influences which social roles we play and how we play them, we consider the life course as a cultural ideal-type that provides a social map through the life stages, and we discuss some of the social processes that connect individuals to the life course and its roles. We then revisit the notion of adult development from the point of view of its social components. The chapter closes with a consideration of aging and its relation to changes in four types of social circumstances: roles, groups, environments, and lifestyles.

Defining Social Aging

In Chapter 1 we saw that individual aging can be defined chronologically, functionally, or in terms of life stages. All three definitions can

be used separately or in conjunction with one another to develop different social conceptions of age and aging. Life stages can be transformed into static *age strata*, which tend to define social age in terms of age-relevant social policies such as voting age or age of eligibility for retirement. Chronological *birth cohorts* can be tied together into social age categories; examples include the baby boom generation, the children of the Vietnam War era, and the children of the Great Depression of the 1930s. Lineal *generations* in the family are yet another way to socially identify a person's age. The term *generation* is also used to refer to age groupings in occupational groups; terms such as "old guard" and "young hard-chargers" call such definitions to mind. Life stages tend to be the most important of the social definitions of age because they are the master timetable to which concepts of life course *role sequences* are tied.

Definitions of physical and psychobiological aging are constrained to some extent by the physical realities of aging, but developmental and social definitions are much more complex. There are many different social definitions of age and aging, and matching the right definition to the issue at hand requires an open and thoughtful stance.

Social Roles

Social roles have great significance because individuals often define themselves in terms of roles, and the places of individuals in society are determined by the specific roles they play. A *role* is the expected or typical behavior associated with a position within the organization of a group. Positions, and the roles that go with them, usually have labels such as *mother, teacher, senator, milling machine operator,* or *volunteer*. Roles are sometimes activities such as artist or gardener that do not necessarily involve interacting with others, but most of the time we behave not as isolated role players but in relation to other role players in **role relationships.** Thus, in the organization of a group, the various positions are related to one another in specified ways *regardless* of who occupies the positions. For example, the ideal is that teachers impart knowledge and skills, assign work to students, and evaluate the results; and students pay attention to instruction, do the work, and aim for a positive evaluation of it. This much we can know in advance without knowing who the actual role players are.

A large part of everyday life consists of human relationships structured at least partially by the social roles and positions of the various actors. But in everyday life the action is much more like improvisational theater, where the characters and dialogue are made up right on stage, than like formal theater, in which the action is predetermined. The following example makes this point clearer.

Suppose you sign up for a course on aging. When you first meet the professor, you both have general ideas of what to expect from each other. For example, you might expect the professor to know a lot about aging and to use that knowledge to structure your educational experience. You might also expect the professor to evaluate your progress. The professor might expect that you have relatively little systematic knowledge about aging but are intelligent and interested in the subject and willing to work to improve your knowledge. These general expectations would serve as the background for your initial interactions, but almost immediately the two of you would begin to incorporate *personal* in-

formation into your expectations for this *specific* relationship. You might discover, for instance, that the professor is relatively easygoing and has a good sense of humor. The teacher might find you conscientious and inquisitive and a good writer. You would each gradually modify and refine your expectations so as to take an increasing amount of personal information into account. The pace of this process would depend on the willingness of each of you to disclose and attach significance to personal information. The longer you keep at this process, the more personalized your relationship might become. Thus, the *duration* of any role relationship is one of its most important features.

In this conception, the social order made up of group structure and the roles that constitute it is a *negotiated order* rather than a rigid, predetermined pattern. Sociologists call this process *role making*. Conflict can arise in groups because some members expect the action to "follow the rules," while other group members expect the rules to evolve along with the action.

Aging Affects Social Roles

When life is filled with relatively new relationships, as in young adulthood, we are uncertain about what to expect from others and what they expect from us. But when life is filled with long-standing relationships, there is a large component of security in knowing what to expect—even if the relationships are not all we might wish them to be. In middle age and later life, most people's role relationships are long-standing and highly personalized.

Age is not a position in the structure of a group.* Along with other individual attributes, such as gender, social class, and race, age is used to determine *eligibility* for various positions, to evaluate the *appropriateness* of various positions for specific people, and to

---

*Others disagree; see Riley, Johnson, and Foner (1972).

modify the expected behavior of an individual in a specific position. Let us look at each factor in turn.

Role Eligibility In adulthood, advanced age occasionally makes people eligible for valued positions such as *retired person,* but more often it makes them ineligible to hold positions they have valued. Whether we are merely *assigned* to the positions we occupy, whether we must *achieve* them, or whether we have *selected* them, we do not simply occupy or take over or retain positions. Positions are usually available to us as a result of our having met certain criteria. In most societies, different positions, rights, duties, privileges, and obligations are set aside for children, adolescents, young adults, the middle-aged, and the old. In our culture, the primary eligibility criteria are health, age, gender, social class, ethnicity, skin color, experience, and educational achievement. These criteria may be gradually modified and personalized in some cases, but overall they govern eligibility in prescribed ways.

A major unresolved research question is the precise position that age occupies in this list. Is age a primary criterion, or are gender and ethnicity more important? Are there situations in which age advances to be a prime criterion and other situations in which it is not very important?

As we pass through life, the field of positions for which we are eligible keeps changing. A young boy can legitimately be a Boy Scout, an elementary school pupil, or unemployed—all roles that would bring disapproval if he held them as an adult. As a young man he can be an automobile driver, a bar patron, and a voter—all roles he cannot occupy as a child. As an older adult he can be retired, a great-grandfather, or member of a senior citizens' center—all roles he cannot play as a young adult. Of course, age also works negatively. For example, age discrimination often prevents older people from holding desirable jobs even if they want one.

Role Appropriateness In addition to affecting eligibility, age also influences ideas about the suitability of particular roles for people of a particular age. For example, Mr. A. began competing in local cross-country motorcycle races at the age of 39. Most of his friends thought he was crazy, and many indicated that *they* would certainly never do anything that dangerous *at his age.* His friends were letting him know that his choice of racing was inappropriate for anyone his age. Interestingly, he was encouraged by other racers of all ages and went on to win a few trophies before he "retired" from competition at the ripe old age (in motocross terms) of 45. These reactions illustrate that the age standards of society at large may not be adopted by all subgroups. And this tendency, in turn, shows why both older people and younger people often restrict their interactions to subgroups of age peers who can be counted on to approve of their role choices.

Role Modification Age also modifies what is expected of people in particular positions. Because of their inexperience, young adults are often dealt with leniently when they begin new adult positions, such as their first full-time job. For instance, young adults sometimes show too much impatience with others on the job. Generally, supervisors tolerate such impatient behavior, attributing it to inexperience rather than a character flaw. The same amount of impatience from a middle-aged worker would not be tolerated as long or as well. Behavioral standards may be different in old age as well. For example, an 86-year-old father is allowed to be more dependent in his relationships with his offspring than is a 56-year-old father.

Role models are people whose role performances are used by others as a guide. By observing the demeanor, skills, and behavior of others in a role that we aspire to play well, we can get a much better mental image of the dynamics involved in effectively playing that role. Obviously, the more potential role models a person can observe, the greater the choice of role-playing styles. For example, people who are retiring today have many more potential role models compared with people who were retiring in 1950.

Cumulative Effects

Positions that adults hold or have held provide various degrees of access to advantages such as prestige, wealth, or influence. Highly rewarded positions can be attained through family background, individual initiative, or both. But despite the rhetoric that ours is an open class system in which one can rise from rags to riches, in reality only a tiny minority of people ever move out of the social class into which they are born.

Growing older in an advantaged social position generally means being able to accumulate personal wealth and prestige. People do not become wealthy, revered, or influential just by becoming old but by occupying an advantaged position for a number of years. Former president Jimmy Carter enjoys a great deal of prestige and influence in his old age, not because he is old but because he was president of the United States and continues to be a skilled conflict mediator. A great many of the richest and most influential Americans are old, but their age is not the reason why they are rich or influential. In fact, their wealth and influence *discounts* their age in that rich and powerful people are much less likely than others to be disqualified from participation purely on account of age. Age disqualification happens mainly to people who are already disadvantaged.

One can also be advantaged by having exceptional skills. Great writers, musicians, artists, therapists, diplomats, and others can use their skills to offset their age and avoid being disqualified from full participation in later life. Artist Pablo Picasso, psychiatrist Carl Jung, artist Grandma Moses, and writer May Sarton never had to worry about age discrimination—their talents remained in demand. Other people find that age often offsets ordinary skills and leads to disqualification.

Although part of the disqualifying character of age is related to erroneous beliefs about the inevitability of age decrements in performance, part of it is related to the scarcity of leadership positions, the desire for younger people to acquire the positions of leadership that are available, and the willingness of older people to give up their leadership responsibilities.

The Cultural Life Course

The term **life course** has several quite different meanings in social gerontology. *Biographical* definitions of the life course focus on the pathways *individuals* take through the life span and how these pathways are influenced by the intersection of life stage and historical eras such as the Great Depression or World War II (Elder 1991). *Statistical* definitions of the life course look at the relative proportions of an *age cohort* who show various patterns of role sequences over time (Rindfuss 1991). These definitions attempt to identify statistical regularities in the role-related behavior of age cohorts, particularly in terms of living arrangements, family, education, and employment. *Cultural* definitions of the life course focus neither on what particular individuals do nor on what most people do in various life stages, but instead focus on *what people are expected to do* as they prepare for, occupy, and relinquish successive life stages.

The cultural life course consists of ideal age-related progressions or sequences of roles and group memberships that individuals are expected to follow as they mature and move through life. Thus, there is an age to go to school, an age to marry, an age to "settle down," and so forth. For example, Settersten (1992) found considerable consensus among adults that people should marry in their mid-20s and that men should be settled in a ca-

Grandma Moses at work at age 88.
Photograph by AP/Wide World Photos

reer by the time they are 30 years old. Life course milestones are most commonly found in relation to family, education, and employment (Gee 1990; Byrd and Breuss 1992).

But in reality the life course is neither simple nor rigidly prescribed. For one thing, various subcultures (whether based on gender, social class, ethnicity, race, or region of the country) tend to develop unique ideas concerning the timing of life course transitions. For example, male automobile workers tend to favor retirement in the mid-50s while self-employed entrepreneurs tend to prefer the late 60s. Life course sequences differ substantially by gender, with women having more potentially conflicting expectations than men, usually revolving around combining employment with child rearing. In addition, even within subcultures, there are often several alternatives. For instance, having made a decision to attend college, a person often has a wide range of types of institutions from which to choose. Thus, like a road map, the abstract concept of the life course in reality is com-

posed of a great many alternative routes to alternative destinations. There are many ways to get from New York to Los Angeles. Most people take the well-traveled interstate routes and a few take the scenic byways, but the destination is the same. Thus, in studying the life course the general outlines of the trip are probably more important than the specific route being followed at a given point in time.

As people grow older, their accumulated decisions about various life course options produce increased differentiation among them (Dannefer 1988). Although very late in life the number of options may diminish somewhat because of social and physical aging, the older population is considerably more differentiated than the young. Yet even with the increasing complexity of the life course with age, certain generally accepted standards serve as a sort of master timetable for the entire population. Although there are many exceptions and variations, most Americans start school, finish school, get married, have children, experience the "empty nest," and retire; and each event occurs within a general chronological age span, an age range during which these events are *supposed* to happen. Most of us spend our lives reasonably on schedule, and when we get off schedule we are motivated to get back on again (Neugarten and Datan 1973). Thus, the cultural life course serves as a global construct that can be used to place a particular life history in context.

Nevertheless, the details of people's life choices are usually much more chaotic than the ideal life course patterns depicted by various subcultures. For example, Rindfuss, Swicegood, and Rosenfeld (1987) followed a national sample of high school graduates for eight years, looking at their role sequences over time in work, education, and household arenas, among others. They found that it took 1,098 separate sequences to describe the eight-year experience of the 6,696 men in the study and 1,827 sequences to capture the life sequences of the 7,095 women. Only about half of the young men pursued the traditional sequence of education followed by work, and the proportion for young women was even lower. Thus, there were more exceptions than adherents to the cultural ideal. In addition, cultural ideals themselves change over time. For example, in the late 1950s, marriage and childbearing in early young adulthood was the middle-class ideal; in the 1980s, the middle-class ideal was to defer both marriage and childbearing until later in young adulthood.

How much of the complexity observed by Rindfuss and his colleagues was the result of subculture and gender and how much was due to "error" in following life course patterns is unclear. However, it is clear that cultural ideals serve as important life course guidelines, even when they cannot be followed. Indeed, considering the complex number of dimensions, such as family, education, occupational preparation, and employment, that must be balanced and coordinated in order to follow "traditional" life course paths,

it is not surprising that there are many exceptions. Another important point is that statistical or demographic definitions of what is "normal" do not take the place of cultural norms. Most people do not know what the statistical norms are, so they could not use them for a guide even if they wanted to.

The various life stages that serve as benchmarks for the life course are made real for the individual in three ways. First, they are related to more specific patterns, such as occupational career and family development. Indeed, the family is a major social context for negotiation and decision making about pathways from one life stage to the next. Second, specific expectations or age norms accompany various life stages. And finally, people are required to make particular types of choices during given phases of the life course.

Dimensions of the Life Course

Figure 6-1 shows, very roughly, how life stages are related to chronological age, and age-graded progress in employment, the family, and education. More dimensions could be added, but the important points are that social institutions such as education, the family, and the economy tend to prescribe their own stage definitions and ideal life courses and that these institution-oriented life courses are related to the various general life stages. For example, tracing lines vertically in Figure 6-1 at age 40 and at age 70 reveals significant expected differences on all of the life course dimensions. At age 40, an adult is expected to have completed her or his education, be advancing on the job, and to be married and rearing children in the home. At age 70, an adult is expected to be retired, live alone with his or her spouse, and have launched the children. There are significant parallels and interactions between and among these dimensions of the life course.

Age Norms

Age norms tell us what people in a given life stage are allowed to do and to be as well as

Figure 6-1 Various dimensions of the life course.

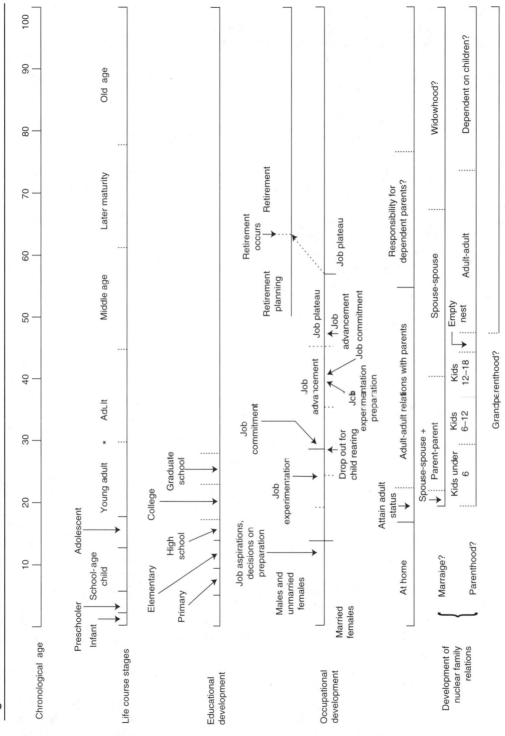

* Dotted lines mean that the timing is discretionary

what they are required to do and to be. Some age norms operate very generally to specify dress, personal appearance, or demeanor in a wide variety of social roles. Other age norms govern approach, entry, incumbency, and exit from social roles, groups, or social situations. Many age norms come down to us through tradition. On the other hand, legal age norms are often the result of compromise and negotiation.

A series of assumptions underlies age norms. These assumptions, often uninformed, concern what people in a given life stage are capable of—not just what they *ought* to do, but what they *can* do. Thus, opportunities are limited for both children and older people because others assume that they are not strong enough, not experienced or educated enough, or not capable of adequately mature adult judgments. For instance, older workers are passed over for training opportunities because "you can't teach an old dog new tricks." Older job applicants are passed over on the assumption that people over 60 have too few years left to work to warrant the investment. By the same token, young adults are passed over because they "don't have enough experience."

Sometimes age norms make useful and valid distinctions. For example, few of us would want to drive automobiles if six-year-olds were allowed to take to the roads. But many of the norms for adolescence and beyond become increasingly difficult to justify because of their essentially arbitrary nature. And the greater the gap between the actual level of individual functioning and the level implied by an age norm, the more likely that the age norm will be seen as unjust.

Various mechanisms secure conformity to age norms. People are taught early in life by their parents, teachers, and peers how to apply age norms to themselves. If they do not conform, then friends, neighbors, and associates can be counted on to apply informal pressures. In the formal realm, regulations bring bureaucratic authority to bear. And finally, laws put the full power of the state behind age norms.

Age consciousness is an important prerequisite for conforming to age norms. Various cues from the environment send age signals to the individual. For example, Karp (1988) studied white professional men and women in their 50s to find out how they conceived of their age and what factors were related to their perceptions. Nearly all respondents mentioned the paradox between how old they felt on the inside and how they appeared to others and to themselves on the outside. They knew their age objectively, yet subjectively they usually felt younger. Karp found seven common reminders of age: body signals, time markers, generational reminders, contextual reminders, mortality reminders, human development reminders, and life course reminders.

Body signals common to people in their 50s included more frequent aches and pains, declines in physical stamina, and less physical flexibility. These experiences located people securely in middle age. For most people, their fiftieth birthday and the hoopla surrounding it was a significant time marker. Generational reminders included experiences such as becoming part of the oldest generation at one's workplace, seeing one's parents' obvious aging, and seeing one's children become parents. Contextual reminders were the most subtle, consisting of experiences of not being fully welcome in certain places and among people of a different age. For example, Karp's respondents reported feeling appreciated at work for their special knowledge of the organization and its history, but at the same time some felt unwelcome at after-work get-togethers. Mortality reminders came from the increased frequency with which people in their 50s learned of deaths of friends and relatives. This knowledge reminded people of their own mortality; they became aware that each year the incidence of death among people their age increases. Human development reminders came from an increased sense of wisdom, of being better able to make judgments, identify appropriate values, and set priorities. Life course reminders consisted of launching children into adulthood, enjoying the period of married life after child rearing, and becoming a grandparent. Thus, people in their 50s had a great deal of information at their disposal that indicated their age.

Decision Demands

The sometimes chaotic nature of the numerous alternatives that various life courses present was mentioned earlier. This chaos is minimized to some degree by age, sex, social class, ethnicity, and so forth. But how do people get into *specific* situations? It is impossible to assign people to each and every niche in a complex, rapidly changing society. **Decision demands*** force the individual to work within the system in order to find a slot in the social organization.

Decision demands require that selections be made from an age-linked field of possibilities. For example, after completing preparatory education, young adults usually enter a period of job experimentation. The field of possibilities expands dramatically immediately after graduation (certification) and continues to expand while the individual gains job experience. But there is an increasing expectation that people find positions of employment into which they will settle; during this period, the field of jobs for which they are eligible may slowly contract. Contraction also occurs as jobs are selected by others of similar age and experience. For many jobs, career tracks are difficult to begin after the age of 45.† For others, it is difficult to break in after 35.

Decision demands tend to be concentrated in young adulthood, between ages 18 and 30. That is, individuals are *required* to make choices and select their "career tracks" in all sorts of areas—education, employment, family, community, voluntary associations, and so on. For example, Rindfuss (1991) examined demographic data and found an extremely high density of life course transitions in young adulthood. This is the life course stage when most people marry, have babies, change residence, divorce and remarry, leave school, and settle into full-time employment. Lifestyles developed before or during middle age tend to persist as long as health and money hold out. Thus middle-aged people who want to switch tracks or get involved in new areas often must confront the fact that the available slots were taken earlier by their age peers.

On the other hand, those middle-aged people who do retrain and switch careers often find that their new careers develop at an accelerated rate compared with those of younger people starting at the same time. This acceleration is probably due to these people's greater maturity and general experience plus the underlying notion that a person's position in an organization should match his or her life stage. More research is needed on this phenomenon.

Thus far we have looked primarily at the social structure of life in human groups and how it is affected by aging. We now turn to how aging influences the various social processes through which individuals are incorporated into that structure: socialization, acculturation, role anticipation, and role adaptation.

Socialization and Acculturation

Socialization is a group of processes through which a group encourages and/or coerces its members to learn, and conform to, its culture—the distinctive way of life of the group and its total human-made environment, including ideas and material products of group life that are handed down from generation to generation. Socialization is society's means

---

*For a more detailed discussion of decision demands, see Atchley (1975).

† The term career track as used here refers to an idealized progression of roles within an ongoing group.

for creating capable participants. **Acculturation** is the actual learning of a culture. Much of what we seek to learn through acculturation is aimed at making us more effective role players. Some social scientists use the term *socialization* to refer to both the action the group initiates toward the individual *and* the individual's response to this action. To the individual, socialization and acculturation are important prerequisites for getting access to whatever society has to offer. If people know and understand the culture and the social system built upon it, they can potentially put the system to use. If they do not understand the way of life of the society in which they live, their lives can be confusing and unpredictable. Thus, people are usually motivated to become acculturated.

For society, socialization produces a measure of continuity from one generation to another. After all, society needs new participants, and socialization is a major process through which society attracts, facilitates, and maintains participation. The efforts of the group to help individuals learn the system range from formal, structured programs in which the group is responsible for the outcome, to unstructured, informal processes in which the individual is responsible for the outcome. For example, families are expected to teach children to speak, and schools are expected to teach children advanced mathematics. On the other hand, adults must generally find out on their own how to buy and insure an automobile.

Age affects what the individual is *expected* to know and what the person *needs* to know to be an effective participant in society. Early in life, the emphasis in socialization is on learning language and customs; then it shifts to preparing for adult social roles. In adulthood, expectations gradually shift toward self-initiated acculturation; that is, adults often recognize their need for knowledge and skills and go after them without waiting to be told. At the same time, fewer publicly supported, formal opportunities exist for socialization in adulthood compared with the opportunities that exist for young people.

Rosow (1974) argued that because norms for older people do not exist, aging is a "roleless role" and there is nothing to socialize the aged *to*. Although this argument may be accurate in terms of formal socialization, its applicability to acculturation is questionable. Older people often have few required contacts with the general community; they associate primarily with people they know and who know them. In addition, most older people continue to function mainly in roles they have occupied for many years. The point is that most roles older people play are roles they have been *socialized to* for a long time. Aging usually affects primarily the nuances of role playing rather than its basic structure.

Most societies pay little attention to their adult members' needs for maintenance or renewal of their knowledge or skills. In this age of the computer, for example, where do middle-aged or older adults go to learn how to use this new technology for dealing with information? As we will see later, this pattern seriously hampers the ability of some older members of society to remain an integral part of the society, particularly in terms of the knowledge and skills required for employment. If older people need information and skills that are necessary for participation in society but that they cannot secure for themselves, then the socialization processes in the society are not adequate.

Much adult acculturation comes from experience. For example, Kohlberg (1973) pointed out that the moral development of young people is mainly cognitive and symbolic. But for the person to develop what Kohlberg called *principled thinking*—"principles to which the society and the self *ought to be committed*"—it is not enough merely to "see" the principles (1973,194). The person must also experience sustained responsibility for the welfare of others and experience having to make irreversible moral choices. A similar process occurs in other areas of life, such as mate selection, career choice, choice of residence, and so on. The adult develops life principles—a philosophy of life—through an interaction between what is known, what is experienced, and his or her personality. And, theoretically at least, the longer one lives, the greater the opportunity for this dialectic to produce refinements in one's approach to living.

For instance, it is one thing to know intellectually that telling lies is wrong, and it is quite another to have experienced the results of having lied. When we are young, it is difficult to believe that all the dos and don'ts we are taught are necessary. As we get older, our experiences give us a fuller understanding of why rules exist. Older people are often more committed to basic moral principles than the young, not because they are rigid but because life has taught them the value of following the rules. They also may be wiser in choosing which rules to challenge.

Role Anticipation and Adaptation

The process of **role anticipation** involves learning the rights, obligations, resources, and outlook of a position one will occupy *in the future*. To the extent that the future position is a general one, it need not represent an unknown. For example, most people understand that in retirement, people are expected to manage their own lives—set goals and form new daily routines. Most also understand that to do so successfully requires financial planning. Through fantasy, it is possible to anticipate what the future will hold, identify potential problems, and make advance preparations. The role changes common to later life can often be anticipated, thus smoothing out the process of transition.

However, many roles we take up in later life have a degree of vagueness that allows some flexibility in playing them but at the same time hinders anticipation. Such roles are not "packaged," as are many roles that young people play. For example, the role of high school student is much more clearly defined than the role of retired person. As a result, older people must often *negotiate* the rights and duties of their positions with significant other people in their environment. Thus, in late adulthood, acculturation—learning or relearning how to function in one's social milieu—depends heavily on the characteristics of the others with whom the older person must negotiate. The attitudes of others concerning who the older person is becoming are probably crucial to the content of acculturation in later life. As images of aging become less stereotyped in American culture, the outcome of these negotiations can be expected to become more positive.

Social roles, even specific ones such as mother or teacher, tend to be defined in terms that allow room for interpretation. **Role adaptation** is a process of fitting role demands to the individual's capabilities. Negotiation—the interpersonal aspect of this adaptation—takes place between role players. The process of tailoring the role to fit the individual and the situation makes it impossible to describe *the* role of *the* grandfather. We can only describe some of the similarities and differences we find when we look at how grandfathers play their roles and what these similarities or differences are related to.

Thus, through socialization, society tells its members that retirement is normal and that it should occur sometime between age 60 and 70. Through acculturation, an individual comes to *agree* that retirement is normal and that it should occur between 60 and 70. Through role anticipation, the individual identifies potential problems associated with retirement and tries to solve them in advance. Once the individual retires, the process of role adaptation involves tailoring the rather vague retirement role to fit the needs of the individual. Part of this process may involve negotiations between spouses about what the retirement role will consist of and how it will influence their life together.

Social Factors in Individual Development

The cultural life course represents a social program for individual development. Life course choices result in the evolution for each individual of a social life structure made up of roles, activities, group memberships, relationships, and environments. The social development of the person takes place in parallel and interaction with psychological development. A person's location in the social world can have profound impact on the life structure he or she develops, but so can that person's aspirations and perceptions.

In Chapter 5, *continuity theory* (Atchley 1989, 1995d) was presented as an evolutionary theory of adult psychological development. This same theoretical perspective can also be used to understand how people interact with their environments to construct their own social development. Continuity theory assumes that people develop patterns of behavior, relationships, and activities that they wish to maintain over time both to achieve their goals and to adapt to changes in the details of everyday living. Continuity theory presumes that people learn from their successes and failures and use this feedback to modify both their goals and the patterns of behavior and relationship that they see as the means for achieving their goals.

In this context, social development is a process of *selective investment* in roles, relationships, activities, and environments. The selection here is based on a combination of personal preferences, life course ideals, age norms, and feedback from the results of life experience. For example, some people have employment that is easily

mastered and offers little opportunity for development; the assembly-line job is an example. People with such jobs are unlikely to expect significant opportunities for advancement and, accordingly, are not apt to put much time and effort into developing their occupational capacity as long as they are willing and able to remain in their jobs. By contrast, others have jobs that both require and offer continuous opportunities for growth and development. Examples here might be physicians, computer engineers, college professors, and information specialists. People in these kinds of jobs tend to invest large amounts of time and energy in developing and maintaining their occupational knowledge and skills.

By the time they have reached middle age, most people have developed an obvious life structure that is not just frozen in time but linked both to the individual's social past and desired social future. The persistence of this life structure results not just from the goals of the individual but from social pressures from others that the individual remain socially recognizable. Another way of saying this is that social continuity increases the predictability of feedback that we get from others.

A central premise of continuity theory is that in adapting to aging, people will attempt to preserve and maintain the long-standing social patterns of living and coping that they identify as being uniquely their own. The theory does not tell us what continuity looks like to a specific individual. To know that, we must know more about the person and his or her social background. How people respond to changes in social circumstances that often accompany aging depends on how that specific change fits into the preexisting life structure of that particular person. It would be easier to work with aging people if we all responded in similar ways to similar situations, but that is simply not the way development works. Let us now look at some of the common shifts and changes that occur in middle and later life.

Aging and Changes in Social Context

Writings about later life often give the impression that the aging individual is firmly in the grip of inexorable physical, psychological, and social processes that allow little room for individual or group initiative or maneuvering. It is easy to lose sight of the fact that, while our abstractions concerning these processes are built up from individual or group experience, the categories we use are often intellectual conveniences or necessities and not concrete elements of the social world. Thus, when we read that having a certain social role does this or that to a person, what is really meant is that certain concrete behavior or concrete ideas are more likely to occur in the lives of people who play a given social role. The term *social role*, then, is merely convenient shorthand for a certain pattern of behavior or particular situation.

An earlier edition of this book stressed role change as the central element of situational change in the later phases of life. Certainly, role changes are important and deserve a good bit of attention. However, roles are not the only important social elements that can change in later life. Moreover, the process of role change itself is governed by how one defines the life course and the age norms associated with it, and by various situational changes.

The term **social situation** is a general one that social scientists have used to emphasize that roles, groups, and norms have little meaning for the individual apart from the particular time and place at which the individual encounters them. The immediate social environment is very important in translating social order in the abstract into concrete reality for the individual. In the following sections we consider changes in roles, groups, and environments that can accompany aging.

Role Changes

The role changes we consider are mentoring, launching the children, retirement, widowhood, dependency, disability, institutionalization, mediating, peer counseling, and participation in activities.

Mentoring *Mentors* are senior members of a group who serve as teachers, sponsors, counselors, guides, and exemplars to younger people, often those just entering the group or moving to a new level within an organization. The mentor fulfills a transitional role by helping young people through the process. A good mentor's prime consideration is helping a member of the upcoming generation fashion effective goals for the future and develop specific plans for achieving them. There is a strong element of generativity in mentoring—concern for the well-being of another—but the mentor also can get enormous satisfaction from seeing his or her skills and knowledge put to good use.

Mentoring can occur in any group, but most discussion of mentoring has concerned the workplace. Some work organizations have formal programs that match senior members of the staff with those just starting out. For example, Dr. S. has served in this role several times with younger faculty members at her university. She meets with them regularly to discuss their career plans, aspirations, fears, and frustrations. At the forefront in her agenda is an attempt to help the person gain perspective on what the job requires and how to produce it at a level that will be both satisfying and rewarding. It is not so much a matter of urging them to do what she did at that stage of her career, but of listening to and supporting their visions and using her knowledge about the workings of the profession and of the particular university to make suggestions about how to achieve their aspirations. She sometimes has more confidence in them than they seem to have in themselves, and she likes to think that her faith in them helps them overcome some of their apprehensions about taking on challenging and ultimately rewarding goals.

The good mentor unselfishly makes all of his or her career experience available to facilitate and support the career development of another. Being a mentor can be one of the most satisfying roles in middle age and later life. Certainly, as Dr. S. looks back over her career, some of her most important contributions have been to the successes of younger colleagues who are now themselves passing the torch to the next generations.

Launching the Children   For those who elect to become parents, launching the children into adulthood generally occurs in middle age. For example, people who married in 1965 and had children could expect on the average to see their last child leave home in 1990, at which time the average husband was just over 47 and the average wife was about two years younger.

The rhetoric of American society has proclaimed parenting to be reserved mainly for females and to be a major focus of their lives. Following this supposition to its logical conclusion leads to the expectation that the "empty nest" would result in a meaningless existence for most mothers. Of course, this perspective is based on *role theory,* which contends that roles become so embedded in our identities that we cannot give them up without doing major damage to our self-concepts. However, researchers who have put this conventional wisdom to the test have found quite a few surprises.

First, the term *empty nest* is somewhat misleading, carrying the connotation that young adults, like fledgling birds, leave the nest once and for all. But many young adult newlyweds visit and telephone their parents often and continue to use their parents' charge accounts, cars, and so on. As education has extended to higher ages, young adults may no longer live at home, but a variety of financial connections may remain. With delayed marriage becoming more common, *getting* the

children to leave home sometimes becomes a bigger issue than *having* them leave.

As we saw earlier, young adulthood is a chaotic time in the life course, filled with finishing school, job experimentation, first marriages, births of first children, divorce and remarriage, and so on. Many parents of young adults are faced with having to cope with young adult children who need to return to the nest for a time to regroup and develop a new plan for living as independent young adults. This can be a stressful time for the parents, especially if they have made plans based on their newfound freedom as empty nesters. Returning home is often no picnic for the young adults either. Having to return home can represent the failure of a dream, and the time needed to get over disappointment can delay the formulation of new plans.

Fiske and Chiriboga (1990) found that the period following the launching of the last child was not stressful. Indeed, middle-aged women who had not yet launched their children reported much less positive self-concepts than those who already had, particularly in terms of those characteristics that influence interpersonal relationships. Dealing with adolescent or young adult children may be the most difficult task a parent can attempt. The youngsters' inconsistent yearnings and capacities for independence sometimes yield awesome unpredictability. Parents feel the need to let their children experiment with assuming responsibility for their own decisions and behavior, but at the same time to supervise them closely enough to avoid potentially drastic errors. This conflict makes parents also somewhat unpredictable. The various conflicting postures create an uneasy situation on both sides and help explain why so few parents feel that they are "a success" at this stage.

Women generally look forward to having their children launched because they see it as a chance for greater freedom and opportunity. This is less true if motherhood has been the focal point in their life and if they have few other interests.

Barber (1980) compared reactions to the empty nest of a small number of fathers and mothers. He found that most parents had both positive and negative reactions to having launched their children into young adulthood, but the kinds of feelings they reported varied significantly by gender. Mothers' negative feelings tended to be based on loss of an important role and uncertainty about whether the adult child was adequately prepared to succeed in adulthood. Mothers' positive feelings tended to be based on a sense of having accomplished their mission and having fulfilled a dream. On the other hand, fathers' negative feelings tended to stem from a feeling of having lost an opportunity—a sense of regret at not having taken advantage of the time when they could have developed relationships with their children. In addition, fathers were more likely than mothers to be surprised at the intensity of their feelings of missing their children. Nevertheless, both fathers and mothers felt that being relieved from parental responsibilities and increased freedom and privacy produced positive feelings, including closer marital relationships, that outweighed the negative. Most also reported improved relationships with their adult children. On balance, then, it appears that fathers and mothers have different reactions to the empty nest and arrive at generally positive responses to it via different routes. Only a small minority of either mothers or fathers have negative responses, and these tend to be linked to feelings of not having done as well as they might have at being a parent.

In summary, then, it would appear that having all of one's children launched into adulthood is not as complete or abrupt as folklore would have it. In addition, this change is one that parents look forward to, prepare for, and do not find especially stressful. Stresses of the empty nest are not really about the empty nest so much as about the difficulties that arise when young adults have to return home.

Part of the controversy over the effect of child launching no doubt results from the fact that researchers have sometimes falsely attributed the consequences of other factors to the empty nest. For example, when the departure of children from the home unmasks an empty marriage, it is the quality of the marriage, not the empty nest, that may lead to divorce. There is no evidence that a solid marital relationship is harmed by an emptied nest. Studies are needed that examine the empty nest as a *transition* and look at men and women both before and after the last child is launched.*

Retirement Retirement is the institutionalized separation of an individual from her or his occupational position, with continuation of income from a retirement pension based on prior years of service. Age is usually a prime consideration in such separation, although years of service and health may also be used as criteria. Usually, a certain latitude is given the individual to choose the point at which he or she retires, sometimes as early as 50 and often as late as 70. At age 65 or over, about 85 percent of men and about 90 percent of women do not have an occupational position.

When people leave their occupations, many changes occur in their lives regardless of whether retirement was voluntary or forced. But as we will see in Chapter 9, retirement means not only giving up a role but also taking up a new role. The self-sufficiency required of retired people is certainly a challenge to most of them. Retirement also influences the other roles an individual plays, although the direction of this influence is unpredictable. For example, some marriages are enhanced by retirement, while others are subverted. Some organizations shun retired members, while others rely heavily on them. To some people, retirement represents a crisis, a world turned upside down. To others,

*See Chapter 8 for more detail.

retirement is merely the realization of a long-awaited and well-planned change. Currently, it appears that at least two-thirds of those who retire do so with no great difficulty either in giving up their job role or taking up the retirement role (Ekerdt 1987).

Retirement also can affect other roles such as spouse, father, or community service organization member. Many retirees go through a short period of uncertainty as they renegotiate the ground rules of various roles to take their retirement into account. Retirement is also covered in much more detail in Chapter 9.

Widowhood   Widowhood is another role change that is common among older people. In 1990, more than half the women and 20 percent of the men 65 or over were widows or widowers, and the proportion widowed increases with age (U.S. Bureau of the Census 1992b).

Widowhood is often a difficult role change. It disrupts the routines of everyday life in terms of time use, space, and activities. For some people, widowhood allows no fulfillment of sexual needs. It also diminishes the possibility of gaining identity through the accomplishments or positions of the spouse. Widowhood also can mark the loss of an important type of intimate interaction based on mutual interests. For example, the widow often replaces the intimate, extensive, and interdependent relationship she had with her husband with several interdependent relationships with other widows.

The role changes that retirement and widowhood bring are not necessarily unwelcome to the individual. Often, people happily give up the responsibilities of work and view retirement as a period of increased fulfillment. Likewise, after the initial shock of grief wears off, some widows enjoy the sense of relief and the freedom from responsibility that come with widowhood. For many, widowhood means a reunion with friends who have been widowed for some time.

Perhaps the most significant factor influencing the adjustment of widows in a given society is the number of other roles available to them. For example, older widows who live in small towns tend to be less isolated and to have more access to new social roles after widowhood compared with women who live in large metropolitan areas.*

Dependency[†]   **Dependency**, becoming dependent on others, is one of the most dreaded role changes a middle-aged or older adult can go through. Older people in particular are aware that the need to be dependent on others is common among those over 85. Whether physical or financial, dependency is a difficult position for most adults to accept. Their concern is understandable. We are taught from birth to strive to become independent and self-sufficient; this value is deeply ingrained in most of us, so it is not surprising that we are hostile to the idea of giving up autonomy and becoming dependent on others.

Dependency is all the more difficult to accept because of the changes it causes in other roles. For example, by necessity some older people are forced to fall back on their adult children. For a number of reasons, this necessity strains the older parent-adult child relationship. Older parents sometimes resist or resent having to become dependent on their adult children; in those cases where it occurs, being treated as a child by one's child is especially irksome. Older parents may become angry and frustrated by changes in the quality of interaction, particularly a loss of equality, that can occur when adult children assume responsibility for older parents' care. The parents may feel guilty because they believe that they should not be dependent. Their children, now middle-aged or older, may likewise resent having to provide[†] for both their own children

---

*See Chapter 12 for more details about widowhood.

†Psychologists use the word "dependency" to refer to a psychological state, but here it refers to a position and role characterized by the necessity of relying on others, either physically or financially.

Being dependent on others for basic tasks of daily living is a difficult role.
Photograph by Martine Frank/Magnum

and their parents or having their retirement disrupted by the demands of their parents' dependency. Yet adult children often feel guilty for harboring such resentment. Finally, the spouses of the adult children may not willingly accept the demands that parents' dependency puts on family resources.

Dependency is made all the more difficult by some of the ideas and language used in our culture to discuss dependency of older parents on their adult children. The expression, "parenting your parents," has a certain ring to it, and it may reflect the frustrations that some adult children feel, but generally it is not an accurate reflection of the relationship. Parenting implies a relationship aimed at the future independence of the initially helpless child. Parents expect steady progress toward independence, although whether they get it is another matter. When adult children assume responsibility for some aspect of their

parent's affairs, they do not usually have the total authority that goes with parenthood before the law; they are generally dealing with a parent who is quite capable of participating as an adult in decision making, and the prognosis is not for increased independence but perhaps the reverse. By using the expression "parenting your parent," adult children are sometimes seeking to legitimate taking an authoritarian approach to the task of meeting their parents' dependency needs.

What makes dependency especially difficult is the set of expectations attached to it: Dependent people in our society are expected to be deferential to their benefactors, to be grateful for what they receive, and to give up the right to lead their own lives. We demand such deference of our growing children, the poor, or any other dependent group. Is it any wonder then that older people, who often have spent more than sixty years being independent adults, rebel at the idea of assuming a dependent position? Interestingly, while many older Americans find themselves having to ask their children for some kind of help at one time or another, they still tend, on balance, to give as much help as they receive.

Disability and Sickness **Disability** is another new role that older people may experience. There are varying degrees of disability, and when it becomes extreme, it usually turns into dependency. But even for the many older people whose disabilities have not reached that extreme, disability still restricts the number of roles they can play, and it also affects other people's reactions. Although only a small minority of older people suffer severe disability, disability is a major influence on the structure of a person's life.

Becker (1994) looked at the relationship between elderly people's conceptions of autonomy and how these concepts were affected by physical impairment. Becker found that regardless of whether impairment was a new experience, whether it had stabilized, or whether it was increasing gradually, elders did not see themselves as frail and they focused their concepts of autonomy on those functions they could still control, however minor they might appear to others.

Laborsky (1994) also drew attention to the tendency to think of disabled elders as something less than full persons. Our cultural predisposition to grant full adult status only to those who are fully able-bodied is a long-standing prejudice that many in the disability movement are trying to call into question. The Americans with Disabilities Act was designed in part to affirm disabled Americans' right to full personhood. Laborsky also drew attention to the problems disability can represent to the continuity of a person's long-standing internal definition of full personhood. So newly disabled elders not only have to contend with a hostile environment but with their own past conceptions of self.

Sickness is similar to disability in that it limits role playing. Good health, or absence of sickness or disability, operates alongside age and sex as a major criterion for eligibility for various positions. Most positions outside the circle of family and friends require a level of activity that is impossible for the severely sick or disabled.

Our society has long recognized that sick people occupy a unique position—the sick role. They are not expected to hold jobs, go to school, or otherwise meet the obligations of other positions, and they are often dependent on others for care. If they play the sick role long enough, they may be permanently excluded from some of their other positions, such as jobholder, officer in voluntary associations, or family breadwinner. Sick people are usually exempt from social responsibilities, not expected to care for themselves, and expected to need medical attention. The more serious the prognosis, the more likely they are to find themselves being treated as dependents.

Whether people are defined as sick partly depends on the seriousness and certainty of the prognosis. Even if they have a functional or physical impairment, individuals are less likely to be allowed to play the sick role if the prognosis is known to not be serious. Yet regardless of the prognosis, sick people are often expected to want to get well, and to actually do so. The importance of aging to sickness and disability is of course that, as we age, the probability of being sick or disabled becomes greater.

The Role of the Institutional Resident

When illness and disability become serious handicaps, many older people become residents in a nursing home or home for the aged. While only about 6 percent of the older population lives in institutional facilities at any one time, those who enter an institution usually face important role changes. In the institutional setting, opportunities for customary activities, contacts with the outside world, and privacy are all less frequent than they are outside. Coupled with the extremely negative attitude of most older people toward living in an institution, these changes create significant obstacles to continuity of role playing and adaptation. Institutional care is considered in detail in Chapters 13 and 14.

Other Important Role Changes

There are also changes with age in roles that continue from middle age into later life. For example, there is a decline in church participation (but not in at-home religious activity). Overall participation in voluntary associations also declines, in terms of both number of memberships held and number of meetings attended. Contacts with friends appear to be maintained until about 75, after which there is a decline.

One community role that seems to be increasing in prevalence among older people is that of *mediator*. In the past decade, many communities have come to realize that the peaceful resolution of interpersonal and intergroup disputes is a major need, if for no other reason than to remove some of the pressure from clogged court dockets. Mediation programs seek to bring people together to resolve their differences. The mediation process involves bringing the disputing parties together with a trained mediator who can help them pinpoint the areas of contention, identify the underlying issues involved, and generate solutions that will genuinely resolve the dispute. In most such programs, the mediator's role is to help people in conflict go through a process in which they listen to one another and come to understand one another so that they can then formulate their own resolution to the dispute. The mediator does not arbitrate the dispute or play judge or take sides. Older adults, with their life experience and with appropriate training, make excellent mediators, and retirees are used extensively in this role in many mediation programs around the country.

Another role that elders are playing in many communities is that of *peer counselor*. Peer counselors are trained to use their own experience with some aspect of life to help others cope with similar circumstances. Peer counseling is being used in areas such as adjustment to widowhood, job loss, disability, and loss of housing.

Changes in Groups

Much is said in later chapters about how older people relate to various types of groups. For now, our interest centers on how groups influence the individual's social situation.

Obviously, when people leave jobs through retirement or become widowed through death, their role changes parallel changes in groups. However, certain role changes do not correspond to changes in position. First, individuals may gain or lose power in a group, even though their position may remain nominally the same. Thus, an older member of a board of directors may experience a change in influence, and a retired executive, having lost an economic power base in the community, may find maintaining roles in community organizations difficult. Such changes are often subtle, but their impact on the individual's social situation can be dramatic. Group memberships sometimes interlock in such a way that a change in relationship to one

group will change the individual's relationship to other groups. For example, Atchley (1976b) found that among retired telephone company men, retirement reduced the number of contacts with friends because interaction was tied to the workplace, but among retired men teachers, retirement increased the contact with friends because it was not restricted to the workplace.

Another important situational issue concerns the fact that certain groups have a finite life span and grow old along with their members. For example, in one small midwestern town, the American Legion Auxiliary was made up of women whose husbands had fought in World War I. In 1971, the organization had five members, all widows, ranging in age from 78 to 84. The group met frequently, and the members were close friends who relied on each other heavily for companionship and exchange of small services. A year later, two of the members died, including one who had been a dominant influence on the group. The remaining members found it difficult to carry on community service functions. But more important, they found it difficult to fill their needs for stimulation and companionship within the group. Because new members were not being recruited, this group eventually disappeared.

In groups that recruit new members from upcoming age cohorts, time can bring another sort of situational change. Older members of voluntary associations sometimes complain that the goals of the organization have changed or that the type of person being recruited is "not my kind of people." These changes mainly reflect cohort differences in values and lifestyles. In such cases, nominal continuity of role playing masks some important changes in social situation.

Changes in Environments

People play social roles in particular places, and how they feel about their role performance is partly influenced by the appropriateness of the environment. Aging can potentially bring change to three types of environment: workplace, housing, and community. How people cope with loss of contact with a familiar environment is neglected in gerontological research. However, it is safe to say that some people develop strong feelings about places and the things in them; for these people, environmental change is an important type of situational change.

Retired people often seem to miss a familiar workplace. Even if the work environment was not pleasant, it was mastered, at least in most cases, and this feeling of mastery is important to many people. Also, social scientists tend to neglect the fact that people often develop a feeling of relationship with machines. For example, boat captains sometimes speak of their vessels as if they were people, complete with quirks and idiosyncrasies. In addition, skills associated with work may not be easily exercised in other settings. Retired salespeople can use their verbal skills anywhere, but retired accountants cannot easily transfer their job skills to other settings.

Few older people change housing in later life, and even fewer change communities. Those who do may experience dramatic situational change. People become attached to various aspects of their dwellings—the door that has to be opened just so, the pencil marks on the closet door that mark a child's growth, the stain on the carpet from a wedding reception. When a person lives in the same place for a long time, the environment grows rich in associations with the past. Of course, feelings about this past may as easily be negative as positive. In addition, the routines of role playing to some extent also depend on the environment. Being a wife in a high-rise apartment building may be different from being a wife in a single-family home.

It takes time to get to know a new community—how to get around, what facilities there are, who the "good" physicians are, and so on. Older people who change communities often must disconnect themselves from innumerable long-standing relationships, and reestablishing oneself in the new community is always difficult, sometimes impossible. More than anything else, the potentially negative impact of migration on the individual's social situation and level of integration into the community keeps older people from changing residences even in the face of strong pressures to do so.

Moving into an institution is sometimes necessary in response to the symptoms of physical aging, and as such it is an age-related change in environment, probably the first that comes to mind. However, at most 25 percent of older people actually experience such a move.*

Lifestyles

The life course, age norms, and decision demands combine with other social factors, such as ascription and achievement, to locate people in their positions, roles, groups, and environments. This view explains a large measure of the *social* pressures on the individual. But what about pressures from within the individual—pressures that come from personal preferences, capacities, needs, wants, and fears?

By observing people, we can see not only their roles, relationships, group memberships, activities, and environments but also how these elements are combined by choices, by selective commitments of attention and energy, and by personal mannerisms into a whole life—a *lifestyle*. Even with the restrictions imposed by social class, gender, ethnicity, and age, there are usually many possibilities for mixing one's interests in work, free-time activity, family, friends, creative pursuits, service to others, and a host of other interests into a personal lifestyle. Yet developing a lifestyle is not purely a personal matter.

---

*For further discussion of housing, community, and institutionalization, see Chapters 13 and 14.

The *values* served by one's lifestyle are determined in part by cultural and subcultural *tradition* and in part by personal *experience*. To a great extent, we can expect lifestyle to show large cohort and period influences. For example, in the 1920s, it was not unusual for people to select a lifestyle in which marriage played no part. In the 1950s, this was practically unheard of. Thus, one era's "traditional" lifestyle can be another's "deviant" lifestyle.

Aging has the potential to alter lifestyle in countless ways. Physical capacities, economic wherewithal, group memberships, activities, and environments are all vulnerable to changes with age. In contrast, aging brings certain freedom from social restraints that could alter lifestyle more *toward* a person's values than away from them, as might result from physical aging. Since nearly all studies of lifestyle in later life are cross-sectional, it is not possible to assess the impact of age on lifestyle changes. All we can know from these studies is what lifestyles predominate at various life stages.

In a study of 142 middle-class men and women in their 70s, Maas and Kuypers (1974) found that just over 40 percent of the men had a lifestyle centered on their families and just over 40 percent had a more solitary lifestyle (all their male respondents were married) that revolved around hobbies and cursory social contacts. The remainder found their lifestyle dominated by ill health and the consequent dissatisfactions. Women in their 70s were more diverse in lifestyle, mainly because half of them were no longer married. Widowed or divorced women (all the women respondents had been married at one time) were much more likely than married women to have job-centered lifestyles (13 percent) or diffuse lifestyles with no clear-cut focus (22 percent). The lifestyles of the remaining women paralleled those of the men: about 40 percent had a lifestyle centered on their husband, if married, or visiting with family and friends, if not; and there were just over 10 percent

each involved in group-centered and illness-centered lifestyles. Interestingly, although a third of the men were employed, none had a job-centered lifestyle.

Williams and Wirths (1965) studied the lifestyles of 168 men and women over 50 (average age, 65). They found six basic lifestyles among their respondents: familism (33 percent), couplehood (20 percent), employment (15 percent), living alone (12 percent), and minimal involvement (7 percent). These results are generally comparable to those of Maas and Kuypers (1974), but Williams and Wirths provided no clear-cut data on gender differences. Williams and Wirths found that 65 percent of their subjects were in the "highest" category with respect to being able to adapt to aging successfully within their lifestyle. This was true for all but one lifestyle—minimal involvement, which showed only one-third in the highly successful category. This finding does not support Cumming and Henry's (1961) hypothesis that individual disengagement is a satisfying and successful way to age.

The tendency for marriage and family to dominate lifestyles in these studies from early gerontology were replicated in more recent studies Keith et al. (1994) conducted in various societies in Africa, Europe, Asia, and North America. However, Bellah et al. (1985) cautioned against assuming that American individualism led to a similar valuing of communal life comparable to other cultures. Bellah and his colleagues spoke of "lifestyle enclaves" to describe retirement communities that brought together elders with similar social, economic, and cultural backgrounds to pursue a common lifestyle. They contended that "the contemporary lifestyle enclave is based on a degree of individual choice that largely frees it from traditional ethnic and religious boundaries" (1985,73). These people were overwhelmingly middle class. But the data showing low migration rates among older Americans would lead us to expect that such lifestyle enclaves would be relatively rare.

At the other end of the economic spectrum, Eckert (1980) studied elder residents of several hotels that had become SROs (single-room occupancy) in a blighted area of San Diego's downtown. These facilities ranged in size from 25 rooms to 325 rooms. There were three basic paths to SRO residence in Eckert's study. *Lifelong loners* consisted mainly of men with seasonal occupations such as resort staff or loggers, and military and maritime workers for whom geographic mobility was a way of life. Most were not lonely and considered themselves interesting nonconformists. The *marginally socially adjusted* consisted of people who had attempted a conventional lifestyle but, usually because of alcohol-related problems, had failed. They were less well-adapted to SRO life and tended to see themselves as "victims" of forces over which they had no control. Many of them had served time in jail or prison. The *late isolates* became isolated in later life. These residents were usually age 75 or older and many of them had outlived their close family members. Their way of adapting was to turn to isolation and alcohol. Many of the women in Eckert's study were late isolates, and most in this category were in deteriorating health.

A large minority of the older residents of these hotels had lived in them for many years, knew one another, and provided social support and services to one another. However, reciprocity, not altruism or filial responsibility, was the basis for the exchange of social support and services. At the same time, these SRO residents were a highly individualistic lot, most of whom resisted depending on others if at all possible. Compared to younger and transient residents, the "permanent" older residents of these SROs tended to get favored treatment from the management, but many also experienced patterns of exploitation and social control by building managers.

Even further down on the ladder of economic lifestyles, Cohen et al. (1988) studied men over 50 who lived in the Bowery, a sixteen-block area in lower Manhattan located between Chinatown and East Greenwich Village. The mean age of these men was 62. Most lived in the Bowery because it was possible to survive there on very low incomes. Cohen and his colleagues interviewed 177 men who lived in flophouses, 18 who lived in apartments, and 86 who lived on the streets. The typical flophouse charged about $90 per month for a 4×7 cubicle containing a bed, a locker, and a small table. Walls were thin partitions about 8 feet high, with chicken wire making up the remaining 2 feet to the ceiling. The chicken wire allowed air to circulate, but it also increased noise and cut down on privacy. Beds were often infested with lice, and mice and roaches were endemic.

For men who lived on the street, the major objective, even ahead of food in some cases, was to find a place to sleep that was safe and free from harassment. They found safety in numbers, often sleeping in twos or in groups on park benches or in abandoned automobiles.

Protection from the elements is crucial. When it rains, men who sleep in the park must run across the street and stand all night under . . . an awning. In cold weather, men find a warm doorway, one that has a radiator; or they might try to get into a cellar. Some men bundle themselves inside a big cardboard box. Though group sleeping provides some security, life on the street can be intensively anxiety-provoking: "I can stretch out but I can't relax my body, lay down and sleep. It's no joke sitting up in a hard bench sleepin'. Everytime somebody passes I'm woke. My nerves is on edge 'cause I've seen people attacked in the park."

(Cohen et al. 1988,62).

Most of the men (59 percent) who lived in the Bowery had been assaulted or robbed during the year prior to the interview.

Half the men had incomes of less than $1,300 per year. Most were eligible for public assistance, but many did not receive it, especially those living on the streets. Although on average the men had worked for twenty years or more, most were not working at the time of the interview. Two-thirds of the men reported that they sometimes went without meals. Local meals programs were a major source of food for all the respondents. Compared with men in the general community, the Bowery men had a significantly higher incidence of respiratory problems, hearing impairments, and hypertension, and a third reported that they had been hospitalized in the past year. Most of the health care was delivered by hospitals and health clinics. Cohen and his colleagues also found that informal networks were very prevalent and necessary for survival on the streets.

Homeless women living in urban public places and carrying their households around with them in shopping bags are often referred to as *shopping bag ladies* (Hand 1983). Hand estimated that 400 to 500 shopping-bag ladies were living in Manhattan, most of them in their 50s and early 60s, who at one time were integrated into society's institutions but found themselves cut off from family, income, and a place to live. Hand reported that many were former mental patients. Their lives revolved around places and things rather than people. They needed places to sleep that were safe, warm, and dry. They needed food, places to be, and something to do. They slept in bus stations, public restrooms, apartment lobbies, subways, and a host of other public places. They welcomed pedestrian traffic for the safety it brought but wanted nothing to do with people. Shopping-bag ladies felt that to protect themselves—to avoid "being locked up and put away"—they had to avoid people. Unlike the homeless men, shopping-bag women did not talk to one another or help one another.

Shopping-bag ladies found places to be in locations open to the public—libraries, churches, waiting rooms of clinics, doctors, lawyers, and so on. They existed there by blending into the surroundings. They tried to look like they belonged and tried to avoid drawing attention to themselves. They ate "other people's leftovers found in garbage cans or still on the plates of self-service cafeterias." Scavenging was a major form of recreation for shopping-bag women. They sorted through the plentiful trash on the streets of Manhattan and collected familiar articles that to them became their own belongings and signified status. Obviously, shopping-bag ladies are even further outside the "system" than the men on skid row. They are unknown to social agencies and do not want to be known by them. They lack reliable access not only to shelter, food, and clothing but to health care as well.

From these various small-scale studies, it is clear that lifestyles vary considerably within the older population. Less clear is the relationship between aging and lifestyle. There is some indication that even middle-class elders tend to simplify their lifestyles past age 80, but much more research is needed on how role changes such as economic dislocation, retirement, child launching, and so forth influence lifestyle continuity and change.

While there are many more questions than answers about the effect of aging on lifestyle, more definitive answers than we already have may not be easy to obtain. The subject of lifestyle is as elusive as that of personality and has been subject to much less development. Yet those researchers who are interested in how "whole lives" change with age cannot avoid the issue of lifestyle.

Summary

Social aging involves society's assigning people to positions and opportunities based on chronological age or life stage. Age is used to determine eligibility, evaluate appropriateness, and modify expectations with respect to various social roles in society. Positions in groups and the roles and relationships that go with them largely define the participation of people in their society. As we begin a new role, we tend to play the role rather formally, but the longer we have been in the role, the more personalized it becomes and the more comfortable we are in it. Many people find aging a socially satisfying process because it increases the duration and therefore the security of their role relationships. Holding an advantaged social position for a long time can also result in an accumulation of advantages such as wealth, prestige, and influence. Holding a position for a long time can also lead to the development of exceptional skills that can be used to offset the disqualifying character of older age as an eligibility criterion.

The cultural life course is an ideal sequence of roles a person is expected to play as he or she moves through various life stages. As such, it represents a program for social development of the individual. There are many versions of the life course, depending on gender, ethnicity, and social class. The time schedule implied by the life course can also change over historical time, so that, for example, in one era the ideal age to marry may be much older than the ideal age in another era. As people grow older, their accumulated decisions about various life course options increasingly differentiate one person from another. The life course determines the opportunities and limits for people in a given life stage, and decision demands require people to make choices from among the options open to them. Generally, the older one becomes within a given life stage, the narrower the range of options left. The age norms used to allocate people to positions and roles are based loosely on ideas about what people of various ages are capable of as well as about what is appropriate for them. The greater the gap between the assumptions about functioning implied by the age norm and the actual level of functioning of the individual to whom the norm is applied, the more likely that the age norm will be seen as unjust.

Continuity theory helps us understand social aging by pointing our attention to the social life structure that the individual develops in interaction with the social environment.

This structure of social relationships, activities, and environments is not merely the passive product of what we are taught. We actively construct it to pursue our goals and create capacity to adapt to change.

Society conveys age norms through socialization, and individuals learn them through acculturation. As age increases, opportunities for formal socialization decrease. As adult experience increases, so does appreciation for the value of following rules. Anticipation allows people to prepare for and to smooth out life transitions. Adaptation and negotiation allow people to fit themselves to new roles, and vice versa.

In Chapter 5, we found that aging can have both positive and negative outcomes, that aging is mainly positive for most people, and that it is unpredictable for any given individual. The material in this chapter further illustrates why this is so. Variety among people increases with age as a result of biological, psychological, and social processes that operate differentially. Child launching, mentoring, retirement, widowhood, dependency, disability, sickness, and institutionalization are some major role changes that can accompany later life. Taken together, they represent very significant changes in the situations that individuals confront as they grow old. Role change can affect group membership, and vice versa. In addition, subtle changes can occur in an individual's relationship to groups even though the person's role may remain outwardly the same. The individual's influence may either grow or decline with age, depending on the group. Lost group memberships may influence relationships with other groups. Sometimes groups grow old and die along with their members. In other cases, the group's orientation shifts as the membership changes, sometimes alienating older members.

Environment is an important, but neglected, aspect of the social situation. Roles are played in specific places, and people can become quite attached to places and things in them. Retirement usually brings a loss of relationship to the workplace. Residential mobility creates a new housing situation and sometimes a new community situation as well. Institutionalization is without a doubt potentially the most dramatic environmental change that people experience in later life.

Lifestyle choices allow individuals an opportunity to synthesize their personal characteristics with their outward social ties. Aging can cause both positive and negative discontinuities in lifestyle. To the extent that aging brings significant physical decrements, lowered income, reduced social contacts, and so forth, it may force the individual to abandon a satisfying lifestyle. On the other hand, to the extent that aging brings freedom from various social obligations, it may allow the individual to abandon an unsatisfying lifestyle. Of course, some people never quite develop a lifestyle that is suitable for them. Most people try to maintain their customary adult lifestyle throughout their adult lives.

7　Personal Adaptation to Aging

Aging brings many types of changes. This chapter discusses how people adapt to these changes and achieve the generally positive adjustment that typifies the experiences of most older people. It begins with a definition of adaptation and a discussion of some general means and goals of adaptation. Then the chapter considers specific ways that individuals adapt to specific changes associated with aging. Finally, the chapter focuses on ways people try to avoid adapting and finishes with a discussion of effective adaptation.

What Is Adaptation?

Adaptation is the process of adjusting to fit a situation or environment. It is actually several processes people use to deal with constant changes encountered in everyday living. Changes that require adaptation can occur either in ourselves or in our situations or environments. Early in our lives we develop basic routine approaches to dealing with change. Each of us can adjust to a certain amount of change nearly automatically, without special effort. However, some changes surpass our capacity for such routine, perhaps unconscious, adjustment and require that we take unusual, nonroutine steps to deal with them.

Coping means contending with or attempting to overcome difficulties. Over time, we develop various skills of thought and behavior—*coping skills*—that we use to grapple with problems we encounter in everyday life.

The particular coping skills of a specific individual may change over time in response to the person's perceptions about their effectiveness. However, it is not uncommon for people to continue to use coping styles that worked in the past but are not very effective in their current situation. For example, many women in current older cohorts learned to defer to their husbands in caring for the household finances. But if they are now widowed, to whom do they defer, and is deferral the best way to approach their new situation? We adjust to aging both through gradual routine adaptation and by mobilizing coping skills to deal with crises. As we saw in earlier chapters, many age changes occur gradually and so can be dealt with routinely. Other changes may be coped with routinely at first, but eventually change may reach a threshold of magnitude that requires major adjustments and a corresponding mobilization of personal and social resources. Still other changes may be both sudden and serious, and require conscious coping and social support. For example, we usually adapt to age changes in appearance gradually and routinely, whereas adapting to sudden and severe disability usually requires all the coping skills and social support we can muster.

To adjust to a change, we must first *acknowledge* that change has occurred. If a person denies that aging has affected personal performance, when in fact it has, then it is difficult to adapt successfully. *Denial* is usually a maladaptive reaction to age changes in that it prevents an internal reconciliation with an external reality. Denial may postpone adaptation temporarily, but it may also make the eventual adjustment more difficult by causing those around the person to lose respect for the individual's capacity to face reality. However, there is a great difference between denial of genuine change and refusal to accept and/or internalize negative age stereotypes.

Once people recognize that their situation has changed, then it is possible for them to go through the process of **accommodation.** That is, they change their behavior to fit their new situation more closely.

What is it about aging that requires adaptation? Many physical, psychological, and social changes that accompany aging alter the individual's circumstances in ways that require some sort of adjustment. Sometimes the adjustments are to positive changes, as when older people accustom themselves to the greater freedom and autonomy of retirement. Sometimes increased physical, financial, or social constraints restrict what people are able to do unless some sort of adjustment is made. Aging can also alter relationships, cause relationships or activities to disappear from a person's lifestyle, or change income or housing needs and resources. When everything possible has been done to compensate for changes, people can still find their lives restricted and must then adapt to a new level of capacity. Let us look now at general strategies people use to adapt.

General Ways to Adapt

Continuity is an important adaptive strategy that is available to most aging people (Atchley 1989).* Continuity of self and personality means that changes can be incorporated into a whole that others still recognize as the same unique individual. Continuity of activities and

---

* See Chapters 5 and 6 for more detail about continuity theory.

environments concentrates the individual's energies in familiar domains of activity where practice can often prevent, offset, or minimize the effects of aging. Continuity of relationships preserves the person's convoy of social support—the core of people who provide social support over a major part of the individual's adult life (Antonucci 1985–1986). Many view continuity of lifestyle and residence as an important way to meet instrumental needs for clothing, food, shelter, and transportation. Continuity of roles and activities is seen as an effective way to maintain one's capacity to meet socioemotional needs for interaction and social support. Continuity of independence and personal effectiveness is seen as a way to maintain self-esteem. Thus, for a wide variety of reasons, continuity is regarded as a central adaptive alternative in coping with many of the changes associated with aging.

Like adaptation in general, continuity can be both internal and external. The term **internal continuity** refers to the persistence of a personal structure of ideas based on memory. The term **external continuity** refers to living in familiar environments and interacting with familiar people. *Continuity does not mean that nothing changes*; it means that new life experiences occur against a solid backdrop of familiar and relatively persistent attributes and processes for both the self and the environment. One of the most frequent findings in gerontology is that continuity overshadows change for most people in midlife and after. This finding is true for internal aspects such as personality and self as well as for external aspects such as relationships, housing, community residence, activities, and lifestyles. Continuity is an adaptive response to both internal and external pressures.

Internal pressures toward continuity come from our basic need for stable viewpoints concerning ourselves and the world we live in that we can use as a basis for anticipating what will happen and deciding how to respond to what does happen. Epstein (1980) pointed out that these constructs are often preconscious and that the individual may not be aware of them as viewpoints or be able to describe them. Once developed, these viewpoints exert pressure toward continuity because of the nature of assimilation. Assimilation requires that new information be reconciled with the old; therefore, the more experience that is contained in memory, the less effect new information can be expected to have, which is why new information does not have the same weight for older people as for youngsters and why it reinforces the appearance of conservatism among elders. The greater the weight of previous evidence, the greater the pressure toward continuity. Thus, the more experienced the person (and usually the older), the higher the likelihood that new information will be interpreted as continuity and thus as not requiring adaptation. The focus of internal continuity is a unique past history; therefore, what seems to be continuity to one person may not to another.

We defend our personal viewpoints, once they are developed, because to us they seem necessary for our security and survival. And the more vulnerable the person sees himself or herself to be, the stronger the internal motivation for continuity, which is one reason why older people in institutions may often think and talk about the past. If their present offers them little opportunity to reinforce their images of themselves as capable and self-reliant, then they may concentrate their attention on maintaining ties to the past. Older people who are involved in current activities in the community are not as likely to focus on the past.

External pressures for continuity come from environmental reinforcement and the demands of the roles we occupy. Environments condition us to fit them, and to the extent that environmental demands remain relatively constant, there is pressure for continuity in our responses. To the extent that older people live in familiar environments, conditioning exerts pressure for continuity.

The people we interact with in the roles we play expect us to remain the same from one time to another. They want us to be predictable, which of course involves maintaining continuity in our actions, appearance, mannerisms, and thoughts. But lest we think that this expectation overly constrains our freedom, remember that we also have internal motives for wanting the same thing. In addition, the constraints of continuity provide us with useful guidelines for deciding how to adapt to change. Obviously, the more relationships a person is involved in, the greater the potential for external pressures for continuity.

Anticipation involves realizing beforehand what is likely to happen and taking action that can minimize or eliminate negative aspects and/or promote or increase positive aspects of change. For example, the entire wellness movement is predicated on the notion that there are identifiable lifestyle factors that can prevent, or delay the onset of, serious illness or disability. Millions of people today are attempting to keep their levels of blood cholesterol low and are engaging in regular exercise in an attempt to lower their chances of having a heart attack. People are controlling their weight to avoid high blood pressure; others are eating foods high in fiber to reduce their risk of colon cancer. Millions have quit smoking to reduce their chances of health problems such as heart disease and emphysema. By looking ahead and taking sensible precautions, millions of people will lead healthier, more vigorous lives as older people.

Financial planning involves anticipating financial needs associated with changes in later life. For example, a main goal of retirement planning is to estimate the financial resources needed to lead one's desired lifestyle in retirement and to develop concrete plans to develop those resources. After going through the process, some people settle for a more modest lifestyle and others decide to save more money for retirement. Others find that they must save more *and* lower their aspirations in order to have a realistic financial plan for retirement. In any case, through retirement financial planning (anticipation), people can enter retirement better prepared to adapt to the realities of the new situation because they have identified financial prerequisites and taken steps to meet them in advance.

Another example of taking steps in advance is planning for long-term care needs, which can improve the adjustment process if and when the need for long-term care arises. For example, some people purchase long-term care insurance that will cover a major portion of the cost of nursing home or in-home personal care should they need it. Others identify an amount of savings that they need (over and above what it will take to finance their retirement income) to pay for long-term care.

Meeting long-term care needs can involve more than just financial planning. There must also be people to provide the services. Many older people who move to full-service retirement communities do so because they looked ahead and saw in their not too distant future a strong possibility that they or their spouse would need assisted living or long-term care. Retirement communities are an attractive option because they typically provide ready access to personal care. This is another good example of anticipation.

Compensation involves taking actions that offset or make up for a loss in function. People do not simply sit idly by and accept whatever the fates may give them in terms of declining functions. Their first steps are usually to think of ways to compensate for negative changes. Compensations can be physical, psychological, or social.

Eyeglasses and hearing aids are the most common examples of compensations for *physical* age changes. Through these devices, many people with sensory deficits are able to function at normal adult levels. In the case of hearing, lipreading and voice training are other compensations that allow people to minimize the effects of impaired hearing on social functioning. In the area of ambulation, hip replacements allow people who would otherwise be severely impaired to walk normally. Those who are not able to medically restore the capacity to move about can benefit from motorized carts and wheelchairs. Environmental compensations also can allow impaired people to continue to function in more or less normal ways. For example, lowering the height of the kitchen sink, stovetop, and counters can allow people in wheelchairs to continue to do their own cooking. Entry areas with ramps to go up and stairs to go down enable people with impaired mobility to continue to get in and out of the home. Small, wall-mounted elevators let people continue to live in two-story homes when they can no longer climb stairs.

Compensation for *psychological* changes can include behavioral strategies. For example, as we saw earlier, people can unconsciously adapt to gradual changes in psychomotor capacities. The older professional manuscript typists Salthouse (1984) studied compensated for their slower reaction times and information processing speeds by reading further ahead in the text than younger typists. As a result of this compensation, they maintained their typing speed and accuracy in spite of substantial decrements in the individual psychological skills required for typing. A more mundane example of psychological compensation is the increased use of reminder notes to compensate for declining memory.

Compensation for *social* losses centers mainly around finding new sources of social support and people with whom to socialize. Here social organizations can play a key role. For example, churches, civic clubs, garden clubs, and the like are excellent sources of new friends. Generally, older people have participated in such organizations for many years. When a close friend dies, compensation can consist of developing a closer relationship with

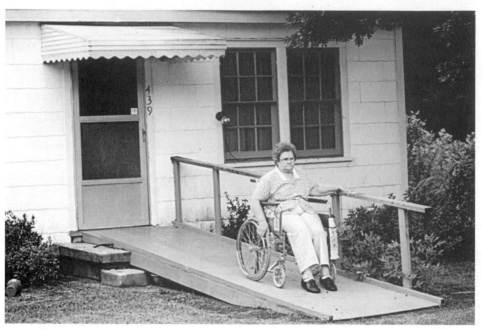

Compensation.
Photograph by Bob Daemmrich/Stock, Boston

a long time associate whom one sees regularly at group meetings.

Compensation can sometimes restore a person to the same level of capability she or he enjoyed before a change occurred, but more often compensation is imperfect; there may be a large degree of restoration, but there is still a sense of loss too. For example, hearing aids generally improve the capacity to hear, but often the quality of that which is heard drops. As a result, even with compensation usually an adaptation has to be made, and the compensations themselves usually require an adaptation. For example, it takes time to get used to bifocal eyeglasses and hearing aids. And even though a person can make new friends to compensate for ones who die or move away, the depth of intimacy and satisfaction is usually less and therefore can create a sense of loss. As one man said, "I'm not lacking for friends. But I miss Millie [his deceased close confidant] so much it hurts, and no one can ever replace her."

Baltes (1993) developed a theory of adaptation centered in part around compensation. He theorized that adaptation to losses of functional capacity was facilitated by a process of selective optimization with compensation. *Selection* is required by the need to focus on high-priority activities in the face of environmental and physical constraints. Priorities are established by preexisting knowledge, skills, and motivations. *Optimization* refers to behavior by individuals to enhance and enrich their capacity to engage in the activities selected. The idea here is that by focusing energy and resources on selected activities, one can still achieve satisfying results, but only if other activities are dropped. *Compensation* is the final element in the theory and refers to the use of both psychological compensations such as mnemonics to aid memory, technological compensations such as hearing aids and eyeglasses, and environmental compensations such as physical alterations to the person's dwelling The point is that people do not simply accept and accommodate to functional declines; they use selection, optimization, and compensation in tandem to create sufficient functional capacity to continue to enjoy life.

Specific Adaptations

In this section we will look at how older people typically adapt to specific changes. We consider how people adjust to changing job circumstances, reduced financial resources, increased prevalence of chronic illness and disability, increased physical dependency, lost roles and activities, declining independence and personal effectiveness, and positive changes in roles and life goals.

Adapting to Changes on the Job

The data on raw age-related declines in functions such as psychomotor performance, memory, and problem-solving capability suggest that aging might have negative effects on job performance. But when objective measures of job performance are used (as opposed to supervisors' ratings of performance, which can be age-biased), job performance of older workers who remain in the workforce is as good or better than that of younger workers. Older workers somehow manage to adapt ways that offset age changes in the components of performance. Some, like Salthouse's (1984) manuscript typists, use cognitive processing strategies, but others may use more complex processes such as selective optimization with compensation (Baltes 1993).

Abraham and Hansson (1995) developed scales that measured each of the elements of selective optimization with compensation as they might be applied on the job. For example, selection might involve focusing on a narrower range of high-priority job tasks, doing more projects that involve group productivity, and cutting out nonessential parts of the job. Optimization might involve paying more attention to maintaining job skills, practicing high-priority job skills, and seek-

ing training to keep up job skills. Compensation might involve being careful to put one's best foot forward at work, to emphasize areas of greatest competence, and to take on tasks that match performance strengths and avoid tasks that involve areas of weaker performance. Using the scales for selection, optimization, and compensation, Abraham and Hansson found that the use of selective optimization with compensation as an adaptation strategy became more important to job competency as age increased. The selective optimization with compensation measures predicted very well which older workers in their study retained their job skills and competence.

Adapting to Less Income

The major way that older people adapt to reduced financial resources is learning to make do with less. Indeed, living within one's means is seen by elders as a major dimension of aging well. As one elderly woman put it, "I was raised and taught all my life, 'Waste not, want not.' We never buy on credit. Never. If we have the money, we buy it. If not, we just do without until we get the money" (A. Day 1991,260). Certainly, a large part of the adjustment to reduced income in retirement is made easier by the fact that employment expenses such as transportation or special clothing are no longer required. However, economizing is still necessary for most retired couples to make ends meet, which the vast majority do. Single retired people often face a very difficult financial situation since retirement income for single individuals averages only about 40 percent of preretirement income. Keith (1993) reported that older women were much more likely than older men to experience chronic financial strain, largely because they were more likely to be single.

Relatively little research has been done on how people cope specifically with reduced incomes. Douglass (1982) studied the impact of the rapid rise in fuel costs for home heating on older people living in Michigan. He found that 40 percent cut down on their expenditures for food and another 16 percent did without needed medical care in order to pay their fuel bills. Older people often have no frills in their budgets that can be cut out in order to deal with unanticipated expenses, so they must dip into the more flexible items in their budget— food and medical care.

Many of us have levels of living well above what we actually need to enjoy satisfying lives. In this respect, retired people can offer the rest of us an example of how to get by on less without sacrificing the enjoyment of living. We need more research on the *processes* people go through in making this adaptation.

Adapting to Chronic Illness and Disability

Significant disability is not part of the life experiences of the majority of people. For most, aging brings gradual changes that do not compromise their customary lifestyles. But in our culture, the concept of full personhood assumes a fully functioning body and mind, and people whose mental or physical capacities fall below the modest threshold required for full adult functioning are often treated as less than fully human (Luborsky 1994). Disabled elders often internalize these negative assessments of disability, which can produce lower self-esteem and an inclination to withdraw into household isolation rather than face the general public's harsh judgments about people with disabilities. Especially when accompanied by incontinence or dementia, disability becomes a "master status" that negates any other potentials the person might have (Herskovits and Mitteness 1994). The "disability movement" is mainly concerned with opposing these culturally and socially constructed negative images of disability in order to restore the right of disabled people to claim full personhood and the access to opportunities and resources that goes along with full personhood.

Thus far, however, the social support structures (social compensations) that have allowed many younger disabled people to escape social marginality, or at least partially neutralize it, have been slow to spread into the world of disabled older people. The disability movement has never been popular among health and social service providers. But that may all change as cohorts of disabled people enter old age with a long-standing attitude of resistance to the negative stereotypes and discrimination applied to disabled people. They may become visible agents of change among disabled elders. On the other hand, older disabled people may just not have the energy needed to mount the needed protest.

Physical disabilities thus not only create hassles with activities of daily living; they also constitute visible sociocultural stigma that constrain social opportunities as well. In addition, the cultural preference to keep disabled people out of sight also limits the extent to which they can become defined as a category with unique needs that require public policy attention.

Sudden and severe disability in later life is most often caused by strokes. Becker (1993) found major upheavals in the lives of most stroke victims. Physical symptoms of stroke often include paralysis, numbness in the affected region of the body, inability to speak or to understand what others have said, and other cognitive impairments. Interestingly, continuity played a significant role in successful adaptation to the disruption caused by strokes. Becker found that trying to reestablish continuity with functioning prior to the stroke was a major goal of therapy in the hospital, and being able to come home was uniformly a symbol of continuity for the sixty-four stroke victims in Becker's study. These stroke victims saw being able to identify continuity in the face of severe discontinuity as an indicator of their ability to persevere in spite of very long odds. In much the same way that self-esteem can be enhanced by moderating one's aspirations to match one's accomplishments, stroke victims gradually revise their expectations for continuity to match their experienced reality during the adjustment period.

In coping with chronic illness and disability, elders are caught in the middle of two contradictory cultural principles: autonomy and institutionalized helping. On the one hand, we believe in the principle that people should solve their own problems whenever possible; but on the other hand, when service providers are involved, we believe that solutions should meet professional standards. Kaufman (1994) illustrated through several case studies that as soon as an older person becomes a "case," professional standards supercede the individual's values and lifestyle choices. For example:

Mrs. M. was an 80-year-old woman who was brought to an assessment program by a neighbor of 15 years. He was concerned that she was increasingly unable to remember her appointments, pay bills, prepare meals, or maintain her personal hygiene.

Following the examination by physician, neurophysiologist, podiatrist and nurse and a meeting with the social worker, the assessment team concluded that Mrs. M was obese, incontinent, visually impaired, had degenerative joint disease, a history of hypertension, anemia, significant gait and balance instability, and severe memory and concentration problems that pointed to progressive dementia.

The team report concluded that Mrs. M. could be safe in her own apartment *only* if: her apartment were professionally cleaned, she accepted a home aide to provide personal care and homemaking services five hours per day, five to seven days per week, she received home-delivered prepared meals, she had a money manager, she got rid of some of her pets, and she attended an adult day health program twice a week to have her physical health monitored, have a meal, and receive physical therapy to improve her walking.

Mrs. M. was congenial during the care planning conference but left without agreeing to receive any of the services suggested.

(Adapted from Kaufman 1994,51–52)

In this and the other cases cited by Kaufman, elders refused services because to accept them required them to compromise lifelong lifestyles and values. Lived experience was transformed by a medical encounter into a problem list that encompassed personal and social behaviors as well as physiological disorders. The medical model of "quality care" was often more invasive than the care management team could imagine and required active, unsolicited behavior modification on the part of the client. To receive services required giving up autonomy.

Chronic illness also can have direct and dramatic effects on the experienced quality of life. Consider a person who begins to periodically suffer attacks of angina, a sharp pain in the chest caused by restricted blood flow to the heart muscle. Angina is a result of coronary artery disease and angina episodes can last from a few minutes to half an hour.

When angina occurs, the whole world is dramatically transformed and placed into question. At virtually any moment the searing pain might [signal] some hidden alarm warning of limits that cannot be breached, warning of the inescapable fact of death. New concerns become relevant, such as which physical activities are tolerable and which are not. Stairs, for example, now seem to have changed their proportion. The angina thus calls into question typical and taken-for-granted experiences of the world and the person's place therein.

(Agich, 1995,149)

Fortunately, the large majority of aging and older adults with multiple chronic conditions experience only minor inconvenience from their symptoms or can control their symptoms with medication. But we should not underestimate the emotional toll that chronic illness can exact.

Adapting to Increased Dependency

Coping with increased physical infirmity and disability cannot be an easy matter for adults who have not only been responsible for themselves but for others as well. A major task is to learn how to accept assistance from others without losing one's self-respect. We know that many older people learn to do so, but very little study has been made of *how* they learn. Lieberman and Tobin (1983) suggested that people who are able to adjust well to increased dependency need two things: (1) sufficient physical and mental resources to mobilize the energy used to adapt, and (2) tough-mindedness and stubbornness about their own worth. For example, in their study, the "nice" older people who moved into institutions did not fare nearly as well as those who were aggressive, irritating, narcissistic, and demanding. In other words, it may take a big ego to accept assistance effectively. Certainly, it must take a sturdy one. Fortunately, as noted earlier, most older people have high self-esteem.

Adapting to Lost Roles or Activities

Gerontologists have paid more attention to the problem of adapting to the loss of roles or activities than they have to the preceding topics. Havighurst (1963) and Rosow (1967) articulated an **activity theory** of aging which held that unless constrained by poor health or disability, older people have the same psychological and social needs that middle-aged people do. They theorized that decreases in social interaction that occur with age are the result of a withdrawal of society from aging people, and that most older people do not want this withdrawal. Therefore, according to activity theory, older people who are aging optimally stay active and resist shrinkage in their social world. They maintain activities of middle age as long as possible and then find substitutes for activities that must be given up. Thus retirees are expected to find substitutes for work and bereaved people are expected to look for new friends and loved ones to replace those who have died. Of course, substitution does not have to be literal. Retirees can find nonjob roles that meet

many of the same needs that jobs did, and widows can find alternative sources of intimacy even if they cannot find new husbands.

Research supports the notion that healthy people who keep active do indeed have higher life satisfaction than those who are inactive or in ill health. But one can be active in later life without maintaining the pace or amount of activity typical of middle age. Ample evidence shows that some of the decline in activity levels with age is the result of a desire by older people to have a more relaxed approach to activities, which usually means a voluntary reduction in the number of social roles being played. To the extent that people see themselves as being too active, they may not find a replacement for lost activities or roles. Thus, the premise in activity theory that older people who are aging optimally maintain a middle-aged activity level and number of social roles is not supported by research on activities (Steinkamp and Kelly 1987). For example, Chambré (1984) found that compared with older people who are still employed, retired people were less likely to do volunteer work. However, there is also some support for the notion that reductions with age in the number of roles are related to reductions in opportunity (Cutler and Hendricks 1990). It is also true that older people's activity levels remain quite high as long as they are able-minded and able-bodied.

People who are involved in several roles and a variety of activities may not need to find *new* roles or activities to substitute for those they lose. They may find it easier to redistribute their time, energy, and emotional commitments among their remaining roles and activities. In an earlier work, I called this the **consolidation** approach (Atchley 1985). The general level of activity that results from this redistribution may be on a par with the preloss level, or it may be somewhat reduced.

For example, retirement commonly involves role loss that for most people can be compensated for through consolidation. When Mr. A. retired, the time he had spent on the job became available for other activities. He now sleeps about an hour longer than he did when he was employed and spends more time reading the morning paper. In good weather, the rest of his morning is spent puttering around the yard. He fixes lunch—usually a sandwich and a beer—and then straightens up the house. Around 3 P.M. he walks to a nearby cafe, has coffee, and talks with friends, some retired and some not. Around 4:30, Mr. A. goes home and starts supper. He has always liked cooking, and since he retired he has more or less taken over the job of preparing evening meals for himself and his wife, who is still employed. Nothing in Mr. A.'s day is new to him. What has changed is how he distributes the time and energy he devotes to activities he has done most of his adult life. As a result, Mr. A.'s home is in the best condition ever, Mrs. A. is delighted to have less housework to do, and Mr. A. is proud of the good adjustment he has made to retirement.

The consolidation approach is available to most people. However, some people have so few roles or activities that when they lose some of them there are not enough remaining to absorb the energies freed by the loss. Unless they can find substitutes, these people are forced to disengage. Consolidation also may not be a satisfactory solution if the lost activity was extremely important to the person and if the remaining activities, although perhaps plentiful, are not meaningful.

Disengagement occurs when people withdraw from roles or activities and reduce their activity level or sense of involvement. Based on their work in Kansas City in the 1950s, Cumming and Henry (1961) theorized that the turning inward typical of aging people produces a natural and normal withdrawal from social roles and activities, an increasing preoccupation with self, and decreasing involvement with others. Individual disengagement was conceived as primarily a psychological process involving withdrawal of interest and commitment. Social withdrawal was a consequence of individual disengagement, coupled with society's withdrawal of opportunities and interest in older people's contributions (societal disengagement).

Disengagement theory caused a flurry of research because, in positing the "normality" of withdrawal, it challenged the conventional wisdom that keeping active was the best way to deal with aging.

Streib and Schneider (1971) suggested that **differential disengagement** was more likely than total disengagement. That is, people may withdraw from some activities but increase or maintain their involvement in others. Differential disengagement is similar to the consolidation approach mentioned earlier. Troll (1971) supported this notion when she talked of disengagement *into* the family, meaning that older people cope with lost roles by increasing their involvement with their families.

Larson, Zuzanek, and Mannell pointed out that

the daily choice between being with others and being alone represents a fundamental sociological alternative. It is a choice between participation in the human community or autonomy from it. It is a choice between sharing one's being—one's thoughts, values, judgments, worries—with others, or solitary cultivation of an independent existence.

(1985,375)

Thus, people seldom are completely engaged nor are they completely disengaged. Rather, they strike a balance between the two states that reflects their own preferences, often mediated by social encouragement or discouragement from others. Compared with people at age 50, at age 80 people tend to be relatively disengaged in terms of their total number of roles and amount of interaction with others. But if you asked these people if they *feel* engaged, they would say yes. This succinctly reflects the subjective quality of engagement or disengagement.

In an interesting study, Larson and his colleagues (1985) studied ninety-two retired adults who carried electronic pagers for one week. The respondents were paged at random times throughout the week and asked to fill out self-reports on their companionship and internal states. The research generated a total of 3,412 reports. The researchers found that the respondents voluntarily spent almost half of their waking hours alone, but being alone was not a negative experience for a large majority of them. When they were alone, the respondents tended to be doing things that required concentration and challenge, which suggested to the researchers that they were highly engaged personally but not *inter*personally. Solitude was not a negative, disengaged state but rather an opportunity for focused thought and absorption. There is no doubt that social support plays an important role in life satisfaction, but this research indicated that these older adults neither needed nor wanted constant companionship. However, for those who lived alone, getting enough companionship did present problems.

Atchley (1995c) studied the impact of disability on level of participation in a wide variety of activities. Most older adults living in the community who became disabled over the sixteen-year course of the study became disabled gradually. In addition, most were able to maintain a reasonably active life by working around their disabilities. If arthritis prevented them from doing handiwork anymore, they adapted by increasing the time they spent doing another of their customary activities. No one adapted to disability by taking up activities entirely new to them. Disengagement was most obvious in situations where the disability involved paralysis or systemic disorders such as emphysema. For example, Mr. L. had been a lawyer until his retirement six years before his most recent response to the longitudinal study. His emphysema severely restricted his activities at that time. He received personal assistance from his wife. He went from having one of the highest activity levels in the study in 1975 to having a very low activity level in 1991. The number of activity areas he never participated in went from one in 1975 to ten in 1991. The only activities he did often in 1991 were reading, being with friends and family, and watching television. He sometimes played cards and occasionally attended social functions. The effects of his illness and the resulting functional limitations and role loss was startling. As late as 1981, he was very positive about himself and his life. But by 1991, his morale was among the lowest in the study and he saw himself as sick, inactive, immobile, unable, dependent, and idle. He saw his life as empty. Obviously, this man did not disengage voluntarily.

Substitution, consolidation, and disengagement do not represent mutually exclusive responses to lost roles or activities. Instead, these concepts often can be blended together when we discuss adaptation to specific losses for specific people.

Consolidation may be the most common outcome because it preserves continuity for both the individual and those around him or her. Clark and Anderson (1967) referred to it as the "path of least resistance." The extent of disengagement may depend very much on how threatened older people feel by their environment. Finally, the balance among consolidation, substitution, and disengagement depends very heavily on health. Good health is a prerequisite for the consolidation and substitution approaches to coping with role and activity losses. People in poor health are much more likely to be forced into disengagement, even in terms of solitary pursuits.

Coping with Threats to Self-Concept and Self-Esteem

Declining independence and increasing vulnerability threaten older people's self-concepts. The major ways that they can cope with these threats are to use defense mechanisms (covered in Chapter 5), develop an acceptance of themselves as aging people, and emphasize their maintained competence.

Older people can use the past to sustain an image of themselves as worthwhile. This is easier to do in a group of friends or in a community of longtime residents because the people the older person interacts with tend to share the same view of the past and may well have participated in his or her life history. For example, Krause and Tram (1989) found that continuing church participation by older African Americans prevented the erosion of self-worth in the face of life stresses. When we interact with older people we do not know, we must keep in mind that their talking about their past is their way of telling us the basis for their self-esteem.

People who are becoming less independent are most likely to use the past in this way, and they are also most likely to have to deal with service providers who are essentially strangers. This fact no doubt has added to the stereotype that "older people live in the past."

By comparing themselves with other older people who are worse off, people can minimize the importance of their dependencies (Heidrich and Ryff 1993). Selective interaction and selective perception allow people to concentrate their attention on people and messages that support their own view of themselves and to avoid those that do not. But these defenses gradually lose their power as the person becomes more and more dependent. At the point where the person becomes bedfast either at home or in an institution, he or she loses the control necessary for defenses such as selective interaction. And the shift in power that accompanies severe dependency means that the older person can no longer influence people to accept relative appreciation or past accomplishments as valid bases for his or her self-concept. This shift may partly explain why the behavior of some older people in institutions appears strange. They are among strangers and in a situation in which their self-concepts are severely threatened, and

they have few resources for defending them. As a result they may be combative, apprehensive, untrusting, and anxious.

Many young people do not understand why some older people get angry when younger people try to help them. They are angry because they believe their image of self-sufficiency is being threatened. Older people revere self-reliance. As a result, they try to do things for themselves as long as possible and to resist letting anyone know that sometimes they just cannot. This is why many older people gradually and knowingly let their homes deteriorate: They do not want people coming in and realizing how helpless they are at times.

It is most important for older people to have some way to maintain a sense of competence. We need to work with them to see how *they* would like to accomplish this goal, and we should be as flexible as possible in accommodating their preferences. It is okay if their views are somewhat unrealistic as long as behavior that endangers themselves or others is not the result. Nevertheless, learning to accept a definition of competence appropriate to their functional capacity is also useful. At some point it is adaptive for a disabled older person to acknowledge that certain activities can no longer be done as successfully as when she or he was younger. This recognition is an important prerequisite for the person's conservation of energy. It is also important for elders to recognize that social norms impose very real limits on what one is allowed to do as an older person. One need not like these limits, only accept their reality. Clark and Anderson (1967) found that people tended to try denying aging and its effects in two ways. The first is by trying to avoid looking old through the use of cosmetics, hair dyes, figure control devices, and so on. Second, people sought to deny aging by attributing their limitations to sickness rather than to aging

Some people seek to deny aging by refusing to go along with rules that exclude them

Most older people have high self-esteem.
Photograph by Phil Borgas/Tony Stone Images, Inc.

from participation and by fighting the system. In itself, this strategy is not bad if the person is indeed still capable. However, it is maladaptive when the individual's level of functioning means that continuation is not a real possibility. For example, Mr. G. had played in the Thursday afternoon men's golf league for over thirty years, and at age 78 he was very proud that he was still capable of playing. But after a hip replacement at age 80, even using a golf cart Mr. G. was no longer able to keep up with the other players. They tried to be gracious and patient, but after several weeks it was clear that waiting for Mr. G. was causing play to take nearly an hour longer each week. When the league secretary asked him to drop out and allow a substitute to play for him, Mr. G. refused. He spoke heatedly with individual league members about the unfairness of the request but got little support since most of the players agreed with the league secretary. By refusing to withdraw, Mr. G. forced the league to exclude him formally. In the process, Mr. G. probably became more dis-

couraged than if he had made the decision himself.

In addition to accepting changes in their concept of activities and roles that are appropriate to the self, older people also find it useful to accept changes in the criteria used to evaluate the self and that serve as the basis for self-esteem. Regardless of their capacities to continue in positions of leadership and to further their achievements, societal disengagement and age discrimination combine to make continuation an unrealistic goal for many older people. Moving toward less emphasis on position and achievement as personal goals is typical of older people, and thus most people seem to accomplish this adaptive change.

But there is probably no easy way to help those older people who resist being realistic about either their capacities or their opportunities, assuming that we can know what their capabilities and opportunities actually are. A certain amount of realism is necessary for adaptation, and sometimes the best thing to do is to tell the truth in as kind a way as possible. It is also essential to be sure that the limits one is trying to get an older person to accept are indeed genuine and not based on one's own age biases.

Most people can adapt to changes without losing their self-esteem because most have never become so disabled that they lose their sense of continuity. Even among people in institutions, a sizable proportion are able to maintain a sense of continuity with their pasts. But people—both in the community and in institutions—who perceive that changes in themselves or in their social situations have produced a negative discontinuity are likely to feel bad about themselves.

Adapting to Positive Changes in Roles and Life Goals

Many age-related changes are challenges that require adaptation. However, role changes such as the empty nest or retirement, or value changes such as a shift in life goals toward less emphasis on other-directed achievement are usually experienced as positive. Positive changes may also require adaptation.

Although both the empty nest and retirement usually result in increased freedom, many people feel awkward at first about exercising their newfound freedom. They are accustomed to setting their everyday goals within a field constrained by their child-rearing or job responsibilities. The vastness of their new, more open field of choices in terms of activity, time, and place may initially seem overwhelming. However, most people quickly realize that the long-standing preferences and life structures they have built through a lifetime of selected investment of time and energy constitute an effective framework for constructing new life routines. Thus, continuity theory (Atchley 1989, 1995d) can be used to understand adaptation to both loss or gain in social roles or activities.

Retired people often experience a shift in life goals. For example, instead of seeking individual achievement, retired individuals may seek to contribute to the achievements of groups devoted to community service. Also, instead of letting employers or occupational cultures define what is important to achieve, retired people often become more inner-directed in their choices of achievement goals. Finally, many people find that material values become less important and that relational and spiritual values become more important as they mature in retirement.

Incorporating new goals or life structures into personal relationships, particularly relationships with the people who constitute the convoy of social support, requires continuous self-disclosure. The person experiencing changes in goals or life routines needs to communicate these changes to others so they can understand why these new directions are important and how they are valid for the person taking them. Such communication is a necessary part of maintaining continuity of social support.

Escape Rather Than Adaptation

Why do some people confront and resolve the conflicts aging poses while others choose various forms of escape, such as isolation, alcohol, drugs, or even suicide? Based on bits and pieces of information on various forms of escape, a picture begins to emerge: People who encounter many losses are prone to seek escape. Although social supports from the community, family, and friends often can lead the person out of escapism, some people perceive their losses as too great to be coped with. Others see the available "solutions" as intolerable compromises. For these people, escape is a preferred alternative. Some people have never learned to cope well with change. These people have retreated in the face of changes throughout adulthood, and escapism in later life is nothing new for them.

Baumeister, Heatherton, and Tice (1994) developed a theory of why some people are able to adapt and others are not. Their theory assumes that adaptation is a form of self-regulation. To adapt, individuals need values or standards they can use to set goals, they must monitor the feedback they get from their life experience in trying to achieve those goals, and they must be able to modify their own behavior or circumstances in response to feedback. The concept of willpower is central to their theory because it is the mechanism through which people stay on track and resist the temptation to give up.

To Baumeister and his colleagues, failure to adapt is a failure of self-regulation. Such failures can occur if people are unable to set realistic goals for adaptation, if they do not pay attention to feedback from experience, or if they lack the willpower to respond to feedback. Failure in adaptation can also occur through misregulation. For example, people can hold inaccurate beliefs about themselves or their environment, which aims them at unrealistic goals that cannot be achieved. The other main form of misregulation is trying to control the uncontrollable. For example, trying to suppress thoughts or moods may be unrealistic, which leads to a sense of failure. In addition, "self-regulation often requires one to adopt a long-range context or higher value or abstract frame of reference. When, instead, attention is narrowly focused in the here and now, the capacity to override impulses or delay gratification or calm unwanted feelings or persist at an unpleasant task is reduced" (1994,244–245). People who repeatedly experience failure to adapt often develop feelings of hopelessness and engage in a variety of self-destructive or self-harming behaviors. That such patterns are relatively rare among elders suggests that people who survive young and middle adulthood have successfully developed the capacity for self-regulation that may be at the heart of adaptation. Baumeister, Heatherton, and Tice's theory was developed to explain why people fail at tasks or let things get out of control. It has not been applied to issues of adaptation to aging, but it could be an excellent framework for research on failed adaptation to a variety of changes associated with aging.

We should not assume that escape is always an irrational choice or the result of repeated failure. For some people, it is the most viable alternative given their values. Some people are able to satisfy their need for escape through withdrawal, alcohol, tranquilizers, or some combination of the three. For others, the pain is too great, and they eventually choose suicide.

If we look at life course patterns of escape as a response to adaptive challenges, rates of alcohol and drug abuse are much lower among older people than among other adults. Although being alone is more common in later life, there is no evidence that intentional isolation as a way of coping is more common among older people than among other adults. Suicide rates are no higher for older women or African-American men than they are for other adults, but among white men suicide rates increase steadily with age. Two factors probably operate to produce lower incidence of escapism among older people. First, alcoholism, drug abuse, and isolation all increase the risk of mortality, and people who use these strategies in early or middle adulthood are less likely to survive into later life. In addition, there is evidence that as a function of adult development, people become less self-destructive. And as retirement and the empty nest reduce levels of conflict in everyday life, the need for escape may diminish.

Isolation

Isolation is not common among elders. For example, in one study only 4 percent of elders had no companions, and only 7 percent lived alone and had no confidants (Chappell and Badger 1989). Bennett (1980) reviewed twenty years of research on social isolation in the older population and found that isolation in later life was connected to the past. Many aged isolates had been isolates their entire adult lives. Others had been outgoing earlier and had become isolates only in later life. Some were voluntary isolates, and others became isolates as a result of changing circumstances, such as widowhood.

Those who use isolation as an adaptive strategy tend to have the freedom to withdraw from interactions and roles they define as unsatisfactory. Retirement, widowhood, and divorce are examples of changes that can free people to choose isolation. Lowenthal and Berkman (1967) found that mental illness tended to lead to social isolation, leading us to expect that people who have difficulty in getting along with others (one symptom of mental illness) would be more likely to withdraw in later life, particularly in reaction to conflicts over their own self-concepts.

Krause (1991a) found that financial strains were both a direct and indirect cause of a lack of social ties among older people. Not only did financial strains directly reduce social ties with others but they also indirectly reduced social ties by making people more distrustful of others. Not surprisingly, financial strains were more common among working-class elders. In addition, fear of crime reduced social ties by making elders more distrustful of others. Krause concluded that stressful events that increase financial strains or distrust of others can actually interfere with coping by reducing the person's access to social support.

Bennett (1980) found that social isolation prior to moving to an institution was associated with difficulty in adjusting to the institution. She concluded that this difficulty was the result of "desocialization," in which the individuals forgot what they had learned earlier in life about conformity to norms and how to get along with others. However, another possible interpretation is that those people chose isolation because they were already having difficulty conforming and getting along with others. In a longitudinal study of adaptation to aging, Atchley (1982a) found that less than 1 percent of older respondents were voluntary isolates who had chosen low levels of social activity, which suggests that isolation is a seldom-used adaptive strategy.

Use of Alcohol

The proportion of older people who use alcohol as a means of coping with life, compared with the proportion who use alcohol for more positive social reasons, is unknown. Whittington (1988) pointed out that very often people who use too much alcohol are attempting to find a chemical solution to a human problem. Maddox (1988) estimated that about 5 percent of older people use alcohol to the point that they are alcohol-dependent or encounter problems such as driving while intoxicated or criticism from others stemming from alcohol-related behavior.

Problem drinking is less common among older people than among younger adults. For example, men under 30 are about four times as likely as men 60 or over to have alcohol problems. In addition, men are about four times more likely than women to have alcohol problems, and as we saw earlier, the proportion of women increases as age cohorts grow older. Maddox also found that trouble connected with alcohol problems tends to be highest among people of lower socioeconomic status. He reported that only a minuscule proportion of people become problem drinkers later in life. Nearly all older people with drinking problems had them before they reached 65. People who are problem drinkers in later life are often responding to depression and boredom, which are sometimes associated with life changes such as widowhood or physical decline (Butler and Lewis 1982; Alexander and Duff 1988).

Kastenbaum (1987) reported that people in general begin to reduce their alcohol consumption in their 40s, and alcohol use steadily declines with age. However, many who stop or reduce their level of alcohol use do so because of poor health. Healthy and active people are more likely to continue their patterns of alcohol use into later life.

Alexander and Duff reported that drinking is more common in leisure-oriented retirement communities, where abstainers account for only 28 percent of the population as compared with 45 percent for the general older population. Part of the reason may be that residents of this type of retirement community often come from an upper-middle-class background in which social drinking is part of the culture. The extent of drinking in full-service retirement communities, which are oriented around a rich service environment, is unknown. However, based on the tendency of people in frail health to abstain, we could expect residents of full-service retirement communities to have a lower-than-average proportion of problem drinkers.

Drug Abuse

David Petersen (1988) reported that nearly all those who use illegal drugs in later life began using them in early adulthood. Indications are that like alcohol use, illegal drug use declines with age. However, estimates of the proportion of older people who use illegal drugs are difficult to make because few studies of this topic have been done and many older drug users have adapted so as to avoid detection better than younger drug users. For example, Capel et al. (1972) studied thirty-eight older narcotics addicts between the ages of 48 and 73 who were living in the community but were unknown to any of the city's drug treatment programs or law enforcement agencies. These addicts had continued to use narcotics as they grew older. Many had switched from heroin to Delaudid, which is a synthetic drug that the addicts reported to be cheaper and easier to get. In addition, they had reduced their daily intake of narcotics by supplementing them with alcohol and barbiturates. Most of these addicts were steadily employed.

Because large proportions of drug users are mainly characteristic of cohorts who entered young adulthood since 1960, the proportion of illegal drug abusers in the older population is likely to increase dramatically over the coming years. For example, Capel and Peppers (1978) estimated that as addicts survive into later life, the number of addicts 60 or over may quadruple, resulting in a substantial problem of illegal drug use among older people.

Aging people can also abuse prescription and over-the-counter drugs. For example, Butler and Lewis (1982) reported that of the top twenty drugs prescribed for older adults, twelve are tranquilizers. Tranquilizers such as Valium and Librium can be addictive; that is, emotional dependence can develop and physical withdrawal symptoms can occur if the drug is withheld. Given that a large number of older people have access to psychoactive drugs such as tranquilizers and sedatives, there is a potential for abuse. However, drug management studies have shown that older people tend to take drugs as prescribed, and when they deviate from the physician's instructions, it tends to be in the direction of taking less than prescribed (Whittington 1988). At this point it appears that intentional overuse of prescription drugs among older people is relatively rare (Whittington 1987).

Suicide

Suicide is a disturbing and fascinating subject for most of us. We have a deep-seated drive to learn why suicides occur, why some people purposefully reject what most of us value highest— life itself. Yet suicide is not a common cause of death at any age. For example, in a cohort of 100,000 men age 65 to 74 in 1986, there were 38 suicides, 1,447 deaths from heart disease, 1,087 deaths from cancer, 185 deaths from strokes, and 81 deaths from pneumonia (U.S. Bureau of the Census 1989c,79). Suicides accounted for less than 1 percent of the deaths among men in this cohort. However, suicide rates for white men are much higher at the older ages than in middle or young adulthood. For example, in 1986, suicide rates fluctuated between 24 and 28 per 100,000 for white men at age 20 to 64, but were much higher at age 65 to 74 (38 per 100,000), age 75 to 84 (59 per 100,000), and age 85 and over (66 per 100,000). For women and African Americans, suicide rates were generally low and did not vary much with age. At this point, we do not know why suicide is strongly correlated with age only for white men and not for women or African-American men. However, the correlation between age and suicide of men is high in all industrialized countries of the world (Atchley 1980a).

Most of the literature on the subject starts by assuming that suicide is more common among older people and then proceeds to speculate about why age is related to suicide. Bereavement, social isolation, failing health, depression, sexual frustrations, and retirement have all been suggested as causes of suicide among older people. However, these factors are usually more common among women than among men, so we cannot use them to explain suicides since male suicide rates at the older ages are several times higher than those for women. Many of the proposed explanations look good on the surface but do not stand up under close examination. For example, it is widely held that suicide is related to retirement. However, if we look at the suicide data, we find that the sharpest increases in suicide rates occur at ages well past the age of retirement.

Menninger (1938) proposed a classification of suicide motives into the wish to kill, the wish to be killed, or the wish to die. Analysis of suicide notes indicates that older people are much more likely than other age groups to see suicide as a release. Their motive is simply to be dead. The individual reasons for this choice are many, and no single or simple explanation can encompass the wide variety of situations that lead to suicide. Nevertheless, certain groups of elders are at high risk. For example, very old men who have recently been widowed, who have re-

cently been given a terminal diagnosis by a physician, or who have experienced rapid physical deterioration are all at high risk. Older people in institutions are high risks for suicide, especially if they receive no visitors. Cognitive impairment also increases suicide risk (Schmid, Manjee, and Shah 1994). Older people who are severely depressed are at a higher risk. Marv Miller (1978) reported that in many cases older people gave clear signals that they were intending to kill themselves, but that they were ignored. Generally, when older people suggest that they might kill themselves, they mean it. And when older people attempt suicide, they usually succeed.

In addition to active suicide, some older people die because they have given up. Failing to continue treatment for illnesses such as heart disease or diabetes, engaging in detrimental or dangerous activities, or neglecting one's health are all indirect ways to kill oneself. No one knows how many older people, or younger people for that matter, die as a result of such actions.

Effective Adaptation

How do we know that a person has managed to cope effectively with aging? Whereas earlier we were concerned with the various means of adapting, here we are concerned with the *outcome* of the adaptive process. There is no consensus about how to define effective adaptation. Some definitions focus on behavioral capacities or social integration. Others are more concerned with the aging person's subjective evaluation of his or her life. For example, the internal adaptive process could be called effective if the person has a high degree of life satisfaction, finds meaning in life, is able to remain relatively autonomous at least psychologically, and is able to maintain a personal sense of life meaning. External adaptation could be defined as effective if the individual maintains social ties and continues to receive rewards for participation. Clearly, these definitions are not the only ones that could be proposed. In addition, the research literature shows consistently that not only are there many outcomes that could be called effective adaptation but also there are many pathways that can lead to effective adaptation (Williams and Wirths 1965; George and Klipp 1991; Maddox 1991).

Life satisfaction is probably the most often-used indicator of effective adaptation to aging. If older people are satisfied with their present and past lives, they are seen as having adapted to aging. Havighurst, Neugarten, and Tobin (1963) identified five components of life satisfaction:

1. *Zest*: showing vitality in several areas of life; being enthusiastic

2. *Resolution and fortitude*: not giving up; taking the good with the bad and making the most of it; accepting responsibility for one's own personal life

3. *Completion*: a feeling of having accomplished what one wanted to

4. *Self-esteem*: thinking of oneself as a person of worth

5. *Outlook*: being optimistic, having hope

The scale they developed to measure these components of life satisfaction has been used in hundreds of studies. The results indicate that a majority of elders have high life satisfaction; that is, on this very general criterion a majority have successfully adapted to aging. Harris and associates (1975,161) found that, when income was controlled, life satisfaction of older people was equal to or greater than that of younger age categories.

For other researchers, *achieving the good life* is a more inclusive concept of effective adaptation. Lawton (1983) developed a four-fold concept of "the good life." According to Lawton, people who experience a high degree of behavioral competence, psychological well-being, perceived quality of life, and positive characteristics of the objective environment are very likely to feel that they have achieved the good life. Lawton defined *behavioral competence* as "the theoretical upper limit of capacity of the individual to function in the areas of biological health, sensation and perception, motor behavior, and cognition." Operationally, having behavioral competence means that the individual is in good health and has the knowledge and skills necessary to function as an independent adult and to interact successfully with others. *Psychological well-being* means that the individual is free of depression, anxiety, or pessimism; rates life as more happy than unhappy over the long run; experiences positive affect or feelings about life at the moment; and sees congruence between desired and attained goals. Operationally, psychological well-being can be measured either as a global construct or in terms of its various components.

Perceived quality of life deals with the individual's assessment of various areas of life such as family and friends, activities, work, income, neighborhood, and housing. *Objective environment* consists of the realities (as opposed to the perceptions) of one's circumstances. For example, people who live in substandard housing, who have incomes below the poverty level, who cannot get access to the health care they need, or who live in areas where it is unsafe to walk on the street are relatively deprived in terms of objective environment compared with people who have comfortable homes, who have enough income to lead the life they want, who are able to get care for any health condition they might have, and who live in areas relatively free of crime. Although the constructs that make up the good life are often correlated, Lawton found that they are independent of one another empirically. Thus, people can have high behavioral competence and low psychological well-being or a high perceived quality of life even though others may see their housing as severely deficient. Indeed, Lawton theorized that "the relative autonomy among sectors is what makes normal human existence possible?" (1983,355). For example, if change in behavioral competence were immediately reflected in all other areas of the good life, then chaotic instability might be the result.

Because psychological well-being is a global subjective assessment of life experience, it is in many ways the linchpin of the good life. Without psychological well-being, it is very difficult to see one's life as good, no matter how well the other components fit our image of the good life. Studies show that well-being generally increases or stays the same with age, and a large majority of the older population report their lives as high in well-being. Lawton found that among older people, lower feelings of well-being were most strongly associated with poor health (.72), followed by low activity level (.42), difficulty performing activities of daily living (.26), dissatisfaction with amount of interaction with friends (.20), dissatisfaction with the physical environment (.20), and lower cognitive capability (19).* This finding of differential contribution to well-being reinforces Lawton's earlier point about the interplay of the various dimensions of the good life.

Lawton's concept of the good life reflects an optimal set of outcomes that most people would probably agree are desirable. However, Minkler (1990) pointed out that by focusing our definitions on the ideal "good life" we deny the possibility of effective adaptation to chronic illness and disability, reduced income, reduced competence, deteriorating housing, and so on. Alice Day (1991) found that maintaining a positive attitude was the mysterious ingredient that differentiated those who had coped well with aging from those who had not. Certainly, favorable economic and social circumstances, physical

health, and social support were related to having a positive attitude, but not in a deterministic way.

Finally, Garfein and Herzog (1995) developed the concept of **robust aging** to refer to elders for whom aging is experienced as a continuation of good physical health, psychological well-being, cognitive competence, and productive activity. In their review of the literature, Garfein and Herzog found only a few studies that had looked at the prevalence of positive experiences of aging, but those that had done so tended to find substantial proportions, even among the oldest-old, who did not fit the negative stereotypes of aging as decline.

To address this issue, Garfein and Herzog used data from a national sample of 1,644 adults age 60 and over to develop scales for four dimensions of robust aging: productive activity, affective status, physical functional capacity, and cognitive capacity. These scales did not focus only on the problematic end of these four functional continuums but included those who were functioning well. *Productive activity* included a variety of task-oriented work such as employment for pay, unpaid work around the house, volunteer work, and helping work such as child care outside one's own home or helping others who suffered from ill health or disability. Robust aging on this dimension was defined as very high if the person engaged in 30 hours or more of such work per week and high at 10 hours per week. *Positive affective status* addressed the issue of positive mood and was measured by scores on a scale of symptoms of depression that included many positively worded items. Respondents were rated as very robust if they showed 1 or fewer depressive symptoms, and robust if they showed 2 to 4 depressive symptoms. (Those who were least robust had 10 to 22 symptoms of depression.) *Positive functional status* was a composite of capacity to perform various physical functions such as being able to do heavy housework, walk up and down stairs, or walk several blocks and frequency of participation

---

* These numbers refer to standardized regression coefficients, which reflect the relative predictive power of various factors. Thus, as a predictor of well-being, health is 30 percent more powerful than activity level.

Table 7-1 The dimensions of robust aging and the percent showing robust aging, by age.

Age Category		High Productive Activity		Positive Affective Status		Positive Functional Status		Positive Cognitive Status	
		Very Robust and Robust, %		Very Robust and Robust, %		Very Robust and Robust, %		Very Robust and Robust, %	
60–69	Very robust	52.8	} 88.6	30.9	} 66.7	14.8	} 78.3	11.5	} 66.0
	Robust	35.8		35.8		63.5		54.5	
70–79	Very robust	28.7	} 76.1	24.2	} 60.3	9.4	} 68.7	8.5	} 62.6
	Robust	47.4		36.1		59.3		54.1	
80+	Very robust	12.5	} 66.5	22.4	} 59.5	5.4	} 53.3	4.5	} 39.7
	Robust	54.0		37.1		47.9		35.2	
N			1644		1634		1644		1644

Source: Computed by the author from data reported by Garfein and Herzog (1995).

in activities such as gardening, sports or exercise, and taking walks. The very robust people on this scale reported participating often in physical activities regardless of their level of physical impairment. The robust people had intermediate levels of participation frequency and were not impaired or only minimally impaired. *Positive cognitive status* was based on a measure of cognitive impairment plus a short sentence completion test. Respondents who performed well on both tests were rated very robust and those who did well on one test but not the other were rated as robust.

Table 7-1 shows the prevalence of robust aging for each of the four dimensions by age in Garfein and Herzog's (1995) sample. The most obvious finding is that a majority of the sample were in the robust or very robust categories for nearly all functional dimensions and age categories. Only for the cognitive dimension among those age 80 or older did the proportion showing robust aging drop below 50 percent. However, the proportion in the combined very robust and robust category dropped with age for all dimensions. The age drop was greatest for positive cognitive status. These data clearly show that most aging people are at the top end of the continuum of physical, psychological, and social func-

tioning even into their 80s. This goes a long way toward explaining why most people are able to adapt to aging with the coping skills they have used throughout middle and later adulthood. Aging is a gentle slope that can be accommodated within their existing values and lifestyles and that lends itself to continuity strategies for maintaining positive physical functioning, psychological well-being, and satisfying relationships and lifestyles.

Summary

Adaptation is both an internal and an external process. Internally, it involves the assimilation of new information about changes in the self and in the environment. Externally, it involves the modification of behavior to accommodate changing demands of the social world. Most adaptation is accomplished through the exercise of routine coping skills to deal with the challenges life presents.

To cope with change, we must acknowledge that change has occurred. Denial is usually maladaptive. Aging requires adaptation because it causes various physical, psychological, and social changes. Three general strategies people use in adapting to aging are continuity, anticipation, and compensation.

Maintaining continuity of personal viewpoints and familiar associations and environments is an adaptive response to both internal and external pressures. Internal pressures for continuity come from the need to make new information consistent with the old. The older the individual, the greater the backlog of information with which new information must be reconciled, and the greater the likelihood of continuity. Individuals who feel threatened by change feel more pressure than others to stick with tried-and-true ways of doing things. External pressures for continuity are exerted by the conditioning we receive from our environments and by the demands of the roles we play. The more consistent the individual's environments and the greater the number and familiarity of the person's relationships with others, the greater the external pressures for continuity. Aging can bring a greater perception of threat, it usually brings a large backlog of experience, and it increases the duration of residence in a given environment and participation in a given set of relationships. Therefore, aging very greatly increases the pressure for continuity as an adaptive strategy. Continuity is also a means for maintaining the capacity to meet instrumental needs, needs for interaction and support, and the need for sources of self-esteem.

Anticipation allows people to identify potential problems and find solutions before problems actually develop. The wellness movement, retirement planning, and the purchase of long-term care insurance are examples of anticipation. Compensation involves taking actions that offset or make up for a loss. People seem to compensate by selectively optimizing certain of their areas of knowledge and skill. Compensations can be physical, psychological, or social. Compensation is often used to adapt, but compensations are seldom perfect; therefore compensation itself often requires adaptation.

Specific adaptations to aging include coping with aging on the job, learning to live on a lower income, learning to live with multiple chronic illnesses and perhaps disability, learning to accept help from others without losing one's self-respect, coping with lost roles and activities, and coping with threats to self-esteem. We do not know very much about how people adjust to lowered incomes. As long as the reduction is less than 50 percent, most people seem to be able to adjust somehow. But when the reduction is greater, financial dependency is likely to result. Older people cope with unexpected demands on their incomes by cutting down on food and needed medical care.

Physical energy and solid self-concepts help people adjust and most older people have both. Nevertheless, because of the high value most people place on self-reliance, there is great potential for both internal and external conflict for older people who experience chronic disease or disability and must rely on others for financial or physical assistance.

Most older people cope with role losses by redirecting their time and attention within the roles and activities that remain. Those who are younger, healthier, and more financially secure are more likely to be able to substitute for lost roles and activities.

Disengagement is sometimes voluntary, particularly for those who see themselves as overinvolved. However, a large amount of disengagement is involuntary withdrawal imposed by poor health or age discrimination.

Threats to self-esteem are dealt with through selective interaction, selective perception, use of the past, and relative appreciation. Sometimes they are dealt with by increasing isolation. To maintain a sense of competence in later life, the aging person also finds it useful to develop a self-concept that incorporates real changes in both capacities and opportunities. By revising the ideal self, people can maintain self-esteem.

Most older people are able to maintain a sense of continuity and meaning throughout later life. But a few experience changes so great that rather than trying to adapt, they escape through isolation, drugs, alcohol, or suicide. However, older adults appear to use alcohol and drugs less than younger adults. There is no evidence that older people are more likely than people of other ages to use intentional isolation as an adaptation. Suicide is the only escape mechanism that occurs at higher rates among older people than among the young, and then only for white men. People with drug and alcohol problems in later life tend to have had the problems when they were younger.

Suicide is not very common, but white men over 75 are much more likely to commit suicide than any other age-sex category in the United States. Factors that expose older people to high suicide risk include severe depression, recent bereavement, terminal diagnosis for a physical ailment, rapid physical deterioration, cognitive impairment, and living in an institution and having no visitors.

Other factors such as retirement are occasionally associated with suicide, but not often enough to represent a trend.

Family and community supports are often successful in helping older people abandon escapism. When older people mention suicide, they are often serious, and if they attempt suicide, they usually succeed on the first try.

The effectiveness of adaptation to aging can be measured through self-ratings of life satisfaction, meaning and purpose in life, and sense of autonomy and continuity. Well over half of older people (about 60 percent) have adapted to aging quite well by these criteria, and perhaps another 25 percent could be said to have adapted adequately. Overall adaptation among older people is not much lower than the level for people under 55. Older people also adapt well in terms of maintaining their participation in social life.

Part 3

Aging in Domains of Everyday Life

Individual aging does not occur in the abstract; it occurs in the context of specific social situations. Part 3 considers separate aspects of this context in detail. Chapter 8 examines the influence of aging on social relationships with family and friends and on the need for care and social support. Chapter 9 deals with employment and retirement. Chapter 10 discusses how aging changes activities and lifestyles, and considers aging in various spheres of activity, such as the workplace, the home, and the community. Chapter 11 examines the interrelationships among aging, religion, and spirituality. Finally, Chapter 12 examines dying, death, grief, and widowhood. Taken as a whole, the five chapters in Part 3 provide concepts and information that can improve our understanding of how aging interacts with everyday social life.

8 Family, Friends, and Social Support

This chapter is about the personal relationships that typically surround people as they grow older. People do not cope with aging in isolation; most often they cope in the company of others who provide social and emotional support and help, and in surroundings that provide a sense of security, connectedness, and belonging. Many of our relationships with others are simply associations that may involve an element of liking but no sense of lasting commitment. Close personal relationships are those in which the individuals have frequent, strong impact on one another in a diverse array of activities and settings over an extended period of time (Berscheid and Peplau 1983). People can develop close personal relationships if they are in frequent and enduring interaction, if they are close to each other in space, and if there is perceived reciprocity in their exchanges over time. Once developed, close relationships can be maintained at a distance and usually can be reactivated relatively easily. Close relationships come in a variety of forms.

Here we will look at how aging influences, and is influenced by, relationships with a spouse or a partner, older parents, adult children, siblings, other relatives, and friends. We also look at how these relationships can come together to form a social support network for an aging individual. In most cases, these relationships are intimate, durable, and highly personal, especially in comparison with other kinds of relationships, such as those between doctor and patient or between social worker and client.

Types of Bonding

Three types of ties hold close relationships together: interdependence, intimacy, and belonging. *Interdependence* brings people together to satisfy their needs better than they could acting alone. For example, division of labor reduces the number of things an individual must do or know how to do. Interdependence can also increase the amount of resources available to the individual. Through relationships, people can pool their knowledge, financial resources, insights, and encouragement. Ties of *intimacy* allow for the exchange of affection, trust, and confidences. Some intimate relationships also allow people to meet their need for sexual expression. Confidants' opinions of the person and her or his past history are important sources of self-respect. Relationships can also be based on the need for *belonging*—the sense of being more than an isolated individual. Relationships of belonging can be a source of companionship, socializing, identity, and safety or security.

These three types of bonds—interdependence, intimacy, and belonging—can exist in combination or independently. Given the many possible combinations, clearly relationships can vary a great deal in both quality and purpose. Also, the quality and purpose of relationships are influenced by such factors as gender, social class, and ethnicity, thus ensuring that relationships do vary a great deal, even within a specific social role, such as adult child or spouse.

Relationships emerge when people *interact* with one another, usually in the process of playing their various social roles. Interdependence is built into many role relationships, but whether intimacy and/or belonging develop depends more on the specifics of the interaction—whether there is equality, self-disclosure, similarity, acceptance, agreement, trust, or cooperation.

Relationships can also be destroyed or diminished. For instance, relationships can be diminished by permanent geographic separation—a moving apart with no likelihood of return. Interdependence bonds can be weakened if one or both of the participants become less able to do their share. Belonging bonds can be weakened by growing feelings of superiority or inferiority or difference in one or both of the participants. And intimacy bonds can be weakened by selfishness, breaks in trust, or withdrawal by one or both participants.

Regardless of the specific role, relationships tend to increase in diversity throughout middle age because of the cumulative effects of individual choices made along the way. In later maturity, the number of long-term relationships is often gradually diminished by deaths of friends and relatives. But even in advanced old age, most people still have an effective network of close relationships.

For convenience, our discussion of relationships looks first at family relationships, and then at friendships and social support. In each case, we are concerned with structural aspects such as availability of kin roles, dynamic aspects such as the nature and quality of relationships, and changes over time, particularly those associated with aging.

Family

The family is perhaps the most basic social institution.* There are a great many different types of family organizations throughout the world; here we concentrate on the pattern that prevails in the United States. Almost everyone is born into a family. Most people spend most of their lives residing with a family group. Most people play several family roles in the course of a lifetime: for example, son, husband, father, grandfather, uncle. For many, the family group is the center of their world, the highest priority in their system of values. Fewer than 5 percent of the older population are without family.

Family researchers often think about the development of family relationships in terms of an ideal-type called the *family life cycle*. The family life cycle begins with marriage—the formation of a couple from two previously unrelated people. Families can develop in many ways. Some develop intact, others split up, and still others represent a uniting of parts from different split families. What family sociologists call the family life cycle is really an ideal representation of the stages in the life of a couple who marry, have children, and stay together until one spouse dies in old age.

The following relatively simple scheme illustrates the concept of family life cycle stages:†

Beginning (couple married 0 to 10 years, without children)

Early childbearing (oldest child under 3)

Preschool children (oldest child over 3 but under 6)

School-age children (oldest child over 5 and under 13)

---

*For a fuller discussion of families in later life, see Bengtson, Rosenthal, and Burton (1990).

†The model presented here is adapted from Rollins and Feldman (1970).

Teenagers (oldest child over 12 and living at home)

Launching children (from oldest child gone to youngest's leaving home)

Middle years (empty nest to retirement)

Retirement (one spouse retired to onset of disability)

Old age (one spouse disabled to death of one spouse)

Of course, since children and grandchildren usually begin their own family cycles, the chain seldom stops. As we will see, usually it is only the last three or four phases of the family cycle that occur in later maturity or old age.

The family that produces the relationships of spouse, parent, grandparent, great-grandparent, and widow(er) is called the *family of procreation;* it is the family within which a person's own reproductive behavior occurs. People also belong to a *family of orientation,* usually the family into which they are born or adopted. Older people very often carry into their later years the roles from the family of orientation (son, daughter, brother, sister) as well as from the family of procreation. There are also family roles, such as cousin, uncle, nephew, brother-in-law, and so on, that derive from the *extended family,* the complex network of kin that parallels both the family of procreation and the family of orientation, usually through a sibling relationship or through the marital bond uniting two separate families of orientation.

Because kinship can be an extremely complex subject, we do not emphasize extended kinship here but concentrate on roles in the family of procreation and the family of orientation as they relate to the effects of aging.

Demographic Factors

Recent demographic trends will continue to have a substantial impact on family structure. Declining mortality has increased the number of generations in most families to the point where five-generation families are becoming common. At the same time, a sharp reduction in the number of children per couple since 1960 has meant a sharp decrease in the number of aunts, uncles, and cousins for young people. Thus, instead of a family with a large number of people in each generation, as was true in the 1950s, the contemporary family has more generations but many fewer people in each generation. Researchers refer to this new form as the vertical or "beanpole" family (Bengtson, Rosenthal, and Burton 1990). For example, Joan S. was born in 1942. She had nineteen aunts and uncles and nearly sixty first cousins. She was her paternal grandmother's twenty-fifth grandchild. Joan has two siblings, one with two children and the other childless. Joan has one child. Her parents are recently great-grandparents, and by the time Joan reaches 65, her parents will probably be great-great-grandparents. Thus, Joan was born into a three-generation family with a large number of people in each generation, but now she is in a four-generation family with fewer people in each generation. As a result, interaction in her family now crosses the lines of several generations rather than being confined primarily to adjacent generations, which was more common in the past.

The development of the vertical family structure also means that people spend much longer in their various family roles. For example, widowhood generally occurs at a much *later* age now than it did in 1960, which means that spouse roles continue later in the life course. Bengston and his colleagues found that in addition, most adults spend many *more years* as grandparents, providing more potential for development of the grandparent–grandchild relationship.

Gender, race, and ethnicity also have a great impact on family structure in later life. For example, as Figure 8-1 shows, marital status is strongly correlated with gender. In 1991, beyond the age of 65, women were

Figure 8-1 Marital status of elders by age and sex: United States, 1991.

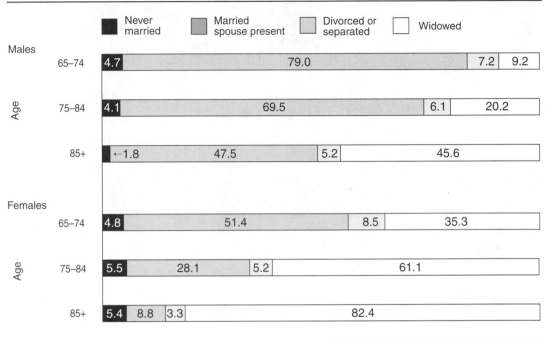

Source: U.S. Bureau of the Census (1992b).

much more likely to be widowed than men, particularly in the oldest cohorts. For example, 26.1 percent more women than men were widowed at age 65 to 74, contrasted with 36.8 percent more at age 85 and over. Put another way, at age 85 and over, nearly half of the men are married and living with their spouses, compared with only 8.8 percent of the women in that age category. The highest proportion of widowed people occurred among African-American women age 85 and over (83.3 percent). Hispanic and African-American older women were also more likely than white women to be divorced. White men were more likely than Hispanic and African-American men to be married.

Social class also has a significant impact on family structure. For example, teenage childbearing is much more common among working-class women. In some cases, several generations of teenage childbearing have produced families in which the grandmother is in her late 20s and the great-grandmother in her 40s. Burton (1985) reported that many young grandmothers did not want to be grandmothers and refused to become substitute parents for their grandchildren. The job of caring for the new baby tended to be left to great-grandmothers, who were also responsible for caring for their adolescent granddaughters as well as their own aging parents.

By contrast, the patterns of late childbearing typical of the upper middle class result in more years between generations (Bengtson, Rosenthal, and Burton 1990). When the interval between generations is thirty years or more, there is a great likelihood that caring for adolescent children and older parents may coincide.

The large increase in marital dissolutions since 1960 also has had a significant effect on family structure. For example, Eggebean and Uhlenberg (1985) found that between 1960 and 1980 there was a dramatic *decrease* in the amount of their lives men spent in house-

Intimacy is an important dimension of older marriages.
Photograph by Frank Siteman/Tony Stone Images, Inc.

holds with young children. In many cases, the reduced interaction with their children is so pronounced that it interferes with the development of intimacy with them, which in turn has profound implications for relationships with those children later in life. In addition, Cherlin and Furstenberg (1985–1986) found that in many cases divorce severely disrupted grandparents' opportunities for interaction with their grandchildren, while in other cases grandparents became more involved with their grandchildren. (We will look at this issue in detail later in the chapter.)

Living arrangements also have a strong impact on the structure and functioning of family relationships. For example, although 80 percent of elders live alone or with their spouses in independent households, nearly 20 percent live in multigenerational family households. For this latter group, the nature of family relationships, especially across generational lines, is undoubtedly qualitatively different from that of elders who do not coreside with other generations. (We will revisit the issue of coresidence at several points later in the chapter.)

As we look at specific family relationships, we will first consider older couples, and then briefly consider how other marital statuses— never-married, widowed, and divorced or separated—influence family relationships. Next, we look at relations with adult offspring, grandchildren, siblings, and other kin.

Couplehood

Most older people are married and living with their spouses in independent households. However, beyond the age of 75, less than one-third of older women are married (see Figure 8-1). Despite the lack of partners for many older women, high marriage and remarriage rates mean that older couples are very prevalent even among older people. The couple relationship tends to be the focal point in married people's lives, especially when the departure of adult children and retirement have increased the amount of contact between spouses and perhaps reduced the number of alternative close relationships. As the average length of life increases, so does the average number of years a couple can expect to live together after the children leave home.

Most middle-aged and older couples have been together since early adulthood, but marriage in later life is not uncommon. For example, in 1987, more than 300,000 adults 45 and over got married, and of these people, over 50,000 were 65 or older. Only 4 percent were first-time marriages. Of brides 45 and over who were remarrying, more than 85 percent wedded men who were in their own age category or an older one, whereas most remarrying grooms 45 or over married women in a younger age category than their own (Vera, Berardo, and Vandiver 1990). Only 12 percent of remarriages involved women who were four or more years older than their husbands. In later life, large age differences between spouses are much more common for men than for women. Large age differences also are more common among nonwhites and people in lower socioeconomic classes (Berardo, Apell, and Berardo 1993).

In looking at the couple relationship and how it changes with age, we consider trends in marital satisfaction, functions of couplehood—including intimacy, sexual intimacy, interdependence, and belonging—and remarriage.

Marital Satisfaction As a concept, marital satisfaction refers to the extent to which a marriage produces contentment and meets expectations. It is usually measured by asking marital partners how satisfied they are with various aspects of marriage such as affection, companionship, sharing of responsibilities, confiding in one another, and caring for one another. Whether the increased emphasis on the couple that comes with launching the children and retirement is a bane or a blessing depends, of course, on the quality of the relationship. A happy and satisfied relationship is a blessing. The couple relationship is a source of great comfort and support as well as the focal point of everyday life, and satisfied couples usually grow closer as the years go by. Such couples often show a high degree of interdependence, particularly in terms of caring for one another in times of illness. The husbands in strong couple relationships are particularly likely to see their wives as indispensable pillars of strength. Happy couples tend to share many activities, and there tends to be greater equality between the partners than is true for unhappy couples. This equal-

ity results from a gradual blurring of sex role boundaries and a decreasing use of gender to define household division of labor (Dobson 1983). Happy couples tend to remain sexually active into later life (Brecher 1984).

Although most couples are generally happy and satisfied, a small proportion are not. For example, Brecher found that among more than 3,000 husbands and wives over the age of 50, less than 15 percent were unhappily married. Higher proportions of unhappily married were found among those who had dependent children living at home, less than very good communication in their marriage, and low enjoyment of sex with their spouses. Among those who reported low enjoyment of sex with their spouses, 32 percent of wives and 38 percent of husbands reported being unhappily married. Of course, these are correlations, and we do not know whether unhappy marriage causes low enjoyment of sex or vice versa. Retirement, income, and education were not related to marital happiness. Stinnet, Carter, and Montgomery (1972) found that lack of mutual interests, differing values, inability to express feelings, lack of companionship, and frequent disagreements were the major troublesome aspects of marriage in later life. Johnson (1985) found that long-term disability was the major factor that brought out negative aspects of marriages. Some of these unhappy marriages end in divorce, although divorce rates are much lower among people over 50 than among those under 30. In 1987, more than 14,000 divorces were granted to people 65 or older.

What happens to marital satisfaction over the life span? From various studies, a consistent picture emerges of high marital satisfaction in early adulthood, a gradual decline in satisfaction through middle age, and a steady increase in marital satisfaction in the period after child launching. This curvilinear pattern has not been found in all studies, but it has been found often enough that it is probably correct. Couples tend to be quite satisfied at the beginning of marriage, but if they have children, marital satisfaction sometimes drops substantially, usually as a result of decreased marital interaction and greater financial pressures connected to having children (White, Booth, and Edwards 1986). In a fifty-year longitudinal study of marriages, Weishaus and Field (1988) found that the U-shaped pattern of marital satisfaction was most common, followed by lifelong positive marital satisfaction.

Gilford and Bengtson (1979) separated marital satisfaction into two components: a positive dimension that included interactions such as working, discussing, and laughing together, and a negative dimension that included criticism, disagreement, and anger. They found that the negative dimension of marriage declined steadily over time and the positive dimension showed the familiar U-shaped pattern. The significant point here is that once the children are launched into adulthood, marital satisfaction tends to steadily increase with age. For example, Brecher (1984) found that the proportion who were happily married increased from 84 percent of respondents in their 50s to 88 percent of those in their 60s, to 91 percent of those in their 70s.

Skolnick (1981) found that "good" marriages were characterized by affectionate and enjoyable interaction between partners. These couples also tended to accept conflict as a normal part of married life. She found that 63 percent of the marriages she studied had improved over time, compared with 14 percent that deteriorated. Things that are important in long-term marriages are being married to someone you like; having a sense of humor; and consensus on life goals, friends, and decision making (Lauer, Lauer, and Kerr 1990). Atchley and Miller (1983) studied both husbands and wives and found that the couples averaged better than 80 percent agreement with respect to values and activities. They concluded that couples who stay together definitely play together.

Swenson, Eskew, and Kolhepp (1981) differentiated couples into *conformist* and *postconformist*. Conformist couples tended to continue into later life the stereotyped sex roles of young and middle adulthood, whereas postconformist couples tended to soften the sex role divisions and accommodate more personality factors into their interactions. They found that if couples change over time, changes occur gradually and only in couples who are ready to see changes in their partners and are open to new information. Conformist couples experienced a decline in expressions of love after the child-rearing stage, whereas the postconformist couples reported more expressions of love in the later than the earlier years.

Atchley and Miller (1983) looked at the relationship between couples' family orientation and their level of life satisfaction. They found that the most satisfied couples were those who valued both their marital relationship and close ties with other family members. Only one couple of the 143 in their study valued marriage but not close family ties. Thus, close ties with other family members is an important contributor to satisfaction within marriage.

For married people, marital satisfaction is a central factor in their overall satisfaction with life. For example, Lee (1978) found that among older married men, only health was a more powerful predictor of their overall life satisfaction than marital satisfaction. For older married women, marital satisfaction was by far the strongest predictor of overall life satisfaction. For both men and women, marital satisfaction was more important than age, education, retirement, or standard of living in predicting life satisfaction. Thus, high marital satisfaction is extremely important to the overall well-being of the individuals in older couples.

Functions of Couplehood   When we look at couples in the period after child rearing, we are struck by the very wide range of functions that couplehood involves. Functions of couplehood include intimacy, sexual intimacy, interdependence, and belonging.

Intimacy in the couple involves mutual affection, regard, and trust. For most couples it also involves sexual intimacy. A wife is often her husband's only confidant. In his review of the literature on intimacy, Wright (1989) found a prevailing lack of mutuality and intimacy in relationships among men, a fact that men regretted. Wright also found substantial evidence that men have difficulty establishing close relationships with anyone but their wives.

Farrell and Rosenberg (1981) found that the wife played a crucial role in validating the middle-aged husband's image of himself as the beloved family patriarch. According to Farrell and Rosenberg, a major problem in most men's lives is the need to be in control at home and at the same time to be loved there. In actuality, the wife more often controls the home situation and, often in collusion with the children, protects the husband by avoiding confrontations that might undermine his belief that he is in charge and has the family's support and respect. Part of the decline in marital satisfaction in middle age may be thus a by-product of the strains involved in supporting what is increasingly a myth of patriarchy. And the upturn in satisfaction after child launching may be related to the fact that, when the children leave, control over them is no longer a power issue dividing the couple and interfering with intimacy.

Sexual intimacy is an integral part of any couple's relationship. In counseling older couples and older people in general, the sexual component of human interaction must be taken into account. Sexuality in middle and later life reflects physical capacity, emotional needs, and social norms. Unfortunately, the simultaneous influence of these various factors has yet to be studied adequately. Studies of sexuality in general suffer from lack of interest among research funding agencies and low esteem among social scientists. As a result, most studies that have looked in detail at sexuality in middle age and later life have used convenience samples, which means that the results are not definitive.

Masters and Johnson (1966) found that male sexual performance declined with age. Levels of sexual tension, ability to achieve an erection, ability to ejaculate, and frequency of masturbation and nocturnal emission all declined with age. These findings were confirmed by Brecher (1984). Starr (1985) concluded that the decline with age in male erectile response was the most limiting factor in the sexual relationships of older people. Brecher reported that impotence was highly associated with health problems such as anti-hypertension medications, diabetes, and prostate surgery, but even among men who have these problems, a substantial proportion report no problems with their sex lives. Hysterectomies and mastectomies among women had no lasting effects on sexual activity. Masters and Johnson (1966) found that older men who had sex frequently in their middle years showed a much less significant decline. In one of the few longitudinal studies of age changes in sexuality, George and Weiler (1981) reported that levels of sexual activity over time remained more stable than cross-sectional results had indicated.

Ade-Ridder (1990) studied 670 married elders concerning their sexual interest and sexual behavior and the relation of the two to marital quality. The respondents averaged 72 years of age and had been married more than forty years on average. The respondents scored high on general sexual interests and feelings, yet most reported frequencies of intercourse that were substantially lower than during the early years of marriage. It appears that carrying sexual interest through to completion in intercourse is not as necessary to marital quality in the later years of marriage as in the earlier years. Most of the couples were in their early 60s when they noticed a decline in sexual behavior, and most of the time dramatic changes were brought on by the husband's declining capacity for physical sexual arousal.

What about the relationship between sexuality and marital quality? Ade-Ridder found that those marriages in which sexual behavior had not noticeably changed tended to be happier than marriages in which there had been a marked decline in sexual behavior. But happier marriages were also typified by higher levels of sexual interest, too.

Brecher (1984) surveyed more than 3,000 husbands and wives over 50 who were recruited through a notice in *Consumer Reports* magazine. He found that 75 percent of the wives and 87 percent of the husbands felt that the sexual side of marriage was important. Of the respondents, 67 percent of the 1,018 wives and 59 percent of the 1,523 husbands felt that frequency of sex was about right; 41 percent of husbands thought that sex was not frequent enough in their marriages, compared with only 18 percent of wives. Of the respondents who felt that frequency of sex was just about right, 95 percent were happily married, compared with 84 percent of those who thought that sex was not frequent enough. Among the small minority who were unhappily married, there was a strong gender difference in the importance of sex: 54 percent of unhappily married wives reported that sex was of little importance, compared with only 16 percent among unhappily married men. This suggests that incompatible views about the role of sex in marriage may be a strong factor in unhappy marriages in middle and later life.

Fear of failure, or "performance anxiety," plays an important role in the tendency of some men to withdraw from sexual activity (McCarthy 1984). Wives sometimes lack insight into this problem and mistakenly feel rejected by their husbands' apparent lack of interest in sex. Counseling can sometimes help alleviate these problems, but many older people are reluctant to talk about these issues, even though most feel that they are very important.

Menopause also has effects on sexuality. Among women, perhaps 30 percent or more experience a decline in sex drive associated with menopause (Sheehy 1992). This is probably related to a drop in testosterone, the same hormone that is responsible for sexual arousal in men. About 50 percent of women stop producing testosterone when they go through menopause. In addition, the anatomical changes in the vagina (for example, thinning skin, less flexibility) can make intercourse painful. As discussed in Chapter 4, hormone replacement therapy can be used to relieve these symptoms, but hormone replacement therapy remains controversial because its negative long-term effects appear to outweigh its benefits.

Aging can also bring some advantages in sexual expression. Starr and Weiner (1981) studied 800 people between the ages of 60 and 91 and found that 36 percent said that sex was better for them than when they were younger. Only 25 percent said it had become worse. Most felt that sex should play a large part in an older couple's life. About 80 percent were sexually active, and a majority had intercourse at least once a week. Most had experienced an increase in the spontaneity of sexual expression as a result of the empty nest and retirement. Older couples viewed being able to spend a long time making love whenever they wanted as a major advantage.

The prevailing myth is that older people are not interested in sex. When Starr and Weiner approached senior center staff for permission to distribute questionnaires to center participants, they encountered resistance because the staff assumed that older people would be offended by questions about sex and embarrassed about their lack of sexuality. In fact, however, older people responded to the questionnaire in substantially higher percentages than in other studies of sex among middle-aged adults.

Sex means much the same thing to older people as it does to others. To the question "What is a good sexual experience?" one of Starr and Weiner's older respondents replied,

"To be really horny with a partner who is just as horny. To take plenty of time and when you can't stand it another minute, make it!" (1981,5). Many adults of any age would probably agree with this definition. Starr and Weiner's respondents liked lots of things about sex: a sense of contributing to the other; feelings of desirability, zest, completion, and relaxation; feeling loved and loving; the comfort of touching and cuddling; and even transcendence. The vast majority had sex lives that were satisfying.

Fortunately, sex is one important area in which older people generally refuse to do what society at large expects them to do. Yet undoubtedly some people withdraw from sexual activity because they feel withdrawal is expected or "normal" or that sex is not something widows should be interested in. It is for such people that the myths and stereotypes about aging and sexuality are most harmful. But even the sexually active may feel the need to hide the fact, for fear of disapproval. This is not a healthy situation. Sex is a powerful expression of intimacy, and its continuation into later life is important for the vast majority of older couples. On the other hand, we do not want to create expectations about the quality of sexuality in later life that are unrealistically high, either.

Interdependence in older couples involves instrumental sharing of housework, income, and other resources. In terms of housework, the picture is not very clear. Early studies found that even employed wives bore a disproportionate share of household chores (Bahr 1973; Ballweg 1967). More recent studies have shown that, with greater employment of wives, the empty nest, and retirement, couples tend to share household responsibilities (Dobson 1983). Others have found a great deal of continuity in gender-differentiated household tasks over the course of the marriage (Brubaker 1985; Brubaker and Hennon 1982; Vinick and Ekerdt 1991). Changing attitudes about "men's work" versus "women's work" around the house may make getting a

clear picture even more difficult. In addition, trends may vary by type of work. Most people, especially men and better-educated women, tend to dislike routine work like housecleaning. Men may try to fall back on their traditional sex role to avoid it, while working women may lobby for hiring a housekeeper. In both cases the idea is to avoid housework. These days, older people seem to enjoy housework more than younger people do. Whether this tendency will change over time remains to be seen.

With time, men apparently increase their involvement in more expressive or creative types of housework, such as cooking and home decorating (Gordon, Gaitz, and Scott 1976), making husbands and wives more equal in these areas in later life than in middle age (McAuley, Jacobs, and Carr 1984). Men seem to be expected to do heavy work and mechanical repairs throughout the life course. Husbands as well as wives tend to increase their involvement with gardening in later life. However, apart from grocery shopping, wives show an unfortunate and persistent lack of knowledge or interest in managing the couple's finances. Although more household activity tends to be shared in retirement, the responsibility for making decisions and taking initiative tends to remain along gender-role lines.

Suitor (1991) looked at satisfaction with the household division of labor across the family life cycle and its relation to perceptions of marital quality. She found that wives' satisfaction with household division of labor followed the now-familiar U-shaped pattern over the family life cycle, but the husbands' satisfaction remained relatively high throughout. Satisfaction with household division of labor was correlated with perceptions of marital quality across the family life cycle. "Satisfaction with the division of labor was more important in explaining marital happiness and conflict than were age, educational attainment, or wife's employment status" (1991,221).

Income sharing is a major dimension of interdependence for most couples. Income adequacy—what the members of the couple have relative to what they feel they need—fluctuates across the life course. Of particular importance is the financial pinch of supporting children. Working-class couples often recover from this by early in middle age, but for middle-class couples the financial pinch can be the worst at that stage, when their children are in college. The need to support older parents financially may arise at the same time that children need financial assistance to get launched into adulthood. The result is more than a pinch; it is a squeeze. And because partners usually share income in middle age, this can lead to conflict. Although retirement usually reduces available income, it normally does not produce a feeling of financial pressure comparable to the strains felt at midlife. Indeed, many couples find themselves more comfortable financially in retirement than at any other time in their married lives, which may also contribute to a rise in marital satisfaction later in life.

Taking care of one another in time of illness is a facet of interdependence that grows much more important with increasing age. Among older people who need assistance in such household matters as personal care, meal preparation, housework, or shopping, the spouse is *the* major provider of such help (Stone, Cafferata, and Sangl 1987). For example, in a large national survey of elders in need of long-term care and living in the community, 76.7 percent of men and 37.9 percent of women under age 75 were being cared for by their spouses, and among married elders receiving care, 93 percent of husbands and 81 percent of wives were being cared for by their spouses (Coward, Horne, and Dwyer 1992). Although assistance is an essential element of continuing to live in an independent household, it can be a severe strain on the spouse, particularly if she or he is also having health problems. Lopata (1973) found that 46 percent of the widows in her study had cared for their husbands at home during their final illness, and nearly 20 percent had done so for over a year. Colleen Johnson (1985) found that in 75 percent of couples in which one spouse took care of the other, both spouses were in poor health. Such health problems not only severely curtail many kinds of marital interactions but also reduce the spouse's freedom of action.*

A major problem can occur when caregiving shifts from interdependence to dependence, when one spouse can no longer reciprocate. The more able spouse may feel resentment at the one-sided flow of attention and energy, and the reduction of equality between spouses may reduce the intimacy between them as well. For example, Hoyert and Seltzer (1992) found that wives caring for husbands were much more likely to experience negative effects of caregiving than were adult daughters caring for parents.

On the other hand, caregiving can provide caregivers with a sense of purpose and competence, and care receivers may feel good about the attention they get from their spouses and the security they feel from having their needs met. Atchley (1992) found that wives with moderate disabilities were more likely than nondisabled wives to be extremely satisfied as opposed to merely satisfied with their marriages. Most spouses do what is needed if they can, and most could use more emotional support, aid, and counseling than they get.

Belonging for the older couple involves individual identification with the couple, sharing of values and perspectives, comfortable interaction and socializing, and a sense of safety and security. A sense of belonging based on shared values or perspectives thrives on agreement. Atchley and Miller (1983) found that older couples averaged 85 percent overlap in personal values and that couples who valued close ties with each other and with

---

*Caregiving by others in the family is considered in more detail later in the chapter.

adult children tended to have high levels of life satisfaction. A major determinant of agreement is potential for conflict. In middle age, most couples have a high potential for conflict—over jobs, money, sex, or children. Child launching, career plateaus (and the withdrawal of personal stakes that usually goes with such plateaus), and greater financial security—all changes common to couples in late middle age—reduce the potential for conflict and increase the fund of agreement that supports a sense of belonging. The rediscovery of one's spouse can reduce sexual conflicts, and retirement can reduce disagreements over the provider function. The reduced size of the household after child launching usually means more socializing and companionship between spouses, which also increases the sense of mutual participation and belonging. Finally, the availability of a spouse to provide caregiving when needed enhances most older people's sense of safety and security. Thus, caregiving supports ties of belonging as well as ties of interdependence.

It should be quite clear from the foregoing that the increase in marital satisfaction that typifies the transition from middle age to later maturity is no accident; it is the result of many forces, the vast majority of which are pulling the couple together toward an increasingly greater appreciation and acceptance of one another. The empty nest and retirement increase opportunity for companionship and decrease potential for conflict, especially for those couples who remain physically healthy.

Though most marriages in middle and later life are strong and satisfying relationships, there can be a dark side to marriage, usually a carryover from marital dysfunction and conflict earlier in the marriage. Mental illness in one or both spouses can poison the relationship. Irreconcilable differences in values and standards can lead to persistent and relentless conflict. Dementia can literally destroy the basis for whatever relationship the marital partners might have had. Spouse abusers

often continue these patterns into later life, and the frustrations of caregiving can trigger spouse abuse even in people with no history of family violence. Rosalie Wolf (1986) reported that 25 percent of elder abuse consisted of long-standing patterns of spouse abuse. These negative aspects of marriage have received very little research attention.

Remarriage A substantial number of older couples are formed in later life. Only three studies have looked at this phenomenon in depth (McKain 1969; Vinick 1979; Burch 1990). In a study of 100 couples who married in later maturity, McKain found the desire for companionship to be by far the most frequently given reason. Previous experience with marriage also predisposed older people to remarry. Few of the couples believed in romantic love, but they were interested in companionship, lasting affection, and regard. As McKain stated:

The role of sex in the lives of these older people extended far beyond love making and coitus; a woman's gentle touch, the perfume on her hair, a word of endearment—all these and many more reminders that he is married help to satisfy a man's urge for the opposite sex. The same is true for the older wife.

(1969,30)

A few older people remarried to allay their anxiety about poor health, and some remarried to avoid having to depend on their children. Many older people tended to select mates who reminded them of a previous spouse. Also, older couples followed the same pattern of *homogamy* (the tendency for people of similar backgrounds to marry) that is found among younger couples.

Using as indicators of marriage success such unobtrusive measures as displays of affection, respect and consideration, obvious enjoyment of each other's company, lack of complaints about each other, and pride in their marriage, McKain found that successful "retirement marriage" was related to several factors. Couples who had known each other well over a period of years before marriage were likely to be successfully married. A surprisingly large number of couples McKain studied were related to each other through previous marriages. Probably the prime reason that long friendship was so strongly related to a successful marriage in later life is that intimate knowledge of the other allowed better matching of interests and favorite activities. Marriages in which interests were not alike were less successful.

Approval of the marriage by children and friends was also important for the success of marriage in later life, according to McKain's study. Apparently, considerable social pressure is exerted against marriage in later life, probably growing partly out of a misguided notion that older people do not *need* to be married and partly out of concern over what will happen to their estates. Older people are very sensitive to this pressure, and encouragement from children and friends is important in overcoming it. Also, a marriage that alienates older people from their families or friends is not likely to be successful.

McKain also found that financial factors were related to successful marriage in later life. If both partners owned homes, success was more likely than if only one or neither did. The importance of dual home ownership was probably symbolic, indicating that each partner brought something equally concrete to the marriage. If both partners had sufficient incomes prior to marriage, they usually had a successful marriage. The arrangements for pooling property or giving it to children were important for predicting marital success because they indicated the priority one partner held in the eyes of the other. It was important for the marriage partner to have first priority on resources, if the marriage was to be successful.

Vinick (1979) found many of the same factors operating in the twenty-four couples she studied. She also found that the time before remarriage was easier for older women than

for older men. Despite their greater financial difficulties, women had more social support than men during their time alone, especially from family. Women also found it easier than men to ease loneliness by keeping busy. Thus, for these reasons men were much more highly motivated to remarry, and the large number of widows in later life made it a much more realistic prospect for them. Courtship did not last long, perhaps because the partners were already familiar with marriage and had a sense of what they were getting into. Family were more likely to approve of the remarriage than friends were, at least initially. After remarriage, interaction with children and friends declined as partners used one another as their primary source of interaction. Most people were quite satisfied with their remarriage.

Burch (1990) looked at age differences in the probability of remarriage for men and women who became widowed or divorced. As Table 8-1 shows, 60 percent of men who became divorced were remarried after five years, compared with only 42 percent among widowers. After fifteen years, over 80 percent of divorced men had remarried, compared with 55 percent of widowers, 46 percent of divorced women, and 38 percent of widows. Among women, the older they were when the marital dissolution occurred, the lower the likelihood of remarriage. For example, among women who became divorced between ages 45 and 54, 39 percent were remarried after fifteen years compared with 69 percent of women who became divorced under age 35. Among women who were widowed at age 45 to 54, only 22 percent had remarried after fifteen years.

Burch pointed out that we do not know very much about the opportunities and motivations for remarriage. The literature generally assumes that everyone wants to be married and that the main factor that influences remarriage is the availability of potential marriage partners. Certainly, demographic factors such as the sex ratio of unmarried people and the tendency of men to marry

Table 8-1 Proportion remarrying for divorced and widowed persons by time since widowhood or divorce, by gender.

Time Since Widowed or Divorced (in Years)	Men		Women	
	Widowhood	Divorce	Widowhood	Divorce
5	.42	.60	.17	.25
10	.50	.70	.27	.35
15	.55	.81	.38	.46

Source: Adapted from Burch (1990).

younger women constrain the remarriage possibilities for women, but there are other possibilities that have not been sufficiently investigated. For example, in some ethnic groups remarriage for widows is frowned upon, which produces social pressure against dating and eventual remarriage. Many women find that being single is a lifestyle that has its rewards, especially if they are pursuing an occupational career. For some of them, singlehood is a satisfying lifestyle that must be sacrificed if they are to remarry, and some are unwilling to do so. These are just two examples of internal and external factors that could influence the opportunities and motivations for remarriage. Much more research is needed on this topic.

Other Types of Couples

Although married and heterosexual couples are the most common types, there are other types of middle-aged and older couples. Nonsexual couples consist mainly of friends who share housing and have a degree of commitment, intimacy, interdependence, and belonging similar to that of most married couples. This type of couple has been mentioned in studies of housing (Hochschild 1973) and in studies of community-dwelling older people (Clark and Anderson 1967), but I could find no studies that dealt with this topic in depth.

Homosexual couples have received slightly more research attention. Bell and Weinberg (1978) described two types of homosexual couples: closed-coupled and open-coupled. The closed-coupled homosexual relationship resembled heterosexual marriage in terms of strong emotional commitment to the relationship. Members of these couples tended to be sexually exclusive. They were the happiest among homosexual couples. This type of relationship is more common among older homosexuals than young adult homosexuals (Berger 1982) and more common among lesbians than gay men.

Open-coupled homosexual relationships were less sexually exclusive and more common among gay men than lesbians. Lipman (1985–1986) reviewed the research on the sharing of household tasks among homosexual couples and found that the traditional heterosexual pattern was not recapitulated. There tended not to be "male role" and "female role" assignments in homosexual couples; assignments tended to be made more specifically and negotiated between partners. Older homosexual couples, like heterosexual couples, tended to be more content in their relationships than middle-aged couples (Berger 1982; Silverstein 1981). But Peacock (1990) found that even among older homosexual men, the norms against sexual exclusivity created a tension that has no parallel for most heterosexual couples. The findings about homosexual couples should be approached with some caution, for as Lipman (1985–1986) pointed out, research on homosexual relationships tends to lack methodological rigor.

Older People Not in Couples

The never-married, widowed, and divorced might be expected to have different patterns of family relationships than older people who are married. Indeed, Atchley, Pignatiello, and Shaw (1979) found that married women had much less contact with kin outside their households than women of other marital statuses.

In 1991, only 5 percent of the older people in the United States had never married. On the surface, one might expect these people to have trouble getting along as single individuals living in independent households. However, apparently because they learn very early in life to cope with aloneness and the need to look after themselves, older single people living alone have developed the autonomy and self-reliance so often required of older people (Johnson and Catalano 1981).

Gubrium (1975) found that never-married older people tended to be lifelong isolates who were not especially lonely in old age. He suggested that these people view isolation as normal. They resent the assumption researchers and service providers often make that relative isolation is necessarily bad and leads to loneliness. This point is a good one to bear in mind.

However, Rubenstein (1987b) called some of Gubrium's generalizations into question. For example, Rubenstein found that most never-married older people in his sample of people living alone had *not* lived alone most of their adult lives. Many lived with others most of their lives; living alone was a late-life occurrence. And while about one-third of the never-married had no close friends, almost none had no family with whom they were close. Most had intimate relationships. Taken together, these data question the notion that most never-married older people are lifelong isolates.

Older people who have never married might be expected to have more contacts with extended kin than those who married (Allen and Pickett 1987; Johnson and Cat-

alano 1981). However, Atchley, Pignatiello, and Shaw (1979) found this contact depends a great deal on social class, at least for women. For example, never-married older women teachers interacted with extended kin significantly more than those who were married. But older telephone operators who had never married had much lower levels of interaction with extended family than did those who were married. In their total interaction patterns, teachers appeared to compensate for being single by disproportionate involvement with relatives, while telephone operators tended to compensate by having relatively more contacts with friends. Single older women teachers had about the same overall level of interaction as married older women teachers, but among older telephone operators, the overall interaction levels of single women were much lower than those of the married women.

Widowhood affects all family relationships. Shortly after the death of the spouse there is usually a flurry of support from family members, but this support is not very enduring. Within a few weeks, widowed persons usually find themselves on their own. Most widowed people live alone, and their major resource in coping with loneliness is most often friends rather than family. Widowed parents must adjust to a new relationship with their children.

In 1991, about 6 percent of the older population was divorced, and about 10,000 older individuals get divorces every year in America. Yet there has been very little research on the impact of divorce on older people or on the impact that being divorced may have on an older person's life. For example, many women who are divorced are effectively deprived of income in later life. In many cases, they are not entitled to private pension benefits. If they are not entitled to retirement benefits in their own right, they are forced into financial dependency. Cooney (1989) found that recently divorced women age 60 or over were much more likely to coreside with an adult child (43 percent) than were recent widows (21 percent). Part of this is probably related to financial factors, but part is also probably related to the emotional devastation that divorce can bring. Divorce of older parents can certainly be expected to have an impact on relations with adult children and other kin (Bulcroft and Bulcroft 1991). This topic is one of the most neglected areas of research in social gerontology.

Uhlenberg and Myers (1981) reported that recent increases in divorce rates are not restricted to young adults. Even though divorce rates among older people are still substantially lower than those for people in their early 20s, there has been an increase in the percentage of people entering later life who are divorced. For example, in 1960, only 2 percent of those 65 to 69 were divorced, compared with over 6 percent in 1991. Uhlenberg, Cooney, and Boyd (1990) predicted that the proportion of elders who were divorced, especially among women, will increase rapidly over the next several decades.

Older Parents and Adult Children

For most older people, particularly those who live alone, their relationships with their adult children are extremely important. There are many myths about how older people and their adult children relate in American society. It is erroneously thought that older people are abandoned by their children, that they seldom see them, that older people in nursing homes have been rejected by their families or have no family, and that older people are mainly takers rather than givers of aid. This section is an overview of what we know about how older people and their adult children tend to relate. It identifies important dimensions of these relationships and provides an idea of what is typical, so that you can put personal observations and experiences into context.

About 80 percent of the older American population has living children. About 10 per-

cent of older Americans have adult children who are also 65 or over. Most older Americans are not isolated from their children. In fact, older people and their adult children have many kinds of relationships, which have been studied in terms of coresidence, family size, residential proximity, frequency of interaction, mutual aid, feelings of affection, and sense of filial duty or obligation.

Coresidence   When older adults live in multigenerational households with their adult children, and sometimes their grandchildren and other kin, their relationships with those adult children are very different from relationships with adult children who live in a separate household. Coward and Cutler (1991) used 1980 census records of more than 47,000 elders to develop the best data we have on coresidence. They found that 19.8 percent of older Americans lived in multigenerational households in 1980. As Table 8-2 shows, about 75 percent of elders who lived in multigenerational households lived in two-generation households, and about 25 percent lived in three-generation households. Of those who lived in two-generation households, a large proportion (61.6 percent) lived with their spouses at age 65 to 74, and, as could be expected, the proportion living with spouses in addition to other family generations dropped sharply with age to only 12.1 percent at age 85 and over. Likewise, the proportion of two-generation households that included parents of the elder also dropped sharply with age, from nearly 10 percent at age 65 to 74 to less than 1 percent at age 85 and over. But it is interesting that even at age 85 and over, there were two-generation households in which the *youngest* generation was 85 or older.

Most elders in two-generation households lived with sons or daughters, which comes as no big surprise, but the age patterns go in opposite directions. The proportion living with sons dropped from 48.7 percent at age 65 to 74 to 36 percent at age 85 and over. The pro-

Table 8-2 Percentage of elders living with specific family members in multigenerational households, by type of household and age of elder: United States, 1980.

Relationship to Elder	Total Sample 65+ N 47,286	Two-Generation Households Age of Elder				Three-Generation Households Age of Elder			
		Subtotal 65+ 35,534	65–74 21,460	75–84 9,720	85+ 4,354	Subtotal 65+ 11,752	65–74 6,658	75–84 3,817	85+ 1,277
Spouse	41.6	46.9	61.6	30.2	12.1	25.4	35.7	14.0	6.2
Parent	5.6	6.2	9.1	1.7	0.8	3.4	5.4	0.7	0.7
Sibling	4.2	5.1	6.2	4.2	2.0	1.4	1.7	0.7	0.7
Son	44.2	45.4	48.7	42.6	36.0	40.1	41.0	39.0	38.8
Son-in-law	12.3	6.7	2.3	10.9	18.6	29.3	25.9	34.2	32.7
Daughter	44.3	39.3	33.8	44.0	55.8	59.6	62.0	57.5	53.9
Daughter-in-law	11.1	5.4	2.1	8.3	15.5	28.3	25.0	32.0	34.8
Grandchild	30.7	8.9	10.0	8.6	4.4	96.3	95.3	98.3	95.5
Other kin	10.0	8.7	8.1	9.8	9.7	13.8	11.3	15.3	22.6
Nonkin	2.7	2.6	2.5	2.8	2.3	2.9	2.9	3.0	3.0

Source: Adapted and abridged from Coward and Cutler (1991).

portion living with daughters increased from 33.8 percent at age 65 to 74 to 55.8 percent at age 85 and over. The proportion of elders living in households that included sons-in-law or daughters-in-law also increased with age of the elder. These data showed more coresidence with sons than is generally recognized in the literature.

When elders lived in two-generation households with grandchildren, they were probably serving as surrogate parents. The age data suggest this, in that the proportion living with grandchildren in two-generation households dropped from 10 percent at age 65 to 74 to 4.4 percent at age 85 and older.

When elders lived in three-generation households, they were much less likely to be living in them with their spouses. Nearly all of these three-generation households involved either a son or a daughter, and the proportion that involved sons-in-law or daughters-in-law was much higher than for the two-generation households. More than 90 percent of these three-generation households also included grandchildren. Interestingly, the proportion of adult daughters in these

households actually dropped as age of the elder increased, from 62 percent at age 65 to 74 to 53.9 percent at age 85 and over. The proportion of sons showed no appreciable pattern, staying at around 40 percent for all age categories of elders. The proportion living in three-generation households with kin other than adult children, siblings, or parents increased significantly with age of the elder, from 11.3 percent at age 65 to 74 to 22.6 percent at age 85 and older. The proportion living with nonkin in multigenerational households was small (2.7 percent) and unrelated to age of the elder.

The census data on coresidence tell us very little about the nature of the relationship between generations; they only tell us that people are living together. If we look at the patterns of coresidence across the life course of the parents, we get a mixed picture of the extent to which dependency of either generation plays an important role in the need to coreside and the types of events that precipitate coresidence. For example, based on data from a 1975 national survey of over 2,000 elders, Crimmins and Ingegneri (1990) found that 56 percent of adult children who lived with their parents had never lived apart from them. Of the 44 percent who had reestablished coresidence, 52 percent had done so to meet the needs of the adult child, 17 percent as a result of the older parent's widowhood, 12 percent as a result of deteriorating health of the older parent, and 6 percent due to declining health of the adult child.

Based on the 1987–1988 national Survey of Families and Households, Aquilino (1990) examined coresidence patterns among nearly 5,000 parents with living children age 21 or older. He found that dependency was not a major factor in explaining coresidence at *any* point in the life course. A large proportion of coresidence was in the household of the older parents, and at all ages older parents were more likely to provide a home for adult children than vice versa. He concluded that only parents with unmarried adult children were at all likely to experience coresidence. Health of the parents was not correlated with coresidence. Daughters were more likely to live with older parents, whereas sons were more likely to live with younger parents.

Among parents age 65 or older and living with adult children, Aquilino found that 75 percent lived in the elder's household. The 25 percent of older parents who lived in the adult child's household were likely to be unmarried and to have low incomes and little education. Among nonmarried older parents, mothers were likely to live with adult children, whereas fathers were not. Older parents who had gone through a marital dissolution and remarriage were much less likely to coreside with adult children. On the other hand, parents whose households included extended kin were much more likely to have adult children also living in the household.

Research of both Crimmins and Ingegneri (1990) and Aquilino (1990) indicate that dependency of elders is not a major cause of coresidence in the general older population. However, data from other studies have shown a strong association between dependency needs of older parents and coresidence. For example, Brackbill and Kitch (1991) looked at the factors that precipitated coresidence in a small sample of fifty-seven adult child–older parent pairs and found that 72 percent of the cases involved health declines among the older parents; 37 percent involved death or incapacity of a previous caregiver; 37 percent involved the adult child and older parent wanting company; and 26 percent involved financial considerations. Crimmins and Ingegneri reported that in a 1984 national sample of over 11,000 elders, 36 percent of elders who lived with adult children did so either due to the older parent's health or the need to share income.

In a large national sample, Cooney (1989) found that recently divorced mothers age 60 and over were much more likely (43 percent) to coreside with an adult child than those who were recently widowed (21 percent). From 25 to 30 percent of long-term older widows or divorcees lived with an adult child, compared with less than 20 percent of married older women. Cooney suggested that the financial repercussions of divorce could be a precipitating factor in coresidence, but that would not explain why a smaller proportion of long-term divorcees live with children, since financial resources of older divorcees are not likely to improve over time. Perhaps a more likely factor is the need for emotional support on the part of older women who have been left by their husbands. In such cases, coresidence may be a temporary situation, but longitudinal data would be needed to evaluate this possibility. Among recently divorced older mothers who lived with an adult child, 55 percent lived with daughters and 45 percent lived with sons, again a departure from the common conception in the literature that older women are unlikely to rely on sons (Lopata 1973).

Chappell (1991a) studied a sample of 400 coresident elders in Canada. She found that their living situations fell into three types:

(1) *dependent* elders who lived in an adult child's household and depended on them for assistance (22 percent); (2) *traditional* intergenerational households in which elders both give and receive assistance and support (56 percent); and (3) *independent* coresidence in which the elder lives with a sister or friend and tends not to expect assistance, but gets it nevertheless (22 percent). In the latter category, older women were more likely to live with sisters than friends, and men were more likely to live with friends than siblings.

Taken together, the results of these varied studies suggest that a large proportion of coresidence consists of lifelong living patterns that are not the result of a crisis on the part of either the older parent or the adult child. Another category consists of adult children who move back in with an older parent in response to problems the adult child is having, particularly divorce, declining health, and financial difficulties. Then there are cases in which the older parent experiences a crisis such as divorce, declining health, or financial difficulty. Finally, there are cases in which older parents and adult children coreside in order to keep one another company and to pool financial resources. We cannot at this point attach firm proportions to these various scenarios, but it is clear that there is no single path to coresidence. In addition, each type of coresident situation could be expected to have different implications for adult child–older parent relationships.

Brackbill and Kitch (1991) looked at the dynamics involved in decisions to coreside as well as decisions to terminate coresidence in cases where caregiving was involved. They interviewed fifty-seven coresident pairs involving an elder and another family member, nearly all of whom were adult children. They also interviewed sixty-one pairs who had previously lived together but were now living apart. They found that the decision to coreside was usually the result of a perceived crisis, most often connected to an elder's need for care, which prompted adult children to

act on their feelings of filial responsibility by proposing that the elder move in with them. The decision to end coresidence was usually connected to a combination of factors: deteriorating functional status of the elder, realization by the caregiver that caregiving is more difficult than anticipated, conflicts for the caregiver between caregiving and employment, excessive dependency or demands on the part of the elder, and conflict between the elder and the caregiver or members of the caregiver's family.

Brackbill and Kitch found Emerson's (1962, 1972) version of *exchange theory* useful in explaining these dynamics. Emerson's exchange theory grew out of the behaviorist school in psychology in which reinforcement (rewarding) of behavior plays a central role in perpetuating interaction between people. In Emerson's view, relationships depend on the capacity of the actors to mutually reward one another with something of value. *Exchange resources* used to reward others can include assistance, money, information, affection, approval, labor, compliance, or property. According to this theory, if one actor has a lower capacity to reward the other, then the actor with the lower exchange resources is assumed to be more *dependent* in the relationship than the other, and the actor with greater exchange resources has more *power* in the relationship. According to Emerson, if exchange relations get too far out of balance, they can become unstable and relationships will disintegrate.

Brackbill and Kitch found these exchange theory concepts helpful in understanding why some coresident situations were congenial and stable and why some were terminated. Elders who continued to live with their adult children had brought more to the exchange than elders who no longer lived with their adult children. Elders who continued to coreside were generally less demanding in terms of assistance; they tended to help the family financially, thus allowing caregivers to cut down on their need for employment; and they tended to have more activities outside the household, which reduced interaction demands from the elder within the household. Elders who no longer lived with their adult children tended to be more demanding and to have less to give. They needed more assistance; they were less likely to help the family financially, which meant that caregiving was more likely to conflict with the need for employment on the part of the caregiver; and they were less likely to have activities outside the household, which increased their need for interaction within the household and increased their potential for conflict within the household over this need. When the elder could bring more of value to the exchange, particularly financial help, the adult child was more likely to feel that her or his lost autonomy, independence, and privacy were counterbalanced. But when the elder brought less to the exchange, feelings of filial responsibility were not enough to prevent a growing feeling on the part of the adult child that caregiving was mostly a losing proposition.

Because their sample was small and nonrepresentative, Brackbill and Kitch's findings must be used with caution, but their research illustrates the value of bringing social theory to bear on the dynamics of social relationships. Certainly, exchange theory shows promise as a framework for future research on coresidence as well as other aspects of relationships between older parents and their adult children.

Family Size The number of children older people have is another important context for adult child–older parent relationships. Obviously, older people with no living children cannot rely on this type of relationship. But what effect does the *number* of adult children have on older adults' interactions with children, their capacity to get assistance from them, or the quality of their relationships with adult children? Uhlenberg and Cooney (1990) looked at mother–child relationships from two perspectives. First, from the point

of view of the adult children, how does the number of siblings influence the probability of coresidence, frequency of interaction, and perceived quality of the relationship with their mother? Second, from the point of view of the mothers, how does the number of adult children influence interaction frequency and perceptions of availability of help and support?

Uhlenberg and Cooney found that in terms of coresidence and interaction frequency, having a larger number of siblings allows for some substitution to occur and, overall, the larger the number of siblings, the lower the proportion living with the mother, visiting her frequently, or communicating with her frequently. Nevertheless, the total amount of visiting and communication for mothers with four or more children was still much greater than for mothers with only one or two children. In addition, the larger the number of siblings, the greater the proportion who perceived their relationship with their mother to be of high quality. For daughters who did not coreside with their mothers, being an only child of a mother in poor health was a strong predictor of seeing the relationship as being of less than high quality. However, even in this type of case, only 25 percent of daughters perceived the quality of relationship as low. On the other hand, daughters who lived with their mothers rarely saw the relationship as anything other than high quality. Generally, less than 20 percent of the respondents felt that they had a poor-quality relationship with their mothers.

The total amount of interaction mothers had with their adult children was directly proportional to their number of adult children. Mothers with just one adult child were in contact at least weekly in 75 percent of cases compared with 88 percent of the time for mothers with four or more children. Face-to-face visits at least weekly were even more influenced by number of children; 46 percent of mothers with one child saw them at least weekly, compared with mothers with four or more children, 79 percent of whom were likely to see at least one of their children each week. The same patterns held for assistance and support. Compared with mothers of single children, mothers with four or more children were more than three times as likely to be receiving help with transportation, house or car repair, or housework; they were also much more likely to see a child as potentially available in case of emergency, financial need, or need for advice. Not surprisingly, proximity was the most important determinant of visiting and communication, but number of children was a more important predictor of who actually received help. We will return to the issue of proximity to children shortly.

Although it represents only one study, Uhlenberg and Cooney's research was based on a large national sample; therefore, we can be reasonably comfortable with their conclusion that large family size has a positive influence on older mother–adult child relationships in terms of quality of relationship, interaction frequency, actual help received by older mothers, and perceived potential help available to them. Large numbers of siblings result in less interaction and helping for each of the siblings, but the net effect is still greater interaction and help from the point of view of the older mothers. Quality of the mother–adult child relationship tended to be seen as high regardless of family size, but high quality was a more prevalent perception among adult children with several siblings.

Hoyert (1991) looked at factors related to the probability that elders receive household or financial aid from adult children. She found that the probability of receiving both kinds of aid increased with the number of children, and this was true for both mothers and fathers. Now let us look at the effect of residential proximity between older parents and their adult children.

Residential Proximity Older people prefer to live near their children, "near" usually being defined functionally in terms of one hour's travel time rather than geographic distance. Estimates of the proportion of elders who have living children and at least one of them living within one hour's travel vary widely, from 51 percent (Mercier, Paulson, and Morris 1989) to 72 percent (Frankel and DeWit 1989).

Frankel and DeWit found that as the geographic distance between older parents and their adult children increased, face-to-face interaction and visiting declined sharply. Visits and phone calls dropped precipitously with increased distance from one to four hours' travel time, with little change beyond four hours' travel. On the other hand, contact by letter and overnight visits increased in direct proportion to distance. However, distance had no relationship to the quality of the adult child–older parent relationship (Uhlenberg and Cooney 1990) or capacity to have important conversations (Frankel and DeWit 1989).

There may be a period in the life course—when parents are middle-aged and adult children are beginning their own households and occupational careers—when the two generations are more geographically distant, especially among those with higher income, more education, and white Anglo-Saxon ethnic background (Uhlenberg and Cooney 1990). But as adult children "settle down" and get older, generations tend to live closer to one another. For example, Speare and Meyer (1988) found that more than 25 percent of movers age 75 or older mentioned kinship as a reason for migration. More research is needed on life course patterns of residential proximity between generations, as opposed to studies that look at proximity only at one point in time.

Interaction Frequency Interaction frequency tends to be high among older parents and their adult children, but it appears to be dropping over time. Crimmins and Ingegneri (1990) reported that the proportion of elders having daily interaction with adult children dropped from 51 percent in 1962 to 43 percent in 1975, to 34 percent in 1984. However, the proportion having contact at least one to two times a week dropped less significantly, from 76 percent in 1962 to 72 percent in 1975, to 63 percent in 1984. Part of this shift is probably the result of improved economic circumstances among older people since 1962. Frankel and DeWit (1989) studied a 1983 Canadian sample and found that only 15 percent of elders saw an adult child daily, but 47 percent saw an adult child at least once a week. Telephone contact was more frequent, with 55 percent of elders reporting at least weekly contact. These data from various sources all show that older people generally have frequent contact with their children.

However, this general trend masks important variations. For example, married older women interact less with their children than do those who are widowed or divorced (Atchley, Pignatiello, and Shaw 1979). Having an unmarried child nearby and having health problems are strong predictors of frequent face-to-face contact (Frankel and DeWit 1989). Daughters are much more likely to keep in frequent contact by phone than are sons.

There also remains the issue of the *nature* of the contacts. Although frequent, intergenerational contacts are often brief encounters to "pass the time of day." If older parents are ill or disabled, the contacts are a way of checking that all is well, and this sort of *monitoring* probably increases with the age of the parent. These frequent contacts mean that adult children are apt to learn quickly if their parents need something or have a problem.

Mutual Aid   Many researchers consider mutual aid the crucial intergenerational dimension. Aid flows in both generational directions and may consist of services such as baby-sitting or housework, information and advice, moral support for various decisions, or money and gifts.

Peterson and Peterson (1988) found that a majority of both generations felt that their intergenerational exchanges were equitable over the long run.

Bengtson, Rosenthal, and Burton (1990) concluded that healthy older people are not primarily dependent recipients of aid; in many cases they are primarily donors. Indeed, recent research has documented substantial aid flows from older parents to adult children (Greenberg and Becker 1988; Hoyert 1991). In general, when adult children have stressful problems, older parents are apt to assume more responsibilities. For example, parents provide much of the care of developmentally disabled adults, and parents provide 85 percent of the family care of deinstitutionalized mentally ill adults (Lefley 1987). When adult children divorce, parents often provide housing and financial assistance, at least for a time. Bankoff (1983) found that parents were by far the most important source of emotional support to women who became widowed in middle age, much more important than children or friends. Widows without strong parental support were more depressed, and no other source of support compensated for the lack of parental support. Bankoff attributed this to the younger widow's need for nurturance, which parents are best able to provide.

Hoyert (1991) looked at intergenerational household and financial aid flows from older parents to adult children and vice versa and found that the balance very much related to coresidence, proximity, age of both the adult child and the older parent, and marital status of both the adult child and the older parent. For parents and adult children who live together, there was a great deal of exchange of household help and pooling of financial resources in both generational directions. For parents and adult children who do not live in the same household, intergenerational aid patterns were considerably different for parents under age 75 to 80 compared with those who were 80 or older. Young-old parents tended to give more aid, and old-old parents tended to receive more aid.

About 33 percent of older parents gave household aid to at least one adult child. Parents between 65 and 74 were the prime givers of household aid, and the recipients tended to be young adults (age 19 to 29), the life stage at which most married and previously married young adults have small children in the household. The parents who gave household aid also tended to be middle income and white. These parents sometimes traveled more than 150 miles to provide household aid, which suggests that there may be differences between day-to-day aid and periodic aid, but this topic has not been studied. Older parents were not likely to give household aid to never-married sons who lived in separate households.

Financial aid to adult children was provided by 32 percent of older parents. Most of those who gave financial aid were married, under age 80, and middle-to-upper income. The adult children aided tended again to be young adults, particularly previously married daughters and sons, and never-married daughters. Divorced or widowed mothers were less likely to provide financial aid, probably because their incomes did not allow it. In terms of financial transfers, Soldo and Hill (1993,199) concluded: "More help flows from parents to children than the reverse . . . Parents rarely receive financial transfers from [adult]children . . . past patterns of transfers tend to repeat themselves, persisting over time rather than exhibiting reciprocity." Older parents who give financial aid to their adult children generally do not expect to be repaid; they expect their adult children to in turn help their own children when they reach adulthood. However, after age 75 to 85, patterns of financial transfer to adult children tend to disappear.

About 52 percent of older parents received household aid from an adult child. Household aid from adult children was positively related to residential proximity, and

household aid to older parents was more likely for blacks than whites. The adult child household aid providers were generally older (age 60 or over), and the parents who received household aid were most likely to be age 80 or older and to have annual incomes of $7,000 or less. Financial aid from adult children went mostly to low-income widowed or divorced mothers who were age 80 or older. Financial aid was also more likely to be provided by blacks than whites.

As a general rule, parents seem to give to their children in one way or another as long as they are able. Looking at older parents' perceptions of the balance of mutual aid, Morgan, Schuster, and Butler (1991) found that parents see themselves as giving more than they receive until age 85 or older. A shift from this pattern usually coincides with a deterioration in the parents' financial or health condition.

Seelbach (1977) found that the more vulnerable older people were, the more aid they expected and received from their adult children. Older respondents who were widowed or divorced women, who had low incomes or were in poor health, expected and received more aid. Blacks tend to receive more aid than whites (Mitchell and Register 1984; Hoyert 1991). The relatively small proportion of older people not in couples who need personal care get most of it from their families (Stone, Cafferata, and Sangl 1987).

Whether or not older parents are satisfied with the frequency of interaction and mutual aid patterns with their adult children seems to depend on what they expect. Kerckhoff (1966) found that older people fell into three types of orientation toward family relationships. Older parents with an extended family orientation (20 percent) expected to live near their children and enjoy considerable mutual aid and affection. Those with a nuclear family orientation (20 percent) expected neither to live near their children nor to be aided by them. And those with a modified extended family orientation (60 percent) believed in mutual aid and affection but took a middle-of-the-road position with regard to interaction frequency and extent of aid and affection. In actual experience, most of Kerckhoff's respondents had relationships that fit the modified extended family orientation. Not surprisingly, older parents with an extended family orientation tended to be disappointed, the middle-of-the-roaders were getting what they expected and were satisfied, and those with a nuclear family orientation were pleasantly surprised at the amount of interaction, aid, and affection between generations.

More recent research supports Kerckhoff's formulation. Markides and Krause (1985–1986) found that older Mexican-American respondents expected more from their children than the children were able or willing to give. On the other hand, Peterson and Peterson (1988) found that older Anglos tended to feel that they got more aid than they expected. Thus, culturally shaped expectations are an important contributor to satisfaction with intergenerational relationships.

Talbott (1990) looked at how widows felt about their patterns of mutual aid with their adult children and found that 51 percent had no negative comments to make about these relationships. Even those who had negative comments still had mostly good things to say. But there was a negative side. Some felt neglected, unappreciated, and dissatisfied with the amount of help they received; were afraid of being a bother to their children; and were too emotionally dependent on their children.

However, the most frequently mentioned problematic aspect of the older parent–adult child relationship had to do with how much aid these widows gave their adult children and their families and how little they perceived they got in return, particularly in terms of easy entree into their child's household. Some adult children apparently took for granted the substantial baby-sitting, transportation, laundry service, meals, and cash they got from their mothers and responded with a marked lack of acknowledgment or appreciation. Why these adult children, who are in a small minority, respond this way is unclear. Talbott concluded that the mothers perceived high levels of contribution to be the price paid for involvement in the adult children's lives and that the adult children resisted letting their mothers too far into the life of their households. The adult children may feel guilty about taking so much from their mothers and try to minimize the issue by ignoring it. They may have a nuclear family orientation in which too much involvement by their mothers is seen as inappropriate. Or they may simply see these patterns as unremarkable continuations of lifelong contributions of parents to their children. In any case, the mothers in these cases feel unwelcome in their children's households despite providing much aid to them. They feel that they irritate their children, and that they need their children emotionally more than the children need their money or services, so the mothers pay the price and make themselves subservient in the relationship.

To Talbott, an unfortunate side effect of this situation is loss of self-esteem and morale among these mothers. The point is that in exchange relationships, both parties have to value the exchange. It seems that in this minority of negative cases, the increased involvement in the child's household desired by the mothers is a price their children are reluctant to pay, even in return for valuable services and money. Nuclear household autonomy is probably valued more highly in these cases than support to the mother or the services she can provide. However, this hypothesis needs to be tested. One would also expect that the mothers who would be most likely to experience this problem would be those with extended family orientations and a poorly developed network of friends. Indeed, as we will see later, contact with friends is more effective in positively influencing elders' feelings of well-being than is contact with adult children per se (Lee and Ishii-Kuntz 1987).

Affection and Regard Proximity, interaction frequency, and mutual aid are important indexes of older parent–adult child relationships. However, *qualitative* aspects such as degree of closeness or strength of feelings (affection or dislike) may be even more revealing. Investigators have only recently started to explore these more intangible aspects of kin relationships.

It is commonly assumed that closeness is synonymous with liking or loving and that distance indicates negative feelings—that we love those relatives we feel close to and hate those we feel distant from. But feelings that run high are rarely only positive or only negative; where love can be found, so can hate. Probably most family relationships ebb and flow.

Most parents and children report positive feelings for each other. For example, Talbott (1990) found that just over half of widows had only good things to say and only a small minority had strong negative feelings about their relationships with their adult children. Hagestad (1984) found that family communications were actively managed to enhance feelings of affection. Most families had a forbidden zone of topics that could potentially lead to conflict, and most families consciously avoided these areas.

Parents remain important to their children throughout the life of the children. When adults of all ages were asked to describe a person, they tended spontaneously to refer to their parents more frequently than to any other person (Troll 1972). The oldest members of Troll's sample, in their 70s and 80s, were still using parents as reference persons.

Johnson and Bursk (1977) found that ratings of their relationship provided by both older parents and one of their adult children were highly congruent. Occasionally, the parents rated their relationship higher than their child had. Both generations felt better about each other when the parents were in good health and able to be financially independent.

Harris and Associates (1975) found that as age increased, so did the percentage of older people who said they felt close enough to their children to talk to them about "things that really bothered them." At age 55 to 65, only 25 percent felt close enough, but at 80 or over, 43 percent felt this way. Glass, Bengston, and Dunham (1986) found that adult children's influence on attitudes of older parents increased with the age of older parents. Older people tend not to disengage from their children, regardless of the amount of affection present in the relationship.

Problems in the Relationship As we saw earlier, adult children often come to their parents for help and support as a result of a variety of problems the adult child is having, including such crises as divorce, drug dependency, and unemployment. Coping with an adult child's problems can have a negative effect on the quality of the relationship and on the morale of both generations. Pillemer and Suitor (1991) looked at the effect of adult children's mental and emotional problems, serious health problems, drinking or alcohol problems, and serious stress on symptoms of depression among the older parents. They found that having an adult child with one or more of these problems was significantly related to the older parent's having more symptoms of depression. Mothers and fathers were equally likely to show this pattern. Next to self-rated health, having an adult child with problems and having an adult child who created tensions and arguments were the strongest predictors of symptoms of depression among parents, even ahead of being widowed or divorced or not having a confidant. These results point up the importance of the quality of the relationship with adult children for the well-being of older parents.

Likewise, conflict with older parents and difficulties in coping with their dependency needs have potential negative effects for the well-being of adult children. Caregiver burden has been studied most often, and will be discussed in detail later. For now, just note that parental dependency has been shown to have negative effects: increased symptoms of depression, lowered self-esteem, and increases in other indicators of stress (Mancini and Blieszner 1989).

Finally, Strawbridge and Wallhagen (1991) found that 40 percent of the adult child, primary parent caregivers they studied were experiencing serious conflict with another family member. Such conflict was most likely to occur between siblings because the sibling was not providing what the primary caregiver thought was a fair share of help. This conflict had a negative effect on feelings of caregiver burden and prevalence of mental health problems.

In recent years, we have come to recognize that most families can transcend enormous difficulties and survive with strong relationships. But we have also learned that child abuse, elder abuse, dysfunctional families, and conflicted family relations are even more common than we thought. Thus far, research in social gerontology is just beginning to focus on these issues as they affect family relationships in later life. For example, Pillemer and Suitor (1991) note that in addition to looking at the simple existence of problems in relationships, we need to consider the duration and severity of the problems. There is much research to be done in this area.

Duty   Feelings of obligation or a sense of duty often underlie relationships between generations. Despite the large sacrifices sometimes required, a very large proportion of adult children attempt to meet their older parents' needs, however extensive. Some of this sense of duty may result from a wish to avoid negative opinions of others, but most seems to come from internalized norms of filial obligation.

Finley, Roberts, and Banahan (1988) studied norms of filial obligation in a probability sample of 667 adults with living older parents. They found that women generally subscribed to norms of filial obligation in higher proportions than men did, but the differences were very small. About 96 percent of their respondents agreed that contact with parents should be maintained even when it is not convenient. More than 80 percent felt that adult children should be willing to share their homes with aging parents. Over 90 percent agreed that in whatever way necessary, adult children should take care of their parents when they are sick. Over 85 percent said that aged parents should be cared for by children rather than social agencies. About 95 percent felt that adult children should help their parents financially when necessary. But over 60 percent disagreed with the notion that adult children should live close to their parents. There was such overwhelming agreement with norms of filial obligation that attempts to look at the effect of factors such as affection, role conflict, and distance were only marginally successful.

Hamon and Blieszner (1990) studied 144 older parent–adult child dyads in terms of their mutual perceptions of filial duty. More than 95 percent of both parents and adult children thought that children should help their parents understand the resources available to them, give them emotional support, and talk over matters of importance. About 86 percent of both parents and adult children felt that adult children should give their parents advice and make an effort to be together on special occasions. However, the adult children were much more likely to think that they should be prepared to make sacrifices to meet their parents' needs than the older parents were. For example, 92 percent of adult children agreed that they should take care of their parents when they are sick, but only 64 percent of the older parents thought they should do so; 84 percent of children thought they should provide financial help if needed, but only 41 percent of parents agreed; and 61 percent of children thought they should adjust their work schedules to help, while only

42 percent of older parents thought so. Indeed, in Hamon and Blieszner's sixteen-item filial responsibility scale, there was no case in which the proportion of elders agreeing with a specific filial expectation was higher than the proportion of adult children subscribing to that expectation. But for eight (half) of the items, the proportion of adult children agreeing with the item was significantly higher than the proportion of older parents agreeing with it.

Thus, we know that most adult children feel obligated to maintain relationships with their older parents and to come to their aid if necessary. We also know that a large proportion conforms to these obligations.

McGrew (1991) studied the process adult daughters went through in responding to the caregiving needs of their older mothers. She found that caregiving was always a response to an *impulse to care*, a predisposition to caregiving that was rooted in women's self-concepts as nurturing and responsible people, their behavioral conditioning to gender roles over their life course, and moral norms of reciprocity and filial responsibility. With regard to reciprocity norms, the reference point was not what the mothers had done for the adult daughter in the past but instead what the daughters thought their mothers would do now for them if they had similar dependency needs. Interestingly, although all of the daughters felt obligated to provide care, they did not want their children to feel a similar obligation. This matches well the findings of Hamon and Blieszner (1990) that norms of self-sacrifice in relation to older parents were held much more by adult children than by older parents. McGrew also found that *guilt* played a major role as a monitor of adherence to filial responsibility, and that guilt avoidance was a major motivator and measure of the success of their striving to be "good" daughters. Thus, filial responsibility does not operate in isolation; it is part of a complex of ideas that serve to define the self, and as such it can be understood as an ex-

ample of the continuity theory of the self. (See Chapter 5.) To the extent that filial responsibility is an internalized part of the ideal self, it serves as a standard against which self-esteem is measured and therefore has great potential motivating power.

Filial responsibility is also expected by society. For example, McGrew concluded that expectations of filial responsibility are an important aspect of society's policies and practices in the area of long-term care. The recent push toward increased use of community-based long-term care under Medicaid is predicated on an assumption that families will provide the bulk of the care needed and that the state can conserve valuable Medicaid resources by relying on family care, with little thought about the full price the families pay. Discharge planners, social workers, and others also often operate on the assumption that adult children have the responsibility for making decisions about their parents' care, even when they know that the children do not understand the issues or alternatives available. This leads us to our next topic, adult children as caregivers.

Adult Children as Caregivers

We use the term *care* broadly here to mean ongoing assistance with a wide variety of functions ranging from periodic chores, housework, and transportation to twenty-four-hour monitoring and help with ambulation, bathing, and eating. As we saw earlier, a large percentage of married older people who need care receive it from their spouses. Adult children are generally considered to be the second major line of defense when ongoing care is needed.

There is considerable diversity in patterns of caregiving by adult children, and no single pattern can be considered typical. Some older parents are cared for mainly by daughters; some are cared for by sons; some are cared for in their own households by unmarried children who have moved in with them; others are cared for in the intergenerational

Table 8-3 Percentage of elders receiving care who are being cared for by adult children, by gender of adult child and gender, age, and marital status of care recipient.

| Care Recipients | Caregivers | | | | Daughter/Son Difference |
| | Daughters | | Sons | | |
	N	%	*N*	%	
Mothers	1,414	41.2	684	19.9	21.3
Age 65–74		32.0		16.4	15.6
75–84		43.5		20.6	23.5
85+		51.6		24.3	27.3
Marital status					
Married		23.7		12.3	11.4
Widowed		51.6		24.6	27.0
Divorced		39.7		28.8	10.9
Fathers	377	20.5	275	15.0	5.5
Age 65–74		15.2		11.6	3.6
75–84		21.5		14.8	6.7
85+		35.2		26.3	8.9
Marital status					
Married		16.2		12.3	3.9
Widowed		43.4		30.3	13.1
Divorced		22.2		13.3	8.9

Source: Adapted from Coward, Horne, and Dwyer (1992)

households of their married children; and still others are cared for by adult children who live independently and travel to the parent's household to provide care. Some older parents are cared for almost exclusively by one adult child; others are cared for by a network of adult children who share responsibilities.

The proportion of older parents cared for by adult children varies widely by gender of the parent, gender of the adult child, and the older parent's age and marital status (see Table 8-3). Overall, about 41 percent of mothers receive care from adult daughters and about 20 percent receive care from sons, whereas about 20 percent of fathers receive care from daughters and 15 percent from sons. This mother–father difference is, of course, due in large part to the interaction of age and marital status on the availability of spousal caregiving. Older men tend to have spouses to care for them even at advanced

ages, whereas older women do not. *About 93 percent of older married men and 80 percent of married women who receive care are cared for by their spouses.*

From Table 8-3 we see that there are major differences between sons and daughters in terms of parent care, with sons less likely to provide care for all categories of parents. However, the gap between the percentage of sons and daughters providing care is smallest with regard to care of fathers under age 85. The greatest proportion of care is given by adult daughters to widowed mothers, but even this common category only accounts for about 33 percent of all care to older women (Coward, Horner, and Dwyer 1992). Nevertheless, because more than half of care to widows age 85 or older is provided by their daughters, the mother–daughter dyad has received the most attention in the research literature on caregiving of older parents by adult children, to the exclusion of caregiving by sons and the wider network of adult children.

In a national sample of caregivers, Stone, Cafferata, and Sangl (1987) found that adult daughters caring for older parents were most likely to help with shopping and transportation (91 percent); household tasks (87 percent); bathing, dressing, and eating (69 percent); handling finances (59 percent); administering medications (57 percent); and assistance with ambulation (44 percent). Adult sons who provided care to older parents were more likely than adult daughters to be involved in providing transportation and help with mobility and less likely to provide other types of care. Nevertheless, at least half of all caregiving sons provided all types of care. On average, adult children provided about four hours of care per day. These findings generally apply to caregivers who do not live in the same household with the older parent. When adult children care for coresident older parents, the intensity and number of care tasks tends to increase.

Under what conditions do adult children assume responsibility for caring for their older parents? In many cases, particularly those involving close residential proximity and frequent contact between generations, responsibility for help with shopping and transportation may be assumed gradually with no definable starting point. But when a major time commitment is made to providing care to an older parent, some form of decision making usually takes place. Usually, in response to some change in the functional status of the older parent, but sometimes in response to social changes such as widowhood, there is a perception that the parent's situation has changed, that a new approach to meeting her or his needs is required, and that this new approach requires ongoing assistance from an adult child (Cicirelli 1992).

McGrew (1991) contended that in most cases, daughters do not go through a complex decision-making process to decide *whether* to provide care but simply act on their impulse to care, a predisposition women have to provide care to family members who need it. However, in deciding *how* best to provide care, daughters seek to balance their open-ended impulse to care against the practical limits of what is possible. They are often at a disadvantage in making good decisions because they often must make decisions in a hurry, they do not know how the situation will progress in terms of their parent's changing need for care, and they have poor information concerning the care alternatives available to their older parent. McGrew's respondents cited many cases in which social workers, hospital discharge planners, and other health service workers failed to provide adult children with the information needed to understand the long-term care alternatives available as well as how to gain access to them. Haste, uncertainty, and lack of information, along with an unchecked impulse to care, can seriously erode the adult child's capacity to make rational decisions about caregiving.

Is the decision to provide care a joint decision between the adult child and the older parent or primarily one made by the adult child? Cicirelli (1992) found that adult daughters generally wanted their mothers to participate in the decision-making process and to retain as much autonomy as possible, but older mothers felt that it was mainly the adult daughter's decision to make. Adult daughters who were most likely to take over decision making were those from working-class backgrounds who had grown up in traditional paternalistic households. Paternalistic decision making was also more likely with regard to health and financial decisions than with decisions about everyday living. Cicirelli concluded that the daughters' beliefs about the appropriate decision-making relationship governed their behavior, regardless of the mother's beliefs, which is itself a strong statement of the relative power of daughters in these situations.

Table 8-4 Types of family support structures.

Social-Emotional Support		Household Support	
		None	*At Least One*
	None	Isolated 12.2%	Obligatory 1.1%
	At Least One	Modified-extended 40.9%	Traditional 45.8%

N = 910
Source: Adapted from Silverstein and Litwak (1993,260).

McGrew (1991) found that most of her respondents spoke in terms of "parenting one's parent." Seltzer (1990) argued persuasively that the legal and social authority relationship in the adult child–older parent situation in no meaningful way compares with the adult parent–minor child authority relationship. The older parent has legal, social, and moral rights that a child simply does not. Nevertheless, the use of the expression, "parenting one's parent," describes a role reversal that many feel occurs when adult children assume responsibility for caregiving. Some of this usage may be an attempt to justify paternalistic decision making, and some of it may simply reflect the newness of parent caregiving in our culture as well as a failure to develop adequate language to describe the actual authority relationship. An additional point of confusion probably comes from the way daughters develop conceptions of the role of caregiver to their older parent. McGrew concluded that adult daughters do not so much reverse their relationship with their older mothers as they *borrow* ideas about how to be a caregiver from their experience in the role of mother. We could expect that the results of borrowing elements of the mother role for application in a caregiving situation might be a different process than simple role reversal implies. This is an excellent topic for further research.

Silverstein and Litwak (1993) developed a fourfold typology of caregiving by looking at the types of support tasks performed by elders' primary adult child helper. Through factor analysis, they divided the tasks on two dimensions: *household tasks* (doing laundry, housekeeping, fixing things, and managing money) and *social-emotional support tasks* (keeping in touch, exchanging gifts, cheering up, and sharing meals). When these two dimensions were cross-classified, they formed four family service structure types, as shown in Table 8-4. They found that 12.2 percent of the 910 elders in their sample were *isolated*, receiving neither social-emotional nor household support from adult children. Only 1.1 percent fit the *obligatory* pattern, in which they received household support only and no social-emotional support. Most elders fit the *modified-extended* model (40.9 percent), in which they received only social-emotional support; or the *traditional* pattern (45.8 percent), in which they received both social-emotional and household support.

Silverstein and Litwak found that elders who were more disabled and were not married were significantly more likely to be in the traditional rather than the modified-extended service structure type. In other words, parents who were alone and disabled mobilized household help, whereas married and nondisabled parents tended to receive social-emotional support only. Other factors related to having a traditional family support structure included being Hispanic, having a large number of adult children, living closer to adult children, and having unmarried adult children.

How effective do adult children feel they are in meeting their older parents' caregiving needs? Noelker and Townsend (1987) found that about two-thirds of their sample of adult child caregivers felt that the family was doing an effective job of caring for their older parents. Having a less impaired older parent, using community support services, having other family and formal support workers who cooperate in caregiving, and having a positive view of the care situation all had a positive influence on perceptions of care effectiveness. Negative perceptions of care effectiveness were most likely when the caregiver felt distressed and overwhelmed and when the older parent was more impaired, which was related to greater difficulty in finding and using community support services. Unfortunately, the more impaired the elder, the more likely he or she was to be reluctant to accept community support service workers as caregivers, which created additional difficulties for the adult child caregivers. Caregivers were particularly likely to see the situation negatively when the parent was perceived to be too dependent, too demanding or critical, or unappreciative of the child's efforts, and when the adult child felt that no matter what she or he did, it would never be enough.

Mui (1995) compared sons and daughters in terms of the amount of strain they felt in the caregiver role. She found that daughters were more likely to experience strain because they were caring for more impaired parents, had fewer resources, and felt that caregiving interfered more with their personal and social lives. Sons who experienced these conditions were just as likely as daughters to report high levels of strain in the caregiver role.

Managing guilt feelings is a major aspect of caregiving for most adult child caregivers. McGrew (1991) reported that none of her respondents had fully avoided feeling guilty about their performance as caregiver. Likewise, Noelker and Townsend (1987) found that avoiding guilt was a major goal in the caregiving process for most caregivers. Why do caregivers, especially daughters, have so much trouble with guilt? Noelker and Townsend

cited three possibilities. First, guilt may be an important reinforcer of norms of filial responsibility. Second, the travails of caregiving may result in declining feelings of affection for the parent, which could precipitate guilt. Third, caregivers may expect themselves to be able to bring the same total commitment and devotion to the care of an older parent that they brought to the care of their children, an impossible task which can lead to guilt. McGrew pointed out that daughters tend to model their caregiving for their parents after their behavior as mothers. Livson (1976) found that most mothers feel guilty about their behavior as mothers, so it would be unsurprising if the same set of unrealistic expectations, carried over into an even less controllable situation, resulted in guilt. More research is needed on the causes and consequences of guilt among caregivers and potential interventions that might blunt guilt's negative impact on the mental health of caregivers.

In summary, caregiving for older parents is becoming a common phase of the family life cycle. Most older adults who cannot be cared for by a spouse can get care from one or more adult children and are satisfied with the quality and quantity of care they get. Most adult children feel that they are doing an effective job of caregiving and are not unduly stressed by having to provide care (Noelker and Townsend 1987). However, as the amount and intensity of care increases, as the age of the adult child caregivers increases, and as difficulty in finding people to help increases, so does the stress connected to caregiving. Accordingly, as our older population ages, the proportion of adult child–older parent caregiving situations that encounter difficulty can be expected to increase.

Other Factors The findings on various dimensions of relationships between adult children and older parents present a generally positive picture of physical nearness, frequent visiting, aid as needed, high mutual regard, and dutiful adult children. However, there are major gender, social-class, and probably ethnic variations. Older women tend to have more intergenerational relationships than older men do. Some researchers contend that older women are more family oriented and thus have more contact and receive more aid, while others find little gender difference. This area needs more study.

Social class seems to have its greatest impact on the flow of mutual aid. Affluent older people tend to continue to aid middle-aged children, while blue-collar older people are more likely to be on the receiving end. However, more research is needed on class differences in family interaction patterns.

Ethnicity is a prominent influence on family life for many Americans. For example, Markides and Krause (1985–1986) reported that intergenerational patterns among Mexican-American families were significantly different from those of Anglo families. As mentioned earlier (see *Mutual Aid*), older Mexican-American respondents expected more than their adult children could or would give; they were also more likely than other groups to be disappointed with what they experienced. Yet the older Mexican Americans were more likely to be involved in strong helping networks with their children. Older Mexican-American women were more likely than Anglos to be heavily relied on by their adult children, which often resulted in high levels of psychological distress. Gibson (1986) reported that although black and white older people tended to turn to family for support in about equal proportions, older blacks did not rely as exclusively on spouses and adult children. Older blacks selected from a more varied array of family helpers. Fictive kin, people who are not legally related to one another but who treat each other as kin, are more common among blacks. For example, in one study among blacks over 60, 47 percent had raised children other than their own and reported equal proportions of fictive "grandchildren" (Bengtson 1985).

Geographic mobility is often presumed to be prevalent in American society and to lead to the breakup of extended families, but research shows that 85 percent of older people live near at least one adult child. Nevertheless, almost half of older people in one study had at least one child living over 150 miles away (Moss, Moss, and Moles 1985). These relationships were maintained more through quality of interaction than quantity. Moss and her coworkers found that frequency of visits and telephone calls declined sharply with increased distance between older parents and their adult children, but the frequency of letters increased. Despite less frequent contacts, close ties between older parents and adult children can be maintained effectively at a distance (Litwak 1981).

Brody (1985) contended that it is becoming normal for middle-aged women to experience being "women in the middle," squeezed between the demands of employment and the dependency needs of both their children and their parents. Data on typical childbearing patterns and the typical age of onset of physical or financial dependency suggest that the average family should not experience such a squeeze. By the time older parents begin to need assistance, the children should be out of the middle generation's household. Indeed, Spitze and Logan (1990) found that the combination of full-time work, active parenting, and caregiving for parents applied to only 8 percent of women age 45 to 49 and to even lower percentages for other age categories ranging from 40 to 65. Even among nonemployed women, only 7 percent were caring for both parents and children at age 45 to 49. Thus, the "women in the middle" thesis does not seem to apply widely.

Retirement of older parents frees them for more visiting with geographically distant children, but it generally reduces their economic resources for doing so. Widowhood tends to increase interaction with children (Atchley, Pignatiello, and Shaw 1979).

We do not know very much about how divorce or separation affects relationships between adult children and older parents. There are a good many aspects that could be studied. Do the children of divorced parents keep in touch with both parents? Is the frequency of contact affected by divorce? Are divorced parents as likely as widowed parents to receive aid from their children? Do older divorced mothers become financially dependent on their children more frequently than do widows? What effect does the timing of the divorce in the adult child's life have on the effect of divorce on older parent–adult child relations? Does divorce affect older parent–adult child relationships for mothers more than for fathers? Many other questions could be added to this list.

The evidence indicates that most older parents understand and comply with the norms for relationships between older parents and adult children. The demands on the older parent are to recognize that the adult children have a right to lead their own lives, not to be too demanding and thereby alienate them, and above all, not to interfere with their normal pursuits. At the same time, the adult child is expected to leave behind the rebellion and emancipation of adolescence and young adulthood and to turn again to the parent, no longer as a child but as a mature adult with a new role and a different, less dependent love, seeing parents for the first time as individuals with their own rights, needs, and limitations and with life histories that to a large extent made the parents the people they are long before the child existed (Blenkner 1965). This type of relationship requires that both older parents and adult children be mature and secure. Therefore, if both sides have adequate physical, mental, and financial resources, then the chances are improved of maintaining a satisfactory relationship.

A growing number of the adult children of older people are themselves over 65. Most people 85 or older in 1985 had living children who were in their 60s. Seccombe (1988)

found that 13 percent of the older men (average age of 69) in the Retirement History Survey were providing financial support to older parents, even though the men themselves had incomes averaging only $10,000 per year in 1979. In these cases, relationships can be strained not only by the financial squeeze in which the older adult children find themselves but also by the increased dependency and disability among the very old.

Siblings and Other Kin

About 80 percent of older people have living siblings. Cicirelli (1985) found that as a result of attrition, the average number of living siblings dropped from three among those in their 60s to one for those in their 80s, although all older cohorts in the study originally had an average of more than four siblings. But even in their 80s, most people had living siblings. Just over half the siblings of older people lived within 100 miles, and about one-fourth lived in the same community. Also, 17 percent of Cicirelli's respondents saw a sibling at least once a week, and another 33 percent saw a sibling at least once a month. Over 50 percent reported feeling extremely close to the sibling with whom they had the most contact, and another 30 percent felt close; only 5 percent did not feel close at all. Men felt closest to sisters, and sister-sister ties were most likely to be extremely close. Several investigators have discerned this general pattern (Bengtson, Rosenthal, and Burton 1990).

Gold (1989, 1990) found that sibling relationships in later life fell into five types: intimate, congenial, loyal, apathetic, and hostile. *Intimate* siblings were extremely close, considered each other as best friends and closest confidants as well as siblings, and accepted one another completely. They would help one another no matter what. Negative interactions seldom occurred and contact was frequent. Among the sibling dyads Gold studied, 14 percent of whites and 20 percent of blacks fell into the intimate category. *Congenial* siblings also felt close, but lacked the empathy shared by intimate siblings. They saw one another as good friends, but named a spouse or adult child as the person to whom they felt the closest. Contact was weekly or monthly rather than daily, and help would be given, but not if it conflicted with obligations to spouse or children. Negative interactions were not a significant part of the relationship. Congenial sibling relationships characterized 30 percent of the sibling dyads among whites and 20 percent among blacks. *Loyal* sibling relationships were based mainly on bonds of belonging to the same family rather than on affection or interdependence. Relationships were structured by family norms rather than by close personalized relationships. Compared with the intimate and congenial types, acceptance and approval were less consistent and intense, and agreement was less important to the relationship. However, even when these siblings disagreed, they would not let the disagreement erode their sibling ties. Loyal sibling relationships typified 55 percent of black sibling dyads compared with 34 percent of white sibling relationships. *Apathetic* sibling relationships were characterized by indifference. There was no sense of affection, belonging, or interdependence, but neither was there strong hostility. These siblings rarely thought about one another. Apathetic relationships were uncommon, accounting for only 11 percent of white dyads and less than 2 percent of black sibling dyads. *Hostile* sibling relationships were typified by strong feelings of resentment, anger, and enmity. These siblings avoided one another and had nothing good to say about their relationship. Unlike the apathetic, who never thought about one another, hostile siblings spent considerable energy denouncing each other. Most of these relationships had been negative for a long time, but some had been precipitated by arguments over inheritances. Hostile relationships were also uncommon, accounting for 11 percent of sibling dyads among whites and 3 percent among blacks.

When Gold (1990) looked at the gender composition of older sibling dyads for whites and blacks, she found that white brother-brother dyads were mostly ties of belonging with few strong emotional components; relationships were either loyal (42 percent) or apathetic (37 percent). Only about 10 percent were hostile, and only 10 percent were either intimate or congenial. But among black brother-brother dyads, 50 percent were loyal, 17 percent were congenial, and 33 percent were intimate. None were apathetic or hostile. Thus, among brother-brother dyads blacks were much more likely than whites to have positive relationships involving emotional attachment as well as a sense of belonging.

Among the older sister-sister dyads, 87 percent were positive relationships, exactly evenly split among intimate, congenial, and loyal types of relationship (29 percent each). About 11 percent were hostile. There were no significant black-white differences among the sister-sister dyads in the proportional distribution among the various types of sibling relationships.

Among the older brother-sister dyads, black sibling relationships were predominately loyal (73 percent); and the remainder were congenial (20 percent) or intimate (7 percent). None were apathetic or hostile. White brother-sister dyads were more widely distributed across the various relationship types, with 11 percent being intimate, 38 percent congenial, 34 percent loyal, 6 percent apathetic, and 11 percent hostile. However, a large proportion (83 percent) consisted of positive relationships.

Gold's (1989, 1990) research showed that in later life sibling relationships come in a number of varieties, that most are positive relationships of belonging which often include a strong component of emotional closeness and affection, that older sisters are more likely to have emotional bonds than older brothers, that older black men are more likely to have emotional ties to their brothers than older white men, and that older black men rarely have apathetic or hostile sibling relationships. Hostile sibling relationships were generally uncommon, constituting only 8 percent of the total.

Cicirelli (1985) reported that most older people felt that they could call on siblings if they needed help. However, the expectation was that spouses and adult children would be the first lines of support; siblings were seen as standby resources should the primary sources be lost or prove inadequate. Siblings are seldom called on to actually provide assistance; their role mainly is to provide a sense of security (Bengtson, Rosenthal, and Burton 1990). For example, Coward, Horne, and Dwyer (1992) found that less than 5 percent of older adults receiving care were being cared for by a sibling.

However, among never-married elders, siblings become much more important as actual care providers. For example, among never-married elders who were receiving care, Coward and his colleagues found that 36 percent of men and 29 percent of women were being cared for by sisters, and 17 percent of men and 11 percent of women were being cared for by brothers. However, the proportion being cared for by siblings declined sharply with age of the care recipient.

After child launching and retirement, older people often resume lapsed relationships with siblings, even over great distances. There is more visiting among siblings in later life than in middle age. Sibling relationships are especially important for the never-married.

Other kin, such as aunts, uncles, nieces, nephews, and cousins, are a reservoir of potential family relationships, but here ties are probably more often voluntaristic and based on individual attractiveness of specific individuals rather than closeness of kinship. Other kin fall in behind siblings as sources of caregiving. Nevertheless, Coward and his colleagues found a significant amount of care being provided by kin other than spouses, adult children, and siblings. Indeed, other kin

ranked third in caregiving behind spouses and adult children and ahead of siblings. Of elders receiving care, 14.3 percent were being cared for by other female kin and 8 percent by other male kin.

As we have considered various types of family relationships, we have found that spouses, adult children, siblings, and other kin can all potentially be involved in caring for older family members. Let us now look at an overall picture with regard to family caregiving.

Family Caregiving

Care by the family is a cultural norm imposed by family values, public opinion, and laws requiring families to provide for care. As we saw earlier, a large majority of adult children believe that they should provide care to their parents, and most do so when care is needed. We also saw that care by other family members is not uncommon. This section addresses several additional questions concerning family caregiving: What is the rationale for laws requiring families to provide care? What is the extent of family caregiving? How do family networks function in making decisions about family care? Does caregiving overburden families, and if so, under what conditions?

Filial Responsibility Laws　If cultural norms require families to care for their older members, why did states feel obliged to enact laws requiring filial responsibility? The primary answer to this question seems to lie in a conflict between public opinion about who should be responsible for paying for care for elders and the financial pressures legislators experience in coping with the increasing cost of programs for elders in an aging society. The public generally believes that government has the obligation to finance care for elders who cannot pay for it themselves (Guilliland and Havir 1990), whereas economically pressured state and federal governments have an interest in shifting as much of this financial responsibility onto the family as possible.

Bulcroft, Van Leynseele, and Borgatta (1989) surveyed the thirty state statutes in place in 1988 that required family support to elders in some form. There was little consistency among the laws in terms of which specific family members were to be held responsible, what kinds of support they were responsible for providing, the agency authorized to enforce the statute, or what sanctions could be imposed to gain compliance. In many cases, these laws seemed designed to allow the state to recover from families public funds spent providing services to older people. For example, in Delaware, the state could require family members to reimburse the Department of Health for the full cost of care provided to an older family member. Financial support was the most common form of support required. Enforcement in most states was problematic; responsibility for enforcement was often not specified in the statutes and enforcement efforts were at best sporadic. Thus, there was little agreement among states about whether to have relative responsibility laws, who they should apply to, what form of support ought to be required, or how they should be enforced. Many states currently without relative responsibility laws had them in the past, but abandoned them because they proved unworkable. At this point filial responsibility laws seem to be mainly a symbolic gesture by governments aimed at exhorting families to do more in terms of financial support for elders who need care in order to relieve some of the financial pressures governments face. As we will see, families already bear an enormous share of the financial cost of long-term care, so such exhortations are not very realistic.

Extent of Family Caregiving As Table 8-5 shows, families are very much involved in caregiving for older adults, and although spouses and adult children predominate, no single family caregiving combination occurs in

Table 8-5 Percentage of older care recipients receiving family care, by relationship of caregiver to the elder, by gender.

Relationship of Caregiver to the Elder	Older Care Recipients	
	Male $N = 1,839$	Female $N = 3,434$
Husband	—	21.6
Wife	67.8[a]	—
Son	15.0	19.9
Daughter	20.5	41.2
Son-in-law	2.7	5.6
Daughter-in-law	4.8	9.7
Brother	1.7	1.6
Sister	3.5	5.7
Other male kin	6.0	9.1
Other female kin	8.5	17.4

[a] Because respondents could report care from more than one person, columns total more than 100 percent.
Source: Adapted from Coward, Horne, and Dwyer (1992).

a majority of cases. Only about 3 percent of elders who receive care in community settings are cared for exclusively by paid caregivers; of the remainder, about 27 percent get a combination of paid care and informal care (U.S. Senate Special Committee on Aging 1992). Family care represents about 85 percent of informal care. Mehdizadeh and Atchley (1992) found that family caregivers spent an average of 7 to 9 hours per day on caregiving, depending on the level of disability of the elder. The sheer volume of this cumulative effort is staggering. The U.S. Senate Special Committee on Aging estimated that more than 27 million days of informal care are provided by families *each week*.

Not only are families providing most of the personnel and time required to provide long-term care to elders in the community, but families also are making the greatest economic contribution, mostly indirectly through the economic value of the uncompensated services they provide. For example, Mehdizadeh and Atchley estimated the economic value of family and other informal care and compared it with the economic value of formal care such as care in nursing homes or in-home care from community agencies. The total economic value was estimated to be $7 billion for the care provided to the 252,000 disabled older adults in Ohio in 1990. Of that amount, 23 percent represented the value of institutional services (mostly nursing home care), 2 percent represented the value of formal community-based services, and 75 percent represented the value of informal care. Elders and their families were paying out-of-pocket for 34 percent of the costs of institutional and formal community services to elders and contributing over 80 percent of the value of informal care. In addition, 26 percent of family caregivers had to reduce their hours of employment because of caregiving, which represented an additional cost of care in the form of lost income amounting to over $680 million in 1990 alone. In this context, the pleas of government for families to do more might seem a bit unreasonable, particularly since federal, state, and local governments combined were paying for less than 25 percent of all services provided. The fact, of course, is that both governments and families make important contributions to providing care to older people, but the public rhetoric that appears in the media tends to ignore the substantial economic value of family care.

Family Caregiving Decisions In many if not most cases, one family member serves as coordinator, case manager, and primary caregiver. Other family members may share in the caregiving tasks and serve as backup, but there is one person who has lead responsibility for dealing with caregiving issues. As long as the elder's needs are modest, the primary caregiver tends to make decisions in consultation with the elder or the elder makes the decisions. When care needs become more demanding, or when crises arise, primary caregivers may become more paternalistic in their decision-making style (Cicirelli 1992), may consult less with other family members, and may become more manipulative (McGrew 1991). As the pressures increase,

practical exigencies tend to overpower the primary caregiver's ideals with regard to how decisions ought to be made.

Caregiving Burden As we might expect, family caregiving can come to be seen by the caregiver as a burden, a weight that is dragging down the quality of life of the caregiver and negatively influencing his or her physical or mental health. Why some caregivers feel this way and others do not is unclear.

Some studies have found that caregivers feel burdened in direct proportion to the amount of care needed and the physical and mental demands of providing it (Miller and Montgomery 1990); others found these factors unrelated (Miller, McFall, and Montgomery 1991; Cattanach and Tebes 1991). Still other studies have found a range of perceptions of caregiver burden within the same levels of care need (Noelker and Townsend 1987). Caregiving situations are complex and so are the people involved in caregiving relationships. Given all the sources of variation, perhaps it is unrealistic to expect that we could arrive at definitive predictions about what causes caregivers to perceive that caregiving is a burden, particularly since the dependent variable in this case is subjective.

Another way to look at this same issue is to look for external manifestations of distress in relation to caregiving. In a particularly interesting study, Hoyert and Seltzer (1992) looked at well-being in a national sample of women consisting of four groups: women who were mothers and/or daughters but who had no caregiving responsibilities; adult daughters caring for a disabled parent; wives caring for disabled husbands; and mothers caring for disabled children. These groups were compared on household performance, employment, participation in activities and groups, marital satisfaction, physical health, and overall well-being. Compared with those not providing care, women involved in family caregiving did more housework, were less likely to be employed, were involved in more organizations, were in poorer health, had lower self-esteem, and were more likely to be depressed. Caregivers who were married were also more likely to see their marriage as in trouble and reported more disagreements with their spouses. The only potentially positive outcome associated with family caregiving was a greater involvement in organizations by caregiving women compared with those who were not doing family caregiving.

The noncaregiving women evaluated their marriages the most positively, reported the best physical health, and were the least likely to be depressed. When Hoyert and Seltzer broke down the caregivers by type, the spouse caregivers had by far the largest incidence of negative outcomes and were the only category to be significantly worse than noncaregivers on all of the outcome measures. Spouse caregivers participated in the fewest social activities, rated their marriages most negatively, and rated their physical and mental health lowest. These negative effects appeared very early in caregiving for spouses. Mothers who were caring for a disabled child were not significantly different from the noncaregivers in terms of outcomes. Adult daughters caring for an older parent were in between in terms of incidence of negative outcomes. However, as the duration of caregiving increased, the incidence of negative outcomes among caregiving daughters increased.

These results led Hoyert and Seltzer to conclude that spouses need support services the most and require them from almost the beginning of caregiving, whereas adult daughters caring for an older parent do not need supportive services as much and do not require them until they have been in the caregiver role for some time. The researchers were impressed with the broad range of negative effects of caregiving, ranging from intrapsychic effects on self-image and mood to social effects on activities, to effects on physical health. These results are similar to those found in a number of other studies, so we can be reasonably confident that caregiving exacts its toll on many of those who provide family care to older people.

Grandparenthood

As a result of declining mortality, more people than ever before are functioning as grandparents. There has been a surge of research on grandparenthood since 1980 (Bengtson and Robertson, 1985). Most people enter grandparenthood in midlife, and many people spend forty years as grandparents (Hagestad 1985). Many become great-grandparents and some great-great-grandparents. However, as Troll (1983) pointed out, whether people become grandparents or not is contingent on their children's decisions. Thus, most people have little authority over the timing of grandparenthood or the number of grandchildren they will have. This section examines the roles grandparents play, styles of grandparenting, and factors influencing grandparenthood.

Grandparental Roles Grandparenthood is a much more indeterminate role than parenthood. There are a few general rules that grandparents can follow, such as the norm of noninterference, which prescribes that under normal circumstances grandparents should not interfere with the way their adult children are bringing up the grandchildren, and the norm of independence, which says that adult generations in the family normally should live independently and autonomously. But these general norms do little to help grandparents get a picture of what they are expected to do specifically in relation to their grandchildren.

Grandparents can serve a variety of functions in relation to their children and grandchildren. In their most passive role, grandparents help families by just *being there* (Bengtson 1985). By being there, grandparents are symbols of family longevity and continuity, and having grandparents available seems to increase feelings of security in younger generations. Grandparents very often serve as *family historians*. Very few families have written histories. Instead, the evolution of the family is conveyed orally through stories told by grandparents and great-grandparents. In times of family crisis, grandparents often serve as *crisis managers*. When crises such as divorce, widowhood, teenage pregnancy, or prolonged unemployment occur in the lives of their children or grandchildren, grandparents often provide substantial assistance to their adult children and become more involved in the daily lives of their grandchildren. For example, Cherlin and Furstenberg (1985–1986) reported that when divorce occurs, the norm of noninterference no longer applies. In their sample, 60 percent of grandparents on the side of the custodial parent said that they provided financial assistance; 30 percent said the grandchild came to live with them, usually accompanied by the custodial parent. Grandparents on the noncustodial side were also involved in substantial assistance, but not in as high proportions. Grandparents often serve as *arbitrators* or negotiators who act to stabilize relationships between grandchildren and their parents (Hagestad 1985).

Grandparents can also serve as *sources of values*. The young adult grandchildren in Roberto and Stroes' (1992) study reported that their family ideals, moral beliefs, and personal identities had been significantly shaped by their grandparents, especially their grandmothers. Indeed, value development was perceived by these young people as being a much more important outcome of having grandparents than doing things with them or even the quality of the relationship, although most had positive attitudes about these aspects as well.

Having grandparents to talk to may be a safety valve for grandchildren who are in conflict with their parents, and grandparents may exert a moderating influence on their adult children's reactions to grandchildren's behavior. Grandparents sometimes serve as *mentors* and *role models* to their young adult grandchildren. Grandparents sometimes become *caretakers* for their grandchildren, par-

ticularly when divorced custodial parents must be employed full-time and grandchildren are young. Finally, grandparents sometimes become surrogate parents, assuming the major responsibility for day-to-day parenting of their grandchildren.

Surrogate parenting by grandparents most often occurs when neither parent is able to fulfill this responsibility. In 1990, 5 percent of children in the United States lived with grandparents, over one million in households where neither of the grandchild's parents were present. This pattern is particularly prevalent among blacks (12 percent of children) compared with whites (3.6 percent). Minkler, Roe, and Price (1992) found that 20 percent of the Headstart students in their Oakland study were being parented by their grandparents. Thus, parenting grandchildren is becoming a significant area of family care and one in which middle-aged and older people are the *providers* rather than the recipients.

What leads grandparents to become surrogate parents? Jendrek (1992) studied a sample of thirty-six grandparents who had formal custody of their grandchildren through a court order, adoption, or guardianship proceeding. Her sample was almost exclusively white, and their ages ranged from 41 to 71, with an average of 56. These grandparents had gained custody of their grandchildren because their adult children had multiple problems that prevented them from parenting their children: emotional problems (73 percent), drug addiction (53 percent), mental problems (48 percent), and alcohol abuse (44 percent). In over half the cases, keeping the children out of foster homes was a consideration for the grandparents. In 62 percent of the cases, the grandparent had offered to take the grandchild; in 29 percent, someone else had suggested it; and in 9 percent, it was just assumed that the grandparents would do it.

Among blacks, drug addiction of adult children has also emerged as a leading cause of grandparents' becoming surrogate parents to their grandchildren. In Burton's (1992) study of black grandparents who were parenting grandchildren, over 40 percent were caring for three or more of their grandchildren. Not only did these grandparents have to deal with the demands of parenting their grandchildren, but they also had to suffer the pain of seeing their children's lives disintegrate as a result of drug use and often deal with the physical effects of drug use during pregnancy on the physical and mental health of the grandchildren they were parenting. They were stressed not only by the demands of parenting, but by the dangers of the neighborhood environment, needs of other kin, conflicts stemming from multiple roles, and lack of time for a personal life. As one grandfather put it, "I got a lot on my back. I take care of my wife who has cancer and my two grandbabies. I chase around after my daughter on the streets trying to make sure she eats, at least" (1992,748). The strains are also obvious in the following: "A 56-year-old woman who reported frequently feeling depressed opened the interview by saying, 'I have three children: my 2-month-old who was born on crack, my 17-month-old who is HIV [positive], and my 83-year-old mother. All three are in diapers'" (Minkler, Roe, and Price 1992,759).

Parenting grandchildren is rewarding to most who do it, but it also has tremendous costs: lost privacy, no time for fun, lost contact with friends, and no time for oneself or one's spouse. Life plans have to be put on hold and routines rearranged. Most who are parenting grandchildren frequently feel worried, tired, drained, and overwhelmed (Jendrek 1992). More than 85 percent of Burton's (1992) respondents reported feeling depressed, and significant proportions reported smoking more, drinking more, and having exacerbated health problems—41 percent reported multiple negative outcomes. Yet grandparents with health problems did not feel that they could let them get in the way. As one 65-year-old great-grandmother who was caring for her 1- and 2-year-old great-grandchildren said, "I've been blessed. Oh, sure, I've got arthritis and I had a bladder infection last week. I have high blood pressure and diabetes, I have sleepless nights, but I'm not concerned about my health. What does worry me lately is this pace I'm trying to keep up" (Minkler, Roe, and Price 1992,757).

Grandparents parenting their grandchildren also needed help. Only 3 percent in Burton's (1992) sample were getting help from other family members, 77 percent reported needing financial help, and most would have welcomed respite care. Jendrek's (1992) respondents also reported difficulty getting health care for their grandchildren. The grandparents' health insurance companies didn't cover grandchildren and the grandparents couldn't qualify for Medicaid.

Grandparenting Styles The various functions that grandparents can perform usually occur in combinations that are influenced by circumstances and negotiations between grandparents and parents. Cherlin and Furstenberg (1985) studied grandparents of teenagers and found that each grandparent had a somewhat unique relationship with each grandchild. Some were detached (26 percent), some passive (29 percent), and a large proportion were active (45 percent). Of the *detached* grandparents, 63 percent lived more than 100 miles from their grandchild, and many had no appreciable relationships with any of their grandchildren. Others showed selective investment: They were detached in their relationship with the teenage grandchildren selected for study, but they focused their energies and emotions on grandchildren who lived nearby or who were their favorites.

Passive grandparents lived near their grandchildren and served symbolic functions, such as being there and as family historian. They considered regular but superficial contact with their grandchildren normal. Cherlin and Furstenberg concluded that the passive grandparents "best fit the popular image of American grandparents: the loving older person who sees the grandchildren fairly often, is ready to provide help in a crisis, but under normal circumstances leaves parenting strictly to the parent" (1985,115). *Active* grandparents not only frequently interacted with their grandchildren, but they also provided support, advice, and sometimes surrogate parenting. Many of the closest relationships with grandchildren were continuations of close relationships that were established in response to family crises (Cherlin and Furstenberg 1986). Neugarten and Weinstein (1964) found that grandparents of young children showed yet another pattern they called *fun-seeker*. Fun-seeker grandparents tended to be middle-aged, and their relationships with their grandchildren centered around recreational activities.

Factors Influencing Grandparenthood

There is an important voluntaristic quality to the grandparent-grandchild relationship. Troll (1971) concluded that "valued grandparent" is an achieved position based on the personal qualities of the grandparent and is not automatically ascribed to all grandparents. Likewise, "valued grandchild" is not automatic either. Some grandparents want to remain distant figures to their grandchildren; others respond to some grandchildren and not others, depending on the grandchildren's personal qualities. Grandparents often have resources grandchildren want and vice versa. Grandparents can bring their grandchildren a sense of history, support and advice, and companionship, and grandchildren can bring their grandparents a sense of the present, love, and companionship (Troll 1986). But both sides must want exchange if the relationship is to persist, especially beyond the grandchild's teenage years.

Nevertheless, being a valued grandparent is not evenly distributed. Hodgson (1992) found that maternal grandmothers were twice as likely as maternal grandfathers or paternal

grandmothers to be picked "closest grandparent" (63 percent likelihood). Paternal grandfathers had the least likelihood of being seen as the closest grandparent.

Other factors that substantially influence the nature of grandparenthood include the timing of grandparenthood and race and ethnicity. Younger grandparents are more likely than grandparents 65 or older to fit the active grandparent pattern. The timing of grandparenthood in relation to other life course events can also make a great difference. Troll gave a personal example:

At the time I acquired the title "grandmother," I was back at the university completing my dissertation, with a new academic job awaiting me. My children were "finding themselves" . . . I had separated from my husband. My father was dying. One whole era of my life was ending and I was beginning a whole new era, becoming interested in many new and exciting pursuits. Thus, at that time, being a grandmother had minimal significance. When my sister became a grandmother for the first time, she was at a point in her life when it made a big difference. . . . After almost a quarter of a century of housekeeping and child rearing, she had not found a niche that absorbed her. . . . My first year as a grandmother was spent revising my dissertation and learning how to be a college teacher. Her first year was spent sewing baby clothes, writing letters to her daughter—the baby's mother—and waiting for new photographs.

(1985,138)

Race and ethnicity also influence the experience of grandparenthood. For example, Bengtson (1985) compared African-American, white, and Mexican-American grandparents. Mexican-American grandparents had many more grandchildren and reported higher levels of contact and more satisfaction with their contacts with grandchildren than African Americans or whites. On the other hand, Burton and Bengtson (1985) found that grandparenthood among young African-American grandmothers was often problematic. They found a pattern of several generations of teenage childbearing, resulting in grandmothers who were in their late 20s when their grandchildren were born. Often these young grandmothers had small children of their own and were unwilling or unable to assume the active grandparent role. By contrast, African-American grandmothers who experienced grandparenthood "on time"—in middle age—showed much more involvement with their grandchildren.

Friends

How do we define friendship? Quite a variety of relationships are lumped together under this label, ranging from close, intense, continuous interactions marked by mutual understanding and concern to cursory sociable contacts over the years. For our purposes, it is useful to call the former contacts "friends" and the latter "acquaintances." Friends are people we know, like, and trust. They are our favored companions and confidants. Acquaintances are people we know and like through social contacts at work, in the neighborhood, or in the community. In either case, these relationships are entirely voluntary. We may have to be civil to our associates at work, for example, but no one can force us to like one another.

Nearly everyone has acquaintances, and having friends is very prevalent in middle age and later life. About 93 percent of Armstrong and Goldsteen's (1990) respondents reported having friends, and the size of their friendship networks ranged from 3 to 36. More than half reported having 10 or more friends.

The boundary between friends and relatives is not always clear-cut. For example, Armstrong and Goldsteen found that many of their respondents included kin in their lists of friends. On the other hand, Mac Rae (1992) found that in situations where people lacked particular family members, especially parents or siblings, they tended to compensate by converting close friends into *fictive kin*.

Factors Influencing Friendship

Friends tend to be selected from among people who are considered social equals. Growing up together, living in the same neighborhood, having similar occupations, having children the same age, having similar interests, and being the same general age and/or the same gender are all factors that can create the sense of similarity and equality necessary for the development of friendship.

Age   Many factors influence the age structure of friendships, including how long one has lived in the same community, the age structure of one's workplace and neighborhood, and how one's main interests relate to age. The friendships of people who live out their lives in the same community are likely to have developed in childhood and to persist into later life. Adults who settle far from where they grew up may find that they select friends from among their neighbors, their coworkers, and their fellow participants in various organizations and activities. How age-segregated these environments are obviously influences the age range of a person's friends.

But even in age-integrated environments, one's position in the life course has a bearing on the age range of one's friends. People at an early age relative to their environment can have friends only their age or older. People at the middle of their environment's age range can have friends of a wide variety of ages, and those at the top end of the age range can have friends only their age or younger. In general, most friends are selected from those of one's own age, broadly defined. Adults generally are not open to age-integrated friendships. Whether this stems from insecurity, age cohort chauvinism, or some other cause is unclear. However, among people in their late 70s and older, there is a tendency to develop more younger friends, perhaps as a shield against attrition (Armstrong and Goldsteen 1990).

Johnson and Troll (1994) studied friendship patterns longitudinally in a sample of elders who were 85 or older at the time they were first interviewed. They found that despite the constraints on friendship stemming from deaths among friends and increasing prevalence of disability, most of their respondents maintained at least one close friendship, three-fourths were in at least weekly contact with a friend, and 45 percent reported making at least one new friend after their 85th birthday. Over the thirty-one-month interval between interviews, disability did take its toll on friendships within the sample. Two factors were especially related to the number and frequency of contact with friends: a social environment that provided easy access to people similar to oneself and a gregarious personality. These oldest-old respondents adapted to the constraints on friendship by relaxing their definition of what constituted a friend to increase the size of the pool of potential friends. Johnson and Troll concluded that the oldest-old do find it difficult to sustain the high quality of friendships that they had enjoyed earlier in their lives.

Gender   The impact of gender on friendship in later life is also unclear. On the one hand, some researchers reported that compared with women, men have fewer friendships, rely on friendships less, and do less to maintain friendships (Wright 1989). Others reported that for most older men, spouses are their only close friends, but older women tend to have close friends other than their husbands (Keith et al. 1984). On the other hand, other researchers found only small differences between genders in terms of friendships (Babchuck 1978–1979). Part of this confusion is no doubt related to imprecise measurement of friendship. The weight of the evidence indicates that men have about as many acquaintances and friends in later life as women do, but men apparently are less likely to confide in their friends, preferring to confide mainly in their wives.

Duration of Friendship   The duration of friendship is also important. Litwak (1989) suggested that we look at duration not only as a quantitative characteristic of friendship but as a qualitative one as well. We expect and get different things from long-term friends than we do from intermediate- and short-term friends. Long-term friends are the only ones with whom older people can reminisce about their childhood or early adulthood. The level of intimacy between long-term friends tends to be high, and changes in circumstances such as retirement or widowhood have less effect on long-term friendships than on other friendships. Intermediate-term friends are with us through particular stages of the life course. For example, friends at work often are an important resource for coping with hassles on the job, but after retirement many of these friendships wither away for lack of contact. Short-term friends help us through immediate changes in roles. For example, when people move into retirement communities, they tend to develop many short-term friendships that provide a sense of welcome and support. But after a year in the community, these relationships are either converted into intermediate-term friendships or fade away.

Approaches to Friendship

Everyone does not approach friendship the same way. Matthews (1986) classified her older respondents into three categories of orientation to friendship: independent, discerning, and acquisitive. The *independent* (20 percent) did not have close friends. They lived mostly in the present and were content to be with people who happened to be available. They were likely to identify themselves as loners who did not really need people. They expected their relationships to consist of superficial ties to acquaintances of convenience. However, they were not likely to be lonely. The *discerning* (13 percent) were very demanding and selective in terms of whom they would accept as friends, and as a result they tended to have many friendly acquaintances but relatively few friends. They were oriented to the past and unlikely to develop new friendships in later life. The *acquisitive* (67 percent) were active people who used their large number of social contacts to develop many friendships. They actively sought new friendships and maintained old ones. They were oriented to the past, the present, and the future.

Not all friends feel the same sense of commitment either. Fair-weather friends can be counted on only as long as times are good. When the going is tough, it is mainly close, intimate friends who can be counted on for help. For example, when older people become disabled, usually only their closest friends can be expected to make the effort required to continue to include them in group leisure activities.

In terms of interdependence, intimacy, and belonging, there is wide variation in what elders expect from their friends. Most expect to have their instrumental needs met through their family, and when friends provide care, most older people are especially gratified. On the other hand, most expect to receive emotional support and companionship from their friends, and if these are not forthcoming, elders tend to be disappointed. But again, the line between family and close friends is often not hard and fast. Many elders feel that family and friends serve similar functions and are equally important (Armstrong and Goldsteen 1990).

Changes in Friendship

Not all aspects of friendship are positive. Rook (1989) pointed out that actions such as breaches of confidence, nonreciprocated affection or self-disclosure, invasions of privacy, and critical or competitive remarks violate implicit norms of trust, respect, reciprocity, and equality and can strain relationships. Rifts between friends not only affect the two parties but their mutual friends as well. Rook found that because negative exchanges between friends were relatively rare among older adults, they were especially bothersome when they did occur. More research is needed on this topic.

Obviously, aging increases the opportunity for long-term friendships. But because mortality rates increase with age, it has been commonly assumed that increased age automatically causes a reduction in the number of friends available. However, this perspective does not allow for the fact that most older people are steadily making new friends. For example, one longitudinal study of friendships found that older women acquired more friends than they lost over a three-year period (R. Adams 1987). Generally, close relationships get closer with age (Blieszner 1989), and older adults continue to put a high premium on maintaining their friendships. Lost friendships tend to be replaced by converting acquaintances with fellow activity participants, neighbors, and former work associates into friendships (R. Adams 1989). Nevertheless, the number of confidants tends to decrease with age. More research is needed on age changes in friendship.

Many changes in friendship are connected more to age-related changes in circumstances than to aging per se. Retirement, widowhood, and health problems are examples of situational changes that can have an impact on friendships. *Retirement* tends to reduce contact with work friends mainly among working class men and women. Middle-class retirees show little effect of retirement on friendship (Allan and Adams 1989). Indeed, Francis (1990) found that social support from a cohort of work friends eased considerably the transition to retirement for a group of middle-class women. *Widowhood* can have a strong effect on friendships. Older widows often increase their reliance on friends as companions and confidants. *Health problems* can reduce the energy available for friendships, but having health problems can also attract new friends. Rebecca Adams (1985–1986) found that older women often felt better when they had a relationship with someone who was worse off than they were. In addition, helping others was a source of respect in the community. As a result, in

Companionship is an important element of being friends.
Photograph by Lawrence Manning/Tony Stone Images, Inc.

Adams's study of women in age-segregated housing, women with physically limiting health conditions had more close local friends and more contact with friends than women with no physical limitations. Litwak (1989) reported that older people who were sick and unmarried were almost twice as likely as those who were married and healthy to have friends who were younger than they were.

We form acquaintances and make friendships from among the people we encounter in our everyday routines. Youngsters develop friendships at school. Young adults develop relationships with coworkers. In a similar way, older adults sometimes develop personal relationships with service providers. For example, Wolfsen, Barker, and Mitteness (1990) studied the relationships of older people living in publicly subsidized housing projects and found that the physicians, apartment managers, and social workers the residents encountered not only provided instrumental, professional services but also were seen as friends or "like family" by a large proportion of the respondents. These elders saw their physicians not just as providing medical services but also as caring about them as people. They described the relationship as good, close, and trusting, and they saw their physicians as someone to talk to, not just about health problems but about life problems in general. The apartment managers and social workers were often seen as people to count on for a wide range of types of help in time of need, such as help with meals or coordinating services, tasks that went far beyond the job description. Thus, even in presumably impersonal relationships with professional service providers, the need to personalize ongoing relationships can lead to a level of self-disclosure that has the characteristics of friendship, and patterns of expectation can develop that are more typical of the informal support network than the specifically delimited professional role.

Eustis and Fischer (1991) found even greater blurring of roles in the relationships between home-care clients and the workers who came into their homes. The role of the home aide is often a vague one that can include elements of companionship as well as instrumental task support. Eustis and Fischer found that two-thirds of the home-care clients in their study and more than half of the home-care workers used friendship terminology to describe the relationship. Often the workers provided extra help, beyond what they were being paid for. This was particularly true for companionship. In addition, clients tended to confide in their home-care workers, especially about family problems, which often put the workers in the middle of family disputes. Because the workers knew so much about the "backstage" aspects of the clients' lives, the element of trust was more important than in many other types of relationships with service providers. The friendship elements of these relationships could and did come into conflict with the formal service element. Workers who worked for agencies often faced different sets of expectations from their supervisors than from their clients. The relationship tended to be easier if the worker was hired directly by the client. On the other hand, clients were often bothered when workers were not interested in personalizing the relationship. This mixing of personal and instrumental elements is probably common in ongoing service relationships, and more research is needed on this issue. In particular, we need to focus on the negotiation processes through which the formal relationships are modified to include the more personalized elements.

Social Support

Thus far we have discussed various kinds of relationships separately, but obviously this is not how most people view them. People tend to look at their close relatives and friends as constituting an informal network that serves their needs, even if each person is not in contact with every other person in that network. Social support networks consist of people we

Social support and social service delivery can be combined.
Photograph by Marianne Gontarz

can count on to provide ongoing assistance, emotional support and affirmation, and information and personal assistance in times of crisis (Cantor 1980). Antonucci (1990) pointed out that both discussion and study of this subject are complicated by the fact that terms like *social support* and **support networks** have a wide diversity of meanings, and she recommended that we not try to attach specialized meanings to these terms or spend too much energy trying to differentiate between them.

The structure of adults' informal support networks usually persists over considerable time. Several longitudinal studies have shown that network size remains relatively stable over as long as fifteen years. Despite the prevalent notion that with advancing age networks decline in size, age differences in size of networks and frequency of contact are small (Antonucci 1990). Social support thus often involves reciprocity over a long period of time. Unlike short-term relationships, which require more immediate exchange, social support relationships are long-term relationships in which reciprocity is assumed to average out over the long run.

Family members tend to provide personal care, while friends are more likely to give emotional support (Antonucci 1990). For example, Coward and his colleagues (1992) found that among older people who received assistance, 85 percent received help from family, 10 percent from friends, about 15 percent from paid household staff, and 13 percent from social service agencies. Most support networks involved a mixture of different types of kin, friends, and formal service providers. Antonucci pointed out that because older people generally do not expect personal care from their friends, they tend to appreciate getting it more than they do care from family, whom they see as obligated to provide care.

Generally, getting social support when needed has positive effects on a person's sense of well-being. Perceived reciprocity over the long run and involvement of friends are especially conducive to the well-being of those receiving assistance. Social support can also reduce the need for institutional care.

An important part of *continuity theory* concerns the creation and maintenance of one's social support network in adulthood. Through years of selective investment, people develop well-tested relationships that can be relied on to meet a variety of needs. These relationships represent an important part of the person's capacity to cope with changes that occur, and elders are generally highly motivated to do what is necessary to preserve their supportive relationships. The core of people who constitute the network tends to remain with the individual through the life course. Kahn and Antonucci (1981) referred to this as the "convoy of social support,"

which is usually made up of close family and all-weather friends.

Social services professionals often view the rise of formal services to older people as evidence of a weakening of family and other informal network supports, but as many investigators have documented, informal support for older people is widespread and provides vastly more assistance to older people than do service agencies. The informal network, through its everyday socializing, can give older people individualized support that stresses continuity of relationships, and can also provide immediate response to crisis situations. Informal support network members can also provide information and help negotiate with formal agencies for services. Formal service agencies, on the other hand, are better able to handle repetitive and uniform tasks such as home health care, home-delivered meals, or homemaker services (Brubaker 1987). Rather than considering support as coming from *either* informal supports *or* agencies, it is better to consider how *both* forms of support can serve important functions. The goal of agencies' assisting networks of family and friends to function more effectively is certainly more in tune with public preferences than is the goal of having agencies take over family functions. By the same token, social support networks can function better if work requiring technical knowledge is done by service agencies.

In addition, as Eustis and Fischer (1991) pointed out, the boundaries between formal service providers and informal support networks are not as clear as we once thought. Especially for elders who have thin family networks, formal service providers are often recruited to become part of the personalized convoy of social support.

Summary

Relationships connect us to one another through our needs for interdependence, affection, and/or belonging. Relationships emerge when people interact with one another, usually in their various social roles. By personalizing role performance, long-term relationships articulate the personalities, resources, and needs of the people involved into a whole that is often greater than the sum of its parts. In this sense, the group is stronger and more capable than any individual acting alone.

Relationships based on interdependence arise because ours is a complex, technologically advanced society in which complete self-sufficiency is virtually impossible. Our need for interdependence is the foundation on which intimacy and belonging are built through communication, agreement, self-disclosure, trust, similarity, and acceptance. Relationships can be weakened by disagreement, disability, selfishness, betrayal, or withdrawal. How well relationships resist negative forces depends greatly on duration. The major long-term relationships for most adults are with spouse or partner, parent, child, relative, friend, associate, and neighbor. These roles are arranged into a family life cycle tied to various life stages.

The reduction in number of children in each generation and the increase in the number of generations in many families will produce a family population pyramid that is tall and slim. One implication of this changing structure may be an increase in caregiving that skips a generation or that involves several generations of caregivers.

Most middle-aged and older people are part of a couple. For those who have children, marital satisfaction tends to reach a low point when the children are in their teens and to improve steadily after the children leave home. One major reason for the rise in marital satisfaction in later life is probably the reduced potential for conflict over jobs and children.

There are many types of couples, and the couple relationship can serve many possible purposes for its members. Nevertheless, satisfied couples in later life tend to have adequate incomes, to share values and activities, to be egalitarian in their approach to household work, to be flexible in terms of traditional sex roles, to be deeply involved with one another psychologically, and to be sexually active. The availability of a spouse or partner for caregiving if needed meets not only the need for someone to depend on but also the need to feel safe and secure. Being part of a couple is *the* major social resource for a large majority of middle-aged and older adults.

Unfortunately, by 70, half of older women are widows, and although they get social support from family and friends, they still must make major adaptations. Whether or not they succeed usually depends on their economic resources and community supports. Older men become widowed too, but most of them remarry since they are greatly outnumbered by unmarried women of their own age and can also marry women younger than themselves. However, the proportion of older men who remain widowers increases steadily with age. Divorce can have a devastating effect on financial resources in later life, especially for older women. Most people remarry for companionship and affection.

Contrary to the popular myth, older people are not isolated from or neglected by their adult children, especially when older parents need assistance. The flow of interaction, aid, and affection tends to go in both directions, with older parents giving as much as they are able. When parents are in later maturity, there is a great deal of independence among generations, with visits spaced weekly or less frequently and focused around socializing. When parents enter old age, they tend to need more assistance and support. Visits or contacts tend to be daily, particularly if the older parent has a serious chronic illness or disability, and there is much more "doing for" at this stage—housework, shopping, and transportation. The older the parents, the more likely they are to see their adult children as confidants.

Most adult children who live in the same household as their parents have never lived apart from them, and most adult children who reestablish coresidence with older parents do so to meet the adult child's needs. Dependency of older parents does not account for a majority of cases of coresidence at any point in the life course. A large proportion of coresidence consists of lifelong patterns that are not the result of a crisis on the part of either the older parents or the adult children.

The number of adult children influences the probability that older parents will visit and talk with their children frequently. Parents with four or more living adult children interact less with each child but have a greater overall amount of interaction than older parents with only one or two children.

Relationships with adult children are greatly influenced by the sex of the parent, social class, and ethnicity. Older women tend to interact more with their adult children than older men do. Obviously, middle-class adult children have greater economic resources with which to aid older parents. Working-class older parents are more likely to assume responsibility for rearing their grandchildren. Compared with other ethnic groups, black families tend to be involved in more mutual aid. Relationships between older parents and adult children are not generally disrupted by migration. However, we do not know how these relationships are affected by such factors as divorce.

Caregiving for older parents is becoming a common phase of the family life cycle. Most older people who cannot be cared for by a spouse can get care from one or more adult children. Most older adults who receive care from adult children are satisfied with the amount and quality of care they receive. Most adult children feel that they are doing an effective job and do not feel overly stressed by having to provide care. However, as the amount of care increases, as the age of the adult child caregivers increases, and as the difficulty in finding people to help increases, so does the stress connected with caregiving.

Relationships with grandchildren, siblings, and other kin tend to be more voluntary and have a less predictable meaning than those with a spouse or adult children. Today, most people who are becoming grandparents can expect to spend over thirty years in that role. Most grandparents are psychologically close to at least one grandchild, but most are not involved in their grandchildren's daily lives. Grandparents serve a wide variety of functions, ranging from just being there to being surrogate parents. In times of family crisis, grandparents must often be crisis managers and provide considerable assistance. They may also mediate family conflicts. The valued grandparent must earn that position.

A large majority of older people have living siblings, and most report that they are close to them. Most older people feel that they could call on their siblings for help, but few do so. Siblings visit with one another more in later life than they did in middle age. Other kin serve as a reservoir of potential relationships. Those relationships that are developed tend to be based on individual attractiveness rather than kinship position.

Friends are a major source of companionship for most older people, especially those living alone. Many friendships carry over from middle age into later life. Most older people are also steadily developing new friendships. Long-term friends are likely to be confidants as well. Short-term friends are important in successful adaptation to changes such as relocation. Rifts between friends rarely occur, but when they do, they are extremely upsetting to most older people. The number of confidants tends to decrease with age.

Support networks are the totality of family and friends that people depend on to meet their social, psychological, and physical needs. Support networks are of vital importance in helping older people adjust to changes such as widowhood and disability. Fortunately, most older people have an informal support system that complements the services available from formal service agencies.

9 Employment and Retirement

Employment and retirement play a big part in the adult life course. When men reach adulthood, they are expected to get a job and to remain employed until they become eligible for retirement. Men are excused from this expectation only if they are or become disabled. The traditional cultural ideal for women is to marry and remain homemakers with no paid employment, but there have always been women who had jobs. For example, women workers were commonplace in the textile industry in the early 1900s, and women were vital to the defense industries during World War II. During the prosperity of the 1950s, many women were able to realize the ideal of being homemakers during their child-rearing period and beyond. But in the 1960s, the women's movement and an expanding economy led more women to stay in the labor market. In the economic climate since 1970, women's employment has become an economic necessity for most middle-class households; more than two-thirds of women are now employed continuously until retirement.

Age plays an important part in the rules governing employment. Age of entry into employment is governed by child labor and compulsory education laws. Likewise, the minimum retirement age is also usually tied to chronological age.* In addition, informal norms about how old a person ought to be in a particular job often disqualify large portions of the adult population because they are considered either too young or too old.

Figure 6-1 (p.139) shows several scenarios linking the life course with occupational development. The idealized sequence of continuous full-time employment depicted in the figure for men and unmarried women is increasingly becoming an expected path for many married mothers as well. Following some basic decisions about education and job preparation, people begin to experiment with various kinds of jobs, usually part-time student employment and summer jobs at first. Early full-time jobs in young adulthood come in an enormous variety, and there is often considerable job changing during this period as young adults sort out their preferences, respond to opportunities, or experience setbacks. But by the time they are in their early 30s, most adults have identified the general area within which they wish to make their living, settled into a pattern of employment and begun to master their occupational field. After age 45 or 50, many adults experience a period of occupational plateau or stability. They are settled into a job that they have well mastered; they do not expect to advance much further if at all; and they receive satisfaction, if not excitement, from routine good

---

*Years of service is sometimes the sole criterion for the minimum age of retirement; and years of service too is related to age. For example, if retirement is permitted after thirty years of service, the lowest possible age of retirement will still be 46 since employment prior to age 16 is unlawful.

performance. At this stage some people become bored and switch careers. Others turn to the family, leisure pursuits, or community involvement for new challenges while they continue to enjoy the security of stable employment. When pension eligibility reaches the point that the amount of retirement income is sufficient to support the household, most people retire.

The *timing* of entry into job paths, job changes, career changes, occupational plateaus, and retirement depends on individual decisions as well as the social environment within which individual decisions are made. For example, the life course of occupational development looks very different for those who do not complete high school compared with those who complete college. But even among those who complete college, there are countless occupational paths that individuals can take, so many in fact that young adults often feel overwhelmed by the range of choices. Most can take comfort in the fact that people can successfully change careers even in midlife.

Midlife Career Changes

The popular press carries frequent accounts and case studies of people who begin new careers in middle age. However, it is difficult to find statistics on how common this phenomenon is. Many employed middle-aged women attend school part-time in order to qualify for new careers. Military personnel often "retire" in middle age in order to take up a new career.

In recent years there has been an increased emphasis on the individual. Self-improvement, self-actualization, and autonomy are some of the values being emphasized in the middle class. Happiness is often tied in popular opinion to being able to change oneself and/or one's circumstances. A certain amount of romanticism is obvious in this approach, and some people subscribe to the be-

lief that if one is unhappy with one's job, spouse, ego, car, stereo, or whatever, then the only and best way to deal with it is to ring out the old and ring in the new. No doubt a certain amount of job change is related to this more general cultural factor. Nonetheless, at this point it is difficult to say how prevalent midlife career change is, what preparations precede such changes, and what consequences they have. Numerous job changes certainly occur, but how these relate to a person's own projections about her or his occupational career is unclear. Obviously, even the idea of a career may apply to only a small portion of the labor force.

Blue-collar workers change jobs in middle age primarily as a result of poor health or being laid off (Parnes 1985). For middle-class workers the situation is more complex: midlife career change has been tied to rejection of the work ethic, personality type, family factors, and achievement motivation. Interestingly, only rarely does midlife career change appear to be motivated by a desire for more money.

Thomas (1977) studied seventy-three middle-class men who had left managerial and professional careers between the ages of 34 and 54, and he found four paths to midlife career change among these men. About 34 percent of his sample were men who had little internal motivation to change careers but were forced into it by external circumstances. Another 26 percent were faced with external pressures to change and had high internal motivation to do so; 23 percent had neither high desire to change nor heavy external pressures. Only 17 percent changed purely because of their own desire. External factors associated with the job career were thus the prime force in these midlife career changes. Career change was not related to rejection of the work ethic, desire for more money or job security, or family factors. There was some indication that a desire for meaningful work was important for most of these men. Thomas's study is noteworthy in that it carefully screened out those who had merely changed jobs and concentrated on those who had genuinely changed careers.

In the 1990s, reductions in the workforce were commonplace in American corporations, because downsizing came to be seen as the quickest way to cut costs. As a result, millions of people in their 40s and 50s were displaced from the labor force. Some of these people were offered retirement pensions or other economic incentives to induce them to voluntarily leave their company. Others' employment was simply terminated. Many of these displaced workers were forced by economic necessity to find new jobs, but so far we have a very poor research picture of what happened to them.

Declining Labor Force Participation in Middle and Later Life

The labor force consists of both people who are employed and those who are unemployed but looking for work. Labor force participation rates are simply the percentage of the population in each age category who are in the labor force. In 1989, almost 95 percent of men between ages 25 and 44 were in the labor force. After age 45, the proportion of men in the labor force declined (see Figure 9-1). For women the pattern is similar, but the peak labor force participation in the 25- to 44-year-old category was just over 75 percent—up substantially from less than 50 percent in 1970.

Table 9-1 shows patterns of labor force participation in later life by single years of age for 1970, 1986, and 1993. Age patterns changed considerably over the first sixteen-year period. For men, the trend toward earlier retirement between 1970 and 1986 is very clear. The proportion in the labor force dropped about 10 points at ages 56 to 58, but after that the drop was precipitous. By age 62, only 53.5 percent were in the labor force in 1986, compared with 73.9 percent in 1970. By 1986, the largest drop was occurring at age 62 rather

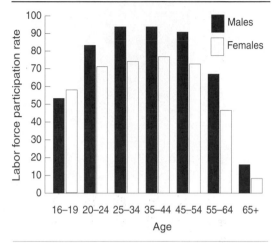

Figure 9-1 Labor force participation rates, by age and sex: United States, 1989.

Source: Based on data from the U.S. Bureau of the Census (1991a).

than 65. But between 1986 and 1993, there was much less change in labor force participation rates for men. The sharpest drop still occurred at 62. More men stayed in the labor force at 63 and 64, but at 65 there was a drop again. There were small drops in participation in the late 50s but most of the age categories showed stable or increasing employment between 1986 and 1993. This suggests that many men were responding to the economic uncertainties of the period by remaining in the labor force, especially at the older ages.

Labor force participation rates for women under 60 steadily increased from 1970 to 1986 to 1993. The age pattern of drop at ages 62 and 65 was comparable to the pattern for men, but more interesting is the fact that between 1986 and 1993 labor force participation rates of women increased for every single age but 74. These data suggest that women were responding to economic uncertainties by remaining in the labor force even more consistently than men.

This complex set of age patterns over time shows clearly that although withdrawals from the labor force cluster around ages 62 and 65, there is considerable variation in labor force participation in late middle age and

Table 9-1   Labor force participation rates, by age and sex: United States, 1970, 1986, and 1993.

AGE	Males			Females		
	1970	*1986*	*1993*	*1970*	*1986*	*1993*
55	91.8	84.1	84.2	52.6	56.3	61.7
56	90.7	80.9	80.1	48.2	54.5	60.6
57	89.1	79.7	77.6	50.0	51.3	56.3
58	87.7	77.2	75.5	47.7	48.4	55.6
59	87.5	73.4	74.5	45.9	46.1	50.5
60	83.9	69.2	69.0	44.0	42.3	45.4
61	81.2	66.2	64.4	38.7	37.8	45.6
62	73.9	53.5	51.5	36.1	31.9	35.4
63	69.4	44.3	46.4	31.9	28.1	31.0
64	64.4	39.4	41.6	28.5	25.2	29.4
65	49.9	30.8	29.9	22.1	18.6	20.6
66	44.7	27.6	30.1	19.3	15.4	16.0
67	39.4	24.0	25.3	16.2	13.5	16.1
68	37.7	20.7	22.2	15.0	12.4	14.2
69	34.0	20.7	19.2	12.9	10.5	11.7
70	30.2	17.1	17.2	12.2	8.4	10.9
71	27.9	15.3	15.7	9.8	7.1	7.5
72	24.8	14.1	15.4	8.6	6.9	8.4
73	22.0	14.5	13.3	7.8	6.1	7.0
74	19.1	12.7	11.4	6.4	5.8	5.2

Source: Adapted from Clark (1988); Current Population Survey data (1993).

later life. The ages of 62 and 65 remain important because at 62 people become eligible for reduced Social Security retirement benefits and at 65 they become eligible for full benefits. Private pensions are often articulated with Social Security and therefore are tied to ages 62 and 65 as well. For example, a private pension might provide $2,000 per month until the retiree reaches age 62 and thereafter provide only the difference between the retiree's monthly Social Security benefit and $2,000.

The Age Discrimination in Employment Act (ADEA) was amended in 1978 to ban mandatory retirement before the age of 70 in most sectors of the economy. Many experts predicted an increase in the proportion of workers staying in the labor force past age

65. But as the data in Table 9-1 show, more men retired before age 70 in 1993 than in 1970. What this means, of course, is that mandatory retirement was not a very important factor influencing labor force participation of older men.

Hayward, Crimmins, and Wray (1994) cautioned against assuming that labor force participation rates are a clean indicator of the extent of retirement. In fact, in the 1970s the duration of long-term jobs began to decline and has continued to do so. In this same period, there has been an increase in postretirement employment. At any point in time, labor force participation rates in later life are a balance between labor force exits due to retirement and labor force entries by retirees returning to the labor force.

For older workers unemployment can be devastating.
Photograph by Andy Sacks/Tony Stone Images, Inc.

Withdrawal from the labor force after age 55 is mainly caused by three factors: employment problems of older workers, disability, and desire for retirement coupled with entitlement to retirement pensions. We will now look at each factor in turn.

Employment Problems of Older Workers

To the U.S. Department of Labor, a "mature worker" or "older worker" is any employee age 40 or over. Mature workers are less likely to become unemployed, but if they do, they encounter difficulty getting another job. Thus, although unemployment rates may be lower for mature workers, unemployment for them has more serious consequences. Once unemployed, mature workers are much more likely than others to remain unemployed long enough to exhaust their unemployment benefits (U.S. Senate Special Committee on Aging 1989a). In addition, many people have been

unemployed so long that they have become *discouraged workers*—they no longer are looking for jobs because they believe it is useless to do so. These people are not even included in unemployment statistics because they are not looking for work. The U.S. Senate Special Committee on Aging reported that in 1984 the number of discouraged older workers was about equal to those who were unemployed and looking for work. Older workers were more than three times more likely than younger workers to be unemployed so long that they became discouraged workers.

Age Discrimination

Age discrimination is at the root of most older workers' employment problems. Age discrimination in employment occurs when, purely on the basis of chronological age or appearance of age, people are denied opportunities. They are not considered for job opportunities, not hired when they are fully qualified for the job, not promoted, not given opportunities for retraining, or not given beneficial transfers. Age discrimination also occurs when management uses age as a criterion for layoffs, transfers to less desirable jobs, reductions in benefits, or reductions in hours worked.

The Age Discrimination in Employment Act of 1967 made illegal the use of age as a criterion in hiring, firing, layoff, working conditions, and referral by employment agencies. The ADEA also prohibited job advertisements reflecting age preferences, age discriminatory pressures by labor unions, and retaliation against workers who assert their rights under the act. As amended in 1986, the ADEA also prohibited mandatory retirement from most jobs. However, because age discrimination has been assumed to stem mainly from ignorant assumptions about the effects of aging on job performance rather than dislike or intolerance of older people per se, prosecutors and the courts have treated age discrimination more leniently than racial discrimination (Eglit 1989). In addition, since 1980 the number of cases reported to the Equal Employment Opportunity Commission has increased dramatically, but the number of prosecutions of ADEA cases by the government has dropped substantially, which indicates the federal government's unwillingness to enforce the ADEA. As a result, most workers have only symbolic protection against age discrimination in employment.

The extent of age discrimination is unknown, but most adults think that it is widespread. For example, 80 percent of Americans agreed that "most employers discriminate against older people and make it difficult for them to find work," and more than one-third of employers believed that age discrimination is common (U.S. Senate Special Committee on Aging 1989a).

Stereotypes about Older Workers
Rosen and Jerdee (1976b) wanted to know how age stereotypes influence managerial decisions. To find out, they presented a group of forty-two business students in their 20s with an "in-basket" exercise involving six memos requiring action by a manager. To discover the effect of age, they referred to the focal person in the memo variously as "younger" or "older"; in addition, four cases included a personnel file with a photograph enclosed. Since each participant received only one version of the six cases, the manipulation of the age variable was unobtrusive. The six cases were as follows:

1. A recently hired shipping room employee who seemed unresponsive to customer calls for service

2. A candidate for promotion to a marketing job that required "fresh solutions to challenging problems" and "a high degree of creative, innovative behavior"

3. A candidate for a position that required not only knowledge of the field but the capability of making "quick judgments under high risk"

4. A request to transfer a worker to a higher-paying but more physically demanding job

5. A request from a production staff employee to attend a conference on "new theories and research relevant to production systems"

6. A request for a decision on whether to terminate or retain a computer programmer whose skills have become obsolete due to changes in computer operations

These cases were based on earlier work (1976a) showing that managers stereotyped older workers as resistant to change, uncreative, cautious and slow to make judgments, lower in physical capacity, uninterested in technological change, and untrainable.

The older shipping room employee was not expected to be able to change his behavior, and most participants thought he should be replaced. The younger employee was expected to be able to change, and most felt that a reprimand would solve the problem. Only 24 percent recommended promotion if they reviewed the older candidate for the marketing job, while 54 percent of those who reviewed the younger candidate recommended promotion, even though both candidates had identical qualifications. The older candidate came off worse for the risk-taking job too. Of those who reviewed the younger applicant, 25 percent recommended selection, as against only 13 percent of those who reviewed the older one. The older worker (age 56) who wanted a transfer to a more physically demanding job was significantly more likely to have his request denied. The older production staff employee was significantly more likely to have his request to attend a conference to update his knowledge turned down. The older computer programmer was less likely to be retrained and more likely to be replaced.

Rosen and Jerdee concluded that the participants' assumptions about the physical, mental, and emotional characteristics of older workers produced managerial decisions that were obviously contrary to the older employees' well-being and career progress. Decisions about hiring, retention, correction, training, and retraining all suffered purely because of the age of the focal person. Rosen and Jerdee cautioned that personnel directors may not behave the way their business students did; but other evidence on employment problems of older workers indicates that they probably do.

A major controversy surrounding the employment of older workers concerns the effect of aging on employability. Earlier we saw that a majority of older people continue to function physically and psychologically at a level well above the minimum needed for most adult performance. But what has research on actual work situations shown? Doering, Rhodes, and Schuster (1983) reviewed the results of over 150 research studies related to aging and work. Six of their major conclusions were:

- The attitudes and work behavior of older workers are generally congruent with effective organizational functioning.

- Job satisfaction is higher among older workers than in other age groups.

- Older workers are more loyal and less likely to leave their current work organization.

- Healthy older workers have lower rates of absenteeism than the young, but unhealthy older workers may have higher rates of absenteeism.

- Compared with the young, older workers are less likely to be injured on the job. However, if injured, older workers take longer to recover and are more likely to be disabled than other age groups.

- Older workers continue to learn.

In the vast majority of the studies Doering and her associates reviewed, "older worker" was operationally defined as someone age 55 or older. They could come to no firm conclusion on the issue of whether older workers perform as well as younger ones; in some studies they did, and in others they did not. The safest assumption to make, until there is proof to the contrary, is that age makes little difference in performance. Park (1992) came to similar conclusions. *Overall, the literature provides no justification for wholesale discrimination against older workers.*

Labor Market Problems

Older workers' labor market problems stem from job dislocations, displacement of homemakers into the labor market, declines in seniority protection, changing benefit costs of older workers, lack of opportunities for part-time employment, and problems of job reentry for retirees who decide to return to employment. Job dislocations occur as a result of plant closings, company reorganizations, mergers, or mass layoffs. For example, between 1979 and

At age 95, this stock broker has been employed by the same firm for 70 years.
Photograph by AP/Wide World Photos

1984, more than 5 million workers lost their jobs as a result of plant closings or declining employment within established industries (U.S. Senate Special Committee on Aging 1989a). Declines in employment were especially pronounced in relatively well-paying manufacturing jobs. From 30 percent of all jobs in the 1960s, manufacturing jobs declined to only 17 percent of all jobs in 1990 (Barlett and Steele 1992). Only 40 percent of those 55 or over who were displaced by this decline were able to find new jobs, compared with over 70 percent of those in their 20s. Parnes and King (1977) found that if experience, education, and technical competence are held constant, older workers tend to be laid off first. Those who suffered job dislocations tended to have had manufacturing jobs, to have worked for small business, to have had no private pension coverage, and to have worked for nonunionized companies. And as we saw earlier, once an older worker loses his or her job, age discrimination makes getting another job difficult (Love and Torrence 1989).

Displaced homemakers can also experience employment problems in middle age. These are women who chose homemaking as a full-time job and who, either through widowhood or divorce, find themselves in need of a paying job. Women who choose to be full-time homemakers generally assume that their economic support, health insurance, and retirement benefits will come as a result of their marriage to an employed man. However, women who become widowed or divorced in middle age can be thrust suddenly into the job market. They often lack marketable job skills. Because homemaking is not looked on as employment, homemakers lack a work history and are ineligible for Social Security or SSI; yet they are old enough to encounter age discrimination in the job market (Sommers and Shields 1979). That displaced homemakers have severe employment problems was recognized by the U.S. Congress in 1978 when displaced homemakers were categorized as "hard-to-employ" under the Comprehensive Employment Training Act.

Older workers have less protection from *seniority* than in the past. Seniority is the length of time a person has been employed by a particular employer. Under the seniority principle, employees with the longest service are laid off last, recalled first, and given first choice of promotional opportunities for which they are qualified. Seniority rights protect older workers with seniority from being discriminated against on the job because the most senior employees are generally older employees. Of course, newly hired older workers have no such protection because they lack seniority. As long as younger employees can look forward to achieving seniority themselves, they tend to view the seniority principle as just.

However, seniority rights are not automatic; they result mainly from civil service regulations and labor union contracts. Gersuny (1987) reported that the proportion of workers covered by seniority protections has dropped substantially as a result of the declines since 1955 in the proportion of union employees within various industries, especially steelworkers, machinists, and autoworkers. Lowered public support for unions, vigorous employer opposition to unions, and the National Labor Relations Board's growing tendency to rule in favor of employers in labor disputes suggest that further erosion of seniority rights is likely.

The changing *costs of employee benefits* for older workers also exert pressures on employers to avoid hiring older workers. There is little evidence that older workers as a category cost more than younger workers, either in absolute terms or in terms of productivity. But in 1982, Congress enacted several changes designed to reduce the amount of funds required by Medicare; these changes directly increased the fringe benefit costs of hiring older workers. If employers offered their employees health insurance, then beginning in 1983 this insurance had to be the primary insurance for older workers, even those who qualified for Medicare by being 65 or older. Private health insurance for people 65 or older is much more expensive than for younger age categories. In addition, if the employee had a spouse 65 or older, the employer's health insurance had to cover that spouse even if she or he were eligible for Medicare. These changes were estimated as having increased the costs of employing older workers by about $500 million per year (U.S. Senate Special Committee on Aging 1989a).

Lack of partial retirement options can also be a problem. Many older employees would like to retire partially. That is, they would like to draw part of their retirement pension but continue working at their same job part-time. However, employers have been very reluctant to develop partial retirement options. Quinn and Burkhauser (1990) reported that only 3 to 7 percent of retirement-age workers have the option of phased retirement. Indeed, since 1980, employers have aggressively used early retirement incentive programs to induce workers to retire early. Part of this action was in response to the need to reduce managerial employment because of the high volume of business mergers, and part was caused by an unfounded belief that getting rid of older employees is good for an organization (Seltzer and Karnes 1988).

Once retired, some people decide that they would like to *return to employment*, but they find that age discrimination often severely restricts their options. Quinn and Burkhauser reviewed the literature on partial retirement and found that about one-third of recent retirees reported being partially retired at some point, but it usually involved taking a new type of job at substantially lower pay. Economic necessity was the main motive for taking this type of employment.

Disability

As age increases in the population, so do rates of job-limiting disability. About 10 per-

Table 9-2 Prevalence of disability by age and sex: United States, 1978.

| | Disabled | | | | | | | | |
| | Severely[a] | | Partially[b] | | Not Disabled[c] | | Total[d] | |
Age/Sex	N	%	N	%	N	%	N	%
Males								
18–44	1,286	3.2	2,520	6.2	36,959	90.6	40,765	100
45–54	1,050	8.9	1,503	12.7	9,240	78.4	11,973	100
55–64	2,240	21.8	1,406	13.7	6,641	64.5	10,287	100
Females								
18–44	1,644	4.0	2,814	6.9	36,434	89.1	40,892	100
45–54	1,418	12.4	1,518	13.3	8,461	74.3	11,397	100
55–64	2,633	28.1	950	10.1	5,801	61.8	9,384	100

[a] Unable to engage in any substantial gainful activity in the national economy
[b] Limited in the amount or kind of work
[c] Numbers in thousands
Source: Computations by the author based on data from Greenblum and Bye (1987).

cent of adult men and women under age 45 have some degree of disability that either precludes or limits employment; this proportion increases to about one-third of the population at age 55 to 64. Past age 45, the proportion unable to work at all increases sharply with age for both men and women, but the increase in prevalence of complete disability is higher among women. By age 55 to 64, nearly one-fourth of the population is job-disabled (Greenblum and Bye 1987). The extent of job-related disability after age 65 is unclear, because most agencies that collect disability data assume that at age 65 an individual who leaves the labor force is retired, not disabled.

The causes of job-related disability also vary with age. McCoy and Weems (1989) found that among those disabled and under age 45, mental conditions, injuries, and nervous system problems were the most prevalent causes. Mental illness and mental retardation accounted for over 60 percent of disabilities of those under age 50 who were drawing SSI disability benefits. Among those disabled and age 50 or older, circulatory disorders, diseases of the bones and joints, and mental conditions were most prevalent.

Using cross-sectional data to infer life course patterns is questionable, but it appears that those who are out of the labor force due to disability in middle age and later consist of a small proportion of lifelong disabled people plus those who have been disabled in midlife by circulatory disorders such as heart disease, hypertension or stroke; by injuries; and by late-onset mental disorders. Iams (1986) found that compared with retired men, disabled men age 60 to 64 were more likely to have been in farming, craft, laborer, and machine operator occupations. However, disability occurred in all types of occupations, and older disabled men were more likely than younger disabled men to have been in executive, managerial, and professional occupations.

Thus, disability is an important factor in the reduction of labor force participation after the age of 45. Indeed, most investigators have found that disability accounts for between one-third and half of labor force withdrawals before the age of 62 (Parnes 1985). Disability in the years before retirement is much more likely among Mexican Americans and African Americans than among whites (Stanford et al. 1991).

Bridges to Retirement

The common conception many of us have of retirement involves working at a career job for a lengthy period of time and then abruptly and completely retiring from employment. But data from over 10,000 men and women who were age 58 to 63 in 1969 and who were resurveyed periodically until 1979 about their labor market experiences revealed a much different reality. Ruhm (1990) found that most people entered career jobs during their late 20s or 30s, but these jobs varied greatly in duration and often ended well before retirement. Only about 53 percent of men and 27 percent of women had held one job for as long as twenty years or more during their working life. Males, whites, and the college educated were more likely to have held long-duration career jobs. Career positions, defined as the longest spell of employment with a single employer, were usually followed by a period of postcareer or bridge employment. By age 60, more than half of the sample had left their career jobs. Only about 45 percent of workers retired from career jobs; 55 percent retired from bridge jobs, and the average duration of bridge jobs was five years. *Bridge jobs* span the time between leaving a career job and full withdrawal from the labor force. Bridge jobs are sometimes part-time, and they are often in a lower occupational classification and lower pay range than the career job. The bridge to retirement is for some a means of making the transition gradual and for others an economic necessity stemming from their need to supplement their job-related pensions until they become eligible for Social Security.

Retirement

For an individual, retirement can have several meanings. Retirement can mean withdrawal from employment, receipt of a retirement pension, subjective identification with the role of retired person, or the stage of the life course that follows employment. Although people can indeed give up such nonpaying positions as board member of a voluntary organization, church elder, or nursing home volunteer, the term *retirement* generally refers to at least partial separation from a position of employment. Retirement is a change in job status from full-time employed person to at least partially retired person, and full retirement usually constitutes a complete withdrawal from employment (Parnes 1989). Ruhm (1990) found that about 45 percent of the more than 10,000 people in the Retirement History Survey sample were partially retired at some time within the ten-year study. He estimated that the average period of partial retirement was five years. By age 70, over 80 percent were fully retired.

Elder and Pavalko (1993) looked at longitudinal data on the transition to retirement for two age cohorts of well-educated men: those born from 1900 to 1909 and those born from 1910 to 1920. The first cohort hit age 65 between 1965 and 1974; the second became 65 between 1975 to 1985. Elder and Pavalko found that nearly half of both cohorts showed a pattern of gradual retirement by reducing time worked. The stereotypical pattern of abrupt withdrawal from the labor force occurred for 37 percent of the earlier cohort but for only 22 percent of the later cohort. This suggests that for well-educated men the stereotype has never dominated and is declining in importance. On the other hand, in the more recent cohort 23 percent showed a pattern of multiple exits and returns to the labor force, compared with only 13 percent for the earlier cohort. This suggests that the transition to retirement is becoming more chaotic from the standpoint of labor statistics.

Note that retirement relates to jobs, not work. In general, people never stop working; they simply retire from positions of employment. Receipt of a retirement pension differentiates retired people from others who are out of the labor force, such as the unemployed or the disabled. Part of the process of retirement involves a shift in personal identity from employed person to retired person. The individual usually does not abandon the former occupational identity but instead adds the identity of retired person. The fact that retirement is a life stage with its own prerequisites for success focuses attention on the need to plan for retirement.

How retirement is *operationally defined* influences our picture of the extent of retirement. Operational definitions convert abstract concepts such as retirement into observable facts. For example, in their national sample of men 60 and over, Parnes and Less (1985) found that 70 percent were receiving a retirement pension, but only 57 percent were working fewer than twenty hours per week and receiving a retirement pension. Thus, the operational definition used made a large difference in the proportion classified as retired. Ekerdt and De Viney (1990) concluded that by collecting data on pension receipt, reduced employment or earnings, and self-definition of retirement, researchers would have the data necessary to use a variety of operational definitions of retirement, depending on the purpose of the study. Their point is that there is no single definition of retirement that fits all purposes.

Conceptually, both receipt of a retirement pension and withdrawal from employment are important and necessary parts of an adequate operational definition of retirement. For example, many former military personnel draw retirement pensions but are employed full-time. In terms of labor force participation, they are not retired. On the other hand, people do not qualify as retired merely by being out of the labor force. For example, there are chronically disabled people who have never been employed; it makes no sense to call them retired. The pension receipt and reduction in employment criteria illustrate important dimensions of retirement: Retirement is an earned position based on service in the labor force, and its main attributes are a reduction in, or cessation of, employment and a shift in the source and usually the amount of individual or family income.

The Link Between People and Jobs

Retirement marks the end of what is generally thought to be a close relationship between the kind of job an individual holds and the kind of lifestyle and livelihood the person enjoys. This relationship results, of course, from the fact that, for most people in the United States, employment is the only fully acceptable way of making a living. Americans tend to identify a job with a style of life, and this way of assessing a person's status is usually correct, for job usually determines income, and income sets limits on lifestyle.

Another factor in the relationship between people and their jobs is the idea that a career can be pursued as an end in itself, purely for the satisfaction of doing a fine job. A related notion is that job careers can give people their greatest opportunities for creativity and in this way can become their entire life. Mills incorporated these ideas into his concept of craftspersonship:

Crafts[person]ship as a fully idealized model of work gratification involves six major features: There is no ulterior motive in work other than the product being made and the processes of its creation. The details of daily work are meaningful because they are not detached in the worker's mind from the product of the work. The worker is free to control his [or her] own working action. The crafts[person] is thus able to learn from his [or her] work; and to use and develop [her or] his capacities and skills in its prosecution. There is no split of work and play, or work and culture. The crafts[person]'s way of livelihood determines and infuses [the person's] entire mode of living.

(1956,220)

Mills also said:

Work may be a mere source of livelihood, or the most significant part of one's inner life; it may be experienced as expiation or as exuberant expression of self; as bounden duty, or as the development of . . . universal [human] nature. Neither love nor hatred of work is inherent in [human beings], or inherent in any given line of work. For work has no intrinsic meaning.

(1956,215)

Understanding retirement thus involves understanding the nature of the links between people and their jobs. It can be said with some certainty that this relationship has changed over the years. In the peasant economy, there was no concept of occupational position or employment. In the days of medieval guilds, the idea of a vocation as a way of life was generally accepted. Not many people lived to become old in those days, but those who did continued to be employed until they grew too feeble or died. Since there were no provisions for gaining income in any other way, people worked as long as possible. The Industrial Revolution, assembly line production, and automation have made some important changes in this picture.

First, our system has become increasingly productive; today, increasingly fewer people are needed to produce the nation's economic output. This increase in productivity was made possible largely by the switch from low-energy power converters such as animals, wind, and water to high-energy converters such as steam, internal combustion, and electricity (Cottrell 1955). Also, the demographic revolution reduced deathrates, particularly in infancy, and prolonged life. Birthrates also declined. The eventual result meant a smaller proportion of young dependents to support, and this decline meant that it was possible to defer income in order to support retirement.

The rise of industrial production systems increased the need for planning and coordination within the economic system. Mechanisms such as the corporation and the bureaucracy were introduced into the economic system and gradually spread throughout society. Govern-

ment bureaucracy grew to become the political counterpart of the economic corporation. Once the state became a large-scale organization, it was capable of pooling the nation's resources and allowing a segment of society to be supported in retirement. It was thus no accident that the first retirement program, inaugurated in 1810, consisted of pensions for civil service workers in England. America's failure to introduce civil service pensions until 1920 can be partly explained by the slow development of our national government.

The rise of industrialization brought with it urbanization and extensive residential mobility. These two factors combined to change the traditional ties to family and local community that had been the source of a great many different kinds of services. These trends resulted in a redefinition of the relationship between the individual and the government. At the same time, the wage system was putting an end to the traditional economic functions of the family. All these trends combined to give individualism more weight in the relationship between the person and society.

The Evolution and Institutionalization of Retirement

Accordingly, the link between people and jobs came to be separated from political and family life. One important result was to reduce the amount of cohesion that people could gain in their lives through their jobs. Another result was that the new relation between the individual and the state set the stage for establishing an institutionalized right to financial support in one's old age, partly as a reward for past contributions to the economic system and partly as a result of deferring income during one's years of employment. Thus, Social Security and employer pensions were established, and with their introduction the institution of retirement emerged in the United States.*

---

*For more detail on the development of retirement as a social institution, see Atchley (1982c).

As the concept of retirement has continued to mature, it has come to embody the idea that, by virtue of a long-term contribution to the growth and prosperity of society, its individual members earn the right to a share of the nation's prosperity in their later years without having to hold a job. For example, Ash (1966) found that between 1951 and 1960, steelworkers shifted substantially in terms of how they justified retirement. In 1951, the majority said that retirement was justified only if the individual was physically unable to continue. By 1960, the majority of steelworkers felt that retirement was justified by years of prior service on the job. The idea that the right must be earned is related to the fact that retirement is reserved only for those who have served the required minimum time on one or more jobs.

Industrialization and the rise of the corporation led to the development of the concept of *job,* as opposed to *craft,* and generated a large economic surplus that could be used to support adults who were not jobholders. The rise of strong national government created the political machinery that could then divert part of the economic surplus to support retired people. The depression provided strong incentive to cut down on the number of older workers competing for scarce jobs. Because routine industrial and bureaucratic jobs are often boring and meaningless to the people who do them, the promise of retirement was also an incentive for people to endure a growing number of repetitive jobs. And gradually the average jobholder has come to accept the idea that people can legitimately live in dignity as adults without holding jobs, provided they have earned the right to do so. All these conditions helped establish retirement as a legitimate institution in American society.

Retirement's main function for the society was and is to keep down the number of persons holding or looking for jobs. Providing a reward for service or supporting people too old or disabled for employment were important but secondary considerations. But retirement was initially sold to the public primarily as a means of supporting people who were physically unable to hold jobs. This linkage was expedient at the time but had the long-run disadvantage of inappropriately connecting the concept of retirement with the concept of functional old age and incapacity for employment. Thus, when 65 was set as the "normal" retirement age, the public made the incorrect logical inference that 65 also meant old age and functional incapacity. It is important to recognize that while retirement was historically linked with old age, this linkage has never had substantial basis in fact. And as the "normal" retirement age has crept downward and the health of the older population has improved, the association of retirement with old age and disability has become even less applicable. But it is also important to remember that the goal of retirement is not to deal with aging workers but to identify a category of people who can legitimately be excluded from a too-large pool of people who need employment.

These long-term cultural trends are important because they have shaped people's concepts of retirement as a legitimate individual choice. In addition, as economic conditions have changed, the role of retirement has become more complex. Retirement began as a measure designed mainly by political and economic leaders to control the labor market, but it has evolved into an individual right to retire that is earned through years of employment and is highly prized by most employees. The evolution of retirement shows that leaders may make policies, but how the people react to these policies is unpredictable. In 1950, no one believed that American workers would ever learn to want retirement, but by 1980, more than 80 percent of adults 50 and over had positive attitudes toward their own retirement (Atchley 1982b).

The Retirement Process

The retirement process begins when people recognize that they will some day be making choices about retirement. They may choose to retire or not. If they do retire, they face many decisions, such as whether to take part-time employment, how much income they will need, whether to move to a new community, and what goals they wish to pursue. In studying individual retirement,* we are interested in how people view their jobs and the prospect of retirement, the preparations people make for retirement, retirement decision making, the retirement event, and the dimensions of retirement as a social role. We are also interested in the effects of retirement and how people adjust to them. Retirement is a complex social institution that is intertwined with the economy, the family, and the life course of individuals. As a result, several different approaches are required for a balanced understanding of individual retirement.

Aging and Attitudes Toward Jobs and Employment

It seems logical that to be positive about retirement, people would first have to become negative about jobs and employment, but this is not the case. People can be positive about both work and retirement at the same time. Doering, Rhodes, and Schuster (1983) reviewed the literature on the relationship between age and attitude toward work and found no age differences in the value placed on having meaningful, fulfilling, and enriching work. Employees of all ages felt that having satisfying work was important. Hanlon (1986) reviewed eighteen studies of commitment to work and found that the proportion with a strong work commitment was

* The role of retirement in the economy and in the politics of aging is discussed in Chapter 3, "History"; Chapter 18, "The Economy"; and Chapter 19, "Politics and Government."

somewhat greater in older age cohorts. He found that a large majority of all age cohorts had a strong commitment to work. In terms of job satisfaction in the twenty-three studies they reviewed, Doering and her colleagues found that job satisfaction generally tended to be higher in older cohorts.

Factors that have been advanced as causes for the increase in job satisfaction in older cohorts include older employees' declining expectations from work, generational differences in initial job satisfaction, higher likelihood of those who are satisfied remaining in a job, and better working conditions for older workers. McNeely (1988b) looked at the relative merits of these potential causes of age differences in a sample of human services workers. He found that older workers had higher job expectations—they were more likely than younger workers to expect intellectually rewarding and stimulating assignments. Over time, successive cohorts had lower mean job satisfaction when they entered the job market than they did later on, which meant that age differences were not the result of generational effects. Those who survived on the job ended up with better working conditions, not the other way around. Thus, McNeely concluded that better working conditions accounted for most of the age difference in job satisfaction. As he pointed out, new, and usually younger, workers start off with the least desirable duties but progress in time to job duties that require more skill, provide better pay and benefits, and permit more autonomy. These improvements over time in the conditions of work produce increased satisfaction on the job.

Lowther, Gill, and Coppard (1985) looked at age differences in sources of job satisfaction among high school teachers. They found that younger teachers received more satisfaction from intrinsic aspects such as the extent to which their jobs allowed them to develop, provided interest, and used their ability to be self-directive. On the other hand, older teachers got more satisfaction from external aspects of their jobs, such as good working conditions, pay and fringe benefits, relationships with their coworkers, and career advancement.

The size of the general increase in job satisfaction masks the fact that a significant minority of workers experience declining job satisfaction, often beginning in their 50s. For example, in Karp's (1987) sample of professionals who were all in their 50s, how men experienced their jobs had shifted significantly after the age of 50, sometimes not for the better. Most professional men saw their 30s as a fast-paced and demanding period during which they made their initial marks on their professions and their 40s as a period of steady advancement during which career building reached its culmination. But by their 50s, some professional men saw their jobs as having become repetitive, boring, and unchallenging, especially compared with the excitement of developing their careers. Many of Karp's respondents said that their jobs had changed very little in the previous ten years. As one physician put it, "There's a certain repetition to my work. A lack of challenge, I'd say. I mean, the practice of medicine doesn't demand that you use all your knowledge. Most things are fairly routine" (1987,218). Those with lessened job satisfaction responded in a variety of ways. Some tried new career options, some "retired" on the job—just took it easy and saw their jobs mainly as a source of income—and others looked to other roles for their satisfactions. Karp divided those who experienced continued satisfaction with their jobs into two major categories: men who had risen to the very top of their professions and stayed there, and men who in their 50s concentrated on passing on to the next generation the wisdom they had gained through their job experiences.

These various studies suggest that at the beginning in an occupation, people are absorbed by the challenge of learning to do the job. Once a solid base of competence has been laid down, keeping current is often merely a by-product of the work itself. At this stage, which often occurs in later middle age, the person turns to extrinsic rewards such as relationships with coworkers or pay increases for new satisfactions. Exercising competence is usually a continuing source of satisfaction, but any increase in satisfaction is related more to extending one's horizons to include other people at work and to include a larger part of the field within which one works.

It is important to remember that being satisfied with one's job is not the same as reluctance to leave it. I have interviewed many people of retirement age who like their jobs very much but at the same time are ready to move on to the next stage: retirement. Now let us look at attitudes toward retirement.

Attitudes Toward Retirement

Attitudes about retirement are generally favorable. In two studies of attitudes toward retirement in people 45 and over (Atchley 1974, 1982a), I found that the concept of retirement had four independent dimensions: activity, physical capacity, emotional evaluation, and moral evaluation. On all four dimensions, attitude toward retirement was very positive, regardless of the respondent's age or gender. The only exception to this very positive picture of retirement was among retired people who would have liked to have continued on the job. Some of these reluctant retirees felt that retirement was unfair and bad. Most adults expect to retire (less than 10 percent do not), and most expect to retire before 65 (Goudy 1981; Ekerdt, Bossé, and Glynn 1985).

Generally, attitude toward retirement is correlated with the degree of expected financial security. About four-fifths of employed adults over 50 expect minimal financial difficulties in retirement, although most expect retirement to significantly reduce their income. No subgroup in the labor force appears to be negative about retirement, but those with expected retirement income needs higher than their expected retirement incomes tend to be less positive, for obvious reasons.

Having a negative attitude toward retirement is uncommon and generally has been found to be unrelated to social characteristics such as age, gender, social class, or occupation (Atchley 1982a; Atchley and Robinson 1982). However, in Karp's (1989) in-depth study of seventy-two professionals in their 50s, more than 35 percent were so invested in their work that the thought of retirement was completely negative. Negativity toward retirement was most likely among professionals who had unfinished agendas on their jobs, had high job satisfaction, expected a substantial negative effect of retirement financially, and were healthy. Negativity was also strongly related to not having a vision or fantasy of what life in retirement would be like. Those who were negative either could not imagine what retired people do or imagined people sitting around with nothing to do or engaging in meaningless leisure activities. Fortunately, most people do not have difficulty imagining what life in retirement would be like and seeing value in what they imagine.

Fletcher and Hansson (1991) developed a scale to measure anxiety about retirement in the preretirement stage. They found that retirement anxiety was not widespread and that it tended to occur mainly in people who generally found social transitions difficult. These tended to be people who were shy, lonely, and lacking in interpersonal skills, and who expected to have little personal control of their lives after retirement. It is important to remember that having a negative attitude toward retirement does not become a self-fulfilling prophecy. I have interviewed many people who were anxious about retirement in the preretirement period but whose experience of retirement was very positive. Also, remember that the vast majority of people have positive attitudes toward retirement.

Retirement Preparation

Since nearly everyone expects to retire, it is useful to examine the planning and preparation that precede retirement. Evans, Ekerdt, and Bossé (1985) looked at how often men talked about retirement with their wives, other relatives, friends, and people on the job and how often they read articles about retirement. Not surprisingly, they found that as retirement drew closer, frequency of informal preparations for retirement increased. However, even fifteen years prior to expected retirement, some men were already talking with others and reading about retirement. Only a small minority of people make formal preparations for retirement, such as preparing a budget or developing written goals, but most have a general idea of what they will do in retirement and a reasonably accurate picture of their retirement income. Only about 5 percent of workers participate in formal retirement planning programs, but most people manage to adjust quite well to retirement without formal preparation (Atchley 1982b). Participants in formal programs tend to be people with relatively high incomes and education who have already been planning for retirement. The main psychological effect of retirement planning programs is to reinforce preexisting positive orientations toward retirement (Atchley, Kunkel, and Adlon 1978).

The retirement income picture has improved substantially since 1970, but at present the most important element of retirement preparation is still financial planning, especially for those who want to have an annual retirement income of more than $35,000 (in 1993 dollars). Below $35,000, a combination of Social Security and employer pensions are likely to be sufficient. Finding the right balance between present spending needs and the need to save for retirement is tricky. Many middle-class people find it difficult to save very much until they are beyond age 50. But once children are launched, many

Effective retirement planning emphasizes individual preferences.
Photograph by Chuck Keeler/Tony Stone Images, Inc.

middle-aged people find that they are able to save substantial amounts. Of course, retiring early is difficult unless one starts saving substantial amounts early or is eligible for a lucrative early retirement pension.

If retirement financial planning focuses on income needs throughout retirement as a life stage, as opposed to just looking at income needs at the beginning of retirement, then additional factors that enter into financial plans include the costs of caring for older parents and long-term care for oneself and/or one's spouse. And as one advances in age as a retired person, more services will need to be purchased, as opposed to performed oneself, to maintain the same style of living. Thus, it is not unreasonable in any way for retirement financial planning to include continued saving in retirement.

Because retirement is a do-it-yourself role in terms of what is expected, it is difficult to prescribe other necessary elements of retirement planning. But many people want to know about how to find part-time employment should they want it, how to develop new life goals, and how to decide where to live. Retirement planning programs are also good opportunities to provide information about issues of aging, such as retaining good health, keeping active, legal affairs, caring for parents, long-term care, and housing, but these concerns are not essentially aspects of retirement planning.

Gradually increasing the time spent in roles other than the job and gradually decreasing employment can be particularly useful preparation for retirement. This tapering off can be accomplished by gradually increasing the length of annual vacations, but a method perhaps more in tune with the goal of practicing retirement is to gradually decrease the number of hours in the workday. This pattern is very common among self-employed professionals such as physicians and lawyers. For example, Quinn (1981) found that among men who retired between the ages of 58 and 63— early retirees—about 60 percent of the self-employed were partially retired, compared with less than one-third of wage and salary workers. As mentioned earlier, employers have been reluctant to offer part-time employment options to employees of retirement age. Thus, employer policies often force retirement to be all or nothing. Bridge jobs at lower pay and status are the main prospect for part-time employment.

Access to formal retirement planning programs is spotty. The American Association of Retired Persons (1986) found that only 23 percent of workers age 40 or over reported that their employers offered retirement planning programs. People who work for large employers are more likely than those who work in small businesses or the self-employed to have an opportunity to participate in such programs. Programs in churches or community centers are sparse. Even among those with access, most employees apparently do not think they need a formal program. For example, in one study, only about 10 percent of employees who had access to retirement planning programs participated, and those who did tended to be those who had engaged in lifelong planning (Campione 1988).

The question of who needs exposure to formal retirement planning is problematic. Gerontologists commonly have contended that low-income employees, especially women and minorities, need access to retirement planning because they are more likely than others to have low retirement incomes (Szinovacz 1982; H. Dennis 1989). However, Social Security replaces a significantly higher proportion of total preretirement income for low-income couples than for moderate-income couples. The average replacement is nearly 60 percent for couples. For single people, Social Security replacement rates are less than 40 percent. Therefore, single people seem to be a prime category of people needing to do retirement financial planning. Another issue relates to the individual's capacity to take advantage of financial planning help. People with low incomes cannot afford retirement income options such as annuities and investments. They rely mainly on Social Security, so their major route to better retirement income is through a better job. Retirement planning programs are not likely to help in this regard.

Middle- and upper-income retirees get a much bigger proportion of their income from pensions and investment income. These people need financial planning to protect their assets and maintain their income. The major problem they have is getting objective investment advice, both with regard to how to distribute their pension contributions—if their plan allows them to select investment options—and how best to diversify their savings. Most financial services providers make their money on commissions from selling specific products and cannot be counted on to provide the most objective advice.

The Decision to Retire

As a process, a rational decision to retire involves three elements: (1) how individuals come to consider retirement, (2) the factors that influence the consideration of the retirement option, and (3) the factors that influence the timing of retirement. Figure 9-2 is a schematic that shows how these elements interact in the decision process. It assumes that the individual is eligible to retire and is not retiring from one full-time job to take another.

Considering Retirement   People mainly come to consider retirement because it is a desired goal. But people can also come to consider retirement through unemployment, reaching a mandatory retirement age, experiencing illness or disability that makes the current job difficult or impossible to perform, family needs, peer pressure to retire, or employer pressure to retire. For example, among new Social Security retirement pension recipients, 56 percent of men and 62 percent of women cited voluntary reasons for retirement. Most men simply wanted to retire, but women also cited family reasons for wanting to retire. Health problems accounted for 60 percent of involuntary retirements, job dislocations for 29 percent, and mandatory retirement for 11 percent (Ozawa and Law 1992).

In all these cases, the individual may or may not consider retirement, depending on his or her felt need to continue employment. If a job separation is involved and the person feels a need to continue employment, then a job search can be expected, and, if successful, retirement will not be considered. However, if there is no felt need to continue employment or if the job search is unsuccessful, retirement will be considered. In these cases, retirement is considered as a result of external pressures.

In the process of considering retirement, two comparisons are crucial. The first is a comparison of expected financial needs in

Figure 9-2 Factors in the decision to retire.

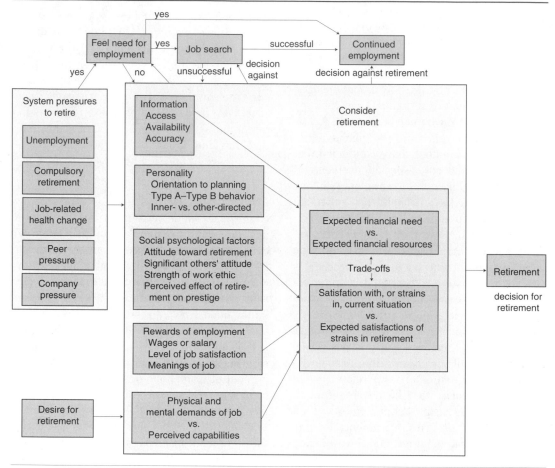

Source: Atchley (1979).

retirement with expected financial resources. The second is a comparison of satisfaction with, or strains in, the current social situation with expected satisfactions or strains associated with retirement. Some trade-offs can be expected to occur between these two comparisons. For example, if the satisfactions of retirement are expected to be very great, then the person may be willing to settle for a lower degree of income adequacy. These two basic comparisons and the trade-offs between them can be influenced by numerous additional factors. Access to information, availability of information, and accuracy of information have an impact on both compar-

isons, as do various personality factors. For example, whether the individual is active or passive in relation to her or his social environment has an influence on the degree to which that person is oriented toward planning, which in turn has a bearing on whether retirement will be considered rationally.

Other factors influence only the comparative satisfactions. Both the economic and noneconomic rewards of employment will have a bearing on the comparison of current satisfaction with expected satisfaction. Particularly relevant here is the meaning the person attributes to employment. Other social psychological factors such as attitude toward retirement, perceived attitudes of others toward retirement, strength of the "work ethic," and perceived effect of retirement on prestige can also influence this comparison. Finally, this comparison is also affected by the individual's perception of his or her capacity to meet the physical and/or mental demands of employment. If retirement compares favorably with continued employment or job search, then the individual will be motivated to retire. The timing of retirement is particularly sensitive to the income comparison. If retirement does not compare favorably, then the person can be expected to continue employment or to engage in a job search.

Of course, the decision to retire and the timing of retirement may not be the result of a rational process. For example, many workers accept the idea that retirement at age 62 or 65 is "normal," and nothing happens in their working life that requires them to think actively about retirement. These people retire when they are "supposed to," without considering that there might be an alternative.

Pressures to Retire   The foregoing ideas about the retirement decision are very general. Now let us look at more specific influences. Both attitudes toward retirement and planning or preparation for it have an impact on the decision to retire. Other important factors include the hiring and retirement policies of employers and the health of the individual.

Employment Problems   First, an individual may decide to retire because she or he cannot find a job. Hiring policies tend to discriminate against older workers. About the only jobs that are not harder to find as one grows older are those that pay badly or for which there is a chronic labor shortage. This pattern has been found in the United States, West Germany, and Japan, and probably applies to most industrialized nations. When asked to defend their hiring policies, employers usually say that older people cannot meet the physical or skill requirements of the jobs. There is apparently no foundation for these allegations; nevertheless, they are acted on as if they were true.

Ruhm (1989) estimated that about one-third of household heads leave the longest job they have held in their lifetime before the age of 55 and reported that job dislocation rates of older workers had increased significantly since 1970. These trends are related to older workers' labor market problems discussed earlier. Following job displacement, older workers are often forced to take bridge jobs at significantly lower pay to provide income until they reach the age or length of service required for eligibility for retirement pensions.

Employer Policies   On the other side of the coin are employer policies that allow employees to generate entitlement to "early" retirement. Most researchers consider retirement before the age of 62 early, but others contend that retirement before 65 should be considered early since 65 is viewed as the "normal" age for retirement. Both definitions stem from the fact that Social Security retirement benefits are geared to age 62 for reduced benefits and age 65 for full benefits. The U.S. General Accounting Office (1986a) reported that the proportion of the population receiving employer-sponsored pensions before age 65 increased substantially between 1973 and 1983.

Employer policies influence individual decisions about retirement mainly through rules for pension eligibility. Such rules include minimum retirement ages, service requirements, and benefit formulas. One common early retirement pattern allows full retirement benefits at age 55 with thirty years of service. For example, Mr. V. worked for a company with such a policy. Because he began working for the company at the age of 28, he satisfied the thirty-year service requirement at the age of 58. At that time he became eligible for an annual pension equal to 60 percent of the average of his highest five years of salary. If he stayed longer, he generated no greater pension entitlement except to the extent that delaying retirement slightly increased the value of his highest five years of salary. Thus, Mr. V. stood to lose substantial pension income if he remained on the job. He decided to retire. Mr. L. had a similar situation, with one important difference. He also was 58 with thirty years of service, but under his company's plan, if he delayed retirement, his benefit increased 2 percent for each additional year of service beyond the thirty. He decided to wait until 65 to retire, at which time, based on thirty-seven years of service, he received 74 percent of the average of his highest five years of salary. Thus, although both companies allowed early retirement, Mr. V.'s company had a policy that penalized employees for not retiring early, whereas Mr. L.'s company had a policy that encouraged delayed retirement. Note that for both companies, full retirement benefits were defined as 60 percent replacement of the average of the highest five years of salary.

Employer-sponsored pension programs vary widely with regard to minimum retirement age, service requirements, and benefit formulas, and even within the same program there are often many options. Therefore, people of retirement age certainly are not created equal in terms of the conditions under which they are eligible to retire. However, in recent years, the propensity of employer policies to encourage early retirement has increased. Some have developed early retirement incentive programs that further encourage early retirement by lowering minimum retirement ages or purchasing additional years of service credit for employees who wish to retire. Some employers develop these programs as a way of coping with the effects of technological changes, mergers, plant closings, and labor force cutbacks. As one personnel director told me:

We have to cut our workforce by 150 employees. Every person I can encourage to retire early means a younger person with a family I don't have to lay off. If I lay off the young guy, his unemployment benefits will eventually run out. But the older guy's retirement pension checks will keep on coming indefinitely.

However, other employers develop early retirement incentive programs simply because they want to get rid of older employees.

Seltzer and Karnes (1988) and Windus (1987) studied university early retirement incentive programs and found that management's main goals were to reduce costs by bringing in lower-cost faculty to replace those who retired, to bring in newly trained faculty, and to get rid of less productive faculty. What actually happened did not completely meet any of these expectations. Retiring older faculty did not save money. Surprisingly, faculty just entering the field were entering at salaries not far below the salaries of faculty on the verge of retirement. When the costs of recruiting were added in, the net effect was no savings. Newly trained faculty did indeed possess more current technical knowledge, but they lacked the know-how involved in working with other people and operating within a university system. On balance, the universities lost as much knowledge as they gained. Finally, faculty who took advantage of the early retirement incentive program were a mixed bag. Some were professors who had stagnated and were just waiting to reach retirement age, but others were among the most productive research faculty in the universities. Even though these early retirement incentive programs usually did not accomplish their goals, most university administrators said that they would offer the program again, illustrating that management decisions are not necessarily rational. Indeed, most administrators who chose to adopt an early retirement incentive program did so based on their personal opinions. In no case among the nineteen universities was a cost-benefit analysis of the expected effects of the program done prior to making the decision.

Compulsory retirement age policies require employees to retire by a specific chronological age. The Age Discrimination in

Employment Act prohibits mandatory retirement age rules in most jobs, but even when such rules were allowed, less than 10 percent of employees were actually forced to retire (Parnes and Nestel 1981; Ruhm 1989). After 1986, mandatory retirement age rules were legal only for those in highly specific occupations, such as commercial airline pilots, law enforcement officers, and commercial bus drivers. More than 85 percent of the general public thinks that mandatory retirement age rules are wrong if the individual wants to continue and is still able to do the job. The legal abolition of mandatory retirement age rules was mainly a highly popular symbolic gesture that outlawed rules that had little practical effect. But as we saw earlier, age discrimination in employment is still very much with us.

Other Pressures   Poor health makes employment difficult and encourages retirement. Parnes and Nestel (1981) found that 28 percent of their national sample of men who retired between ages 62 and 64 did so because of poor health. Sally Sherman (1985) studied new retirees under Social Security and found that the proportion retiring because of poor health dropped from 54 percent in 1968 to 29 percent in 1982. No doubt variations in working conditions, such as the physical demands of the job, are important determinants of the extent to which poor health actually affects the retirement decision. Colsher, Dorfman, and Wallace (1988) found that among men, heart attacks, cancer, and diabetes were the health conditions that most often caused retirement. Among women, hypertension was far and away the most common cause. Not only do these conditions in and of themselves make employment difficult, but the stress connected with having serious illnesses and the medications used to treat them can also impair job performance.

Undoubtedly, an individual's decision to retire is also influenced by the informal norms of the work situation. For example, most professionals are probably discouraged

from retiring by their colleagues' attitudes. Since retirement tends to be viewed negatively among top executives, to retire is to buck the system. On the other hand, routine production and clerical workers are generally favorable toward retirement and indeed tend to retire early. It would be a mistake to assume that two individuals anxious to retire face the same decision in these two quite distinct occupational areas.

Attitudes of friends and family also probably play a role in the retirement decision. One man's children may want him to retire; another's may not. One woman's husband may want her to retire; another's may not. One man lives in a neighborhood where retirement is sneered at; another lives in a leisure-community where retirement is the rule rather than the exception. If we retire, then the characteristics our friends and family impute to retired people will be imputed to us. All these factors may encourage or discourage retirement.

Employers sometimes exert pressure on employees to retire. They may do so by suggestion or by job transfers or job reclassifications that substantially alter the employee's working conditions. There are countless ways that employers can apply pressure to retire, but as yet the proportion of older workers who experience such pressures is unknown. However, the fact that three-fourths of complaints of age discrimination in employment concern conditions of employment other than hiring or firing suggests that employer pressures to retire may be common.

Timing   For most people, the retirement decision is more a matter of when than if; 9 out of 10 people plan to retire, and the small proportion who do not will find it easier to stay on as long as they are willing to work full-time and able to do the job. The timing of retirement depends mainly on personal preference, job history, pension rules, and health. As we saw earlier, most people prefer to retire at the minimum age of eligibility for what they perceive to be adequate retirement benefits. The early retirement bias of most employer pension plans coincides with this preference. However, pension rules can cause people to delay retirement in order to gain access to more adequate pensions, either by coercive methods such as service requirements or incentives such as benefit increases for delayed retirement. People with interrupted work histories can find themselves having to delay retirement in order to meet service requirements.

Women, much more than men, are apt to find the timing of their retirement influenced by their marital status. Married women are more likely to retire early because they rely on their husbands' pension eligibility and husbands are generally older (Atchley 1982a). On the other hand, single women are more likely to have to stay on past 65 in order to meet service requirements and because the pensions they are entitled to are inadequate as a result of their having had relatively low-paying jobs. This factor will decline in importance in the future because women today are less likely to have interrupted work histories, but the gender gap in pay and access to employer pension coverage still persists.

As we just saw, poor health promotes early retirement but has diminished as a retirement decision factor in recent years. Nevertheless, there are mature workers whose health is poor, but not poor enough to qualify them for disability benefits. Early retirement with reduced pension benefits is their only real choice.

People generally are able to time their retirement to match their expectations. Quinn and Burkhauser (1994) reported that 65 percent of men and 55 percent of women expected to be able to retire at their desired retirement age. Only about 10 percent expected to have to retire earlier than they wanted to; but 32 percent of women and 25 percent of men expected to have to work longer than they wanted to. Ekerdt, Vinick, and Bossé (1989) found that 66 percent of the men in their longitudinal sample accurately predicted their year of retirement.

Who Does Not Retire?

In a national longitudinal study of the labor force behavior of men, Parnes and Summers (1994) were able to look at the characteristics of men who remained employed from their 50s until their 70s or 80s. Of the more than 5,000 men who responded to the study when they were age 45 to 59, about 20 percent of the 69 to 74 age category were still in the labor force twenty-five years later and about 12 percent of those 75 and over were still employed. About one-third were employed full-time, mainly in the 69 to 74 category.

What factors predicted continued employment into later life? Continuing good health was an obvious prerequisite. But attitudes were also important. Those who stayed in the labor force had a long-standing negative attitude toward retirement. That is, they expressed negative attitudes toward retirement long before they reached retirement age. Those who remained employed also had higher education. But they also had relatively low incomes, which suggests that need for income is also an important motivator. Parnes and Summers speculated that cohorts entering traditional retirement age categories in the future will have better health, more education, and less retirement income security, and therefore, a greater proportion of elders may reject retirement.

Retirement as an Event

When retirement occurs at a specific predesignated time, it may become an occasion for a ceremony to mark the end of employment and the beginning of retirement, or the passage from one role to another. Unfortunately, we do not know very much about retirement ceremonies or the part they play in the transition from employment to retirement. Certainly, the stereotypical "gold watch" ceremony has never been part of most workers' retirement experience, and as retirement has become increasingly more desired and commonplace, the austerity and mock honor of the stereotyped ceremony seem even more out of place.

The most notable characteristic of retirement events is their complete lack of standardization (compared with ritualistic events such as graduations or weddings). Retirement ceremonies range from highly personalized, informal affairs involving current coworkers only to all-day tributes and speeches involving colleagues from around the country. Sometimes the occasion is a dinner party for family members as well as coworkers; others are late afternoon events at the workplace. Most occasions are arranged by the employer, but some are planned by friends at work. Some are mass ceremonies honoring several people approaching retirement; others are for a particular individual. People whose job performance was highly respected tend to be singled out, whereas ordinary performers tend to be honored as part of a group. When speeches are made, they tend to focus on acknowledgment for past contributions, with much less emphasis on the transition or the "new career" as a retired person. However, informal conversations usually display plentiful expressions of envy and good wishes. The atmosphere is seldom negative.

We do not know how common retirement ceremonies are, but it is probably safe to say that they are most likely for those whose upcoming retirement is known well in advance. Those who retire unexpectedly or gradually or because of ill health or unemployment would seem less likely to experience a ceremony.

Retirement as a Role

After the retirement event, be it formal or not, the individual is expected to assume the role of "retired person." Gerontologists have had a lengthy debate over the nature of the retirement role and, in fact, over whether such a role even exists. Most people agree that there is a "retired person" position that individuals enter when they retire. The dis-

agreement comes when specifying the role associated with that position. The term *role* can refer to either the culturally transmitted norms governing the rights and duties associated with a position in society or a relationship between holders of complementary positions.

The rights of a retired person include the right to economic support without holding a job (but at the same time without the stigma of being regarded as dependent on society, as in the case of the unemployed); the right to autonomy in managing time and other resources; and often more specific rights associated with the former job, such as membership in an employer-sponsored retiree organization or access to employer health facilities.

The retirement role also involves duties, those things expected of a retired person. The foremost duty is to avoid full-time employment. This expectation is backed up by the provision in Social Security of a benefit deduction for retired people with annual earnings from employment that exceed the allowable amount. Retired people are also expected to carry over into retirement their skills, experience, knowledge, and identity with the jobs or positions they had. They are sometimes expected to provide free services to community projects or organizations. In addition, retired people are expected to assume responsibility for setting their own goals. For a great many people, retirement adds a lot of decision-making responsibility, which, in addition to the continuing responsibilities of parenthood, friendship, and so forth, means that not all the time formerly spent on a job is available for new pursuits.

Retired people also have the duty to live within their incomes. Although some people fail to live up to this expectation—usually because their retirement incomes are too low—the expectation is there, nevertheless. Retired people are expected to avoid becoming economically dependent on either their families or the community.

As a relationship, the retirement role connects retired people to those who are still employed, either in a particular trade, craft, or profession or in a particular organization. The crux of the relationship is that both the retired person and the person still on the job identify themselves in terms of the same occupation or work organization. In this sense the position of retired person is similar to the position of alumnus or alumna. The role of retired person is usually defined in terms that are flexible and qualitative rather than concrete and instrumental.* It was probably the absence of the instrumental element that led many investigators to view retirement as "roleless role" and therefore as an inevitable problem for the retired person. In this view, retirement creates a gap that only a new instrumental or "functional" role could fill.

The Busy Ethic†   The work ethic is a general value that justifies self-respect in return for initiative, industriousness, self-discipline, and commitment to seeing tasks through to completion. The busy ethic in similar fashion justifies leisure activities in retirement, protects retired people against implications that they are no longer able to perform, and gives definition to the retirement role (Ekerdt 1986). It is not a new instrumental role that gives meaning to retirement, but rather the continuation of the abstract ethic of work into retirement. However, in the case of retirement, work is defined not in terms of effort within society's economic system but in terms of effort toward worthwhile goals of the individual's own choosing.

All retired people get asked questions such as, "What are you doing with yourself these days?" That these questions will be asked impels the retired person to develop activities that indicate serious effort toward worthwhile

goals. Activities such as home maintenance and renovation, gardening, maintaining and using artistic talents, household finance, housework, volunteer work, part-time jobs, and a host of others take planning, commitment, and effort and are continuing testimony to the individual's capacity to work and engage in life. Even for those who prefer to focus their retirement lifestyles around leisure, the necessary tasks of maintaining self and household provide enough tasks to justify continuation of an image of self as productive enough to justify leisure. For example, Hooker and Ventis (1984) found that men with a strong work ethic who were not able to perceive their retirement activities as being useful had lower life satisfaction in retirement. Fortunately, an overwhelming majority of their respondents felt that their retirement activities were useful.

The busy ethic is reinforced by lifelong individual socialization and acculturation, social pressures from members of the social support network, and the media. Once ingrained, the need to feel that one is accomplishing something does not easily disappear. In retirement, this learned disposition is merely applied to the wide variety of nonemployment areas of life. Performing worthwhile tasks in retirement is as important to the individual's life satisfaction as being employed is during the labor force years. Members of the social support network understand what is important to the retired individual, and they usually can be counted on to affirm the individual's definition of meaningful retirement goals. The media reinforce the busy ethic by featuring retired people who are engaging in worthwhile activities, such as helping children learn to read, spearheading a local recycling effort, or preserving folk art. Keeping responsibly busy is the cornerstone of the busy ethic, which in turn is an important part of the retirement role.

Phases of Retirement

It is useful to consider the various phases through which the retirement role is ap-

---

*Jobs also have important qualitative aspects, but these are seldom emphasized in "job descriptions" or in discussions about jobs, which is perhaps a major oversight.

† This discussion was suggested by Ekerdt (1986).

proached, taken on, and relinquished.* The following phases are not inevitable responses to retirement. Instead, they are concepts that can be used to organize ideas about the issues people face in taking up, playing, and relinquishing the retirement role.

Preretirement The preretirement period includes a remote phase, in which retirement is vaguely seen as something positive that will happen someday, and a near phase, in which individuals orient themselves toward a specific retirement date. Two important things often happen during the near phase of preretirement. First, people begin to gear themselves for separation from their jobs and the social situations in which they carried them out. They may adopt a "short-timer's attitude." They may begin to see their jobs as more burdensome (Ekerdt and De Viney 1993). They may begin to notice subtle differences in how others view them. Second, people often develop fairly detailed fantasies of what they think their retirement will be like. These fantasies may turn out to be quite accurate pictures of the future, or they may be totally unrealistic. If realistic, they can serve as a "dry run" that smooths the transition into retirement by identifying issues that require advanced decision making. But if the fantasies are unrealistic, they may thwart a smooth transition by setting up detailed but unrealistic expectations. Worries about retirement revolve around income and health, with little concern about missing the job (Atchley, Kunkel, and Adlon 1978). When retirement occurs, people begin by taking one of three paths, each of which corresponds to a phase of retirement: the honeymoon, immediate routine, and R and R phases.

Honeymoon The retirement event is sometimes followed by a rather euphoric period in which the individual tries to "do all the things I never had time for before." A retired person going through this honeymoon phase typically says, "What do I do with my time? Why, I've never been so busy!" Extended travel is common in the honeymoon phase.

Some people do not go through a honeymoon phase. For one thing, a honeymoon requires a positive orientation. It also requires money, a scarce commodity for some older people. Most people cannot keep up the hectic pace of the honeymoon indefinitely, and they soon settle into a retirement routine.

Immediate Retirement Routine The nature of the retirement routine is important. If the individual is able to settle into a routine that provides a satisfying life, then that routine will probably stabilize. People whose off-the-job lives were full prior to retirement are often able to settle into a retirement routine fairly easily. These people have already made their choices among activities and groups earlier in life; all that remains is to realign time in relation to those choices.

Rest and Relaxation Following retirement, many people go through a period of low activity that is in marked contrast to the very active honeymoon phase. Atchley (1982a) called this the R and R phase to emphasize the temporary nature of the decline in activity. In a cohort of 168 people Atchley followed for four years, activity levels went down following retirement but returned to preretirement levels after three years of retirement. After a long period of having been employed, many people apparently welcome a period of "taking it easy." But after sufficient rest and relaxation, and perhaps a lengthy life assessment, they begin to get restless and at that point begin to pursue their planned retirement activities.

Disenchantment Some people do not find it so easy to adjust to retirement. For a small

---

*These phases of retirement were identified through a series of empirical studies of retirement (Atchley 1967: Cottrell and Atchley 1969; Atchley 1974; Atchley 1982a).

number of people, once the honeymoon is over and life begins to slow down, or if retirement has not turned out as expected, there is a period of letdown, disenchantment, or even depression. During the honeymoon period the retired person lives out the preretirement fantasy. The more unrealistic the preretirement fantasy turns out to be, the more likely the individual is to experience a feeling of emptiness and disenchantment. The failure of the fantasy actually represents the collapse of a structure of choices, and what is depressing to the individual is that she or he must start over again to restructure life in retirement. Disenchantment can also result when the situation one expects in retirement is disrupted. Death of a spouse is the most common such difficulty. What proportion of people become disenchanted is not known. However, cross-sectional surveys have found less than 10 percent at any one point in time who were disenchanted with retirement (Atchley 1976b,110). In Atchley's longitudinal study, no one became truly disenchanted, which suggests that disenchantment may in fact be rare. Those who became less positive toward their life in retirement did so as a result of disability, not difficulty in adjusting to retirement (Atchley 1994).

Reorientation   A period of reorientation sometimes occurs during the R and R period or among the few retired people who become disenchanted with retirement. During reorientation, individuals "take stock" and "get their life together." This process involves using their life experiences to develop a realistic view of alternatives within their particular set of resources. Reorientation also involves exploring new avenues of involvement. Very few people elect to become hermits in retirement; most want to remain involved with the world around them.

Groups in the community sometimes help people reorient themselves toward retirement. For example, many people become involved in senior center activities for the first time during this phase. But for the most part, people are on their own during the reorientation phase, seeking help most often from family and close friends. The goal of this reorientation process is creating a set of realistic choices that establishes a structure and a routine for living in retirement with at least a modicum of satisfaction.

Retirement Routine   People develop a set of criteria for dealing routinely with change. People with stable retirement lifestyles have well-developed criteria for making choices. These allow them to deal with life in a reasonably comfortable, predictable fashion. Life may be busy, and certainly it may have exciting moments, but for the most part it is stable and satisfying. Many people pass into this phase directly from employment; others first go through the honeymoon or R and R phases; others reach it only after a painful reassessment of personal goals; others never reach it. People with a satisfying retirement routine have mastered the retirement role. They know what is expected of them, and they know their capabilities and limitations. They are self-sufficient adults, going their own way, keeping usefully busy, and managing their own affairs.

Termination of Retirement   Some people reach a point at which the retirement role becomes irrelevant to their lives. Some return to a job, but more often the retirement role is overshadowed by illness and disability. When an individual is no longer capable of engaging in major activities such as housework, self-care, and the like, the retirement role is displaced by the disabled role as the primary organizing factor in the individual's life. This change is based on the loss of able-bodied or able-minded status and the loss of independence, both of which are necessary for playing the retirement role adequately.

The increasing dependence of old age usually comes so gradually that the retirement role can be given up in stages. Only with severe disability does independent choice become so limited that the dignity of the retirement role is diminished.

Timing of Phases Ekerdt, Bossé, and Levkoff (1985) studied patterns of life satisfaction in a sample of men at six-month intervals following retirement. They found that life satisfaction tended to be highest in the first six months following retirement, which gave some support to the idea of a honeymoon phase. They also found a significant dip in life satisfaction in the six months following the first anniversary of retirement; they interpreted this finding as mild support for the disenchantment phase. However, like Atchley (1982a), they did not observe actual negative life satisfaction, merely a reduction in the degree to which life satisfaction was positive.

Because there is no universal point of retirement and each retirement is highly individualized, it is difficult to tie the phases of retirement to a chronological age or a specific time period. Rather, the phases of retirement form a conceptual framework that can be used to understand the issues and possibilities connected with approaching, playing, and giving up the retirement role. This framework *does not* represent an inevitable sequence that everyone must go through.

Effects of Retirement on Individuals

In addition to affecting the individual's other roles, retirement and the loss of a job have various effects on the individual and his or her situation. Retirement is widely thought to have an adverse effect on health. Everyone seems to know people who carefully planned for retirement only to get sick and die shortly after leaving their jobs. However, the crucial question is whether people retire because they are sick or whether they are sick because they retire. If people retire because they are sick, then it is not surprising that some of them remain ill or die.

The decisive test of the impact of retirement on health is to compare health following retirement with health just preceding retirement. Using data from a large longitudinal study of people both before and after retirement, Streib and Schneider (1971) concluded that health declines were associated with age but not with retirement. That is, retired people were no more likely to be sick than people of the same age who were still on the job. In fact, unskilled workers showed a slight improvement in health following retirement. Haynes, McMichael, and Tyroler (1978) studied nearly 4,000 rubber tire workers before and after retirement and found that preretirement health status was the only significant predictor of mortality within the five years after retirement. Iams and McCoy (1991) also found that withdrawal from the labor force because of poor health was the major predictor of mortality.

Ekerdt et al. (1983) used longitudinal data from the Veterans Administration Normative Aging Study to examine the relation between retirement and health. After excluding men who retired because of illness or disability, they found no difference in health change between those who retired and those who did not. This study is noteworthy because it used physical examinations to corroborate earlier studies of self-reports of health. Herzog, House, and Morgan (1991) also found no relationship between employment status or hours of employment and physical or mental health. They did find that people who felt in control of their own participation or nonparticipation in the labor force were healthier than those who were involuntarily either in or out of the labor force.

Ekerdt, Bossé, and LoCastro (1983) went even further. They examined the possibility that retirement might actually *improve* health among those who retire because of ill health. They found that retirement did not improve the level of preretirement self-reported health, but it did reduce perceived health-connected role strains. This led the investigators to suggest that retirement can improve functional health without improving health level by reducing the level of demands placed on the individual.

Midanik et al. (1995) did a longitudinal study of the influence of retirement on health behaviors. They found that compared with others the same age who stayed on the job, the retirees had lower stress levels and were more likely to engage in regular exercise. Thus, one effect of retirement may be to give people time to take better care of themselves. Atchley (1994) found that walking for exercise was much more common among retirees than among those the same age who were still employed.

Why does the notion persist that retirement harms health, despite the fact that over thirty years of research has found no support for the idea? Ekerdt (1987) provided an insightful analysis. First, retirement is seen as a major life event, and big events are likely to be seen as causes of health problems; no one looks deeper to see the complex and cumulative set of factors that influence health changes. Whenever someone dies just following retirement, the death reinforces the association between retirement and death in people's minds. It does not matter that the association may be spurious, that poor health causes retirement rather than vice versa. If people can imagine an association, it seems plausible and is often taken as true without examination. Second, exaggerating the perils of retirement is also a way to affirm the value of employment. And if to show his zest for work, someone such as Chrysler's former chairman of the board, Lee Iacocca, says, "Some guys in this business slow down, retire and take it easy. A couple of months later, they're dead" (Ekerdt 1987,455), people tend to believe it. Among gerontologists, the fact that retirement has no obvious effects on health is hard to accept because it challenges theories in which many people have a stake. For example, the theory of stressful life events (Holmes and Rahe 1967) presumes

that retirement must have harmful effects because retirement was a life event people *expected* would be stressful. For advocates of this theory, retirement's lack of impact on health is difficult to accept because such acceptance undermines a basic premise of the theory. In the final analysis, Ekerdt concluded that people continue to believe the myth that retirement harms health because they want to. We need myths because they help us make sense out of our experiences and reinforce our beliefs, and sometimes we hold onto our myths with irrational tenacity.

It is also commonly assumed that retirement increases the incidence of mental health problems. Again, research provides little support for this idea. In fact, when Crowley (1985) compared the mental health of retired men with men the same age who were still working, she found that those who retired for nonhealth reasons scored higher than men who were still employed on measures of positive feelings and lower on measures of negative feelings. However, men who retired for health reasons were much more likely than voluntary retirees or employed men to score lower on happiness and higher on negative feelings. Thus, poor physical health was more likely to precede retirement than to be a consequence of it, and poor health, rather than retirement, correlated with negative feelings. Studies have shown no systematic link between retirement and depression. Palmore et al. (1985) looked at several longitudinal data sets and found little support for the idea that retirement has a negative effect on life satisfaction or morale. In the research on suicide, there is no indication that suicide risk is increased by retirement (Stenback 1980). Atchley (forthcoming) looked at the impact of retirement on self-esteem and found no effect. Bossé et al. (1991) found that their recently retired subjects rated retirement as the least stressful on a list of thirty-one life events; only 30 percent found retirement at all stressful; and retirement hassles were rated as less stressful than work-related hassles. Taken as a whole, the research results firmly contradict the conventional wisdom about retirement's negative effects on mental health. Yet the myth persists.

Howard et al. (1986) found that retirement actually improved mental health for men who exhibited a type A behavior pattern in their preretirement period. Type A behavior is characterized by feeling pressed for time, being easily aroused to anger or hostility, extreme competitiveness, impatience, and striving for achievement. In excess, these traits, especially explosive hostility, are thought to increase the risk of heart attack. Howard et al. found that men who had shown type A behavior just prior to retirement tended to lose these traits, and after one year of retirement, they were similar to other retirees.

Financial distress can also have a negative effect on the experience of retirement. Women, minorities, and relatively older retirees are all more likely to experience inadequate income in retirement, which in turn can affect opportunities to play the retirement role. Grad (1990a) found that income changes can begin with lowered salary and wages just before retirement, involve substantial reductions in income at the time of full retirement, and extend over several years. This pattern is related to the large proportion of people who experience bridge employment, as discussed earlier in this chapter. Logue (1991) found that because of gender differences in work history, occupation, and economic sector, African-American women were particularly vulnerable to financial stress early in retirement.

Another area in which retirement is widely thought to have a negative effect is *social participation*. Social participation is assumed to be tied to support from a job role; loss of the job role presumably hinders participation. Only in a relatively small minority of cases does retirement produce a drop in participation and consequent loneliness and isolation. And if the effects of widowhood are controlled, the proportion in this category is substantially reduced.

Thus, it is widowhood in retirement, not retirement itself, that tends to produce isolation. George Rosenberg (1970) found that retirement is more likely to produce social isolation among working-class men than among the middle class. However, surveys done since the retirement income improvements of the mid-1970s indicate that social isolation is rarely caused by retirement (Parnes 1985).

Simpson, Back, and McKinney (1966) found that many patterns of involvement supported by jobs persisted into retirement. They found that having a higher-status occupation and an orderly work career were as crucial for involvement during retirement as at earlier ages. They also found that if social involvement did not develop prior to retirement, it was unlikely to be initiated after retirement. Finally, Simpson and her colleagues found that retirement itself was not responsible for lack of involvement among semiskilled retired people and some middle-status workers; instead, it was their work histories that had not allowed these people to become integrated into society. Simpson and her coworkers stressed particularly the role of financial security in providing support for participation in society. Atchley (1994) observed many of thesesame patterns.

Bossé et al. (1990) found that although retirees tended to have slightly fewer contacts with coworker friends than still-employed respondents, there was no difference in the qualitative social support that retirees got from their work-related friends. Bossé et al. (1993) also reported that there were no changes in the quantity of social support for men before and after retirement. They also reported that most workers carried close friendships at work over into retirement.

Most people in retirement continue to do the same *kinds* of things they did when they were working. About one-third increase their level of nonjob-related role activity in order to fill the time created by retirement. About one-fifth of the retired population experiences a decrease in amount of activity. However, gains or losses in activity level are a relative matter. For someone uninvolved prior to retirement, leaving the job can result in an increase in the number of activities and still leave gaps of unfilled, unsatisfying time. On the other hand, for an overinvolved professional, retirement may reduce the net amount of activity but at the same time bring it down to a level more suitable to the person's capabilities and desires.

A great deal of attention has been paid to the impact of retirement on *leisure participation*. According to one school of thought, leisure cannot legitimately be engaged in full time by adults in Western nations without bringing on an identity crisis for the individual and a social stigma of implied inability to perform (Miller 1965). At one time, and in some cultures more than others, this set of assumptions may have been widely applicable. In fact, much of the retirement research done in the United States in the early 1950s supports such a view. However, there is growing evidence that in recent years both retired people and society at large have viewed the relatively greater freedom of retirement as an earned privilege and opportunity (Atchley 1982b; Ekerdt 1986).

Bossé and Ekerdt (1981) compared retirees' perceptions of leisure involvement with those men of the same age who had continued employment. They found that, contrary to expectations, retired men saw themselves as only slightly more involved in leisure activities than their employed counterparts. Morgan, Parnes, and Less (1985) found no differences between retired men and those still employed in terms of types of leisure activities, and retired men reported spending 47 percent more time at leisure activities compared with those who were still employed. Finally, in a longitudinal study of retirement, people were found to engage in the same leisure activities to about the same extent both before and after retirement (Kunkel 1989). Taken together, these studies indicate that leisure participation is as great or greater in retirement as before.

The amount of increase in leisure following retirement is probably not as great as we might think because activities such as housework, home maintenance, personal care, and so on still absorb a large share of time. Moss and Lawton (1982) studied how retired people spent "yesterday," which in their study was always a weekday. They divided activities into obligatory activities such as housework or personal care and discretionary activities such as exercise, hobbies, or watching television. They found that their respondents spent about 5.5 hours per day on obligatory activities and 10.5 hours on discretionary activities. Since the respondents were substantially involved in obligatory activities, the opportunity for more involvement with leisure may have been limited.

Situational Consequences

Much of the preceding discussion focused on individual responses to retirement in general. We now discuss three specific situational changes associated with retirement: income changes, changes in residence, and changing family structure.

In Chapter 13, we will deal with income sources and strategies in detail. In the present context, our concern is the *social impact* of the income changes that accompany retirement. Income in retirement is usually about 60 percent of what it was prior to retirement. McConnel and Deljavan (1983) found that even with the drop in income at retirement, the typical retired household did not feel economically strapped; it was no more likely than the employed household to spend more income on necessities or to supplement regular income with savings.

Based on her analysis of a series of longitudinal studies, Holden (1989) concluded that once individuals have weathered the initial effects of retirement on income, economic resources do indeed hold up. Cross-sectional studies had suggested that longevity in later life produced a sharp decrement in financial well-being, but longitudinal studies found a much smaller longevity effect. Holden cautioned, however, that the data she analyzed were collected during a period in which economic circumstances of retired people were improving. Whether retirement income will continue to hold up in the future remains to be seen.*

Change of residence is closely associated with retirement in the minds of many, but the overwhelming majority of those who retire do not relocate. The most significant impact of retirement is the economic incentive to find less expensive housing; even then, such moves are usually made within the same general locality. The common notion that retirement inevitably brings on migration is a myth. However, in the future, as better-educated cohorts enter retirement, the trend to move at retirement may increase somewhat, although it probably will never become prevalent.

Retirement also affects *marriages.* As we saw earlier, marriage is a central part of life for older people who are married. Retirement usually changes a household's organization with respect to both time and space. Employment dramatically influences the rhythm of household routines. Retirement allows more flexible schedules, spouses may do more activities together, and both spouses often spend more time around the house.

In terms of the timing of retirement, Atchley and Miller (1983) surveyed retired couples and found that in 45 percent of cases, both members of the couple were retired from employment, 24 percent were traditional couples in which the husband was retired and the wife had been a nonemployed housewife; in 16 percent of the cases, the husband was employed but the wife was

retired; and in 15 percent of cases, the husband was retired and the wife was employed. Thus, in almost one-third of the cases, only one member of the couple was retired, and wives were as likely as husbands to be the still-employed member of the household. How these varying patterns influence couples' space-time routines would be an interesting subject for study. We did find, however, that these differences in timing did not significantly affect the overall life satisfaction of retired couples.

Most studies have found that marital satisfaction in retirement is quite high. For example, in Atchley and Miller's sample, 70 percent of retired couples were extremely satisfied with their marriages, 27 percent were satisfied, and only 3 percent were dissatisfied. There were no significant differences in marital satisfaction between retired couples and couples the same age who were not retired. We also looked at longitudinal data for those who retired during our six-year study and found no significant difference in marital satisfaction from preretirement to retirement. We concluded that retirement had no measurable effect on marital satisfaction.

Some studies have found that retirement improved marital satisfaction. For example, Vinick and Ekerdt (1989) found that 60 percent of the husbands and wives they interviewed during the first year of their retirement said that the quality of their lives was better than before retirement; only 10 percent said it was worse. Other studies have found that retirement changes marriages for the worse. For example, Keating and Cole (1980) found in their Canadian sample that a substantial minority of couples had worse relationships in retirement, mainly resulting from lack of communication between spouses about what they expected from each other in retirement. In the problem couples, wives tended to assume that they had to plan activities to keep their husbands busy, and husbands did not feel that they needed such help.

---

* The future of retirement income is discussed in Chapter 13, "Income and Housing."

Earlier studies found significant class differences in the impact of retirement on couples' marital satisfaction. For example, Kerckoff (1966) found that across all occupational levels, men were more involved in household tasks after retirement. In middle-class couples, this involvement enhanced marital satisfaction, while in working-class couples it led to conflict. Heyman and Jeffers (1968) also reported lower levels of marital satisfaction among wives of working-class retirees. However, more recent work has not found such class differences (Atchley and Miller 1983).

Most couples experience a minor realignment of household tasks after retirement. Brubaker and Hennon (1982) looked at the impact of retirement on household tasks such as cooking meals, washing dishes, mowing grass, doing laundry, car maintenance, arranging family social events, cleaning house, and shopping. They found that compared with the preretirement period, retired husbands participated more in household tasks, but mainly in those activities traditionally associated with male sex roles: mowing grass and car maintenance. There was a small increase in joint responsibility for tasks such as washing dishes and cleaning house and a large increase in joint responsibility for shopping. There was a substantial gap between what wives expected and what they got, especially regarding joint responsibility for washing dishes and cleaning house. Brubaker and Hennon found that working-class women were more comfortable sharing tasks than middle-class women were.

Vinick and Ekerdt (1989) found that recently retired men attached top priority to home maintenance and renovation projects. Some had saved up projects specifically to provide useful work in retirement. They found that although women continued to retain responsibility for most of the household tasks, they appreciated the help they got from their husbands.

After retirement, spouses often have more opportunity to do things together. Vinick and Ekerdt found that about half the couples in their study reported an increase in leisure activities they did together without the company of others; 40 percent of wives reported that they had reduced personal activities to spend more time doing things with their husbands. Only a few wives were unhappy with this change.

Another retirement myth is that wives do not like having their husbands around the house after retirement. Expressions such as "getting under foot" and "sitting around with time on his hands" are often used to articulate this myth. It is also illustrated in such familiar jokes as, "I married him for better or for worse, but not for lunch!" and "After he retired I found that I had twice the husband and half the paycheck." This myth presumes that the household is the wife's domain and that she sees the retired husband as invading her turf. Studies from the 1950s found some support for this perspective (Kerckhoff 1966), but more recent studies have shown it to be less applicable. For example, Vinick and Ekerdt (1989) found some examples of this kind of feeling, but it was mild and more than offset by appreciation of the sharing of household tasks and the increase in companionate activities. Part of the reason that the myth does not fit the reality is because the myth is based on a traditional household structure in which the man has been employed and the woman has been a nonemployed housewife. But as we saw earlier, today only about one-fourth of retired couples fit this pattern; most couples have both been employed. In addition, launching the children causes a realignment of household time and space routines (Atchley and Miller 1983), which means that dual-earner couples grapple with some of the issues of living together in a two-person household well in advance of retirement.

Interestingly, Vinick and Ekerdt (1989) found that wives received some benefits from having their husbands become more familiar with the demanding nature of household work. When husbands were around to see the tedious nature of many household tasks, they were likely to encourage their wives to develop more outside interests. In our longitudinal study (Atchley and Miller 1983), we interviewed a couple whose adjustment to retirement illustrates this point. The couple had been retired for four years. At the beginning, he helped her with many household tasks such as house cleaning and laundry, but he saw quickly that these tasks were taking time that they could be using more profitably. They talked it over and decided to hire a housekeeper twice a week, freeing them to enroll in college courses on a nearby campus. Both said that this new exposure to learning had revitalized their lives.

Some couples find that their retirement is spoiled by factors that interfere with their expected retirement lifestyles. Widowhood usually devastates a person's retirement plans. Disability interferes with a person's capacity to lead the independent life necessary for playing the retirement role. Although most spouses face their partner's disability with courage, no doubt disability significantly alters what is possible in retirement. Sometimes couples' retirements are spoiled by having to take adult children and grandchildren into their households. Having small children in the household is not consistent with the laid-back, uncomplicated life that most people want in retirement. Finally, caring for older parents may sharply curtail the freedom that for many people is the most important aspect of retirement. Most people cope well with these realities, but they still experience a sense of loss relative to their concept of what their life in retirement might have been.

Personal Adjustment

As public acceptance of retirement has grown and as the systems that provide retirement income have matured, adjustment to retirement has become easier. Cottrell and Atchley (1969) found that as many as 30 percent of retirees in the 1960s reported they would never adjust to retirement. By the 1970s, this proportion had dropped to less than 10 percent (Parnes and Nestel 1981).

Heidbreder (1972) examined differences between white-collar and blue-collar adjustment to early retirement. While she found that an overwhelming majority were satisfied with retirement, she also found that adjustment problems were concentrated among former blue-collar workers who had low incomes, poor health, and little education. Anecdotal reports suggest that middle-class highly job-absorbed "workaholics" are also likely to have trouble adjusting to retirement, but research has yet to support this idea. The problem may be that this is a rare pattern and must be studied using methods designed to detect rare patterns.

Adjustment to retirement is greatly enhanced by sufficient income, the ability to give up one's job gracefully, and good health. In addition, adjustment seems to be smoothest when situational changes other than loss of job are at a minimum. In other words, if people's notions about the retirement role are realistic, then factors that upset their ability to live out their retirement ambitions also hinder their ability to adjust smoothly to retirement.

Any type of role behavior is at least partly the result of negotiations between the role player and the other role players to whom his or her behavior relates. Thus, retirement also changes the set of people with whom one negotiates. At work one negotiates one's work role behavior with one's peers, superiors, subordinates, and audiences. One's family and friends are involved only on the periphery of work role negotiations. In retirement, the people associated with one's former work drop out of the picture almost entirely (except for friends who also were work associates), and one negotiates primarily with family and friends in order to translate the general demands of retirement into particular behavior. Most people seem able to make these transitions reasonably smoothly. A major influence on this transition is a commonly held stereotype of the retired person. The fact that both the retired person and the people with whom she or he interacts share at least to some extent a common idea of the nature of the retirement role gives everyone a place to begin the interaction. Very quickly, however, the individual retired person learns how to personalize this interaction. Thus, the vagueness of both the retirement role and the stereotype allows retired individuals to negotiate definitions of the retirement role that fit their particular situations and are consistent with their own personal goals.

Differences by Race and Gender

Because the capacity to retire is tied closely to occupational level and work history, it should not be surprising that racial and gender discrimination in the labor force produce differences in patterns of retirement. As we will see in Chapter 17, "Social Inequality," African Americans are still concentrated at the lower ends of the occupational and income distributions. In many cases, African Americans of both genders experience sporadic work histories because they are employed on the fringes of the economy, where jobs are insecure. Gibson (1987) looked at data from a national sample of African Americans 55 and over who were not employed. She found that 62 percent defined themselves as retired and 38 percent did not. African Americans not in the labor force who did not see themselves as retired accurately used the pension receipt and labor force withdrawal criteria to make this determination. They tended to be people who had never had a full-time job, were poorly educated, were born in rural areas, and viewed disability as the main reason for their not being able to work full-time. They were much less likely than the retired to be receiving any form of retirement benefits and much more likely to be receiving disability benefits and welfare. Thus, the tenuous access to jobs among African Americans in the United States also means tenuous access to the systems that create eligibility to retire.

Parnes and Less (1985) found that in most respects the labor market and retirement experiences of black and white men in the fifteen-year National Longitudinal surveys showed more similarities than differences. For example, about 7 percent of both African Americans and whites withdrew from the labor force as a result of labor market problems. However, of those who retired between 1976 and 1981, 48 percent of African Americans retired for health reasons compared with only 31 percent of whites. Parnes (1985) concluded that when the effects of health, marital status, and age were controlled, racial differences in retirement were quite small. Not surprisingly, the retirement income picture was less promising for African Americans than for whites. For example, the average Social Security retirement benefit of whites in 1990 was $7,400, compared with $6,000 for African Americans.

Belgrave (1988) compared retirement among African-American and white women. She found that African-American women were considerably more likely than white women to have worked continuously most of their adult lives. Because of their greater average number of years of service in the labor force, African-American women were more likely to be eligible to retire. However, lifelong attachment to work and greater probability of widowhood resulted in more than half of African-American women at age 62 to 64 being in the labor force, compared with one-third of white women that age. Health factors were strongly related to early retirement for women of both races, but retirement induced by labor market problems was much more common among African-American women.

The issue of women's retirement has received a great deal of attention since 1970. In recent studies where retirement for women was compared with retirement for men, women have been found to be more influenced by marital status in terms of the timing of their retirement (Atchley 1982a), with married women and men being more likely to retire early and nonmarried women being more likely to retire late, mainly for economic reasons. Atchley found that retirement generally had no negative effects on life satisfaction for retirees of either gender. Women tended to be more positive about retirement, although both genders were quite positive. Women's activity levels tended to drop slightly more after retirement than men's, but both genders tended to remain quite active. Poor health reduced activity levels more for retired women than for men.

Pension entitlement is the single largest gender difference in retirement. Women's discontinuous work histories and higher likelihood of part-time employment decrease the probability that they will become entitled to adequate retirement pensions (O'Rand and Henretta 1982). Belgrave (1988) found that retired women had monthly pension benefits only 60 percent as large as those men received. However, both genders were likely to rate their retirement incomes as adequate.

Hayward, Grady, and McLaughlin (1988b) looked at national data on gender differences in withdrawal from the labor force and found that during the period from 1972 to 1980, women's patterns of employment and retirement had become similar to those of men in terms of overall increases in the extent of job changing and early retirement. Like men, women were likely to retire earlier from jobs characterized by low status, low earnings, and lack of seniority protection.

Summary

Employment and retirement are important organizing roles in life. Most people are employed throughout adulthood until about the age of 55, when participation in the labor force begins to drop. Peak labor force participation rates are about 95 percent for men and 70 percent for women. Most people retire before age 65, and retirement is often sudden. In the 1980s there was a significant rise in the prevalence of bridge employment (usually at lower pay and less generous benefits) between leaving career jobs and full retirement. On the other hand, the trend toward early retirement has accelerated in recent years.

Withdrawal from the labor force after 55 is mainly due to desire for retirement, employment problems, and disability. Since 1980, desire for retirement has surpassed disability as the major reason middle-aged and older adults leave the labor force. Employment problems are caused mainly by age discrimination. In the labor force, those 40 and over are considered older workers. Older workers are widely stereotyped as less effective and more expensive. As a result, they often experience age discrimination in hiring, retention, and promotion. Early retirement incentive programs try to lure older workers from the labor market. Those who lose their jobs often find new ones, but usually at lower pay. Job dislocations from corporate mergers and layoffs in declining industries increased in the 1980s.

Protections for older workers against age discrimination declined in the 1980s. Age discrimination in employment is illegal. The Age Discrimination in Employment Act (ADEA) prohibits the use of age as a criterion in hiring, firing, layoffs, promotions, and working conditions. However, the Equal Employment Opportunity Commission, which is responsible for enforcing the ADEA, has severely cut back enforcement. The decline of labor union power in the 1980s has also reduced dramatically the number of older workers protected by seniority rules on their jobs.

As the age of a population cohort increases, so does the prevalence of job-limiting disability. By age 55 to 64, about one-fourth of the population has disabilities that limit capacity for employment. Most employment-limiting disabilities are caused by heart disease, hypertension, injuries, and mental disorders. Disability accounts for a large proportion of withdrawals from the labor force before age 62.

Retirement can refer to withdrawal from employment, receipt of retirement income, identification with the retirement role, or a stage of life that follows employment. Operationally, retirement is best defined by both a substantial withdrawal from the labor force and receipt of retirement benefits. Most people in American society view retirement as a position that is earned through many years of participation in the labor force.

As the American economy evolved into a large-scale social system dominated by fast-changing labor markets, intricate division of labor, impersonal personnel policies, and bureaucratic rules, vocations gave way to jobs—specific positions of employment requiring a circumscribed commitment from the employee. In this system, retirement began as a personnel policy designed to tie workers to jobs and to reduce unemployment by providing a systematic way to phase out older workers. However, employees quickly saw the economic freedom and lifestyle autonomy retirement offered, viewing it as an attractive and justified reward for years of service.

The process of retirement involves thinking about retirement beforehand, considering retirement, making decisions about retirement lifestyles, and adjusting to retirement. A large majority of people have very positive attitudes toward retirement, seeing it as active, healthy, fun, and right. People do not have to become disenchanted with holding a job to be attracted to retirement. Most people have thought, at least in general, about what life in retirement might look like for them, but only about 5 percent participate in formal retirement planning programs. Retirement planning involves informing people about decisions they need to make as well as helping them get information and develop skills necessary to make sound choices. Information is usually more effective than counseling as an approach to retirement preparation.

People mainly consider retiring because retirement is a desired goal. However, prolonged unemployment, employer policies, and disability also can make employed people consider retirement. In a rational retirement decision-making process, people will generally retire if they feel that they will have enough retirement income and if they anticipate that their life satisfaction will be at least as good as when they were employed. Many considerations go into making these two evaluations. Employment problems tend to push people out of the labor force, while employer retirement incentive programs attract people to retirement. Many pension provisions encourage retirement at the minimum age for full benefits by penalizing workers economically for delaying retirement. Poor health often pushes people out of the labor force, while partial retirement policies encourage workers to stay on. Of course, there are people who do not consciously decide to retire; they simply assume that at some age people are supposed to retire, and they do so.

Most people retire at the minimum age for what they see as adequate retirement benefits. Employer pension plans generally encourage this action. However, service requirements can combine with interrupted work histories to force people to stay in the labor force long enough to amass the needed service credits. This is most likely to happen to women who have spent time out of the labor force because of child rearing. Disability is a leading cause of very early retirement, but its influence has diminished in recent years.

As an event, retirement ceremonies are only just beginning to deal with the opportunities and obligations attached to the retirement role. People who occupy the position of retired person are expected to maintain a certain amount of continuity in areas of life not related to the job, to continue being the same person as before. Retired people are expected to avoid full-time employment, to set their own life goals, and to avoid becoming economically dependent. In return, retired people have more personal freedom than perhaps any other category of adults. The busy ethic is a cornerstone of the retirement role. By having a number of worthwhile goals to structure life in retirement, the retired person demonstrates continuing capacity for productive endeavor and legitimates the leisure of retirement.

The retirement process can involve several phases. In the preretirement period, attitudes toward retirement tend to be positive, and people develop general ideas about what their life in retirement will be like and the social and economic resources required. Immediately following retirement, any of three phases may occur: a highly active honeymoon period, an inactive rest and relaxation period, or movement straight into a retirement routine. Most people settle into a satisfying retirement routine within a year after retirement. The successful retirement routine is characterized by a firm set of criteria for making day-to-day decisions and structuring time. Some people become disenchanted with retirement and need to rethink their life goals, but depression as a result of disenchantment with retirement is rare. Finally, life changes such as disability can end the life stage in which the role of retired person is the major organizing force in an individual's life.

Despite the widely held myth that retirement harms health, the overwhelming weight of the research evidence shows that retirement has no negative impact on either physical or mental health. Health declines can be traced to aging but not to retirement. In terms of social adjustment, retirement has no negative impact on life satisfaction, self-esteem, or morale. Activity patterns tend to be carried over into retirement; most people continue the same activities after retirement that they did before. Economic resources tend to hold up over time in retirement, at least for those who retired in the 1970s or earlier.

Retirement has only minor effects on marriage. Marital satisfaction of retired couples tends to be quite high on average, about the same as for employed couples of the same age. Most couples report minor realignments of household tasks in retirement, with wives retaining responsibility for most housekeeping tasks but with husbands providing more help than when they were employed. Women who do not enjoy having their husbands around the house are in a small minority.

Widowhood, disability, having to take adult children and grandchildren into the household, and having to provide regular care to older parents are all factors that can prevent retirement from turning out as people wished it had.

Because the capacity to retire is very closely tied to a continuous employment history, older women and older members of minority groups tend to have problems generating entitlement to adequate retirement pensions. At the same time, these people are also likely to have employment problems that make earning a living difficult. For such people, the dignity of retirement is difficult to attain.

10 ▌ Activities and Lifestyles

People like rhythm in their lives, and the ebb and flow of activities provide it. Each of us has a familiar routine for approaching the day. Our routines can also vary on holidays, on weekends, in the summer months, and on special occasions. These routines give a comfortable predictability to life, especially if they are satisfying ones. Our routines can be highly structured or not, but they are constructed of expected and/or preferred activities.

Aging changes activities in many ways. Role changes in later life, such as retirement or launching one's children into adulthood, change the amount of time that must be given to the job and to child rearing. Changes in physical functioning can impose limits on what we are able to do physically and may eventually become serious enough to restrict our activities to an institutional environment. Age discrimination can also limit the quality and quantity of activities available to us. Thus, age-related role changes may free us to concentrate our efforts on activities of our own choosing, while declining physical capacity or financial resources and age discrimination may constrict our choices. The balance between freedom and constriction that an individual experiences is greatly influenced by the person's position in the social hierarchy. People who are well-to-do and powerful are not likely to find their activities seriously limited because they are likely to remain in good health and to have the economic resources necessary to enjoy the freedom of retirement. However, for those at the lower end of the social ladder, inadequate incomes and serious health limitations are more likely, and so is age discrimination. Most people's experience is somewhere between these two extremes.

When aging becomes obvious, the gradual attrition of activities can be quite distressing. The case of Mrs. A. is not necessarily typical in its particulars, but it illustrates the process. Mrs. A. was first interviewed at the age of 70. At that time she lived alone in a comfortable apartment. Her daily round of activities was not a complex one. She spent a good deal of time keeping her apartment and herself comfortably neat and clean. She took frequent walks to nearby stores to pass the time of day and make purchases. She spent an hour or more per day just talking with friends. On Thursdays, she usually drove to a nearby city to attend the program at a senior center. Afterward she would often visit with friends. Her major pastimes were reading, watching television, talking with friends, and housekeeping.

However, Mrs. A.'s most meaningful activities involved visiting and talking with her daughter and her grandchildren. Although these contacts were not always smooth ones, because of generational differences in values, the level of mutual affection and respect was an absolutely essential element of Mrs. A.'s life. Although Mrs. A. had several close friends, they were not able to provide her with a feeling of belonging. Whatever her pattern of activities, Mrs. A. needed enough interaction with her family to maintain her sense of belonging to a family.

By 74, Mrs. A.'s situation had changed remarkably. She increasingly had found the stairs to her apartment difficult to negotiate, especially with packages, laundry, or groceries. As a result, she had moved to a first-floor apartment in another neighborhood. This move relieved the problem of the stairs, but it cost Mrs. A. her contacts with neighbors and shopkeepers in the area she had left. These contacts were no longer a convenient by-product of everyday living; to see these people became a special project. About this same time she began having difficulty driving at night and had to give up her visits to the senior center and to her friends in the nearby city. She still talked frequently with friends, but such contact was more likely to be by phone rather than face to face. Mrs. A. had had to curtail her reading because of eye trouble; she could still read, but not as much. Her eye specialist said nothing could be done about the problem. She seemed to feel that she was watching about all the television she could stand. To add to her distress, Mrs. A. was finding caring for herself and her apartment more difficult. She resented the fact that she could not vacuum under her furniture without being "out of commission" for two days afterward. She seemed particularly angry at having to give up tub baths in favor of showers because she could no longer easily get up and down in the tub. Thus, Mrs. A. not only had to relinquish activities, but she had to substitute less preferred options for several of those she retained.

The big gap in Mrs. A.'s altered activity pattern was contact with people and the outside world. She lost the contacts in her old neighborhood, at the senior center, and in the nearby city. She compensated somewhat by talking to old friends and making new friends in her new neighborhood. But her decreased physical capacity made it difficult for her to get out into the community. Mrs. A. turned to her daughter for more interaction and for assistance in getting out and about, especially when she was feeling bad. So far these increased demands have been met, but Mrs. A. is fearful lest she jeopardize the most meaningful relationships in her life.

As Mrs. A.'s case illustrates, changes in activity patterns as a result of aging are not simply a matter of dropping activities and sometimes substituting others. Lifelong preferences, emotional reactions, and social relationships are usually involved. This area is not an easy one to research, especially if the research is to follow people through time adequately. Nevertheless, such research is essential if we are to have an authentic understanding of the dynamics of aging.

In this chapter, we will look first at concepts that can be used to describe and evaluate activities, and then consider some general factors that influence the mix of activities in middle age and later life. Next, we look at how changes often connected with aging influence activities. Finally, we deal with some of the age-related factors that influence activities in specific environments such as the workplace, the community, and the home. Thus, we provide an overview of the forces influencing activities and the context in which activities occur. This understanding is vital if we are to understand how aging people are oriented toward their activities.

Concepts about Activities

How do we spend our time? This question implies that as long as we exist we are *doing* something. To explain what we are doing, we use concepts such as work and leisure to differentiate various types of activities. In its narrow sense, *work* is activity for which we receive payment, but in its broader sense, work is any expenditure of physical or mental effort toward the accomplishment of something. Thus, we can differentiate between *paid* and *unpaid* work. Unpaid work consists of productive activities such as housework, home maintenance, caring for one's own children, caring for older parents, or doing volunteer work through community organizations such as churches, nursing homes, or hospitals.

In its narrow sense, *leisure* is time not spent at work or in compulsory activities. In its broader sense, the concept of leisure refers to activities that are to a large extent freely chosen for their own sake rather than in response to social pressures (J. Kelly 1987a). John Kelly differentiated leisure into *play,* which refers to spontaneous enjoyable activity, and *recreation,* which is organized and goal-directed activity aimed at attaining personal and social benefits.

Kelly (1983) also made a useful distinction between *core* activities, which tend to persist throughout the adult life course, and *balance* activities, which are more sensitive to changing roles and opportunities. Core activities are readily accessible and include household conversation, reading, watching television, walking, shopping, and other informal interaction with family and friends (Kelly, Steinkamp, and Kelly 1987). Balance activities are related more to personal identity, life course stages, and roles. As a result, balance activities such as outdoor recreation, travel, and sports change more over the life course than core activities such as family leisure or social activities.

These concepts are designed primarily to sensitize us to relevant aspects of activity, not to be a definitive taxonomy. For example, spontaneous moments of play can occur within the context of both work and recreation. Paid work may be chosen because it offers freedom and is satisfying in and of itself. Recreation such as arts and crafts can contain elements of unpaid work. Thus, concepts such as paid or unpaid work, leisure, play, and recreation often are not mutually exclusive. As a result, a rigid classification of activities as either work or leisure ignores the fact that there are elements of leisure in both work and free-time activities and elements of work in many leisure pursuits. Apart from pay, jobs and leisure activities had the same potential meanings for people.

Continuity and Activities

Throughout this book, we have repeatedly seen examples of how the continuity principle influences a person's vision of his or her past, present, and future. Patterns of activity over the adult life course are perhaps the most obvious example of the continuity principle. Remember, continuity theory (Atchley 1989) says that in making choices in life, people are attracted to past views of the self, pathways they have used in the past, coping strategies that have been successful, ways of thinking that have been effective, people who have been supportive and helpful, and environments that have met the need for security and predictability. People are also drawn back repeatedly to activities that have, at best, proven satisfying, or, at worst, the least punishing.

Part of the continuity of self that people seek is based on activities. The extent to which people base their identities on activities varies widely. Some people strongly define themselves in terms of what they do or have done, and others are more concerned with the qualities they bring to everything they do. But in the social world, we are evaluated by others both for what we do and how we do it.

The relationship of activities to identity can also change over time. For example, Fiske and Chiriboga (1990) characterized statements people made about themselves into several categories: physical characteristics such as eye color or gender; roles such as employee or mother; ideological stances such as Republican or Presbyterian; activities such as sailing or theater going; physical capacities such as strong or healthy; attitudinal qualities such as being outgoing, happy-go-lucky, or responsible; interpersonal skills such as sensitivity or domination; and emotional qualities such as hostility or warmth.

Continuity of activities is important to many elders.
Photograph by Marianne Gontarz

The statements young adults made about themselves tended to focus on intrapersonal and interpersonal characteristics. The young adults were in the midst of trying to establish their identities in relation to other people. They described themselves in terms of personal qualities such as being enterprising, intelligent, knowledgeable, uncertain, or somewhat egotistical. They tended *not* to see themselves in terms of roles or activities. By contrast, adults who had just launched their children into adulthood tended to focus on activities, roles, and memberships in their statements about themselves. Respondents who had completed the transition to retirement were similar to those who were middle-aged, but added a heightened concern with spirituality as a personal quality.

Fiske and Chiriboga concluded that this evolution of identity seemed completely appropriate. "For people who are still struggling with who they are and where they are going, for example, a focus on inner concerns and descriptors seems quite fitting. Similarly, as the years stretch on and the self becomes established, it can now be characterized by givens: the activities and interests that have stabilized now define the person. An interest in cars or music may not appear particularly descriptive of a 20-year-old, for example, but to a 60-year-old with a forty-year history of interest in cars or music, this interest may now represent a central theme in life" (1990,68).

Thus, young adults often have a rich array of activities, but they do not know yet whether they want to define themselves in terms of activities and if so, which ones. But by middle age, most adults have selected an array of activities and roles that they are willing to stake their identity on. Once activities become a part of identity, then motivation for continuity in activities can be expected to increase and remain high.

Activity domains are the settings where activities take place; they include workplaces, community organizations, households, public places, recreational facilities, and so on. Activity domains are not valued equally. Each person has his or her own most valued activity domains, and makes decisions about investment of time and energy based on those values. For example, some people are heavily invested in their church, and the demands of this activity domain take precedence. Other people spend every discretionary moment honing and practicing a craft. Some people are very focused on one activity domain, and put most of their time and energy into it. Others take a more balanced orientation, and are invested in having the knowledge and skill needed to engage fruitfully in a wide range of activity domains. Such people may be accomplished at their jobs, involved in several community organizations, devoted to their families, and skilled

at a number of individual activities such as art, crafts, or music.

By the time most people reach middle age, they have developed a clear, identifiable orientation toward activity domains. And once formed, these orientations tend to persist over time. In addition, within a particular domain people arrange their preferences for specific activities into a hierarchy. For example, a person may have placed a high priority on producing art over most of her or his adult life; within the domain of art, the person's preferences may have been toward line drawing and painting watercolors, and he or she may have accumulated considerable skill and accomplishment in these areas. If that person were to experience an increase in time available for art, she or he might decide to explore new areas within the domain, such as learning to make block prints, which involves a combination of drawing and woodcarving. In this case, the individual would be working in a domain of known skill and success, and at the same time developing a new area of skills and growth potential within that general domain. This is a dynamic view of continuity in which people can exhibit continuity of activity domains and try new activities at the same time.

As people age, they often gradually reduce the number of activities they engage in and the time they spend on each activity. Carstensen (1991) suggested that such reductions in activity result from a lifelong process of selection, not disengagement from life. For example, when she looked at longitudinal data on interpersonal interaction, Carstensen found that contact with acquaintances began to decline early in adulthood and continued to go down through middle age. By contrast, contact with close friends went up during the same period. Thus, once selection of a circle of close friends is made, then contacts tend to be concentrated within that domain and contact with more peripheral acquaintances declines. This decline is the result of selection, not withdrawal.

Activities in Middle Adulthood

The range of activities that occupy human time and attention is practically infinite. Nevertheless, we can predict with some assurance that most middle-aged adults will have jobs, will spend time at home, will be involved in community activities to some extent, and will do things with their families. Within these general areas or spheres of participation, the factors that influence the choice of *particular* activities include the meanings attached to activities and how these meanings are related to the individual's personal goals, to the person's sense of competence for various activities, and to gender, social class, and ethnic norms about the desirability of various activities.

The Meaning of Activity

The same activity can mean quite different things to different people. Activities can be a source of personal identity, personal development, sensory experience, prestige or status, new experience, peace and quiet, fun and joy, feelings of accomplishment, or something to look forward to. Activities can also meet goals such as making money, being with people, getting the "vital juices" flowing, serving others, passing time, exercising competence, or finding escape.

Any activity can be rated as to how much a given individual attaches these various meanings to that activity. For example, writing a book could be a source of personal development, a way to focus creativity, a source of prestige or status, a way to be of service to others, a way to exercise competence, a source of feelings of accomplishment, and a way to make money—not necessarily in that order. It is important to see that any activity can have many meanings for an individual. To understand how an activity fits into an individual's life, we must know the meaning of that activity for that person *and* how that activity (given its mean-

ing) fits into the person's life goals. We cannot assume that we know the meaning of an activity for a particular person—we have to find it out (Atchley 1993b).

Activity Competence

Although there are plentiful opportunities to enjoy a wide variety of activities in adulthood, people need *skills* and *knowledge* to take advantage of them. For example, older people—particularly the less educated—are reluctant to engage in activities such as art, music, handiwork, or writing. And this reluctance is at least partly caused by the older person's feelings of incompetence in such activities.

Critics of education have attacked this apparent deficiency in our orientation toward education. Contemporary education, it is said, devotes anywhere from 80 to 90 percent of the students' time for 12 to 19 years to teaching them how to fill jobs, but makes little effort to prepare them for life outside the job. As the noted publisher Norman Cousins said:

I contend that science tends to lengthen life, and education tends to shorten it; that science has the effect of freeing [people] for leisure, and that education has the effect of deflecting [them] from the enjoyment of living.

(1968,20)

The point is that to open up the full range of possible activities requires some training.

The predominant pattern of stability that activity patterns show over the last half of the life span attests to the fact that activity competence created in early or middle adulthood can be maintained into later life. However, the explosive growth of programs that teach new activity skills to elders, reviewed later in this chapter, also shows that activity skills can be developed at any age. However, basic skills such as literacy are prerequisites for learning. Lambing (1972) found, for example, that lack of literacy seriously limited the leisure activities available to older lower-class

African Americans in Florida, a situation that these respondents recognized and wished to correct.

Sinnott (1994) presented an indictment of educational institutions' failure to address the needs of adult learners, not only with regard to activity competence but work competence as well. Although cognitive researchers generally find age-related declines in specific cognitive functions, the success of middle-aged and older adults in everyday learning situations suggests that the types of activity skills aging adults are interested in learning are well within their capacities. But programs addressing those interests are in very short supply. We will come back to this topic later in the chapter when we discuss education.

Gender, Social Class, and Ethnicity

Although many choices of activities are based on individual preferences conditioned by experience, many others are constrained by what is considered "normal" for a person of a given gender, social class, or ethnic group. Gender roles have received much scholarly attention over the past two decades. Many activity differences between men and women are not the result of biological differences, but instead are the result of differences in the education and training that boys and men and girls and women receive. Paid work roles differ substantially by gender. Despite three decades of social agitation for equal employment opportunities for women, employed women are still concentrated in helping professions such as nursing and teaching, and in clerical and secretarial work. As we saw earlier, men are more likely than women to be employed. In nonpaid work, women spend more time than men doing housework and providing child care, whereas men spend more time doing home repairs and improvements (Herzog et al. 1989). In leisure activities, men are more likely to be involved in exercise, sport, and outdoor recreation, whereas women are more likely to be involved in cultural activities such as attending plays, concerts, or art exhibits, and in social and home-based activities (Kelly, Steinkamp, and Kelly 1986).

Over time there have been changes in gender role definitions. For example, in the 1950s very few young women engaged in competitive sports, but today it is common for young women to compete in basketball, volleyball, track and field, swimming, tennis, gymnastics, and a host of other sports. Most public schools and colleges have girls' and women's teams in several sports. In addition, today more girls and women have opportunities in school to learn mechanical arts and crafts such as automobile repair and carpentry, and more boys and men have opportunities to learn skills such as cooking and sewing than in the 1950s. How these shifts in activity patterns will affect future cohorts of older people is uncertain. However, even with these changes, traditional gender roles still exert strong influences over activity patterns.

Substantial social-class differences also exist in preferences for activities. For example, while upper middle class people may watch television about as much as others do, they tend to watch different programs. A large portion of the audience of the Public Broadcasting System (PBS) is made up of the upper middle class, while the audience for all-sports networks tends to be mainly working class. Voluntary associations, sports, reading, and gardening are much more common among the upper middle class; the middle-middle class is more inclined toward crafts and television; and members of the working class are more likely to spend their time visiting with neighbors and kin. Social class also has obvious economic effects on the range of possible activities. Traveling, entertaining, going to concerts or plays, playing golf or tennis, and dining out require more money than watching television, talking with neighbors, or puttering around the yard. Thus, some social-class differences in activities are the result of differences in what people are taught to prefer, and part is due to differences in financial capacity.

We do not know very much about ethnic differences in activity patterns, but certainly they exist. For example, blacks participate more than whites in church activities, but less in other types of activity (Parnes and Less 1983). People of Italian or Mexican descent are more likely than people of English or Scandinavian descent to spend time with their families.

Thus, gender, social class, and ethnic differences in socialization all can affect activity meanings, development of activity competence, and the value placed on various types of activities. The result is wide variation in adult activity patterns. We now turn to some of the changes that commonly occur with age and how they influence activities.

Aging and Changes in Activities

The life course exerts significant pressure to make choices about jobs and family. Most remaining choices of activities result from preferences that have been learned by trying the alternatives in vogue at the time of one's early adulthood. Once adult activity patterns are established, they tend to persist. Yet aging and changes associated with it cause activity patterns to change—if not in the type of activity, at least in the amount of it.

Herzog et al. (1989) studied age differences in annual hours spent at both paid and unpaid work. They found that almost all the differences between age groups could be accounted for by differences in paid work and child care. In other areas of productive activity, such as housework, household maintenance, and volunteer work, age differences were generally minor. Gender produced more differences than age did. (See Figure 10-1.) However, those age 75 and over reported lower involvement in unpaid work than did other middle-aged and older age categories.

Figure 10-1 Mean annual hours of paid and unpaid work, by sex and age.

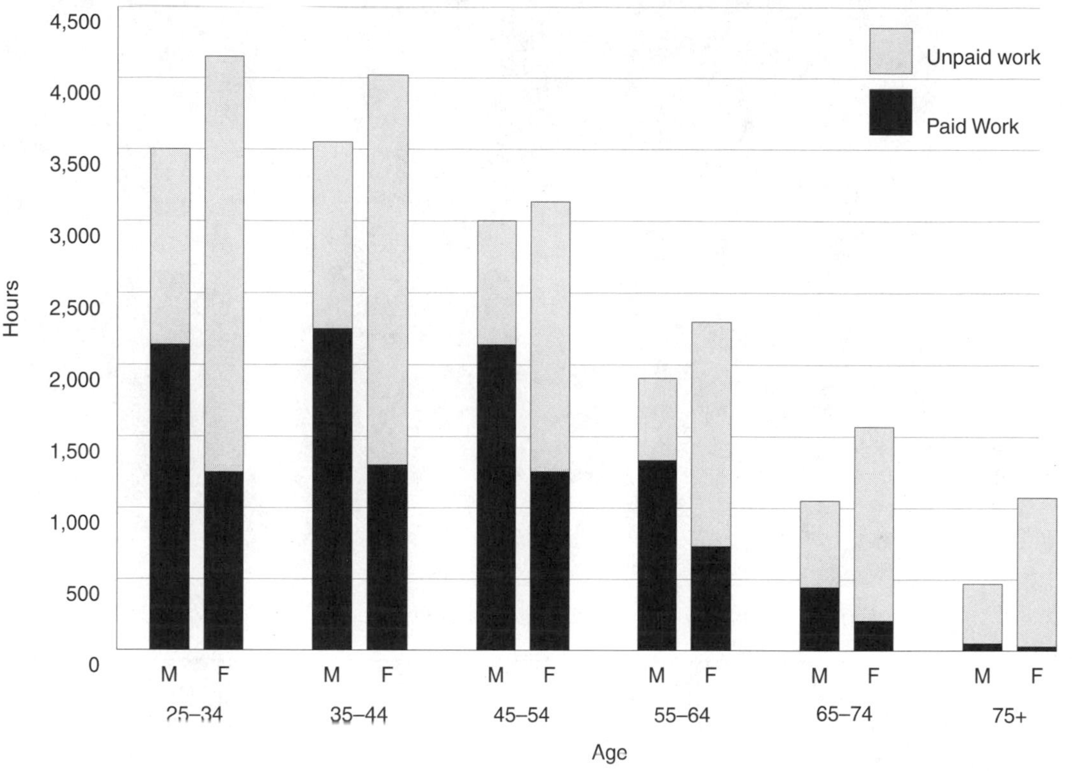

Source: Adapted from Herzog et al. (1989).

In leisure activities, Kelly and his colleagues (1986) found that age differences in some areas were pronounced, with significantly less involvement of older cohorts in exercise, sports, and outdoor recreation. However, for social activities and participation in community organizations, age differences were very small.

In my longitudinal study of a panel of adults who were 50 or over in 1975, I found a number of different patterns in terms of activity stability and change from middle age to later maturity to old age (Atchley 1994). Involvement in many activities remained stable from age 50 to age 85; these activities included travel, spending time with family and friends, reading, and watching television. Some activities were dropped in later matu-

rity (age 60 to 74); these included involvement in professional organizations (related to retirement) and attending sporting events. However, the greatest number of drops in activity were observed after age 75; these activities included gardening, political organization involvement, participation in exercise, attending social events, and attending church (this pattern was found among men only). These patterns of declining activity were partly a result of an increasing prevalence of disability after age 75, but the patterns of decline still remained in milder form even for those with no disability. These findings suggest that elders simplify their activity patterns in response to changes in social networks and declining energy as well as to disability.

Reading is a favorite solitary activity among older people.
Photograph by Marianne Gontarz

Life course factors that account for the age patterns of activity include the completion of child rearing, retirement, and widowhood. Factors associated with aging that influence activity patterns include physical aging, increasing frailty, a move to congregate housing, and institutionalization.

Completion of Child Rearing

For most people, life after launching the children does not seem to require a large emotional adjustment. However, it probably does mean some adjustments in the use of time. In the middle class, lost interaction with children tends to be replaced by increased interaction with one's spouse. Reduced demand for service work in the household is offset by employment for most women. Social circles that brought parents together as a by-product of having to do things for their children, such as parent-teacher organizations and carpools, are likely to disappear. Apart from more travel and eating out, leisure activities probably do not change very

much after the empty nest. This area is much in need of research.

Retirement

Despite the seeming inevitability of substantial change in activities, retirement produces little feeling of discontinuity in activities for a large majority of people. Part of this is due to the fact that most adults have considerable amounts of time for activities in the preretirement period and have a well-developed repertoire of familiar activities to engage in in retirement. Part is due to the expansion in productive household and volunteer work in retirement. Part is due to the continuity of familiar activities and relationships. John Kelly (1993) concluded that at least in the early years of retirement, most people continue to do the same activities they did before retirement, but with timetables revised to fill some of the time freed from employment. He also concluded that this continuity is tied to the self and retirement does not cause people to suddenly or dramatically redefine themselves. Cutler and Hendricks (1990) concluded that social participation was not adversely affected by retirement.

When people retire, they increase the amount of time they spend at both obligatory and leisure activities. Parnes and Less (1983) found no differences between male retirees and nonretirees in types of activities, but, as expected, retired men spent nearly 50 percent more time at each activity. Many retirees find that properly taking care of domestic, financial, and social affairs occupies much more time than they were able or willing to devote to such activities prior to retirement (Atchley 1976b). As one woman put it, "Now that I'm retired, I feel guilty if I put off answering letters. I sometimes spend a whole morning on one letter!" About 25 percent of those who retire experience an overall decrease in activity. But this decrease is sometimes welcome. For some, particularly those who retire for health reasons, it is a relief not to have to keep up the pace of the preretirement period.

Apart from the impact of retirement on activity *level,* there is the question of retirement's effect on the *meaning* of activity. According to one school, adults in Western societies cannot engage full-time in leisure activities without provoking an identity crisis. The thesis is that the job identity mediates all other activities and that without it other activities cannot provide the person with an identity (Miller 1965). This view was very prevalent in the 1950s, and it received modest support from research being done at that time. However, more recent research shows that job identity carries over into retirement and that in general people view an increase in the amount of leisure activity in retirement as an earned privilege and opportunity (Atchley 1982b). The fact that only a small proportion of people take up entirely new activities following retirement does not appear to be the result of an identity crisis. Ekerdt (1986) pointed out that in retirement unpaid productive activities give legitimacy to leisure in much the same way that paid work did earlier.

Retirement also tends to diminish financial resources, which in turn can reduce activity level. Financial security plays a crucial role in providing support for participation in various activities. Kelly (1993) found that poverty creates a barrier to getting involved in new activities in the general community, especially for those who formerly had middle-class incomes. For example, Fischer, Mueller, and Cooper (1991) found that low income was a major factor that inhibited participation in volunteer work by older people.

Widowhood

Widowhood usually causes fundamental changes in the core of activities—day-to-day conversations, shared meals, and activities in and around the household. Where once there was a longtime friend and confidant to talk to and share experiences with, now there is no one. Although most widowed people adjust to this major shift in lifestyle, the result is often less human interaction and sense of belonging. In addition, many couples participate together in leisure activities outside the home. Many studies report that loneliness is a prevalent aspect of widowhood, especially during the bereavement period (Kalish 1985). Part of this loneliness stems from missing the presence of the spouse, and part stems from no longer having a ready-made companion for activities.

The impact of widowhood on activities has not been studied systematically. Instead we have bits and pieces of information generated by studies focusing primarily on other aspects of widowhood or bereavement. For example, there have been several reports of the effects of widowhood on participation in voluntary associations (Ferraro 1984). In general, the effects appear to be small, with a great deal of continuity in participation. Small decreases in participation sometimes occurred for men following bereavement.

Physical Aging

Physical change has two types of influences on activities. The first is a series of changes that move individuals from more active pursuits to more sedentary ones. The second is the constraining effects of serious illness and disability. Gradually, as the human body ages, it becomes less capable of high rates of physical output, and recovery from strenuous activity takes longer. This decline probably accounts for some movement away from strenuous activities in middle age. However, people who have remained in peak physical condition can probably continue these activities into later maturity. Perhaps more important influences on strenuous activity come from life course factors and age norms. Middle-aged people who are highly involved in job, family, and community responsibilities may find it difficult to allot time to stay at a level of physical conditioning that allows strenuous activity without discomfort. In addition, there is a general expectation (age norm) that after about age 35 or 40, people should be exempt from pressures to involve themselves in strenuous activities. It will be interesting to see what effect the current exercise movement has on attitudes about strenuous activity in middle age and later maturity. On the one hand, the stress on lifelong exercise promotes a healthier cardiovascular system, stronger bones, and greater physical flexibility; on the other hand, the increase in fitness activity-related injuries of the joints, particularly the knees, suggests that exercise needs to be approached with caution and moderation much like any other prescription.

The relatively sedentary activity patterns of most middle-aged Americans are well within the physical capabilities of later maturity. Only when symptoms of disabling illness or old age appear must there usually be a change in activity patterns. How disability and aging change activities is a highly individualized matter, depending on the mix of activities prior to physical change, the specific physical changes that occur, and the individual's capacity to adapt to them. Generally, the narrower the range of activities in middle age, the more vulnerable the individual is to physical change.

Increasing Disability*

The activity patterns of middle age tend to persist into later life, but gradually, if people become increasingly disabled, there is a constriction of activity. This constriction is minor in the early years of later life, but by the time a cohort reaches 90, the survivors tend to lead life very close to home. The household assumes greater importance as a center of action, and going out into the community may require physical assistance from another person.

Thus, a large majority of people 65 to 74 lead lives that frequently take them out into the community. Among those 75 to 84 is a growing proportion whose disabilities limit their capacity to be active outside the home, and an increasing proportion lives in nursing homes or other congregate housing with assisted living. Those 85 or older who still are able to move about normally in the community are a minority. This constriction of activities does not occur in all cases, nor should it be considered a negative result of aging. Some older people attain remarkable insights while leading a simple life. Again, it is not only the outer appearance that determines the quality of life but the inner vitality as well (Gadow 1983).

A Move to Congregate Housing

Although only a tiny proportion of older people live in congregate housing, a move to congregate housing generally increases the opportunity for activity, and activity levels usually increase for the mover. This increase is not simply a short-term flurry of activity in

*The ideas in this section were suggested by John Kelly.

response to a new environment, but a genuine change that is maintained over time. In a well-done longitudinal study of this subject, Carp (1978–1979) found that compared with older people who did not choose to move to an activity-rich congregate housing environment, people who did move reported increased activity levels. Increases in activity took place both in "regular responsibility" activities, such as lobby receptionist, senior center responsibilities, and church work, and in pastimes such as club meetings, table games, and visiting. Over the eight-year period of the study, movers maintained their increased activity levels even though they averaged 72 years of age when the study began. On the other hand, comparison respondents in the community showed a decline in activity over this same period. People who move to congregate housing with extensive activity programs are self-selected, and certainly it would be incorrect to assume that a high level of activity is good for everyone. Perhaps more important is the degree of fit between the person's desire for activity and the opportunities present in that person's living environment. Carp's findings show that for people who want a high activity level, congregate housing that offers plentiful opportunities for activity can successfully meet this need.

Institutionalization

Institutionalization by its very nature could be expected to reduce activity because it cuts people off from their daily contacts in the community. And institutionalization is usually the result of disability that curtails activity. Yet little research has been done on before-and-after changes in activity resulting from institutionalization. We do not know for a comparable disability if the activity reductions due to disability occur more frequently in institutions than in the community. We do not know how institutionalization influences the person's desire for activity or how discontinuity between wants and possibilities for activity in the institution influences adjust-

ment to the institution. Whether the disability that led to institutionalization is mental or physical undoubtedly influences activity. All these topics deserve further research.

Spheres of Activity

All the common life changes that accompany aging can potentially influence activity patterns. In addition, age discrimination and societal disengagement do not operate with the same force in all areas of life. Accordingly, we now examine how aging influences activity within various domains: on the job, in community organizations, and at home. We then look briefly at special activities of older people.

The Job

Aging affects employment in several ways. Aging can cause job-related disabilities, it can cause people to be denied the opportunity to work even when they are able-bodied, and it can increase the motivation to leave the workforce in favor of the freedom of retirement. Whether people continue to hold a job into later life depends on several factors. First, there must be an opportunity to continue. Second, the person must want to continue. Reasons for wanting to continue might include to supplement insufficient retirement income, to get satisfactions that can be gained only on the job, or to maintain associations with others that would be lost if the person retired. Third, the person must be physically able to continue. Unless the person can continue in the same job past the normal retirement age, employment is tough to come by in later life. As we saw earlier, age discrimination in employment is common. In addition, most people who reach retirement age feel they have served their time and deserve to be allowed to retire. The end result is that jobs are a major activity for a small minority of older people, especially after age 70.

We should not confuse retirement with stopping work. Work is the expenditure of energy toward a goal. Able-bodied adults never stop work; they simply stop working toward economically motivated goals in a position of employment. They become "self-employed" in that they are free to decide for themselves what goals to pursue. Indeed, life satisfaction in retirement is very much tied to having numerous personal goals (Atchley 1982a).

Community Organizations

Community organizations are groups that develop around a collective desire to achieve some purpose or pursue some interest. Most communities have churches, political parties, labor unions, veterans' groups, fraternal organizations, community service groups such as the Kiwanis and the Rotary clubs, professional associations, and parent-teacher organizations. Most communities also have hobby and garden groups, groups related to sports, and other special-purpose groups. In this section, we will look first at how aging affects participation in community organizations in general, then at participation in later life, and then at participation in churches, political groups, volunteer work, and education. Last, we will examine barriers to participation.

Participation in General   Stephen Cutler (1976) reported that if the effects of socioeconomic status were controlled, then age produced little or no change in organizational participation from middle age into later life. He found that young adults tended to have low levels of community participation and middle-aged and older people had high levels.

Cutler and Hughes (1982) also examined age and sex patterns in membership in various types of organizations (see Table 10-1). For both men and women, affiliation with church-related groups was by far the most common, with women being more involved than men. This pattern persisted across the entire age range. For men, involvement with sports-related, labor, and professional orga-

nizations peaked in middle age and was lower for older age categories. Involvement with fraternal, veterans', and service groups was generally consistent across the age categories. For women, involvement in sports-related groups diminished across the age categories, while involvement with professional, hobby or garden, and discussion or study groups remained relatively stable. Note that although women were less involved than men in professional groups in middle age, after the age of 55 their involvement paralleled that of men. This is contrary to the stereotype and indicates a greater interest in career among women. Except for church-related groups, participation in voluntary associations was less common among women than men. For most organizations other than churches, only about half the people who claim membership in organizations actually participate (Kelly 1983).

Participation in Later Life   Changes such as residential mobility, retirement, and widowhood have no predictable, consistent impact on participation in community organizations. For example, depending on the situation, retirement has been found to increase, decrease, or produce no change in participation. In a longitudinal study, Cutler (1977) found that people with either increased or stable participation far outnumbered those whose participation declined. However, poor health or dwindling financial resources—factors that are particularly prevalent among the working class—have a predictable dampening effect on participation. The impact of declining health is direct and obvious. Eroded financial resources have a subtler impact. Because community groups are almost always nonprofit, they must be subsidized, usually by members' contributions. Older people on tight budgets can be forced out of participation by the embarrassment of being unable to contribute.

For those who do continue to participate, community organizations take up a great deal

Table 10-1 Percentage of people belonging to various types of associations, by age and sex: United States, 1978–1980.

Type of Association	Age (Years)				
	35–44	45–54	55–64	65–74	75+
Men					
Church-related	29%	32%	33%	34%	42%
Sports-related	29	18	18	10	11
Fraternal	15	16	25	24	27
Labor unions	26	25	24	17	21
Veterans' groups	7	18	29	10	18
Professional	19	19	18	7	7
Service	14	16	10	9	12
Hobby or garden	10	7	6	7	3
Women					
Church-related	43%	47%	38%	43%	51%
Sports-related	18	17	9	3	2
Fraternal	7	9	9	13	11
Labor unions	11	6	7	6	1
Veterans' groups	1	6	5	6	5
Professional	14	12	9	5	6
Service	10	13	8	10	4
Hobby or garden	11	9	9	11	14

Source: Adapted from Cutler and Hughes (1982).

of time. Yet as they age, many people experience declining satisfaction from participation as they see groups "letting in a different sort of people" or "having changed so much." Others are self-conscious of their age and would rather "leave it to the youngsters." Still others have done every job in the organization two or three times and see little opportunity for continued growth in further participation. Thus, older members are often subtly "squeezed out."

We see that aging in and of itself has no predictable effect on overall participation in community organizations. Changes that do occur in participation are most likely to be associated with poor health, inadequate income, and transportation problems. We now look at specific types of organizations, beginning with churches.

Participation in Churches* Churches are the most common type of community organization membership for older people. Membership in churches also tends to be higher at the older ages, especially after 75. Leadership positions in churches also tend to be concentrated among older people. Participation in religious organizations declines among older people at a much slower rate than does participation in other types of organizations (Cutler and Hendricks 1990).

Some churches set up senior centers or clubs for older people; others allow such groups to use their facilities. The churches are also becoming interested in housing programs, particularly retirement homes for

*See Chapter 11, "Religion and Spirituality," for a more detailed discussion of this topic.

middle-income older people. They are beginning to see housing programs as a legitimate service and not merely as an act of charity.

Because the proportion of older people in the average congregation is higher than the proportion in the total population, it is not too surprising that churches have begun feeling some pressure from their older members concerning housing problems. At this point, however, church programs for older people are few and far between. Although 80 percent of the Presbyterian churches, for example, report special social groups for older people (including age-segregated Sunday school classes), only two-thirds report any type of educational program specifically for older people, and very few have employment, homemaker, or health services.

Participation in Politics   There is a significant increase with age in political activity, such as working for the party in a local vote, signing petitions, and belonging to political groups. Older people represent the same percentage in party organizations that they do in the general population, but party membership has declined sharply among young adults over the past two decades. Older party members who remain involved in politics are influential because experience carries more weight in politics. In politics, the older person can play the role of sage. In fact, the word "politic" means "wise" or "shrewd." Political prowess is still something one learns mainly from experience rather than from a book or a professional school.

However, in my sixteen-year longitudinal study of adults who were age 50 and over at the beginning of the study, I found that after age 70, most people begin to withdraw from political activities, and the proportion who never participate increases substantially. Large proportions of elders continue to vote, but they gradually reduce other types of political participation.

Participation in Volunteer Work   Herzog et al. (1989) looked at the age pattern in vol-

unteer work and found that the proportion doing volunteer work increased from age 25 to age 45, at which point about 51 percent of men and 59 percent of women were involved, a significant gender difference. After 45, gender differences disappeared, and the proportion participating dropped from 45 percent at age 45 to 54 to about 26 percent at age 75 and over. However, for those people who did volunteer, the average annual hours of volunteer work was similar across the adult life course and there was no gender difference. Those who did volunteer work in later life tended to be people with a history of volunteerism earlier in life, which again illustrates the continuity principle.

Retired elders are often discussed as a great sea of untapped volunteers. However, Fischer and her colleagues (1991) pointed out that, depending on how volunteer work is defined, elders are already engaging in a substantial amount of volunteering and how much increase could be expected is questionable. For example, in their Minnesota study, Fischer and her coworkers found that nearly 60 percent of elders provided help to their families, most often in the form of caring for grandchildren. Is this volunteer work? The answer is not altogether clear, but what is clear is that people who have ongoing responsibility for caring for someone else probably have constraints on how much other volunteering they can do. What about people who provide help to their neighbors? Just over 40 percent of older Minnesotans provided this type of help, usually in the form of transportation or visiting. Is this volunteer work? Or is volunteer work only that which one does as part of an organized effort? In that case, 52 percent of elders in the study qualified as doing volunteer work, mostly organized by churches. When all kinds of help to others was combined into a general category of unpaid help to others, *only 17 percent of older people in the Minnesota sample were not doing any type of voluntary service.* The mean amount of time devoted to this service was about 14 hours per month.

Less than 10 percent donated more than 40 hours of service per month.

In the Minnesota study, the volunteers who worked for organizations (52 percent) tended to have higher incomes, to have more education, to be married, to have no disabilities, and to be younger. Even at age 75 to 84, 47 percent of the sample were still doing volunteer work, but among those 85 and older, the proportion dropped to 23 percent, still a substantial proportion.

The Older Americans Act* has provided funding and organizational support for the development of several national programs and transportation services to facilitate local participation.

The *Retired Senior Volunteer Program (RSVP)* offers people age 60 and over the opportunity to do volunteer service to meet community needs. RSVP agencies place volunteers in schools, hospitals, libraries, courts, day-care centers, nursing homes, and a host of other organizations. RSVP programs provide transportation to and from the place of service.

The *Service Corps of Retired Executives (SCORE)* offers retired businesspeople an opportunity to help owners of small businesses and managers of community organizations who are having management problems. Since 1965, over 200,000 businesses have received help from SCORE. Volunteers receive no pay but are reimbursed for out-of-pocket expenses.

The *Senior Companion Program* offers a small stipend to older people who help adults with special needs, such as the handicapped and the disabled.

A program called *Green Thumb,* sponsored by the National Farmers Union in twenty-four states, provides part-time employment in conservation, beautification, and community improvement in rural areas and existing community service agencies.

The U.S. Department of Labor has three programs that offer older people part-time *employment* as aides in a variety of community agencies, including child-care centers, vocational training programs, building security, clerical service, and homemaker services. The *Senior Aides* program is administered by the National Council of Senior Citizens; *Senior Community Services Aides* is sponsored by the National Council on the Aging; and *Senior Community Aides* is sponsored by the American Association of Retired Persons (AARP).

The success of these programs illustrates that older people can be quite effective in both volunteer and paid positions. For the time being, however, we can expect volunteer opportunities to outnumber opportunities for part-time employment. A major obstacle to the effective use of older volunteers has been an unwillingness to assign them to responsible, meaningful positions on an ongoing basis. The result is a vicious circle: Because volunteers are assigned to menial tasks, they get bored or frustrated and quit. Because they quit, administrators are reluctant to put volunteers in anything other than nonessential jobs.

Studies have shown that older volunteers can be counted on to perform well on an ongoing basis (Chambré 1993), particularly if the agency placing volunteers adheres to a few simple guidelines. First, be flexible in matching the volunteer's background to assigned tasks. If the agency takes a broad perspective, useful work can be found for almost anyone. Second, *volunteers must be trained.* All too often agency personnel place unprepared volunteers in an unfamiliar setting. Then the volunteer's difficulty confirms the myth that you cannot expect good work from volunteers. Third, the volunteer should be offered a variety of placement options. Some volunteers prefer to do familiar things; others want to do anything but familiar things. Fourth, volunteers should get personal attention from the placement agency. There

---

*See Chapter 15, "Social Services," for a detailed discussion of the Older Americans Act.

should be people (perhaps volunteers) who follow up on absences and who are willing to listen to the volunteers' compliments, complaints, and experiences. Public recognition from the community is an important reward for voluntary service. Finally, transportation to and from the placement should be provided (Sainer and Zander 1971).

Older people are often reluctant to commit themselves to a rigid schedule of volunteer work because such schedules interfere with the freedom and autonomy that are hallmarks of the retirement role. For this reason, the proportion of older people who are willing to participate is likely to remain small unless ways are found to make use of short-term contributions. Many older people who are not interested in making ongoing commitments are willing to donate time and energy to one-time events or programs that last a few weeks at most.

Education   Education is the systematic imparting of knowledge or skill. We typically think of education as taking place in schools, but it actually occurs in a wide variety of other settings, such as community centers, churches, libraries, senior centers, and workplaces. Formally organized education is our collective attempt to bring together important ideas and viewpoints, skilled teachers, and motivated students so we can expand the capabilities of our people to participate in all levels of society effectively and satisfyingly. However, as a society, we are not highly committed to providing formal education to adults.

In 1988, state and local governments spent $243 billion on education in the United States. But, our public education systems are heavily oriented toward youth and preparation for jobs. In addition, the proportion of government resources going to education fell from 40 percent in 1970 to 35 percent in 1983 to 29 percent in 1988, which means that the fiscal climate has not been good for expanding public educational programs for

anyone, especially for middle-aged or older people. Except for the offerings of public colleges and universities, the little adult public education that does exist tends to focus on adult basic education, mainly literacy. Very little public education is oriented toward developing or expanding the knowledge and skills middle-aged and older adults need for employment, service in the community, caring for themselves or their families, or enjoying life.

Our approach to education involves an intense dose early in life with no organized effort to continuously update knowledge and skills throughout adulthood. But education is not like a measles immunization that, once given, lasts a lifetime (Birren 1989). We need regular doses of new information if our knowledge is to remain up-to-date. In this sense, our life course ideal, which concentrates education in our first twenty-five years, is out-of-date and needs to be revised to include regular updating of knowledge and skills. The notion that we can concentrate all our effort and financial resources on young people and let adults fend for themselves in terms of educational needs simply does not fit the realities of an aging society. Today's world is one of rapidly paced social change combined with an explosion of information. Frequent and systematic exposure throughout adulthood to various knowledge bases and ongoing attention to maintaining skills, especially in relation to technology, are vital to our capacity to provide useful opportunities for middle-aged and older adults. They need this to participate in the work of society, whether their work is through employment, volunteer work, or family caregiving.

The U.S. Senate Special Committee on Aging (1991) has estimated that as much as 35 percent of older adults are functionally illiterate. Although among elders the proportion of high school graduates has increased substantially in recent years, there is a large minority with very little education, especially among minority and rural populations. Thus, we need two varieties of effort: *basic* education to provide literacy in language and mathematics, and *lifelong* education designed to enhance and update existing knowledge.

The Adult Education Act was designed to provide basic literacy skills and to encourage adults to complete high school. Programs were supposed to be directed toward adults who had not completed high school and who were not enrolled in adult education, a category composed of more than 30 percent elders. Yet less than 5 percent of people typically served by these programs are age 60 or over (U.S. Senate Special Committee on Aging 1991).

The record for lifelong education is not much better. De Crow (1975) found that older adults were interested in a variety of educational programs, but only about one-third of the available offerings were through conventional educational institutions. Most were provided by cooperative agricultural extension services, libraries, senior centers, and recreation centers.

Most public lifelong education is done by colleges and universities. However, there has been some question as to how effectively older people can be integrated into college classrooms with the traditional 18- to 22-year-old college undergraduate populations. Kay et al. (1983) conducted an interesting experiment in which older students were integrated into a combination writing and literature course for second-semester first-year students. They found that students in their 60s acted as catalysts and that compared with classes with traditional college-aged students only, the intergenerational classes discussed more and laughed more. The older students stimulated productive and positive interaction. In addition, compared with the control classes, younger students in the intergenerational classes emerged with more positive attitudes toward their own aging and toward intergenerational learning. Finally, the students in the intergenerational classes did as well or better on their essay examinations.

These findings match my own observations. In my seminar on adult development and social change, I regularly have a few middle-aged and older students mixed in with fifteen or more students of traditional college age. These older students, most of whom have experienced careers and child rearing, offer insights that younger students value very much. The older students are especially valuable in stimulating discussion in the early days of the seminar, when members of the class are still getting to know one another.

In recent years, community colleges increasingly have promoted their course offerings among middle-aged and older adults. Over 90 percent of the population lives within commuting distance of a community college, which should make for easy access to education. But in 1989 less than 25 percent of the more than 1,200 community colleges in the United States currently offered programs designed to meet the needs of middle-aged or older students (Ventura-Merkel 1991).

Among the courses designed for older adults, exercise, arts and crafts, and retirement financial planning were the most commonly offered. Of course, older adults also take courses along with other adults of all ages. General courses that attracted the greatest participation among older enrollees included arts and crafts (42 percent), literacy training (40 percent), technological and computer literacy (37 percent), humanities and social sciences (35 percent), participation in the performing arts (31 percent), career counseling (30 percent), and exercise (30 percent).

This varied mixture indicates that elders have multiple motives for being involved in lifelong education. Some are interested in personal issues such as fitness and managing their finances; others are seeking job training and placement; others want to keep their knowledge current on issues ranging from the latest computer technology to political and civic affairs; some participate in crafts and in the performing arts; and still others are interested in learning for its own sake through the humanities and social sciences.

Private enterprise has also slowly awakened to the market for education among retirees. In the travel tour business, for example, an increasing share of tours marketed to mature adults are being billed as "study tours."

Elderhostel is perhaps the fastest-growing educational program aimed at people 60 or over. Elderhostel is a national nonprofit corporation that coordinates short-term programs at institutions of higher education. In keeping with the hostel concept (simple accommodations and modest cost), most Elderhostel participants often stay in college dormitories and eat in college dining facilities. Scholarships are available for low-income participants. The usual program lasts for a week, during which the Elderhostel participant will have three courses from which to choose. The participant may take any combination, or all three. Courses are not for credit; there are no exams or grades. No previous knowledge or study is required. The recreational facilities and resources of the host institution are also available to Elderhostelers during their stay.

In 1991, for an average cost of about $275 per week per person (lodging and meals included), Elderhostel programs were provided to 450,000 people at over 1,800 educational institutions spread throughout all of the United States, all Canadian provinces, and 45 foreign countries. This was up from just 14,000 participants and 235 institutions in 1979. The range of course offerings at any time is mind-boggling, ranging from writing family history to silt painting in the arroyos of the Southwest. Elderhostel is an exciting new movement and seems in tune with the individualism and humanism that currently form the moral backdrop for retirement lifestyles. Its success shows clearly that older Americans are very much interested in learning if the right kind of opportunities exist.

Senior Net is another interesting example of education for elders. Founded in 1986, Senior Net's goals include teaching computer skills to older adults, offering opportunities for older people to communicate with one another through an on-line computer network, and conducting research on the use of technology by older adults (Middleton 1991). In 1994, Senior Net had more than sixty sites where members could learn to use computers. The organization has several publications aimed at increasing the use of computers by elders. In addition, many older people already have substantial computer skills, and they work as volunteer teachers and also participate at home in Senior Net On-Line, the computer network. All it takes is an inexpensive personal computer, a modem, and a phone line. Members of Senior Net use their computers in a variety of ways: to keep in touch with their families via e-mail, to play board games via the Internet, and to subscribe to interest groups. Interest groups include a mind-boggling variety of topics such as widows supporting widows, gardening, genealogy, World War II experiences, and preventive health. By 1994, Senior Net alone had a membership of over 13,000. The success of Senior Net, and of senior-oriented forums and bulletin boards on other on-line services such as Compu-Serve and Prodigy, contradicts the stereotype that older people are afraid of new technology and are unwilling to use it. What elders often lack is access to opportunities to learn to use new technology.

Barriers Four major barriers currently constrain the participation of aging people in education: the heterogeneity of the aging population, inexperience with education, sparse availability of adult education, and insufficient quality of available programs. As we have seen repeatedly throughout this book, the aging population is extraordinarily diverse. When we consider the full range of differences in interests and skills that could potentially be

Senior Net trains older adults to use electronic networks.
Photograph by Todd Powell/Tony Stone Images, Inc.

related to differences in age, gender, educational background, current or former occupation, ethnicity, and so on, the difficulty of planning educational programming for an aging population is clear. But the difficulty of the task can become an excuse for inaction. We know that basic literacy is important both to the society and its older citizens, yet we have not aimed our literacy programs effectively at aging or older people. We know that aging and older people are interested in job training and placement, but offerings in this area are sparse, especially programs that consider the special problems associated with confronting age discrimination in employment. There is no doubt that many types of lifelong education are needed, but at the same time, the priorities are reasonably clear. What is lacking is the will to act on them.

Lack of literacy is obviously a barrier to participation in educational programs. But even among those who have basic literacy skills, images of school and learning may be so outdated that the potential value of education is unclear. Because most readers of this book are currently involved in education, it is probably difficult to imagine what it would be like to have never had any formal education or to have been away from school for 30 years or more. Yet many older adults have not recently experienced the enabling power of information or knowledge and do not see the potential linkage between the quality of their everyday lives and the quality of the information they have about the world around them.

Aging adults cannot avail themselves of programs that do not exist. Lack of availability is probably the major barrier to participation in education. As we saw, only a small minority of public schools and community colleges have programs designed for elders. Participation by elders in general adult education is low, and the meager resources and effort devoted to public adult education in general translates into sparse availability of adult education of all kinds.

Finally, to attract adults, educational offerings have to be high in quality. To be high quality, education aimed at older adults must take into account the special needs of older learners, such as providing sufficient lighting, using flexible procedures for providing feedback on student performance, being sensitive to the pace of instruction, avoiding time pressures and promoting self-paced learning, using multisensory messages, and being sensitive to hearing and vision problems (Heimstra 1991). In summarizing a review of what makes for successful programs for elders, Kelly (1993) concluded that quality was a major determinant: quality of leadership, quality of the setting, and quality of the students enhance one another.

The Home

The tendency to view activities in terms of roles has led to an artificial separation of the study of activities. Work, home, community, and other environments are viewed separately and studied by different researchers with different concepts and research agendas. Even those who look at activities at home tend to study specific types of activities rather than the entire array of at-home activities.

Table 10-2 is a classification of at-home activities with examples. The examples are not exclusive, in the sense that a particular activity could fit into different categories, depending on its meaning to the individual. For example, if a person gardens mainly to sup-

plement the family food budget and does not enjoy it very much, then gardening is mainly a household chore rather than an appreciation of or contact with nature. Cooking can be either a chore or a creative act. The examples have been placed in the categories that fit the meaning of the activity to most middle-aged or older adults most of the time.

The amount of activity centered on the home depends on the amount of time spent there, which even before retirement tends to increase after age 50. Part of this increase is due to a loss of roles in environments outside the household, and part is due to a growing preference for in-home activities. Physical limits restrict activities to the home for only a small proportion of adults, including those over 75.

Moss and Lawton (1982) studied time budgets in a sample of older people living independently in the community. The average age of the sample was 76; 57 percent of the sample was female. They found that the respondents occupied their days with a diversity of activities. (See Table 10-3.) Obligatory activities, such as shopping and housework, took up 22.7 percent of the day; discretionary activities, such as socializing and watching television, occupied 39.2 percent of the day; and sleep accounted for 32 percent. Although watching television was the most common discretionary activity, it occupied

Table 10-2 Types of at-home activities.

Type of Activity	Examples
Work	Housework, home maintenance, household chores
Relaxation	Resting, sleeping, mediating
Diversion	Television watching, listening to music, reading
Personal development	Studying, learning, practicing, inquiring, tending to personal care, worshiping
Creativity and problem solving	Crafts, cooking, planning, handiwork, arts
Sensory gratification	Engaging in sexual activity, eating, drinking, smoking
Socializing	Discussing, arguing, entertaining, gossiping
Appreciation of nature	Walking, gardening, bird-watching

Source: Adapted and modified from Gordon, Gaitz, and Scott (1976).

Table 10-3 Percentage of time independent older people spend on various activities.

Type of Activity		Percent of Time Spent
Obligatory:	22.7	
Personal/sick care		3.4
Eating		5.3
Shopping		1.3
Housework		5.8
Cooking		5.1
Helping others		0.6
Discretionary:	39.2	
Family interaction		4.0
Friend interaction		3.1
Religious activity		0.8
Reading		4.3
Listening to radio		2.1
Watching television		14.3
Recreation		2.8
Rest and relaxation		7.7
Unaccounted time	1.7	
Travel	2.6	
Sleep	32.0	
Total	98.2	

Source: Adapted from Moss and Lawton (1982).

only a little over three hours per day, no more than is average for all adults.

Special Activities

Everyday activities in workplaces, communities, and homes tend to have a rhythm designed for the long haul. There are usually times that are more exciting than others, but in general, events progress on a more or less even keel. These everyday rhythms are punctuated by special events: vacations, holidays, community festivals, and the like.

The opportunity to enjoy extended travel is a major special event that differentiates later maturity from most other life stages. Older people do not necessarily travel more than people of other ages, but in many cases they travel greater distances and they travel

while young people are in school. As mentioned in our discussion of the family, older people often travel to renew relationships with siblings and to visit children and grandchildren. The freedom and financial resources to enjoy travel are often mentioned as major advantages of being retired.

Older generations in families also often find themselves playing host to children, grandchildren, and extended kin during holidays. For most this is a welcome opportunity to be with family, but dealing with a house full of people for several days at a time can be a substantial amount of work too. Activities during winter or summer family holidays offer a considerable departure from normal routine, which is usually seen as a needed break. But for some people, the interruption of normal routine and the hubbub involved are stressful. We do not know how aging affects perceptions and experiences of family holidays; this would be a good topic for research.

What Activities Are Desirable for Older People?

Middle-class professionals often judge the activities of all older adults in terms of their appropriateness for middle-aged, middle-class people. Thus, the literature often implies that work is better than other kinds of activity and that group-centered activities are "better" for people than solitary activities. From the viewpoints of older people, this evaluation may not be justified. Because activity patterns established in early adulthood tend to persist and because what is fashionable in the way of activities varies from time to time and between social groups, it seems safe to assume that no single standard can be used to determine the "adequacy" of activity patterns among mature adults.

The tendency to assume that only leisure activities or employment can give meaning to life has caused many researchers to overlook the potential importance of mundane activities such as personal care, housekeeping, cooking, shopping, tinkering, and puttering (Lawton 1978). In later life, these activities reflect the continued ability to be independent in the face of age, whereas earlier the ability to do them is taken for granted. Likewise, there is a common conception that inactivity is detrimental. In massive doses, inactivity is undoubtedly harmful, but the general statement overlooks the benefits of contemplation. It is important to find out what is happening inside a person during physically inactive but wakeful times rather than simply to assume the result is negative. For example, reading for pleasure may appear to be a passive activity, but for the reader it can be an involving, exciting activity.

In an interesting study, Larson, Zuzanek, and Mannell (1985) recruited older participants (median age of 68) who were willing to carry electronic pagers for a week and fill out self-reports of what they were doing when they received signals from the pager. Signals were sent between 8 A.M. and 10 P.M., with one signal occurring at random within each two-hour block of time throughout the week. In all, they collected 3,412 self-reports about the objective circumstances and subjective states of their respondents at random moments in their lives.

The researchers were interested in the relationship between being alone, activity, and subjective well-being. They found that 48 percent of the time, the respondents were alone when paged, and that far more time was spent alone than with any single person. However, for the married respondents, spending substantial parts of their days alone was not a negative experience but instead was related to feeling energized. Being alone was a chance for focused thought and concentration, and was not seen as a condition of loneliness. For the unmarried, the situation was different. Whereas married elders were likely to feel involved when they were alone, unmarried older people were likely to

feel bored and to wish they were doing something else. Although the respondents were not on the unhappy end of the scale under any circumstances, the unmarried respondents were less happy and cheerful when alone than the married respondents, and this was especially true in the evenings, when the unmarried respondents were much more likely to be alone.

The implication of these findings is that the objective state of aloneness cannot be evaluated apart from its social context and subjective results. For people who live with a spouse, being alone is experienced as an opportunity to be free of the demands of the relationship and to concentrate one's energies on activities of one's own choosing. For people who live alone and who may therefore feel the need for more interaction, being alone can be experienced as deprivation. For these people, solitary activity cannot substitute for interaction. So is solitary activity bad for older people? There is no single answer.

Summary

Activities structure people's lives, and for many people, satisfaction with life depends on having a satisfying and meaningful round of activities.

Activities can be categorized in a vast number of ways. Social activities provide the context within which human relationships are developed, nurtured, and maintained. Some important dimensions of an activity are its intensity, the frequency with which it recurs in a person's life, the number of different environments in which the activity occurs, the skills and the knowledge required to do the activity, the extent to which the activity is obligatory, the capacity of the activity to provide a sense of leisure, the meaning(s) of the activity for a particular individual and for people in general, whether the activity can be solitary or requires others, and how costly the activity is.

The knowledge and skills needed for a varied set of activities are generally developed in early or middle adulthood and are maintained into later life. If people want to develop new activities in later life, they can sometimes get assistance from organizations such as senior centers, continuing education institutions, and voluntary organizations. But most older people prefer to be on their own if they want to take up new activities.

It is important not to impose our own notions of which activities "ought to be" satisfying to older people. For example, mundane activities around the house are meaningful to many older people—much more than many people who work with elders would expect.

Core activities generally show a great deal of stability in later life, even with the substantial environmental changes that can occur as a result of child launching and retirement. People tend to cope with constriction and lost activities by becoming increasingly involved in the activities that remain. Most people have a wide variety of activities that can absorb even important losses. However, some people have a limited set of alternatives or place a very high value on a lost activity and so are distressed by changes in activities.

As a prevalent activity, employment begins to fade in importance in the early 50s, although much of the decline prior to 60 is related to problems such as ill health or unemployment rather than personal choice. After age 65, the overwhelming majority of retirements are voluntary.

Participation in community organizations tends to be highest in middle and later life. Transportation problems, poor health, and lack of money are the major factors hampering participation for older people. Religious groups are the most common voluntary associations with which middle-aged or older people are involved. Churches generally welcome the participation of older people, but do little actively to promote it. Apart from churches, no other type of voluntary organization involves participation by as much as one-third of the middle-aged or older population. Older people who have been involved in local politics for many years usually find that their skills and experience give them a good deal of political influence. Volunteer work by older people has expanded dramatically in the past decade, although its exact prevalence is as yet unknown.

Activity patterns are highly individualized and highly stable across the life course. Nevertheless, options are sometimes limited in later life by physical, financial, or transportation factors. Personality, family, and social class values can narrow the range of options even further. Social activities remain important personal goals for most people throughout the life course, but with increasing age, the circle of people with whom one can enjoy social activities becomes smaller. It is probably fortunate that, as they age, people tend to gravitate from obligatory activities to discretionary activities, from activities outside the home to at-home activities, and from social activities to solitary activities. This general movement can create flexibility and freedom for the individual, provided it is what the individual wants, which in most cases it is.

11 ■ Religion and Spirituality

Spirituality and religion are important issues that influence opinions, knowledge, personal goals, behavior, and lifestyles for most middle-aged and older people. Religion and spirituality also significantly influence both mental and physical health in later life.

In its essence, *spirituality* is a holistic inner domain within which people deal with existential issues such as the meaning of life, the existence of God, the causes of and cures for human suffering, and the meaning of death. Spirituality deals simultaneously with cognitive, experiential, emotional, and motivational aspects of these concerns. Ultimately, each human being confronts these questions alone, but the inner experience of spirituality and socially constructed religion interact and for most people are inseparable.

Religion is a social institution that attempts to give each new generation an opportunity to benefit from the religious knowledge and spiritual insights of those who came before. People create (most believe with divine help) religious beliefs, philosophies, and rules for living a good and worthy life that serve as guides for individuals as well as set moral and ethical standards for religious communities. Throughout the world there are literally hundreds of religions, and many religions consider their specific answers to the great existential questions of life to be the only ones acceptable. And because disputes over these questions cannot be decided by science or reason alone, there is no way to resolve conflicts between groups who

do not grant religious freedom to one another. Thus, religion can be a cause for hope and comfort against suffering within a given religious context. But at the community, national, or international level, religion can also be a cause of suffering and despair as a result of religious conflict.

The importance of religion and spirituality varies widely across individuals, religious traditions, cultures, and nations. On the one hand, some of us are born into families that are centered around religion—filled with religious discussion, religious music, religious art, religious holidays, and devotional religious practices. Life literally revolves around religious definitions of both the self, the group, and every social situation. Early life socialization emphasizes the learning of religious worldviews, language, beliefs, and practices, both in the home and in religious institutions variously called churches, synagogues, temples, mosques, and so forth. The individual's developing inner sense of spirituality is conditioned by the language and concepts used in specific religious socialization. Choices of schools, friends, and potential marriage partners are often heavily influenced by religious considerations.

For example, the old-order Amish stand out as a group because their religious beliefs require them to reject many of the changes associated with modernization, and their religion so infuses their cultural life together that Amish culture has evolved along a very different path from "mainstream" American culture. Many orthodox, traditionalist, or fundamentalist variants of the major faith groups, such as Christian, Jew, Muslim, or Buddhist, require their adherents to put their religion at the center of their lifestyle. This results in their appearing "different" from those who compartmentalize religion into the private sector of their lives and whose behavior and dress conform to the secular culture. For instance, traditional Muslim women wear loose robes and veils that cover all but their eyes, Orthodox Jewish men wear yarmulkes (skullcaps), and Catholic parochial schoolchildren usually wear uniforms that differentiate them from other children.

On the other hand, others are born into families where religion is given little or no emphasis. Instead of including religious doctrine or philosophy as a way of resolving questions about the meaning of life and death, definitions of right and wrong, or questions of value, people who reject religion decide these issues on pragmatic and/or scientific grounds by the use of reason and dialogue. Those who do not use religion to answer fundamental metaphysical questions often base their values on ideals such as honesty, courage, justice, fidelity, and respect for one another that have been handed down through the centuries. Morality is decided through democratic dialogues based on experience rather than by reference to divine authority. These people attempt to be principled and responsible but nonreligious.

In the United States, most people grow up in families and in communities where the emphasis on religion is somewhere between these two extremes. They identify themselves as members of one of the major faith groups, attend religious services regularly, subscribe to the basic beliefs of their religious tradition, and pass on their religious orientation to their children. But their religion does not cause them to stand out from the general population.

Aging can affect both the experience of spirituality and connections with religious or-

ganizations. In this chapter, we will look at how aging relates to participation in organized religion, informal religious practices, and subjective elements of religion and spirituality. We then consider various theories of individual spiritual development in adulthood. Next, we look at the effects religion and spirituality have on individuals' health and psychological well-being and how these effects change with age. The chapter concludes with a discussion of specific research needs in the area of religion, spirituality, and aging.

Compared with other areas of social gerontology, religion and spirituality have been neglected topics. Until the late 1980s, well-done studies that included more than rudimentary measures of religiosity were rare.* Many early researchers in social, behavioral, and health sciences saw a fundamental conflict between scientific and religious worldviews, which led them to define religion as a topic that could not be studied scientifically and to therefore ignore it. Indeed, many scientific researchers were and are actively hostile toward religion as a topic of research (Levin 1994b). In addition, social gerontology began its rapid growth during the 1960s, a decade in which religion took a back seat to a new secular morality in public discourse, especially on university campuses. Except at a few sectarian institutions, faculty interested in studying religion were seen as outside the mainstream and poor candidates for tenure and promotion (Sherrill and Larson 1994). The resulting body of research was often of poor quality (Levin 1994a). Fortunately for us, the literature on religion, spirituality, and aging has grown rapidly and improved greatly in quality. By 1995 there were a number of very good sources (see Kimble et al. 1995; Levin 1994a; Koenig 1995a, 1995b; Thomas and Eisenhandler 1994).

---

*See Maves (1960) for an excellent early review of the literature on religion and aging.

Concepts and Language

The concepts and language used to study religion and spirituality are not well organized or agreed-upon. Most scholars acknowledge the importance of distinguishing between the behavior and ideas connected to membership in organized religion, the behavior and ideas associated with individual, private expressions of spirituality, and the subjective, experiential aspects of spirituality. However, many scholars use terms such as *religion, religiosity, religiousness, religious involvement,* and *spirituality* as if they were interchangeable ways to describe the same thing (Koenig 1995a). But equating religiousness, religiosity, religious affiliation, church attendance, or adherence to specific religious beliefs ignores the great religious diversity in America and the multiplicity of sometimes conflicting ways of expressing religious commitment contained within the hundreds of religious subgroups in our society. Thus, to study religion and spirituality systematically, distinguishing among these various terms is both necessary and useful.

As mentioned, *religion* can be regarded as a social institution organized around specific human needs—to know the meaning of life, to cope with the reality of suffering and death, and to have a concept of a moral life. All religions attempt to meet these needs. Organized religions vary in complexity from the very large and formal hierarchy of the Roman Catholic church, through the loose confederation of congregations that make up Protestant religious denominations, to unique sects whose organizations are fluid and whose major beliefs are shared only within their relatively small group. Thus, equating religiosity with religious affiliation, church attendance, or adherence to a specific set of beliefs ignores the enormous religious variability in America.

Most of the high-quality research on the relation of aging to religion and spirituality comes from the United States, and most of this research has used concepts and measures that reflect a Judeo-Christian cultural tradition (Koenig 1995a; Bianchi 1992; Kimble et al. 1995). However, the United States is extremely diverse in terms of religious beliefs and practices. In 1990, about 55 percent of Americans classified themselves as Protestants, another 25 percent as Catholics, and 2 percent as Jews; so about 82 percent associated themselves in general terms with a Jewish or Christian biblical tradition. However, about 9 percent said they were Muslims, Hindus, Buddhists or other religions, and about 9 percent did not identify themselves with any religion (U.S. Bureau of the Census 1992h). Among young adults, the proportion who express religious affiliations other than Christian or Jewish is growing as is the proportion who express no religious affiliation, which means that the domination of the Judeo-Christian cultural tradition may lessen in future cohorts entering the older population.

Further diversity comes from the variation of perspectives within the major faith groups. For example, although the Roman Catholic church had by far the largest membership, there were nine other variations of Catholicism that reported membership (U.S. Bureau of the Census 1992h). And although collectively the Protestant category was the largest, there were more than twenty denominations reported within this category. In addition, religions such as Unitarian-Universalist, Mormon, and Jehovah's Witnesses are included in the data on Protestants although these religions are not primarily based on Protestant theology. Finally, there is great diversity of religious belief and practice even within the relatively small Jewish population.

Each of this multitude of religions has its unique ideas that represent a collective response to questions about the existence and nature of a deity, about the relation of human beings to God or the universe, about the nature of death, about the existence and nature of an afterlife, and about what is required and forbidden in terms of lifestyle and behavior. Religious groups also vary considerably in the nature and frequency of rituals and the degree to which they hold their members accountable for conforming strictly to religious doctrine, which in turn influences the relationship between religious beliefs and behavior. Thus, knowledge of the proportion of people of various ages who report religious group membership tells us very little about the extent to which religion significantly influences their lives.

The degree to which people espouse and act on the beliefs of their religion is much more relevant information for our purposes. *Religiosity, religiousness,* or *religious involvement* are terms for a multidimensional concept often used to describe the extent to which people identify with a religion, have religious knowledge, espouse religious beliefs, experience religious emotions, and are exposed to the influence of religion in their lives (Glock and Stark 1965; Krause 1993). These concepts also usually differentiate between participation in organized religious activities, private religious behavior not done in an organizational setting, and a subjective level that involves faith and commitment to religion or mystical religious experience (Krause 1993; Moody 1995).

Spirituality is usually thought of as a personal quality or experience. Koenig pointed out that most research on religion and aging is based on a model in which "there is a spiritual dimension to humans which is separate from the biological, psychological, and interpersonal dimensions, but which heavily influences each one of these and in turn may be influenced by them" (1995a,10). Thibault (1995) defined *spirit* as a generic, nonreligious term for the *energy,* similar to the concept of a psychobiological drive, that impels people to seek answers to ultimate questions. She defined spirituality as a person's style of seeking, finding, creating, or using personal meaning in a context that includes the entire universe. Spirituality is more often considered a continuing quest than a final destination.

Spirituality is a broad concept that itself can have several dimensions. From the point of view of personal construct psychology (G. Kelly 1955), spirituality stems from a personal cognitive integration of the multiple dimensions of religiosity previously cited. But from a philosophical point of view, spirituality is not simply a personal cognitive construction; it is also a holistic integration of body, mind, and spirit into an orientation toward ultimate concerns that transcends personal conceptions (Moody 1995; Tornstam 1994). This more mystical view of spirituality is much harder to measure than beliefs, knowledge, or behavior because mystical experience is largely nonverbal and intuitive.*

Contemporary measures of religious involvement or religiosity attempt to collect information on most of the dimensions already cited, but seldom have individual investigators used extensive multidimensional indicators within the context of a single study. Figure 11-1 lists several significant dimensions of religiosity, religiousness, or religious involvement and provides examples of indicators that could be used to collect data on various dimensions. There are several scales available to measure different aspects of religious behavior and beliefs (Koenig 1995a). For example, the multidimensional Springfield Religiosity Schedule (Koenig, Kvole, and Ferrel 1988) contains 306 items, probably more than could be included in most surveys; but it is an excellent resource for researchers.

In the next section, we will look at how organized religion relates to aging or older people.

Organized Religion

A large majority of elders have had a life-long identification with their religion and are

---

*Mysticism is defined in more detail later, in the section, "Subjective Elements of Religion and Spirituality."

Figure   11-1   Concepts and indicators of religion and spirituality.

Concept	Examples of Indicators:*
External Manifestations	
Religious affiliation	Checklist of denominations (for example, Protestant, Catholic, Jewish, other, none)
Formal religious behavior	Participation in religious services, ceremonies, or rituals
Informal religious behavior	Reading sacred scriptures Private prayer or meditation
Religion as life organizer	Religion provides guidance for everyday life choices
Internal Manifestations	
Knowledge	History of one's own religion and its basic tenets
Faith	Belief in a personal afterlife
Emotions	Being filled with a spiritual joy Being afraid of God's disapproval or anger
Commitment/Motivation	Religion or spirituality as an enduring priority
Mystical	Feeling God's actual presence
Transcendence	Experiencing the unity of all the universe
Consequences	
Use in coping	Prayer helps to cope with suffering
Well-being	Relationship with God reduces loneliness

* Specific measures usually assess the frequency with which and/or the extent or degree to which a statement fits a respondent's experience.

long-standing members of their local congregations.* Membership in religious organizations is more prevalent among those in their 50s and 60s than in other age categories. Older members represent a substantial proportion of most congregations, especially in the long-standing religious denominations, but congregations vary greatly in the extent to which they encourage leadership by elders, seek out elders to do the work of the organization, or have programs aimed at the needs of elders. There are also substantial racial and gender differences in the respon-

siveness of organized religions. African American churches generally respond much more effectively to the needs of older church members than do churches attended exclusively by whites. Older women, even though they represent a large majority of church attenders, often find that patriarchally organized church organizations are insensitive to their needs.

Most religious clergy begin their careers under the assumption that worship services and pastoral support are as available to older people as they are to others, and for the able-minded and able-bodied older population this may be true. But through contacts with shut-ins, hospitalized elders, and older people who live in nursing homes, most local clergy learn firsthand that disabled or terminally ill elders have spiritual needs and interests that differ from the needs of the general congregation (Koenig 1995a). In addition, many clergy

*Local religious groups vary greatly in how they refer to themselves. Most Protestant denominations refer to the people affiliated with the local church as a congregation, local Quaker groups are called meetings, Catholics participate in parishes, and so forth. Here we use the term *congregation* generically to refer to the people who make up religious organizations at the local level.

come to realize that elders in general are interested in a contemplative, inner religious journey that is not easily accommodated within the traditional structure of religious activities. Although most clergy spend a majority of their time working with young people and youth programs, Longino and Kitson (1976) found them to be quite open to working with elders.

Many congregations emphasize programs for children and youth. After all, religious socialization of the young is a major function that must be performed in order to carry on the religious tradition. But many congregations also have practices that unintentionally push elders out of responsible positions. For example, people responsible for scheduling may be insensitive to elders' reluctance to drive at night. By scheduling events and committee meetings at night, elders may be excluded.

Religious organizations, and clergy in particular, are not immune to the ageism that permeates the general culture. Many congregations give a low priority to special programs of outreach to older members. Sheehan, Wilson, and Marella (1988) found that congregations in which responsibility for meeting their members' social needs was a major goal were much more likely than other congregations to sponsor programs for older members. Interestingly, having a high proportion of elders in the congregation was unrelated to the offering of programs for elders.

Because the proportion of elders in the average congregation is high and often growing, it is not surprising that denominations have begun to feel pressure to develop programs aimed at older members. At this point, however, it appears that the most common program consists of special social groups for older members of the congregation. Some denominations have developed "older adult ministries" staffed by retired clergy and older lay ministers, who serve institutionalized elders and homebound elders living in the community. Programs aimed at fostering deeper spiritual development among elders are scarce. While most congregations are willing to passively accept the participation of older people, few place high priority on actively soliciting the participation of well elders much less those who are ill, disabled, or isolated.

Despite this somewhat discouraging general picture, the spread of several innovative programs suggests that change is under way (Huber 1995). For example, in 1972 Dr. Elbert Cole began the Shepherd's Center movement. The Shepherd's Center model focuses on elders' providing services to elders. Shepherd's Centers primarily involve middle-class older people (Seeber 1995). Space is usually donated by a local religious organization, financial support usually comes from several different congregations, and participation is open to all local congregations and to people of all faiths. Shepherd's Centers serve a delimited geographic area and have few paid staff. Leadership is vested in the older people who are center members. Shepherd's Center programs depend primarily on older participants to do research, plan, and deliver their services. These services, particularly education and in-home services, are aimed at needs of older people. Shepherd's Centers avoid duplicating already available services and instead seek to fill high-priority service gaps. Shepherd's Centers of America, located in Kansas City, helps elders throughout the country start new centers. As of 1995, there were nearly one hundred Shepherd's Centers spread over twenty-six states. Total membership was about 110,000 and about 15,000 elders received services in their homes from Shepherd's Center volunteers.

The Spiritual Eldering Institute, begun in 1986 by Rabbi Zalman Schacter-Shalomi in Philadelphia, is an innovative program designed to bring elders of all faiths together to do the inner work needed to deepen their spirituality and to advocate for a revitalization of the roles elders have traditionally played in the spiritual life of the community. Hundreds of elders all over the nation have attended Spiritual Eldering workshops offered by the institute, and the institute also trains elders to offer such workshops.

The entry into later adulthood of the cohorts who were part of the 1960s "consciousness movement" has sparked widespread interest in "Conscious Aging"—a program initially sponsored in 1992 by the Omega Institute for Holistic Studies in Rinebeck, New York, and centered around the inner spiritual journey and how it interacts with aging. In addition to sessions on physical, mental, and spiritual wellness, special sessions in Omega's weekend Conscious Aging programs focus on the effect of participants' inner spirituality on their experience of aging, and the effect of aging on their experience of the inner spiritual journey. Ram Dass, whose book *Be Here Now* (1971) heavily influenced the consciousness movement, has played a leading role, along with Zalman Schacter-Shalomi and several "enlightened" gerontologists, in convening the Conscious Aging meetings.

Most of these innovative programs were carried forward by elders with vision who persisted in spite of obstacles thrown up by traditional religious organizations. As the demand increases for programs focused on the spiritual needs of elders, traditional congregations will either begin to offer such programs or perhaps find themselves competing with new types of religious organizations. For example, Thibault (1995) proposed the development of "spiritual care communities" within local congregations. In these communities, elders would support and nurture one another and communicate honestly about their spiritual lives—they would become spiritual companions. Thibault recommended that these communities be based on a contemplative model rather than social role activity or social advocacy models. Thomas

(1994a) also saw value in approaches that focused on development of transcendent vantage points toward worldly concerns rather than trying to force elders back into the workaday world. Thomas emphasized the importance of this transcendent wisdom that develops among some elders as a spiritual resource for the community.

Attending Religious Services

About half of the general population attends organized religious services regularly (twice a month or more). Catholics are more likely than Protestants or Jews to attend regularly, but this disparity is diminishing. A higher proportion of women than men attend frequently. Attendance is positively related to income, education, and length of residence in the community (Cutler and Hendricks 1990). Age is thus only one of several factors that relate to differences in frequency of attending religious services. In cross-sectional research, done at single points in time, there is a steady increase in attendance from the late teens to a peak attendance of about 60 percent in the late 50s or early 60s. After that, attendance shows a slight decline, which has been attributed to increased prevalence of disability, ill health, and transportation problems (Koenig 1995a; Levin 1994b). Among older people, there is also an increase in the proportion never attending religious services.

In an ambitious study, Levin and his colleagues (1994) used four large national data sets to examine gender differences and black-white differences among older Americans in church attendance. A number of other indicators of religiousness, such as religious reading or feeling near to God were also used. Only church attendance and religious affiliation were asked in all the surveys. They found a consistent gender difference across all surveys, with older women attending church more often than older men among both blacks and whites. They also found that blacks attended church in significantly larger proportions than whites in all four surveys. By contrast, religious affiliation did not show consistent differences, partially because very high proportions of both genders and blacks and whites in the older population listed a religious affiliation—there was not much variability in religious affiliation in three of the four surveys.

In a sixteen-year longitudinal study, Atchley (1994) found that church attendance peaked at later ages than in cross-sectional studies. Members of his longitudinal panel reached their peak prevalence and frequency of attending religious services in later maturity—in their 60s and early 70s. But after age 75, there was a steady decline in frequency of attendance and a steady increase in the proportion never attending. These declines were larger for those with disability, but even those with no disability showed a decline in frequency of attendance. However, some elders responded to disability by increasing their frequency of attending religious services. Without longitudinal data, the latter finding could not have been possible, which stresses the need for more longitudinal research if we are to understand age changes in religion, spirituality, and aging.

Williams (1994) pointed out that although attendance is by far the most often used indicator of participation in organized religion, other important aspects of participation such as financial support, holding leadership positions, and doing volunteer work have been neglected in most research on religion and aging. In addition, religious services vary enormously in their content and form, and little research has been done on age changes in perceptions, cognitions, or emotions experienced during religious services. Rituals, music, sermons, and other aspects of religious participation probably have separable subjective effects on participants. Achenbaum (1995) examined Jewish and Christian age-based rituals and found no examples of ceremonies that affirmed the value of elders to

religious congregations or communities. These topics are much in need of research.

Informal Religious Behavior

Informal religious behavior consists of activities such as scripture reading, prayer, devotions, and meditation that are done at home or in very small, informal groups. Data on these aspects of religious behavior are scarce. Levin, Taylor, and Chatters (1994) found that listening to religious programs and religious reading were common among older people and that elders were slightly more likely to listen to religious programs on radio or television than to engage in religious reading. Among both blacks and whites, women were much more likely than men to do religious reading.

Koenig et al. (1993) studied 1,299 community-dwelling people age 60 or over who lived in central North Carolina. They reported that 56 percent of their respondents engaged either in reading scripture or in prayer several times per week and 32 percent listened to religious radio or watched religious television programs several times per week. Koenig (1995a) compared the religious behavior of older people in three settings: a nursing home, a geriatric clinic, and a senior center. Nursing home residents showed a significantly higher proportion who prayed at least daily (79 percent) and read scripture at least daily (41 percent). However, more than 60 percent in all groups reported praying at least once each day. Peacock and Paloma (1991) looked at cross-sectional age differences in types of prayer. Among those who prayed, the older respondents were more likely than younger respondents to engage in meditative prayer and less likely to engage in petitionary or ritual prayer. These findings suggest that informal religious behavior is very common indeed among older people and that the contemplative and meditative aspects of informal religious behavior increase in importance with age. However, much more research is needed to assess these possibilities.

Older people read scripture for inspiration and for comfort, to keep their minds occupied with spiritual thoughts. They engage in prayer and especially meditation or contemplation because these activities are soothing and supportive self-care for the mind as well as the soul. Their positive experiences of the results of meditation and prayer reinforce the value of these activities (Koenig 1995a). In addition, the everyday lives of older people often contain less time pressure and more opportunity and freedom to engage in meditation and prayer.

Subjective Elements of Religion and Spirituality

Subjective aspects of religion and spirituality include religious identification, religious attitudes, values, beliefs, and knowledge, religious conversions, and mystical experience. These aspects have received much less research attention than religious affiliation or church attendance.

Those who have a religious affiliation also tend to attach subjective value to religion. Atchley (1994) found that over the sixteen-year duration of his Ohio longitudinal study of adults who were 50 and over at the beginning of the study, over 70 percent maintained consistently that "being a religious person" was an important personal goal. Koenig (1995a) reported that more than 80 percent of the older respondents in his North Carolina study said that their religious faith was the most important influence in their lives. Levin et al. (1994) found that the importance of religion was high among older people, but they also found the familiar gender and race patterns, with women and blacks both scoring higher on the importance of religion. These findings suggest that the subjective importance of religion is high across a number of subgroups within the older population.

Religious beliefs, attitudes, and values tend to be specific, and represent a topic that has seldom been studied in relation to aging. Krause (1993) divided religious beliefs into three categories: relationship with God, relationships with other humans, and beliefs about the devil, sin, and hell. A large majority of older Americans believe in a God that watches over them and answers their prayers (Koenig 1995a). Most elders in Koenig's North Carolina study also believed the devil existed and in the literal interpretation of biblical miracles, but we should be careful about generalizing from this regional population, which is mostly evangelical Protestant, to other regions of the country that are much more diverse in terms of types of religious beliefs. About 80 percent of the older respondents in the 1990 General Social Survey believed in an afterlife, and most who believed in an afterlife thought that their personal consciousness would survive death. Data on religious beliefs and attitudes are contained in various scales of religiosity, but so far little systematic analysis has been done that looks at the effects of holding various specific beliefs.

It is important to look not only at the existence of religious beliefs but also at their content. Much of the research on religion and aging has treated religious belief as a generic characteristic, but beliefs are always specific. It matters a great deal whether people believe that long life is a gift from God that indicates moral worth or whether disability is punishment for sin. However, most research looking at specific religious beliefs has been confined to people of a specific faith tradition, and comparative effects of differences in religious beliefs have not been systematically studied. In addition, religious beliefs differ in the extent to which they interact with or supersede other beliefs, and this aspect of religious belief needs study as well.

Faith means accepting beliefs as true without proof and being willing to act on them. There is no way to prove scientifically that God does or doesn't exist. Those who believe

Prayer is important to most older people.
Photograph by Larry Kolvoorn/The Image Works

in a personal God do so as part of their faith. Koenig (1995a) argued that mature faith results from spiritual development that in most cases takes time and life experience. As a result, he postulated that aging increases the prevalence of mature faith. We will return to this question shortly when we consider various perspectives on spiritual development.

Religious conversions have two basic meanings. The first refers to a dramatic upheaval in a person's belief system such that an existing structure is literally destroyed or significantly modified and replaced by a new belief system. In the prototypical conversion experience of this type, a sinner suddenly "sees the light" and becomes devoutly religious in both beliefs and behavior. However, Bianchi (1992) pointed out that the destruction of religious faith can also occur suddenly,

often in response to intense doubt and/or profound loss. A second type of conversion experience is a sudden increase in the quality of religious commitment and does not imply a revolution in beliefs. In this second type of conversion experience, people who have believed in their religion in their minds suddenly experience a certainty of belief that extends to their hearts and souls as well. Both types of conversions have been observed in older people.

It was once thought that religious conversions occurred mainly among adolescents, but recent research has indicated that conversion experiences definitely occur in later life. For example, Koenig (1995a) found that 27 percent of the 1,011 men in his Veterans Administration clinic sample reported having a conversion experience at some time in their lives, and 31 percent of these conversion experiences occurred at age 50 or older. This means that about 8 percent of all reported conversion experiences occurred in later life. Koenig focused on conversions in which people turned to religious faith for comfort in times of life crisis. Research is needed that also allows for the possibility that people turn away from religion in response to life crises. For example, the death of one's child is a life crisis that often precipitates a profound challenge to faith in God, and for some this doubt eventually results in their conversion to nonbelief in God. We have little indication of how common this type of experience is, especially among aging people.

Mysticism is the ultimate subjective religious experience.* Mysticism refers to a variety of spiritual disciplines aimed at union with the divine through deep meditative or contemplative practices. It also involves belief in the existence of realities central to life that lie outside the capabilities of sensory or cognitive apprehension but that are directly accessible through intuition. Mysticism's common core of meaning includes "a detachment from a superficial experience of life in favor of a deeper reality hitherto unknown; dissolving of barriers between the self and the world; and a powerful sense of certainty and existential security that gives meaning to everything" (Moody 1995,87–88). For most people, mystical experiences are intermittent, not constant. Most sacred traditions contain texts in which mystics of that tradition attempt to point the way inward for interested seekers. For example, influential writings by Christian mystics include those of St. John of the Cross, Meister Eckhart, Hildegard of Bingen, George Fox, and Thomas Merton.

Because mystical experience cannot easily be studied by conventional research methods

and in many ways challenges both religious and scientific worldviews, research on mystical aspects of spirituality and aging is very recent. As Moody says, "Taking mysticism and aging seriously would mean a very far-reaching reassessment of the possible meaning of old age. One can think of old age as a kind of 'natural monastery' in which earlier roles, attachments, and pleasures [lose significance]"(1995,96).

The mystical journey is a journey inward, one that Atchley (1995b) argued was made easier by aging. Many elders experience a mellowing of physical drives and lead simpler, less hectic lives conducive to contemplation, a key process in inner discovery. Instead of trying to live an old age based on lifestyles required of young or middle-aged adults, many elders intentionally simplify their lifestyles and turn their attentions inward to focus on their experience of spirituality. Meditation, scripture reading, and prayer are activities that support this spiritual journey. We can infer that mystical experiences are probably more common among elders than in other age categories by observing the large proportion among the old-old who have an absolute sense of certainty about their spirituality. These elders do not stand out; rather, their main characteristics are their quietness, inwardness, empathy and kindness toward others, forgivingness, and gratitude for having life (Moody 1995).

Attempts to measure mysticism are rare. Levin (1993) analyzed data from the 1988 General Social Survey, a probability sample of adults age 18 and over, which asked "How often have you felt as though you were very close to a powerful, spiritual force that seemed to lift you out of yourself?" About 32 percent of respondents reported having such experiences at least once. There were no age differences in the proportion who reported ever having a mystical experience or in the frequency of these experiences. Unfortunately, the very nature of mystical experiences makes them difficult to verbalize,

---

*This section was heavily influenced by the excellent discussion in Moody (1995).

so it is difficult to confirm that those who say they have had mystical experiences have indeed done so (Thomas and Cooper 1978).

Mysticism involves transcending the conventional personal worldview from which most of us interpret the world around us. Tornstam (1994) developed an operational concept of transcendence that involved an increasing sense of unity with the universe, looking at a greater span of time, a reduced fear of death, an increased affinity with past and future generations, a decreased interest in material things, a decreased interest in superfluous social interaction, a decreased self-centeredness, and an increase in time spent in meditation. Tornstam developed a ten-item scale that he used to study the prevalence of transcendence in a large sample of Danes ranging in age from 74 to 100. A majority of his respondents agreed with items such as "Today I feel that the border between life and death is less striking compared to when I was 50"; "Today I feel to a higher degree, how unimportant an individual life is, in comparison with the continuing life as such"; "Today I take myself less seriously than earlier"; and "Today material things mean less, compared to when I was 50." Within this older population, age was unrelated to having transcendent perceptions, but nearly all of the respondents were already in old age. Tornstam also found that respondents who were high in transcendence were much less likely to rely on social activity for life satisfaction. Transcendence involves a change in the focus of attention in which internal desires and fears as well as current values and events of the social world assume less importance in comparison with a very long-range view of life and peaceful acceptance of the personal self's impermanence. Such shifts in worldview can be seen developing in later life in the autobiographical writings of influential scholars such as Ludwig Wittgenstein, Alfred North Whitehead, Sigmund Freud, Carl Jung, and Albert Einstein.

Conceptions of Spiritual Development

In this section, we will consider several theoretical conceptions of how spirituality or religious faith develops. Most of these conceptions posit a positive relationship between age and spiritual development.

James Fowler (1981,3) defined faith as "the dynamic patterns by which we find life meaning." Fowler's (1981, 1991) theory of faith development conceived of spiritual development in adulthood as having three stages: *individuative-reflective faith,* which could begin in the early 20s, in which the self begins to turn away from external sources of spiritual authority toward the development of an internal moral and spiritual orientation that has meaning for the individual; *conjunctive faith*, beginning in midlife or later and involving acceptance of paradox and ambiguity, a deepening sense of understanding, disillusionment with the overreliance on logic and rational thought typical in the individuative-reflective stage, and a more open attitude toward religious traditions other than one's own; and finally *universalizing faith*, occurring in late life and involving a rare willingness to give up oneself and one's life to make spiritual values a reality on this earth. Exemplars of people in the latter stage include Mahatma Gandhi, Thomas Merton, and Mother Teresa. But Fowler felt that most adults probably never progressed beyond individuative-reflective faith, even in very old age. But, Fowler's theory has been criticized for relying too much on cognitive concepts of spirituality (Koenig 1995a).

Koenig proposed a theory of faith development based on the presumption that development is more continuous and aimed at achieving the developmental goal of "mature faith." To attain mature faith, commitment must be freely entered into. In mature faith, spirituality is central to all of life, which involves the ability to commit to this goal,

deemphasize other potential central values, and develop the self-discipline required to keep spirituality in the center of consciousness and action. Mature faith involves believing without doubt the absolute truth underlying one's religious beliefs and spiritual life. People move toward mature faith by making religion and a religiously guided life their ultimate concern and primary motivation (1995a,113). Koenig asserted that life experience is conducive to the development of mature faith, that faith is more important to older people than to younger people, and that elders have probably had to overcome more tests of their faith over their greater life experience. But his theory contains no elements to explain the mechanisms through which aging might lead to a greater prevalence of mature faith.

Continuity theory (Atchley 1989, 1995a) might be helpful here. Continuity theory presumes that adults have goals for developmental direction that guide the evolution of their inner structure of ideas, including their religious belief systems, and that shape their lifestyles. The direction of evolution depends on the person's goals for development of spirituality and religious behavior. If the goal is to deepen spirituality, the implications are quite different from if the goal is simply to maintain an existing sense of spirituality. If people aim toward mature faith as their prime goal for developmental direction, then continuity theory points to internal mechanisms through which faith could be expected to evolve toward maturity. Of course, continuity theory could equally well be used to describe what would happen if a mystical path were chosen as a goal for developmental direction (Atchley 1995b). Continuity of religious or spiritual identity may become a more important goal for those who find that they have been immersed during middle adulthood in the secular aspects of their identity. Religious beliefs, self, and behavior may also be increasingly reinforced as vital sources of life meaning over a lifetime of experience.

Thomas (1994b) pointed out that doubt can serve as an important stimulus for spiritual development. Indeed, confronting and resolving doubt—enduring and triumphing over the "long dark night of the soul"—may be the mechanism that leads people from what Fowler (1981) called individuative-reflective faith toward what Koenig (1995a) called mature faith.

The evolution of a life philosophy can be seen as a primary outcome of spiritual development. Through spiritual development, people can evolve increasingly deeper and more reflective answers to the questions of what gives life meaning and what makes for a moral and worthy life. Many elders speak about this evolution using the metaphor of an inner journey (Cole 1992). They speak about spiritual awakening, which is for most a subtle kind of conversion experience, and they often speak about their spiritual journey as a learning process that focuses increasingly on transcendence.

For example, in 1994 and 1995, the Omega Institute sponsored several "councils of elders," where people age 60 or over were invited to meditate on two main questions: How has aging affected your spiritual journey? and How has your spiritual journey affected your experience of aging? Admittedly, the 250 elders who attended these three meetings were not a random cross-section of the older population. They were mainly middle-class people who had consciously been dealing with spiritual issues for some time and therefore had a better vocabulary for talking about these issues that many elders do. Here are a few examples of what they had to say:*

"I looked inward out of desperation. I wanted to seek something better. I've been on a twenty-

---

*The statements in this section were made by people age 60 or older who attended the Conscious Aging programs offered by the Omega Institute in 1994 and 1995, at which the author served as unofficial recorder.

year quest for purpose and clarity, and life gets better as I age."

"I was proud of being an 'active ager,' but then a bunch of my friends died, some quite a bit younger than me. I could imagine myself dead. . . . What will be important after I'm dead? . . . Silence is important to me now."

"At first, taking care of my mother focused on the hassles and the difficulty of the job, but gradually I came to see that it is a spiritual journey."

"I was awakened by a minister who introduced me to the spiritual life. I cannot imagine growing older now without God in my inner life."

"After a lifetime of searching for spiritual roots, at 65 I went to massage school. People thought I was crazy, but I had to do it. Now I give the gift of touch. I don't give advice."

"We should all have a teacher. My teacher was my late wife. She saw the divine in me and led me to see it. Through her I gained a sense of spiritual self that is strong within me."

These statements indicate that spiritual awakening can be gradual or sudden, stimulated by loss or inner need, self-generated, or inspired by others. There are apparently many pathways. Maves (1986) believed that elders are drawn by their awakened spirit to simplify their lives, accept who they are, heal bitter memories and forgive old enemies in order to liberate themselves from their personal agendas so they can reach out to others and can see their lives in the context of eternity.

Here are some comments elders made about their experience of their inner spiritual journey:

"Knowing for certain that I will die, I make better judgments about what to do in my life. I don't want to die watching television."

"In my old age, my mother continues to push me to grow. As she gets more and more difficult, the spiritual fire gets hotter."

"I watched my interests shift from having everything, to doing everything, to just being here. I focus on my religious path, and as I follow the path, I become a teacher."

"I find myself sinking more deeply into my memories and finding so much beautiful space."

"I feel lucky to have a strong belief that we do not end with this life. This frees me from fear of death and makes me want to take advantage of potential experiences and to evolve as much as I can."

"My inner journey has freed me from judgments about my body and about others."

"We deepen with age, but it is more difficult to express this depth. To the young it may seem passive, but there may be another language, language with more silent attention, more meditation, that can expose this depth."

"Peace comes from within. Creating peace means being peace, recognizing oneness and allowing it to be expressed through how you live your life. We need role models of peace."

"Life is not always wonderful. Life is like a dance—full of twists, turns, ups, downs, but all in a pattern that it takes a while to see."

These comments indicate that experiences of the inner spiritual journey are far from uniform, but they often involve experiences of the value of silence and inner strength and usually point toward "mature faith," "spiritual development," or "transcendence."

Although the results of spiritual development are nearly always expressed in positive terms, inner spiritual growth often involves overcoming emotions such as fear, hurt, anger, or grief.

"To grow spiritually, you have to embrace death as your own, as a natural part of your life, and a part that inspires fear sometimes."

"I was just going along, aging and feeling bad about it. Then my husband died. I watched my feelings closely and watched my thinking. Gradually my attitude changed. I learned to release the things tethering me."

"I've just gotten a diagnosis, and things don't look good for me. But I can take off my mask and be a model for others of letting go, getting gracefully off the stage."

"Losses, deaths, seen as beauty, can be seen as extensions of our capacity to give ourselves *every day* to each other: the little kind gestures, living with loving kindness in everyday things."

"My inner journey involved turning my attention away from a lifestyle of work, family, and social achievement that dominated me in middle age. I still do many of those things, but they are not the main goals of my life anymore. My main goal now is to be fully present in as many moments of my life as I can."

"I was very angry and sad about the rejection I felt from others at work when I reached my 50s. But gradually I realized that I had been relying too much on their approval. Eventually, I learned to rely on my own private relationship with God."

Although there are may be many pathways to spiritual awakening and development, enough people report having such experiences to merit serious study. In addition, many report that spiritual growth arises out of personal experience and that aging has had a positive effect on the depth of their spiritual life. For example, Markides and Martin (1983) found that more than half of the older Hispanics in their San Antonio study felt that their faith and their feeling of closeness to God were the greatest they had ever been, much stronger than in the past. But at this point we are only just beginning to develop the language and research tools needed to give serious study to these topics.

Many of the elders whose comments are cited obviously saw benefit in having religious beliefs and being involved in an inner spiritual journey. The next section will examine more systematically research results on the effects of religion and spirituality.

Effects of Religion and Spirituality

Membership in a religious congregation and religious beliefs both have numerous potential benefits, particularly for mental and physical health. Participation in religious or-

ganizations can provide social support and a sense of belonging and security. Congregations often function as social networks that can increase members' social support resources for coping with both crisis situations and ongoing challenges such as the need for assistance. However, little systematic research has been done on the effects of participation in religious organizations on the experience of aging in the general population. This topic is much in need of research.

The role of the church for African-American elders has received more research attention. Religious affiliation is diverse among African Americans: 52 percent reported themselves to be Baptist, 12 percent Methodist, and 6 percent Roman Catholic; 20 percent were affiliated with thirty-five other religious groups, and 10 percent gave no religious affiliation (Chatters and Taylor 1994). Going back to the early days of American social science (Du Bois 1899), African-American churches have been recognized as central institutions within their communities. Drake and Cayton (1962) detailed the great extent to which African-American churches in Chicago served as multipurpose organizations for African Americans of all social classes. Billingsley (1968) noted the historical importance of churches to rural African Americans. More recently, Lincoln and Mamiya (1990) presented a well-developed, dynamic model of the functioning of African-American churches. Churches were historically the only institution African-Americans were allowed to control, both in the rural South and in northern cities. Churches were a main vehicle for exercising leadership in African-American communities, and they tended to be dynamic organizations that adapted to changes in both local and national social and economic conditions.

However, churches in African-American communities are divided along a number of dimensions. For example, some are ritualistic, others are evangelistic; some emphasize other-worldly concerns, others stress this-

For older African Americans, churches are an integral part of the social network.
Photograph by Owen Franken/Stock, Boston

worldly concerns; some are charismatic groups, others are bureaucratic organizations; some emphasize resistance to racial prejudice and discrimination, while others focus on accommodation. Nevertheless, African-American churches have been very "instrumental in the development of the black self-help tradition (that is, mutual aid societies) and in providing the institutional foundation for educational, civic, and commercial endeavors within black communities" (Chatters and Taylor 1994,201).

Taylor and Chatters (1986a, 1986b) found a complex picture of church aid to older African Americans. Those who were church members and had adult children received more church aid than those who were childless. This suggests that adult children played a role in mobilizing church aid, which consisted mostly of advice, encouragement, and help in time of sickness. For African-American elders receiving assistance in the community, churches were an integral part of a social network that also included adult children, close friends, and extended kin. Maldonado (1995) reported that churches play a similar role in the lives of older Mexican Americans.

Churches can also serve as important intermediaries between older individuals and their families and community service organizations. For example, ministers often can intercede with social service and health-care organizations to improve communications and responsiveness. However, very little research attention has been paid to this role. We do not know how common or how effective such action is.

The normative systems of organized religions can require healthy lifestyles that prohibit smoking, drinking, and drug use and can emphasize stress reduction through meditation and prayer. Levin (1994b) reported that adherents of behaviorally strict religions, such as Mormon or Seventh Day Adventist, have better health, less illness, and greater longevity than other religious groups and people with no religious affiliation.

Although churches are important resources for a number of reasons, most research has focused on religion as an individual characteristic (Levin 1994a). This research has looked primarily at the effect of religious beliefs and behavior on physical health, mental health, and psychosocial coping.

Levin, Chatters, and Taylor (1995) proposed a model in which organizational, nonorganizational, and subjective religiosity were presumed to directly influence both health and life satisfaction and to indirectly influence life satisfaction through the influence of health on life satisfaction. Using data from the National Survey of Black Americans, which provided multiple measures for each of their theoretical constructs, Levin and his colleagues found that, for African Americans, their model generally fit the data very well. The only links that did not receive support were those between nonorganizational religiosity and life satisfaction and between subjective religiosity and health status. They also tested this model separately for three age groups and found that it applied equally well to young, middle-aged, and older adults. This latter finding suggests that the relation between various dimensions of religiosity and health and life satisfaction holds across the adult life course.

Levin (1994b) reviewed over 250 studies examining the relationship between religious involvement (however measured) and a wide variety of health conditions. Regardless of the measures of religious involvement used, the greater the degree of religiousness, the better the health and the less of whatever illness was being studied. As individuals aged, both formal and informal religious participation were associated with better health, happiness, and life satisfaction. Controlling for other factors such as gender, race, and social class did not eliminate the positive association between religious involvement and life satisfaction.

Even when health was controlled, the strong link between religiosity and subjective well-being remained. However, much of this research was rudimentary, and how religious involvement produces these positive benefits is unclear. For example, religious involvement is associated with healthier lifestyles, greater availability of social support, lower feelings of stress, and greater self-confidence and self-efficacy. Religious involvement probably affects physical health through a number of these and other mechanisms, but much more research is needed to sort out these potential relationships.

Ellison (1994) proposed four basic reasons that religious involvement might be expected to have a salutary effect on mental health: (1) by reducing the risk of stressors such as divorce or job strains, (2) by providing a cognitive framework within which stressors might seem less threatening, (3) by encouraging greater objective and subjective social support, and (4) by enhancing psychological resources such as self-esteem. This is a promising theory that could be used to study the relation between religious involvement and mental health conditions such as depression.

Koenig (1995a) reported that religious involvement was in fact related to good mental health. For example, he found that greater religiousness was associated with lower prevalence of anxiety or depression, less fear of death and better coping with grief. Payne and McFadden's (1994) discussion of loneliness versus solitude is an excellent example of dynamics through which religiosity may affect emotional health. Although many people consider being alone an undesirable state, Payne and McFadden pointed out that aloneness is not the same as loneliness. Loneliness is psychological distress associated with being alone, and it stems from a predisposition toward being with others that begins in infancy. Some people feel emptiness when they are alone, others feel sadness, fear, or anger. On the other hand, Payne and McFadden follow theologian Paul Tillich in defining solitude as "the glory of being alone." Solitude can indeed be glorious for people who feel a stronger spiritual presence when they are alone in contemplation or prayer. To the extent that the greater opportunities for aloneness that come with age are accompanied by a spiritual attitude toward solitude, we might expect lower prevalence of loneliness among elders in response to aloneness compared with that in other adult age categories. Empirically, loneliness is much less commonly experienced among elders than among young adults per unit of time spent alone (Malatesta and Kalnok 1984). However, Payne and McFadden's formulation of why loneliness is less common among elders, no matter how plausible, still remains a theory to be tested by research.

There is ample evidence that religion can be an enormous source of comfort and a great resource for coping with illness, loss, and suffering at any age, but especially in later life (Levin 1994a; Koenig 1995a; Kimble et al. 1995). Koenig pointed out that questions about religious coping are susceptible to social desirability bias if asked directly. About 90 percent of Americans agree with a general statement that religion helps older people to cope. But only about 40 percent of older people cite religion spontaneously in response to open-ended questions such as, "What enables you to cope with life? What keeps you going?" But religion is by far the most commonly cited coping mechanism (Pargament, Van Haitsma, and Ensing 1995).

Koenig (1995a) developed a scale of religious coping and used it to study the relation between age and various aspects of religious coping in a large sample of adult men either under 40 or 65 and over who were admitted to the Durham, North Carolina, Veterans Administration Hospital. He found that men age 70 or older were more likely than other age categories to spontaneously mention religion as a means of coping with the illness or disability that brought them to the hospital. The use of religious coping was more common among men who were older, held fundamentalist or evangelical religious beliefs, were African American, and were living alone. The more serious their physical or neurological condition, the more likely the men were to use religious coping. Koenig found an increase with age in religion's importance, but he did not find that nonbelievers suddenly turned to religion as a way of coping with illness. Rather, illness and religious coping were perceived by the respondents as strengthening their already-existing religious faith.

Koenig's study illustrates the role of religion in psychological coping, but Ellison (1994) pointed out that pastoral counseling and social support from church members are both important resources for identifying coping options and connecting with institutional sources of help. However, these aspects of religion's possible relation to coping have received little attention from researchers.

Religion does not always have positive results for coping. Pargament and his colleagues (1995) pointed out that religious zeal can interfere with interpersonal relations; some religions believe that illness or disability is caused by sin; and some elders find that their congregations are not much interested in helping them. For example, Tobin (1991) described a case in which a homebound older woman felt abandoned by her congregation. Mrs. C. was 81, a widow living alone, and had health problems that severely restricted her mobility and left her homebound. She was unhappy that the people she knew through her synagogue had not visited her. "I don't know many people, but I did belong to the sisterhood, and they know I'm sick," she said (1991,125).

Research Issues

The survey of the literature on the relationships among religion, spirituality, and aging in this chapter contains both good news and bad. The good news is that there has been a sharp increase in published research on religion, spirituality, and aging since 1990. These findings allow us to begin to map the conceptual territory on these topics. But the bad news is that there are numerous gaps in our knowledge. This concluding section looks at these research issues in more detail.

Good research depends on having solid concepts that can guide the accumulation of descriptive data, and well-developed theories that can serve as working explanations for the multitude of relationships between aging and various aspects of religion and spirituality. We have already mentioned Fowler's (1981) and Koenig's (1995a) theories of faith development, the promise of Atchley's continuity theory (1995a) to help us understand the persistence of religious behavior and spiritual concerns and identity over the adult life course, and Ellison's (1994) theory of how religious involvement might mediate the impact of stress on mental health. In psychology, researchers are beginning to apply attachment theory, which was developed initially to explain interpersonal attachment, to explain attachment to religion and spirituality. Thus, research on religion, spirituality, and aging is beginning to move in the very positive direction of forming the well-developed concepts and theories needed to guide high-quality research.

Most research has found a steady increase across successive adult age cohorts in the strength of religious beliefs, knowledge about religion, and reliance on religion in everyday coping (Koenig 1995a; Courtenay et al. 1992). However, few of the theories about how and why this increase occurs have been empirically tested, and nearly all of the findings are based on cross-sectional research. Longitudinal panel and cohort studies are needed to help sort out age, period, and cohort effects.

Measures of religious ideas and behavior need to be improved. Although research has begun to move beyond religious affiliation and frequency of church attendance as measures, religion and spirituality remain difficult areas to measure, and the development of valid and reliable scales that can be applied across religious groups should be a primary goal of research in this area. At the same time, survey methods cannot address all the topics of interest, especially the integration of religion and spirituality into the whole of life experience and the nature and effects of mystical experience. These topics may be better studied by qualitative and interpretive methods (Weiland 1995).

Thus far, most research on religion, spirituality, and aging has tended to look at undifferentiated age patterns. But the cross-cutting variables of gender, ethnicity, and social class operate to produce diversity in this area of life as much as in any other. For example, the few studies that have looked at gender differences have shown that women have consistently higher religiosity than men over the entire life course, even controlling for education, marital status, and urban-rural residence (Levin and Taylor 1993; Atchley 1994). Yet organized religions tend to retain traditional patriarchal structures that may not serve women well, especially in later life (Larsen 1995; Knutsen 1995). Much more research in needed in this area.

As discussed earlier in this chapter, older African Americans and Mexican Americans both experience religion as a central social institution in their communities and rely heavily on religious coping. But we know almost nothing about the role of religion and spirituality in relation to aging in other ethnic groups. The impact of ethnic diversity on cultural conceptions of religion and spirituality can be substantial. Native Americans and recent immigrant groups from Central and South America, the Caribbean, Asia, and the Middle East have been especially neglected.

Even within ethnic groups, the effects of the great diversity of religious beliefs and practices on the relationship between religion and aging are poorly understood. For example, there are substantial differences among the Protestant denominations in religious beliefs, views of religious development, and attitude toward aging. In addition, the number of Muslims, Hindus, and Buddhists is increasing in America, and these religions are practiced by large numbers of people in developing countries. We know that these religions differ from the Judeo-Christian cultural tradition in terms of their conceptions of the Almighty and their views of spiritual development and aging (Thursby 1992), but we know virtually nothing about internal diversity within these groups or how these differences play out in ethnic communities in the United States.

Various religions are not evenly distributed throughout the United States. Although we do not want to stereotype the complex relation of religion to regional culture, it is probably safe to say that communitarian Christian religions such as the Presbyterians, Pilgrims, and Puritans had a strong influence on the evolution of New England culture (Cole 1992), whereas more individualistic and evangelical forms of Christianity were stronger influences in the South. Even within the same state, there are large differences in religious microclimates. For example, Catholic influence is much stronger in Cincinnati than in Columbus, Ohio. The interaction between the prevalence of specific religions and the evolution of regional subcultures has received very sparse attention, but certainly the religious climate in the San Francisco Bay area is very different from the religious climate in rural North Carolina, the site of much of Koenig's (1995a) research. We know virtually nothing about how these issues intersect with aging.

Thus, even though a good beginning has been made toward adequate empirical and conceptual mapping and theory development in the area of religion, spirituality, and aging, this area still remains perhaps the most neglected area of social gerontology. New research programs are under way (Levin 1994b; Koenig 1995a) that may begin to establish firmer findings and fill research gaps, but many more researchers are needed to improve our knowledge of an area that is very important to a large proportion of aging people.

Summary

Spirituality is an inner domain that refers to an individual's holistic quest to locate his or her place in the universe—to confront ultimate concerns such as why we are here, how we got here, and what happens to us when we die. Organized religion attempts to pass on to new generations either a set of beliefs about these ultimate concerns or a program for arriving at individual answers to these spiritual questions. Mystical spiritual disciplines attempt to lead those who practice them to directly experience the divine. Religiosity, religiousness, and religious involvement are the extent to which individuals exert energy and pay attention to religion and/or spirituality. All these dimensions are thought to increase in importance with age, and research generally supports this contention. In particular, strength of religious beliefs, knowledge about religion, and reliance on religion in everyday coping increase steadily with age over the adult life course.

There is enormous variability in both spirituality and religion at all adult ages, but there is growing interest in how aging influences spirituality and religion and vice versa. Most aging people have long-standing attachments to a religious faith and are involved in a local congregation. Church membership peaks in the 60s, but churches have been slow to develop programs to serve the unique spiritual needs of elders, even though the number of elders in most congregations is increasing rapidly. In addition, most religious organizations encourage older members to continue to participate, but are not interested in recruiting elders to do responsible work in the organization. New programs such as the Shepherd's Centers, the Spiritual Eldering Institute, and the Omega Institute's programs on Conscious Aging illustrate constructive alternatives to the usual lack of programs aimed at elders' needs within most congregations.

Longitudinal data on church attendance shows a curvilinear pattern with age. Attendance peaks in the 60s and declines thereafter. Attendance also shows persistent gender and black-white differences, with women and African Americans tending to attend church more across the life course. We know very little about how aging affects the experiences people derive from attending church. Informal religious practices, such as scripture reading, prayer, and meditation, become more important with increasing age and are especially important to disabled elders.

Over 90 percent of adults express a religious preference, and a large proportion of older adults believe that their faith has grown stronger over time. Nearly one-third of older people report having a mystical experience at least once in their lives, and about one-third report having a conversion experience. There is some evidence that the increased depth of faith in later life is also accompanied by a more transcendent quality to faith, in which faith seems to become more universal and less personal.

There are several theories of why aging might cause a deepening of faith. Most are extensions of existing adult development theories. Fowler's theory assumes that faith is more individualistic and specific in young adulthood and moves in the direction of growing certainty, greater acceptance of other faith traditions, and transcendence of worldly aspects of religion. Koenig's theory posits that maturing of faith involves placing more emphasis on religion and spirituality and having the self-discipline to maintain this focus. Spiritual development may also involve a continuation of deepening spirituality as a conscious goal for individual developmental direction. In addition, the need to overcome doubt may be an important stimulus for spiritual development. We need to be cautious in applying these concepts and theories, however, because none of them has yet been adequately tested in research.

Elders often speak of their spiritual development in terms of their spiritual awakening and the nature of their spiritual journey. Events associated with aging, particularly deaths and illnesses, sometimes stimulate spiritual awakening, and aging can influence the experience of the spiritual journey through such factors as reducing the number of work or family distractions, providing more time for contemplation, and providing the freedom to travel to seek out spiritual teachers. But the nature of the spiritual journey can also profoundly affect the experience of aging.

Involvement in organized religion and subjective religiosity both appear to have positive effects on physical and mental well-being and longevity. For African Americans, it is well documented that church members work alongside family to provide help to older people. Churches probably provide helping networks more generally, but this topic has not been systematically studied for the population as a whole. People affiliated with religions that prohibit tobacco, alcohol, and drug consumption tend to be healthier and live longer than others. As people age, both formal and informal religious participation are associated with better health and life satisfaction.

In general, the greater the degree of subjective religiousness, the better people's health and subjective well-being. The greater the religiousness, the lower the prevalence of anxiety, fear of death, and loneliness. Highly religious people also cope better with grief. Religious beliefs and orientations are the most prevalent resource for coping with negative aspects of life, especially in later life. The more serious the problem, the more likely that people will use religious coping. However, religiosity can also be maladaptive if it isolates elders from others, if it construes "negative" aging as resulting from sin, or if elders seek support from their congregation and do not get it.

Many of the findings covered in this chapter need much more research to firmly establish them. Fortunately, the volume of high-quality research on religion, spirituality, and aging is increasing rapidly. Nevertheless, more work is needed to develop better measures. In addition, qualitative research is needed to amplify the relationships discovered in survey research. Our most important research needs in this area require exploring gender and ethnic differences and differences across various religions in terms of age patterns of religiosity and the effects of religiousness.

12 Dying, Death, Bereavement, and Widowhood

In the nineteenth century, death was a common occurrence at all stages of life. Uhlenberg (1969) studied deaths of native-born Massachusetts women born in 1830. He found that 36 percent died before they reached the age of 20 and another 12 percent died while giving birth. Nine percent were widows before they were 55, and only 20 percent survived to the age of 55 as a member of a married couple.

By 1920, mortality was concentrated much more in the late stages of life. Uhlenberg found that of Massachusetts women born in 1920, only 10 percent died before reaching 20, while only 4 percent died giving birth. There was no change in the proportion who were widowed by age 55 (9 percent), but the proportion who survived to 55 as part of a couple had increased from 20 to 57 percent.

In addition to the increasing concentration of death among the old since the nineteenth century, there has also been a sizable change in *where* people die. In 1900, a majority of older Americans died at home, but by 1980, 50 percent died in hospitals, another 25 percent died in nursing homes, and 25 percent died in private homes (Marshall 1980).

These trends are important for our understanding of how people view death. In earlier times, death was a more "normal" part of everyday life at all life stages. People died in everyday environments, and physical death was something that most people had witnessed directly by the time they reached adulthood. Today, most deaths occur among old people and in institutional environments, where family and friends are often prevented from actually seeing the death or from seeing the dead body until a mortician has restored it to a lifelike appearance. Given these circumstances, it should not be surprising that death has little reality for the young. Nor should it be surprising that older people are concerned about where they will die since institutional environments are designed more to prevent death than to be good places to die.

The concentration of death among older people has probably influenced what we think and how we feel about death, particularly in terms of seeing death that comes before later life as unfair or unjust. The conquering of infectious diseases and the advent of high-technology medicine have given us an illusion of control over death. The idea that death is a constant possibility has been replaced by the idea that death happens only to the old. The association of death with aging has no doubt contributed to ageism—prejudice against older people—because people feel that if they can deny aging they can deny death. But none of this has changed the true inevitability of physical death.

As far as we know, humans are the only beings who are aware that they will eventually die. Reactions to this knowledge range from denial to resistance to acceptance, depending on culture and circumstances. Through the eons, interest in death as a major part of life's drama has almost always been high, but until recently in our culture death was not a legitimate topic for open discussion. But that has changed. It is now possible, at least for some, to discuss feelings about death openly and to see death as a transition that can be accomplished better in a context of understanding and open communication than in a context of ignorance and awkward silence.

Among well-educated people, age is mildly related to awareness that life is finite. For example, about 25 percent of 30-year-olds with high school or better education report that they often think about the uncertainty of their own lives, and this proportion changes very little through the age of 60. However, about 50 percent of well-educated people 60 or over often think about this uncertainty. Interestingly, among people with less than nine years of education, nearly half often think about life's uncertain length at all ages (Marshall 1980,100). Most people who are 85 or older expect to die within five years, whereas only 7 percent of those age 65 to 74 have this expectation. Perceptions of the amount of time remaining to a person are also affected by the ages at which one's parents died.

Marshall found that when people perceive that the amount of life remaining is finite, they often go through a process of legitimizing their biography. In young adulthood, this is a prospective process of creating biography. In later life, the process more often involves reminiscence designed to make sense out of the life a person has led and is similar to the process Erikson, Erikson, and Kivnick (1986) spoke of as leading to ego integrity. Marshall found that this process resembles the writing of one's autobiography, that it often begins in the 60s, and that it involves images of continuity for most people.

In this chapter, we will examine various definitions needed to understand dying and death, what death means to most people and to older people in particular, the process of dying, care for dying people, the conse-quences of death for the survivors, and life as a widow or a widower.

Defining Death

Death can be defined as a process of transition that starts with *dying* and ends with *being dead* (Kalish 1985). For practical purposes, a dying person is one identified as having a condition from which no recovery can be expected. Dying is thus the period during which the organism loses its viability. The term **dying trajectory** refers to the rate of decline in functioning leading to death, and it is used to estimate the time frame within which dying will take place. The word *death* can also be defined as a point at which a person *becomes physically dead*. When we say that someone died yesterday, we are not referring to the entire dying process but to its final result.

What moment in time a person becomes physically dead was once considered an easy practical question to resolve. Recently, however, it has become possible to stimulate artificially both breathing and heartbeat. As a result, there is currently a huge legal tangle over the issue of when a person is physically dead. For example, a Harvard Medical School committee proposed four criteria to be met before certifying death: (1) unreceptivity and unresponsivity, (2) no movements or breathing, (3) no reflexes, and (4) flat electroencephalogram, which indicates that no brain activity remains. Not only would all these criteria have to be met, but the tests would be made again twenty-four hours later to confirm death. Someone who met these criteria would almost certainly be actually dead. However, it is unlikely that such cumbersome procedures will become widely used. Instead, no reflexes, no breathing, and no heartbeat will probably continue as the primary operational definition of death.

Death can also be a social process. People are socially dead when we no longer treat them as individuals but as if they were unthinking, unfeeling objects. Social death has occurred when people talk *about* the dying person rather than *to* the dying person even when the dying person is capable of hearing and understanding what is being said.

To *deny* death is to believe that people continue to be able to experience—to see, hear, think, and feel emotions—after their physical death. Physical death is undeniable, but whether the interconnected thoughts and feelings we call the mind cease to exist at the point of physical death is arguable. Belief in an afterlife, belief in reincarnation, and belief in ghosts, spirits, angels, or demons are all ways of denying both the death of the ability to experience and an end to the integrity of personal existence.

Some people view the practice of displaying the dead cosmetically "restored" to resemble the living as a form of denial too. If the dead are made up to appear that they are "just sleeping," then it may be easier to recall their former capacities as well. However, there is no evidence that viewing the body hastens the resolution of grief (Marshall 1980), and the expensive casket, embalming, and cosmetology involved add substantial—and some would say unnecessary—expense to the cost of burial. However, Kastenbaum (1992) argued that seeing the dead body gives young children vital sensory information they need to assimilate the reality of death.

The Meaning of Death

How people approach their own deaths and the deaths of others depends to some extent on what death means to them—and death can have many meanings. Some people see death as an ugly, punitive extinction of life. Others see it as God's will or as a beautiful, rewarding transition to a new and better type of life. Some see death as hateful destruction; others see it as a welcome release. A crucial point is that death's meaning at any given time and in any given culture is mainly a

social creation. For example, even in the United States, the meaning death has for a Creole, a Navaho, a Lutheran, a Catholic, a Muslim, a Jew, a Buddhist, or a Baptist could be expected to vary systematically.

Ariès (1981) traced the cultural meaning of death in Western thought from the Middle Ages to the present. There is not sufficient space in this book to summarize adequately the myriad and complex patterns of thought Ariès discovered about death, only to sketch some main points. He found that before the year 1200, people were resigned to death as the collective destiny of our species. Death was to be neither glorified nor avoided. From 1200 to 1700, people began to be concerned with the self and its death. During this period, writings stressed the art of dying, the religious concept of a final personal reckoning with God, and the personal meaning of death. Also during this period, personal tombs and epitaphs assumed significance. From 1700 to the late 1900s, people began to think in the abstract about the difference between the "nonnatural" death and the "beautiful and edifying" death—death as punishment as opposed to death as natural transition. Death was glorified and romanticized. Around the turn of the twentieth century, according to Ariès, death became culturally invisible. Physicians and hospitals assumed control over dying, both death and mourning became private, and people were encouraged to deny death and place their faith in medicine. Thus there have been wide swings in the cultural meanings attached to death. Currently, there is a swing away from the invisible death toward the notion of "death with dignity."

Back (1971) examined the meaning of death through ratings of various metaphors about what death is like. He found little age variation among respondents who were over 45 concerning the meaning of death, but he did find a significant gender difference. Women tended to be accepting of death, to see it as a peaceful thing. Death was most like a compassionate mother or an understanding doctor. Men tended to see death as an antagonist—a grinning butcher or a hangman. Thus, women tended to stress death as a release, whereas men tended to stress death as unwanted destruction.

The most frequent reaction to the idea of death is commonly thought to be fear. However, as Marshall (1980) pointed out, we need not assume that a person either is afraid of death or not. It is quite possible that the same person could be afraid of death in some respects and not afraid of it in others. Also, fear of death probably fluctuates over time. Numerous studies have found that older people do not appear to be extremely fearful of death. In fact, older people express fewer fears of death than young people do (Kastenbaum 1992). There also appears to be no increase in fear of death among older people with terminal diagnoses (Kastenbaum 1969). Kalish (1976) attributed the lower prevalence of death fears in the face of a higher prevalence of death to several factors: (1) older people see their lives as having fewer prospects for the future and less value, (2) older people who live past the age when they expected to die have a sense of living on "borrowed time," and (3) dealing with the deaths of friends can help socialize older people toward acceptance of their own deaths.

In addition, Kalish (1985) reported that fear of death was also related to religiosity. The most consistent finding was that people who are strongly religious showed few death fears, as did affirmed atheists. Uncertain and sporadically religious people showed the most death fears (Koenig 1995c).

In a sample of older women, Quinn and Reznikoff (1985) studied the relationship between death anxiety and psychological characteristics such as purposefulness, goal direction, and experience of time as rushing forward. They found that compared with older women with low death anxiety, those who measured high in death anxiety had less

sense of purpose in life, saw time as marching relentlessly on, felt harassed by time pressures, and saw less continuity in their lives. These relationships held even when general level of anxiety and propensity to give socially desirable answers were statistically controlled. Thus, being anxious about death had a profound effect on the existential quality of older women's lives.

Apparently, when people know that they are approaching death, they see little point in continuing to do things for purely social reasons. In addition, if it is widely known that the person is dying, continued participation may be awkward because many people do not feel comfortable around the dying. Those who "come back from the dead" often experience extreme social isolation as they are avoided by people who have already mentally "said goodbye" to them. For example, a few years ago a colleague of mine had a severe heart attack and was not expected to live—but he did. He shared with me the anguish he felt at being avoided by people he had known for many years. And those who did interact with him felt ill at ease. "It's like I'm not supposed to be here anymore. And now that I am, people don't know where I fit in anymore or how to relate to me," he said.

Although death is generally accepted with little fear among older people, it seems reasonable to assume that there are some who would not necessarily react in this way. Less acceptance and more fear of death might be expected among older people who have dependent children, disabled spouses, career goals yet to be achieved, or socially crucial positions. We know, for example, that legislators are often old. How does the prevalence of death fears among older legislators compare with that of the general older population? This and other questions along these lines need more research.

De Vries, Bluck, and Birren (1993) found that young, middle-aged, and older respondents alike differentiated death as an event from dying as process. Kastenbaum (1992) concluded that older people worry more about the circumstances of their dying than about being dead.

Dealing with Dying

Dealing with dying involves understanding what is expected of the dying, the stages dying people go through, the needs and tasks of the dying, the conditions under which people should be permitted to die, and factors in the care of the dying.

The Role of the Dying Person

When a person is diagnosed as terminally ill, that person is assigned to the social role of "dying person." The context of the role depends a great deal on the age of the dying person and the expected length of his or her dying trajectory. People with short dying trajectories do not spend much time in the role of dying person, regardless of age. However, age heavily influences whether or not a person is defined as terminal. For example, Sudnow (1967) found that in hospital emergency rooms, older people were more likely than the young to be routinely defined as dead, with no attempt to revive them.

Among those with long dying trajectories, age is an important element in determining what is expected of them. Young people are expected to fight death, to try to finish business, and to cram as much experience as possible into their remaining time. In short, we expect young people to be active and antagonistic about dying. Older people who are dying are expected to show more passive acceptance. Older people in the dying-person role are less likely than the young to see a need to change their lifestyles or day-to-day goals. Older people are also less apt to be concerned about caring for dependents or causing grief to others (Kalish 1985).

Older people are more apt than the young to find that the role of dying person also means having less control over their own lives. All dying people find that family members and medical personnel (usually with good intentions) take away their free choices, but this interference is particularly likely for older people.

Finally, all people in the dying role are expected to cope with their impending deaths, but older people are allowed less leeway than the young in expressing anger and frustration about their own deaths.

Stages of Dying

A great deal has been written about the stages people pass through in the process of dying. Kübler-Ross (1969) proposed five stages in the dying process: (1) denial, (2) anger, (3) bargaining (such as asking God to postpone death in exchange for good behavior), (4) depression and sense of loss, and (5) acceptance. These stages are probably not progressive, and they probably overlap. Kalish (1976) suggested that although denial may be more common in the early stages of dying and acceptance more common toward the end, there may be considerable fluctuation and movement in the person's orientation. Kalish also reported that many people have mistakenly treated Kübler-Ross's stages of dying as an inevitable progression. As a result, dying people have been chided for not progressing through their own dying on schedule and have been made to feel guilty for not having accomplished the various tasks, such as acceptance.

Marshall (1980) pointed out that Kübler-Ross's formulation begins with the assumption that the first reaction to one's own impending death is denial. In the relatively young people Kübler-Ross studied, this may have been the case. However, Weisman (1972) studied dying people who averaged about 60 years of age and found that acceptance of death without denial was indeed possible. Kalish (1985) reported that there is little research support for Kübler-Ross's stages as anything more than some common reactions to one's impending death. They do not appear to be a natural progression; not everyone passes through them. In addition, other reactions, such as hope, curiosity, apathy, and relief are not included in the Kübler-Ross formulation. Kalish concluded that death, like most individual life experiences, is highly personalized and not easy to generalize. Thus, although Kübler-Ross's stages may have given people who work with the dying a framework to use, it may also have deflected them from recognizing the individuality of the dying person.

Awareness that one's death may not be far off can have substantial effects on the choices one makes. Sill (1980) found that as the perceived length of time before death decreased, activity decreased. In fact, perceived length of time left was a much more powerful predictor of activity level than either age or physical incapacity.

No doubt a good bit of the anxiety about death results not so much from the idea of death itself but from uncertainty about the manner in which death will occur. We do not usually know what will cause death, what the circumstances will be, or when and where death will occur. Although it is impossible to predict on an individual basis, age is related to what one is liable to die from and under what circumstances. Beyond the age of 45, as age increases, the probability of dying from heart disease or cancer rather than other causes increases dramatically, and the probability of dying from cirrhosis of the liver declines.

Marshall (1980) reviewed the literature on where older people die. He concluded that although only about 5 percent of the older population live in nursing homes at any one time, about 25 percent of older people who die in a given year die in a nursing home. Another 50 percent die in hospitals, and 25 percent die at home or in the community. As more older people live alone, the probability of dying at home alone also increases. Bradshaw, Clifton, and Kennedy (1978) found that of 181 people found dead at home

in the English city of York, two-thirds were over 70, and three-fourths had died suddenly. The significance of these facts rests in the gap between what people prefer as an environment in which to die and what people get. Most people prefer to die at home, in the presence of close family and friends. Most people actually die in a hospital. Although many hospitals do what they can to accommodate family and friends of the dying, the hospital's bureaucratic rules and procedures may take precedence. In nursing homes, the environment is less restrictive, but again family and friends may be seen as a nuisance to those who provide the dying person with "bed and body" care. These facts have given rise to the hospice movement, which is discussed later.

Needs and Tasks of the Dying Person

Kalish (1985) outlined a number of needs that dying people have. First, dying people have the same need for food, clothing, shelter, rest, and warmth that we all have. In addition, they often have the need for freedom from pain. They also need reassurance that their close friends and family will not abandon them to die alone. Weisman (1972) proposed that the dying person needs *safe conduct*, or cautious and skillful guidance and support through peril and the unknown. Weisman believed that if health-care professionals themselves cannot provide safe conduct, then they should ensure that someone else does. Dying people need to maintain their sense of self and self-esteem. As we saw earlier, close friends and family are those most likely to be able to relate to the dying person in terms of her or his lifelong interests and accomplishments.

Some people have a need to deny that they are dying. Weisman differentiated denial into two types: (1) brittle denial, which amounts to repression of the facts and failure to assimilate their implications, and (2) adaptive denial, in which the individual is aware of the facts and their implications but elects to focus attention on what remains rather than what is being lost. Most dying people have a need to know that they are dying. Whether they are told directly or infer it from behavior of those around them, most terminally ill people know that they are dying. But access to this information varies considerably across environments. Some practitioners believe in directly telling the patient and others do not.

Tasks that the dying person often must deal with include completing unfinished business, such as getting insurance and financial paperwork in order; dealing with medical care needs, especially decisions about alternative forms of treatment; scheduling time and preserving energy resources; and arranging for after death—things such as distribution of possessions, funeral arrangements, and making a will. Kalish and Reynolds (1981) reported that the only death preparations that a majority of people over 60 had made was taking out life insurance (66 percent). Forty-four percent had paid for a cemetery plot and arranged for someone to handle affairs after their death. Only just over one-third had made out a will, and 24 percent had made funeral arrangements.

Permitting People to Die

As for when death occurs, except for sudden death, a prime issue is when a person should be permitted to die. There is no clear consensus among older people about whether "heroic" measures should be used to keep them alive or whether death should be speeded in hopeless cases involving great pain. However, many older people are very aware of the financial costs of their care and may prefer to die rather than have their families incur the expense of keeping them alive a little longer. These issues are apt to remain controversial because in our society there is as yet no clear-cut moral ethic that dying persons, their families, and service providers can fall back on. Kalish (1985) reported that many states have passed laws allowing individuals to make a "living will" specifying the conditions under which they would want life-support procedures withheld or withdrawn. Most require that the person already be diagnosed as terminally ill and also be mentally capable of understanding the implications of his or her choice.

Care of the Dying

In caring for the dying, it is important to remember that most dying people fear being abandoned, humiliated, and lonely at the end of their lives (Weisman 1972). Thus, encouraging the maintenance of close personal relationships is an important aspect of the social care of dying people. Kalish (1985,268–70) suggested several guidelines for caring relationships with the dying. First, do not reinvent the wheel. Read up on what others have found to be effective. Second, what are your feelings about death and about aging? How might they affect your relationship with dying older people? Third, be honest with yourself about what you are able to give to the relationship. Caring for the dying can be physically and emotionally exhausting. Fourth, be willing to try to just *be* there rather than always feeling the need to be *doing* something. Many dying people know that there is little tangible that can be done for them. They may not even need to be talked to, just accompanied. Fifth, pace yourself to avoid burnout. Sixth, do not try to impose your philosophy of death and dying on dying people; let them have their own.

Deciding who should be told about a terminal diagnosis has a significant impact on close relationships. Kalish (1976) concluded that it is difficult for the patient to maintain such relationships unless the topic of death can be openly discussed. Thus, when dying people are kept in the dark about the nature of their condition, they are denied an opportunity to resolve the issues surrounding death in the company of those they love and who love them. By the time the prognosis becomes obvious, the dying person may no longer be capable of meaningful discussion. In evaluating the question of who should know, Weisman suggested that "to be informed about a diagnosis, especially a serious diagnosis, is to be fortified, not undermined" (1972,17). Yet some people do not want to be told they are dying, and others keep their prognosis secret.

Hospices represent a way to organize the efforts of health services providers and families around the goals of eliminating pain associated with terminal illness and allowing terminally ill people to die with dignity. In the ideal, a multidisciplinary hospice staff seeks to care not only for dying people but for their families as well. An additional goal is to let people die at home or in a homelike environment.

The first well-known hospice, St. Christopher's, was started in London in 1967. The primary goal at St. Christopher's was to make the patient free of pain and any memory or fear of pain. Besides providing freedom from pain, the staff provided comfort and companionship to their patients. Families, including children, were free to visit any time (except Mondays, when families were given a day off and did not have to feel guilty about not visiting). Families were also encouraged to help with the patient's care. Patients often went back and forth between the hospice and home. The median length of stay at St. Christopher's was two to three weeks; about half the patients returned home to die, with dignity and without pain, after a ten-day session with the hospice staff (Saunders 1976).

Interest in the hospice movement has grown steadily in the United States. In 1977, there were about 50 hospices in various stages of development, in 1980 about 400, and by 1983 the number had risen to over 1,200. Hospices do not all follow the model set out by Saunders. Davidson (1979) found four basic models in operation. Hospital-based programs take advantage of existing physical and administrative structures. The major disadvantage of this approach is that it is not easy to operate nontraditional programs such as hospices within traditional medical facilities. For example, the hospice philosophy is that freedom from pain is more important than the possibility of addiction to pain-relieving drugs, and this is in direct contradiction to the general policies of most hospitals. A second model is limited to home-care programs. These programs may or may not have formal ties to inpatient facilities that will allow continuity of care from staff and volunteers. Third, free-standing facilities offer both inpatient and home care. Inpatient care is often provided in a more homelike environment than is found in a hospital. A variant of this third model is the hospice operated in conjunction with a nursing home. The fourth model is essentially self-help, with volunteer professionals and lay counselors providing services at very modest cost. As yet, which of these models is most effective has not been systematically evaluated.

Studies that have compared hospice patients with other terminally ill patients have found the hospice patients to be more mobile, to rate their pain as less severe, and to see their physicians and nurses as more accessible. Spouses of hospice patients were also less anxious and spent more time visiting than spouses of other terminally ill patients (Parkes 1975). More recently, Mor and Masterson-Allen (1987) reported the results of an extensive review of the research literature on hospice outcomes. They found that hospices provided as good or better symptom control and emotional support as traditional hospital care of the dying, and the cost of hospice care, especially if home-delivered, was less than that for traditional hospital care. However, only those who had substantial family support at home achieved these financial savings.

The hospice movement was given a big boost in 1982, when the U.S. Congress made hospice benefits available to all persons who were eligible for Medicare Part A and who had a life expectancy of six months or less. These benefits covered routine home care, continuous home care, inpatient respite care, and general inpatient care, and included nursing services, physicians' services, medical social services, counseling, physical therapy, home health services, homemaker services, and medical supplies.

Levy (1989), in an insightful analysis, pointed out that the hospice concept was part of a "death with dignity" social reform movement within medicine itself. It was not imposed on medicine from without. The death-with-dignity philosophy provided both a humanistic ideology and a set of pragmatic concerns that rallied leadership and support from among the people who had to deal with dying patients within the medical system. With the institutionalization of hospice through its recognition by Medicare, Levy saw inevitable challenges to the basic tenets that gave rise to reform in the first place. For example, hospices still serve only a tiny proportion of people who die each year. Most of those served are cancer patients. The institutionalization of hospice means increased pressure to serve new groups, bringing new challenges. For example, AIDS patients often do not have the family support upon which the hospice concept relies, and they are often a financial liability because they have already exhausted their insurance benefits. Care for the oldest-old (those age 95 or older) presents these same challenges.

Medicare recognition of hospices has already affected the structure and services provided. Hospice program administrative staff are tempted to select patients from among those who best fit the Medicare reimbursement profile. Medicare reimbursement requires hospice personnel to keep detailed records of every action taken in relation to the patient and to follow complex Medicare regulations governing standards of care, hardly the relaxed atmosphere Dr. Saunders had in mind when she founded St. Christopher's. Hospices may not want to follow the Medicare guidelines because they want to adhere to the original concepts of the hospice movement, but they find themselves in a bind because if they do not qualify for Medicare, they may find it difficult to compete for clients with programs that have the Medicare "seal of approval" (Levy 1989). Thus, for Levy, the question is not will the hospice concept continue to grow and prosper, but whether there will be anything of the death-with-dignity philosophy left in it. It remains to be seen whether the humanistic goals of St. Christopher's Hospice will be overshadowed by a concept of hospice as merely a cheaper way to care for terminally ill people.

Bereavement

Bereavement is getting over another person's death, a process that may be finished quickly or may never be finished. In Lopata's (1973) study of widows, 48 percent said they were over their husband's death within a year, while 20 percent said they had never gotten over it and did not expect to. Individual bereavement takes three forms: physical, emotional, and intellectual. Some common physical reactions to grief include shortness of breath, frequent sighing, tightness in the chest, feelings of emptiness in the abdomen, loss of energy, lack of muscular strength, and stomach upset. Emotional reactions to bereavement include anger, depression, anxiety, and preoccupation with thoughts of the deceased. These reactions are particularly common in the period immediately following a death; they generally diminish with time. The mortality rates of widowed people are slightly higher than those of the married.

The intellectual side of bereavement consists of what Lopata called the "purification" of the memory of the deceased. In this process, the negative characteristics of the person who died are stripped away, leaving

Tending a gravesite is a form of ongoing connection to the person who died.
Photograph by Frank Siteman/Tony Stone Images, Inc.

only a positive, idealized memory. Somehow we think it wrong to speak ill of the dead. Lopata reported that even women who hated their husbands thought the statement "My husband was an unusually good man" was true. The content of obituaries and memorial services also attests to the results of this process. The idealization of the dead has positive value in that it satisfies the survivor's need to believe that the dead person's life had meaning. But it can have serious negative consequences for the future of the bereaved widow or widower because it can interfere with the formation of new intimate relationships.

Glick, Weiss, and Parkes (1974) found that men and women reacted somewhat differently to bereavement. Men more often responded to their spouse's death in terms of having lost part of themselves, while women responded in terms of having been deserted, abandoned, and left to fend for themselves. Men found it more difficult and less desirable to express grief, and they accepted the reality of death somewhat more quickly than women did. On the other hand, men found it more difficult to work during bereavement than women did.

Thompson et al. (1991) studied widows and widowers for a period of thirty months following the spouse's death. Compared with a sample who had not experienced bereavement, both widows and widowers showed significantly higher incidence of depression and psychopathology at two months after the loss, but by twelve months these differences between bereaved and nonbereaved groups had disappeared. However, on measures of grief, results indicated that both widows and widowers were still experiencing significant grief for at least thirty months following bereavement.

People usually do not have to go through bereavement alone. Others can help the individual through bereavement in various ways. At the beginning, bereaved people become exempt from certain responsibilities: They are not expected to go to their jobs; family and friends help with cooking and caring for dependents; older women often find their adult children making decisions for them. But social supports to the bereaved person are temporary; people are expected to reengage the social world within a few weeks at most.

Some researchers have speculated that some of the grieving process can take place during the terminal stages of illness. This process has been called anticipatory grief. However, Hill, Thompson. and Gallagher (1988) compared adjustment to widowhood among older widows who expected their husbands to die with those who had no advance warning and found no difference. They speculated that this lack of difference was because women in their 60s and older view widowhood as an on-time, generalized expectation. There may not be any warning, but older married women realize that widowhood is a common experience among women of their age.

Death of a Spouse

Widowhood is often thought to be something that happens primarily to women, yet in 1985 about one-third of the male population over age 75 was widowed. It is reasonable to assume that death of one's spouse has different effects on women than on men; therefore we first examine the role of widow and then look at the role of widower. We then discuss the controversy over whether coping with the death of one's spouse is harder for women or for men.

Being a Widow

The role of widow in American society is a long-term role primarily for older women. Young widows can play the widow role for only a short time and then are considered single rather than widowed. Younger widows feel stigmatized by widowhood because they

are so much in the minority, but older widows see widowhood as more normal because, even as young as age 65, 36 percent of women are widows. The prevalence of widowhood in later life combines with low rates of remarriage to produce for women a more definite social position for the older widow.

Yet the role of older widow is a vague one. Ties with the husband's family are usually drastically reduced by widowhood. The position of older widow primarily labels a woman as a member of a social category with its own salient characteristics. For instance, older widows are supposed to be interested in keeping the memory of their husbands alive. They are not supposed to be interested in men and are expected to do things with other widows or with their children. Thus, being an older widow says more about the appropriate social environment for activity rather than the activity itself.

Being a widow changes many women's basis of self-identity. For traditionally oriented women, the role of wife is central to their lives, structuring their lives not only in their households but also on the job. In answering the question "Who am I?" these women often put "wife of" at the top of their list. In addition to loss of a central role, widowhood often also causes the loss of the person best able to support the woman's concept of herself in terms of her personal qualities. If a woman's husband knows her better than anyone else and is her best friend and confidant, his opinions may be very important in supporting her view of herself as a good person. I have encountered older widows who after more than ten years of widowhood still "consulted" their dead husbands about whether they were "doing the right thing" by referring to the husband's values. How these women cope with the identity crisis that widowhood brings depends to a large extent on whether they base their identity on roles, on personal qualities, or on possessions. Role-oriented women often take a job or increase their investment in a job they already have. They may also become

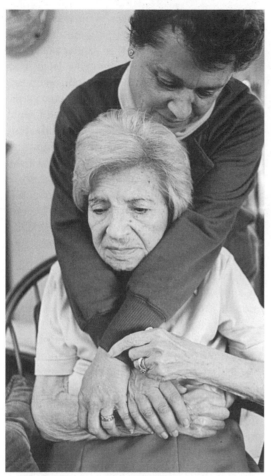

Long-term social support can help older widows cope with bereavement.
Photograph by Marianne Gontarz

more involved in various organizations. Those who need confirmation of their personal qualities may become more involved with friends and family. Those who are primarily acquirers base their identity on things rather than on people. For such women, an adjustment in self-concept is required only if widowhood brings a substantial change in acquisition power. Unfortunately, many widows find that their level of living declines markedly. Of course, all these orientations may exist in the same person.

Widowhood obviously carries great potential for an identity crisis. However, that po-

tential remains unrealized far more than might be imagined. The ability of older women to maintain their conceptions of self as "Mrs. John Doe" means that their memories can preserve a continued identity. Having other widows around helps legitimize this continuity. Children are also important reinforcers of a continued identity.

However, women vary a great deal in the extent to which the role of wife is central to their identity. For many women, the role of mother supersedes the role of wife, and after the children grow up and leave, the role of wife becomes an empty prospect. Other women have resented the traditional role of wife because of its subordinate status. Others have never developed a close, intimate relationship with their husbands. For any or all of these reasons, some widows do not wish to preserve the wife identity. These widows must then negotiate with family and friends to gain acceptance in their own right rather than as a wife. This renegotiation of social identity is a particularly necessary prelude to remarriage. Thus, for some widows, the identity problems brought on by widowhood are more external and social than internal and psychological.

Loneliness is generally thought to be a particularly prevalent problem among widows. Widows most often miss their husbands as persons and as partners in activities (Kalish 1985). While no doubt much of the loneliness stems from the absence of a long-standing and important relationship, some of it results from economic factors. Widowhood means poverty for most working-class women, which translates into lower social participation outside the home (Atchley 1975). The influence of economic factors on loneliness in widowhood deserves more research attention than it has received thus far. But not all studies show loneliness to be such a problem in widowhood. Kunkel (1979) reported that only one-fourth of her small-town sample of widows felt lonely a lot, which suggests that there may be important differences in reactions to widowhood between those who live in large urban areas and those who live in small towns. Helena Lopata (1973) and Zena Blau (1961) found high levels of loneliness among urban widows, while Robert Atchley (1975) and Suzanne Kunkel (1979) found low levels of loneliness in medium-sized cities and small towns. Both Atchley (1975) and Morgan (1976) suggested that there is an economic component to loneliness in widowhood, and Kunkel (1979) pointed out that economic conditions for widows have greatly improved in recent years. This may also play a part in the lower prevalence of loneliness in more recent studies.

However, it would be a mistake to equate aloneness with loneliness. Many widows quickly grow accustomed to living alone, and more than half continue to live alone. And as they become more involved with friendship groups of older widows, they tend to miss a partner in activities less. In residential areas with a high concentration of older widows, loneliness is much less prevalent.

The social disruption widowhood causes depends largely on the number of role relationships affected by the spouse's death. Middle-class widows are particularly likely to have viewed themselves as part of a team and to see their involvement in a wide variety of roles as having been impaired by widowhood. Such feelings are particularly common in cases where the couple operated a family-owned farm or business.

Widowhood has the most immediate impact on family roles. When older women become widows, they usually lose their contacts with in-laws, especially if their children are grown. However, within the reduced kinship network, the average amount of interaction with each available family member is higher than before widowhood (Morgan 1984). Contacts with children usually increase for a time, but those widows who move in with their children do so as a last resort. There are two basic reasons why wid-

ows seem to prefer "intimacy at a distance." First, they do not wish to become embroiled in conflict over managing the flow of household activity, and after being in charge it is hard for widows to accept a subordinate position in another woman's house, especially a daughter-in-law's. Second, they do not want to be involved in the dilemmas of rearing grandchildren. They feel they have done their work, raised their children, and deserve the rest. Gratton and Haber (1993) pointed out that intimacy at a distance is a relatively recent trend made possible by more adequate Social Security benefits. In the past, older widows often had no source of income and were forced to live with family. This was especially true for farm widows.

The value of social support is indicated by what can happen to those who do not have it. Goldberg, Comstock, and Harlow (1988) interviewed 115 widows six months after bereavement. They found that the main predictors of need for the 20 percent who reported that they needed help with emotional problems were recent disability, having few friends, and not feeling close to one's children.

Many communities have programs to help widowed people get the kind of social support they need. For example, in 1985, the American Association of Retired Persons sponsored over 175 Widowed Persons Service programs nationwide (Bressler 1985–1986). The Widowed Persons Service is an interesting model because it uses a peer counseling approach in which trained volunteers who themselves have been widowed for at least eighteen months offer support, someone to listen, and referral to community service programs. Newly widowed people enter the program through outreach phone calls and letters from staff of the Widowed Persons Service. When new programs are established, trainers from the national office work with a local advisory committee to identify and train volunteers to administer the organization as well as to provide peer counseling. Training manuals and written materials showing how to deal with various issues ensure that each new program can learn from the national experience in offering these programs and does not waste energy and volunteer time "reinventing the wheel."

David Morgan (1989) used focus groups of widows to examine the question "What sorts of things made it easier for you to deal with your widowhood and what sorts of things made it harder for you?" Of the things mentioned, 50 percent were positive, 40 percent negative, and 10 percent neutral. Contrary to the presumption that family and friends mainly provide social support, Morgan found that family members were more likely to do things that made the adjustment harder (45 percent) than were friends (37 percent). Not only that, the actions that caused problems for the widows were perceived to be intentional rather than unintentional more often among sons (67 percent of the time) compared with daughters (50 percent) or other family members (47 percent). The chief problem seemed to be that family, especially sons, expected the widows to get over their grief and get on with their lives more quickly than the widows felt was possible or desirable. Thompson et al. (1991) found that the experience of grief persisted for at least two-and-a-half years following the death of a spouse, so the expectation that widows should be able to dispense with grief and get on with it after a few months is probably unrealistic. Yet family, friends, and people in general seem to widely hold this unrealistic expectation.

The impact of widowhood on friendship largely depends on the proportion of the widow's friends who are also widows (Hong and Duff 1994; van den Hoonaard 1994). If a widow is one of the first in her group of friends to become widowed, she may find that her friends feel awkward talking about death and grief, not wanting to face what in all likelihood is their own future. If friendship groups had consisted mainly of couples, then the widow may be included for a time, but she will probably feel out of place. The widow may also encounter jealousy on the part of still-married friends. On the other hand, if the widow is one of the last to become widowed in a group of friends, then she may find great comfort among friends who identify very well with the problems of grief and widowhood. As a group of women friends grows older, those who are still married sometimes feel somewhat "left out" because their widowed friends do many things as a group that they do not feel free to leave their husbands in order to do. For these people, widowhood brings the compensation of being among old friends again.

Churches and voluntary associations offer avenues for increased contact with people. Church groups often present opportunities for increased involvement that do not hinge on having a spouse. The same is true of voluntary associations oriented around interests (as opposed to purely social clubs). However, in heterosexual groups, widows are apt to encounter the stigma of being the lone woman (Van den Hoonaard 1994). It is no accident that many widows are drawn more to women's groups than to heterosexual groups. Again, the age at which widowhood begins is important in determining its impact on community involvement.

Throughout the discussion of the impact of widowhood, the importance of age has cropped up again and again. The younger the widow, the more problems she faces; the older the widow, the more "normal" widowhood is considered to be and the more supports are available from family, friends, and the community at large to help her cope with widowhood. Older widows appear to adjust better than younger widows (Kalish 1985).

The impact of widowhood also varies considerably, depending on the widow's social class. Middle-class women tend to have balanced their roles between being a wife and companion to their husbands and being mothers. The loss of comradeship triggers considerable trauma at the spouse's death. As a result, middle-class women tend to have difficulty dealing with grief. However, middle-class women also tend to be broadly engaged. They have a number of friends and belong to various organizations. They have many personal resources for dealing with life as widows. They usually have a secure income, are well educated, and often have job skills and careers. In contrast, working-class women tend to emphasize the mother role more than the wife role. Consequently, they may experience less trauma associated with grief. But working-class women have fewer friends, fewer associations, less money, and fewer of the personal resources that make for an adequate long-term adjustment to widowhood. Working-class widows are thus much more likely to be isolated and lonely than middle-class widows.

Class differences are particularly pronounced among African Americans. Working-class black women tend to become widowed at much earlier ages than working-class whites. Lopata (1973) found that overt hostility between the sexes was more prevalent among African Americans, so widows sanctified the husband's memory less. As a result, widowhood resulted in even less emotional trauma among working-class blacks than among working-class whites. This subject needs further research.

There are also considerable ethnic differences in the impact of widowhood. Foreign-born widows are much more likely than others to have had a traditional marriage, which, as we saw earlier, can entail greater identity problems for widows. In addition, widowhood brings more potential for family conflict among the foreign-born. Many foreign-born older women were reared in cultural traditions that offer widows a great degree of involvement with extended kin. To the extent that extended kin do not share this orientation, there is room for a greater gap between what the older foreign-born widow expects from her family and what she gets. A

similar pattern prevails among older widows reared in the Appalachian tradition.

Being a Widower

The impact of widowhood on older men has received little attention. The literature on this subject is long on speculation and short on systematic research. Nevertheless, as a stimulus for further research, it is important to outline both what little we know and what we need to learn about being a widower.

The role of widower is probably even more vaguely defined than that of widow. Because there are relatively few widowers in any community until after the age of 75, the status of widower does not solidify groups of older men as that of widow does groups of older women. But older widowers are expected to preserve the memories of their wives and not show an interest in women. Indications are that many widowers adhere to the former expectation but ignore the latter.

Because the male role traditionally emphasizes other roles in addition to the role of husband, widowers are probably not as apt as widows to encounter an identity crisis caused by loss of the spouse *role*. But men are more likely than women to see their spouse as an important part of themselves (Glick, Weiss, and Parkes 1974). In addition, older men are less likely than women to have a confidant other than their spouse. Thus, both widows and widowers are likely to have problems because the spouse was a significant other.

How older men cope with widowhood's impact on their identity also probably depends, as older women's reactions do, on how the lost relationships fit into personal goal structures. Despite current stereotypes concerning men's overinvolvement with their jobs, there is little evidence that widowhood is any less devastating for men than for women. In fact, widowhood is very likely to wreck a man's concept of life in retirement. Likewise, there is little basis for assuming that marriage is less important to older men than it is to older women.

There is no apparent difference in the extent to which older widows and widowers experience loneliness (Atchley 1975). However, this finding may be an artifact of the higher average age of widowers. Were age to be controlled in research, older widowers might turn out to be less lonely than older widows.

Thus far, there has been little study of the impact of widowhood on men's roles outside the household. Widowers have more difficulty with work during the grief period (Glick, Weiss, and Parkes 1974). It also appears that widowers are more cut off from their families than widows (Troll 1971). Widowhood tends to increase contacts with friends among middle-class widowers and to decrease them among lower-class widowers (Atchley 1975). Quite likely the large surplus of widows inhibits widowers in developing new roles in terms of community participation. Particularly at senior centers and in retirement communities, widowers tend to be embarrassed by the competition among the widows for their attention. They also feel pressured by widows who constantly try to "do" for them.

Very little has been written about age, social class, racial, or ethnic variations in the impact of widowhood on older men. However, some of the variations noted for widows no doubt apply to men as well. This area is greatly in need of research.

Comparisons Between Widows and Widowers

There is currently some disagreement over whether widowhood is more difficult for older women or for older men. Age is an important compounding factor in comparing the experience of being a widow with that of being a widower. Widowhood occurs in old age for most men, while it occurs in later maturity for most women. It is important therefore to compare widowers with widows *of a similar age* in order to cancel out the impact of age. Much of the research that has been done to date has not controlled for age.

Atchley (1975) compared various dimensions of widowhood among 72 widowers and 233 widows, all 70 to 79 years old. The widows were significantly more likely than the widowers to suffer high anxiety, although even then only 15 percent of the widows had experienced high anxiety. The widowers were more likely to have a high level of anomie, but even the widowers had only 20 percent with a high level of anomie. The widowers were more likely than the widows to increase participation in organizations and increase contacts with friends. Sex differences in response to widowhood were greater among working-class respondents; here the widows were more likely than the widowers to be isolated and have inadequate incomes. Atchley concluded that the working-class widows were considerably worse off than the middle-class widows and widowers in both social classes. The key seemed to be their inadequate incomes. Income inadequacy in turn produced lowered social participation and loneliness.

Kunkel (1979) studied widows and widowers over age 50 in a small town. She found that the vast majority of both men and women adjusted well to widowhood. High morale, good health, adequate incomes, high levels of self-confidence, and positive attitudes toward retirement typified both widows and widowers. Age had no effect on adjustment to widowhood. Length of time widowed generally had little affect on adjustment. Widows were slightly more likely to have physical symptoms of stress, but these symptoms did not affect either physical health or morale. Widowers had lower levels of interaction with family and friends *and* were more satisfied with their level of interaction. It appears that widows *expected* a high level of interaction, usually got it, and wanted still more, especially among those widowed longer than one year. Widowers expected little, got more than they expected, and were satisfied.

Gallagher et al. (1983) looked at gender differences in mental health among widowed people. Compared with still-married controls, the widowed people in their study showed much greater psychological distress two months following bereavement, but there was no gender difference. Only a small percentage of recently bereaved older people showed symptoms of serious depression. Likewise, Scott and Kivett (1985) found no relationship between gender and morale among widowed people. Finally, Lund, Caserta, and Dimond (1986) concluded that widows' and widowers' bereavement adjustments were characterized by more similarities than differences. Indeed, they found no significant gender differences in feelings, coping behaviors, or eventual adjustment to bereavement.

There is no consistent support for the idea that men are less able to live alone, have more difficulty finding alternate sources of intimacy, or are more likely to be lonely. However, in terms of objective circumstances, compared with older widows, older widowers generally have better financial resources and greater opportunities for remarriage.

Summary

Death touches the lives of older people more often than it touches people of other age categories. The process of dying, the meaning of death, bereavement, and widowhood are all more salient to older people than to younger people, at least in industrialized societies. Death is both an end state and a physical and social process. In a given year, about 6 percent of the older population will die. Yet more and more people are not only surviving *to* old age but *in* old age as well. Due to variations in deathrates, women vastly outnumber men; fewer blacks than nonblacks survive to old age; married people have higher rates of survival than others; and poor people die at higher rates than others.

Men tend to see death as an antagonist, while women tend to see death as merciful. In general, older people fear death less than do other age categories. However, more research is needed on acceptance of death among specific categories of older people who are still highly integrated into society.

For older adults, the role of dying person demands passive acceptance of death. Older people are less apt than the young to change their lifestyle when they learn that they are dying. Older people are also apt to find that being a dying person means having less independent control over one's life.

People go through various stages in dying, but there is no set sequence or necessary progression. Denial, anger, bargaining, depression and sense of loss, and acceptance are all common reactions to dying—and dying persons often experience more than one at the same time. Acceptance of death does not mean a wish to die; it is merely an acknowledgment of an inevitability. Where and when a person dies is a matter of values. Most people prefer to die at home. However, at what point in the process of dying a person should be allowed to die is still a subject of controversy.

Two crucial aspects of care for the dying are maintaining personal relationships and reassuring the dying person that he or she will not be abandoned or humiliated. A lengthy dying trajectory allows survivors to grieve in advance and sometimes softens the final blow.

Bereavement is the process of dealing with someone's death. It usually encompasses symptoms of physical, psychological, and emotional stress. In addition, bereavement usually results in a sanctification of the memory of the person who died. People usually receive extra social support during bereavement, but only for a short time.

Being a widow is a long-term role primarily for older women. The prevalence of widowhood in later life combines with low prospects for remarriage to create a "community of widows." For women with traditional marriages, widowhood creates identity problems: It removes a valued role that is hard to regain, and it takes away the person who was most able to confirm the older woman's conception of herself. Some of the potential for crisis is softened by the fact that society allows older widows to maintain an identity based on having been someone's wife in the past. However, not all women have a high investment in the role of wife. For these people, the impact of widowhood on identity may be minimal.

Loneliness was and perhaps still is prevalent among urban widows, but recent studies of widows in middle-sized cities and small towns have found less loneliness. Widows tend to miss having someone around the house and someone to go places with. However, aloneness does not necessarily mean loneliness. Many widows adjust very quickly to living alone.

The impact of widowhood on family relations is ambiguous. Children respond to the widow's needs in times of illness, but the widow enters the household of a married child only as a last resort. "Intimacy at a distance" is the preferred pattern. Extended kin are not usually important sources of interaction for widows.

If a woman is one of the first of a group of friends to become widowed, she tends to be cut off from friends, but, if she is one of the last, widowed friends can be an important source of support and interaction. Generally, the younger the widow, the more problems she faces, and the older the widow, the more "normal" widowhood is considered.

Widowhood's impact seems to differ according to social class. On the average, working-class widows adjust better during bereavement but later have economic problems that make widowhood difficult. On the other hand, middle-class widows encounter more problems with bereavement, but because of their greater personal and economic resources they make a better long-term adjustment to widowhood. Class differences are especially great among widows who are African American. Owing to their traditional values, foreign-born widows may also have more difficulty.

Widowers have many of the same problems that widows face: vague role definition, identity problems, and loneliness. However, in general, older people of both genders appear to adjust well to widowhood, although it may take as long as a year or two.

Part 4

 # Aging Affects Needs and Resources

A great deal of what we experience in life is shaped by the resources available to us: How much *income* we have affects what we are able to do, the kinds of clothing we wear, where we live, whether we have cars and what kind, what food we eat, and even whether we are able to afford medical care. The quality of our *housing* and where it is located affect not only our comfort and enjoyment but also how secure and safe we feel. These needs are discussed in Chapter 13. Preventive and remedial *health care* affect our physical capacities and morale. Ready availability of *personal care* influences whether we can continue to live at home or whether we need to live in a more assistive environment. Physical health in turn strongly influences the nature and amount of our activities. These topics are examined in Chapter 14. In Chapter 15, the final chapter of Part 4, we will look at community social services. The kind, number, and location of *community facilities and services* affect our choice of activities and assistance. *Transportation* becomes an important necessity if we are to take advantage of the full range of activities in most communities.

Because income, housing, health care, personal care, transportation, and community facilities and services enable people to participate, we call them *instrumental* factors. They are the means by which people pursue an incredible variety of goals, and they greatly affect our sense of independence.

Accordingly, to fully understand aging requires that we know how aging affects our instrumental needs as well as the resources available to meet these needs. As with physical and mental aging, instrumental needs and resources vary widely among people, and aging has neither a uniform nor a predictable influence. In general, aging does not alter instrumental needs or resources enough to change lifestyles appreciably.

However, certain categories of older people are more likely to encounter problems. In each chapter, we look at needs and resources; typical problems; the types of older people most likely to have those problems; what the individual, the family, and the community can currently do about them; and what changes could be made to our social system that might reduce the number of older people who have problems.

People's resources are not merely the result of individual or family decision making—they are greatly influenced by the *social systems* we have created to provide income security, health-care financing, housing assistance, transportation, and a host of other programs and services. Most of our public programs for older Americans were created in response to large populations of older people with severe problems that did not respond to private solutions. Public programs are thus an important part of the social situation within which people adapt to aging. In this part of the book we look at the impact of these systems on the resources of individuals and families. The significant interplay between public programs and people's life chances is a recurring theme throughout this book.

13 ■ Income and Housing

Having sufficient money to live on and a place to live are the most basic necessities of life in modern urban societies. Because very few of us grow all our own food, make our own clothes, construct and maintain our own homes, and so forth, access to disposable income is essential. In addition, housing, or lack of it, heavily influences our location in the community both geographically and socially. In this chapter, we will examine how aging affects access to these basic necessities.

Income*

Income consists of the amount of money, or in-kind goods or services, that a person or household receives during a period of time. Most people receive a relatively stable monthly income, but some experience wide fluctuations in income from month to month or from year to year. Money income can be generated from a very wide variety of sources, and the range of variation in income is enormous.

To understand the income issues related to aging, we need to look first at how aging influences the distribution of income, then we can examine income needs and income sources. We then can look at income issues relating to aging.

---

*For more details on income, see Schulz (1995).

Income Distribution

The distribution of income is strongly related to age. As Figure 13-1 shows, median income increases sharply from age 15 to 24, peaks at age 45 to 54, and then drops sharply to its lowest level at age 75 and over.* The median income for the older population is less than half that of the population age 45 to 54. Indeed, young adult households are reasonably evenly distributed across the income categories, and some have very high incomes; middle-aged households are concentrated in the higher-income categories, and older households are concentrated in the low-income categories.

Figure 13-2 compares the 1990 income distribution for persons age 45 to 54 with the distribution for persons age 65 and over. It shows quite clearly that the distribution of income is spread relatively evenly across the income categories for the middle-aged 45 to 54 age category, whereas for the older population, the distribution is skewed very much toward the lowest income categories, with 68 percent of the older population concentrated in the lowest four income categories.

---

*The median income splits the income distribution for each age category into two equal halves, one-half above the median and one-half below. For a variable such as income, which can have extremely high or low values, the median is a more useful statistic than the mean or average because the mean is much more sensitive than the median to extreme values.

Figure 13-1 Median household income by age; 1991.

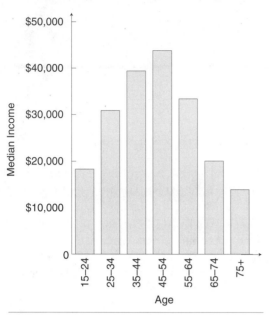

Source: U.S Bureau of the Census (1992e,1).

However, income patterns by age are difficult to summarize because even within the older population income differs substantially by employment status, gender, marital status, and race and ethnicity. The nonemployed, female, nonmarried, and minority older populations all have significantly lower incomes than their counterparts. Households with more than one person can vary even more than individuals. This picture is complicated even further because income can come from an enormous variety of sources, and these sources, particularly earnings from employment and retirement pensions, change with age.

Income Needs

Aging affects income needs in several ways. On the one hand, retirement reduces the amount of money needed for expenses connected with employment—transportation, special clothing, meals away from home, and

Figure 13-2 Distribution of money income of persons, selected ages: United States, 1990.

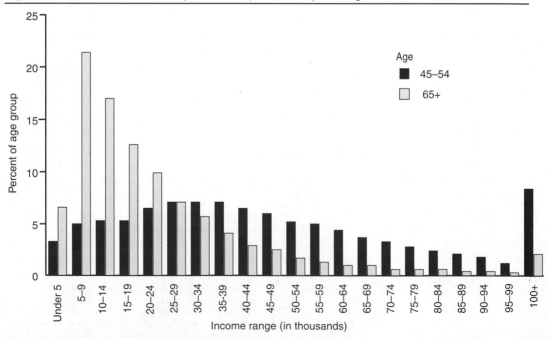

Source: U.S. Bureau of the Census (1991b).

so on. Launching children into adulthood reduces day-to-day expenses such as allowances, food, clothing, entertainment, or vacation expenses. On the other hand, increased physical frailty increases the need to buy services that formerly could have been provided by oneself or one's spouse. For example, after age 75, hanging drapes, washing windows, house painting, and heavy yard work are all more likely to be hired out rather than done by older individuals themselves.

Let us now look at three ways of estimating income needs: the poverty level, expenditure budgets for people at different levels of income, and the extent to which retirement benefits replace preretirement earnings.

Poverty Level The *poverty level*, established by the Social Security Administration in 1964 and revised in 1969 and 1980, is often used as an index of income need. The poverty level is based on an estimate of the cost of items in the Department of Agriculture's least costly nutritionally adequate food plan, which is weighted for the number of persons in the household. A 1955 survey by the Department of Agriculture found that households of three or more spent about one-third of their incomes on food. Therefore, the poverty level was set at three times the cost of the economy food plan. Obviously, the resulting estimate of the poverty level is quite arbitrary because it makes no allowance for variations in the cost of food throughout the country. In addition, the proportion of total income spent on food probably fluctuates according to local costs for food, housing, transportation, health care, and so forth.

In 1991, the official poverty level was $8,241 ($686.75 per month) for older couples and $7,086 ($590.50 per month) for older individuals. This was an austere definition of minimum income, especially in our larger urban areas where rents are high. Accordingly, it is useful to look not just at the poverty level but also at the proportion near poverty. Table 13-1 shows the proportions at or near the poverty level, by age, for 1991. Extreme poverty, income less than 50 percent of the poverty level, is concentrated among the young, especially children under 18. Income at or below the poverty level is most prevalent among people under 35 and

Table 13-1 Ratio of income to poverty level, by age: 1991.

	Ratio of Income to Poverty Level			
Age	Under .50 (%)	Under 1.00 (%)	Under 1.25 (%)	Under 1.50 (%)
Under 18	9.8	21.8	27.4	32.8
18–24	7.5	16.9	22.0	27.6
25–34	5.4	13.1	17.3	22.0
35–44	3.4	9.1	12.4	16.2
45–54	2.9	8.0	10.7	13.5
55–59	3.2	9.6	12.9	16.5
60–64	3.1	10.6	14.4	18.6
65–74	2.2	10.6	16.8	22.8
75+	2.3	15.0	24.2	32.9
65+	2.2	12.4	19.7	26.8
Total population	5.6	14.2	18.9	23.8

Source: U.S. Bureau of the Census (1992f).

age 75 and over. The pattern for income below 150 percent of the poverty level is the inverse of Figure 13-1, with the percentage below the threshold high in the youngest age category, dropping sharply to reach its lowest point at age 45 to 54, and then increasing steadily to age 75 and over. Note that at the income threshold of 150 percent of the poverty level or below, children under 18 and elders age 75 and over experience income deprivation in almost identical proportions.

The linkage between income and labor force participation is clear in these data. People at the beginning of their involvement with the labor force are much more likely to live in poverty than those in the midst of their working lives, and those who are retired are more likely to be in poverty or near poverty the longer they have been out of the labor force.

As a result of improvements in Social Security pensions in the 1970s, the proportion of elders living in poverty dropped from 24.5 percent in 1970 to 12.4 percent in 1991. Much has been made of the comparison between the declining proportion of older people in poverty and the increasing proportion of children in poverty. The argument is usually posed to suggest that the old have benefited at the expense of the young. However, this argument ignores the causes of poverty. Poverty among the old is primarily caused by inadequate retirement or survivors' pensions. Most older people now in poverty were economically productive during most of their adulthood, but because they had low-income jobs or were married to people with low-income jobs, their retirement or survivors' pensions are not adequate to provide income much above the poverty level. Increases in Social Security over the past twenty years have raised many older people above the poverty level, but not by very much.

Poverty among the young is caused mainly by having parents who are unemployed or who have very low-income jobs. Three-fourths of children in poverty are African American. Employment problems of the parents of African-American children are not caused by increases in Social Security benefits to older people.

It is also useful to remember that poverty is not a permanent condition. If we look at longitudinal data, the proportion of children who escape poverty is substantially higher than that of older people. Also, the longer people are in poverty, the less likely they are to escape poverty, especially if they are older. For example, older people whose incomes dropped below the poverty level were found to have a 42 percent chance of freeing themselves from poverty in their first year, but after four years of living in poverty, the probability of escaping dropped to 5 percent (Coe 1988).

Poverty is much more prevalent at the older ages within the older population (Radner 1993). For example, 26.3 percent of the older population as a whole had 1990 incomes at or below 150 percent of the poverty level. But as Table 13-2 shows, near-poverty income is much more prevalent at age 85 and older, especially for women in general (nearly 45 percent) and widowed women living alone (about 60 percent). Thus, there are subgroups within the older population that have extremely high proportions living at or very near the poverty level. This certainly contradicts the recent media stereotype of elders as an affluent population.

Expenditure Budgets Expenditure budgets take a more sophisticated approach to income needs, building from data on actual consumer expenditures for people at different levels of income. For example, the Bureau of Labor Statistics (BLS) developed budgets for retired couples (see Table 13-3). In 1991, the low budget was about 120 percent of the poverty level. The low-budget total of $9,845 was very close to the median income of people age 75 and older and reflected a very modest level of living in terms of the costs of housing, household maintenance (for example, utilities and

Table 13-2 Percentage of persons with incomes below 150 percent of the poverty level, by selected characteristics: United States, 1990.

	Age 65+	Age 65–69	Age 85+
Men	19.0	14.3	29.5
Women	31.5	22.3	44.9
Widowed, living alone	50.6	44.4	59.8

Source: Adapted from Radner (1993).

Table 13-3 Annual budgets for a retired couple at two levels of living: 1991.

	Low-Budget Total, $9,485	High-Budget Total, $20,542
By component:		
Food	2,974	4,962
Housing	3,239	7,228
Transportation	753	2,670
Clothing	332	857
Personal items	269	578
Medical care	1,479	1,496
Entertainment and recreation	375	1,228
Other[a]	424	1,523

[a] Includes gifts, contributions, and insurance
Source: Estimates by the author.

repairs), household furnishings, food, clothing, transportation, medical and dental care, personal items (for example, laundry, dry cleaning, and grooming supplies), entertainment and recreation, insurance, and miscellaneous other expenses. Personal income taxes were not included. It was assumed that the couple would be self-supporting, living in an urban area, in reasonably good health, and able to care for themselves. The low budget assumed almost no replacement of clothing or home furnishings; very little home maintenance, recreation, entertainment, and insurance; and a subsistence level for food. The high-budget figure of $20,542 was very close to the median income for persons age 65 to 74 and approximated a modest middle-class level of living. It allowed for such items as replacing a winter coat or television set every five years and modest amounts for travel, eating out, insurance, and personal care. The high budget allowed more than double the low budget in most budget categories.

The BLS also found significant variation across major metropolitan areas in the costs of various budget items. For example, in 1981 the national average low-budget figure for housing (including household maintenance) was $2,377 ($198 per month). But the low-budget housing cost for renters varied from $4,445 in Anchorage to $3,375 in Seattle to $1,902 in Atlanta. Costs for most other budget categories also fluctuated widely. Of course, these differences in living costs meant that the income needs of older households differed considerably from one part of the country to another. Thus, income need is not easy to estimate.

Schulz (1995) reported that expenditure patterns shifted with age within the older population. Compared with those age 65 to 74, older Americans age 75 and older spent a greater proportion of their income on health care, utilities, and charitable contributions and less on transportation, insurance, and clothing.

Earnings Replacement Earnings replacement methods of estimating income needs attempt to determine the extent to which retirement income must replace preretirement income in order to allow continuity of lifestyle. Schulz (1995) estimated that preretirement lifestyles can be maintained with a retirement income of 65 to 75 percent of preretirement income. However, couples at lower income levels may need to replace a larger proportion of their preretirement income. For example, Gibby and Dorer (1988) estimated that couples with preretirement incomes of $15,000 or less would need to replace 77 to 85 percent of their preretirement incomes to maintain the same level of living.

Both the poverty level and the BLS low budget attempt to estimate minimum income needs for objective necessities of life. But a middle-class person's necessities may be a poor person's luxuries. Most people judge the adequacy of their incomes in relation to what they think they need. Older people generally tend to think that their incomes are adequate even when their incomes are well below what others might think they need. For example, in an unpublished community study of older adults, I found in 1991 that 64 percent of those with incomes of $5,000 or less thought that their incomes were adequate, even though they were surviving on incomes several thousand dollars below the official poverty line.

Thus, the subjective element is very important, not just in estimating income needs but in assessing the adequacy of a person's income. We will return to this topic later when we consider income security as a concept. Now let us turn our attention to sources of income and how they are related to age.

Income Sources

Aging changes the available sources of income. Older people have access to retirement income and financially valuable services and facilities specifically reserved for older people. But, aging also reduces the prevalence of employment and the accompanying earnings. For our discussion, we divide income sources into direct income sources, which deliver cash income that can be used for any purpose, and indirect income sources, which have economic value only if an individual uses subsidized programs.

Direct Income Sources   Direct sources of income include earnings from employment; Social Security benefits of various kinds; retirement pensions; income from property such as savings, securities, or real estate; and public benefits such as Supplemental Security Income (SSI), public assistance, or workers' compensation. There are so many direct sources of income that no particular pattern can be called typical. However, by looking at the proportion of the income of older households that comes from various sources, we can get some notion of the relative contribution of each source. Table 13-4 shows the relative contribution of various direct income sources to the aggregate income of all older households. Social Security is the primary income source for the older population as a whole, followed by asset income, earnings, and pensions. Means-tested welfare benefits such as public assistance and SSI represent only a minor source of income to the general older population.

Table 13-4 also shows the relative contribution of various direct income sources by level of income. Not surprisingly, these data show that Social Security and other public benefits are the mainstays of income for elder households at the bottom of the income range. As we go up the income levels, earnings and property income increase in importance, whereas the proportion of income coming from Social Security decreases. The importance of pensions increases, but not perhaps as much as we might expect; it peaks in the middle income categories and declines slightly for the highest income level. Overall, the data in Table 13-4 show that the older population is quite heterogeneous in terms of sources of direct income. Now let us look at specific direct income sources.

In 1990, about 37 percent of older households with heads age 65 to 74 had some income from earnings. The proportion with earnings dropped sharply with age, to 16 percent at age 75 to 84 and 8 percent at age 85 and over (Waldrop 1992). The median income from earnings was extremely modest, ranging from $8,160 at age 65 to 74 to $6,808 at age 85 and over. Nevertheless, those with earnings tended to have incomes about double those of elders who relied primarily on Social Security. About half of those who were employed needed earnings to supplement inadequate **pensions.**

Retirement pensions come in two basic varieties: Social Security benefits and **employer**

Table 13-4 Income sources, by income level, for older households: 1989.

| Income Source | All Older Households | Income Quintile* | | | | |
		1	2	3	4	5
Mean income	$14,991	$4,020	$7,046	$10,717	$16,488	$36,676
Income Source				**Percent of Aggregate Income**		
Earnings[a]	22.1	1.8	5.1	11.1	18.4	27.8
Social Security[b]	35.0	77.8	73.5	56.6	37.9	17.3
Property[c]	24.2	4.2	8.7	15.6	24.4	39.0
Pension[d]	15.8	1.9	6.6	13.4	17.4	14.6
Other[e]	2.9	14.2	6.0	3.4	2.0	1.3

* Quintiles divide a frequency distribution into five parts, each containing an equal number of cases (people in this instance).
[a] Wages, salaries, self-employment income.
[b] Railroad Retirement benefits, retired worker benefits, spouse benefits, and survivors' benefits.
[c] Interest, dividends, rents, royalties, estate and trust income.
[d] Government and private pensions and annuities.
[e] Supplemental Security Income (SSI), public assistance, unemployment benefits, and workers' compensation.
Source: Adapted from Radner (1991).

pensions. *Social Security benefits* can be retirement benefits paid to covered workers or *survivors' benefits* paid to the surviving spouses of covered workers. Social Security retirement benefits are available to those who have reached age 62 and have worked on covered jobs for a specified minimum period of time (usually ten years). Social Security retirement benefits also carry a 50 percent spouse benefit, which means that retired couples receive 150 percent of the pension entitlement amount of the covered worker. Either husbands or wives can be the covered worker, depending on who qualifies for the highest benefit, but as we might expect, among couples receiving a spouse benefit, 90 percent of covered workers are men.

The standard retirement age under Social Security is 65. Those who retire younger receive a benefit that is reduced by an amount that roughly compensates for the greater amount of time over which benefits will be received. Those who retire after 65 receive an increased benefit that roughly compensates for the shorter time that benefits must be paid.

More than 90 percent of older Americans receive Social Security retired worker or spouse benefits, and more than 75 percent of

new beneficiaries receive reduced benefits as a result of having retired prior to age 65. In 1993, average annual Social Security benefits were $13,272 for retired couples, $7,836 for retired individuals, and $7,296 for surviving spouses (U. S. Social Security Administration 1993). These very modest benefits replaced about 55 percent of earnings in the last year of employment for retired couples and only about 37 percent in earnings for single individuals. The purchasing power of Social Security retirement pensions has increased about 60 percent since 1965, but the increased Social Security payroll taxes to accomplish this increase in pensions have gone to eliminate poverty, not to provide lavish pensions.

Eligibility for *job-related retirement pensions* is established by working for an employer, being in a union, or being in an occupational group that has an institutionalized mechanism for making financial contributions into a pension plan on behalf of the employee. In some cases the employer makes contributions into the pension plan in place of wages or salary. In others, employees must make retirement pension plan contributions out of current income. In 1988, 41 percent of wage and salary workers in the private sector had job-related retirement pension coverage in addition to Social Security; this included 46 percent of men and 35 percent of women (Woods 1989). On the other hand, over 80 percent of state and local government employees and nearly all federal employees had job-related retirement pension coverage. People receiving job-related retirement pensions tend to have had high earnings and thus to qualify for high Social Security benefits. The result is a large and growing disparity between the incomes of those who have multiple pensions and the incomes of those who are entitled only to Social Security, with dual pensioners averaging incomes almost twice as high as those with Social Security alone (Schulz 1995).

Assets such as income-generating property, savings, and investments are often presumed to be a major source of income for older people, but this is true only for the small proportion of very well-to-do elders. In fact, few people find themselves actually able to accumulate the amounts of savings or investments required to provide significant income in their later years. About 70 percent of elders have interest-bearing savings, with an average annual income from savings of less than $2,000 per year; about 20 percent have income from dividends on stocks and bonds, with an average income of just over $1,000 per year; and about 10 percent have income from rents, royalties, or trusts, with an average income of less than $3,000 per year (Waldrop 1992).

Home equity is by far the most valuable asset that older people tend to have. About 60 percent of older couples and 30 percent of older people living alone own their homes free and clear. As Figure 13-3 shows, the financial assets of people of all ages consist mainly of home equity. In 1980, over 80 percent of married older people lived in an owned home (Chevan 1987), which increased discretionary income by about 5 percent (Smeeding 1990b). Liquid assets are a modest source of income for most retired people

(Radner 1990). For example, in 1995 a relatively generous annual yield on assets was 7 percent, which would provide an income of about $145 per month on assets of $25,000 (about the median amount of assets for the older population).

According to the life course theory of savings behavior (Ando and Modigliani 1963), people save during their employment years and dissave during retirement. However, numerous researchers have found that people who developed the habit of saving continue to save in retirement (Stoller and Stoller 1987). Most older people recognize that having a cushion of savings is an important protection against fluctuating financial needs. Unexpected replacement of a furnace, the need for a new car, or high medical bills are typical unplanned expenditures that make it prudent to have a contingency reserve. In addition, most older people recognize that the need for long-term care can require substantial resources, and they are reluctant to spend their contingency funds. Thus, many older people try to live on their income from assets, pensions, and Social Security without tapping into the principal of their savings.

Because home equity is a prevalent asset among older people, finding ways to convert home equity into income without requiring the older person to vacate the home has aroused considerable interest. *Home equity conversion* is the expression used to describe such programs. There have been a number of experimental home equity conversion projects since 1980, and for older people who are nearing the end of life, these projects are a way of generating the income needed to remain at home. Kenny and Belling (1987) provided a useful case example. Mrs. F. was a 78-year-old widow with no children. She was in a hospital recovering from a leg amputation due to cancer. She was scheduled to enter a county nursing home, but she was also referred to a home equity conversion program counselor. Mrs. F. wanted most to return to her own home, but

Figure 13-3 Median net worth of households, by age of householder: 1988.

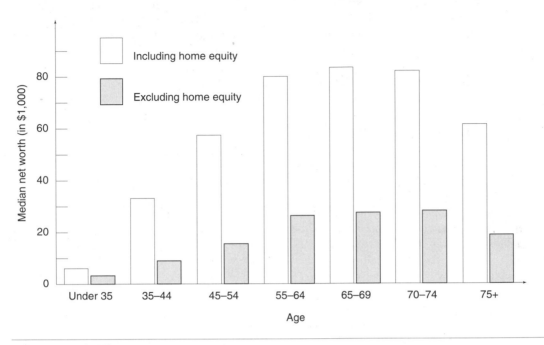

Source: Radner (1992).

her income of $470 a month could not provide the needed live-in care. She elected to secure a reverse annuity mortgage of $88,000 on her home valued at $110,000, which would give her an income of $1,400 per month for four years while she continued to live in the home, enough to cover a live-in caregiver and additional household and health-care expenses. Mrs. F. died at home seven months later, and her heirs repaid the modest loan balance from the sale of the home. Had Mrs. F. lived beyond four years, she would have been compelled to enter a nursing home at that time, but she was faced with admission anyway, and the home equity conversion program allowed her the option of staying at home for four more years.

Because home equity conversion plans are complex, older people and their families usually need considerable counseling to fully understand the advantages and liabilities. To prevent home equity conversion programs from being exploitative, it is important that the counselors who work with the older person be considered service workers rather than salespeople.

Despite the fact that home equity conversion can work, many obstacles are likely to limit the number undertaken (Kenny and Belling 1987). For home equity conversion to significantly augment income, the term of the loan usually must be three to four years. For most well older people, this term is far too short. The prospect of mortgaging one's home is a psychological barrier that only the desperate seem willing to pass. Those older people who need income the most are least likely to own a home of sufficient value to interest lenders. Further, the staff time and skills required to deal in home equity conversions are beyond the capacities of most lenders. The prospect of having to evict older people who refuse to leave at the end of their loan term further discourages lenders. Thus, most counseling in this area is currently funded by public agencies and private foundations.

Accessory apartment conversion has also been proposed as a way of transforming existing housing equity into income. In this approach, elders are able to borrow funds at below-market interest rates to convert part of an owner-occupied home into an apartment for which they can receive additional income in the form of rent. These programs have been tried in California, Maine, Minnesota, and Connecticut (Retsinas and Retsinas 1991). The rationale rests on three assumptions: that older people want to increase their incomes, that many of the houses older people occupy are larger than they need, and that having renters in close proximity might create informal supports for frail elders. But as Retsinas and Retsinas pointed out, interest in this alternative has been slight to nonexistent because it requires that the older couple or individual become borrower, remodeler, and landlord, which may be very intimidating roles for elders with no prior experience. And while close-by informal support is a desirable goal, it is unpredictable coming from renters, and loss of privacy may cancel out the benefits in the minds of many elders. Finally, until the mortgage is paid off, the income from the accessory apartment is meager. Thus far, practical methods of transforming home equity into income have remained illusive.

Supplemental Security Income (SSI) is a federal-state program of public assistance to needy older people. In 1993, SSI guaranteed every older American $434 per month regardless of prior work history ($625 for older couples). Note that this level of income is below the official poverty level. Here is how SSI works. Suppose an older person receives a Social Security retirement pension of $160 per month, has no other source of income, and has assets apart from home equity of less than $2,000. To determine the SSI amount, $20 of Social Security income is excluded, giving an income for SSI purposes of $140 per month. This $140 is then subtracted from the guarantee of $434 to yield an SSI benefit of $294 per month. About 1.5 million older people were receiving SSI benefits averaging about $174 per month from this program in 1990. In addition, forty-two states provided supplements averaging $140 per month to the national SSI guarantee. To qualify for SSI, all applicants must demonstrate that their incomes from other sources fall below a prescribed minimum; this condition of eligibility is called a *means test*. For example, to qualify for SSI in 1993 an older person had to have an income of less than $424 per month and assets of less than $2,000.

The number of older people receiving SSI peaked at 2.3 million in 1975 and has declined substantially since, mainly because of improvements in Social Security benefits. However, about 40 percent of those who are eligible for SSI do not apply for benefits, primarily because they see the means test as stigmatizing (Schulz 1992). Drazga, Upp, and Reno (1982) found that a larger proportion of eligible elders apply for SSI in states that supplement the federal SSI guaranteed amounts.

Indirect Income Sources Indirect sources of income include many public and private programs that serve older people directly and also help indirectly by supplementing incomes. Federal programs that give indirect financial aid to older people include low-rent public housing and rent subsidy programs, Medicare, Medicaid, the food stamp program, and meals programs for elders. Many states give property tax exemptions to people age 65 and over. In addition, senior centers and transportation programs often provide services free or at reduced prices.

Indirect income increases the real incomes of older Americans. The U.S. Bureau of the Census (1981) collected data on receipt of means-tested noncash benefits among the noninstitutionalized population. They looked particularly at food stamps, government-assisted housing, and Medicaid. While elders represented about 11 percent of the noninstitutionalized population, they were 17 percent of those receiving food stamps, 34 percent of those living in government-assisted housing, and 33 percent of those living in independent households and receiving Medicaid. But although elders made up a large proportion of those who received noncash benefits, only a small proportion of all older people actually received such benefits. For example, only 6 percent of older people received food stamps, about 7 percent lived in government-assisted housing, and 16 percent received Medicaid.

It is probably safe to say that as a result of eligibility criteria, most indirect income sources help the poor rather than the middle class. The major exception is Medicare hospital insurance, which in 1989 provided an average indirect income of $1,689 to each enrollee, which included most of the older population (U.S. Social Security Administration 1991a). Older people also benefit indirectly from Medicare's supplemental medical insurance, which partially covers physicians' charges, but they must pay a premium for this coverage. Indirect income sources such as public housing and rental assistance are much more readily available to the urban than the rural older population.

Two hypothetical case histories illustrate the variety of income sources for older people. Mr. B. worked as an insurance analyst for the same company for forty-five years. He retired at 65, in very sound financial shape for retirement. Having earned a high salary during his working life and having paid the maximum Social Security taxes, he was eligible for a monthly Social Security retirement benefit of $757 for himself plus an additional $378 for his wife. In addition, Mr. B. drew a private pension of $690 per month from his former employer and had an annuity income of $150 per month. His total income from all these sources was $1,975 per month, or $23,700 per year, which represented 65.8 percent of Mr. B.'s income in the year preceding retirement. And only a little over $10,000 of his retirement income was tax-

able, compared with $36,000 of his preretirement income. Mr. and Mrs. B. both reported that retirement had not affected their capacity to afford to continue their preretirement lifestyle into retirement.

In contrast, Mr. J. had not been so fortunate in his work life. Working for various employers, mostly small businesses, he had done unskilled work in shipping or stocking. His last job had been as "stock boy" in a liquor store, working six days a week unloading cases of liquor, wine, and beer and shelving them. He was paid the minimum wage. Mr. J. had been unemployed many times throughout his working life, although never for an extended period. At retirement, Mr. J.'s Social Security benefit was only $182 per month, despite the fact that he had worked nearly fifty years. His wife received her own $122-per-month Social Security pension. Having no other income, they applied for and received $93 per month in SSI. They also moved into rent-subsidized housing for older people, where their monthly rent was fixed at $99 (25 percent of their monthly income), saving them $38 per month over what they had paid for their previous apartment and leaving them $3,576 per year for all their other expenses. Combined, their benefits replaced only 38 percent of their earnings the year before they retired. Mr. J. reported having a lot of trouble making ends meet.

Who Has Income Problems?

Income problems come in two basic varieties: absolute and relative. People with absolute income problems do not have enough income to provide basic food, clothing, and shelter. For example, the increasing homeless population in the United States certainly has absolute income problems by any definition. Because all older people in America are at least hypothetically eligible for Supplemental Security Income, no older person should fall below $5,000 per year in income, which is about $1,000 below the official poverty level.

Nevertheless, in 1991 there were more than one million older households in America trying to subsist on incomes of less than $5,000 per year.

Relative income problems have more to do with: (1) a lack of *income security,* the sense that one's customary standard of living is protected from the winds of circumstance; (2) the experience of declining income and level of living that can accompany life transitions such as widowhood or divorce; or (3) the persistence into later life of the lifelong results of gender-related or racial and ethnic biases in access to jobs that produce adequate retirement income.

People at all ages and income levels can experience threats to their income security, but older people may be especially vulnerable. Because of age discrimination in employment, older people often cannot cope with declining income by getting a better job, working more hours, or working more than one job. Elders depend on the nation's pension systems, Social Security, and investment markets to operate effectively.

For example, if we look at an upper-income older household with an annual retirement income of $85,000, about $16,000 would come from Social Security retirement benefits, about $45,000 would come from a job-related retirement pension, and the remaining $24,000 would come from income generated by assets—investments, real estate, and savings. The job-specific retirement pension is likely to be fixed at a set amount, not indexed to increase with inflation but also not subject to falling with interest rates either. Fixed income is very vulnerable to inflation; double-digit inflation can cut purchasing power of a nonindexed pension in half in less than a decade. Social Security income is indexed, and whenever inflation exceeds 3 percent, benefits are adjusted upward to maintain purchasing power. Social Security income is thus a reasonably secure income source, and because it is the only secure element of the retirement income mix, it

is not surprising that retired people get nervous when policy makers propose to deal with the federal budget deficit by means-testing Social Security retirement benefits or by decreasing Medicare coverage.

The security of the substantial private investment part of retirement income depends entirely on the types of investments chosen. Income from relatively secure investment vehicles such as federally insured certificates of deposit are vulnerable to falling interest rates. In the face of falling interest rates, to maintain income retirees can be forced into riskier choices that increase their chances of losing their money, which of course is the ultimate in income insecurity. For example, 60 percent of the investors who lost their money in the collapse of Lincoln Savings and Loan Association in the late 1980s were people over age 60 who were trying to maintain their investment income (U.S. House Committee on Banking, Finance, and Urban Affairs 1989). The need to maintain investment income, coupled with inadequate information on the reliability of various forms of investment, makes elders especially vulnerable to a wide variety of investment frauds and mismanagement.

Middle-income elders rely on these same sources of retirement income, but in different proportions. For example, a household with a retirement income of $35,000 per year is likely to receive about $14,000 in Social Security retirement benefits, about $15,000 from a job-specific pension, and only about $6,000 in investment income. Thus, because Social Security is a larger proportion of their income, their vulnerability to both inflation and falling interest rates is proportionally less.

Low-income older people receive their incomes mainly from Social Security and SSI. Both of these sources are indexed for inflation and not subject to decreases with falling interest rates. Thus, in terms of income security, low-income retired people are not as vulnerable as middle- and upper-income retired people to income insecurity coming from inflation or falling financial markets. Low-income elders may have predictable incomes, but unfortunately they are more likely to be confronting absolute income problems. Their *sources* of income are secure, but the *amount* of income is not sufficient to provide basic necessities.

Retirees, widows, the divorced, and the lifelong poor are most likely to experience income problems in later life. Among retirees, those who have worked for large employers are quite likely to have private pensions in addition to relatively high Social Security benefits. They are most likely also to have fringe benefits such as medical and dental insurance coverage that continue into retirement. Income security for these people is tied very closely to the rate at which retirement income replaces preretirement earnings. About 40 percent of couples achieve the 68 to 72 percent replacement rate necessary to maintain the level of living they experienced in the year before retirement, and most of them achieve this favorable replacement rate through job-related pensions in addition to Social Security (Grad 1990b).

On the other hand, those who worked for small businesses are least likely to have private pensions; most likely to have earned less for the same work, with correspondingly lower Social Security benefits; and most likely to lack substantial fringe benefits before or after retirement. Because Social Security replaces only about 40 percent of income for these people, they are likely to experience a substantial decline in level of living at retirement (Grad 1990a).

Smeeding (1990b) called attention to the serious income vulnerability of the 20 percent of all elders (40 percent of older single people living alone) whose incomes fall between 100 percent and 200 percent of the poverty level. Smeeding called them *'Tweeners* because they are caught in the middle: their incomes are not low enough to qualify for means-tested SSI and yet are not high enough to provide

genuine financial security. 'Tweeners are especially likely to lack Supplemental Medical Insurance under Medicare because they cannot afford the premiums; to have no resources to pay for long-term care should they need it; to rely on Social Security as their only income source; and to spend a larger-than-average proportion of their income on housing. Smeeding concluded that in many ways, the 'Tweeners are "the least well-off and most economically insecure persons" (1990a, 371).

More than two-thirds of the 'Tweeners are women. Women are more likely than men to have worked sporadically, part time, for small businesses that lack private pensions, and in jobs not covered by union bargaining for pensions and fringe benefits. Thus, because retirement income in the United States is so closely tied to type of job and level of earnings over a lengthy working life, women are more likely than men to be poor in retirement or to fall into the 'Tweener category.

The very old are also likely to find their retirement incomes inadequate because, although there have been cost-of-living increases in Social Security benefits, private pensions usually do not rise to match inflation. Workers who are very old today have Social Security pension entitlement based on average wages much lower than today's, even when adjusted for inflation.

Finally, retired minority group members are likely to have inadequate retirement incomes for many of the same reasons that women's retirement incomes are too low: working for employers who provide no private pension, interrupted employment over the working life, lower average wages even for the same work, and so on.

Widows are more likely than married women to encounter income problems in later life. First, many women find they are too young to collect survivors' benefits at the time they become widows. About 13 percent of women in their 50s are widows, yet widows do not qualify for survivors' benefits under Social Security until the age of 60. Second, even if the widow qualifies, household Social Security income will go down by one-third from its level before her spouse's death. Unfortunately, fixed expenses such as rent and automobile payments do not go down when household size drops from two to one. If the husband's Social Security pension was already marginal, then widowhood can bring financial disaster. Bound et al. (1991) found that widowhood caused an 18 percent drop in level of living and pushed 10 percent of widows into poverty.

Fethke (1989) also called attention to the effects of marital dissolution on income. She found that people who divorced in later life were unlikely to be able to establish the economic base needed to maintain their lifestyles into retirement. She projected that as the proportion of divorced elders increases in the future, divorce may emerge as a new major cause of poverty among elders. Divorce causes income problems because even if pension and asset income are divided equally, the amount now has to support two households instead of one.

The lifelong poor are also likely to have income problems in later life because most poor people have no access to private pensions. Their low-wage averages generate low Social Security benefits, even if they remained employed nearly all their adult lives. However, many poor people encounter no drastic change in their level of living. In fact, with SSI, food stamps, and low-rent housing for elders, some lifelong poor find that their situation in old age actually improves, although it is hardly a standard they would choose.

What Can Be Done?

When older people encounter income problems, there are only a few courses of action they can take. They can get jobs, they can apply for SSI and take advantage of programs such as low-rent housing or food stamps (provided their incomes are low enough),

they can move in with family, or they can ask their families for financial support.

The prospects for getting a job in later life are not good, especially for those already likely to have inadequate retirement incomes. Although they may have difficulty sustaining their former middle-class lifestyles, their incomes are usually still high enough to disqualify them for income assistance programs. Most older people believe it is not their children's responsibility to provide them with an adequate income. Social Security is a social insurance program designed to protect people from loss of income, and because a very large proportion of elders have contributed to this program, they see providing adequate income protection as a government responsibility (Harris and Associates 1981). As we have seen, the national income guarantee for elders provided by SSI is just under the poverty level, hardly a standard to be proud of. Older people are very reluctant to ask their children for money, and the end result is that thousands of older Americans find there is virtually nothing they can do if they feel their incomes are inadequate.

In 1995, Congress considered a number of policy changes that would further erode income security for older Americans, especially those at the bottom of the income distribution. Welfare reforms proposed decreasing federal support to states through block grants for programs such as SSI and Medicaid. Proposals were also made to decrease retirement benefits under Social Security and to reduce federal Medicare commitments. These changes would represent a major step backward in terms of income security for older Americans, especially the poor and the near poor.

Housing*

Housing is an important source of continuity in the lives of most middle-aged and older

---

*For more details on housing, see Golant (1992).

people. After age 35, Americans do not change residence often, partly because of home ownership but mostly because of the feeling of comfort and ease that comes with living in a familiar environment. As we saw earlier, one way that people cope with changes in their physical and mental capacities with age is to concentrate time and attention on long-standing patterns of activity (Atchley 1989). And a large measure of this continuity comes through having lived in the same dwelling for a long time. The majority of older Americans have lived in their present dwellings for more than twenty years. Golant (1992) reported that the homes older people owned were typically small and had numerous defects. Nevertheless, the older occupants subjectively felt that their houses fully met their needs, and few wanted to move to another house.

Housing is more than a place to live. It can be a symbol of independence, a focal point for family gatherings, a source of pleasant memories, and a link to the neighborhood and surrounding community. Remaining in one's customary abode can become more important as people grow older because one's home is a major symbol of one's standing in the community. Its appearance can be a source of self-esteem and also a focus of meaningful goals.

However, not all elders have strong and positive relationships with their homes and their neighborhoods. Just as aging individuals change and adapt, so too do their homes and their environments. Homes that at one time seemed ideal may now seem to require an inordinate amount of time and money to keep up. Neighborhoods that were once vital and energetic communities may now resemble bombed-out sections of the city. Retirement lifestyles may not be feasible within the geographic context of one's customary dwelling.

Aging can change housing needs in a variety of ways (Golant 1992). People can respond to these changes by aging in place or by relocating. Most housing choices involve trade-offs among values, and in later life, trade-offs between security and autonomy are particularly likely to require compromise. To get the security of meeting needs for personal care, elders often must give up at least some measure of autonomy. Ideally, housing choices match lifestyle preferences but at the same time acknowledge and accommodate constraints stemming from changes in income, marital status, or functional status.

Aging in place involves remaining in one's customary household over the long term. Aging in place is often a preferred course of action, but is not always the easiest. First, aging in one's familiar household may involve paying for more space than an older couple or an older individual may need. Although most older couples enjoy the increased space created by the empty nest, eventually, as real incomes decline or health expenses increase, maintaining the family home may become a financial burden. Second, elders who become disabled often need to make environmental modifications to preserve their mobility. They may also need a variety of services, including home-delivered meals or personal care from inside or outside the household, either from family members or paid helpers. For example, some adult children coreside with elder parents to provide the care the parent needs. Thus, aging in place is not necessarily the path of least resistance. Indeed, as dependency needs increase, aging in place may become an extremely expensive alternative that is difficult to arrange.

Part of the difficulty of aging in place results from a failure of the housing industry to adopt universal design principles that would allow any housing constructed to be used by adults with a wide range of functional capability. Most housing being constructed today still has stairs at the entries, narrow doorways that will not accommodate wheelchairs, bathroom fixtures that do not provide secure handholds, and counter heights that are not

Retirement community facilities are often designed for easy use by disabled elders.
Photograph by Gordon Baer/Black Star

adjustable. These changes would not be expensive in the long run if universally adopted by architects and contractors. But the building industry has all but ignored these important considerations.

Disability can also bring the need for more in-home services. Almost 20 percent of older people who live alone need help with tasks of daily living. For these people, access to in-home services is related to their capacity to continue to live in an independent household. In communities that have home-aid programs, utilization is usually high. For example, the Elderly Services and Passport programs of Cincinnati provided regular in-home services to more than 1,800 older people in 1995, with more than 250 on their waiting lists. Over the past decade, despite substantial increases in availability of services, demand has continued to outstrip supply.

In 1981, the federal government enacted legislation that allowed states to use Medicaid funds to provide home care for older people who would otherwise qualify for Medicaid-funded nursing home care. By 1987, Medicaid home-care programs were available in forty-one states, but even by 1990, home care under Medicaid accounted for only 3 percent of Medicaid spending (U.S. Senate Special Committee on Aging 1992). In addition, the Older Americans Act and Social Services Block Grants also provide support for home care to low-income elders. Many local communities also are providing tax funds to provide in-home services for low- and middle-income elders. As a result of this stimulus, private-pay home-care programs also grew in number. Over the coming years, more and more older Americans will find that they are able to secure home-care services that extend the period in which they are able to remain in their homes.

Relocation

Relocation is not common among older people at any given point in time. But each year, some elders relocate to pursue a specific retirement lifestyle, and still others move in response to widowhood or increasing levels of disability. Some elders decide to move in response to problems with their dwelling or with their neighborhood, and these are usually local moves. When making a move, housing choices depend on elders' preferences and constraints in terms of geographic destinations, their housing needs, and their financial and functional capabilities.

Let us look at a case in which disability prompted the need to relocate. Mrs. M. was a 72-year-old widow with no children or other family living in the area. She lived alone in an older apartment building. Her neighbors had noticed that she had become increasingly disoriented. She often did not recognize her neighbors or remember their names, even though she had known them for several years. The situation came to a head when Mrs. M. forgot a pot on her stove, which resulted in a great deal of smoke (but fortunately no fire). A protective services caseworker was called to investigate the situation.*

After a thorough medical checkup, it was determined that Mrs. M. was probably suffering from dementia and would be unable to continue to live unassisted on her income of $434 per month (a small Social Security pension plus SSI). There were several choices theoretically available to the caseworker. Adult Foster Care caseworkers could monitor Mrs. M. while she continued to live in her apartment, she could move to a private boarding home or nursing home, or she could move to the county home for indigent older people. None of these solutions was ideal. Mrs. M. was too mentally impaired for in-home Adult Foster Care. With her meager income, no private boarding home would take her. She qualified for Medicaid nursing home care, but there was a waiting list for spaces.

The county home for the aged also had a waiting list. Eventually, it was decided to monitor Mrs. M. at home until she could get into the county home. This short-term solution placed a heavy burden on the Adult Foster Care program, but it eventually worked. Mrs. M. moved into the county home.

Mrs. G.'s case involved moving in response to widowhood. Mrs. G. had lived with her husband in a small town in the Great Lakes region. Their modest home was paid for, and even with the reduction in income following her husband's death, Mrs. G. could still afford to remain there. However, she did not drive and had relied on her husband for transportation. At 82, she did not think she would learn to drive. In addition, she needed help preparing meals and in ambulating around her home. After a few months, even with frequent visits from her adult children, Mrs. G. began to feel a prisoner in her own home and a burden to her children. She missed being with other people her own age. One of her friends told her about Mapletree, a small assisted living facility that was being built in the town. It consisted of 30 one-bedroom apartments, with bath and cooking facilities, plus large common spaces, and a communal dining room and kitchen. Personal care was available, depending on the individual residents' needs. The monthly fees were very affordable for Mrs. G. Three of her friends were planning to move there. She got information about the new facility and discussed it with her children. They wanted her to stay in the family home, but she eventually persuaded them that she needed to be in an apartment building with her friends. Mrs. G. sold her home and was one of the first to move into the new assisted living facility.

Mr. and Mrs. L. relocated in response to a number of factors. First, they were both in their late 60s and eligible to retire from their respective professions. They lived in the Northeast and Mrs. L.'s arthritis was increasingly affected by the cold winters. Their careers had required them to live in a large

---

*Protective services are discussed in more detail in Chapter 15, "Social Services."

metropolitan area, but their preference was to live in a small town. They owned a large apartment that could be sold to provide them with more than enough money to relocate and purchase a very nice home in most regions of the country. They enjoyed doing things together, especially hiking and bird-watching. Mrs. L. was a talented amateur watercolorist who specialized in nature scenes. Mr. L. liked photography, and while Mrs. L. sketched, he wandered around shooting photos. They also liked the arts, so they wanted to live where they would have access to theater and music. They scouted several small college towns in the Appalachian Trail region of Virginia and Tennessee and eventually moved to one in Virginia. They liked their new community, felt that they had made the right decision, and especially enjoyed frequent visits from their children and new grandchild.

Groger (1994) developed a model of relocation decisions that focused on three major contingent questions: whether to move, where to move, and how to accomplish the move. Each of these areas usually involves the older couple's or person's personal perceptions and preferences as well as input from family members. She pointed out that although mentally competent older people are legally entitled to decide for themselves whether, where, and how to move, these decisions often involve intense interpersonal conflicts as the older and younger generations manipulate, persuade, intimidate, and coerce one another. In other cases, the processes are more consensual negotiations as family members discuss, propose alternatives, agree on a course of action, and reassure one another. Groger also pointed out that relocation does not simply involve decisions about whether to move, where, and how, but also decisions about the selection of significant possessions to be taken, belongings to be disposed of, and often what to do about the apartment or house left behind. Although her theoretical model was designed to map decisions concerning nursing home

entry, it could be applied to other types of relocation decisions as well.

Housing Alternatives

Table 13-5 describes a general continuum of housing alternatives for older people. Over 90 percent of older people live in fully independent housing such as detached single-family homes, duplexes, condominiums, manufactured homes, or apartments. Semi-independent housing consists of fully independent housing units in which the residents have access to support services. For example, in some continuing care retirement communities, the dwelling units are fully independent cottages or apartments, but the residents have ready access to meal service, housekeeping, laundry, and other support services, usually for an additional fee.

Assisted living is a relatively new and rapidly growing form of semi-independent housing (Weiss and Applebaum 1994). In assisted living facilities, individuals or couples live in independent apartments with full bath and toilet facilities, kitchen space, lockable doors, temperature controls for heating and cooling the unit, and the residents' own furniture. The buildings usually include dining rooms, group cooking facilities, laundry, and living rooms. In addition to housing, assisted living facilities usually offer four types of services: hotel-type services such as housekeeping, meals, and laundry; personal care services such as help with walking, bathing, dressing, or grooming; routine nursing services such as catheter maintenance or skin care; and special care such as service coordination, monitoring, or behavior management.

A key component of the assisted living concept involves disabled elders in deciding what their needs are and what package of services are necessary to meet those needs. Customized care plans are developed in close consultation with autonomous residents who, unless they are cognitively disabled, have the authority to direct their own care. Unlike

Table 13-5 Levels of housing, by degree of independence.

Housing Type	Significant Criteria
Independent household:	
Fully independent	Household is self-contained, self-sufficient; residents do 90 percent or more of cooking and household chores.
Semi-independent	Household is self-contained but not entirely self-sufficient; may require some assistance with cooking and household chores (for example, an independent household augmented by Meals on Wheels, Homemaker Services, or Adult Foster Care).
Group Housing:	
Congregate housing	Household may still be self-contained but is less self-sufficient; cooking and household tasks are often incorporated into the housing unit. (Common examples are the full-service retirement community and the assisted living facility).
Personal care home	Resident unit is neither self-contained nor self-sufficient; help is given in getting about, personal care, grooming, and so forth. Cooking and household tasks are usually done by paid staff. (A common type is the group home or adult care facility).
Nursing home	Resident units are neither self-contained nor self-sufficient; total care, including health, personal, and household functions, is provided. (A common example is the skilled nursing facility).

nursing homes, which are often required by regulatory agencies to provide care in a standardized fashion, assisted living facilities deliver services to elders in independent households that are constructed to facilitate the delivery of long-term care services. Organizationally, they are structured to enable resident preferences to be a primary guide concerning which services will be provided and how often. Unlike custodial facilities, which assume full responsibility for resident care and as a result tend to be structured in their approach to risk management, assisted living facility managers are freer to negotiate care plans in which the resident assumes some of the risk should the care plan not be successful.

Homesharing can also be a semi-independent housing alternative in cases where a frail older homeowner provides housing and perhaps a small stipend or charges minimal rent in exchange for monitoring and caregiving (Danigelis and Fengler 1990). For example, a frail elder who was afraid to live alone provided a room and kitchen privileges to a college student each year in return for companionship and monitoring. In another case,

a man who was just barely able to manage his household because he was frail and his wife was in a nursing home was matched with an older woman who was willing to do cleaning, cooking, provide companionship, and pay $150 per month to live in his home. Home sharing is not for everyone. It involves being willing to lose a certain amount of privacy; negotiate a new living arrangement, often with a stranger; and develop agreements about ground rules.

About 6 percent of elders live in *group housing* of all types, but as age increases, so does the probability of living in group housing. About 25 percent of those 85 or over live in group housing.

In older cities, many older people live in residential hotels—often called SROs because they consist of single-room occupancies. SROs are usually just group housing, with no formal service component. The stereotypical SRO is a small, run-down hotel catering almost exclusively to elders. However, in their study of SROs in Manhattan, Cohen and Sokolovsky (1980) found that at least one-third were larger hotels (110 rooms or more) of relatively better quality. Erikson and

Eckert's (1977) study in San Diego also reported a few SROs that catered to middle-class clients. Thus, SROs cater to a wider socioeconomic category of urban elders than has been generally assumed. Residents of SROs have also been stereotyped as social isolates, "loners" who lack the support systems found among the elderly at large. However, Cohen and Sokolovsky found that although most SRO residents had indeed lived alone for many years and had fewer contacts with others compared with other urban elders, they were far from isolated, and their contacts both within and outside the SRO were important sources of support. In recent years, central cities have been expanding, resulting in the demolition of many old hotels, which has drastically reduced the supply of SRO housing, especially for the poor. For example, Singelakis (1990) reported that from 1980 to 1988 alone, SRO housing in Manhattan declined by 51 percent.

Many older people who live in group housing are similar to Mrs. M., whose case was cited earlier. They need housing that provides room and board, personal care, and supervision, but they do not need nursing care. *Board and care homes, group homes,* or *adult care facilities* are terms used to describe congregate housing designed to provide shelter, food, monitoring, and a minimal degree of personal care to people who are not able to manage independently but who also do not need the level of care provided by nursing homes. Middle-income older people can find such housing with relative ease, but the 'Tweeners usually cannot afford open-market housing of this type and the poor have trouble because Medicaid financing is not available for group housing and SSI usually does not cover the cost of adequate care in group quarters. Congressional hearings have periodically documented unsafe and unsanitary conditions in group homes, and in 1976 Congress passed a law requiring states to set standards for admissions, safety, sanitation, and protection of civil rights for group

homes where SSI recipients lived or were likely to live (Reichstein and Bergofsky 1983). Nevertheless, because there is a lack of funding for enforcement and, more important, for adequate services in group homes, the standards the states generally developed have been loose and ineffective. Reichstein and Bergofsky recommended the development of a model set of standards that all states can adopt. But until an effective means to pay for group home living for the elderly poor is developed, the situation cannot be expected to improve very much. Indeed, Benjamin and Newcomer (1986) found that the single strongest predictor of the supply of group home space within a state was state supplementation of SSI for residents of group homes.

Despite the poor image the public and policy makers tend to have of group homes, most research studies have concluded that the majority of these facilities provide needed and important long-term care services in a homelike environment at a relatively low cost (Applebaum and Ritchey 1992). Eckert and Lyon (1991) found that two-thirds of older residents of group homes needed help with activities of daily living and more than half were confused to at least a mild extent. The median income of the residents was less than $3,000. Applebaum and Ritchey found that most group homes were small, with about 75 percent having six or fewer residents. Cost of living in group homes was modest, with 61 percent charging less than $700 per month. As states struggle to find ways of providing facility-based long-term care that is less expensive than nursing home care, the number and size of group homes may well increase, but so will the challenges of monitoring and controlling the quality of care provided (Applebaum 1992).

Nursing facilities are usually the most restrictive housing alternative available for older people. Nursing facilities usually consist of residential rooms, often shared, that are neither self-contained nor self-sufficient.

Women caring for women is very common in long-term care.
Photograph by Marianne Gontarz

Nursing facilities are very hospital-like in their construction and appearance, and the approach to personal care, health care, meals, and activities tends to be institutional and only marginally flexible. In 1995, there were more than 25,000 nursing homes in the United States, up from 11,000 in 1976. Nearly 85 percent of their 1.5 million residents were 65 or older, and over half were age 80 or older. Thus, it is an accurate perception that nursing homes are primarily housing for the very old. In addition, nursing facility residents tend to be quite disabled, with most being impaired on at least three activities of daily living.

Part of the dread elders experience at the prospect of entering a nursing home is based on an outdated picture of it as the "last home for the aged." However, in recent years the high cost of nursing facility care, coupled with the development of less costly and less restrictive alternatives to nursing home care, resulted in a declining length of stay. Many nursing home entrants now stay only a short time to finish recuperating from a hospital episode or to complete a program of rehabilitation, and then return to their community household. However, there are still many nursing facilities throughout the country that are oriented primarily around providing long-term personal care for severely disabled old people.

Thirty years ago, nursing homes in America probably deserved their negative image as depressing warehouses for the afflicted. Today, that image is largely inaccurate. Although problems have by no means been eliminated, the quality of care provided in most nursing homes has greatly improved. A good part of that improvement resulted from two facts: (1) Medicaid finances most nursing home care of older people, and (2) facilities must conform to federal standards for care quality in order to be reimbursed by Medicaid. Another large part of care-quality improvement has come from the steady development of professional standards of care within the long-term care industry.

Relocation Problems Within Group Housing

Even those congregate facilities that cater to middle- and upper-income older people are not without their problems. One of the most difficult problems is developing humane procedures for dealing with residents who become disabled to the point of no longer fitting the independent or semi-independent nature of the facility. Take, for example, the York Houses in Philadelphia. They were established in the early 1960s, and the initial residents were generally independent. As expected, as the residents aged, their needs for assistance increased. In addition, as the average level of functioning among the residents declined, the facilities attracted fewer fully independent older people. If they continued in this direction, the York Houses would eventually become nursing homes, a use for which the facilities were not well suited. To maintain their status as semi-independent housing, York Houses would have to develop "skills and sensitivity to counsel tenants for whom transfer might be sought" (Lawton, Greenbaum, and Liebowitz 1980). As of 1995, the York Houses were still temporizing by providing a great deal of in-home service. Very few of the new residents were under 80.

Bernstein (1982) studied the process of transferring residents from independent living in eighty-seven federally assisted housing projects. She found that project managers seldom had to ask tenants to leave; usually the tenant or the tenant's family recognized the need for a more supportive setting and took responsibility for finding alternative housing. When the tenant suffered severe illness, needed frequent medication monitoring, or was bedridden, a voluntary move was most likely. Mental decline, emotional problems, being a safety hazard to oneself or others, or having a problem with alcohol or drugs were problems less likely to be acknowledged by either tenants or their families and more likely to require more active management intervention. Most facilities relaxed their retention criteria and allowed tenants more problems than a new applicant would be allowed. However, the higher the concentration of older people in the facility, the stricter the retention policy tended to be. In addition, on-site administrators, who would have to deal with complaints about inequity, tended to be stricter than absentee management staff. Most facilities have only very generally stated policies and a great deal of flexibility in deciding when a tenant should be asked to leave. However, much public housing for elders is relatively new, and as the older population ages, the resolution of tenant retention issues is going to become more problematic.

Relocation issues are made even more difficult by the lack of alternative housing arrangements. It is difficult to tell tenants they have to move when they are frail and there is literally nowhere for them to go. This problem, of course, relates to a lack of semi-independent housing, board and care facilities, or personal care homes in most communities.

As a result, the main housing alternative for many older people who must leave independent housing is a nursing home, which older people tend to view very negatively. Although many older people recognize that a nursing home may be best for people who cannot care for themselves, most fear the loss of independence and privacy that moving to one can bring and also fear that their children will desert them once they are in a "home." Despite these fears, many older people choose to move into nursing homes.

But even when the image of the nursing home is not a problem, dread of relocation may be. The research on the effects of relocation has not produced consistent answers as to whether relocation increases the probability of death or worsens illness. However, it is clear that many older people are upset at the prospect of relocation. Borup (1981) suggested a few general reminders that can help ease the transition:

- Relocation stress is usually temporary.

- Unwilling movers are most likely to experience stress.

- Stress can be reduced by making the new environment predictable to the mover.

- Keeping tenants and their families informed tends to reduce stress.

Residents of nursing facilities also often move from one room to another for a variety of reasons, some of which may be beneficial (such as reducing conflict between roommates) and some of which may not (such as movement to a special care unit). However, little research has been done on this phenomenon (Mirotznik and Lombardi 1995).

Retirement Housing

Retirement can bring an opportunity for people to move away from housing or a community to which they were tied not by preference but by employment and into one more in tune with their ideals. Some choose planned **retirement communities;** others move to communities not actually planned for retirement but that simply attract large numbers of retirees.*

Retirement communities' residents tend to be homogeneous with respect to social class, race, and ethnic backgrounds. And although there are retirement communities for people at most points on the social spectrum, the majority draw residents from the more affluent social classes (Streib et al. 1984b; Somers and Spears 1992). Movement to retirement communities also seems to self-select those older people who are amenable to life in age-homogeneous settings. As a result, morale in retirement communities is quite high. Conversely, there are indications that many older people who remain in age-integrated communities would not be at all happy in a retirement community.

The homogeneity of retirement communities fosters frequent social interaction and the formation of new friendships. Because retirement communities draw heavily from former managers and professionals, their members are often quite accustomed to moving and have developed the skills needed for making new friends. For the vast majority, the move does not bring increased isolation from their children. Although life in a planned retirement community may not be for everyone, there is no evidence that it is detrimental for those who do choose it.

The lower-cost equivalent to the more affluent retirement community is the mobile home park for older adults. Early-on, Johnson (1971) concluded that the mobile home park is a mixed blessing for retired residents. Most older people who lived in mobile home parks wanted to live in a controlled environment without having to pay a high price for it.

Slightly authoritarian in outlook, justifiably fearful of urban violence, and anxious to maintain all the visible outward signs appropriate to "decent people," the retired working-class mobile home resident wants to live in a park that is neat, attractive, quiet, safe, all-white, and friendly.

(1971,174)

These people were willing to pay a park owner to control the neighborhood, but they complained bitterly if the controls infringed too much on them. All too often, the cost of living in a mobile home park also included considerable economic exploitation by park owners. Nevertheless, the popularity of such parks in many areas of the country attests to the felt need for living areas that provide a safe and secure environment for a community made up largely of older adults.

---

*Retired people can be referred to as "retirers" or "retirees." Since most people retire on their own initiative, "retirer" is probably more appropriate but harder to say than "retiree."

Golant (1992) reported that by the 1990s, the term "manufactured homes" had replaced the old concept of the singlewide mobile home. Many retirement communities today are made up of "doublewide" manufactured homes that provide an average of 2,000 square feet of living space. These homes are often placed on permanent foundations on relatively large lots that are landscaped and located in attractive surroundings. In most respects, these homes are not distinguishable from conventionally constructed homes. Since 1976, federal standards for manufactured homes have increased quality to the point that, if located on a permanent foundation, they are comparable to conventionally constructed homes, especially in terms of susceptibility to fire or wind damage.

The advantage of manufactured homes is that they cost 20 to 30 percent less than a comparable site-built home (Golant 1992). However, they also depreciate more than conventionally constructed homes. In addition, elders who move to retirement communities made up of manufactured homes may not own the land on which the home is located. Many communities lease the lot to the homeowner. In such cases, if the land is sold, homeowners can find themselves faced with costs well in excess of $10,000 to move their homes. Nevertheless, manufactured retirement residences remain a popular alternative, especially among those who maintain residences in two parts of the country.*

Longino (1981) reported on a study of three types of retirement housing: planned, **full-service retirement communities;** planned public housing for elders (nearly all of whom are retired); and **de facto retirement communities** in certain regions of the country that result from large-scale migration of retired people to small towns. Those who moved to planned, full-service retirement

communities tended to be older (mean age 76), upper-income individuals who were oriented around the availability of housekeeping, meals, and personal care services. Those who moved to de facto retirement communities tended to be younger (mean age 68), upper-income couples who were oriented toward independence and enjoyment of nature. Movers to public housing tended to be low-income older women (mean age 78) from the immediate area who wanted to be out of an unsafe neighborhood and to take advantage of the financial relief provided by subsidized housing. Longino's study supported the idea that older people select housing that meets their needs. He predicted that those who moved to planned communities probably would not move again because they had selected housing suitable to the needs of the old. However, he predicted that those who lived in de facto retirement communities would resemble people on an extended vacation who, when needs for assistance increased substantially, could be expected to return to their area of origin to be near family.

Retirement Nomads

As an extension of the working-class camping culture of the 1960s, the 1980s saw the emergence of a population of young-old retirees who lived full-time in motor homes and travel trailers. Hartwigsen and Null (1989 1990) estimated that at least 350,000 mature adults had adopted a nomadic lifestyle and lived exclusively in their recreational vehicles (RVs). Hartwigsen and Null's sample of RV nomads ranged in age from 45 to 81, with a mean of 63 for men and 61 for women. Over 90 percent were married couples. These people did not see themselves as being on vacation; they had no other residence to return to since most sold their homes to purchase their recreational vehicles. They traveled about 12,000 miles a year on average, staying roughly a week or two at each stop. They enjoyed sightseeing and natural settings. Most had incomes in the $15,000 to $30,000

---

*See Golant (1992) for more details on retirement communities made up of manufactured homes.

range, which meant that they had to watch their expenses closely. They tended to travel a circuit of campgrounds that have amenities such as electrical, water, and waste hookups; swimming pools; and laundry facilities.

To the respondents, a major advantage of the RV lifestyle was affordability. Prices of fully self-contained, furnished RVs started at $40,000, and their construction requires minimal upkeep. Another advantage was variety and flexibility. Didn't like the neighbors? Moving was easy. Many also used this flexibility to visit family scattered around the country. Being able to live most of the time in places with good weather was also seen as a major advantage. Many retirement nomads traveled in small caravans consisting of a few couples. There was an established culture around this nomadic lifestyle, complete with instructional literature on how to deal with issues such as mail and banking.

The main disadvantages of the RV lifestyle predictably revolved around health issues. Getting health care in an unfamiliar community can be difficult, and the onset of long-term care needs usually signaled the need for a new lifestyle. Nearly 70 percent of Hartwigsen and Null's respondents expected to give up their nomadic retirement eventually; and most expected to move into a detached home or mobile home in a community they had already identified.

Types of Housing Problems

Many older people cannot find housing that is both suitable and affordable. There is a shortage of subsidized low-rent housing for elders, particularly outside the major metropolitan areas. For example, Golant (1992) reported that although the numbers in the older poor population have increased, the number of subsidized rental units has declined; the income threshold needed to qualify for rental subsidy was lowered to well below the poverty level; and the share of rent older poor people were required to pay was increased from 25 percent to 30 percent of income. All of these trends meant less availability of housing for low-income elders. For the middle class, the situation is not much better. For instance, Mr. and Mrs. N. wanted to sell their home and move into an apartment, but could not find a buyer because their home was in a neighborhood going through a transition from white to African American. In addition, the available apartments in their community either were too expensive or were cheap and shabby, with virtually nothing in the middle. They decided to stay in their home, at least for a while, even though it had much more space than they needed or wanted to take care of.

The majority of older people probably do not want to move and do not need in-home services yet. However, they do need to maintain their homes. Because their homes tend to be older and more dilapidated than the housing of the young, more maintenance is required, a problem that translates into a need for funds to pay for maintenance and the people to do it (Reschovsky and Newman 1991). In 1979, the state of Ohio began offering home maintenance and repair services to older people in rural areas. In the counties that instituted such programs, the demand for services was nearly four times the capacity. This finding indicates that a main reason that older people do not maintain their homes is that they just cannot afford to. Reschovsky and Newman came to a similar conclusion.

Another important factor involved sponsorship. The construction industry, especially in renovation and repair, is composed mostly of small-scale entrepreneurs and companies that often have not been in business long. Various confidence rackets involving home improvements have been widely publicized. As a result, older people are not sure with whom to deal or whom to trust. With government-sponsored programs, older people were more willing to allow workers in their homes.

Finding temporary or emergency housing during the transition from independent living

Homeless elders sometimes live in their automobiles.
Photograph by Robert E. Daemmrich/Tony Stone Images, Inc.

to a congregate living facility can be a major challenge. Often there are waiting lists for the better group quarters, and there are precious few places where one can live temporarily and receive the needed support services while waiting. The same type of problem can arise for people who lose their housing as a result of fires, eviction due to their apartment building being converted into another use, and so on. The problem is finding somewhere to live while trying to make alternative living arrangements. Mellinger (1989) discovered that a program which found temporary emergency shelter in private homes for up to four weeks while displaced elders tried to find new housing arrangements resulted in a lower likelihood of institutionalization.

Although reliable statistics on homelessness do not exist, it is clear that the number of homeless is large and that it increased dramatically throughout the 1980s. Based on a sample of people who used shelters and meals programs in twenty cities, the Urban Institute

estimated that in 1988 there were at least 600,000 homeless Americans. Estimates of the proportion of homeless who are 60 or over range from 15 to 27 percent (U.S. Senate Special Committee on Aging 1989a). Thus, the number of older homeless Americans could be conservatively estimated at 100,000 in 1988. The causes of homelessness among older people include deinstitutionalization of mentally disabled older people, dramatic increases in rents, destruction of SROs, and a shrinking supply of low-income housing.

Who Has Housing Problems?

Older people who are most likely to have housing problems include the frail, the disabled, rural dwellers, people with incomes just above the poverty level, and the poor. The frail and the disabled are apt to find that they can get neither the in-home services they need to remain independent nor suitable group housing that can meet their needs for assistance within their capacity to pay. Middle-income older people face a severely limited housing market in their price range, so they are apt to be locked into their current housing. Low-income elders find that there is a shortage of low-rent housing for older people. At the root of all three problems is financing; our system for financing housing development relies heavily on private investment, and housing for middle- and low-income elders does not offer a potential rate of return on investment that is attractive to investors. Although government incentives have been used in the past to increase the supply of middle- and low-income housing, these programs were scaled back drastically first in the 1980s and again in the 1990s. In addition, the rental housing markets of many major metropolitan areas have undergone profound changes. As landlords sold apartment housing to condominium associations, many older residents found themselves unable to stay in their existing housing. To find affordable housing many were forced to relocate to other cities. However, there has been little research on the extent of this pattern, but as Golant (1992) reported, problem-free housing alternatives are the exception and not the rule.

Public policy has important effects on the housing problems of older Americans. First, the federal government has backed away almost entirely from low-rent housing construction and mortgage guarantee programs aimed at increasing the availability of low-income housing for elders. In addition, federal commitments to rental subsidy programs for very low income elders have also been scaled back considerably. Every national effort to create more units of affordable housing for older Americans has been eliminated or scaled back, and these programs have not been replaced by state or local programs. Second, a number of federal changes in response to fair housing and handicap legislation have required that more units of housing formerly earmarked for elders be made available to "nonelderly" poor people (Filinson

1993), which has further reduced the public housing available for older Americans.

Given that the private sector has not responded to the housing needs of middle- and low-income elders and that government has abandoned its effort to increase the amount of affordable housing, we can expect that housing problems among elders will become increasingly widespread.

Summary

In industrial societies, most people rely for income on earnings during their working years and on pensions and savings in retirement. Adequacy of income greatly influences quality of life because monthly income is the vehicle through which other instrumental life needs such as housing, food, clothing, transportation, and so on are met. For most people, income peaks during the later years of employment, usually age 45 to 54, and tends to decline thereafter, especially after retirement. Despite the fact that Social Security benefits are indexed to rise with the cost of consumer goods, they do not keep pace with general increases in level of living. As a result, the level of living that Social Security retirement pensions supports tends to gradually decrease with age. In addition, most employer pensions are gradually eroded by inflation. Most elders own their homes free and clear, but otherwise they have very modest savings (under $20,000 on average) and tend to conserve savings for emergencies rather than use them to meet everyday expenses. Within the older population, expenses for health and home maintenance tend to increase with age. The end result is that most elders enter retirement with incomes only modestly above the poverty level, and as each cohort ages, the proportion whose incomes fall below the poverty line increases.

There are also significant categories within the older population that have severe income problems, especially widowed or divorced older women living alone and minorities. Regardless of how income needs are measured—poverty status, expenditure budgets, or preretirement earnings replacement—older women, especially minority women, living alone are very likely to have increasingly inadequate incomes as they grow older.

The main direct income sources for the older population as a whole are Social Security retirement benefits, asset income, earnings from employment, and pensions. But these data are somewhat misleading because income sources vary considerably at each level of living. Low-income elders tend to rely on Social Security and Supplemental Security Income. Middle-income elders rely on Social Security retirement benefits, employer pensions, and some earnings and savings income. High-income elders rely mainly on earnings and savings income, with employer pensions playing a less important role and Social Security benefits accounting for only a very small percentage of income. As a significant income source, earnings all but disappear by age 85. Only a very small proportion of elders receive income from means-tested sources such as Supplemental Security Income or other public assistance. Improved Social Security benefits have been mainly responsible for the declining proportion of elders on SSI since 1975. Nevertheless, in 1991, about one million older households were living at or below the poverty level.

Various experiments designed to help elders use home equity to finance current consumption, such as home equity conversions and accessory apartment conversions, have not been widely successful, except among those whose homes are worth $250,000 or more.

The pension and savings systems that were created to provide retirement income are structured in such a way as to preserve into retirement the same patterns of income inequality that are associated with gender and minority status differences in earnings from employment. The privileged upper-middle class labor force is predominately made up of white men, and most of them gain entitlement to generous retirement benefits from tax-sheltered savings and employer pensions and severance packages. These fortunate individuals also enjoy above average Social Security retirement pensions. Minority men are more likely to have to get by on employer pensions and average Social Security pensions. Women are more likely to end up relying almost entirely on very modest Social Security survivors' or retirement pensions.

Income security depends on having a predictable source and amount of income. As a result of age discrimination in employment, most elders have only limited opportunity to adjust to changing economic circumstances by increasing their income from job-based earnings. Most sources of income in later life, including Social Security retirement benefits, can be eroded by economic changes such as increases in the general level of living, inflation, and declining interest rates. In addition, as a result of declining access to generous employer pensions since 1980, upcoming cohorts will enter retirement with much less pension wealth than those who entered retirement in the 1980s.

In 1995, Congress considered a number of sweeping changes to federal programs that would lower incomes of older Americans by taxing and reducing Social Security benefits and at the same time increase the share that older people would have to pay out-of-pocket in order to gain access to health care through Medicare. These changes would increase the proportion of older people below the poverty line and increase substantially the number qualifying for income assistance. However, there may be no funding to provide public assistance for the growing number of elders who would need it. These changes would represent a major step backward in income security for older Americans, especially those with middle and lower incomes.

Housing is obviously linked to income. It takes income to rent, purchase, or maintain housing. Level of income strongly influences the affordability of various housing options, which in turn have a great deal to do with the geographic and social context within which aging takes place.

When people experience changes in their housing needs as they age, they have two primary choices. They can age in place or they can move. Most elders prefer to age in place and in fact do so, but aging people who experience increasing levels of physical and mental disability find that this is often difficult to do. Aging in place often requires extensive and expensive home renovations because private housing lacks universal design principles that are accessible to disabled people. Also, aging in place can require the purchase of increasingly expensive in-home personal care services that can eventually outstrip income and overburden the informal caregivers who usually supplement formal caregiving.

Relocation can be caused by increased disability, a desire to live in a different climate, a desire to live closer to adult children, or a need to escape neighborhood deterioration and increasing crime. Most moves probably involve negotiations within families about whether to move, where to move, and how to accomplish the move.

Over 90 percent of elders live in independent households, about 6 percent live in nursing homes and homes for the aged, and another 3 percent live in a variety of group quarters such as board and care homes and single-room occupancy hotels. Assisted living, which combines the householder autonomy of independent households with easy access to services, is the fastest-growing housing option for disabled older adults and their spouses.

The large variety of housing options for elders includes free-standing homes, duplexes, condominiums, manufactured homes, housing subdivisions restricted to older residents, assisted living apartments, continuing care retirement communities, rent-subsidized housing for low-income elders, board and care homes, single-room occupancy hotels, and nomadic residence in recreational vehicles ranging from modest trailers to palatial motor homes. But despite this vast array of possibilities, most middle- and low-income elders find their housing choices very limited. There is a particularly strong need for emergency housing for elders who lose their housing due to fires, eviction from apartments converted to another use, or other misfortune. At least 100,000 older Americans are homeless at any given point in time.

Housing problems stem at least in part from our heavy reliance on private enterprise to construct middle- and low income housing. The federal government has backed away from almost all housing construction and is gradually terminating its contracts for rent subsidy for elders, which has been a significant incentive to private developers. Without government rent and loan guarantees, developers are reluctant to build new housing or renovate older housing stock for middle- and low-income older renters. Expect housing problems of older Americans to increase significantly in the near future.

14 ■ Health Care and Long-Term Care

Most older adults do not suffer from serious health problems or from disabilities that result in a need for personal care. But with aging, an increasing proportion experiences multiple chronic conditions, periodic acute episodes connected to chronic health conditions, and multiple functional impairments that require personal care. In the first part of this chapter, we examine the need for health care, how it is affected by age, and how the nature of the health-care system influences access to health care and its adequacy. Then we look at the need for long-term care, how it is affected by aging, and how adequacy and access to long-term care are influenced by our systems for providing long-term care. We then examine various models of care delivery and financing and their influence on the availability, cost, and quality of care. We finish the chapter with a look at the effects of advocacy, regulation, and planning on the availability, cost, and quality of health care and long-term care.

Obviously, health care and long-term care are connected by the underlying physical and mental conditions that give rise to the need for care. But we have two very different systems for delivering *health care*, which emphasizes short-term hospital and physician care, and for delivering *long-term care*, which usually emphasizes ongoing personal care more than health care.

Health Care

Good health is a major factor enabling people to enjoy life. As we saw in Chapter 4, most older people are not seriously limited by illness or disability. But aging does increase the proportion of people who have chronic diseases that require effective management. Adequate health care can enable older people to maintain good health and avoid worsening health. Preventive health care can reduce the probability of chronic disease and disability, treatment can often reverse negative effects of chronic disease, and rehabilitation can help people restore lost functions as well as compensate for unrestorable functions.

Health-Care Needs

In many cases, people manage their own health care. They identify behaviors that offer the promise of maintaining good health or preventing disease and incorporate them into their lifestyles. When they experience symptoms of illness, people often respond first by treating themselves with bed rest or over-the-counter medications; only if these actions are ineffective do they seek formal health-care services.

Preventive self-care involves risk avoidance, establishment of good health habits, health monitoring, and information seeking. Risk-avoidance behavior includes using automobile seat belts, installing smoke detectors in the home, and using safety equipment such as grab rails in the bathtub. Health-promoting habits include regular physical and mental exercise, stress-reduction behavior such as meditation, avoiding tobacco, attention to diet, maintenance of proper weight, and adequate sleep. Monitoring blood pressure, body weight, and blood cholesterol and regular self-examination of the body to detect quickly conditions that may need treatment also promote good health. Information seeking includes reading articles and watching television programs about health; discussing health issues with friends; and asking questions of physicians, pharmacists, and other health-care providers. In a national sample, Bausell (1986) found that older people were more likely than other adults to engage in preventive health behaviors, especially watching their diets and having regular blood pressure checks. However, only 23 percent of older adults and 36 percent of other adults exercised at least three times a week, only about 25 percent of all adults maintained proper weight, and only 20 percent wore seat belts.

Self-care is an inclusive term that involves taking actions to promote wellness or prevent disease or disability, monitoring bodily states, changing activities or environments to compensate for health changes, diagnosing possible causes of physical or mental symptoms, consulting with family and friends about the meaning of symptoms, identifying possible treatments, engaging in various forms of self-treatment, deciding when professional help is needed, and blending professional and self-treatment.

Self-treatment is common. Between 70 and 90 percent of acute conditions such as common colds are self-treated (Chappell 1987). Most drugstores carry numerous preparations for self-treatment of aches and pain, colds, allergies, constipation, insomnia, indigestion, hemorrhoids, skin rashes and itching, and so on. Indeed, many chronic ailments such as minor eczema or arthritis pain can be effectively treated with over-the-counter medications. Kart and Engler (1995,435) reported that "self-evaluation of symptoms and self-treatment are the basic and predominant form of primary health care." Self-treatment is far more common than care prescribed by physicians or other health professionals (Eisenberg et al. 1993).

Norburn et al. (1995) found that more than 75 percent of elders reported having changed their behavior as a form of self-care. Among elders, common conditions that are seen as most appropriate for self-care, without physician consultation, include dry skin, pain or stiffness in joints, leg cramps at night, feeling cold much of the time, feeling anxious much of the time, loss of appetite, difficulty sleeping, and frequent indigestion (Kart and Engler 1994, 1995). Many elders attribute some of their physical or mental symptoms of dysfunction to normal aging and, consequently, do nothing about them or wait for nature to take its course (Stoller 1993).

Many people use self-treatment in combination with physician-recommended care and usually do not tell their physicians about their self-care measures. Elders who feel that they can control their own health are the most likely to actively use self-care as their first line of health promotion and disease prevention and treatment. Dill et al. (1995) reported that people develop self-care strategies based on their experiences throughout life and tend to maintain them into later life, as we might expect from continuity theory.

When people are sick, family and friends often help provide care at home and assist the ill person in recognizing symptoms, making diagnoses, and deciding when expert care is needed. However, social support networks are not always a positive resource. Different members of the social network can provide inconsistent or contradictory advice, which can lead to procrastination or conflict.

In adulthood, as age increases, so does the severity of chronic and acute illnesses and the probability of seeking the care of a physician. In 1993, the average number of physician visits per person was 3.9 at age 45 to 64 compared with 5.7 at age 75 and over (National Center for Health Statistics 1994a). Older people generally have to go to the physician's office; house calls fell from 23 percent of all physician visits in 1959 to less than 3 percent in 1982.

Home health care from registered and licensed practical nurses has increased dramatically in recent years. For example, Medicare home health visits increased from 11 million in 1975 to 38 million in 1986 (U.S. Senate Special Committee on Aging 1989a). This increase reflects increased demand for home health services, coupled with a perception that Medicare costs could be partially contained by liberalizing access to home-delivered skilled nursing benefits. Many states also have substantially increased availability of Medicaid for in-home health and personal care services to older people needing a nursing home level of care.

Older people are also more likely than the young to use short-stay hospital services. Because the number of older people has been growing rapidly, the number of admissions of older people to hospitals grew by more than 50 percent from 1965 to 1986, but the age-adjusted rates of hospitalization changed very little. In 1986, older people (12 percent of the population) accounted for 31 percent of all hospital stays (U.S. Senate Special Committee on Aging 1989b). Hospital admission rates increase markedly with age, with those over 85 having rates double those for people 65 to 74. However, since 1983, the average length of hospital stays by older people has decreased because of the Medicare prospective payment system, which limits the number of days' reimbursement for various conditions requiring hospitalization.

Utilization of nursing homes is also much higher among older people than among those in other age categories. For example, in 1985 there were 1.3 million older people living in nursing homes, compared with 173,000 people under 65 (National Center for Health Statistics 1989b). Nursing home residence rates rose from 1.3 percent at age 65 to 74 to 22 percent at age 85 and over. Shapiro and Tate (1988) studied which older people were most at risk of institutionalization. They found that those 85 and over who had no spouse living with them, who had been hos-

pitalized recently, who lived in retirement housing, who had one or more ADL* limitation, and who had mental problems had a 62 percent probability of entering a nursing home within thirty months of the interview. Among those 85 and over who lived with a spouse, who had not been hospitalized, who did not live in retirement housing, who had no ADL limitation, and who had no mental problems, the probability of entering a nursing home within thirty months was only 4 percent. Thus *disability, not age, is the main predictor of nursing home residence.* Nevertheless, at all levels of disability there are about three disabled older people living in the community for every disabled older person living in a nursing home (U.S. Senate Special Committee on Aging 1989).

As noted earlier, many middle-aged and older people have more than one chronic condition requiring long-term treatment. As we saw earlier, long-term care has both personal care and health care components. Long-term health care includes the long-term management of illness and disability (Koff 1982). Thus, long-term health care is possible in both community and institutional settings. However, the specialization of medicine discourages the coordination required to effectively manage multiple chronic conditions in the community. As a result, an older person with several different conditions may be going to two or three doctors, often with no communication among the physicians. Effective long-term care in the community requires **case management,** which involves looking at all of a person's needs and resources and developing a coordinated treatment plan that includes a variety of services, including health care.

---

* Activities of daily living, such as bathing, dressing, and feeding oneself. These tasks are generally used to determine the need for a nursing home level of care, which often qualifies a person for home care reimbursed by Medicaid.

The older population is the heaviest user of physicians' services, hospitals, personal care services, home health services, and nursing homes. As a result, the number and types of health-care services available in a community have a stronger impact on the quality of life for the older population than for those in other age categories.

As age increases, so does the probability that a person will face catastrophic out-of-pocket costs connected to health care and/or long-term care. For example, Liu, Perozek, and Manton (1993) used data from the 1982 and 1984 National Long-Term Care Surveys to document the proportion of elders who faced catastrophic health and long-term care costs. They found that among community-dwelling older people, the proportion experiencing private care costs exceeding 20 percent of their income increased substantially with age, from 21 percent at age 65 to 74 to 34 percent at age 75 to 84 to 41 percent at age 85 and over. Among those living in nursing homes, the out-of-pocket cost of care exceeded 30 percent of income for 75 percent of residents.

Types of Health-Care Problems

Older people are likely to find that the kind of care they need is unavailable, and the kind ·of care they are offered may be inappropriate. No one may be willing to manage the long-term treatment of their multiple chronic conditions. They may be plagued by problems resulting from interactions among multiple medications used to treat multiple chronic conditions. Most older people also find that our system for financing health care is inadequate.

Mrs. C.'s case illustrates the problem. Mrs. C. was 76 and lived alone. She was in good health generally, but had come down with the flu. She tried to manage on her own, but reached the point where she could not get out of bed to fix meals or go to the bathroom. She called her doctor and instead got the receptionist, who insisted that Mrs. C. come into the doctor's office. When Mrs. C.

replied that she could not get out of bed, much less come to the office, the receptionist instructed her to call the emergency squad and go to the hospital emergency room because "that's what emergency rooms are for."

Mrs. C. did as instructed, albeit more than slightly embarrassed to be carried out of her building on a stretcher "all on account of the measly flu." After a two-hour wait at the hospital, she was seen by the emergency room physician, a complete stranger who knew nothing of her medical history, which included two chronic conditions and five regular medications. He diagnosed flu and told her to go home and stay in bed. Almost in tears, she told him about being unable to care for herself at home, to which he said that there was nothing to be done about it because she was not sick enough to require hospitalization, and Medicare would not pay for it.

Mrs. C. could not afford to pay $1,225 per day to finance her own hospital stay, yet she needed around-the-clock care for a few days. In addition, nursing homes usually do not admit short-term residents, and the health care financing system makes no provision for personal care. In desperation, Mrs. C.'s sister, who lived more than 200 miles away, came for a few days, during which time Mrs. C. recovered. What would have happened if Mrs. C. had not had family?

Mrs. C.'s problem was partially a result of the way care is financed, but more important perhaps is the underlying philosophy that pervades our entire health-care system about what illness is and how it should be treated. The **medical model** assumes that there are only two kinds of people to be treated: those who are seriously ill and need hospitalization, and those who are up and about and can go to a doctor's office or clinic. Obviously, incapacitation is a matter of degree, and our system for treatment needs to recognize this fact.

In October 1983, in an attempt to control costs, Medicare implemented a prospective payment system for hospital charges based on Diagnostic Related Groups (DRGs). Medicare hospital reimbursement was tied to the average length of stay and the hospital procedures indicated in the DRG category into which the patient's illness was classified. The DRG system was derived from twenty-three major diagnostic categories that represent the major organs of the body. The major categories were then divided into 467 DRGs, which were differentiated by principal diagnosis, types of procedures usually performed, secondary diagnosis, discharge disposition, and age. DRGs are medically meaningful in that they attempt to group illnesses and procedures that fall together naturally in the practice of medicine.

However, within a given DRG, the resources and hospital days required by an array of cases may vary considerably. Horn et al. (1985) found that because the DRGs do not take severity of illness into account, reimbursement to hospitals that get a high proportion of severe cases may fall short by as much as 35 percent. Obviously, if the prospective payment system is not covering the costs of care, hospitals may be motivated to reduce lengths of stay and limit procedures to those for which expenses will be reimbursed. DRGs often result in older patients being released from hospitals "quicker and sicker," resulting in higher nursing home and home health agency use and heavier reliance on family caregiving (National Center for Health Statistics 1989b). Fischer and Eustis reported the following case studies:

[A daughter about her mother:]

they only had her up one day before they sent her home. That day they gave her light broth and soft Jello and she was not ready to go home. She was so weak. They hadn't walked her for a week. She barely sat up and it was very hard.

(Did they tell you why she was going home?)

Well, just because Medicare doesn't want to pay. They'd just like to shove them out.

[Another daughter about her mother:]

Then, you see, she was no longer on the Medicare after 9 days in the hospital because she was eating and eliminating on her own. No matter that she was in terrible pain and under heavy Demerol and couldn't get out of bed—out you go.

[A son about his mother:]

They almost sent her home half dead because she didn't qualify for Medicare. . . . They scared the living daylights out of me about when they were going to kick her out on a Thursday and they told me that they would discharge her at 8 P.M. and I said there was no way we could handle it. She's sick.

(1988,385)

Fischer and Eustis found that since the implementation of DRGs, the need for both family caregiving and formal home care has expanded. They learned that families were having to manage many of the concrete hospital discharge planning issues for older people, who often were too ill to do it themselves. Unless there is someone at the hospital to fight for proper treatment, older people can be discharged too soon, and their family must usually arrange any in-home services needed. Family monitoring of home care is also an important aspect of ensuring that needed services are provided. Thus, uncompensated managerial services by family are a major reason why the Medicare program saves money through DRGs; if the cost of this time were included, the savings would be considerably less.

Although an estimated 15 percent of the older population needs mental health services to help them cope with depression or chronic illness, only 2 to 4 percent of the clients of psychiatric clinics, clinical psychologists, and community mental health centers are older people. However, rates of service are not uniformly low and are strongly related to the service strategy mental health service providers take. For example, community mental health centers that make no special efforts to serve older clients but simply expect them to come to the center for service typically have very low percentages of older clients. On the other hand, centers that develop services specifically for older people, such as home visits, programming in senior centers and nursing homes, and support groups made up of older people, have much higher percentages of older clients (Spore and Atchley 1990; Lebowitz, Light, and Bailey 1987).

Finally, drug management problems are very common in the older population. Although they represent just over 12 percent of the population, elders account for over 30 percent of prescription drug sales (Doone 1991). Hale et al. (1987) found that 77 percent of elders in their longitudinal study regularly took prescription or over-the-counter medications. Drugs are a major component in the treatment of most acute and chronic conditions common to older people. What drugs to use and in what amounts depend on several factors: how much of the drug is absorbed how fast from the site of administration, how efficiently the drug is distributed, how quickly and completely the drug is metabolized, and how quickly the drug or its by-products are eliminated from the body (Kayne 1976). Aging changes all these factors, "often resulting in greater concentrations of the drug at its site of activity or a longer persistence of drug activity" in older people (1976,437). For example, because kidney functioning declines sharply with age, drugs that are eliminated by the kidneys must be given in lower dosages to older people to avoid accumulation of the drug to the point where it produces adverse effects. Moreover, many older people have multiple chronic conditions for which they are taking several drugs.

Establishing proper dosages and avoiding adverse drug reactions or adverse interactions among drugs require initial metabolic information, history, and monitoring that is beyond the capacity of medicine as it is currently organized. In many cases, the physi-cian does not know how age affects appropriate dosage. Sometimes such information is simply not available. In addition, individuals vary considerably in their reactions to drugs. As a result, establishing proper dosages is usually a trial-and-error process in which a great many errors are made and only sometimes detected.

Even if proper dosage can be established, elders do not always take their medications as prescribed. Doone (1991) found that about 25 percent of her older respondents did not adhere to the dosage or timing prescribed by the physician. Most did not know the possible side effects of the drugs they were taking.

About 5 percent of all hospitalizations occur as a result of adverse drug reactions, and this percentage is higher for people over 60. In addition, 10 to 30 percent of those who are hospitalized have adverse drug reactions in the hospital, and such reactions occur more often in older patients than in others (Kayne 1976). These data probably reflect only a small portion of the actual medical problems that result from the use of inappropriate drugs, incorrect drugs, excess dosage, or drug interactions. In addition, mental problems such as depression or confusion are common results of adverse drug reactions. Unfortunately, these symptoms are all too often attributed to "senility" in older patients.

The costs of health care often represent a substantial problem. Although, as we will see, various forms of public and private health programs pay some of the costs of health care provided in hospitals, nursing homes, physicians' offices, and clinics, elders use more health services than other segments of the population, and also spend more out-of-pocket. Whereas price inflation in general has dropped substantially from double digits in the late 1970s to only around 3 percent in 1993, inflation in health-care prices continued in the double-digit range throughout the 1980s and early 1990s and shows no signs of abating. Increases in costs of hospital care, physicians' services, and prescription drugs accounted for 70 percent of overall health-care inflation from 1985 to 1993. The result is growing pressure on resources to pay for health care. Government and employers alike are cutting back the services they finance, and their costs are still increasing faster than they can budget funds to cover them. Accordingly, more and more health-care costs are being shifted to the consumer in the form of lost health insurance coverage and higher deductibles and copayments.

Prescription drug costs, which are largely uncovered by Medicare or private insurance, have increased at an especially high rate of inflation (U.S. Senate Special Committee on Aging 1992). Among older people, 65 percent must pay the full cost of their medications, which averaged $818 for 1992. From 1977 to 1987, annual drug expenditures among older people increased by over 230 percent, compared with general inflation of about 45 percent during the same period.

Problems with Financing Health Care

Medicare and Medicaid   Medicare is a program of health insurance for older Americans that is administered by the Social Security Administration. It consists of two parts: hospital insurance (Part A) and supplementary medical insurance (Part B). Nearly all older individuals in the United States are eligible for hospital coverage.

In 1993, after the patient paid a *deductible* of $676, hospital insurance paid the full cost of up to 60 days of hospitalization for any illness. From 61 to 90 days, the patient paid a *copayment* of $169 per day. In addition, a "lifetime reserve" of 60 days' coverage could be used for illness that required hospitalization for longer than 90 days, but the copayment increased to $338 per day during coverage under the lifetime reserve. Hospital

insurance also covered up to 100 days per calendar year in an approved skilled nursing facility, provided that entry into the nursing facility was preceded by a stay of at least 3 days in a hospital. Medicare covered in full the first 20 days in the skilled nursing facility, and after that the patient paid a copayment of $84.50 per day. In 1993, in the extreme case in which all Medicare benefits were used to the maximum, the older person's deductible and copayment liabilities would have amounted to $32,110 for hospital and skilled nursing services for less than six months' care. And this figure does not include excess physicians' charges, copayments for physicians' services, or medications not provided in a hospital.

Medicare hospital insurance also covers skilled nursing care at home if it is certified by a physician as medically necessary. Hospital insurance pays 80 percent of the cost of equipment, such as hospital beds or lift chairs, that is certified by a physician as a necessary part of medical treatment.

Supplementary medical insurance is an optional part of Medicare that, for a 1993 monthly premium of $36.30, covered services from health-care professionals, approved diagnostic tests, necessary medical devices, outpatient hospital services, and laboratory services. In 1993, after an initial deductible of $100, Medicare supplementary medical insurance paid 80 percent of "reasonable" charges. The client was responsible for charges above the predetermined Medicare rate, and about half of physicians charged more than what Medicare defines as "reasonable" (U.S. Senate Special Committee on Aging 1992). Medical checkups, prescription drugs, eyeglasses or contact lenses, dentures, and hearing aids were not covered.

Table 14-1 shows the increases in Medicare copayments and deductibles from 1985 to 1993. Note that while the deductibles and costs per day of Medicare services to older people increased by 69 percent across the board, Social Security cost of living increases

Table 14-1 Changes in Medicare copayments and deductibles and average Social Security benefits: 1985–1993.

	1985	1990	1993	Percent Increase 1985–1993
Hospital deductible (for each episode)	$400	$592	$676	69
Hospital copayment (per day)	100	148	169	69
Lifetime reserve copayment (per day)	200	296	338	69
Nursing home copayment	50	74	84.50	69
Social Security benefits (average per day)	16.87	19.81	21.47	27

Source: Data compiled by the author from records of the Social Security Administration.

amounted to only 27 percent during this same period.

Medicare has covered almost all people 65 or over since July 1, 1966. But Medicare coverage is not automatic. To qualify, older people must have participated in Social Security-covered employment or be eligible by being a spouse, survivor, or former spouse of a covered worker. Initially, demand for in-hospital services rose, along with the demand for supporting medical services. Since then, age-specific rates of hospital use have leveled off, and the average length of each hospital stay has actually dropped slightly. The use of ambulatory physician services has remained fairly stable throughout the period since 1966. Per-capita expenditures under Medicare have more than tripled since the program began, mainly because of the increases in what hospitals charged the program for covered services. A large proportion of the funds are spent on behalf of a relatively small number of people with serious illnesses. About 20 percent of the insured population use no covered service in a given year. Another 20 percent are hospitalized. Of this latter category, one-fourth are hospitalized more than once in the year. The bulk of physician costs arise from care for hospitalized patients.

Medicaid is a comprehensive health care program designed to provide health care to the poor, regardless of age. It is administered by local human services departments using federal and state funds and following federal guidelines and regulations. Medicaid pays for everything that Medicare does, as well as for many other services, including drugs, eyeglasses, and long-term care in licensed nursing homes. Medicaid also pays for home care in most states and group home care in a few states.

Health Insurance Gap Fillers *Private health insurance* has great potential to help defray the costs of both acute and long-term care, but most of this potential is unrealized. In 1991, about 65 percent of people covered by Medicare had private insurance to supplement their Medicare benefits (U.S. Senate Special Committee on Aging 1992). Most of those with private coverage were middle- and upper-income elders. About half purchased individual policies and half received coverage as part of their retirement benefits. However, nearly all these policies are designed primarily to handle copayments and deductibles connected with services Medicare covers, and they are of no help in paying for acute care or long-term care not covered by Medicare. Thus, the so-called "gap fillers" fill only minor gaps and leave some very major ones unfilled, especially the costs of long-term hospitalization and prescription drugs, and the need to finance long-term care.

In 1987, of the 8 percent of older people's personal health-care costs funded by private health-care insurance, nearly all was provided under basic "medigap" policies designed to cover only hospital deductibles and copayments for Medicare, Part A. Medigap policies generally provide no protection against major "excess" expenses for physicians' services or expenses for services not covered by Medicare such as dental care or eye care. If an older person received a service and Medicare reimbursed for it, then these medigap policies covered any deductibles or copayments.

More extensive medigap policies cover the 20 percent copayment for physicians' services under Medicare, Part B, but few older people have such policies, probably because of the added expense. Even fewer older people buy medigap policies covering physician charges above the Medicare "allowable" limit, and more than half of physicians charge more than Medicare defines as allowable. Some medigap policies also cover prescription drugs, care provided in foreign countries, or care in skilled nursing facilities not certified by Medicare.

Consumer Reports (1989) found that a basic package consisting of Medicare, Part B (premiums of $383 paid by the insured) and medigap coverage for hospital and physicians' fees not covered by Medicare ($987 annual premium) would cost a 65-year-old woman $1,370 per year. Private medigap rates rose with age. Thus, many older people are simply priced out of the medigap market.

To be competitive, insurance companies offer such a variety of medigap policies that older consumers are faced with a bewildering array of choices. To get the fullest array of benefits possible, some older people have several medigap policies, but most of the money they spend on these policies goes to useless duplication of hospital coverage. Unscrupulous insurance agents sometimes sell additional medigap policies to older people by overstating policy provisions. For example, *Consumer Reports* found that the misleading claim agents most often made was that their medigap policy covered 100 per-

cent of physicians' charges not covered by Medicare, Part B, when in fact their policies covered considerably less.

Long-Term Care Policies Since 1982, private insurance companies have been writing long-term care insurance policies. Most policies guarantee to pay a fixed amount per day for covered services rather than to pay for all needed services. In this way, companies attempt to limit their liability in an arena where inflation in costs appears out of control. The most desirable long-term care policies cover home care as well as care in long-term care facilities such as homes for the aged or nursing homes. In a study of Ohio retired teachers, I found that most who bought long-term care insurance elected to insure for $50 per day should they need nursing home care. All recognized that $50 per day would not cover the complete costs, even at 1988 rates. They planned to supplement long-term care insurance benefits with their retirement income, savings, and home equity if necessary.

Long-term care insurance has evolved rapidly over the past ten years. Sound long-term care coverage is now readily available for those who can afford it. Desirable long-term care policy features include: no prior hospitalization requirement, because many people do not require hospitalization prior to needing long-term care; provision for at least three years of coverage, which would cover 80 percent of nursing home stays; no exclusion of illnesses such as Alzheimer's disease or heart disease, which are leading causes of the need for long-term care; coverage for care in a variety of locations, including skilled nursing, intermediate care, and custodial facilities as well as care at home; benefits that rise automatically with inflation; premiums that do not rise over time but are set based on age at the time of purchase; automatic renewal; and benefits that begin within 3 to 6 months of the determination of eligibility. Eligibility for home-care benefits is usually established through a physician's certification or a determination interview conducted by a long-term care case manager. Inability to perform at least two activities of daily living, such as bathing, ambulation, or dressing, is the most often-used criterion for home-care eligibility.

However, long-term care insurance is expensive for those who wait until later life to buy it. For example, under the plan offered through the Ohio State Teachers Retirement System in 1988, a person age 65 who purchased modest coverage for $50 per day of nursing home care paid an annual premium of $731.40. But a 70-year-old who purchased the same plan paid an annual premium of $1,164.60, and an 80-year-old paid $2,796.60 annually. Remember that the median annual income of older individuals in 1987 was only a little over $8,000. It is easy to see how Rivlin and Weiner (1988) concluded that no more than 12 percent of the older population could be expected to be able to afford private long-term care insurance.

A major obstacle to the development of long-term care insurance is the widespread misperception that Medicare and private health insurance policies that supplement Medicare cover long-term care, when in fact they do not. In a study conducted by the American Association of Retired Persons (AARP), over one-third of the respondents mistakenly thought that their private supplemental health insurance covered long-term care (Goldberg-Alberts 1986). People must be educated concerning their need for private long-term care insurance; otherwise the cost will continue to be prohibitive for most Americans because the base of insurees will remain small.

The lack of standardized definitions for types of care and appropriate providers is another important obstacle to the development of long-term care insurance. Clear definitions for skilled nursing care, assisted living, and personal care as well as home care, home health care, day care, respite care, and hospice care are required not only to determine eligibility for benefits but also to accumulate the actuarial data essential to set accurate premium rates.

Development is also being slowed by a perception in the insurance industry that the existence of Medicaid coverage of long-term care reduces the demand for private coverage. Goldberg-Alberts (1986) concluded that the demand for private long-term care insurance could be increased by educating the public about the liabilities of relying on Medicaid, such as asset depletion and limited choice and access. To qualify for Medicaid funding of long-term care, individuals must have very low incomes and limited assets, which means that they are supposed to first liquidate most of their assets and use them to pay for long-term care. In addition, states are limiting the number of long-term care cases that they are able to fund through Medicaid, which means that even though they may be eligible for Medicaid, some people will have to go on a waiting list. Even if Medicaid funding is approved, they may have fewer choices of providers.

As a result of these obstacles, long-term care insurance has been slow to develop, but over the past decade this situation has steadily improved as more people became aware of their need for coverage and more private companies solved the technical problems involved in offering effective protection. However, both public policy analysts and insurance industry decision makers have been pessimistic about the potential of long-term care insurance to be a significant source of funding for long-term care. But this pessimism is based on a series of faulty assumptions. For example, a study by the Brookings Institution concluded that private long-term care insurance is likely to remain too expensive to be of help to most older people (Rivlin and Weiner 1988). Insurance company analysts point to low rates of purchase at older ages as an indicator that demand for long-term care insurance is too low to provide significant financing of long-term care.

On the other hand, Atchley and Dorfman (1994) found that middle-aged, middle-income people are the most likely to buy long-term care insurance. High-income people in their 70s or older did not buy long-term care insurance because they had the resources to self-insure and the premiums for long-term care insurance were very high at ages above 70. But if people bought long-term care insurance in their late 40s or early 50s, the premiums were low, did not increase with age, and allowed the purchase of a benefit package that could significantly slow down the depletion of savings and preserve choice, this would become a significant financing alternative. This study points out the importance of basing our ideas of what is feasible on actual behavior rather than on guesses, no matter how educated.

Who Has Health-Care Problems?

Although virtually everyone in the United States with a serious chronic condition has some sort of problem finding appropriate and effective care or paying for it, certain subgroups of the older population are more likely than others to have problems. The very old (85 and over) are especially likely to be frail. Often the adult children who might care for these frail elders are themselves older people with limited physical and financial resources. The very old tend to have multiple serious conditions and are much more likely to be disabled by them. Older members of certain minority groups are also more likely to have health problems, particularly African Americans and Native Americans. Disabling chronic illness tends to occur at younger ages among these groups than among whites.

What Needs to Be Done?

In the area of health care it is literally true that an ounce of prevention is worth a pound of cure. Yet our approach to providing and financing care does not reflect this fact. At the very least we need a much greater effort to educate the public about the value of prevention, to educate health-care professionals about the most effective measures, and to perform the research needed to advance our knowledge about effective prevention.

But we must also acknowledge that the major problem in prevention is not financing or lack of knowledge or lack of prevention services—it is a lack of motivation. A majority of the American population probably knows that it is unhealthy to smoke, to be overweight, to drink too much alcohol, and to be sedentary. Yet cigarette sales have increased every year since the surgeon general reported over thirty years ago that smoking was linked to lung cancer and heart disease. The average weight of the American people has gone up steadily, as has the proportion of people who are obese, despite widespread publicity linking being overweight to heart disease, hypertension, and a host of other serious health problems. Despite the popularity of exercise right now, millions of adults live very sedentary lives.

In rehabilitation, we should disconnect the financing of rehabilitation from the prospects for employment. People should not be denied needed rehabilitation services simply because they are retired. We also need to do a better job of training family members to aid in the rehabilitative process because they are the ones most likely to provide care at home. In treatment, we pointed earlier to the need for a continuum of services. In-home services, outpatient clinics, residential care, and long-term care are all in short supply right now.

We also need to fill in the gaps in our current systems for financing health care. For example, right now Medicare does not pay for prescription drugs, eyeglasses, dentures, or hearing aids. Medicare also does not recognize the fact that long-term care can indeed be long-term. As the system operates now, older people, including couples in which only one member needs long-term care, are forced to become paupers to qualify for the only system that provides funds for truly long-term care: Medicaid.

Long-Term Care

When the general public hears about **long-term care**, the image that most often comes to mind is health care provided in nursing facilities. But long-term care is a much broader concept. It involves providing ongoing assistance to people with chronic illnesses or disabilities in a wide variety of settings by a wide variety of types of caregivers. Some of this assistance may require health care and may be provided in nursing homes, but most long-term care consists of personal care provided in the person's independent household, most often by family members.

In contrast with *health care* such as catheter maintenance, care of pressure sores on the skin, injections, or tube feeding, *personal care* consists of help with everyday activities such as bathing, getting in and out of bed, using the toilet, grocery shopping, doing laundry and housework, and preparing meals. Disabled older people usually need considerable personal care. As we saw in Chapter 4, physical disability is not a part of life for a large majority of older people, but for the significant minority of older people who are disabled, the need for assistance is a central organizing feature of their lifestyles. About 6 percent of older people live in nursing homes or homes for the aged, and these people tend to have extensive needs for personal care and ready access to it. Perhaps as much as 12 percent of the older population living in the community are equally impaired and in need of personal care.

The proportion of noninstitutionalized older people who need personal care increases sharply with age. As Table 14-2 shows, the proportion who have moderate needs for personal care increases from 5.4 percent at age 65 to 69 to 26.4 percent at age 90 to 94; and the proportion with severe needs for personal care increases from 4.2 percent at age 65 to 69 to 52.2 percent at age 95 and over. Thus, the proportion of elders with little or no need

Table 14-2 Need for personal care, by age, 1990.

| Age | Need for Personal Care | | |
	Little or None	Moderate[a]	Severe[b]
65–69	90.4	5.4	4.2
70–74	88.6	6.5	4.9
75–79	82.6	9.4	8.0
80–84	71.9	14.4	13.7
85–90	56.6	19.7	23.7
90–94	37.9	26.4	35.7
95+	26.6	21.2	52.2

[a] Needs assistance with at least *one* of the following activities of daily living (ADLs): eating, getting in or out of bed or chair, using the toilet, dressing or bathing; or *two* of the following instrumental activities of daily living (IADLs): walking, shopping, preparing meals, light housekeeping, using the phone, or using transportation.
[b] Cognitively impaired or needs assistance with at least *two* ADLs.
Source: Adapted from Kunkel and Applebaum (1989).

for personal care drops from 90.4 percent at age 65 to 69 to only 26.6 percent at age 95 and older (Kunkel and Applebaum 1989). As the older population ages, and the advanced age categories increase rapidly in size, so too will the need for personal care among elders living in the community.

As we saw in Chapter 8, family members are the major source of personal care for older people, but paid help is also important. In the 1984 Long Term Care Survey, 74 percent of care was provided by unpaid caregivers (mostly family), 21 percent by a combination of paid and unpaid caregivers, and 5 percent by paid caregivers alone (Rabin and Stockton 1987). Stone, Cafferata, and Sangl (1987) found that most care provided by paid caregivers was in tandem with family care. Remember that here we are looking at only the minority of older people who need personal care.

Litwak (1985) argued that families are the primary providers of personal care because it involves knowing the person and having a long-term commitment to the individual's well-being. Nursing homes, homes for the aged, or home-care agencies are well equipped to provide routine services such as housekeeping or giving medications, but in areas such as bathing, cooking, grocery shopping, or leisure activities, family and friends may be more effective personal caregivers because they are familiar with the older person's preferences. When these functions are assumed by organizations, there is a necessary element of standardization that results in a greater impersonal element to care, which explains why most older people prefer family care to even the best institutional care. However, many older people who live in full-service retirement communities and nursing homes do so to protect their families from the burden of caregiving. And the demands are substantial: the average family caregiver spends four hours a day, seven days a week providing care (Stone, Cafferata, and Sangl 1987).

The 1980s saw substantial growth of social service agencies and private businesses that offered personal care. Part of this growth was stimulated by changes in Medicaid regulations that for the first time allowed reimbursement for home-delivered personal care. Another stimulus was the aging of the older population combined with improved financial resources that allowed more elders to afford private-pay personal care. Nancy Kane (1989) reported that 75 percent of formal home care consisted of unskilled homemaker or home aide personal care. Estimates of total sales of personal care ranged from $3 to $4 billion in 1985 alone. More than half of the cost of home-delivered personal care was private pay. The entry of publicly traded corporations such as Upjohn and Kelly into the home care-market indicated that perceived demand was high. However, cost-containment efforts by Medicaid tended to keep profit margins low, and by the end of the 1980s it appeared doubtful whether investors would continue to support corporations remaining in the home-care business. We will return to this topic shortly.

Who Has Problems?

People can have trouble meeting their personal care needs if they do not have family and friends who can help; if their care needs are so severe that they require around-the-clock supervision; if their caregivers are stressed from having to meet other obligations; if they lack sufficient income to pay for needed help; or if needed services are not available in the community or are not acceptable to the client because the services are unreliable or of poor quality.

Middle- or low-income older people with no family or friends to help may find that they are able to get paid help. But most communities lack formal organizations that can supply people to provide personal care, and locating individual paid helpers who are reliable and can be trusted not to exploit the care recipient is difficult and time consuming at best. At worst, disabled older people seek help, but instead become victims of abusive and exploitative strangers. Leading home-care programs are implementing training and monitoring programs to ensure that the care needed is actually being provided and at a satisfactory level of performance (Phillips et al. 1989). When family work in tandem with paid helpers, exploitation and poor quality of care are less likely to go unnoticed.

Wallace (1990) found that the conspicuous gaps in availability of personal care occurred for respite care, transportation, seven-day home-delivered meals, and live-in help. Most clients were reluctant to complain about services they desperately needed, so the complaints received could be considered only the tip of the iceberg. The most common complaints about formal care involved unreliability (workers failing to come when they were supposed to or failing to provide the agreed-upon services); poor supervision; having to transport workers; poor coordination among case managers, discharge planners, and home workers; and problems with reimbursement, mostly from Medicaid. Worker turnover, racial prejudice, intimidation of clients by workers, and fear of theft were also major barriers to the acceptance of formal services.

People who cannot get the needed paid help in their own home and who have no family to provide it have reduced options. If the individual has the economic resources to pay for personal care and space is available, then moving to an adult care facility, assisted living facility or nursing home can provide ready access to personal care. However, nursing homes are expensive and adult care facilities and assisted living facilities for middle-income people are in short supply in most parts of the country.

For low-income elders, Medicaid is the last resort for financing congregate care, and many states fund only nursing home care through Medicaid; care in group homes is not covered. However, states are increasingly funding assisted living through Medicaid. Not all nursing homes participate in Medicaid, and those that do often have long waiting lists, leaving a sizable number of indigent and disabled older people living in the community who need personal care but are unable to get it through either home or institutional care.

Although theoretically states are required to provide Medicaid financing for long-term care services to those who qualify, the steps that many states have taken to contain rapidly escalating Medicaid budgets have had the effect of limiting the number of qualified elders who actually receive services. For example, many states have instituted moratoriums on the construction of nursing home spaces. If no new spaces can be added and the number of people in need of nursing home care is increasing, someone has to do without. In addition, once the state's budget allocation for nursing home or home care under Medicaid has been exhausted, most states stop taking new clients, which of course results in waiting lists and unserved elders. For example, Crystal (1982) found that about half the applicants approved for Medicaid nursing home services never got off the waiting lists.

The type and severity of the older person's disability also have a strong impact on the amount and type of personal care available. Disabled older people with no prognosis of recovery, those whose care is physically demanding, and those needing constant care are most likely to exhaust the coping resources of their family caregivers. For example, Silliman and Sternberg (1988) found that care of older people with Alzheimer's disease involved a constant level of supervision that interfered with caregivers' continuing their customary social and leisure activities, and this situation tended to produce high feelings of burden. Compared with the caregivers of stroke victims, who often have the prospect of some recovery, the caregivers of older people with Alzheimer's disease, who generally only get worse, were much more likely to be depressed. Caregivers of people with hip fractures were often depleted by the physical demands of frequently having to lift the care recipient. All these various types of caregivers reported negative effects on their own physical and emotional health. Despite these difficulties, most families continue to provide care, even at the expense of their own health and lifestyles. Use of home-care agencies and respite care to supplement family caregiving are important ways to lighten the workload and let the family extend the time over which it is able to provide care. Without this additional help, families must turn to institutional care sooner. When family members move to group homes or nursing homes, family and friends still remain major resources of companionship and care monitoring.

Problems can also arise when caregivers have obligations to other family members and to jobs. Stone, Cafferata, and Sangl (1987) reported that more than one-fourth of adult child caregivers also have children of their own still living at home. In some cases older children are important resources for spreading the demands of caregiving, but in other cases the demands of young children compete with those of older parents for the attentions of the middle generation.

Stone and her colleagues (1987) also found that most adult child caregivers were employed. Caregiving often caused problems on the job, and many caregivers reported having to take shortcuts or make adjustments at work to juggle the demands of work and caregiving. About 9 percent even quit their jobs to meet caregiving obligations. Obviously, not all caregivers can afford to leave their jobs, and most jobs allow only minor flexibility for scheduling jobs around caregiving. When the demands of caregiving seriously conflict with jobs, getting the personal care they need from family becomes more problematic for disabled older people.

What Can Be Done?

As the older population grows older, the proportion living in the community who need personal care will increase and the severity of needs will increase as well, especially among those 85 or older. As the prevalence of home-care programs to provide personal care increases, disabled individuals and their families may have more resources to supplement family caregiving efforts, depending on how fast the number of providers increases in relation to the increasing demand. Increasing the number of these partnerships is crucial to our being able to meet the personal care needs of an increasing older population. At the same time, long-term care facilities will also continue to be a vital resource for meeting the personal care needs of the severely disabled because when people require twenty-four-hour monitoring, care in the home is very difficult.

Nancy Kane (1989) concluded that the outlook for private business to play a major role in home-delivered personal care was not good. She found that many corporate providers were losing money, particularly in the personal care segment of the home-care market. There were several factors that con-

tributed to this problem. Medicaid cost-containment efforts tended to set rates for personal care workers at the minimum wage, restrict sharply the number of hours per home visit, not allow travel time to be included in the cost of service, and cap overhead, thereby constraining agencies that might otherwise provide better training, fringe benefits, or try to expand their market base through advertising. When charges for publicly funded services are artificially low, private-pay charges can also experience downward pressure. Hourly wage rates for home-delivered personal care were 20 percent lower than personal care wages in nursing homes, which created problems in recruiting and retaining home-care workers; this in turn increased costs and reduced profits.

In the private economy, investment seeks the highest possible return. Kane concluded that the low profit margins typical of the home-care market will probably prevent a rapid expansion by large-scale employers into its personal care segment. This means, of course, that communities will be forced to rely more on underfunded public agencies and smaller-scale entrepreneurial organizations, which tend to be more transitory and more difficult to monitor for care quality.

Family care will continue to be financed mainly by families, but it is important to expand the base of people who have insurance coverage against the risk of needing personal care. Some long-term care insurance policies that came on the market in the late 1980s provide benefits for those who need personal care in the home. These policies are expensive if purchased after age 75, but they offer older people with sufficient incomes important protection against depletion of assets. As group long-term care policies become available at younger ages, more people will be able to afford long-term care insurance premiums. Medicaid funding for home care is also increasing, which will also be a way of financing home care to supplement family care of older people whose incomes are low enough to qualify them for Medicaid. However, whether the expansion of Medicaid home-care funding is keeping pace with the increase in numbers of elders qualified for it is unclear.

Health and Personal Care Service Integration

Although health services and personal care services to older adults are sometimes thought of as discrete entities, especially by providers of specific services, actual service needs are a continuum, ranging from intensive, high-technology care in hospital settings to telephone reassurance services to older people living in independent households. The specific combinations of health and personal care services needed by people with acute health problems such as pneumonia are very different from those needed by people with several activities of daily living disabilities and multiple chronic health conditions that are being managed at home. The rapid growth of the older population needing long-term care has coupled with increased availability of home-care benefits to dramatically increase the complexity of the mix of providers involved in delivering health and social services.

The availability and operation of health and personal care services to older people largely depend on how these services function as societal institutions—how they are organized, funded, and regulated. Accordingly, here we deal with concepts used to structure health and personal care services delivery, the types of organizations that provide such services, organizations that attempt to influence service policies, and the means for financing and regulating health and personal care.

The health care and long-term care institutions in the United States are a very loosely organized conglomeration of individual and organizational service providers, suppliers, professional and trade associations, and government agencies. The people involved range from nurses' aides to surgeons, from government bureaucrats to private entrepreneurs,

from local to national officials. The organizations range from storefront clinics to huge medical centers, from small proprietary nursing homes to long-term care corporations controlling thousands of nursing home spaces, from individual home care providers to large home care corporations.

Financing comes from a wide variety of sources. Any attempt to discuss such a complex system is bound to be difficult. Nevertheless, it is important that students of gerontology have some grasp of the basic societal issues that influence health services to older people.

Models of Service

Three basic models are used to develop and operate health and personal care services: the medical model, the social model, and the holistic model. Right now, the medical model is dominant, at least in part because governmental and private insurance payment systems favor it for both health services and personal care services. The **medical model** is oriented around the treatment of disease and injury, and services are controlled primarily by physicians. The settings in which services dominated by the medical model are most often provided are physicians' offices, outpatient clinics, acute care hospitals, mental hospitals, community mental health centers, Veterans Administration hospitals, and many nursing homes. The dominant medical view of treatment favors surgical repair, chemical treatment through drugs or diet, and rehabilitation through exercise, physical therapy, or prostheses. Although many physicians are quite humanistic in their orientation toward patients, they are trained primarily in medical intervention, with little attention to the social components of care, such as counseling, education, activity therapy, and long-term care management.

The **social model** of care assumes that medical services are only part of the total array of services needed to promote good health or deal adequately with disease or injury.

In addition to medical interventions, patients and their families may need education in order to comply with the prescribed medical treatment, counseling that can help minimize the deleterious effects of anxiety or depression on recovery or adaptation, monitoring for changes that may require a new approach to treatment, or other services, such as personal care, that might help prevent further deterioration of the individual's condition. In the social model, physicians are only one among an array of professionals that includes social workers, nurses, physical therapists, and case managers as well. Ideally, a variety of professionals would be involved in developing a treatment plan. The social model is favored in many long-term care facilities, personal care homes for the aged, and community-based long-term care systems. However, its usefulness has been severely hampered by the fact that many nonmedical services, even if ordered by a physician, are not covered by Medicare, Medicaid, and most private health insurance plans.

The social model is often used by nonprofit long-term care facilities and facilities that serve private-pay* residents. In addition, home-delivered long-term care has become more common as a result of changes in federal regulations that allow limited use of federal funds to pay for a variety of community-based services.

The **holistic model** of health care emphasizes education and responsibility for self and is aimed at *high-level wellness*—a positive state of well-being, not merely the absence of disease. Prevention of disease and injury is a major thrust of the holistic approach. Prepaid group health systems—often called *health maintenance organizations (HMOs)*—have built-in incentives for prevention because the

system receives the same amount of money from each participant regardless of the type of service provided, and prevention is generally much cheaper than treatment. Accordingly, other than a few HMOs that have been active in promoting prevention through education and self-responsibility, there have been few organizational advocates of the holistic approach, and virtually no governmental or private insurance reimbursement is available for holistic health either in the form of education or prevention.

In addition to the medical, social, and holistic models of care, there are also alternative models of how best to provide personal care in the community. Simon-Rusinowitz and Hofland (1993) pointed out that the philosophy guiding personal care services in the disability community is very different from that guiding home-delivered personal care to older people by most agencies. The *home-care model*, used mainly to serve older clients, is based on the medical model of care and tends to emphasize "avoidance of nursing home placement, case management services, and public regulation of providers to control quality of care" (1993, 161). Agency-based home care views clients as dependent persons who need professional help to develop, implement, and monitor care plans. Older people in this system who want to direct their own care usually find that it is not allowed. This model is based on an assumption that older people who need long-term care are incapable of directing their own care.

The *personal assistance model* has its roots in "the independent living movement, which began in the early 1970s among working-age disabled persons. Borrowing from the civil rights movement, the independent living philosophy holds that consumers should control the choice of services best suited to their needs" (Simon-Rusinowitz and Hofland 1993,161). The personal assistance service model is broader than the home-care model. In addition to assistance with IADL and ADL

---

* Residents who pay for all of their own care, either in cash or through private insurance, are called *private pay* to distinguish them from residents who receive *public third-party* benefits such as Medicare or Medicaid.

tasks, personal assistance aims to maintain the person's participation in the wider community. For example, personal assistants, who help with ambulation and other personal care tasks, travel with the client out into the community to maintain the client's capacity to be a self-directed community participant. Services are user-directed whenever possible. Clients supervise their own personal care and receive public funds they use to hire, train, and pay their personal care attendants. Most of these services are funded through Medicaid personal care option programs and Social Security's Social Services Block Grants.

Certainly, many disabled elders living in the community are fully capable of directing their own personal care using the personal assistance model, and a few community long-term care programs serving older people are beginning to experiment with this alternative. The biggest obstacle to the use of this model with older populations is Medicaid regulations that prohibit its use with older clients.

Patient and Client Flow

The operation of the health-care "system" is best seen through the way patients—persons under medical treatment—and clients—persons to whom professional services are being rendered—are categorized and steered through the maze of organizations providing health care. There is no single, readily identifiable point of entry. People enter the system through physicians' offices, outpatient clinics, hospital emergency rooms, and nursing homes. Of course, they may be referred there by a variety of sources: family, minister, physician, social worker, landlord, and so forth. If a person is able to manage treatment on his or her own at home, then outpatient services are most likely to be used. If a person requires admission to an acute care hospital—a facility oriented around high-technology medicine and relatively rapid recovery—then a hospital tends to be selected on the basis of the patient's ability to pay.

Private hospitals tend to serve mainly private-pay patients and those who have third-party medical care insurance such as Blue Cross/Blue Shield. Public hospitals tend to receive the poor, whose care is most often financed by Medicaid. If a person requires long-term care—one or more services, provided on a sustained basis, enabling individuals with chronically impaired functional capacities to be maintained at maximum levels of health and well-being (Brody 1977)—then the person may be routed to an agency that provides in-home services, if such services are available and the client has a way to pay for them. If in-home long-term care services are not available or are inappropriate, then placement in a long-term care facility may be sought.

Very often the job of identifying possible institutional placements is taken on by hospital discharge planners, who are responsible for seeing that the needed posthospital care is arranged for. Sometimes family members are involved in placement decisions, but often they are not made aware that they might have more than one choice. Private-pay clients are relatively easy to place because they can afford the more expensive long-term care facilities. But those who must rely on Medicaid to finance their care, especially those requiring extensive care, are often difficult to place. Private-sector nursing homes often limit their proportion of Medicaid patients because Medicaid reimbursement sometimes does not fully cover the costs of care. In addition, public nursing homes often have a waiting list.

The philosophy of care differs significantly between acute care settings and long-term care programs. In acute care settings, the medical model is dominant, with the idea being to treat the immediate condition and discharge the patient as soon as possible. In long-term care settings, the goal is to manage one or more chronic conditions over an extended period. In long term care, several specialists may be involved in managing a variety of chronic conditions, and acute episodes connected with them, perhaps over a period of years. While much long-term care is controlled by physicians using a medical model, there are programs that employ case managers to monitor the client's condition and coordinate the activities of a variety of health and social service professionals, and there are still other programs in which decisions about care are made mainly by long-term care administrators and directors of nursing. These variations do not exhaust the possibilities, but they do suggest the variety of ways in which long-term care is organized and delivered.

Funding sources exert a major influence on the ways that care can be organized and delivered. Rules and regulations of funding agencies have had a major impact on the kinds of care available, the quality of care provided, and the kinds of professionals who provide services. The emphasis has been on services provided by acute care hospitals, services using the medical model, and services provided by physicians and nurses, all of which have tended to raise the financial cost of providing care, especially long-term care.

As we move toward more community-based and more self-directed long-term care, the problems associated with establishing eligibility for long-term care service and targeting reimbursement from both government and private long-term care insurance increase. In 1995, most long-term care programs used some combination of ADL and cognitive impairment to determine eligibility. In addition, publicly funded programs usually contained operational processes designed to direct public resources to the neediest among the needy. Public policy makers are especially concerned about the cutoff points used for eligibility, because they want to serve clients who have genuine needs, but they are constrained by a growing number of disabled elders and stable or shrinking pools of tax dollars that can be used to finance services. Spector and Kemper (1994) pointed out that policy makers face difficult issues in their attempts to create policies

that direct scarce government resources to those who most need them. Any set of criteria that are easy to administer, such as two ADLs or cognitive impairment, misclassifies some proportion of clients. Some people with cognitive impairment do not need heavy care, for example. Spector and Kemper recommended liberal eligibility criteria coupled with a careful benefit determination process. The liberal eligibility criteria would make more heavy-care clients eligible, and the careful benefit determination process would limit the amount of resources used by light-care clients. They also recommended ongoing processes to verify that services are being directed to the clients most in need.

Spector and Kemper also pointed out that need is not easy to define, either. Is the needy population those with the greatest need for long-term care services, those with the least capacity to get services from family, those least able to pay for services, those most at risk of serious outcomes if they do not receive services, or some combination of all of these definitions of need? Currently, the underlying definition of need is unstated in most programs.

Financing Health-Care Services

The funds for health-care services to older people are provided by older people and their families, Medicare, Medicaid, third-party health insurance plans, and private philanthropy. Since each source tends to operate a separate accounting system, it is difficult to pin down precisely the relative importance of the various sources. For example, many church-related homes for older people use church funds to underwrite cost of care that exceeds what Medicaid will pay for needy residents. However, such expenditures do not usually become part of our national statistics on the costs of long-term care. Likewise, contributions to the financing of health care of elders by their families are difficult to estimate because data on these contributions are not collected. In addition, funding sources vary considerably by type of care. For example, in 1989, nursing homes received only 3.5 percent of the funds Medicare spent, whereas 31 percent of Medicaid dollars went for nursing home care (U.S. Bureau of the Census 1991a).

Although the public generally thinks that Medicare covers most of the cost of health care for older people, the fact is that no single source of funding bears most of the responsibility. For example, in 1987, per-capita personal health-care expenditures in the older population were $5,235. Of this amount, 41 percent was paid by elders or their families (8 percent through private health insurance), 41 percent came from Medicare, 11 percent was paid by Medicaid, 6 percent came from other government sources such as the Veterans Administration, and 1 percent came from private philanthropy (U.S. Senate Special Committee on Aging 1990).

Nevertheless, between them, Medicare and Medicaid provide well over half the funds used to provide health care for older people. In addition, because Medicare and Medicaid reimburse health providers only for specifically defined categories of people and specifically defined services, Medicare and Medicaid policies have had an enormous effect on the structure and operation of health-care delivery to older people. To understand this impact, let us examine each program in detail.

Medicare

Medicare hospital insurance (Part A) is geared to hospital episodes, called "benefit periods." The benefit period begins when an insured person enters a hospital and ends when that person has not been in a hospital or skilled nursing facility for 60 consecutive days. During a benefit period, Medicare will reimburse up to 90 days of hospital care and 100 days of skilled nursing care. To be eligible for reimbursement, skilled nursing care must immediately follow at least 3 days of hospital care. Each covered individual has a

"lifetime reserve" of 60 extra days' care reimbursement, which can be used if more than 90 days of hospitalization are required.

Medicare Part B, which covers physicians' services, medical therapy, diagnostic tests, surgical dressings, prostheses, ambulance service, home health care, and rural health clinic services, can be used to cover hospital costs not covered by Part A, but excludes eyeglasses, hearing aids, dental services, physical examinations, immunizations, and outpatient psychiatric care exceeding $250 in cost. It also excludes prescription drugs except when administered in-hospital.

Medicare spent $517 per covered person in 1975, $1,142 in 1980, and $2,700 in 1989 (U.S. Social Security Administration 1991c). This dramatic increase has been blamed on "overutilization" of health services by older people. However, over 90 percent of this increase can be traced to inflation in hospital costs alone. Thus it was runaway inflation in health-care costs that produced impending problems in Medicare financing, not irresponsible use of services by elders.

Although Medicare ostensibly covers long-term care, in fact it funds very little long-term care of older people. Only 3.5 percent of the $60.5 billion Medicare spent on care for older people in 1986 went to provide long-term care in nursing homes or at home (U.S. Social Security Administration 1988). Part of the reason for this is a logical contradiction in the regulations: Many individuals requiring long-term care are not sick enough to qualify for three days of hospital care, yet if they do not qualify for hospital care, they cannot qualify for nursing care. Another reason is the "episode" assumption built into Medicare reimbursement; that is, the assumption that illness occurs in episodes. But long-term care is mainly continuous, not episodic. Medicare provides less than six months' long-term care coverage, even for those who manage to qualify for Medicare long-term care reimbursement. Thus, our national health insurance program for older people provides virtually no protection for the vast majority of middle-class older Americans in the event they need long-term care.

Medicaid

Medicaid is a federal and state program that uses general revenues to fund health care for the poor. Eligibility for Medicaid is tied closely to economic status. To be eligible, an individual must have not only a low income but also few assets that can be converted into income. The two major categories of eligibility for Medicaid are the categorically needy and the medically needy. *Categorically needy* older people are those who receive SSI or who would be eligible for SSI if they applied for it. States have the option of adopting an eligibility standard for Medicaid that is lower than the income and assets standards for SSI. In such states, to qualify for Medicaid, SSI-related individuals must be allowed to "spend down" to the more restrictive asset or income level. The *medically needy* are people who ordinarily would not qualify for SSI but who have incurred medical expenses Medicaid would normally cover and whose incomes are below eligibility level, usually lower than the eligibility criteria for SSI. Not all states offer Medicaid to the medically needy, and those that do use complicated formulas to define spending-down and eligibility criteria. Eustis, Greenberg, and Patten reported that

> the considerable choice in program eligibility has led to wide variations in coverage across the states. Some states have chosen the federal SSI definition for eligibility. Others have liberalized cash assistance benefits and extended Medicaid eligibility to more individuals. Finally, some 25 states have more restrictive requirements. . . . These states are generally more concerned with client assets than with income.
>
> (1984,122)

The 1982 Tax Equity and Fiscal Responsibility Act allowed states to place liens on the property of Medicaid recipients, although not if the home was occupied by a spouse, a disabled or dependent child, or a sibling who had lived in the home over a year and had equity in the home.

Medicaid is not a program directed primarily at older people but at providing health-care financing for all Americans who are poor. In 1987, only 14 percent of the 23 million recipients of Medicaid were 65 or over. However, 36 percent of Medicaid funds were used to pay for health services to older people, reflecting the fact that 80 percent of Medicaid expenditures for long-term care went to provide care for older people, mostly in nursing homes. Less than 10 percent of Medicaid long-term care expenditures went to provide home care, and 56 percent of Medicaid-funded home care was provided in just five states: Florida, New York, California, Illinois, and Oregon (U.S. Senate Special Committee on Aging 1988).

Thus, both Medicare and Medicaid stress payment for care in institutions and provide very little support to noninstitutional alternatives. Indeed, both programs have regulations that greatly restrict the use of home care, adult foster care, and group homes. Part of the reason for this is that legislators and government officials fear that the financial costs of more liberalized noninstitutional care would be too high. Another factor may be that mechanisms for policing the adequacy, financial accountability, and safety of hospital and nursing home care are already in place, while the means to police other alternatives would in many cases have to be created, also tending to increase their cost. And while it is often assumed that noninstitutional care is less expensive than institutional care, studies have shown this may not be true in some cases (U.S. General Accounting Office 1982). Finally, the providers of institutional care are well-organized to protect their interest in funding for institutional care, whereas those who might provide alternative care are not.

Despite these difficulties, rapidly rising Medicaid costs caused Congress, in 1981, to approve limited use of Medicaid funds to finance community-based services. The intent was to give states authority to develop community-based long-term care programs in cases where home-delivered services would be less expensive than nursing home care (Applebaum, Atchley, and Austin 1987). States were required to apply to the secretary of Health and Human Services for a waiver of the usual requirement that Medicaid funding be used only for facility-based services. Services that could be provided under these waivers included case management; homemaker, home health, and personal care services; adult day care; and respite care. The very existence of Medicaid home-care reimbursement encouraged an increase in the number of organizations providing home care, but mainly in states with Medicaid home-care waiver programs. By the end of 1986, more than one hundred community-based programs were operating in forty-one states as a result of the Medicaid waiver process (Applebaum, Atchley, and Austin 1987), and by 1995 all states were operating at least limited home-care programs under Medicaid waivers.

Nevertheless, growth of these programs has been slow, partly because of controversy over whether community-based programs save money and whether they actually keep people out of nursing homes (Mathematica Policy Research 1986). Perhaps these are the wrong questions. Applebaum, Atchley, and Austin found that care plans for Ohio's Medicaid home-care waiver program cost on average about 30 percent of the cost to Medicaid of intermediate care in a nursing home. However, many of the home-care clients would have refused to enter a nursing home, so the cost of their care was an increased Medicaid cost, not a saving. In addition, costs for home care of clients with no family to provide free care approached the cost of nursing home care. Thus, Medicaid home care does not always substitute for nursing home care, and although home care

is usually much less expensive than nursing home care, this is not always the case. On average, Medicaid home care can meet the needs of about three older clients for the amount of public funds needed to care for one older person in a nursing facility. Home care may not produce net savings to Medicaid, but it sharply increases the number of older people whose long-term care needs can be met through Medicaid.

Limitations in funding, opposition from nursing home lobbyists, and questions about quality control have also slowed home-care expansion. However, considering the financial pressures on Medicaid as the main government vehicle for financing long-term care and the preference of the general population for at-home care rather than care in nursing homes, the number of community-based programs has increased dramatically and will continue to do so for the foreseeable future.

Private Insurance

Two very different varieties of private insurance policies address older people's health and long-term care financing needs: policies that supplement Medicare hospital and physician benefits and policies that cover long-term care.

Private health care insurance is generally available to cover deductibles and copayments and sometimes drug costs, but it is usually too expensive for those elders who need it the most. Those whose incomes are just above the federal poverty level are especially less likely to be able to afford private health-care insurance. Long-term care insurance is much less widely used than health-care insurance, but it is growing in availability and adequacy. If properly marketed to younger, middle-income people, long-term care insurance could eventually provide significant levels of long-term care financing. Recognizing that their state budgets are being consumed by nursing home costs under Medicaid, several states offer incentives to individuals who buy long-term care insurance. For example, those who buy long-term care insurance may be exempted from Medicaid spend-down provisions so that when their long-term care insurance benefits run out, they immediately become eligible for Medicaid without having to meet spend-down criteria.

Gaps in Financing for Health Care

By now it should be obvious that there are substantial gaps in our systems for financing health care for older people. For one, there are many aspects of care Medicare does not cover; dental care, eye care, hearing aids, and preventive services are particularly important exclusions. In addition, overly restrictive criteria keep Medicare support for long-term care very low. Although Medicaid ostensibly covers many of the areas Medicare does not cover, Medicaid is not available to most older Americans. And even those who are eligible for Medicaid are subject to restrictions on availability of important services. For example, the oral health of many Medicaid patients in nursing homes is atrocious. Still, although older people get over 80 percent of the Medicaid support for nursing care, they receive only 6 percent of the Medicaid payments for dental services, and not because of lower need for service—elders in general need more dental care than younger people. Still, most states have adopted Medicaid rules that exclude reimbursement to older people for all but the most severe dental problems, exclusions that do not apply to younger Medicaid recipients. Finally, as we have seen, neither Medicare nor private insurance provides much protection against the need to finance long-term care; although Medicaid does, it is unavailable to most middle-class people, and it provides too little funding for noninstitutional alternatives.

We pay a big price for our entrepreneurial approach to the provision of health care in the United States. Himmelstein (1990) provided an interesting contrast between the administrative costs of piecemeal care in the United States and the same care in Canada, which has a national health program. In the United States, services provided to each client must be documented in excruciating detail, to provide the paperwork needed to bill separately all the various third parties such as Medicare, Medicaid, and private health insurance carriers as well as the client. The administrative costs of this approach to reimbursement amounted to 25 percent of the cost of care in 1988. In Canada, providers are given an annual budget from the provincial government, and costs are justified in the aggregate. Clients are given service, and there is no need to keep track of the costs of specific supplies used or costs of services given to a specific patient. The administrative costs of this system amount to 4 percent of the total cost of care, and in Canada everyone is eligible to receive care, not just those who can pay or who are eligible for third-party reimbursement.

The Structure of Advocacy

As we have seen, the health-care system is structured in favor of reimbursement for services provided in hospitals and nursing homes, mainly because of the *institutional bias* in the laws and regulations that provide federal funding. To understand why this bias exists and persists, we must consider what groups are involved in developing and maintaining health policy and which ones have the most power. The groups involved include federal and state governmental agencies, state and national professional and trade associations, and organizations in the "aging network."

Although federal and state governmental bodies must develop and administer legislation that provides structure and funding for health care, they do not perform these functions in a vacuum. Health policies and regulations are subjected to a review process, during which the advocates of various interest groups have an opportunity to become informed about legislative and administrative proposals and to address their concerns about them. In the area of health policies toward older people, one might expect reactions from professional associations such as the American Medical Association; trade associations such as the American Association of Homes and Services for the Aging (an organization representing the interests of non-profit homes), the American Health Care Association (representing the interests of for-profit nursing homes), or the American Hospital Association; third-party insurers such as Blue Cross/Blue Shield; and advocates for the interests of older people such as AARP (a large membership organization of older people). While not exhaustive, these examples indicate the diversity of organized efforts to influence the development and implementation of health-care policies toward older people.

A detailed analysis of how these various interest groups have influenced specific legislation or regulations is far beyond the scope of this book. Nevertheless, it is obvious from the results that more than other advocates, physician-directed advocates such as the AMA, the American Hospital Association, and Blue Cross/Blue Shield seem to have been more influential in creating the system, largely due to the resources available to these organizations, whose members can afford to pay for first-class lobbying efforts in their behalf.

The general American tendency to consider physicians *the* authority on health care has also tended to give physicians a greater voice in these matters than other equally well-informed professions such as nursing, social work, long-term care administration, and gerontology. Many groups have useful contributions to make to the development of health care for older people, and some practitioners in nursing, social work, occupational therapy, gerontology, and other professions have training and experience in health aspects of aging that far exceed the average physician's. Physicians tend to favor institutional forms of health care because that is

what they learn in medical school. Physicians tend to do their internships and residencies in hospitals, not in wellness programs, home health agencies, or nursing homes. Thus, the medical model has tended to dominate the physician's thinking about how health care for older people should be provided. In addition, physicians tend to affiliate with hospitals and not with wellness programs, nursing homes, home health agencies, or homes for elders, which increases their willingness to advocate for hospitals but not for other types of health-care organizations. This goes a long way toward explaining Medicare bias toward hospital care. The medical-model bias in Medicaid arose mainly from the use of Medicare as a model for organizing Medicaid (Eustis, Greenberg, and Patten 1984).

There is nothing immoral about physicians' organizing to pursue their own interests. The potential for harm comes from the fact that those who oppose the physicians' positions are seen as having less credibility. If the situation is to become more balanced, then governmental officials, legislators, and the general public must learn that no single group's perspectives can suffice in planning and developing programs to serve a population as diverse as the older population in a field of service as diverse as health and long-term care.

Dunlop reported that the physician-directed organizations cited previously were very instrumental in the 1950s in redefining the hospital's primary concern as acute care and in defining the medically dominated nursing home as

the specialized setting for the delivery of long-term care for the elderly in this country. Virtually no one questioned the medical model employed for their development both by Congress and the Public Health Service. That model is easily discernible, first in the standards applied to facilities [built with federal assistance] as early as 1954, and then in those applied to all Medicare and Medicaid certified homes.

(1979,101)

A classic textbook for long-term care administrators follows the medical model strictly. All decisions about care are to be made by physicians, and the description of the medical director position is practically identical to that of hospital chief of staff (Miller and Barry 1979). This model is unrealistic for a majority of nursing homes because it is difficult to get physicians to even visit such facilities, much less find one to serve half-time or more as medical director. As a result, administrators and nursing staff are faced with regulations reflecting a medical model and physicians who in general are reluctant to get involved in long-term care.

In the late 1980s and early 1990s, several trends in health-care advocacy emerged that began to sharply undercut the hegemony health-care lobbyists had enjoyed for decades (Hill and Hinkley 1991). First, reorganizations within Congress increased the number and autonomy of subcommittees, which increased the number of members of Congress lobbyists had to communicate with. Second, the preoccupation of the administration and Congress with budget development and deficit reduction resulted in a wide range of issues being dealt with in the budget reconciliation process rather than in separate pieces of legislation. Because budget legislation is often placed on a "fast track" through Congress, hearings are more limited and opportunities for lobbyists to review legislation prior to providing their reactions are rare. Finally, health-care interest groups no longer present a united front to Congress even within narrow sectors. For example, the American Medical Association no longer represents even a majority of physicians, and rival groups such as Physicians for National Health Insurance have organized opposition to the AMA's traditional protectionist positions. Important splits have also occurred among hospital advocates and health insurance advocates as well. Health-care advocates have also not been helped by the failure of voluntary efforts at cost control among hospitals and the growing public perception that physicians' groups are more interested in protecting physicians' incomes than in improving the public's access to adequate health care. The result has been a decline in the credibility of health-care advocacy organizations. At this point it is unclear who will fill this policy vacuum.

Regulation of Health and Long-Term Care

The quality and cost of health and long-term care are regulated in several ways: licensing of people to provide care and of facilities in which care is provided, certification and accreditation of facilities, and periodic review of facilities for compliance with standards of quality and cost of care (Eustis, Greenberg, and Patten 1984).

Regulating the Quality of Care

All states require physicians, registered nurses, practical nurses, nursing home administrators, dentists, physical therapists, and the like to be *licensed,* which means that to perform professional services in the state, an individual practitioner must pass a licensure examination and have the appropriate training. In addition, state laws govern licensure of hospitals and nursing homes and specify the conditions facilities must meet to qualify for licensure; without licensure, facilities cannot operate as hospitals or nursing homes. Licensure laws set standards in areas such as staff qualifications; the number and type of staff required in relation to the number of residents; medical supervision; use of restraints; required equipment and supplies; dispensing of medications, food, and nutrition; space requirements; fire protection; and physical plant features such as dining rooms, plumbing, water supply, sanitation procedures, and building maintenance.

Certification ensures that an agency or facility is eligible for reimbursement from Medicare or Medicaid. When Medicare was

The quality of health care is not easy to measure.
Photograph by Dan Smetzer/Tony Stone Images, Inc.

initiated in 1965, standards and conditions of participation were developed for hospitals and hospital-like nursing homes called "extended-care facilities." Medicaid standards were developed for a less intensive form of nursing care in facilities called "skilled nursing facilities." In 1971, Medicaid standards were expanded to include care in nursing homes called "intermediate-care facilities." The major difference between skilled and intermediate care is the seriousness of the resident's condition and the type of personnel needed to provide care. Skilled nursing care is practiced predominantly by registered nurses, and intermediate nursing care is provided mainly by licensed practical nurses. In 1992, both categories of facilities were combined and are now called "nursing facilities."

Setting federal standards is the responsibility of the Health Care Financing Administration, but responsibility for administering the process rests with the states, which must do all the fieldwork and inspections as well as the paperwork. In many states the certification process is identical to state licensure (Eustis, Greenberg, and Patten 1984).

Accreditation is generally handled by organizations sponsored by professional or trade associations, not by governmental agencies. For example, hospitals, long-term care facilities, and home health-care agencies can be accredited by the Joint Commission on Accreditation of Health Care Organizations (JCAHO). JCAHO accreditation for hospitals indicates that they exceed state licensure standards and are automatically certified for Medicare and Medicaid in most states. Accreditation for long-term care facilities or home health agencies is not as likely to result in automatic Medicare or Medicaid certification.

Responsibility for enforcement of standards of licensure and certification generally falls to the states. State inspectors periodically check hospitals and nursing homes for *compliance* with state licensure laws as well as for compliance with Medicare and Medicaid Conditions of Participation, a detailed set of standards for patient or resident care and facilities, management, and staffing of the program. There is also review by Utilization and Quality Control Peer Review Organizations (PROs). PROs focus mainly on cost containment in hospital care (Quinn 1987); they review records of patients whose care is being financed at least in part by federal benefits. PROs review such things as validity of diagnoses, ensure that Medicare patients have not been admitted unnecessarily, and check that patients have actually received needed treatment. In some states, PROs also screen prospective patients prior to admission to certify their eligibility for government benefits.

State licensure laws and federal conditions of participation coupled with state inspections and PRO review would seem adequate to ensure that standards for quality of care are being met. However, Butler (1980) pointed out that many nursing homes are cited repeatedly for the same violations, and certification or licensure is revoked only under the most extreme circumstances. Several factors interfere with the enforcement of quality of care standards. First, little Medicare or Medicaid funding is available for enforcement. Second, governmental agencies are sensitive to charges of heavy-handed "overregulation" of private business. Finally, the legal constraints on enforcement have rendered enforcement very difficult. In most states, the only action that can be taken against facilities that violate the law is to revoke their operating licenses. In a few states, fines are also possible. However, both measures require lengthy court proceedings, so states have been reluctant to use them. In addition, many locales have a shortage of nursing home spaces, so closing down one home would merely increase pressure on those remaining. As a consequence, enforcement officials are often reduced to relying on harassment through frequent inspections and citation of deficiencies as their only usable enforcement tool. In fact, the U.S. General Accounting Office (1987b) found that decertification was a rare occurrence.

These problems led Congress in 1987 to enact legislation authorizing "intermediate sanctions" against Medicare or Medicaid care providers. Such sanctions included denial of payment for new Medicare or Medicaid admissions, civil fines for each day of noncompliance, appointment of temporary management for the facility, and emergency authority to close the facility and transfer its residents (U.S. Senate Special Committee on Aging 1988). Whether these added sanctions have resulted in better quality of care is unclear at this point, but at least now regulators have more tools to work with.

Regulating the Cost of Care

Inflation in health-care costs has run considerably above general inflation since 1970, and although the rate of increase moderated somewhat in the mid-1980s, health-care inflation began to accelerate rapidly again in 1987. Figure 14-1 shows actual and projected trends in health-care expenditures from 1986 to 2000. These projections reveal an increase in health-care costs from 11 percent of gross national product in 1987 to 15 percent by the year 2000. This kind of cost pressure on our national economic resources has had several important effects. First, it has increased the costs of broadening access to health care to everyone and thus reduced the political chances of achieving this important objective. Second, it has led to the development of a number of interest groups, including groups of physicians and major corporations, that are pushing for an overhaul in the health-care delivery system in order to make possible a national health system available to all Americans. Third, it has spawned efforts at every level of government to regulate the cost of care.

Federal and state governments have attempted to regulate the rate of growth in Medicare and Medicaid costs through changing the eligibility standards, the definitions of services covered, and the reimbursement methods.

States have seldom lowered the actual dollar values of the income and assets criteria used to establish eligibility for Medicaid. But by not increasing these ceilings to account for inflation, most states have actually increased substantially the number of poor who do not qualify for Medicaid. Newcomer and Harrington (1983) reported that since 1977, Medicaid population growth has not kept pace with the growth of the population in poverty and that by 1981 about half the population with incomes below the federal poverty level were ineligible for Medicaid.

Under federal legislation, states are allowed to establish "utilization controls" on

Figure 14-1 Actual and projected national health spending: 1986–2000.

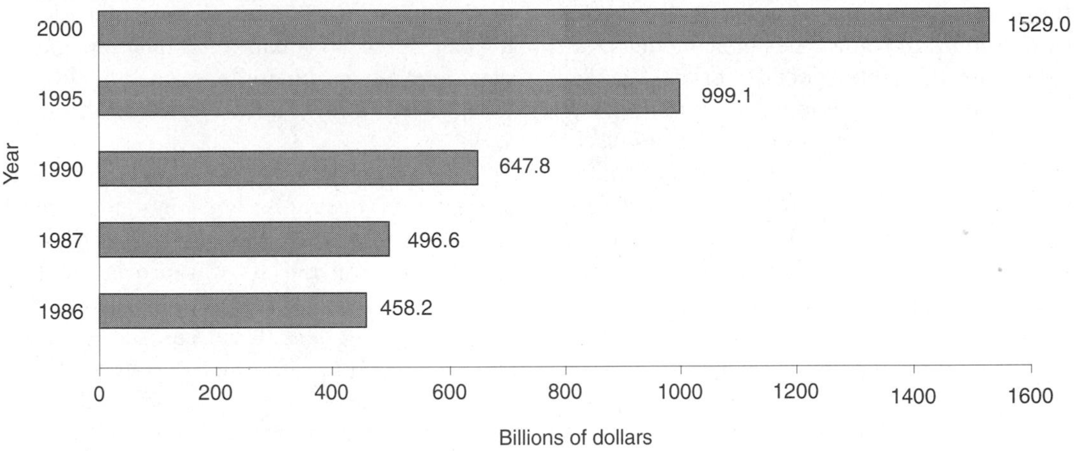

Year	Billions of dollars
2000	1529.0
1995	999.1
1990	647.8
1987	496.6
1986	458.2

Source: U.S. Senate Special Committee on Aging (1988).

services that must be provided under Medicaid. For example, half the states limit the number of days of hospitalization and the number of physician visits that Medicaid will reimburse in a given year. More than two-thirds of the states require approval by the state prior to admission to a nursing facility if Medicaid reimbursement is involved. Although such measures may have resulted in some savings, they have tended to create as many problems as they have solved. For example, arbitrary limits on the number of hospital days obviously have no relationship to needed lengths of care. Likewise, prior authorization is a moot point for nursing home residents who have been Medicare or private-pay patients prior to qualifying for Medicaid.

Reimbursement formulas are another way that states can attempt to control costs. These formulas fall into two general classes: those that incorporate some recognition of differences in costs among facilities and those that attempt to establish standard rates for service independent of variations in cost to the facility. *Cost-based reimbursement* provides little incentive for efficiency. The institution's allowable costs are simply passed on to Medicaid. Indeed, these systems provide

incentives to inflate cost figures and, therefore, require careful cost monitoring to prevent abuse. On the plus side, cost-based reimbursement provides no incentives to limit the quality of care. *Standardized reimbursement* is easier to administer and provides incentives for providers to be cost conscious. However, standardized rates tend not to cover the costs of high-quality care and thus tend to lower the quality of care (Eustis, Greenberg, and Patten 1984). Standardized reimbursement also does not allow for the wide variation in costs of land and building construction that can occur within a state.

Whatever means we take to contain the costs of health care must address the costs to all categories of care recipients. For example, if Medicaid constrains reimbursement, providers may simply pass the unreimbursed cost of care on to private-pay patients or clients. It is safe to say that the methods employed thus far have been relatively ineffective and that health-care cost containment will remain a major issue in the coming years. And although we have seen that there is need for expansion of the services our health-care financing systems cover, such expansion is most unlikely as long as costs keep escalating at such a high rate.

Conventional economic theory holds that the best way to lower prices is to increase competition. In some types of economic markets, this principle indeed works. The decline in prices for personal computers in the early 1990s is a good example. *But in health-care services, increased competition actually increases costs.* How can this be? Most people do not select health care based on price. If Mr. J. needs a hip joint replacement, he does not shop around to find the cheapest price for the procedure. He tries to identify who among orthopedic surgeons in his area does the best job of replacing hip joints. If Mr. J. can afford it, price is secondary to the perceived quality of the service. To be competitive in health-care services means to offer the highest-quality services, which are usually not the cheapest services. In order to attract the best practitioners, who get a large share of the business, hospitals need the latest equipment, up-to-date facilities, and skilled support staff, all of which drive up overhead costs. The more competition there is for their services, the higher the compensation skilled practitioners can demand. The most skilled practitioners are usually in tremendous demand, and they tend to increase their fees as one way of limiting the number of potential patients who come to them. Thus, for several complex reasons, competition in health-care services tends to increase prices rather than reduce them. For this reason, looking to increased competition as a way of controlling health care costs will probably not bring the expected payoff.

Health Planning

We know that the older population is growing older and that the need for health services is going to increase dramatically as a result. To cope with the increase in demand, we need planning to ensure that adequate facilities, personnel, and financing are available. Unfortunately, the history of health planning in the United States over the past two decades gives little cause to be optimistic that the needed planning will be done. Instead, we will probably continue our piecemeal and uncoordinated approach to the development and delivery of health-care services.

The infusion of federal funding for the construction of health facilities in underserved areas took place in 1946 under the Hill-Burton Act. Included in the act was provision for a planning process. In 1974, the National Health Planning and Resources Development Act provided funds to establish state and local planning agencies charged with assessing area health needs, setting priorities, identifying populations or geographic areas most in need, and controlling health costs. Unfortunately, these agencies were established at a time of high inflation in health costs, and controlling health costs soon overshadowed planning as their primary function.

Local planning agencies' control over health costs was exercised indirectly by limiting the construction of new facilities. In most states, to build new hospital or nursing home facilities, organizations were required to obtain "certificates of need" (CONs) certifying a need for the facility. The major goal of certificate of need review was to prevent the use of federal funds to finance "excess capacity, unnecessary duplication of services, . . . and maldistributed health services" (Benjamin and Lindeman 1983,213). Although such reviews were developed mainly in reaction to growing costs for hospital care, CON review was also applied to the construction of long-term care facilities in many states. The result was a restriction of competition in the long-term care field that defeated many efforts to enforce standards of care quality.

Since 1981, federal funding for health planning has been severely cut, and in many states local health planning agencies have been abolished. Benjamin and Lindeman concluded:

Any analysis of health planning in the 1980s is likely to have an unhappy ending. What began as a grand experiment in state and local planning has fallen victim to excessive expectations, inadequate authority, the intractability of medical care costs, and finally to a substantial decline in federal funding.

(1983,222)

Thus we are in great need of health planning in order to cope with an aging society, and what planning organizations we have had have been curtailed or eliminated.

Summary

Although most older people are healthy and active, the need for health care and personal care assistance is strongly age-related, and a large majority of health care and long-term care spending goes to the older population, wherein the needs are by far the greatest. Aging increases the number of chronic conditions people experience. Most aging people care for minor chronic conditions themselves. Self-care is the most common type of primary care, much more frequent than care by physicians or health clinics. Older people are more likely than other population age categories to engage in preventive health practices.

Despite the strong prevalence of self-care among elders, older people are more likely than other population age categories to visit physicians, use hospital facilities and services, use community-based long-term care services, and reside in nursing homes. The oldest-old are especially likely to experience catastrophic out-of-pocket expenses connected with health care and long-term care.

Health-care problems common among aging people include unavailability of needed services either because they cannot afford them or the services do not exist where they live, premature release from hospitals as a result of DRGs, problems stemming from the difficulties of managing multiple medications used to treat multiple chronic health conditions, and problems paying for needed care.

Medicare, Medicaid, private health and long-term care insurance, and family savings are the primary sources used to pay for health care and long-term care. Only the very wealthy escape problems associated with our fragmented systems for financing care. Medicare requires considerable out-of-pocket participation by beneficiaries in the form of deductibles and copayments. Thus, some people covered by Medicare cannot afford to use it. In addition, many needed services are not covered by Medicare, especially the need for long-term care either at home or in a nursing home. Medicaid theoretically covers indigent older people for both health care and long-term care, but in reality Medicaid funds are limited and many eligible elders are forced to wait, sometimes for years, to receive Medicaid-reimbursed services. Private health insurance is too expensive for many middle- and low-income elders and tends to cover only Medicare deductibles and copayments. It seldom covers service gaps in Medicare. Long-term care insurance is becoming more adequate, available, and affordable, especially if purchased before age 60; but so far the proportions buying long-term care insurance have remained small.

Much of the chronic disease and disability currently associated with aging could be prevented or substantially delayed by greater use of primary prevention. However, we have yet to discover how to motivate a majority of the public to engage in preventive health practices such as regular exercise, dietary monitoring, and smoking cessation. Rehabilitation services are too often connected to the possibility of employment, which limits their applicability in an older population.

Long-term care involves more emphasis on personal care services such as help with bathing, dressing, and ambulation than on health care such as catheter care or home kidney dialysis. About 75 percent of all long-term care services are provided in the community by informal helpers, mostly spouses, adult children and friends. Paid home-care workers often work in tandem with informal care providers. Indeed, the relatively low cost of home-based long-term care service plans is contingent on the availability of informal helpers to provide much of the hands-on personal care. When families work alongside formal service providers, formal service quality is higher and potential for elder abuse or neglect is lower.

However, informal caregivers can be overwhelmed by the demands of caregiving, especially for elders who are incontinent or who suffer from dementia. Respite care, adult day care, and other service programs can substantially lighten the load on families, but as the condition of the elder becomes more demanding, institutional options can become necessary both to provide adequate care to the elder and to preserve the mental and physical health of the family caregivers. Most home-care programs will not provide care for elders who genuinely need twenty-four-hour nursing home care.

Although the availability of home-based long-term care has increased substantially since 1985, most of the providers are still in the not-for-profit and small-business sectors. Because most individuals and public programs provide only modest pay for home care, most large corporate providers who have entered the home-care market have not stayed long. This industry is simply not yet profitable enough to attract large-scale private investment.

Several competing models are used to organize and deliver health care and personal care. In the health-care arena, the medical model reigns supreme, with most funding going to hospitals and physicians who provide acute care or care connected with acute episodes of chronic disease. Even funding for home care is dominated by home health care ostensibly directed by physicians rather than by home-delivered personal care directed by the client or by a case manager. But because long-term care mostly involves personal care, there is more pressure to use a more inclu-

sive social model of care in home- and community-based long-term care programs. Holistic care models, which stress prevention and rehabilitation, are primarily confined to a few health maintenance organizations. Within the area of personal care, the home-care model, which sees elders as relatively helpless cases to be managed, currently dominates care financed by Medicaid and Social Services Block Grant funds. But there is a growing challenge to consider the personal assistance model, in which elders receive the funds to arrange and pay for the personal and health care they decide they need.

Elders who need a variety of health care and long-term care services do not find it easy to move through the disconnected system we now have. Often the acute care system and the long-term care systems are unaware of what the other is doing, which often leaves the elderly clients angry, frustrated, and inappropriately cared for. Lack of coordination can also be frustrating for care providers. For example, home-care workers can arrive to provide services only to discover that the client is in the hospital, but no one bothered to inform the home-care agency.

But by far the biggest problem in providing adequate health and long-term care services to older Americans is the runaway inflation that has plagued the health-care sector for the past several years. As long as we cannot bring health-care inflation under control, everyone, including older people, finds themselves without health coverage, paying more out-of-pocket, and worrying more about their ability to get needed health care in the future. The problems we have now with financing Medicare and Medicaid can be traced back, not to overutilization of services by older people or even the rapid growth of the older population, but rather to the astronomical rates of health-care inflation for everyone.

Part of the high costs of health care results from the institutional bias built into our methods for reimbursing health-care services. For example, Medicare only pays for care in hospitals or following a hospital stay. Likewise, Medicaid emphasizes reimbursement for care in nursing homes. In both cases, less expensive alternatives exist, are effective and would free public resources to cover more services or more people.

This institutional bias did not emerge out of thin air. It was the result of extensive lobbying by hospital associations, the American Medical Association, and nursing home trade associations. These organizations spend substantial sums to support the election campaigns of legislators, and in return they often have been given a greater influence than citizen's groups in shaping health care and long-term care policy. However, the budget crises facing both federal and state governments is forcing legislators to look for less expensive ways to provide health care, and this pressure is gradually altering the way we do business in both health care and long-term care.

A variety of mechanisms, such as certification, licensure, and reimbursement formulas, are used to control the quality and cost of care, but none of them is likely to work very well as long as there is a lack of commitment to enforcement of care standards or cost controls. In the political climate of 1995, which emphasized getting the government out of the control business, it is highly likely that quality monitoring and cost-containment efforts will become even less effective than in the past. Economic conservatives are banking on competition to lower costs of health and long-term care, but in these areas, consumers do not shop for the lowest-cost care, they shop for the most effective care. As a result, current policy makers are probably overestimating the savings that can be achieved through relatively unrestrained competition. Unfortunately, we are likely to be left to conclude that some sort of price controls are needed, but price controls are anathema to economic conservatives. How we handle this thorny set of public policy issues will have a profound effect on the future availability of health care and long-term care for older Americans.

Thus far, cost-containment regulations have succeeded in reducing the number of people being served and increasing the portion of care that elders themselves must pay for, but they have had little effect on inflation in care costs.

15 ■ Community Social Services

The community, or subcommunity in larger areas, is an important focal point in most people's lives. People are born, reared, educated, married, housed, fed, healed, mourned, and buried in the local community. Work, play, love, politics, fellowship, and self-discovery most often occur in the context of the local community. The picture of society that people gain is substantially influenced by the extent to which the city or neighborhood in which they live is a unified community. Because older people are usually long-term community residents, they are especially likely to see the community as the locus of life's most salient moments. Thus, how services to elders are organized within communities is extremely important. As federal and state governments feel pressured to cut spending, communities are being called on to provide more local funding for all forms of social services, including services to older residents.

The Older Americans Comprehensive Service Amendments of 1973 created a new community organization, the **Area Agency on Aging (AAA).** Along with significant increases in federal funds to local programs for older Americans, the Comprehensive Service Amendments (through the AAA) placed new priorities on coordination of services and on planning. In 1993, there were about 670 AAAs, each charged with developing plans for a comprehensive and coordinated network of services to older people and offering services in the areas of information and referral, escort, transportation, and outreach. The AAA concept emphasizes flexibility in uniting particular sets of local organizations to meet the needs of local older people, therefore allowing for local and regional variations in resources and service needs. At the same time, the AAA concept seeks to make a minimum set of services available to all older people. After the AAAs were created in 1974, services to older people improved dramatically in most parts of the nation.

AAAs are important focal points around which local fund-raising efforts can be organized. For example, in 1993, the AAA in the Cincinnati, Ohio, area organized a successful tax levy campaign to provide funding for home-care services to elders whose incomes were above the poverty threshold used to qualify for Medicaid-funded home care. As a result of this initiative, over one thousand more elders were able to secure home care than could be accommodated by the Medicaid home care program, also administered by the AAA.

Community Facilities That Serve Older People

Community facilities are organized service centers—stores, banks, churches, doctors' offices, hospitals, and so forth. There is usually a mixture of public and private facilities in any given community. Taietz (1975) intensively studied community facilities serving older people in 144 New York communities. He found that specialized facilities tended to be present in communities with a high degree of complexity and specialization. Only the most rudimentary facilities were available in a majority of the 144 communities surveyed, and facilities were particularly lacking in rural areas.

Because most Americans live in urban areas, it is easy to forget that the majority of communities are in rural areas. More than 80 percent of the land area in the United States is rural, nonmetropolitan land. The impact of this fact on the prevalence of facilities is a topic very worthy of study.

Senior centers are widespread in the United States and are the most common community facilities aimed at serving older people. Krout, Cutler, and Coward (1990) estimated that there were more than 10,000 senior centers serving at least 5 million older people. The quality of senior center facilities ranges from relatively new, spacious buildings constructed specifically for senior center use to shabby homes converted for program use. In the Older Americans Act, multipurpose senior centers have for many years been identified as a preferred focal point for the coordination of services to older people in the community. Having a visible facility aids in this function.

Services

*S*ocial services consist of a broad range of often unrelated programs that revolve around a general goal of helping people get the things they need. The range includes family services, senior centers, the Foster Grandparent program, Talking Books, meals programs, employment services, and protective services. Since communities vary widely with regard to the number and range of such programs, and program titles vary as well, instead of describing the typical community, we examine the types of programs commonly found in communities.

Variety of Services *Meals* programs either take food to people or bring people to food.

Meals on Wheels programs deliver hot meals to older people in their own homes. Congregate meals programs bring older people to a central site for meals. In recent years, the congregate meals approach has been gaining favor because it provides fellowship as well as more opportunities to tie in with other social service programs at the congregate meal sites. Many senior centers have congregate meals programs.

Information and referral services provide a bridge between people with needs and appropriate service agencies. Area Agencies on Aging are responsible for these programs in most areas. Many areas have directories of services for older people.

Visitor programs offer contact between older shut-ins and the outside world, and often serve institutionalized older people as well as those living in their own homes. These programs are usually staffed by volunteers, and older people often serve as visitors.

Outreach services seek out older people in need of services, who often become known to the program via relatives or neighbors. Outreach workers contact older people and refer them to appropriate agencies. Sometimes outreach workers make agency contacts on behalf of older people.

Telephone reassurance services give isolated older people a point of contact and a sense of continuity. In an ideal phone reassurance program, the people working on the phones are well trained in referral and yet also know the older person through regular telephone visits. Regular calls from the program assure older people that someone cares about them and will be checking on their welfare regularly. Volunteers are often used for telephone reassurance.

Employment services seek to connect older workers with jobs. Sometimes such services maintain a file of retired people from various occupations who are available for short-term or part-time employment.

Homemaker services provide household support services to semi-independent older people living in their own homes. In addition to the usual housekeeping chores, such as cleaning, shopping, and laundry, some programs also offer home maintenance and food preparation services. Homemaker services need to be suited to the older person's level of impairment. Highly impaired people need a variety of housekeeping services on a daily basis, while others may need only occasional specialized services.

Income counseling programs help people get maximum use of their income resources by making sure they are aware of all possible sources of income and by helping them make the transition from a preretirement to a postretirement budget. Income counseling often includes such things as how to buy consumer goods at the lowest prices, how to take advantage of seasonal sales, how to avoid the cost of credit buying, how to form consumer cooperatives, how to save on rent or get into low-rent public housing, how to save on automobile insurance, how to save on building repairs, and so on. *Money management* programs help elders with such tasks as bill paying, banking, and safeguarding their savings. Although older people highly value these types of assistance, very few communities offer them.

Most large communities have *senior centers*, which usually take the form of private nonprofit corporations. Often such centers constitute the sole community attempt to offer recreational and educational programs for older people. The small percentage of older people who use them (usually under 15 percent) would seem to indicate that senior centers do not constitute a focal point for recreation and education among most older people. Of course, many older people who might otherwise use a senior center's facilities are prevented from doing so by difficulties of access—poor transportation service, disability, and so on. Yet even in communities where a concerted effort has been made to give older people access to senior centers, only a small minority have taken advantage of them.

No one is sure why most older Americans do not participate in Senior Centers. However, continuity of lifelong involvement in churches and other community organizations probably meets the needs of many who wish to maintain community involvement.

The average multipurpose senior center tends to adopt a flexible program—one that offers informal companionship, community services, self-government within the center, and a wide variety of other possible features. Membership in senior centers tends to be drawn from a wide area rather than a single neighborhood. Initial membership is often related to a major life change such as retirement or widowhood, and joiners tend to be healthy and ambulatory. Members of senior centers do not seem to be different from other older people.

Adult day care primarily provides daytime services to older people who are being cared for in the community by their families and who are unable to be left alone. Services range from meals and custodial care to physical therapy, rehabilitation, and psychotherapy. Programs funded by Medicaid tend to emphasize medical and health services, while those funded under social services programs tend to emphasize activities and meals (Rathbone-McCuan and Coward 1985). Adult day care allows families to be employed while still assuming primary responsibility for the care of an older family member.

Respite care is temporary personal and nursing care to disabled elders. The purpose of respite care is to provide temporary relief for informal caregivers who are caring for older people who live in the community. By providing respite care, communities can help prevent caregiver burnout and allow impaired older people to remain longer in their independent households.

Home care and home health care were discussed in the previous chapter. These services, which usually employ *case management* approaches, are designed to deliver health and personal care to older people living in the community.

Protective services take responsibility for older people who can no longer be responsible for themselves. Everyone has heard of aging, confused hermits who have outlived their families and are starving because they have hidden their Social Security checks and cannot recall where they put them. Take, for example, the case of Mr. M., who had been complaining to a variety of agencies about Mrs. S.—an 88-year-old woman whose house was adjacent to his—and finally reached the Senior Information Center. Mrs. S. a recluse, had harassed the M.'s by making loud noises through their bedroom window, banging her porch door, and calling obscene things to their children playing in the yard. Her house was run down, and weeds and bushes had overgrown the garden. Neighbors had photographed her feeding rats in her backyard. They described her as resembling a fairy-tale witch—with flowing gray hair and a long, dirty skirt, living in silent seclusion. Her only relative was a niece, who for the last several years had refused to become involved. Mrs. S.'s Social Security income was very low.

A home call from the Senior Information Center caseworker confirmed the grim picture. Mrs. S., suspicious at first, refused to let the caseworker into the house but came out on the porch to talk. She tried to hide a dirty, stained slip, and there was a strong odor of urine about her. She denied having any problems other than those caused by the "gangsters" next door, and she revealed a paranoid trend in her thinking. As she talked, she became almost friendly, joked about getting a miniskirt, and assured the caseworker that there was nothing to worry about. Later, a city sanitation inspector went through the house, saw one rat and evidence of rats' presence throughout the house. The odor of urine and feces was strong. The inspector also

Social Services can include counseling and case management.
Photograph by Marianne Gontarz

found about fifty wine bottles, which added to the picture. Finally, Mr. M., angered over the continued harassment, nailed shut the back door of her house. Fortunately, another neighbor realized the danger in case of fire and removed the nails (adapted from Ross 1968,50).

Protective services are used by the community when it becomes clear from the behavior of older people that they are mentally incapable of caring for themselves and their interests, as in the case of Mrs. S. It is estimated that about 1 in every 20 older people needs some form of protective service, a proportion that can be expected to increase as the proportion of older people over age 75 increases. The typical response to someone like Mrs. S. has been to put the individual in an institution; however, this solution is coming under increased scrutiny. Many older people now in institutions could be left at home with a minimum of help in securing the support services already available to them. The most difficult problem often is to find someone who will initiate action. In many states, the law assumes that a relative will initiate the proceedings. In the absence of a relative, no one is willing to take responsibility. Laws need to be changed to more clearly pinpoint that responsibility.

Ultimately, protective services are outreach services rather than deskbound ones. Legal intervention is usually a last resort, used only when all other alternatives have been exhausted. Agencies must often seek out people who might not even want the service at all. It is obvious that while protective services must be a part of any community service program for older people, assigning responsibility for such services is a difficult task.

Nursing home ombudsman programs monitor care in nursing homes, investigate complaints from residents or family members, assist residents in making their case, and serve as advocates for the rights of nursing home residents (Nelson, Huber, and Walter

1995). Ombudspersons also create and maintain communication links between residents, nursing home administrators and staff, and nursing home regulators. Some ombudsman programs use a collaborative model, which assumes that just being there to monitor activity and advocate for resident rights in a conciliatory way will improve the quality of nursing home care. This approach works when the nursing home is open to this kind of participation (Litwin and Monk 1987). But in many cases, the nursing home is defensive about complaints and regulatory agencies may be less inclined to investigate complaints. In these cases, a more assertive ombudsman process may be needed. The adversarial model of ombudsman program operation stresses advocacy more than conciliation. In this model, the ombudsperson encourages residents to file official complaints with regulatory agencies when they feel there is a basis for complaint, follows through to insure that the regulatory agency investigates complaints, reviews results of the investigation, and requests further investigation if necessary.

Nelson and his colleagues (1995) studied a large sample of Oregon nursing homes: forty-six that had been assigned a trained volunteer ombudsperson for at least four hours per week and twenty facilities that did not have one. They found that the presence of an activist ombudsperson resulted in a significantly greater number of official resident abuse complaints, a greater number that were substantiated by investigation, and more letters of reprimand in resident abuse cases. Presence of an adversarial ombudsperson did not reduce the incidence of nursing home survey deficiencies or increase the number of civil penalties. Thus, the adversarial approach appears to result in greater regulatory control in cases of resident abuse.

Elder Abuse Programs   *Elder abuse prevention and intervention* is aimed at one of the most tragic circumstances that can befall older Americans. In recent years, the topic of *elder abuse* has received widespread attention by legislative committees and the media. In addition, this topic has received increasing research attention. Hickey and Douglass (1981) classified neglect and abuse in the order of prevalence and seriousness as perceived by a sample of service providers. *Passive neglect*, leaving an older person alone, isolated, or forgotten, was seen as the most common and least serious. *Verbal or emotional abuse* occurs in situations in which an older person is humiliated, frightened, insulted, threatened, or treated as infantile and was seen as the next most common type of abuse. *Active neglect* involves forced confinement or isolation or withholding of food or medication and was seen as being not very common. Actual *physical abuse,* hitting, slapping, or physical restraint, was seen as the least common type of mistreatment.

Research on abuse and neglect is difficult because many of the concepts are vague and open to multiple interpretations. For example, what seems insulting or humiliating to one person may not to another. Pedrick-Cornell and Gelles (1982) surveyed the research on elder abuse and concluded that "the extent and incidence of abuse of the elderly is still unknown." In addition, although elder abuse is widely thought to be associated with high levels of impairment on the part of the older person, Pedrick-Cornell and Gelles found that these claims had almost no empirical evidence to support them. In a later study, however, Wolf and Pillemer (1989) found that among reported cases of abuse and neglect, 75 percent of the victims had experienced recent declines in physical health and 40 percent were mentally disoriented.

In one of the few careful studies of elder abuse, Pillemer (1985) studied a sample of forty-two older people who had been physically abused and forty-two nonabused older people who were matched by gender and living arrangements with an abused elder. He found no support for the hypothesis that dependency of the elder leads to abuse. Compared with the control group, the abused elders in his study were *younger, less* physically impaired, *less* physically dependent, and *less* financially dependent. Pillemer then turned his attention to the characteristics of the abusers and found, surprisingly, that abusers were very likely to be *dependent on the older person* for housing and financial assistance. Indeed, only one-third of the abusers were financially independent of the abused elder.

When Pillemer classified the abuse cases according to dependency, he found that dependency was not involved or was ambiguous in eight cases (half of which involved longstanding patterns of spouse abuse); the abused elder was dependent in seven cases (five involving wives becoming dependent on their husbands); and in the remaining twenty-seven cases the abuser (usually an adult son or daughter) was dependent on the abused elder. A number of the abusive adult children were mentally impaired. Pillemer found that the main reason given for staying with an abusive dependent relative was a perception of having little choice. Responses such as "Where else is she going to go?" and "You can't throw your children out," were common.

Pillemer and Finkelhor (1988) studied abuse and neglect among noninstitutionalized older people living in the Boston area. The Commonwealth of Massachusetts requires each municipality to survey the population and publish a listing of all residents of each dwelling unit. Potential respondents were randomly selected from this listing, which included birthdates. Of the 2,020 older people interviewed, just over 3 percent had experienced abuse or neglect: 2 percent physical violence, 1 percent chronic verbal aggression, and 0.4 percent neglect in the form of withheld aid in activities of daily living. These findings contradict the perceptions of service providers, who saw physical violence as least prevalent (Hickey and Douglass 1981).

Although maltreatment was not widespread in terms of the proportion of the population involved, the number of cases was large. For example, based on their findings, Pillemer and Finkelhor estimated that at least 8,600 cases of maltreatment occurred annually in the Boston area alone.

Abused elders tended to be men who lived with their spouse or with their spouse and an adult child, to be married, and to be in poor health. Abuse was not related to minority status, age, or socioeconomic status. Spouse abuse was the major form of physical violence. Most elders who are abused live with someone, and most older people who live with someone live with their spouse. Thus, 58 percent of people who abused older people were spouses, compared with 24 percent by adult children. Men were twice as likely to be abused as women; however, abuse against women was more serious. For example, only 6 percent of physically abused men said they suffered injuries, compared with 57 percent of physically abused women. Thus, the image of the abused elder as being a very old, single, very impaired female who is excessively dependent on an adult child can be seriously called into question.

Wolf and Pillemer (1989) looked at the characteristics of the offenders in reported cases of abuse and neglect. They found that 72 percent were age 40 or over, men outnumbered women by 2 to 1, and 75 percent lived with the victim. About 40 percent of the offenders had a history of mental health problems and alcohol abuse, and one-third were experiencing chronic financial problems. Perpetrators tended to be dependent on the elder financially but to be independent otherwise. They also tended to have unrealistic expectations concerning how much the elders could do for themselves and to see the elders' requests for aid as unnecessary and unreasonable. Sons were the most common offenders (28 percent), followed by daughters (18 percent), and husbands (18 percent). When these data are compared with the results of Pillemer and Finkelhor's (1988) survey of the total population, it seems probable that many cases of spouse abuse go unreported.

Abuse is sometimes a two-way street. Steinmetz (1988) pointed out that disabled elders use a variety of abusive tactics in their attempts to maintain control over their caregivers. Elders whose relationships with their children have been typified by verbal and physical abuse over the duration of the relationship could be expected to continue these patterns even in a caregiving situation. Steinmetz reported that more than one-third of the elders in her study yelled at their caregivers and/or cried as a way of gaining sympathy or control; two-thirds used guilt to manipulate their caregivers; and about 18 percent were physically violent toward their caregivers. The reciprocal nature of abuse deserves further research.

Cases of abuse and neglect of older people are highly dramatized by the media because they offend very central American values, respect and consideration for elders. They arouse the fears of older people because there is no way to be positive that elder abuse will not happen to them.

Wolf and Pillemer (1994) reported on several innovative, grant-funded social service programs designed to address elder abuse. In San Francisco, a multiprofessional team made up of representatives from case management, family counseling, mental health services, geriatric medicine, law enforcement, civil law, financial management, and adult protective services met monthly to consider two new elder abuse cases and follow up on at least two others. Community agencies referred cases to the team to clarify individual agency roles, receive advice in handling nontypical cases, seek solutions in difficult cases, resolve disputes between agencies about aspects of the case, or get legal or medical advice not available to them otherwise. The main objective of the team was to produce a care plan that was comprehensive. The team produced a written summary of the problems in each case and the interventions recommended. It also monitored the cases to ascertain if recommendations were followed. Most cases were successfully resolved, but not all. "Particularly difficult were those in which the victim's mental capacity was in question or in which an abuser, a mentally ill adult child, required treatment that was not readily available" (1994,127).

In Wisconsin, the Senior Advocacy Volunteer Program used volunteers to provide elders experiencing or threatened with abuse or neglect with a well-trained partner who would serve as a link to direct services, as their advocate, and as someone to whom they could turn for emotional support. The program staff matched trained volunteers with clients based on the seriousness of the case and the background of the volunteer.

In New York City, a victim support group was established at Mount Sinai Medical Center for men and women age 60 or older who had been identified by various agencies as victims of abuse or neglect by a family member. Before entering the support group, each client's situation was assessed by staff and individual needs were addressed. The victim support groups were limited to ten members, facilitated by the project director and a social worker, and met in ten-session cycles. Participants reported that the group "eased their sense of isolation, buffered their feelings of victimization, and served as a 'family' for them." Participants also were much more aware of their options and felt more control over their situations (1994,128).

Wolf and Pillemer pointed out that these various solutions were practical in large urban areas where there was a sufficient client base when they had the full support of local adult protective services systems as well as the cooperation of the many human services agencies that make up an "elder abuse network" in local communities. The greatest challenge these programs faced was finding long-term funding. Wolf and Pillemer were especially impressed with the potential for replication of these low-cost, flexible, and innovative programs.

Transportation *Transportation* is an important everyday need in America. Aging changes our transportation needs and resources in several ways. For some, declining sight impairs the ability to drive an automobile. Lowered income forces still others to give up their cars or to use them sparingly. The increase with age in the proportion of physically impaired people also raises the proportion who need specialized transportation.

In this country, we have been experiencing a long-term trend away from public transportation. Its use is declining in all but our largest cities. No longer a paying proposition, and usually requiring tax subsidies, public transportation is used most for travel to and from work. The needs of the older population have not been considered in most decisions about public transportation.

Older people fall into two categories with regard to transportation: those who can use present facilities and those who cannot. Those with no transportation problems tend to be the ones who can afford to own and operate their own cars, about 46 percent of older people. For these people, public transportation is something to be used when it snows or only for long trips.

Older people with transportation problems fall into three groups: (1) those who could use existing public transportation but cannot afford it, (2) those who for whatever reason need to be picked up from and returned directly to their homes, and (3) those who live in areas with no public transportation.

Cost is an important factor. In the United States roughly 10 million older people are hampered by the cost of transportation often because they cannot afford a car, bus fares are beyond their means, or cab fares are too high. For these people, lack of adequate, inexpensive transportation is one of the most important limitations on their independence and activity.

Many elders depend on public transportation.
Photograph by Bill Aron/Tony Stone Images, Inc.

Older people who live in suburban and rural areas often have no access to public buses or taxi service. About 8 million older people in rural areas lack adequate transportation, and the growing population of older suburban residents is also likely to lack alternatives to private automobile transportation.

Among older people with transportation problems, most are still able to get to the doctor, dentist, and grocery store. But many do not get out to see their friends and relatives or go to church or find recreation. Although they manage to keep alive, they are unable to do the things that give meaning to life. And when they can get out, it usually must be at someone else's convenience. Older people's pride and dignity often prevent them from relying on friends and relatives for transportation. What they need is a dependable transportation system at prices they can afford.

A comprehensive transportation plan for older people might consist of four elements. First, low-income older people could receive fare reductions or discounts on all public transportation, including interstate transport. Second, public subsidies could be provided for adequate scheduling and routing of existing public transportation. Third, taxi fares for the disabled or infirm could be subsidized. Fourth, funds could be allocated to senior centers to purchase and equip vehicles for use in transporting older people, particularly in rural areas lacking public transportation.

In the 1970s, it became clear that transportation was a major issue in access to facilities and services. As a result, most programs funded by the Older Americans Act developed specialized transportation services. In 1988, over 7 million older people received transportation services funded through the Older Americans Act (U.S. Senate Special Committee on Aging 1989a). However, since 1980, the federal government has steadily pressed the states to assume more responsibility for funding transportation programs; therefore, continuation of funding for transportation programs aimed at the needs of older people is far from assured.

The Organization and Financing of Social Services

Social services assist people in improving their level of functioning or reducing their difficulties in securing adequate income, health care, housing, transportation, or social participation. Such assistance can take many forms: financial aid, information and referral, counseling, education, physical assistance, chore services, in-kind assistance such as rent subsidy or free meals, and a host of others.

Like health care, social services are provided by a wide variety of types of personnel and types of organizations, with funds from a variety of sources. At the national level, various federal departments include social services among their responsibilities. However, federal funds for social services to older people are provided mainly through the programs of the Older Americans Act and through Social Services Block Grants to state and local governments.

State Units on Aging and state departments of human services administer both state and federal social services funds at the state level, and local Area Agencies on Aging and human services departments administer them at the local level. Local United Way organizations also provide local funding for social services, as do many local philanthropic foundations. In addition, many social services agencies do independent local fund-raising. As a result, it is impossible to identify the precise amounts of funding for social services to elders at all levels.

However, given that more than half of all funding for social services to older people comes from federal sources and that the one-third of social service dollars that state governments provide are tied to participation in federal programs, it seems safe to conclude that the structure and operations of federal social service programs greatly influence local efforts to provide social services to older people. Accordingly, we look at the Older Americans Act and Social Services Block Grants, the two major federal funding vehicles.

The Older Americans Act

The Older Americans Act of 1965* identified a series of broad objectives aimed at improving the lives of older Americans; facilitated the agencies responsible for funding, planning, coordinating, and monitoring social services to older people; and provided funding for specific social services such as multipurpose senior centers and community service employment. In 1972, the act was amended to create a national nutrition program for elders and to authorize grants to public and nonprofit agencies for the development of congregate meal services for older people. In 1973, the act was amended to create a network of Area Agencies on Aging, which were given the major responsibility for planning, coordinating, and advocating local programs that would benefit older people. In 1973, amendments created a special authorization for home-delivered meals for elders and required states to establish statewide nursing home ombudsman programs. During its first decade, Congress repeatedly supported the Older Americans Act, and it gradually developed into a major social services resource for agencies providing social services to older people.

As amended in 1981, the Older Americans Act has six sections, called *titles*. Separate appropriations are usually made for each title and sometimes for separate components within titles.

Title I outlines ten broad objectives for older Americans, those 60 and over, including:

1. adequate income

2. physical and mental health

3. suitable housing

---

* Material in this section is taken from the U.S. Senate Special Committee on Aging (1988).

4. full restorative services for those needing institutional care

5. employment without age discrimination

6. retirement in health, honor, and dignity

7. participation in civic, cultural, and recreational activities

8. efficient community services

9. benefits from research designed to sustain and improve health and happiness

10. freedom to plan and manage their lives

These very general goals are for the entire society, and they are addressed through many efforts at the federal, state, and local levels. However, the Older Americans Act clearly articulates a national policy of joint federal, state, community, and family responsibility for the well-being of elders.

Title II established the Administration on Aging (AoA) as the principal agency for carrying out the purposes of the Older Americans Act and for administering the various programs authorized by it. Located within the Office of Human Development Services, AoA is directed by the U.S. Commissioner on Aging, who is appointed by the president and confirmed by the Senate. AoA was intended to have high visibility in the executive branch of government and to serve as an effective advocate on all matters related to the field of aging.

Title II also authorized the federal Council on Aging, a fifteen-member body whose members are appointed by the president. Members of the council cannot be federal employees, and at least five of them must be older individuals. The council was intended to serve as an advisory and advocacy body to the president and the commissioner on aging, to inform the public about matters related to aging, and to provide a forum for discussion of problems and needs of the aging.

Title III authorized grants to state agencies on aging for the development of a compre-

hensive and coordinated delivery system of supportive social services and senior centers, congregate nutritional services, and home-delivered nutrition services. To qualify for funds, the state agency was required to divide the state into separate geographic areas, called Planning and Service Areas (PSAs), and establish AAAs charged with developing generally comparable delivery systems within each PSA. AAAs were responsible for coordinating existing services and fostering the expansion and development of community services for older people. This organizational structure was intended to form an "aging network" made up of AoA, State Units on Aging, AAAs, other public and private agencies, and local service providers and to provide a continuum of services to older people.

Title III required AAAs to allocate "an adequate proportion" of their funds to three categories of social services: access services such as transportation, outreach, and information and referral; in-home services such as homemaker, home health, telephone reassurance, and friendly visiting; and legal services. In addition, AAAs were allowed to contract for ombudsman services, counseling, case management, health screening, employment services, crime prevention, victim assistance, and volunteer service opportunity programs. Funds for these services were authorized under Title III-B.

Title III also authorized grants to establish and/or operate congregate and home-delivered meals programs for persons 60 or over and their spouses of any age. Grants could go only to public and nonprofit organizations. Participants could be asked to pay what they felt they could afford, with the proceeds used to increase the number of meals served. Funds for meals programs came from Title III-C.

In 1987, the Older Americans Act was amended to add Title III-D, which provided specific authorization for nonmedical home care for frail older people. Services that could be provided under this new program included homemaker and home health aides, visiting and telephone reassurance, chore and minor maintenance services, and adult day care as a respite for family caregivers. Other additions to Title III included health promotion activities, elder abuse prevention, and outreach to older people potentially eligible for SSI, food stamps, or Medicaid.

The 1987 amendments also strengthened the long-term care ombudsman program, which investigates citizen complaints and solves problems on behalf of older nursing home residents and their families, in two important ways. First, states were required to enact provisions providing ombudspersons with immunity from lawsuits connected to the good faith performance of their duties and to provide legal counsel if required. Second, retaliation against residents or others who complain was prohibited.

Title IV authorized support for training, education, research, demonstration, and evaluation projects that would add knowledge that could be used to improve the effectiveness and efficiency of Older Americans Act-sponsored programs. Part A provided funds to recruit and train personnel for the field of aging. Part B provided research and demonstration funds for projects in a variety of areas, such as long-term care, rural transportation, and mental health. Title IV-B provided most of the applied research funding for gerontology.

Title V established the Senior Community Service Employment Program within the Department of Labor, which created part-time public service employment positions at the minimum wage for people 55 and older who had incomes no higher than 125 percent of the poverty level. Positions created under Title V included aides in schools, libraries, hospitals, and social service agencies. This program is managed by state Units on Aging and eight national contractors, including the National Council on Aging, the National Council of Senior Citizens, and AARP.

Title VI promotes the development and delivery of services to older Native Americans through grants to tribal organizations.

Older Americans Act programs are administered through a complex of organizations that in 1995 involved 57 State Units on Aging, about 670 AAAs, nearly 7,000 senior centers, and more than 18,000 other local supportive and nutritional service providers, including 13,000 nutrition sites. Throughout the 1980s and early 1990s, funding for the Older Americans Act remained stable at about $1 billion while the number of services that could be provided dramatically increased as did the numbers of older people. Older Americans Act funding remains a minuscule part of the total federal budget, but locally it plays an important role by providing an organizational infrastructure that serves as a magnet for funding from state governments, local governments, and philanthropic agencies.

Social Services Block Grants

Before 1981, federal funds for social services to the poor were provided by Title XX to the Social Security Act, which targeted specific populations to be served, such as those on SSI or Medicaid. In 1981, Congress created less restrictive "block grants" for social services but at the same time reduced the federal appropriations for social services by 20 percent. A total of $2.5 billion was authorized for fiscal year 1983. Social Services Block Grant funds are funneled to local agencies through state and local human services departments.

Appropriations for Social Services Block Grants (SSBGs) held steady at around $2.5 billion from 1981 to 1991. Gaberlavage (1987) surveyed states to determine the proportion of SSBG funds being used to provide services to older people: forty-seven states were using an average of 18 percent of their SSBG funds to provide social services to older people. The most frequently provided services were home care, adult protective services, adult day care, transportation, and nutrition services. Most states saw these services as preventive, designed to maintain older people in independent households. Although there has been a consistent level of funding for

SSBGs, there has been a sharp drop in the number of clients served because services have been focused on older people with very low incomes and who therefore need more services. This lowering of income eligibility levels for SSBG programs has reduced access to in-home care for many older people.

The U.S. Senate Special Committee on Aging (1988,338) concluded its examination of SSBG programs with some sobering thoughts:

It seems clear that while funding for the SSBG has remained relatively constant, there is a strong potential for fierce competition among competing recipient groups. Increasing social service needs along with declining support dollars portend a trend of continuing political struggle between the interests of the elderly indigent and those of indigent mothers and children. In the coming years, the fiscal squeeze in social service programs could have massive political reverberations for Congress, the Administration, and State governments as policy makers contend with issues of access and equity in the allocation of scarce resources.

As the federal government seeks to lower its funding commitments to social programs by using block grants for a larger array of programs, these kinds of problems can be expected to become more widespread.

The Older Americans Act, Social Services Block Grants, United Way, private philanthropy, and local foundations together may have provided as much as $2.2 billion for social services to older people in 1988 (an educated guess). In addition, about $1.5 billion from Medicare and Medicaid went for social services. Thus, a total of perhaps $3.7 billion was available from all sources, amounting to about $130 per older person per year. Contrast this with the $65 billion for Medicare and $9 billion to the older population from Medicaid—a total of about $2,500 per older person per year. From this it should be obvious that although many of older people's needs in 1988 involved in-home services of a nonnursing nature, such as homemaker or meals services, 98 percent of the federal dollars were going to health care. Given that health care and social services support one another, our health efforts would probably be more effective if the social services activities were not so severely underfunded. Ironically, social services to the older population tend to be much better coordinated than health services. Perhaps coordination is improved when it is required and when the amount of money to be made is insufficient to attract competing, large organizations. This would be a good topic for further research.

The increasing demand for services, cuts in funding for social services, and the projected growth in the older population are all fueling a debate over whether social services to elders should continue to be made available to all older people, as they are now in most cases, or whether we should go further toward means-tested or needs-tested programs, as is now the case for those receiving SSI and Medicaid-related social services. Indeed, since 1990 there has been a concerted effort to target Older Americans Act funding to the most economically disadvantaged older Americans. But an old axiom states, "Services that serve only the poor tend to be poor services." When the middle class is not served by government programs, the political will to finance services tends to diminish, resulting in a spiraling inability to sustain programs designed to meet needs concentrated only among the poor. The debate over further exclusion of the middle class from Older Americans Act programs will surely intensify in the coming years and should be a fertile field for research.

Summary

The local community is where social services to older people are delivered. The availability and adequacy of services varies considerably across communities, with urban communities generally offering a greater variety of services than rural communities. The creation of Area Agencies on Aging greatly improved access to services for older Americans throughout the country, but the failure of federal and state funding for these services to keep pace with the growing older population since 1980 has meant a greater reliance on local funding, either through tax levies or local philanthropy. This in turn has led to greater variability in services across communities.

Senior centers are the most common community facility serving elders. There are about 10,000 multipurpose senior centers throughout the country, and most of them provide at least minimal service coordination within local communities.

Services that are available to older people in many communities include information and referral services, outreach, transportation, visitor programs, telephone reassurance programs, senior employment services, income counseling and money management programs, home health and home care services, protective services, nursing home ombudsman services, adult day care, and respite care. These services are often coordinated by senior centers, Area Agencies on Aging, or case management agencies.

Elder abuse programs are generally aimed at intervening to stop abuse and neglect of older people, usually by family members. Although the proportion of elders who experience abuse or neglect is estimated at a seemingly small 3 percent, this represents nearly one million elders nationally. Elder abuse and neglect occurs at all economic and social levels. Abusers are usually spouses or adult children. Adult child abusers are often alcohol or drug-dependent or mentally ill people who live with their parents and are dependent on them. Most service programs aim at intervention through multiprofessional teams, interagency cooperation, advocacy programs, victim support groups, and better education of social workers to deal with elder abuse.

Transportation becomes increasingly problematic as age increases because our society is organized spatially on the premise that everyone has access to a private auto-

mobile everywhere but in the largest cities, where public transportation is still available. However, because disability increases with age, the proportion who can use private automobiles declines with age, and the need for specialized door-to-door transportation increases with age. Most communities have developed specialized transportation systems as part of their social service programs for older people. These systems provide access to programs such as senior centers and basic transportation to grocery stores and doctors' offices, but they often ignore the transportation needed to remain active in other recreational and social activities.

Organization and financing of social services to older people has traditionally benefited from strong leadership from national government and from national voluntary organizations such as the National Council of Senior Citizens and the National Council on the Aging. The Older Americans Act of 1965 has had a particularly strong effect on the organization and operation of social service programs. In addition to articulating a national commitment to providing for the needs of America's elders, the Older Americans Act provided national visibility and developed a national network of state and local agencies responsible for planning and coordinating a comprehensive variety of supportive services for older people. Services funded and coordinated through Older Americans Act–supported State Units on Aging and local Area Agencies on Aging included information and referral, outreach, transportation, meals programs, in-home personal care, legal services, nursing home ombudsman programs, and a host of others. The Older Americans Act also provided funds to train aging network staff, to provide gerontology training in higher education, and to conduct research and demonstration programs aimed at improving services to older Americans. Older Americans Act programs also included senior employment programs for low-income elders. Although its mission is comprehensive, funding for the Older Americans Act social service programs remained at a modest $1 billion from 1980 to 1995. As a result, Older Americans Act services have increasingly been targeted to low-income elders.

Social Security Block Grants provide home care, adult day care, and nutrition services targeted to elders with very low incomes. Block granting generally has been used to reduce federal funding commitments and often produces conflict between various needy groups for a smaller pool of state and federal resources. Because state participation in social service programs is usually geared to match federal grants, when federal dollars diminish, so do state dollars. States often respond to these trends by passing responsibility for maintaining programs to local communities, which then have to either raise local taxes or cut services.

Ninety-eight percent of federal service dollars go to health care and personal care, which ignores completely the enormous value of social service programs in preventing or delaying dependency among older people. But given the antigovernment climate of the mid-1990s, redressing this imbalance is unlikely to be given a high priority.

As the older population itself ages, the types of services needed increase in number. The number of elders is growing rapidly at all ages, but especially at the oldest ages. Yet budget cuts are being planned in all forms of health and social services. These trends will undoubtedly increase pressures to further ration health and social services to older people. It is quite likely that elders at most economic levels will find it increasingly difficult to get the services they need and/or to pay for them. We will return to the subject of rationing in Chapter 16, when we discuss ethical issues in aging.

Part 5

Aging and Society

Part 5 focuses on society at large and its institutions rather than on aging individuals or local situations. Chapter 16 deals with aging in contemporary American society and culture. This chapter focuses on how aging is treated within American culture, particularly in its language, attitudes, values, beliefs, and stereotypes; in the mass media; in education; and in the arts. It also discusses the underpinnings of age prejudice, age discrimination, societal disengagement, and age stratification. It concludes with a discussion of ethical issues related to aging.

Chapter 17 documents the carryover of social inequalities, particularly in job opportunities, based on race, ethnicity, social class, and gender, into later life and the effects of this carryover, particularly in terms of income inequality. Chapter 18 deals with how the economy responds to an aging society; the impact of economic structure on income in later life, the economic functions of retirement, the present and future of various societal retirement income strategies, the minimal effect of an aging population on society's economic problems, aging people as an economic market, and economic exploitation of older people.

Chapter 19 covers the political structure of America, the political behavior of aging and older people, the degree to which elders can exert political influence and how, the influence of political trends on the politics of age, the concept of social insurance, the politics of Social Security, lessons from the controversial Medicare Catastrophic Coverage Act, and government responses to issues of aging.

As we will see, most of aging and older people's problems arise from categorical treatment based not on the effects of individual aging but on biases and mistaken assumptions built into our social institutions.

16 ▮ Aging in Contemporary American Society and Culture

In this chapter, we will examine what aging means in contemporary American society and culture. We start with a general discussion of society and culture to set the stage, then look at cultural aspects of aging such as ideas, language, and aging as presented in various socializing media such as television, books, and newspapers and magazines. We then consider the sources and dynamics of age prejudice and age discrimination, society's disengagement from aging and older people, and the nature of relations between various age strata in the society. We conclude the chapter with a discussion of ethics as a framework within which various conflicts about aging can be examined.

The Nature of Society and Culture

A *society* is a relatively self-sustaining and autonomous group of people who share a common and at least somewhat distinctive culture, have a feeling of belonging, see themselves as a distinguishable entity, are united by social relationships, and live in a particular territory. Societies are special types of groups in that they have a comprehensive social system that includes all the basic social institutions required to meet human needs. Institutions include such spheres as the economy, government, religion, the family, the military, science, health care, education, social welfare, sports, entertainment, and so on. Autonomy is the key to differentiating the society from smaller social units. For example, the national society generally has the authority to impose legal and administrative rules on states and communities within the nation. Thus, when we speak of society in this chapter, we refer to the geographic territory, population, social arrangements, and culture that represent the United States of America. This chapter deals with general responses to aging and older people by the society and culture as a whole. In the chapters that follow, we will pay special attention to how aging is dealt with in specific social institutions such as the economy, politics, and government.

When democratic societies are interactive groups in which most of the people know one another and communicate with one another regularly, discussion tends to be the dominant mode of communication, and opinions offered can be immediately and effectively considered, answered, incorporated, or rejected. When authorities propose action, the public can engage in direct dialogue about it and provide immediate democratic feedback. Members of the public thus feel empowered by their opportunity to participate in the process of transmitting, maintaining, and reshaping their society and culture.

We live in a different type of society, one in which beliefs, values, and attitudes are shaped by impersonal educational institutions, television, radio, newspapers, and magazines more than by face-to-face discussion. This type of society is called *mass society* because the public is relatively passive in the communication process, and public beliefs and opinions tend to be influenced primarily by information produced by centers of economic and political power. The mass society is dominated by large social organizations: government bureaucracies, corporations, educational establishments, organized religion, professional organizations, advocacy groups, labor unions, and so on. Feedback from the democratic public consists less of give-and-take discussion and more of passive opinion polling. On the surface, opinion polling seems to be a reasonable way to include the public in the decision-making process because the samples used in polls are usually selected to be representative of the population. However, the people conducting the polls decide what questions are asked and how the questions are worded. Both of these features have a great impact on the results of polls. Decisions in the mass society tend to be dominated more by representatives of organized interest groups than by public opinion because polls are often inconsistent and do not provide specific direction. In a pluralistic and democratic mass society, preventing the public from becoming alienated as a result of their passive role in decision making is a major challenge.

Robert Bellah and his colleagues (1985, 1991) pointed out the importance of having an underlying moral framework to guide the action of social institutions in a mass society. Moral frameworks consist of a limited number of basic principles of right and wrong that are used to organize, modify, criticize, and evaluate various goals that society could pursue as well as the means for achieving them. From its very beginning, the moral framework underlying American culture has contained a basic conflict over the proper means for achieving our basic goals of peace, prosperity, freedom, and justice. On the one hand, we believe that the best way to achieve these goals is through preserving the freedom and autonomy of individuals, a "free market" economy, and a procedural political state in which a few impersonal rules are sufficient to protect the rights of individuals. This dimension of our public morality assumes that the best government is the least government, that social institutions should be kept small and local, and that the public good emerges out of individuals operating independently to pursue their own self-interests. On the other hand, we believe that the public good can only be identified through a democratic process of communication and that individual well-being can best be protected and enhanced by an activist government that manages the national economy, protects individuals from exploitation and abuse, and promotes well-being through institutionalized programs such as public health, universal education, unemployment insurance, Social Security, and so on. Resolving the conflict between these two opposing visions of how to achieve the good society involves continuous balancing because there is value in both perspectives. Rational, utilitarian individualism has fueled substantial material progress, but at the same time our commitment to the use of democratic processes to identify and pursue the common good has so far prevented our individualism from descending into a grim war of each against all others.

Many of the ethical dilemmas we experience in relation to aging have their roots in this conflict between utilitarian individualism and our need for a collective vision of the common good. For example, utilitarian individualism has been used to justify making health-care policy on the basis of economic cost-benefit analysis. People using this perspective have argued that we cannot afford to provide needed health care to older people because doing so would place an unfair economic burden on the younger population. Using principles such as community responsibility for the well-being of all of our people, others argue that access to affordable and adequate health care should be a basic human right for all people in our society and that we all must bear our fair share of the cost of providing this care. These people argue that age, like gender or race, is not a fair criterion for denying people access to health care.

Culture is the way of life of a people that is handed down from generation to generation. Culture influences people's language and thought, their habitual ways of meeting their needs, and their physical and social environments. Culture provides the *context* within which life at a particular time and place is led. Culture is a collection of ideas about good and bad, true and untrue, beautiful and ugly, just and unjust, free and unfree, fair and unfair, and equal and unequal. Culture also contains our repository of ideas about how we produce these various outcomes. Finally, culture contains the collective conceptions of the relative desirability of various goals and means that frame our cultural priorities.

Ideas about Aging

Culture consists mainly of *ideas* and their manifestations. All the material artifacts of culture stem from ideas. Ideas are embedded in language, which is our most often-used means of communicating. Ideas can also be classified into various categories such as values, norms, beliefs, attitudes, and stereotypes. Although they tend to be interrelated, these various types of ideas serve slightly different purposes. *Values* tell us the relative desirability of various goals, *norms* tell us what we can and cannot do in pursuing our goals, *beliefs* tell us the nature of reality in which we make choices and act. *Attitudes* are positive or negative predispositions toward people, objects, or ideas. *Stereotypes* are composites of beliefs and attitudes about categories of people, which we use to define what types of people we are dealing with when we have little personal information about them.

Language

As we saw in Chapter 3, historically Americans have had a great deal of ambivalence in their thinking about aging. Because aging has both positive and negative aspects, this ambivalence is understandable. Somewhat harder to understand, though, is the fact that the English language as it is spoken in contemporary America contains few terms for aging or older people that are neutral in meaning. Most of the words we use, such as *the aged, elders,* and *older people* are rejected as terms people would like to have applied to them (Barbato and Feezel 1987). Instead, most of us prefer euphemistic expressions such as *senior, senior citizen, mature American,* or *retired person* (Harris and Associates 1981). However, most people do not want to be referred to by any of these labels.

No doubt part of the problem stems from a long history of negative connotations attached to the word *old.* For most of us, being old does not simply mean that we have existed for a relatively long time; being old also carries the connotation that we exhibit the characteristics typifying someone in the old-age life stage. As we saw in Chapter 1, old age is characterized conceptually by extreme frailty. Given that most people under 85 are not frail, it should not surprise us that most "older" people refuse to answer to that label.

In one survey (Barbato and Feezel 1987), *mature American* and *retired person* were seen as the most positive labels, followed by *senior citizen* and *golden ager.* Labels such as *elderly, aged, old-timer,* and *old folks* were somewhat negative synonyms (as rated by survey respondents), and terms such as *biddy* and *fogy* were seen as definitely negative. On the other hand, gerontologists seem to prefer *elderly, older,* and *aged,* in that order, in the titles of their articles. Obviously we gerontologists are using a different language from the one used by the general public.

Having given the matter a great deal of thought, I believe that *elder* is the most neutral, noneuphemistic noun we can apply to someone who is chronologically age 65 or older. As a noun, elder means a *comparatively* older person. The adjective form is *elderly*. The term *elder* stresses the period after late middle age without implying physical or mental decline and may also carry a connotation of influence in the context of the family or the community (Morris 1982). *Older person*, though ostensibly general, strongly stresses advanced years of age and so applies more to those in their late 70s or older. *Aged* implies infirmity, which is not descriptive of the population age 65 and over as a whole.

Nuessel (1982) identified over seventy terms, such as *coot, geezer, fossil,* and *fogy,* which refer to older people in a negative way. Terms such as *elder statesman* or *gray champion* are seen in a positive light, but they refer to specific older people in specific roles, not to older people in general.

The oversupply of negative language about old age may be in part responsible for the widespread perception that American culture is overwhelmingly prejudiced against aging and elders. Yet if we go deeper into the ideas that help us organize our thoughts, we find a much more mixed picture about how Americans actually feel about aging. The double-edged nature of aging is definitely reflected in our ideas. Part of this apparent contradiction no doubt comes from the fact that we seldom deal with specific people as only "older"; we deal with them as older neighbors, grandmothers, friends, church members, and the like. When we are dealing with specific elders in specific social roles, age is seldom seen as a negative trait. But when we deal with elders *as a social category,* especially in terms of social policy, we tend to do so in terms of negative images. Let us now systematically examine values, beliefs, stereotypes, and attitudes in terms of how they are shaped by and shape our ideas about aging.

Values

One key cultural issue concerns how aging fits into the general values of a society. Rokeach (1973, 1978) did what is perhaps the most comprehensive work on personal values in American society. Rokeach's approach recognized that values concern not only the desirability of various outcomes or goals but also the means for achieving them as well. Accordingly, he developed two scales, one that asked people to rank various *outcomes* and another that asked people to rank various personal *instrumental qualities* in terms of their importance as guiding principles in their lives. Despite the common notion that men and women are socialized to have different values, there was a great deal of overlap between men and women at both the top and bottom of the value hierarchies for both outcome values and instrumental quality values. For both men and women, peace, family security, freedom, happiness, and self-respect were the most desired outcomes, and social recognition, pleasure, and an exciting life the least desired. In pursuing these outcomes, honesty, ambition, responsibility, broad-mindedness, courage, and forgivingness were the most desirable personal qualities; being logical, intellectual, obedient, and imaginative were valued the least.

Does aging pose a threat to any of these personal values? Maybe. Individuals who value family security, freedom of choice, and contentment may believe these outcomes to be threatened by their own aging, the aging of family members, or the aging of society in general. Their own aging may threaten family security, freedom, and happiness by reducing their economic resources or by reducing their level of physical self-sufficiency. These values also may be threatened if older family members are forced to become dependent. And to the extent that taxes go up as a result of demands for increased services to older people, the reduced income that results may be seen as a threat to family security, freedom, and

In later life, "secondary values" such as friend-ship are valued more than "primary values" such as accomplishment.
Photograph by Marianne Gontarz

happiness. The fact that aging can potentially affect highly desired values in several different ways no doubt contributes to a general sense of uneasiness about aging. Even if we acknowledge that most people never become poor or highly dependent, that most families do not have to handle severe dependency of older members, and that American tax burdens are still quite low compared with other industrial societies, the *possibility* of negative outcomes probably has an effect on how aging and older people are viewed, and this effect is likely to be negative.

In addition to the values that individuals use as guiding principles, there are values that guide the operation of organizations, such as growth of production and profits,

efficiency, and conformity. Aging is sometimes seen as a threat to these values as well. If aging reduced people's capacities for productivity and efficiency (which, as we have seen, it usually does *not*), then growth of production and profits (and in turn wages and benefits) would be jeopardized by an aging workforce.

Aging also seems to affect personal values. In his cross-sectional sample, Rokeach (1973) found that a comfortable life, true friendship, and being forgiving and cheerful were more important to older cohorts than to younger ones. Accomplishment, wisdom, responsibility, happiness, and being loving were less important for older cohorts. There were few age differences in the value of family security, self-respect, social recognition, honesty, ambition, and helpfulness. Equality of opportunity was more important to those in their 20s and those in their 70s than to those in between, perhaps because age discrimination operates at both ends of the adult age continuum. Likewise, independence is more salient to the young and the old than to those in between because for those people it is more problematic.

If these age differences reflect the effects of social maturation and are not simply cohort differences in socialization, then aging brings shifts in values that are adaptive for older people. There is indeed some movement toward what Clark and Anderson (1967) called "secondary values," such as friendship and being forgiving, and away from values more amenable to the marketplace, such as accomplishment and responsibility. However, what is striking is not the differences in values across age categories but the similarities. The only values (out of Rokeach's 36) that showed large age differences were a comfortable life and wisdom (with wisdom more highly valued by young adults and comfortable life more highly valued by elders). The other values showed either modest age differences or none at all.

Rokeach (1978) found that over time value changes were common among people in their

20s and less common among those in their 30s. Beyond the 30s, value changes were "few in number and could easily have arisen by chance" (1978,140). The older cohorts he surveyed were more settled in their values.

In his study of grandparents, parents, and grandchildren in three-generational families, Bengtson (1989) looked at changes in values over the period between 1971 to 1985. He found the greatest intergenerational differences in terms of individualism. The young adult grandchildren were much higher in individualism, followed by the parents and grandparents in that order. Young adults were also the only category to show an increase in individualism from 1971 to 1985. Bengtson attributed this increase to a greater susceptibility of the young adults to the individualist values that dominated the 1980s. On the other hand, the grandparents were significantly more likely to espouse humanist and collectivist values, parents were in between, and the young adult grandchildren were the least humanist and collectivist. Although all three generations increased in their support for collectivist values from 1971 to 1985, the generational differences remained. And although all three generations showed a decline in humanist values over the period, the grandparents remained significantly more humanistic, the parents remained in between, and the young adult grandchildren were the least humanistic. These results showed that period effects were operating to produce changes in values over time, while cohort effects were operating to maintain generational differences.

Beliefs

Beliefs are ideas about what is true. Sometimes beliefs are based on systematic knowledge and facts, but often they are simply *assumed* to be true, with no serious attempt to refer to facts. For example, although research over the past thirty years has consistently shown that families do not abandon older members in need of aid, the cultural belief that many vulnerable older people are abandoned persists. Retirement research has shown over and over that retirement has no negative effect on mental or physical health in the overwhelming majority of cases, yet the cultural belief that retirement is harmful to individuals persists. This book is filled with examples of common beliefs about aging that have not held up when tested by researchers, illustrating an important point about beliefs: *Being widely held has nothing to do with whether or not beliefs are actually true.* For instance, in a 1994 national sample, 61 percent of the general public believed that most older people were lonely, but only 6 percent of older people in the sample said loneliness was a problem for them personally (Speas and Obenshain 1995).

Unfortunately, as often as not when tested, our cultural beliefs about the causes and consequences of aging turn out to be either inaccurate or misleading. For example, it has been widely reported that suicide rates increase after the age of 65, which implies that entering the older population increases the risk of suicide. This is both true and misleading. It is true that suicide rates increase with age after 65, but it is misleading to imply that before 65 suicide rates do not increase. In fact, suicide rates increase with age steadily after the age of 15, and there is no change in the rate of increase at 65 or any other older age. In addition, this age pattern is completely caused by increasing suicide rates among white men; women and African Americans have different age patterns of suicide. Thus, the general statement that suicide rates increase after 65 oversimplifies the facts to the point of being very misleading.

Beliefs are used to make inferences and draw conclusions. To the extent that our beliefs are faulty, so too will be our conclusions. For example, Table 16-1 compares what a national sample of adults aged 18 to 64 *believed* were problems most older people had with the actual proportions of older people having those problems. A majority of adults under 65

Table 16-1 Beliefs about older people's problems versus their actual experiences: 1994.

Problem	Percentage of Public 18 to 64 Who Attribute the Problem to "Most People Over 65"	Percentage of People 65 and Older Actually Experiencing the Problem
Fear of crime	71	37
Inadaquate income	64	12
Loneliness	61	6
Insufficient medical care	61	11
Not feeling needed	57	8
Poor health	57	15
Sparse job opportunities	47	5
Lack of transportation	45	4
Poor housing	38	3
Number	983	202

Source: Adapted from Speas and Obenshain (1995).

believed that fear of crime, inadequate income, loneliness, insufficient medical care, not feeling needed, and poor health were problems for most people age 65 or over, but the actual proportion of elders reporting these problems was 37 percent for fear of crime and 15 percent or less for the other problems assumed to be rampant among elders. The widespread cultural image of older people as beset with numerous problems is inaccurate. And although this inaccurate image may help garner support for programs serving older people, it also perpetuates the image of later life as an unattractive life stage. But we should not use the data on low prevalence of these various problems to argue that older people therefore do not need programs such as those funded by Social Security, Medicare, or the Older Americans Act. Indeed, *the existence of these programs is an essential reason why the prevalence of problems such as low income, inadequate health care, or lack of transportation is so low.*

Beliefs are difficult to study because they tend to be specific. For example, Tuckman

and Lorge (1953) asked people if they agreed or disagreed with more than 150 specific beliefs about older people, such as "Older people are set in their ways," "Older people spoil their grandchildren," "Older people walk slowly," "Older people worry about their health," "Older people are forgetful," "Older people are able to learn new things," and "Older people can manage their own affairs." This research showed that even in the 1950s, before the rapid increase in older people and information about them, beliefs about older people were as often positive as negative. Compared with research results, sometimes these beliefs were accurate, sometimes inaccurate, and sometimes their accuracy or inaccuracy had not been determined. Negative beliefs were more likely than positive beliefs to be inaccurate. Accurate negative beliefs tended to reflect facts about physical aging.

Stereotypes

Stereotypes are composites of beliefs we attribute to categories of people. Through the psychological process of stimulus generalization, stereotypes allow us to treat widely differing individuals as members of a single category. In this process, stereotypes never capture diversity because they are intended to reflect central tendency. All stereotypes are culture-specific in that the ideas used to categorize people reflect the value hierarchies contained in the culture of the group creating the stereotype. Some stereotypes are essentially accurate descriptions of characteristics within a social category, and others are inaccurate and disparaging composites.

Stereotypes of older people in general differ from stereotypes of older workers, older volunteers, and so on. Schmidt and Boland (1986) found that older people are characterized according to three separate kinds of stereotypes: physical characteristics, negative social characteristics, and positive social characteristics. The physical stereotype of older people included gray hair, wrinkled skin, false teeth, hearing impediment, poor eyesight, gnarled hands, and hair loss. The negative social stereotype included negative affect characteristics such as feeling lonely, sad, bored, or neglected; mild impairment such as moving slowly or being forgetful; vulnerabilities such as fear of crime, living on low incomes, or being poor drivers; severe impairments such as needing nursing care, being dependent on family, or having dementia; shrewish characteristics such as complaining, demanding, being selfish, or having no sense of humor; and inflexible characteristics such as living in the past, finding it difficult to change, or being old-fashioned. The positive social stereotype included conservative characteristics such as patriotism, being Republican, or not liking handouts; patriarchal/matriarchal traits such as living life through the children; exemplary characteristics such as being capable, wise, useful, understanding, happy, alert, healthy, active, or generous, liking to be around young people, or giving social support to others; and sagelike characteristics such as being intelligent or interesting, knowing a great deal, being loving, being concerned about the future, or telling stories about the past.

Hummert and her colleagues (1994, 1995) found an even greater number of stereotypes. Based on adjective checklist data from a sample of young, middle-aged, and older adults, Hummert et al. found eleven different stereotypes of elderly people, six negative and five positive. However, these eleven stereotypes fit roughly into Schmidt and Boland's physical characteristics, positive social characteristics, and negative social characteristics typology. Physical stereotypes differentiated the *severely impaired*, who were typified as being incompetent, feeble, and slow-thinking, from the *mildly impaired*, who were characterized as tired, fragile, and slow-moving. The mildly impaired stereotype was used only by middle-aged and older respondents.

The five positive social stereotypes included: *Golden Ager,** described by twenty-four positive adjectives such as lively, adventurous, active, productive, capable, sexual, interesting, sociable, and independent; *Perfect Grandparent*, seen as kind, loving, family-oriented, wise, knowledgeable, intelligent, and generous; *John Wayne Conservative*, characterized as nostalgic, patriotic, conservative, religious, and retired; *Activist*, seen as political, health conscious, and liberal; and *Small-Town Neighbor,* viewed as old-fashioned, quiet, conservative, tough, and frugal. The latter two positive stereotypes were used only by older respondents.

The four negative social stereotypes included: the *Despondent*, typified as sad, afraid, hopeless, neglected, and lonely; *Shrew/Curmudgeons,* seen as ill-tempered, complaining, demanding, inflexible, selfish, nosy and prejudiced; *Elitists*, characterized as demanding, snobbish, wary, and prejudiced (this stereotype was only used by older respondents); and the *Vulnerable*, seen as afraid, worried, victimized, wary, hypochondriac, bored, and sedentary (this stereotype was only used by younger adults). Note that elders use more stereotypes about older people than do middle-aged or younger adults.

These results confirmed that there is no single overarching stereotype of a "typical elder." Instead, there are numerous stereotypes of older people based on appearance, behavior, and capabilities. Specific combinations of characteristics that are called up to form a stereotype probably vary with the reason for needing a stereotype. Sometimes we need stereotypes in order to begin interactions with strangers. For example, a sales clerk dealing with an elder in a department store can use the specific individual's appearance

---

* Hummert et al. (1994) called this category *Golden Ager*, but a better term might have been *Vital Elder*, which avoids the euphemistic description of old age as a "golden age" when most people do not actually consider it "golden" at all.

and behavior to locate that specific person within the complex matrix of stereotypical characteristics. If the person is attractive, smiling, and alert and shows no obvious impairment, then the clerk most likely will summon up a positive social characteristics stereotype such as Golden Ager or Perfect Grandparent. On the other hand, if the person has a dour expression and makes comments about service not being what it used to be, then the clerk might summon up a negative social characteristics stereotype such as Shrew/Curmudgeon. At other times, decisions need to be made about how to treat a social category, without reference to specific people. For instance, a legislative aide working on a report for a congresswoman who supports home care for impaired older people is apt to use a negative social characteristics stereotype that emphasizes the frailty and vulnerability of older people. On the other hand, a personnel director who wants to institute a flexible retirement policy to allow more older workers to stay on part-time after retirement age is apt to use a positive social characteristics stereotype of elders that emphasizes exemplary characteristics such as alertness, capability, and experience, such as the Golden Ager. The point here is that we do not simply *have* stereotypes, we *construct* and *use* stereotypes to fit our purposes.

Hummert et al. (1995) found age differences in the age ranges to which various stereotypes would be applied. For example, young adults applied the very positive Golden Ager stereotype only to elders in their 60s, not to adults in their 70s or older. Middle-aged respondents applied the Golden Ager stereotype to people in their 60s and 70s, and older respondents applied it to people in their 60s, 70s, and 80s or older, although in decreasing prevalence. All age groups saw both the Shrew/Curmudgeon stereotype and the John Wayne Conservative stereotype as applying more to people in their 70s than to people in their 60s or to people in their 80s or beyond. All age categories of respondents

saw the negative stereotypes of mildly or severely impaired, despondent, or vulnerable as being much more applicable to people in their 80s or older than to people in their 60s or 70s. Again, these findings suggest considerable complexity in how age-related stereotypes are constructed and used.

Unfortunately, based on appearance alone, young people generally call up negative stereotypes when dealing with strangers. For example, Levin (1988) used pictures of the same man at the ages of 25, 52, and 73 to elicit ratings from college students on characteristics such as activity, competence, intelligence, reliability, energy, flexibility, and memory. On each of these dimensions, purely based on appearance, the students rated the same man at age 73 much more negatively than at ages 52 or 25.

Fortunately, most people do not act on appearances alone but make at least some attempt to ascertain personal characteristics. In this process, the attribute most likely to activate a negative stereotype is disability. Braithwaite, Gibson, and Holman (1985–1986) presented high school students with four short vignettes in which older people were described as active, alert, memory-impaired, and physically disabled. Only 15 percent of the students rated the active older person negatively, compared with 56 percent for the physically disabled older person. Not only does disability create a negative image, but it is associated with sharply reduced expectations of what disabled people are capable of regarding growth, continued development, and decision making. Elias Cohen (1988) pointed out that the widespread image of disabled elders as passive and homogeneous leads to paternalistic stereotyping by medical and social services personnel, which in turn severely restricts any potential for autonomy that disabled older people might have.

Stereotypes about the capacities of older workers compared with their younger counterparts are particularly relevant to the situation of older people in the labor force.

Rosen and Jerdee (1976a) found that prospective business managers rated older workers (age 60) lower than younger workers (age 30) on productivity, efficiency, motivation, ability to work under pressure, and creativity. Older workers were also viewed as more accident-prone and as having less potential. Not only were they seen as not being as eager, but they also were seen as unreceptive to new ideas, less capable of learning, less adaptable, and less versatile. Older workers were also viewed as more rigid and dogmatic. On the plus side, older workers were seen as more reliable, dependable, and trustworthy and as less likely to miss work for personal reasons. No age differences were found in the ratings of interpersonal skills.

Obviously, the stereotype applied to older workers operates to their disadvantage. And it is mostly inaccurate. Older workers are *not* consistently less productive or efficient. Generally, the research shows inconsistent age differences on job-relevant factors.

In a Canadian sample of college students and older people, Bassili and Reil (1981) found that older people were stereotyped by young and old alike as conservative, traditional, and present-oriented. But they also found that older men described as "former engineers" were stereotyped almost identically with younger men who were currently engineers. In addition, Levin and Levin (1981) found that upper-socioeconomic status offset the negative image of old age. Thus, age is but one of many characteristics used for stereotyping. Note that negative stereotypes can also sometimes have positive consequences. For example, Kearl (1982) cited the negative stereotype as a reference point many elderly people use to perceive themselves as relatively advantaged.

Finally, Schonfield (1982) pointed out that the proportion believing a given stereotype to be true does not necessarily reveal all we might want to know. He went further and asked what proportion of older people fit each of several stereotypes. Surprisingly, Schonfield found that although most of his respondents believed several stereotypes about older people were true, it was common for them to believe also that there were numerous exceptions to the stereotypes. When he coupled belief that a stereotype was true with belief that no more than 20 percent of older people were exceptions in order to isolate those who "stereotyped" elders, Schonfield found that at most 20 percent could be classified as stereotyping older people and that rarely did the same individual qualify as a stereotyper for more than 3 of the 10 stereotypes presented.

Attitudes toward people are predispositions we have to be attracted to certain types of people and repelled by other types. These attitudes are usually closely tied to values, beliefs, and stereotypes. In the early gerontology literature, attitudes toward aging and older people were found to reflect the basic ambivalence our culture expresses, some attitudes were positive and some were negative. From an extensive literature review, McTavish (1971) concluded that only about 20 to 30 percent of adults have more negative than positive attitudes toward older adults. But attitudes are difficult to study because they are extremely specific likes and dislikes. As a result, attitudes about aging and older people have received little study in the 1980s and 1990s.

Aging as Portrayed in Mass Media

People also learn what to think about aging through media portrayals.

Television Television is the most important mass medium in American society. Americans watch television more than three hours a day on the average, and television is the leading pastime of middle-aged and older Americans (Moss and Lawton 1982; J. Kelly 1993). How television treats the topic of aging varies by the type of message, and in a medium as varied and complex as American

Television can challenge or confirm our stereotypes about aging.
Photograph by AP/Wide World Photos

television, it is difficult to generalize about how aging is portrayed. In our discussion we consider separately how aging is treated in Saturday morning cartoons, daytime serials, prime-time series, news and documentary programs, and commercials.

Bishop and Krause (1984) studied the depiction of aging in Saturday morning cartoons. They found that 95 percent of the characteristics attributed to children in these programs were positive, compared with 52 percent positive characteristics for both adult and older primary characters. Thus, the contrast was between adults of all ages and children. Older primary characters, although much less numerous (25) than other adults (177), were not portrayed any more negatively than were adults in general.

Elliott (1984) studied aging in daytime television serials—usually called soap operas or soaps. Thirteen daytime serials were selected for the study, which took place between July and November of 1979. Each

program was monitored by four trained raters for twenty consecutive episodes. Raters judged characters to be "older" based on their appearance, family roles, retirement, self-disclosed age, or direct age attribution by other characters. For example, appearance factors associated with "older" characters included gray hair; wrinkled skin; a high, reedy voice; and bent posture.

Elliott found that 71 percent of male continuing characters were concentrated in the 20- to 50-year-old range. Slightly under 10 percent were "older," a modest underrepresentation. Female continuing characters were even more age-concentrated, with 60 percent being judged as under 40. Only 7 percent were "older," a significant underrepresentation. Only 3 of 404 continuing characters (0.7 percent) were judged as 70 or over.

Characters were also rated as to their behaviors. Older male characters most often listened, followed by providing information, giving directions, and nurturing. Older women characters also mostly listened, followed by nurturing, providing information, and giving directions. There were more similarities than differences between older men and older women characters' behavior.

Researchers who have studied the treatment of aging in nighttime television series have concentrated on the proportion of older characters and how these older characters are portrayed. Estimates of the proportion of older characters range from 1.5 to 13 percent. Jeffreys-Fox (1977) found that 75 percent of older characters were men. The weight of the evidence seems to indicate that about 10 percent of characters are older—which is close to their prevalence in the population. Of course, the main reason for such varied estimates is that raters have to guess television characters' ages, and this is apparently a haphazard method of assessing age. There is also wide variation in the findings about how older characters are portrayed. Petersen (1973) found that 60 percent of older characters were presented in a favorable light. On the other hand, Aronoff (1974) found that older characters were portrayed as "bad guys," prone to failure, and generally unhappy. Jeffreys-Fox (1977) found that older characters were rated as less attractive, sociable, warm, and intelligent compared with younger characters. He found no age difference in rated happiness. These findings are also quite sensitive to the cues raters use to select "elderly" characters and to the raters' own biases about older people.

Research into television viewing has generally used *content analysis,* a technique in which *samples* of program content are rated by judges according to a set of preexisting categories. If the researcher is not prepared to find positive outcomes, this is often reflected in the categories and the research results. In addition, sampling program content often does not capture the *context* within which characterization occurs.

I took a somewhat different approach in looking at how aging was treated in television

series. In March 1979, there were 45 continuing series appearing on three networks. For each of these series, the *continuing characters* were rated on whether they were teenagers or younger, young adults (20 to 35), middle-aged (35 to 55), mature adults (55 to 70), or older adults (70 or over). Series were also rated on whether intergenerational themes were emphasized and, if so, whether they involved older people. About 12 percent of the continuing characters in series were judged as being over 55, an underrepresentation. Of the 45 series, 24 portrayed intergenerational themes, and 8 involved mature or older adults.

There appeared to be a trend in television drama toward recognition of adult child–older parent relationships as a dramatic issue viewers could identify with. In addition, adults of widely varied ages were being shown working together in a wide variety of occupational settings (radio and television, law enforcement, small business, the military, the press, and education). Also, when mature adults were shown working with young and middle-aged adults, the mature adults usually supplied the experience and leadership while the young adults supplied the enthusiasm, eagerness, and physical prowess. Mature and even older adults were shown as capable of running and using physical strength if necessary, as interested in romance, and as capable human beings. Intergenerational relations in families were portrayed as conflict-ridden situations in which love and family ties generally prevailed.

Dail (1988) examined the portrayal in 1984 of adults approximately 55 or older in prime-time series that centered around family life. Older characters' behavior was rated as positive, neutral, or negative with regard to cognitive functioning, physical functioning, health status, sociability, and personality. Ratings were done for 2,012 incidents involving older male characters and 1,341 incidents involving older women. The results showed that older characters were portrayed in overwhelmingly positive terms. Positive portrayals occurred 98 percent of the time for cognitive functioning and sociability. Ratings for physical functioning, health status, and personality were positive more than 90 percent of the time. The only departure from this very positive picture was for men about 55—positive ratings occurred only about 70 percent of the time, which Dail suggested reflected an image of midlife as more problematic for men.

Mundorf and Brownell (1990) looked at how age differences affected choices of favorite entertainment programs on television. In general, members of age categories showed a preference for programs containing lead characters generally close to their age. For elders, the top two entertainment programs were *Murder, She Wrote* and *Golden Girls*, both of which involved several continuing characters past middle age. For young adult college students, the top two were *The Cosby Show* and *Cheers*, both of which involved continuing young adult characters but not as lead characters. Interestingly, the older respondents rated both *Cosby* and *Cheers* in their top five. However, the young adults' top five was rounded out by *LA Law*, *Who's the Boss*, and *Growing Pains*, all shows involving young adult characters.

When asked which characters in entertainment programs were their favorites, the respondents tended to identify lead characters close to their own age-gender category. Young men preferred Sam Malone of *Cheers*, young women chose Whitley of *A Different World*, older men identified Bill Cosby, and older women preferred Angela Lansbury of *Murder, She Wrote*. This pattern suggests that vicarious identification is important in preferences for continuing characters on television. Interestingly, the young respondents referred to their preferences by the character's name in the program whereas the older respondents named the actor, not the character he or she played.

What emerges with respect to contemporary television series is a picture of a medium responding to changing times. In dramatic series, intergenerational cooperation and conflict resolution are presented as ideals. In comedy series, the humor in intergenerational situations is exposed, presumably with some cathartic effects. And these portrayals are not unrealistic. Older and younger people can and do work well together, and families do love each other and manage to transcend their conflicts.

In general, television seems to be moving toward more representative and accurate portrayals of mature adults in continuing series. This trend is noteworthy because continuing characters can be developed in real depth. What may start out as a relatively stereotyped older character can eventually become a realistic one—a person who can be charming one time and cranky another, depending on the situation or circumstances.

Television's treatment of aging in news programming is another matter. Network news, local news, and television tabloid shows thrive on sensationalism. Emphasis on sensationalism and video voyeurism, once excluded from network nightly news programming, has become part of television news at every level. There is nothing sensational about individuals and families who are effectively coping with the trivial conflicts of everyday life. As a result, news programming ignores typical middle-aged or older people, along with most other ordinary people. Those older people who are given attention in the news tend to be those with "a problem" that can be a springboard for human interest or commentary. Two main themes in human interest are "Can you believe this?" and "Ain't it awful!" Needless to say, the portrayal of older people in this context is seldom positive.

On the other hand, public affairs and talk shows generally present aging in a favorable light (Harris and Feinberg 1977). Older people on public affairs and talk shows tend to be influential business leaders or politicians, or respected actors or creative artists. The 1980s also saw the development of retrospective programs in which celebrated older public figures and performing artists were recognized for their many years of contribution. These programs indicate that one's past does count, at least for people of high achievement.

Commercials also present a mixed picture of aging. Francher (1973) found that the majority of commercials present young, attractive people, and that over half of television commercials were "action-oriented." Harris and Feinberg (1977) found that the proportion of older characters presented in commercials generally matched the proportion of older people in the population. Older characters in commercials were less likely to be physically active and more likely to have health problems, but none were portrayed as physically disabled or incapacitated. However, older characters were substantially overrepresented in commercials for health aids and totally absent from commercials for clothing, appliances, cars, and cleaning products.

Aging in Feature Films In recent years, filmmakers have turned their attentions toward issues of aging and have often produced entertaining and enlightening visions of aging. For example, in 1981, *On Golden Pond*, which starred Henry Fonda and Katharine Hepburn, won three Academy Awards. It depicted the resilience of a couple in their late 70s as they coped with the anxieties of chronic disease, early signs of dementia, and family problems that provided comic relief but also a sense of the importance of older people to families. In particular, it showed how a "new" teenage stepgrandson could be profoundly affected by the acceptance and nurturing provided by these frail but vital old people. This was but one of nearly one hundred feature films produced since 1980 that incorporated middle-aged or older people as significant characters either dealing with the realities of aging (*Driving Miss Daisy, Fried Green Tomatoes, Camilla, Nobody's Fool, The Trip to Bountiful,* and *Strangers in Good Company*) or coping with intergenerational issues (*Terms of Endearment, Roommates,*

Miami Rhapsody, and *How to Make an American Quilt*). In general, films have treated aging and intergenerational issues with respect and sensitivity, even in comedies like *Harold and Maude, Being There, Grumpy Old Men,* or *Don Juan de Marco.*

Of course, there were also some films that missed the mark. For example, the theme of the popular film *Cocoon* was that if given a chance, older people would rather go to a planet they had never seen before than grow old here on Earth.

Children's Books General orientations toward aging and older people can also be studied by looking at how these topics are treated in the educational system. After all, schools are the primary mechanism society uses to exert social pressure on young people to adopt the attitudes, values, beliefs, and norms that make up their culture. This pressure is more often subtle and indirect rather than heavy-handed. For example, children's books seldom say "do this" or "don't do that" or "think this but not that." Instead they most often present the child with an attractive young character, a role model, and simply describe how heroes and heroines ideally cope with various types of situations.

Several studies have analyzed the content of children's books. Old characters are common in children's books, but they tend to be underdeveloped and peripheral (Peterson and Eden 1977; Peterson and Karnes 1976). The older characters in children's books are portrayed in a consistently positive way (Robin 1977), and illustrations tend to present attractive, healthy older people (Storck and Cutler 1977). Seltzer and Atchley (1971) surveyed children's books from the past 100 years and concluded that there was little change over time in the evaluation of old people and that, if there is a negative attitude or evaluation of elders, it is not to be found in children's books. Ansello (1977) found few hostile or negative characterizations of older characters. However, Ansello argued that older characters ultimately come off negatively not in content but by virtue of their blandness; older characters were poorly developed, and old age seemed boring.

Writers try to write for an audience. Fiction writers in particular try to develop central characters with whom readers can identify and then place the characters in conflict situations that readers can understand. Given these constraints, it is hardly surprising that few older characters are central figures in children's literature. The main characters in children's literature tend to be children, and their interactions with adults are governed by role relationships that do not depend on having an in-depth understanding of adult characters. Thus, the characters of school principal, camp counselor, grandparent, and so forth often can be sketchy. What seems more relevant is that to the extent that children's books specify *ideal* ways of perceiving older people, the ideal presented tends to be a positive one.

Lloyd Alexander's five-book series, *The Chronicles of Prydain*, published between 1964 and 1968, represents a good example of even and rich treatment of elders in books aimed at young people. Older characters are described in physical detail and given a wide variety of nonstereotyped roles to play in these books. There is an old wizard, an old shepherd, a middle-aged warrior, an old wise man, an older bumpkin, three magical and ancient sisters, a middle-aged farm couple, an older evil enchanter, an old great poet, an old master weaver, and an older master pot maker.

The description of Dwyvach, the weaver-woman, illustrates Alexander's masterful combination of accurate detail about physical aging's effects on the body with a positive portrayal of the skills and wisdom that older people can possess. In the midst of his sojourn, Taran Wanderer, the young male lead character in the series, approaches a cottage:

Taran saw a bent old woman cloaked in gray beckoning him to the hearth. Her long hair was white as the wool on the distaff hanging from her belt of plaited cords. Below her short-girth robe, her bony shins looked thin and hard as spindles. A web of wrinkles covered her face, her cheeks were withered; but for all her years she gave no sign of frailty, as though time had only toughened and seasoned her; and her gray eyes were sharp and bright as a pair of new needles.

(1967,228–229)

Taran for the first time noticed a high loom standing like a giant harp of a thousand strings in a corner of the cottage. Around it were stacked bobbins of thread of all colors; from the rafters dangled skeins of yarn, hanks of wool and flax; on the walls hung lengths of finished fabrics, some of bright hue and simple design, others of subtler craftsmanship and patterns more difficult to follow. Taran gazed astonished at the endless variety, then turned to the weaver-woman of Gwenith.

"This calls for skill beyond anything I know," he said admiringly. "How is such work done?"

"How done?" The weaver-woman chuckled. "It would take me more breath to tell than you have ears to listen. But if you look, you shall see."

So saying, she hobbled to the loom, climbed to the bench in front of it, and with surprising vigor began plying the shuttle back and forth, all the while working her feet on the treadles below, hardly pausing to glance at her handiwork.

(1967,229–230)

Taran is so impressed with the weaving that he decides to learn the craft. He wants to work on a new cloak, which he sorely needs.

The aged weaver-woman had not only a tart tongue but a keen eye. Nothing escaped her; she spied the smallest knot, speck, or flaw, and brought Taran's attention to it with a sharp rap from her distaff to his knuckles. But what smarted Taran more than the distaff was to learn that Dwyvach, despite her years, could work faster, longer, and harder than he himself. At the end of each day Taran's eyes were bleary, his fingers raw, and his head nodded wearily; yet the old weaver-woman was bright and spry as if the day had scarce begun.

(1967,231)

But before he could learn to weave, Taran had first to learn to comb and card raw wool, spin yarn, fill bobbins, and dye the yarn. Only then could he sit at the loom and begin to weave a new cloak. But his lessons were not over. The weaver-woman taught him the value of patience and the importance of forethought. When Taran left the cottage to resume his search for his role in life, it was with a fine new cloak that he had woven himself and would never be able to look at without recalling every thread and with a deep respect for the skill and generosity of the old weaver-woman.

This is but one story among the many in Alexander's series that illustrate over and over the important role that mentors and role models can play in the forming of a new young adult. Alexander's books are unusual in their use of many and varied older characters in this role.

Alexander's *Chronicles of Prydain* series received the Notable Book Award from the American Library Association, and the last book in the series, *The High King*, won the Newbery Award. Over 2 million copies of the books have been sold, and today libraries all over the United States are circulating multiple copies to avid young readers of both genders. These books represent idealism at its best, and the ideals presented in them portray the diversity that is aging in a fair, balanced, and positive light.

Print Journalism   Less attention has been paid to the treatment of aging in the press and in popular magazines. Some of the comments made about television news probably apply to newspapers, but research is needed to establish how newspapers differ from television news. Buckholz and Bynum (1982) examined 1,703 newspaper articles from 1978 and found that 56 percent presented a neutral picture of the aged, 30 percent a positive picture, and only 14 percent a negative picture. However, Gene Cohen (1994) used the term *journalistic elder abuse* to refer to the casual recycling of pejorative labels for older people. For example, he reported that in

1991 and 1992 the term "greedy geezers" was used in sixty-eight separate newspaper articles, mostly syndicated to multiple newspapers. Journalists are often too quick to accept news releases from interest groups seeking to pin a variety of the country's problems on Social Security, Medicare, or Medicaid for older people. In the tradition of C. Wright Mills (1959), one could suggest that lobbyists, news release writers, corporate public relations personnel, legislative staff people, and elite journalists tend to come from elite families, attend the same elite colleges, circulate among these jobs, share a general middle-of-the-road political worldview, and accept the greedy geezer premise—that elders are wealthy and selfish and do not deserve or need government benefits—on faith, without any felt need to determine the facts. This would be an interesting thesis to investigate.

Those who have negative things to say about programs that benefit older people seem to have an easy time getting media opportunities. For example, an organization called "Lead or Leave"—with a paid staff of two—enjoyed a brief but well-publicized two-year career that saw its invective against Social Security receive front-page national newspaper coverage and space in every national news magazine.

Fiction, Poetry, and the Arts   Schuerman, Eden, and Peterson (1977) examined fiction in nine women's magazines and found that older characters were more likely to be portrayed positively than negatively. Sohngen (1977, 1981) analyzed the *experience* of old age as depicted in more than one hundred contemporary novels. She felt it significant that so many novels dealt with the actual experience of growing old (compared with simply presenting well-developed older characters). The demographic emphasis was on white middle-class characters. Nearly half of these novels used life reviews as a technique for examining the experience of aging.

Otherwise, the content was quite varied but tended to center on gerontological issues such as institutionalization, retirement, isolation, segregated living, or intergenerational relations. Sohngen was impressed with these books. In general, she found that they used language freshly, avoided soap opera clichés, and presented vivid characters in an entertaining, readable, and sometimes moving way. Loughman (1980) found similar positive portrayals.

Rooke (1992) contended that the prevalence of fiction about aging and old age is on the rise for both literary and sociological reasons. Curiosity about aging is growing stronger in both readers and writers, as they grow older. And as society ages, the market for fiction dealing with aging has strengthened and the supply of writers able to accommodate the demand has increased, too. She sees the "novel of completion" as a developing new literary genre in which writers and readers grapple with issues such as the conflict between the desire to hold on and the desire to let go; the need to sum up life, even if only temporarily; and the need to describe what is good and satisfying about old age, which can be difficult to capture in words.

Smith (1992,236) argued that if poems about aging are not just isolated artifacts but taken together are collective representations of the human experience, then aging poets may touch our deepest perceptions about life, death, despair, and joy in the last stages of life. In some poetry written in later life, there is a transcendental quality that may evoke a mystical understanding of the meaning of all of life, including aging. Winkler (1992) found that art contains graphic portrayals of grotesque old age, wise old age, and an old age that is transcendent in the face of death. Kastenbaum (1992) concluded that many famous composers who lived to grow old wrote some of their most beautiful yet simplest melodies in old age. They achieved a clarity and summation in their art form similar to the novel of completion.

Thus, various expressive art forms—fiction, poetry, art, and music—can facilitate the quest for meaning that stands at the center of a person's capacity to adapt to aging. And these portrayals not only make meaning for the artists who create them but can stimulate the development of a sense of meaning in others. These are indeed valuable cultural resources.

Much research is needed before we have a solid picture of how the mass media treat aging. Some scholars see a very negative emphasis; others do not. However, it seems that media portrayals of aging individuals are moving in a positive direction; we hope this is simply an accurate response to the changing age structure of American society. On the other hand, there is a disturbing trend in television and print journalism to accept negative stereotypes of older people as a selfish and burdensome population category.

Harris and Associates (1975, 1981) studied how older people *perceived* the treatment of aging in the media. In 1975, just under two-thirds thought that television, newspapers, and magazines gave a fair picture of what older people are like. Older people felt that television treated older people with respect, made them look like important members of their families, wise, and successful. They did not feel that television portrayed older people as narrow-minded or old-fashioned, meddlesome, sick or incapacitated, useless, or unattractive.

By 1981, this picture had changed. Less than half of the older people in Harris and Associates' national sample thought that television and magazines gave a fair picture of older people. Interestingly, most who thought that the media presented a distorted picture felt that the media made older people look better than they really are. They felt this way possibly because the older people in this survey also tended to buy into the stereotype of older people as beset by numerous problems.

Age Prejudice
and Discrimination

*P*rejudice is an unfavorable attitude toward a category of persons based on negative traits assumed to apply uniformly to all members of the category. *Age prejudice,* or **ageism,** is a negative attitude or disposition toward aging and older people based on the belief that aging makes people unattractive, unintelligent, asexual, unemployable, and mentally incompetent (Comfort 1976). It may be that only one-fourth or less of the general public endorses this extreme and inaccurate view, but most Americans probably subscribe to some erroneous beliefs about aging and have at least a mild degree of prejudice against aging and older people. This notion is borne out by the research reviewed so far.

If people kept their prejudices to themselves, no harm would come from them. But people act on their prejudices. Discrimination is some sort of negative treatment that is unjustly applied to members of a category of people. *Age discrimination* is treating people in some unjustly negative manner because of their chronological age or their appearance of age and for no other reason. Age discrimination occurs when human beings are avoided or excluded from everyday activities because they are "the wrong age." In 1981, 80 percent of adults believed that most employers discriminate against older people and make it difficult for them to find work, and 61 percent of employers agreed with this assessment (U.S. Senate Special Committee on Aging 1986).

Older people sometimes must intrude into various spheres of daily life in order to make people aware that they have something to offer. Most older people are not willing to fight for this recognition, and as a result there is a great deal of age segregation in activities and interactions. Only in the family do older people usually escape this sort of informal age discrimination.

Equally important is the impact of age discrimination on opportunities for beginning or continuing participation in various organizations. As we saw in Chapter 9, job discrimination makes it more difficult for older workers to continue in the labor force or find jobs if unemployed. The stigma of implied inability and the resulting discrimination sometimes extend past paying jobs and into volunteer jobs and other types of participation as well. Organizations especially for older adults offer an alternative to those who have been rejected by organizations in the "mainstream," but at the cost of age segregation. While many older adults prefer associating mainly with their age peers, those who do not often find themselves without choices. Research on age discrimination in organizational settings other than jobs is greatly needed.

Age discrimination can also take the form of unequal treatment by public agencies. A report from the U.S. Commission on Civil Rights (1977) found that age discrimination was present in numerous federally funded programs—community mental health centers, legal services, vocational rehabilitation, social services to low-income individuals and families, employment and training services, the food stamp program, Medicaid, and vocational education. And this problem existed in all regions of the country. People in the oldest cohorts were most likely to experience discrimination from public agencies. In addition, age discrimination was often compounded by discrimination on the basis of race, sex, national origin, or handicap status.

The Commission on Civil Rights concluded that much of this discrimination stems from a narrow interpretation of the goals of legislation. For example, community health centers generally interpret "preventive health care" as applying only to children and adolescents. Directors of employment programs see their most appropriate clients as males age 22 to 44. Even age 22 is too old as far as some job-training programs are concerned. The commission also found that state legislatures sometimes convert federal programs designed to serve all Americans into categorical programs aimed at specific age groups. For example, the state of Missouri passed a strong child abuse and neglect law—a worthy goal; but the state provided no funds for its implementation. Instead, federal funds for social services to everyone were earmarked to support the child abuse program, and as a result most cities in the state discontinued their adult protective services programs.

In many cases where states or local governments are responsible for defining the population eligible for federal programs, age discrimination results. For example, several states excluded older people from vocational rehabilitation programs because they were not of "employable age." Age discrimination sometimes occurs when services are provided under contract with agencies that limit the ages of people they will serve. For example, a general social services contract with a child welfare agency is unlikely to result in social services to older adults. The commission also found that outreach efforts are often aimed at specific age groups, which lessens the probability that other age groups will find out about programs for which they are eligible. The commission also concluded that general age discrimination in the public and private job market was an important underlying factor in age discrimination in employment, training, and vocational rehabilitation programs. As long as older people are denied jobs, agencies see little value in preparing them for jobs.

Exchange Theory and Age Discrimination

Exchange theory assumes that people try to maximize their rewards and minimize their costs in their interactions with others.*

---

* Exchange theory essentially begins with Homans (1961) and was later elaborated by Blau (1961) and Mulkay (1971). The major attempts to use exchange theory in social gerontology were made by Dowd (1975, 1980).

Rewards can be defined in material or nonmaterial terms, and they include such factors as assistance, money, property, information, affection, approval, labor, skill, respect, compliance, and conformity. Costs can be defined as the loss of any of these rewards. Because its proponents use such terms as "exchange rates" and "return on investment," exchange theory initially seems to be a rational equivalent to the well-known but more emotionally oriented pleasure-pain principle. But the exchange process involves elements that are not necessarily either rational or conscious. And as we will see, classical exchange theory is better at helping us understand impersonal interactions in organizational settings than other types of interaction. It also works better when applied to transitory or new relationships rather than to long-standing ones. Exchange theory may be less helpful in understanding complex long-term relationships with family or friends.

Dowd (1975, 1980) advanced the major argument concerning how aging affects exchange. It starts with the familiar assertions that people want to *profit* from social interaction and that profit consists of a *perception* that the rewards coming from the interchange outweigh the costs. The ability to profit from an exchange depends on the exchange resources that the actors bring to the exchange. When resources are reasonably equal, then a mutually satisfying interdependence may emerge. But if one of the actors has substantially fewer exchange resources, then the actor's ability to profit from the interchange can be sharply restricted.

The actual exchange resources that a person has may not be as important as the resources the person is *assumed* to have. And this is where age comes in. Along with other personal characteristics such as gender, ethnicity, and social class, age influences the resources we are *presumed* to have, particularly in terms of information, skills, and ability to do physical labor. Older people are often assumed to have less up-to-date information, obsolete skills, and inadequate physical strength or endurance.

Thus, ageism is converted into age discrimination at least partially because of its influence on our assumptions about exchange resources. And many of these assumptions have been around so long that people are not even conscious of them. Their validity is taken for granted not only by people who deal with older adults but also by older adults themselves, at least as applied to other older adults (but usually not to oneself).

The presumed reduction with age in exchange resources is acted on as if it were real, which results in a power differential that puts older people at a distinct disadvantage. And because they are at a disadvantage, elders are relatively powerless actors who are forced into a position of compliance and dependence because they are seen as having nothing of value that they can withhold in order to get better treatment.

There is no doubt that this scenario describes quite accurately the situation that confronts many older adults who are looking for a new job, trying to become involved in community organizations, or trying to get service from a bureaucracy.

Societal Disengagement

Social institutions outlive the people who compose them, and thus most social institutions are constantly phasing young people in and older people out. However, this pattern is most common in *occupational* positions associated with various institutions such as the economy, education, politics, government, religion, health, and the arts. This sort of analysis is less applicable to more private institutions, such as the family, marriage, or friendship (Ward 1984). The processes through which society loses interest in and no longer seeks older individuals' efforts or involvement is called **societal disengagement.**

As part of an overall theory of disengagement, Cumming and Henry (1961) suggested that societal disengagement is half of the "inevitable" withdrawal of older people and society from each other. From the societal point of view, older people may no longer be sought out for leadership in organizations, their employers may no longer desire their labor, their unions may no longer be interested in their financial problems, and their government may no longer be responsive to their needs. Societal disengagement is probably motivated by the same stereotypes that fuel age prejudice and discrimination, but Cumming and Henry's theory of societal disengagement is based on a different theoretical logic.

Societal disengagement theory is based on *functionalist* logic, which holds that social patterns are invented to meet specific needs. Societal disengagement theory posited that because older people are closer to death and death would disrupt the smooth operation of organizations or groups, then society needed a mechanism for preventing this disruption and societal disengagement emerged to provide for this need.

The explanation offered by societal disengagement theory does not stand up to inspection. Most of the withdrawal of interest in the participation of older people in society occurs long before the ages at which the probability of mortality is high. For example, people begin to feel pressures to retire in their 50s, but large proportions who survive to reach their 50s will not die until their 80s or later. In addition, organizations are quite able to survive constant turnover in personnel, so the "disruption" argument seems specious. Indeed, because it is difficult to identify any social need that requires societal disengagement, a functional argument is difficult to make. So if societal disengagement did not arise to meet a functional need, then how do we explain it?

In my opinion, societal disengagement is *generalized age discrimination.* Many people directly experience age discrimination in specific social role contexts. They are not hired for jobs they are obviously qualified for. They are not even thought of as being interested in promotions. They are not called on to serve as volunteers in their churches or in community service organizations. But most aging and older people are unaware that based on age stereotypes and age prejudice they have, *as a category*, been *generally* excluded from the pool of desirable eligibles for any type of useful work, paid or unpaid. Often societal disengagement is unintended and even unrecognized by society, but it remains an important reality for older adults. More research is needed to uncover the extent and dynamics of societal disengagement.

Atrophy of Opportunity

For those who use society as a point of departure, a major criticism of societal disengagement theory is that it does not give enough weight to the role of the socially determined situation, particularly the structure of opportunity. Let us review studies that have looked at the relationship between opportunity structures and individual patterns of involvement.

Carp (1968) studied the effect of moving from substandard housing and socially isolating or interpersonally stressful situations to a new apartment house having within it a senior center. Prior to the move, the 204 respondents were assessed in terms of their degree of involvement in three separate roles: paid work, volunteer work, and leisure pastimes. Opportunities for all three were very limited in the premove situations of the study sample. "Special effort was necessary in order to participate in any of the three. Expenditure of this effort was assumed to express a strong need for involvement" (1968,185).

Following the move, opportunities for volunteer work and leisure pastimes were expanded, but opportunities for paid work remained about the same. Carp predicted that those who had expended the effort on leisure or volunteer work in the antagonistic

premove setting would be happier and better adjusted in an environment that facilitated continued engagement. She also predicted that those who had worked for pay in the premove setting would be no different in terms of satisfaction or adjustment after the move. Her results fully supported her predictions. People who had been involved in leisure pursuits or volunteer work prior to entering the apartment building tended to be happier, more popular, and better adjusted than the other tenants. On the other hand, those who had worked were not significantly different from the other tenants. Carp interpreted her findings as supporting the idea that the greater the congruence between the person's desires for continued engagement and the opportunities for such engagement offered by the situation, the higher the degree of adjustment and satisfaction. Carp (1975) restudied these same people eight years later and found that these relationships still held.

Implicit in these conclusions is the idea that many people want to remain engaged and that a main obstacle preventing them from doing so is societal disengagement, which creates a situation in which continued engagement is difficult. To assess the impact of societal disengagement directly, Roman and Taietz (1967) studied an occupational role in which continued engagement after retirement is allowed—the role of the "emeritus professor." Unlike most organizations, some American colleges and universities, instead of removing retired faculty from the organization, make available a formalized, postretirement position with a flexible role, whose definition is a function of the individual's preretirement position as well as of the individual's own choice of postretirement activity. The important point here is that this system allows the *opportunity* for role continuity between full-time employment and retirement.

In Roman and Taietz's study, the amount of continuity possible varied. The research professors had the most, since they could generally continue to get research grants through the university. Those involved in teaching, public service, or administration still had opportunities for involvement, but their emeritus role was quite different from their preretirement role. They often ended up writing books, consulting, or becoming administrators. In no case, however, was the continuity complete. The emeritus professor always gave up a measure of involvement. This situation was excellent for studying individual disengagement since societal disengagement exerts less pressure.

Roman and Taietz assumed that disengagement was the product of *particular* social systems, not of systems in general, and that opportunity structures would greatly influence the individual's exposure to societal disengagement. Roman and Taietz predicted "that a significant proportion of emeritus professors would remain engaged, and that those allowed role continuity would exhibit a higher degree of continued engagement than those required to adopt new roles" (1967,149). They found that 41 percent of the emeritus professors were still engaged within the same university, 13 percent had taken employment in their profession elsewhere, 24 percent were in bad health, and 22 percent were disengaged from both the university and their profession. If these percentages are recomputed leaving out the group in bad health, for whom no determination of voluntary disengagement is possible, and combining both categories of those still engaged, the pattern shows 71 percent still engaged and 29 percent disengaged. In addition, they found that those emeritus professors who had had a research role were still engaged significantly more often than the others. Thus, all of Roman and Taietz's predictions were supported by their findings. These data suggest that societal disengagement is very much a product of the opportunity for continued engagement. The fact that the organization provided a continuing role after retirement allowed 71 percent of the healthy people to

remain engaged, whereas in many occupations the percent allowed to remain engaged would have been near zero. This finding is all the more revealing given that almost half of Roman and Taietz's sample was over 75 years of age.

The point is that the external aspects of aging and the socially structured situation can have a far greater influence than personal desires on the social withdrawal of older people. A recurrent theme throughout our examination of how older people fit into a modern industrial society is that the difficulties many older people face are brought on less by withdrawal on their own part than by decisions, often consciously made, on the part of others to exclude them from the mainstream of society. In a social system built around production, rationality, and efficiency, it is often more expedient to write off older people than to expend the energy required to create a "citizen emeritus" role.

Age Stratification

The theory of **age stratification** focuses on how the population is divided into *age strata* such as youth, adulthood, and old age; how inequalities, differences, segregation, or conflict between age strata influence *age relations* (interactions within and between age strata); and how *age grading* operates to sort people into age strata and to channel them through age-graded roles and opportunity structures (Riley 1987).

Aging causes the movement of age cohorts—*cohort flow*—through the system of age strata, like a cork rising in a bottle. Each broad age stratum implies through its *age norms* the capacities, attitudes, and contributions of its members in comparison to the capacities, attitudes, and contributions of other age strata.

In the United States, the study of age stratification is complicated because the number of age strata varies among society's major institutions (polity, government, economy, family, education, religion, military, sports, and so forth) and because age grading is clearer in some institutions than in others.

Age stratification is probably easiest to discuss in relation to the economy, where legal definitions and administrative rules set relatively clear boundaries for youth, adulthood, and later life. For example, except for youngsters who work in family businesses, people under 16 are required to be in school and are legally barred from the labor force. Adults 16 and over who are not in school or full-time homemakers are expected to be employed until they reach the age of eligibility for retirement. Although retirement is permissive in the sense that people are by and large no longer required to retire and forcing them to retire is illegal, the financial systems we have created to provide retirement income serve as strong incentives to induce retirement, and nearly all older adults eventually retire.

When we look at our systems for providing income to people without jobs, our priorities clearly focus on support to people who have society's permission to be out of the labor market. For example, if we look at federal welfare programs that provide cash assistance, 32 percent of all expenditures in 1990 went to support children under 18 through Aid to Families with Dependent Children, 36 percent went to indigent older poor people and disabled children and adults through SSI, and 12 percent went to needy elderly veterans (U.S. Bureau of the Census 1991a,358). Thus, 80 percent of federal cash assistance went to support people who were in age strata where getting a job to provide one's income is either prohibited or not expected.

As we saw in Chapter 3, the exclusion of elders from the labor force was tied in large part to the use of high-energy technology, new ways of organizing work, and age prejudice. The multitude of retirement systems we now have were *responses* to the effects of age discrimination, namely, poverty among older

people. And once these income systems existed, we could then discriminate against and disengage from older people with a clear conscience, as long as we did not look too carefully at the illusion that because retirement income is available age discrimination does not hurt anyone economically. As we saw in Chapter 13, our income security systems function well for many older people, but there are also many for whom age discrimination in employment translates into economic hardship.

As an age stratum, older Americans are partially protected from the harsh treatment in age relations that might logically follow from age prejudice by the fact that family values still operate strongly within our general cultural ethic. The tradition of respect for and care of elders has been around for a long time. Who knows, by now it may have worked its way into our genetic program. Collectively, we could not face ourselves if we deprived our grandparents and great-grandparents of the dignity that comes from self-sufficiency. Thus, we have institutionalized income and health-care support for older Americans in order to allow retirement to become a period of personal fulfillment free of the stigma of unemployment or unemployability. Indeed, most individuals enjoy retirement a great deal, but our reasons for creating retirement in the first place stemmed from negative stereotypes about the capabilities of older people, not a very noble base on which to build one of society's more important institutions.

The theory of age stratification is based on the concept of birth cohorts as significant reference points for understanding age relations. The salience of birth cohort begins with the age grading in education, is carried forward by the tendency toward peer group friendships that is maintained through middle adulthood, and emerges again in later adulthood through legal definitions that use chronological age to determine eligibility for a variety of benefits and services.

But whether the members of a birth cohort, however defined, come to identify with one another in terms of common interests is a crucial question for any theory of age relations. Because without a conscious identity with one another, how can an age category take action in relation to another age category? For example, Americans born during the baby boom are often assumed to constitute a generation, a collection of birth cohorts that share common interests. But just what interests do "baby boomers" share? People who were born in 1947 grew up in a very different world than those who were born in 1967. They are also divided by social class, gender, religion, political attitudes, ethnicity, and so forth. On close examination, the concept of "the baby boomers" as a meaningful social entity appears to be a reification, a concept that is taken for real but has no actual substance. When we look for this presumably cohesive "generation" in terms of common attitudes or behavior, it disappears.

What is more likely is that age relations stem from the development of social class consciousness among some members of birth cohorts. For example, when Social Security tax rates were only 1 percent and applied to only the first few thousand dollars of annual income, upper-middle-class younger people did not object to paying these taxes or call elders "greedy geezers." But in 1983, the Social Security tax rate was scheduled to increase gradually to over 7.5 percent for employees and over 15 percent for the self-employed. And these taxes applied to income up to just over $50,000, not merely to a minor fraction of income. In this context, it is not surprising that upper-middle-class young adults found themselves resonating to the "greedy geezer" message. Young adults in this class were noticeably affected by the sharp increase in the base to which Social Security taxes applied because they were in the stage of the life course where all social classes find it harder to make ends meet. And they were having to cope with a tax their parents had experienced as only a trivial inconvenience, not a hardship. These upper-middle-class young adults rarely took into account that, due to income tax reform in 1983, their income taxes were substantially lower than those paid by their parents. The point is that age relations may be activated by a perception of common economic self-interests among members of a social class group within one age stratum, which creates the potential to mobilize consciousness of the advantages of opposing policies that seem to advantage members of other age strata. Whether this potential consciousness-raising is realized depends to a great extent on whether an organized effort is made to accomplish it.

This explanation, based on *political economy theory*, seems to work better than the theory of age stratification to explain the "generational equity" conflict that pitted interests of the upper middle class against the interests of older Americans. In the late 1980s, largely in response to changes in Social Security tax rates, an organization named Americans for Generational Equity called for sweeping changes to remedy the "generational inequities" in the Social Security system. There is always value in debating social policy, but here the "debate" was not really about generational equity so much as the *preservation of class inequity*. The tactic of painting the older generations as "greedy geezers" who were mortgaging the future of the nation's children in order to live a life of luxury was not designed to foster rational debate about Social Security but rather to discredit a system that had begun to seriously tap the discretionary incomes of the upper middle class for the first time. The young adult members of this class may have been the hardest hit and the most vocal, but upper-income Americans of all ages, along with financial services institutions, provided significant financing to the movement, which suggests that it was more a class issue than a generational issue. But, Americans do not like to acknowledge social class interests as a causal force in our politics. Ageism makes age a more palatable explanation.

In the United States, age relations between older adults as an age stratum and other age strata reflect the basic paradox of aging. Aging can bring increased skill, perspective, or wisdom, and it can bring dementia or physical disability. Because the variety of capacities in the older age stratum is quite large, there is probably no way that categorical norms could be created that could result in equitable age relations. Indeed, with the aging of the older population, there is some indication of the development of at least the young-old and the old-old as separate age strata *within* the older population. Fortunately, age strata are not very relevant to the everyday lives of most older people. Living mainly in familiar environments and interacting mainly with people who know them as individuals, older people are only occasionally faced with having to interact with people who treat them as members of an age stratum (Ward 1984). Nevertheless, such interactions are likely to be unpredictable because of our societal ambivalence about aging.

Ethics, Law, and Aging

If we all shared common definitions of what is good, true, and beautiful and what is right and wrong, then we would not need ethics, laws, or regulations. But humans are fallible creatures who often disagree with respect to what is true, what is just, and what is right. We need ways to resolve our disputes, and ethics, laws and regulations all serve as guides that are often effective in helping us deal with our everyday conflicts and dilemmas. *Ethics* refers to the culturally given body of principles used to decide moral issues of what is right and good. Ethics are usually based on underlying abstract values such as honesty and fairness. *Law* combines ethics

with procedures, placing them into a written record that can be used to resolve disputes or to enforce prohibitions against actions seen as wrong. *Administrative regulations* are procedures designed to accomplish objectives that are defined as moral or good or set down in law.

Ethics is not a static field. Social change insures that we have an endless supply of new situations not covered by our existing ethics. *Ethical inquiry* is a field of study and practice devoted to examining the basis for exploring the issues and identifying ethical courses of action.

Because the aging of society has brought about so many new situations, interest in ethical issues of aging is growing rapidly.* Ethicists work to identify ethical practices in such diverse areas of gerontology as the family, social services, and health and long-term care service delivery settings. We will illustrate the role of ethics in identifying ethical issues and resolving ethical disputes by looking at biomedical ethics.†

Biomedical Ethics

Some of the most difficult issues in gerontology concern allocation of medical care and resources, when to interfere with the natural process of dying, how to define the client or patient's proper role in the medical decision-making process, and the conflict between individual rights to self-determination and professional standards of proper care.

Resource allocation discussions arise because it appears to some people that we do not have the resources to provide all the health care needed by all the people. If we accept this premise, and by no means all health services scholars do, then some means must be found to fairly determine

---

* For more on ethics, see Moody (1992a, 1992b), Jecker (1991), or Dubler (1994).

† This section was heavily influenced by Moody (1992a).

who will be provided health care and who will not. Age has been involved in discussions of this issue in a number of ways. For example, Callahan (1987) argued that people over 75 had had their opportunity for a good life and should be able to claim no more than palliative care to ease their process of dying. More extreme forms of this argument asserted that elders' had a duty to die and make way for new generations. Others argued that older people should not be given high-cost medical interventions because they have relatively few years of life remaining for society to get a return on its investment. Still others argued that too much money is spent on high-cost care for terminally ill elders and that we needed to focus on cheaper ways of dying. Some of these assumptions, such as that elders receive a disproportionate share of their health care as useless prolongations of dying, have been found to be inaccurate (Cohen 1994).

The problem with all of these arguments is that, as we have demonstrated repeatedly in this book, older people are not a homogeneous category. For some elders it makes no sense to provide costly heart bypass operations because they are likely to die soon from something else. For others, bypass operations might result in another twenty-five years of healthy and productive life. And on what grounds could we say that a year of life is less valuable at 80 than at 45? On what grounds could we assert that *as a category* older people are less valuable than other age categories? The problem here is obvious. When society has a demonstrable prejudice against older people, the selection of elders as a category to receive less health care is more likely to be based on prejudice than on practical or ethical considerations.

I have suggested several times that if we genuinely believe that we must ration health care, then we should do it by lottery. A lottery would be fair; everyone would have an equal chance to get scarce health-care services and unconscious societal prejudices

could not enter in. But I suspect that if all Americans were at risk of being denied health care, the amount of health care we thought we could afford would instantly increase. Certainly, the case has yet to be made that wholesale denial of care to older people can be justified by our cultural ethics.

Decisions about when to allow life to end have become increasingly complex. As medical technology increases in its capacity to bring people back from death, so do the dilemmas of deciding which people to bring back and which ones to let go. Because death is concentrated in the older population, these issues are often played out in the context of an older patient and his or her family. *Advanced directives* were developed to allow physicians to know their patients' desires with regard to resuscitation efforts and to provide grounds for issuing DNR (do not resuscitate) orders to guide health-care facility staff in cases where cardiac arrest occurs. Without an advance directive, family often are asked to provide guidance, and if they cannot agree on a course of action, the health-care providers may be forced to resuscitate a terminally ill person repeatedly.

Voluntary death and physician assisted suicide are two very volatile issues in the ethics of death and dying. People vary widely on whether people should be free to end their own lives when they rationally choose to do so, and whether physicians should be allowed to provide people with humane means of ending their own lives.

Principles of client or patient *autonomy and informed consent* have become a very important part of the move toward greater emphasis on assisted living and home-delivered long-term care. The concept of *negotiated risk*, giving the resident/client/patient an opportunity to assume mutually acceptable risks in return for less strict procedures or monitoring by service providers, is a cornerstone of care planning in many assisted living and home-care programs. In general, negotiated risk lowers costs for providers and increases choices for clients. However, these issues are often difficult to address in cases where the older person is cognitively impaired, and the power differential between the client and the provider raises the potential for abuse by providers.

Principles of client autonomy also can conflict with *professional standards and regulations* concerning what constitutes care of adequate quality. Kaufman (1994) reported that elders often refuse services that require them to conform to what they see as invasive standards of care providers. Few programs have mechanisms for mediating such disputes, and even if they did, many service providers are prohibited from providing anything other than a full array of "approved" services.

These examples of ethical issues and dilemmas in the field of aging are but a small sampling of the many areas of ambiguity and uncertainty that surround the process of helping aging people. Ethical concerns are sometimes resolved when laws require issues to be handled in a certain way. State advanced directive laws and the federal Resident Self-Determination Act of 1987, which guaranteed participation in care plan decision making to nursing home residents, are examples of laws that attempt to create uniform standards for deciding ethical issues, but gray areas always remain. For example, how do providers create autonomy for demented residents? Recent approaches to ethical decision making have focused on processes through which fair and ethical decisions can be made as well as on uniform standards, which are increasingly difficult to develop for such a heterogeneous population.

Summary

General social responses to aging exist in the ideas of a culture, in the media and educational materials through which people learn these ideas, and in the attitudes and actions that result. America appears to be genuinely ambivalent about aging. While most people seem to want to think well of older people, aging poses both real and imagined threats to important social values. Family security, freedom of choice, and general happiness are all vulnerable to the ups and downs of one's own aging, the aging of family members for whom one may be responsible, and the aging of society in general. In addition, what is generally *believed* to be true about physical and mental aging makes it *appear to be* a threat to societal values such as productivity and efficiency. Yet our general beliefs and stereotypes about aging are not predictably negative or inaccurate; sometimes they are and sometimes they are not. More important perhaps is the set of negative and inaccurate beliefs about older people as jobholders or productive contributors that underpins widespread age discrimination, not only in employment but in other areas of life as well.

People seem to dislike both the idea of aging and the people who experience it. This is true throughout the life cycle. And this dislike seems to result from the public's association of aging with unpleasant outcomes such as illness, unattractiveness, and inability. On the other hand, people also seem to like the idea that wisdom, warmth, and goodness increase with age. Such is the nature of ambivalence. People hold a wide variety of positive and negative stereotypes about aging and older people, and these stereotypes are used to organize thought and behavior in specific situations rather than being applied across the board.

Contemporary children's books and adult novels generally portray aging in a positive and humanistic light. Children are presented with the ideal that older people are to be treated with respect. Adults are presented with older characters who represent a full range of human qualities. There is also evidence that contemporary television drama and feature films are moving toward a recognition of multigenerational relationships as a natural and interesting part of life. Older characters are often well-developed, successful people who can serve as models for either being or interacting with older adults, particularly in family or job situations. More than 90 percent of older characters are portrayed positively. However, television's treatment of older people in news programming tends to present the inaccurate view of later life as a life stage beset with serious problems, and this tends to reinforce the prevalent view of later life as an undesirable life stage. Journalists are susceptible to imitative repetition of disparaging labels for older people and faulty arguments that oversimplify issues affecting programs such as Social Security, Medicare, and Medicaid.

Our ambivalent beliefs and attitudes about aging and the perception of aging as a threat to important values are translated into age prejudice, or ageism, which in turn feeds age discrimination. Age discrimination imposes negative outcomes on older people just because they are older. Some of it is subtle, as when older people are ignored or avoided in interaction or social planning. It is more direct where people are denied participation because of their age. Direct age discrimination is especially prevalent in the occupational sphere but also appears to operate in volunteer work and other areas too. Age discrimination also occurs when public agencies fail to orient public programs meant for the general population as much toward older people as they do toward younger people.

Societal disengagement is institutionalized age discrimination that results in the withdrawal of interest in the contributions or involvement of older people. It is not a mutually satisfying process but one that is imposed on older people by the withdrawal of opportunities to participate. Like age discrimination, the more general process of societal disengagement rests not so much on the functional needs of society as on its prejudices.

Age strata are created by age norms, which govern access to positions within social institutions such as the economy. Our age relations within the economy are ambivalent. On the one hand, based on age prejudice we institutionalized the removal of older adults from the labor force; on the other hand, we created retirement income systems that transformed retirement into a reward. Society's treatment of the older age stratum is softened by centuries-old family values of respect and caring for elders. However, it is also hardened by the rise in self-centered individualism and social class interest among some of today's young adults.

As the aging population continues to diversify, the potential for ethical, legal, and administrative ambiguities increases as well. Important areas of ethical consideration in the field of aging include resource allocation within and among generations, making decisions on death and dying, and how to balance disabled elders' need to continue to be involved in decisions about their lives with professional concerns about care quality. Recent efforts have focused on how to create processes for making ethical decisions in specific situations as well as creating universal standards embodied in law.

17 ▮ Social Inequality

All known societies treat categories of people differently; such treatment is usually based on the values of the dominant culture. And as with other elements of culture, those social inequalities are passed on from one generation to the next. In the previous chapters we considered a variety of inequalities people encountered in the past and continue to encounter as a result of growing older. To a great extent, the experience of aging is also mediated by other sources of inequality: social class, race, ethnicity, and gender. This chapter focuses on these sources of inequality and their impact on aging.

Social Class

"I don't go to senior center events because the others who go there are not my class of people. We have no interests in common." This quote from a retired woman executive reflects the essence of social class. *Social classes* are *categories* of people considered to have certain social, economic, educational, occupational, or cultural characteristics in common. Furthermore, social classes are arranged in a *hierarchy* of relative social desirability. Social classes are not organized groups; they are aggregates of people who would recognize one another as overall social peers and who would be recognized as such by others.

The study of social class in American society is difficult because a number of factors affect social class. First, ours is an open class system in which people may move up or down. Although most people live their entire lives in the social class into which they were born, there is enough upward and downward mobility to complicate the study of social class in life course or intergenerational perspectives. Second, the criteria used to categorize people into social classes vary between social classes. For example, family history may confer little status in the working class, whereas among the rich it may be of crucial importance in defining the individual's peers. Social class criteria also vary by size of community. In large urban areas, "objective" characteristics such as occupation, education, income, and place of residence generally carry a great deal of weight. "Appearances" is another important factor influencing secondary social relationships with service personnel, bureaucrats, shopkeepers, and the like. This reliance on attributes is an obvious outgrowth of the relative anonymity of large cities. In medium-sized cities or smaller communities, criteria for class assignment tend to focus on behavior as well as attributes. Thus, in long-standing neighborhoods and small communities, a person's or family's past history of performance is usually an important dimension of social class assignment by the community at large. There are also regional differences in class criteria. Finally, there are class criteria specific to certain life course stages. For example, there is a class of retired people in American society in which peers are defined by their affirmation of leisure as the focus of their lifestyle and by having the financial resources to pursue an expensive leisure lifestyle dominated by lavish homes and exclusive country clubs.

Class Differences in Aging

There are no clearly defined points that separate various social classes in American society. The terms *upper class, middle class,* and *lower class* refer to stereotyped composites of lifestyles, educational backgrounds, family values, occupations, housing, and so forth. In fact, the attributes and behavior used to rank others in relation to an individual form a continuous distribution over which a very large number of distinctions can be made. Yet it is still possible to discuss general categories if we keep in mind that in every case the boundaries of the categories are fuzzy.

The *upper class* is typified by extraordinary wealth, and distinctions within the upper class center on family history and values. Upper-class children are educated in private schools with other upper-class children. They attend the "best" private universities. Their occupational careers prepare them for their "proper" place as leaders of the nation's corporate, military, financial, political, educational, and legal institutions. Upper-class people lead very private lives in circles largely unknown to the general public. For example, every year *Forbes* magazine publishes a list of the 100 richest Americans. Most members of the general public do not recognize more than one-fourth of the names on this list. In addition, there are many very rich Americans who avoid being placed on the *Forbes* list.

Upper-class people have access to wealth and power. Our tax structure allows Americans to amass great fortunes and pass them on from generation to generation (G. Nelson 1983). The fortunes of upper-class people do not depend on their having jobs, and upper-class people carry their fortunes into later life. In fact, most of the very rich (billionaires) are over 65. As a result, older upper-class family members have wealth and the power that comes with wealth. This privileged position means continuity of power within the family, generally better health and vigor, and much less likelihood of facing dependency in old age. Nevertheless, no amount of money can eliminate aging; it can only delay it. Eventually, even upper-class people experience a decline in physical functioning and erosion of the social environment as their friends and associates die. Because the upper class tends to insulate itself rather severely from others, developing new relationships in old age may be more difficult. In addition, physical aging in the upper class may produce much greater change in lifestyle than in the middle class. These topics deserve more research.

The *middle class* in America is a large and heterogeneous category. Jobs and educational attainment are the measures most often used for sorting out various subcategories of the middle class. At the top are college-educated people with professional and managerial jobs. The middle area consists of well-educated white-collar workers who push the nation's paper and market its goods. The lower reaches of the middle class consist of routine clerical and skilled blue-collar and service workers who may or may not be well educated, whose jobs may involve "dirty" work, but whose incomes are often more than sufficient to allow them a middle-class lifestyle. Middle-class lifestyles emphasize both "getting ahead" in the educational and occupational sphere and diverse activity patterns that often revolve around the family. Middle-class jobs and values usually result in good health and adequate financial resources in retirement. And family values in the middle class usually ensure prompt response to dependency needs of older family members.

The *working class* is made up of people whose livelihood is or was generated by semiskilled and unskilled blue-collar and service jobs. Working-class elders often have not finished high school, and their jobs are or were more physically demanding and precarious than those of the middle class. Working-class lifestyles emphasize avoiding slipping downward more than getting ahead, and activity patterns tend to separate men and women. Working-class jobs carry more health and disability hazards and are less likely than middle-class jobs to produce adequate retirement benefits. Family values of the working class support the idea of caring for older family members, although the financial capabilities to do so are limited. Working-class people often have to rely on public assistance to supplement inadequate retirement income.

The *poor* are those who live on incomes from the least desirable types of jobs—for example, dishwasher, presser in a cleaning shop, farm laborer, or private household worker. Many poor people never have much success at getting or holding a good job. They have neither the education nor the family upbringing necessary to get steady jobs. Lifestyles emphasize survival, and older family members' old age benefits are sometimes a crucial source of family income. In fact, the availability of SSI and other indirect income assistance can mean a substantial *improvement* in level of living for those poor people who survive to become eligible for them. Yet even though aging may improve their financial situation, the elderly poor are apt to find themselves physically less able to take advantage of these opportunities than their working-class counterparts.

Social class has often been used as a variable in gerontological research. Studies of retirement, widowhood, adaptation to aging, and numerous other topics have found that social class makes a difference. For example, compared with the working class and the poor, the middle class enters later life with better health, more financial resources, more activities, better housing, and fewer

worries.* Accordingly, middle-class people seem to cope better with just about every life course change in later life, particularly in the long run. Comparatively, middle-class people have good marriages that tend to get better as a result of child launching and retirement. This is less often true for the working class and poor. Middle-class people generally adjust better and more quickly than working-class people to widowhood. Both loss of independence and institutionalization are less likely for middle-class people than for the working class and the poor. Thus, to a great extent the optimistic picture of aging presented in earlier chapters is a result of the fact that most Americans are middle class. But for those who are not, the picture is much less rosy. A lifetime of poverty often translates into poor health and poor living conditions in later life. Poor working conditions earlier in life can lead to greater rates of disability in later life. Semiskilled and unskilled jobs provide meager financial resources not only during working life but in retirement as well. To the extent that adequate income and good health are prerequisites for a satisfying life in the later years, the working class and the poor are at a distinct disadvantage. Their jobs are unlikely to entitle them to adequate retirement income and may expose them to a greater risk of premature physical aging and disability.

From a research perspective, the foregoing general view is well supported. We need to know more about precisely where in the occupational structure the high risk begins. Jobs that produce adequate salaries and wages do not necessarily produce adequate retirement income, although there is considerable overlap. For example, a computer repair specialist may earn a salary sufficient to support a middle-class lifestyle, but may work for a small business with no pension plan. The result

---

* Source references for the social class findings are in the other chapters of this book.

could be a retirement income that does not support a middle-class lifestyle. We also need to know more about illness and disability histories in various occupations. For example, people who work in construction jobs are much more likely to enter later life with disability than people who work at clerical jobs.

Because social class position tends to persist across generations, not only are the lower-class older person's physical, financial, and social resources for coping with life more limited than those of their middle-class agemates, but the adult children of lower-class older people are also substantially less likely to have financial resources sufficient to aid their older parents.

Many think that the improvements in Social Security benefits since the early 1970s have reduced class differences among older Americans, but in a carefully controlled analysis, Crystal and Shea (1990) found that income inequality in the older population was actually higher than in the general adult population. Elders who were helped most by the improvements to Social Security were by and large not middle-class but instead working-class elders whose incomes now hover just above the poverty level. On the other hand, those who were helped most by tax-break-subsidized private pensions and annuities tended to be in the highest income categories. Thus, the occupational inequalities of the working years were carried over into retirement by Social Security benefits tied to lifetime earnings. In addition, the mildly redistributive effects of Social Security benefit formulas designed to provide slightly higher earnings replacement to low-income workers were more than offset by the tendency of our tax incentive systems to encourage multiple retirement income sources for workers in the top 20 percent of the preretirement earnings range. We examine this topic in more detail in Chapter 18, "The Economy." Thus, our approaches to retirement income, taken as a whole, increase rather than reduce the distance between social classes.

Pampel and Hardy (1994) looked at the predictors of men's income before and after retirement and found that the economic advantages of being white, urban, highly educated, and having high occupational status carried over into retirement. Because it is tied to level of earnings during the working years, Social Security mimics market wages to produce wide differences in retirement income. Another way of saying this is that our systems for providing income in retirement are constructed in a manner that preserves or even increases class differences in income.

Social class position is central to later life because it is central to life in general. The social class into which one is born or adopted determines the family context of early socialization. One's values, attitudes, beliefs, lifestyle, and opportunities early in life depend largely on the social class position of one's parents. Social class also has a great influence on role models. Imitation is an important social process. Children try to be like their parents (although they often do not like to admit it). They also look to people around them for ideas about what their options are with regard to jobs, marriage, and other aspects of lifestyle. Middle-class children have access to a different set of role models than lower-class children do. Those people who manage to rise out of lower-class beginnings are predominantly those whose families had middle-class values and made sacrifices to create opportunities for their children (Billingsley 1968).

There is a great deal of intergenerational inertia in the social class system, particularly in the pool of jobs and lifestyles available to family members. Educational, occupational, and marital decisions made early in the life course are heavily influenced by the perceptions and social pressures that grow out of the social position of one's family. For the children of the working class and of the poor, the working people around them are laborers, waitresses, dry cleaning employees, equipment operators, truck drivers, and the like. Early in life they learn that employment for the people around them is hard and unpredictable. These children may want to be astronauts, but they expect to have to settle for considerably less. And when they settle for job and family responsibilities before or soon after high school graduation, they have made decisions that are hard to reverse and that greatly influence the physical and financial resources with which they approach middle age and later life. Not only that, but less education also usually means fewer skills for enjoying life off the job.

Dimensions of Disadvantage

The class differences discussed resulted from a dominant culture based on an amalgam of values that have their roots in the Western European cultures from which the early white immigrants came—England, Scotland, Wales, Germany, France, and Spain. Later large streams of white migrants also came from Ireland, Italy, Poland, and Russia. It is important to note that the high natural increase (excess of births over deaths) of the early Americans added more to the population of the United States than immigration did, so the dominant Euro-American culture evolved in the context of overwhelmingly superior numbers.

Women are disadvantaged even as members of the dominant white majority, because the European cultures that formed the base of America's dominant Euro-American culture contained strong elements of sexism that have continued to be part of the evolving American culture. Other disadvantaged groups suffer because they are the "wrong" color, speak the "wrong" language, or practice an "alien" culture, all from the ethnocentric point of view of the dominant Euro-American culture, of course. In the next sections, we will consider race and ethnicity as dimensions of disadvantage. We conclude the chapter with a discussion of how gender-based disadvantages cut across all the other dimensions of social structure.

Because the United States is a "work society"—a society in which eligibility for major benefits such as income, health care, and protection against misfortune are derived from a person's or family's relationship to the occupational structure—we focus on employment and income as the major outcomes of our systems that create social inequality. These are by no means the only types of inequality that disadvantaged categories of Americans experience, but they are the easiest to document and among the most important.

Race

Race is ostensibly a biological characteristic, but because so much interracial reproduction has occurred over the centuries, race is as much a matter of social definition as biology. As defined by the U.S. Bureau of the Census, the major racial groups in the United States are Caucasian (white), African American (black), Asian American, and Native American. There are also a number of less prevalent but separate racial categories such as Pacific Islander, Aleut, and Eskimo.

In 1990, over 90 percent of the older population of the United States was white. Another 8 percent was black, and the remaining 2 percent was made up of Asian Americans (1.4 percent),* Native Americans (0.4 percent), and a smattering of other races. Perhaps because the United States is so predominantly white, white Americans tend to view persons of other races unfavorably.

Racism is a complex of attitudes and discriminatory behavior patterns based on the notion that one racial category is inherently superior to another. In American society the white majority behaves in general as if it

---

* Asian Americans are those of Japanese, Chinese, Filipino, Korean, or Hawaiian descent.

were superior to most other racial categories, but particularly to African Americans, Asian Americans, and Native Americans.

African Americans

In 1990, the 2.5 million older African Americans represented 8 percent of the total older population and 8.3 percent of the African-American population. They were by far the largest racial minority in the older population. The visibility of the African-American population is heightened by its concentration in central cities as a result of residential segregation.

Racial discrimination also produced a situation in which the African-American population is concentrated in the less desirable jobs within the blue-collar category (see Tables 17-1 and 17-2). For example, in 1994 among employed African-American men age 55 to 59, 47.7 percent were working as laborers, nonprecision equipment operators, janitors, and helpers and at other menial jobs, compared with 23.3 percent in the same types of jobs among white men of the same age. In 1994 among African-American women age 55 to 59,

49.1 percent were working as maids, food-service helpers, and nurses' aides and at other menial jobs, compared with 22.4 percent of white women of the same age. In addition, unemployment rates of African Americans are much higher than those for whites.

The occupational situation for African Americans has improved substantially since the 1960s. African-American men on the eve of retirement are about as likely as younger men to have middle-class jobs. For example, in Table 17-3, if we compare workers age 30 to 34 ("settled" adults) with those 60 to 64 (on the eve of retirement), 32.5 percent of young males had middle-class jobs compared with 30.3 percent of older males. However, age differences were still pronounced for women. For example, in 1994, 62.6 percent of young women had middle-class jobs compared with only 39.2 percent of women on the eve of retirement. A much greater proportion of younger women were working as teachers, social workers, nurses, and retail managers compared with women on the eve of retirement. Women approaching retirement age

Table 17-1 Occupational distribution by percent of employed males age 55 to 59 by race and ethnicity: United States, 1994.

	Total	White	African[a] American	Native[a] American	Asian	Other Races	Non-Hispanic	Hispanic
No. employed	3,798,457	3,379,674	298,112	13,538	88,304	18,820	3,546,373	252,085
Managerial and professional	32.2	33.3	18.8	31.6	37.9	17.0	33.4	15.9
Technical, sales, and administrative support	18.7	19.2	13.7	9.3	20.4	1.5	19.0	13.8
Service	7.5	6.5	16.2	10.5	15.6	22.5	7.3	11.2
Farm, forestry, and fishing	4.8	5.0	2.7	7.0	2.2	11.2	4.5	9.8
Precision production, craft and repair	18.8	19.2	17.1	17.8	13.6	12.1	18.7	20.6
Operators and laborers	18.0	16.8	31.5	23.9	1.2	35.8	17.2	28.6
Total	100.0	100.0	100.0	100.1[b]	99.9	100.1	100.1	99.9

[a] The U.S. Bureau of the Census uses the categories *black* and *American Indian*.
[b] Columns may not add to 100% due to rounding.

Table 17-2 Occupational distribution by percent of employed females age 55 to 59 by race and ethnicity: United States, 1994.

	Total	White	African[a] American	Native[a] American	Asian	Other Races	Non-Hispanic	Hispanic
No. employed	3,199,951	2,774,329	311,391	16,131	80,422	17,678	3,045,790	154,161
Managerial and professional	28.6	29.9	18.0	47.8	25.1	17.5	29.3	15.4
Technical, sales, and administrative support	41.3	43.1	28.1	21.6	39.5	21.0	41.8	30.9
Service	17.1	15.1	33.7	22.8	16.8	39.5	16.3	34.0
Farm, forestry, and fishing	1.7	1.9	0.9	0.0	0.3	0.0	1.8	0.5
Precision production, craft and repair	2.8	2.7	3.9	0.5	3.6	0.0	2.7	6.5
Operators and laborers	8.4	7.3	15.4	7.3	14.8	22.1	8.2	12.7
Total	99.9[b]	100.0	100.0	100.0	100.1	100.1	100.1	100.0

[a] The U.S. Bureau of the Census uses the categories *black* and *American Indian*.
[b] Columns may not add to 100% due to rounding.

Table 17-3 Occupational distribution by percent of employed African-American[a] workers of selected ages, by sex: United States, 1994.

	Age 30–34		Age 60–64	
	Male	Female	Male	Female
No. employed	986,902	997,132	159,525	195,687
Managerial and professional	15.0	22.8	18.6	17.3
Technical, sales, and administrative support	17.5	39.8	11.7	21.9
Service	19.4	21.4	21.1	46.3
Farm, forestry, and fishing	2.6	0.5	4.6	0.0
Precision production, craft, and repair	12.6	3.9	11.7	1.4
Operators and laborers	32.9	11.7	32.5	13.1
Total	100.0	100.1[b]	100.2	100.0

[a] The U.S. Bureau of the Census uses the category *black*.
[b] Columns may not add to 100% due to rounding.
Source: Tabulated from the 1994 Current Population Survey.

were much more likely to be janitors, laborers, or maids. This cohort difference in the occupational distribution of women is the result of historical job discrimination. Older African-American women are still paying for it, particularly in terms of their ability to generate adequate retirement income.

A main effect of racism toward African Americans has been to restrict the number able to get jobs that would support a middle-class lifestyle. For example, in the 1920s, college-educated African Americans had to settle for jobs as porters on the nation's trains (Drake and Cayton 1962). Yet many African-American families adapted in ways that allowed them to retain middle-class values. It is commonly assumed that stifling of opportunities gradually erodes values, but this certainly has not been the case for African Americans historically. Even in the face of job discrimi-

African Americans are concentrated in less de-sirable blue-collar jobs.
Photograph by Bruce Davidson/Magnum

more like older middle-class whites not only in terms of values but also in terms of education, health, and financial resources.

O'Hare (1989) reported that African-American families with annual incomes of $50,000 or more in constant 1987 dollars grew from 212,000 in 1967 to 764,000 in 1987. These households were similar to any other well-off group. They were likely to be "middle-aged, married, relatively well-educated and to own their own homes" (1989,25). Compared with whites, upper-income blacks were more likely to be 35 to 44 years old and less likely to be 65 or older. The recency of upper-income status for many African Americans was indicated by much lower levels of assets, including homeowning. Nevertheless, these members of the African-American middle class can expect to enjoy many of the advantages middle-class whites do in terms of retirement income and good health.

On the other side of the coin, however, more than 2 million working-age African Americans entered poverty between 1978 and 1983 alone. O'Hare (1985) reported that the poverty rate for African Americans was 36 percent in 1983, compared with 12 percent among whites. In 1983, 45 percent of African-American children were living in poverty households, compared with 17 percent of white children. Duncan (1984) reported that during the 1970s there was considerable movement in and out of poverty, with only about 1 to 3 percent in poverty over more than an eight-year period. However, among those in long-term poverty, more than 60 percent were African American. Most African Americans in poverty are there because they are unable to get jobs or better-paying jobs.

The outlook for poor African Americans as they grow older is bleak. As we will see in Chapter 18, the United States' economy is struggling to maintain sufficient jobs for the middle class, let alone create new jobs for the minority poor. We have not found an effective way to deliver adequate education and

nation and low incomes, many African-American families raised their children to value education in itself and to live by the middle-class values of thrift, hard work, sacrifice, and getting ahead. Liebow (1967) found these values to be alive and well even among the black working poor. While a smaller proportion of African Americans than whites are middle class, there has been an African-American middle class for many decades (DuBois 1915), and many African Americans were well-prepared to take advantage of new opportunities that grew out of the Civil Rights Act of 1965 (Kronus 1971). As a result, future cohorts of older middle-class blacks will be

Table 17-4 Poverty rates[a] for white, African-American,[b] and Hispanic persons, by age and sex: United States,1992.

	White		African American		Hispanic Origin	
	Men	Women	Men	Women	Men	Women
Age			**Persons in Married Couples**			
65+	5.4	4.9	20.5	18.4	14.4	10.8
65–74	4.9	4.6	17.2	18.9	14.2	10.3
75+	6.3	5.7	29.4	16.8	14.6	12.1
			Unmarried Individuals			
65+	14.5	23.8	44.4	57.5	39.9	50.7
65–74	14.8	20.6	43.5	53.5	—	46.5
75+	14.2	26.2	45.8	62.5	—	55.7

[a] The poverty rate is the percent of a population category that has income below a government-set poverty line.
[b] The U.S. Bureau of the Census uses the category *black*.
Source: U.S. Bureau of the Census (1993).

job placement to unemployed young African Americans. And as long as there are few jobs for inner-city African Americans even if they complete high school, we cannot expect great gains for the African-American poor.

Although poverty rates for older African Americans in general went down from 48 percent in poverty in 1970 to 34 percent in poverty in 1990, there are significant subgroups within the older African-American population that are at extreme risk of poverty. For example, Table 17-4 shows that poverty rates for older African-American couples are three to four times higher than the rates for white couples the same age. For older African-American individuals, the risk of poverty increases with age, especially for women; 62.5 percent of African-American women age 75 and over and living alone had incomes below the poverty line.

Older African Americans also suffer the effects of numerous inaccurate stereotypes. It is commonly assumed that all African Americans are lower class, that most African-American families are headed by women, that older African Americans die substantially earlier than older whites, and that older African Americans are mostly dependent on welfare. However, as Table 17-3

Table 17-5 Life expectancy, by age, sex, and color: United States, 1990.

	Males		Females	
Age	White	Black	White	Black
50–54	26.7	22.5	31.6	28.2
55–59	22.5	19.0	27.2	24.2
60–64	18.7	15.9	23.0	20.5
65–69	15.2	13.2	19.1	17.2
70–74	12.1	10.7	15.4	14.1
75–79	9.4	8.6	12.0	11.2
80–84	7.1	6.7	9.0	8.6
85+	5.2	5.0	6.4	6.3

Source: National Center for Health Statistics (1995).

shows, a substantial portion of older African Americans in 1994 were indeed middle class. "The black experience" in American society is definitely mediated by social class, and so is the experience of African-American aging.

Because life expectancy at birth is about five years higher for whites than African Americans, it is commonly assumed that older African Americans die sooner than

Table 17-6 Income sources of persons 65 and over, by race and Hispanic origin: United States, 1992.

Income Source	White			African American[a]			Hispanic Origin		
	N[b]	%	Median Income	N	%	Median Income	N	%	Median Income
Earnings (employment)	4,113	15.1	$16,132	344	13.2	$12,564	147	12.7	$14,759
Social Security retirement	25,461	93.3	$6,751	2,345	89.7	$5,492	969	83.8	$5,716
SSI	1,103	4.0	$2,263	500	19.1	$1,999	249	21.5	$3,031
Pensions	9,423	34.5	$8,327	522	20.0	$7,016	224	19.4	$7,624
Interest	19,438	71.2	$2,997	624	23.9	$1,655	416	36.0	$1,829
Number with income	27,291			2,615			1,156		

[a] The U.S. Bureau of the Census uses the category *black*.
[b] In thousands.

older whites do, on the average. In fact, however, because higher mortality forces operate on the African-American population at younger ages, older African Americans are relatively robust people whose life expectancy is close to that for older whites (Markides 1989). Table 17-5 shows that black-white differences in life expectancy are just over three years at age 50 to 54, but gradually decrease with age.

There is no support for the notion that older African Americans suffer substantially higher mortality rates than whites at the oldest ages. But they do have a higher incidence of illness and disability. For example, Markides and Mindel (1987) reported that older African Americans were more likely than whites to suffer from hypertension, but were like whites with respect to prevalence of heart disease, cancer, and strokes—the main causes of death among older people. However, about 19 percent of older African Americans had serious physical limitations of mobility or activity, compared to about 14 percent of older whites.

In terms of income, older African Americans had incomes in 1992 that averaged about 60 percent of those for older whites (U.S. Bureau of the Census 1994). This represented a substantial improvement

over the situation in the mid-1960s due to increased pension coverage and improvements in Social Security. The uneven occupational histories of older African Americans because of racial job discrimination were reflected in the fact that much smaller proportions got private pensions compared with older whites. In 1992, Social Security, earnings, interest, pensions, and SSI were the major income sources for older African Americans. For older whites the major sources were Social Security, interest, pensions, and earnings. Only 4 percent of whites were receiving SSI. (See Table 17-6.) Note also that interest income was a major income source for about 71 percent of whites, but provided income to only about 24 percent of African Americans. Since nearly 90 percent of older African Americans drew Social Security retirement benefits and less than 20 percent drew SSI, the stereotype of older African Americans as mainly dependent on welfare is degradingly inaccurate.

In summary, the racial discrimination that has typified the treatment of African Americans for many decades has concentrated older African Americans in low-paying jobs. These conditions continue into later life in the form of lower Social Security benefits, fewer private pensions, and more prevalent health problems. There is evidence that the situation of African Americans is improving, which should mean less disadvantage for upcoming cohorts when they reach later life. However, parity with older whites is still nowhere in sight.

Asian and Pacific Island Americans

In 1990, there were nearly 450,000 older Americans of Asian—primarily Japanese, Chinese, and Filipino—and Hawaiian and other Pacific island races. Although the white majority tends to see these various categories of Asian and Pacific island Americans as similar, they differ substantially in physical appearance, language, customs, background of involvement in American life, and social class structure. Some Asian ethnic and racial groups have been able to overcome the occupational disadvantages of racial discrimination. For example, older Japanese Americans are quite similar to older whites in terms of the occupations they held. A bigger proportion of older Chinese Americans held jobs as professionals or managers, compared with older whites. But at the other end of the occupational range, a bigger proportion of older Filipino and Chinese Americans than older whites held menial jobs. Older Japanese Americans tend to have a median income near that of older whites, and all Asian races tend to be better off than older African Americans. (See Table 17-7.) Yet the wide variations within the older Asian and Pacific island category mean that a substantial proportion tends to be financially comfortable and a substantial proportion tends to be financially poor.

Older Asian and Pacific island Americans have unique language, dietary, and cultural problems that make the current mix of services for older people more difficult for them to use, especially in regions of the country away from the Pacific coast states. Older Asian Americans have much higher proportions of men than the general population be-

Table 17-7 Median income of persons age 65 and over, by sex, race, and Hispanic status, 1994.

Race	Male	Female
White	$14,998	$8,368
African American*	8,750	6,000
Native American*	9,240	5,320
Asian American	10,057	6,327
Other races	6,542	7,013
Non-Hispanic	14,667	8,306
Hispanic	9,000	5,592

* The U.S. Bureau of the Census uses the categories *black* and *American Indian.*
Source: Tabulated from the 1994 Current Population Survey.

cause immigration laws that allowed men to enter as laborers early in this century did not allow entry of women and children.

Native Americans

Older Native Americans are the most disadvantaged among the disadvantaged. The situation of Native Americans, or American Indians, is clouded by numerous issues, not the least of which is difficulty in gathering information about those who do not live on reservations. According to the Bureau of the Census, there were about 116,000 older Native Americans in the United States in 1990, of which about 30 percent lived on tribal lands. About 53 percent of Native Americans lived in rural areas, compared with 25 percent of the white population. Because Native Americans tend to live a rural life, they are less likely to have had paid jobs, and those who live in cities tend to have jobs similar to those held by low-income African Americans. As a result Native Americans have the lowest median income among disadvantaged minorities in the older population.

Older Native Americans on tribal lands face conditions that are substantially different from the older population in general. They lead an agricultural lifestyle that has changed much more slowly than society as a whole. Native Americans share in the general un-

Life has changed very slowly for older Kickapoo Indians in Texas.
Photograph by AP/Wide World Photos

availability of federal programs for older people in rural areas. Housing and sanitation on Indian reservations are the most substandard in America.

Conditions on tribal lands also include poor roads that must be traveled great distances. The low incomes of most residents mean that the vehicles they use are often not in the best of condition. As a result, older Native Americans, especially men, have death rates from motor vehicle accidents five times higher than the national average (Markides and Mindel 1987). In addition, Native Americans are particularly more susceptible than the general population to obesity, diabetes, cirrhosis of the liver, and alcoholism. Diabetes is especially likely to lead to disability among older Native Americans because untreated diabetes can lead to the need to amputate feet and legs and to low vision (Markides and Mindel 1987).

In cities, the older Native-American poor suffer disadvantages similar to those low-income African Americans experience. They have had low paying jobs, little education, and little preparation for life in urban America. And like Asian Americans, they often have language problems. These problems of older Native Americans are likely to persist. Younger generations of Native Americans are not making the same strides as other minorities.

Ethnicity

The cultural and ethnic differences among the various racial minorities in American society influence the differences discussed in the previous section. Within each racial category, there are often social class, religious, and regional *subcultures*, and the probability that individuals will experience racial discrimination, especially in education and employment, is highly contingent on these ethnic differences. Thus, there are degrees of ethnic disadvantage even among the racially disadvantaged. Hispanic Americans constitute the largest minority that experiences disadvantage mostly as a result of culture rather than race.

Hispanic Americans

In 1990, about 3.7 percent of America's older population was of Spanish heritage—mainly Mexican, Puerto Rican, and Cuban. Although Hispanic Americans can classify themselves as any race, most classify themselves as white. While African Americans have experienced discrimination based on their skin color, Hispanic Americans have mainly experienced discrimination because of their language and culture. However, different cultural groups within the Hispanic population have different perspectives on the extent to which they are a racial as well as an ethnic minority. For example, people of Cuban and Puerto Rican descent are likely to classify themselves in terms of white and black racial characteristics. But Mexican Americans in the Southwest and in California often identify themselves as "brown," reflecting their identity with an intermingling of Spanish and Native-American populations, and to see a racial as well as ethnic character to the discrimination they experience.

Succeeding generations of white Hispanic Americans can adopt the language of the dominant white majority, but adopting skin color is another matter. As a result, older Hispanic Americans are substantially better off than older African Americans (see Tables 17-1, 17-2, and 17-6). Their occupational histories and incomes are generally better than those of African Americans. Yet Hispanic Americans still have a substantial distance to go before they will be on a par with non-Hispanic whites.

Unemployment just before retirement is particularly prevalent among Hispanic Americans. In addition, like older Asian Americans, Hispanic elders make up a diverse category. The Hispanic elderly poor share with other minority poor the problems of little education, high illiteracy, high incidence of disability, language barriers to participation in society, and cynicism toward the effectiveness of government programs. As Table 17-4 shows, Hispanic women age 75 and over have a poverty rate above 50 percent. However, substantial numbers of Hispanic-American older people have the resources to live middle-class lifestyles in retirement. In addition, Mexican Americans in the Southwest, Puerto Ricans in the Northeast and the Great Lakes region, and Cubans in Florida are quite different in terms of culture of origin and the economic and social situation in their area of settlement in the United States. Generally, Cuban Americans are more advantaged than Mexican Americans, and Mexican Americans are more advantaged than Puerto Ricans.

Thousands of Mexicans live in the United States illegally, yet very little information is available on aging among illegal aliens. Given the difficulties of doing research on this population, we are not likely to fill this gap in our knowledge soon.

Gender

Women constitute one of America's most oppressed social categories. *Sexism*—the belief that women and men are inherently different in overall capability and that each sex is fit only for certain jobs—has resulted in a unique position for women in relation to jobs. To many, a woman's place is in the home. As we saw in Chapter 9, substantial proportions of older women have been nonemployed homemakers, and while the percentage of employed women has dramatically increased just since 1970, there are still substantial proportions who will be nonemployed homemakers throughout their adult lives. These women are completely dependent on their husbands' lifetime earnings and the retirement income those earnings can generate, and they are *quite* vulnerable to the effects of divorce and widowhood. If they become divorced after age 35, they will find it difficult to work long enough to get adequate Social Security pensions on their own. The system is stacked against people not employed throughout adulthood since Social Security benefits are tied to average earnings over as many as thirty-five years. For example, suppose a person is in a category in which earnings are averaged over thirty-five years and that person has worked only twenty-five years, with average earnings of $10,000 a year. The Social Security Administration would divide the total earnings of $250,000 by 35, yielding a pension tied to average earnings of $7,143 rather than the actual $10,000 average over the twenty-five years worked.

In addition, women who become divorced in middle age often find themselves with no marketable skills, and as a result they either have to find the resources to go back to school or settle for a low-paying job with little or no chance for advancement and little likelihood of generating an adequate retirement pension.

Widows who have been nonemployed homemakers over their adult lives are less likely to find themselves out in the cold economically, but it can happen. For example, most pension systems allow spouses to waive

survivors' benefit. In such cases, widows find that they have lost not only their husbands but a large chunk of their incomes as well. Many women like caring for house, husband, and children rather than participating in the job market. But by selecting this option they take a sizable gamble that the marriage will stay intact long enough to give them financial security in their later years.

Women who do seek employment often find themselves channeled into "women's work." (See Table 17-8.) In 1994, two-thirds of women age 55 to 59 held retailing, clerical and administrative support, and service jobs. Those in professions tended to be in teaching or nursing. These jobs are precisely the ones that fit the stereotype of "women's work." The general overview of women's occupations has changed very little since 1940 (Baxandall et al. 1976). Even if a move toward greater gender equality should take place in the job market, it is unlikely to affect the position of middle-aged and mature women for some time to come.

Of the top ten women's jobs, only teachers and nurses have a high probability of getting

Table 17-8 Most common jobs among employed women of selected ages: United States, 1994.

Age 30–34		Age 55–59	
Job	Number Employed (in thousands)	Job	Number Employed (in thousands)
Manager/administrator	664	Office clerk	497
Office clerk	633	Secretary/typist	283
Secretary/typist	546	Manager/administrator	277
Sales worker	370	Teacher	211
Teacher	351	Sales worker	171
Nurse, registered	270	Janitor/house cleaner	149
Food-service worker	234	Hairdresser/cosmetologist	124
Janitor/house cleaner	222	Nurse, registered	115
Cashier	198	Nurse aide	101
Nurse aid	194	Food-service worker	91
Percent of all employed: 66%		Percent of all employed: 76%	

Source: Data compiled from the 1994 Current Population Survey.

a private pension in addition to Social Security. In addition, women's earnings average only about 60 percent as high as men's earnings at the same occupational and educational levels (see Table 17-9). These various factors mean that retirement incomes of women average only about 55 percent as high as those men enjoy. And for older African-American or Hispanic women, poverty is the rule rather than the exception.

Multiple Jeopardy

Thus far we have been concerned primarily with how various types of social inequality, taken separately, affect aging. But many older Americans are simultaneously members of several disadvantaged groups. For example, compare an upper-class white older man and a lower-class black older woman, and imagine their lives as they grow older. Beyond such obvious extremes, however, it is difficult to get an idea of just how various minority characteristics interact (Markides 1983).

One way to examine the issue is to look at how education (social class related), race and ethnicity, and gender simultaneously affect earnings. In general, lower education and being African American are decided disadvantages. However, *the greatest disadvantage is being a woman.* The gender difference in median income is much greater than income differences associated with differences in education or minority status (see Table 17-9).

These social inequalities reviewed in this chapter result from actions taken on the basis of people's social class, race, ethnicity, and gender. And based on the preceding chapter, we must add age to the list. Inequality becomes injustice when the basis for the inequality has no validity. A large proportion of the social inequality in the United States rests on categorical assumptions—ageism, racism, ethnocentrism, and sexism—that are at least as much a *result* of our system for linking people with jobs as they are a justification for it.

Table 17-9 Median earnings of year-round, full-time workers at age 55–64, by education, race, and ethnicity, by gender: United States, 1993.

	Males	Females
Education		
Less than 9th grade	$22,067	$12,759
9–12 (no diploma)	28,612	15,203
High school graduate	30,739	19,701
Some college (no degree)	38,709	25,945
Associate degree	38,990	25,456
Bachelor's degree or more	56,416	37,599
Race and ethnicity		
White	31,737	22,423
African American[a]	22,942	20,299
Hispanic	20,312	17,743
Total	31,012	22,167

[a] The U.S. Bureau of the Census uses the category *black*.
Source: Tabulated from the 1994 Current Population Survey.

The quality of life in the later years depends heavily on health and financial resources. In our society both are closely tied to jobs, so we must conclude that while they may have a bearing on the availability of housing and services, *the primary impact of social class, race, ethnicity, and gender on aging is their impact on access to education and jobs* and on the opportunities those jobs allow or the limitations they impose. Thus, the inequalities we observe among people in later life are for the most part created by the social structural arrangements that sort and select people within the field of education and that link people to segments of the labor market.

Social Structure and Life Chances

Gender, social class, race and ethnicity all influence how young people are perceived and processed by educational institutions (Dannefer 1992). All these dimensions also influence the exchange resources attributed to various job applicants, candidates for training or promotion, and candidates for reductions in force. There is a great deal of inertia over the life course. People who start life in disadvantaged positions tend to remain in disadvantaged positions, mainly because our systems for delivering income in retirement are designed to replicate the structure of social inequality during the working years. Only a very small proportion of people born to disadvantaged parents are able to improve their relative socioeconomic position over the course of their adult lives.

Summary

Social class affects aging by influencing the attitudes, beliefs, and values people use to make life course choices and by influencing life course opportunities, particularly in terms of education and jobs. People whose social class backgrounds lead to middle-class jobs or higher approach aging with much greater resources—knowledge, good health, adequate retirement income—compared with the working class and the poor. The positive picture of individual aging presented earlier is primarily a middle-class picture because most older Americans are middle class. On the other hand, many of the problematic aspects of aging are concentrated among the working class and the poor.

Racial discrimination has concentrated African Americans disproportionately in low-paying jobs and in substandard housing; this fact applies more to older African Americans than to African Americans in general. Compared with older whites, older African Americans have lower Social Security benefits, fewer private pensions, and greater incidence of illness and disability.

Older Native Americans face an even worse situation than older African Americans. Excluded from participation in American society and heavily concentrated in rural areas and on tribal lands, older Native Americans are much less likely than other older people to have access to services. Compared with older whites, older Native Americans are much less likely to have had middle-class jobs and much more likely to have inadequate incomes and poor health.

The picture for older Asian Americans is mixed, although all groups show some negative effects of racism. Japanese-American older people have had jobs that closely parallel those of whites and as a result have retirement incomes closer to those of whites than any other racial category. There is great diversity among older Chinese Americans in terms of jobs and retirement income. Filipino-American older people are more likely to have had low-paying jobs and thus low retirement incomes. Despite their lower incomes, older Asian Americans tend to be in better health than older whites.

The Hispanic population is quite diverse also. Older Hispanic Americans tend to be better off than older African Americans but not as well off as older non-Hispanic whites or older Asian Americans in terms of health and retirement income. Very little is known about aging among Mexicans living illegally in the United States.

Of the categories of people who experience discrimination in American society, women experience the greatest inequity. Women who opt to be homemakers are quite vulnerable economically to the breakup of their marriages via divorce or widowhood. Those who are employed are concentrated in "women's work," which tends to be low-paying and not covered by private pensions. Even in areas where women and men work at similar jobs, women are often paid less. As a result, retirement incomes of women are only about 55 percent as high as those for men.

Multiple jeopardy increases the probability of having poor health and inadequate income. Being a woman is the greatest disadvantage, followed by having less than high school education (being working-class) and by being African American.

Social inequality has a great influence on aging through its effect on jobs and lifetime earnings and their consequent impact on health and retirement income in later life.

18 ▌ The Economy

The complex system we call the economy is a hodgepodge of organizations and activities that extract raw materials and food from the physical environment, transform them into consumer goods, convert people's knowledge and energies into services, and distribute these goods and services to people. The amount of goods and services the economy produces provides subsistence for the general population plus an economic surplus—the value of goods and services over and above what is required to sustain those who do the producing. In the United States, the economic surplus is distributed among various groups: owners get profits; workers get salaries, wages, and fringe benefits; retirees get earned pensions; governments get taxes on wages, salaries, profits, and pensions; those who are poor, disabled, or unemployed get financial support from governments; and the public gets facilities and services from governments (roads, fire and police protection, national parks, public health programs, national defense, and so on).

Economic Ideology

How an economy responds to an aging population is to some extent shaped by the fundamental organization and underlying values of the economy. All economies produce goods and services; they differ mainly in how the benefits of that production are distributed. The American economy is basically capitalist, which means that one of its primary goals is private accumulation of wealth (capital), which can then be invested in ownership of the organizations that produce and distribute goods and services. Additional goals include a drive for constant growth and "progress," limitation of government's share of profits by limiting taxation, and individual wealth as a "measure" of the person.

Underlying our capitalist system is a free-market ideology that assumes that all people have equal access to the means of gaining wealth, the free market is the fairest way to distribute goods and services, poverty or need is the result of individual weakness, the family and the private sector are the "proper" agents to respond to need, and the way to minimize demand on public services to the disadvantaged is by making these services punitive and stigmatizing.

The free-market economic ideology sees government's major functions as supporting the private sector by keeping economic growth a prime goal; minimizing regulations or restrictions on production and distribution; providing incentives such as tax breaks for those who accumulate capital; minimizing the amount of goods and services produced by nonprofit organizations, including governments; and financially underwriting production costs through such mechanisms as tax breaks, military research and development, health research and development, and government loan guarantees. Regardless of how one feels about the appropriateness of this ideology, it is a fundamental basis for our economic system and its supporting political structure.

However, as we saw in Chapter 3, many saw the Great Depression of the 1930s as a failure of the capitalist free market to provide adequately for the needs of the public as a whole. The "Roosevelt revolution" marked the entrance of government into major areas of the economy: public works, public health, education, occupational health and safety, environmental conservation, and social services.

Americans also espouse a humanitarian ideology in which the alleviation of human suffering is a major goal. Thus, both government involvement in promoting the well-being of the people and the humanitarian principles of service to humankind offer ideological tools to those who oppose the free-market ideology. This opposition has produced a countervailing ideology that assumes that, left unchecked, the free market leads to exploitation of the powerless and disregard for the public good. To provide the needed checks, government must regulate economic activity to prevent exploitation such as child labor or unsafe working conditions, and government must tax the profits from economic activity in order to provide public benefits such as public health, highways, national defense, unemployment insurance, and public education. The dialectic between these two opposing viewpoints has led to a complex web of laws and regulations that constrains the economic system but cannot be said to actually control it.

Economic Structure

The U.S. economy can be divided into three major sectors: a core private sector, a peripheral private sector, and a government sector. The *core private sector* is made up of the largest corporations and financial institutions—organizations that use high-energy technology, economies of scale, and noncompetitive pricing to achieve high out-

put per worker and relatively high profits. People who work for core private sector organizations tend to be unionized, to have relatively high wages and fringe benefits, and to be covered by both private pensions and Social Security. The *peripheral private sector* is made up of relatively small organizations that tend to be labor-intensive, to have relatively low output per worker, and to have lower profit margins. Workers in the peripheral private sector tend to have relatively low wages and fringe benefits and to lack private pensions. Although these workers pay payroll taxes that generate eligibility for Social Security pensions, the level of benefits reflects their generally lower career earnings. In addition, workers in the peripheral sector who are covered by private pensions are less likely to get those pensions because labor turnover is high in the peripheral sector and workers often must work at least ten years for the same employer to qualify for a pension.

In 1990, there were 127 million people in the labor force of the United States. Of this number, about 94 million (74 percent of the labor force) worked for establishments in the private sector—about 30 million in the core segment and 64 million in the peripheral segment (U.S. Bureau of the Census 1991a). The remainder of the labor force was made up of the armed services (1.6 million), the unemployed (6.9 million), the self-employed (9 million), and those employed by government (17.6 million).

In 1990, the *government sector* was made up of 17.6 million people working for federal, state, and local governments and they represented 14.9 percent of the labor force. The nearly 3 million federal employees had relatively stable employment and received salaries, benefits, and pensions similar to those in the core private sector. Most federal employees did not participate in the Social Security system until 1983, at which time new federal employees began to contribute and were covered. The nearly 15 million state and local government employees were a more variable lot. Some had relatively high wages and fringe benefits, pension and Social Security coverage, and stable employment; others had relatively low wages, government pension coverage but no Social Security, and insecure employment. About 30 percent of government workers then were unionized, and they have tended to have done better in recent years at securing better wages and pensions. Many local government pensions are unfunded, which means they must be paid for out of current revenues, which in turn means that, in the future, these pensions may not be very secure.

Thus, how well a worker fares in the American economy in terms of wages, fringe benefits, and retirement income depends to a very important degree on the sector of the economy in which the person works. Incidentally, work in the three sectors of the economy is not evenly distributed across all segments of the population. For example, workers in the core private sector are predominantly white males. Low-income minority and women workers are more likely to be employed in the peripheral private sector. Middle- and high-income African-American workers are more likely to be employed by government.

We are interested in both how the nature of the American economy affects its ability to deliver income, goods, and services to the older population and how retirement and aging affect the operation of the economy. We consider first the role that retirement plays in the economy through controlling unemployment and encouraging capital accumulation. We then look at how effective the economy is at providing retirement income for older people. We then examine its ability to deliver goods and services. We then consider the emergence of the aging as a new consumer group. The chapter concludes with an examination of economic exploitation of elders by unscrupulous and criminal business practices.

The Economic Functions of Retirement

People commonly assume that retirement and the pension systems that support it were created to provide for the welfare of the older population. But although income security for older people played a part in the development of Social Security, the major functions of retirement for the American economy have mainly revolved around controlling the size of the labor force. Private pensions were initially developed to tie workers to particular employers and to cut down on labor mobility. Social Security was fostered by a need to *control unemployment* as much as by a desire to improve the lot of the older poor. Retirement still performs these functions today. Private pensions are most prevalent in the core private sector, where workers are highly skilled and labor mobility is less desirable. Social Security has tended to become more liberal during times of high unemployment. When unemployment is high or when organizations must make cutbacks in personnel, then encouraging workers near retirement to get out of the labor force is often one of the first actions taken.

As pointed out in Chapter 3, our high-energy economy simply does not need every adult to be employed in order to produce all the goods and services we need. And our economic surplus is more than sufficient to generate retirement incomes. Retirement benefits the entire society by keeping the number of people in the labor force closer to the number of jobs available. Without retirement, the unemployment rate would be much higher than it is today.

Retirement is also an incentive for workers to put up with less than satisfying jobs by setting a limit on how long a bad job situation has to be endured. Employers also see it as a way to phase out less effective workers, and unions view it as a way to create new jobs and opportunities for promotion. Thus, for workers, unions, and employers, retirement serves different but generally positive functions.

Retirement as a life stage and the corresponding need to create resources to finance it have played a major role in *capital formation* in the American economy since World War II. By 1992, retirement trusts, usually managed by banks or insurance companies, had over $2.5 trillion invested in corporate securities, which represented nearly half of the market value of all corporate securities listed on the New York Stock Exchange (U.S. Bureau of the Census 1993).

Drucker (1976) called this trend "pension fund socialism" since a great deal of the ownership of American business is in the hands of workers in the form of pension entitlements. However appealing this concept may be to some, in fact these huge pension funds are not managed to maximize pension income but to fulfill what the financial decision makers see as the need for capital investment. Olson (1982) found that compared with other large investors such as mutual funds, pension funds' performance was extremely poor. In fact, between 1962 and 1978, 87 percent of more than one hundred pension funds studied showed well-below-average returns compared with the performance of the stock market as a whole (1982,111). Apparently, because pension funds often do not have to increase benefits to keep up with inflation, managers do not put a high priority on the funds' getting their share of the capital growth available through the stock market. This issue is further complicated by the fact that employers' choices of pension fund trustees are sometimes not based on pension fund performance at all. Olson pointed out that a corporation may select a bank to manage its pension fund in return for prime rate loans and other benefits. Thus, the contributions of workers to pension funds constitute a very important source of capital for the American economy, but these assets have not always been managed with the workers' best interests in mind.

In addition to providing enormous amounts of capital to American business, in 1992 private and government employer pension funds held more than $2.4 trillion in federal, state, and local government securities as well (U.S. Bureau of the Census 1993). Social Security trust funds held another $794 billion in special U.S. Treasury bonds. Thus, the financial resources laid aside to finance retirement constitute a major source of capital not only to private enterprise but to local school districts, state construction projects, and the like, as well as to federal budget expenditures.

The Economics of Retirement Income*

Older people commonly are thought to constitute an economically dependent category, and it is often held that the aging of the population constitutes an economic "problem" because the older population is growing faster than the younger population, who presumably must "support" elders. But if we examine the sources of income for older people, we see a different picture.

Sources of Retirement Income

Although we tend to think of the older population as being retired, a significant minority is employed. In 1993, at ages 65 to 69, 25 percent of men and 16 percent of women were employed, and, at 70, 11 percent of men and 5 percent of women were employed (U.S. Bureau of the Census 1994). Earnings accounted for about 17 percent of the aggregate income of older people (Schulz 1995). For most who were employed, earnings from employment supplemented retirement income and many worked part-time. Thus, elders continued to make important contributions through their

---

* For more details on the economics of retirement income, see Schulz (1995).

labor in the economy, and the money they received from employment certainly could not be labeled "economic dependency."

Private- and public-employer pensions earned in connection with previous employment accounted for 19 percent of the aggregate income of older people in 1992. In most cases, employee contributions were deducted over the working life of the individual, and employer contributions were made in lieu of salary or other benefits. Thus, private pension income could hardly be called economic dependency either.

Income from property, in the form of rents, dividends, and interest, constituted 21 percent of the 1992 aggregate income of elders. Income from this source is concentrated among the most well-to-do older people and is the result of private savings, so it could not be classified as economic dependency either.

Thus, *about 57 percent* of the aggregate income of elders came from sources that clearly did not represent economic dependency in any sense of the word.

Social Security benefits represented 39 percent of the aggregate income of older people in 1992. Since this income was financed with revenues from current payroll taxes, it might appear that it represented economic dependency. However, there is another way to look at this issue. Anyone eligible for Social Security retirement benefits has helped support the system financially during his or her working years. No cash fund accumulated for the individual to be used later to pay a pension, mainly because when Social Security was established business interests did not want the government controlling a large fund of investment capital. Instead, the transaction was in the form of a social contract between the U.S. government and the Social Security taxpayer: In return for taxes paid at the going rate, based on current earnings, the individual became *eligible for a pension* based on the average level of earnings subject to Social Security taxes over a large

number of years of employment. The entire society incurred an obligation to pay these pensions.

However, payroll taxes are not the only way to meet Social Security pension obligations; they simply represent the current *concept* of how best to meet them. When someone pays off a debt to a bank, we do not call the bank "economically dependent." Likewise, it is misleading to refer to paying off our national obligations to retirees as "economic dependency." These pensions are IOUs being called in under the terms of an agreement.

This kind of confusion has led to frequent reference in the media to the "fact" that a large percentage of federal expenditures are going to "support" older people. However, if we examine federal expenditures more closely, we can see how misleading this statement is.

Table 18-1 is a breakdown of federal expenditures for fiscal year 1993. Money set aside to cover our IOUs to older people and their dependents or survivors consisted of Social Security ($304.7 billion) and Medicare ($132.8 billion). In all, these *earned* benefits totaled $437.5 billion and represented nearly 30 percent of federal expenditures. However, only $41 billion (9 percent) of these expenditures (subsidies for Supplementary Medical Insurance, which covered mostly physicians' charges) were financed with general revenues. The remainder was financed from payroll taxes, premiums paid by Medicare recipients, and interest on Social Security trust fund reserves. Thus, very little of the federal budget deficit was connected with either Social Security or Medicare.

Public assistance to "dependent" older people amounted to $32.8 billion, or 3.1 percent of the federal general revenue expenditures, which is where the much-discussed federal budget deficit comes from. In 1993, less than 4 percent of the aggregate income of the older population came from "unearned" sources such as Supplementary

Table 18-1 U.S. federal expenditures affecting older people: 1993.

Type of Expenditure	Amount (in $ billions)	Percent of Federal Expenditures
Earned benefits:		
Social Security	304.7	20.6
Medicare	132.8	9.0
	437.5	29.6
Public assistance:		
Supplemental Security Income	3.5	
Medicaid	18.6	
Food stamps	.6	
Subsidized rental housing	9.5	
Low-income energy assistance	.6	
	32.8	3.1
Social services:		
Older American Act programs	.8	
Senior employment programs	.3	
Social Services Block Grants	.4	
	1.5	0.1

Source: U.S. Bureau of the Census (1993).

Security Income, Medicaid, or food stamps. Thus, the vast majority of federal expenditures going to older people consisted of benefits earned by participation in Social Security through working at a covered job, and it is misleading to label these benefits as "support" or "dependency."

In the early 1970s there was little question that older people had earned their Social Security retirement pensions, but by the 1980s a "me generation" of the baby-boom "yuppies," perhaps a minority but vocal and well-connected in the media nevertheless, had begun to see the older population as "greedy geezers" who used their political clout to pry ever-increasing and undeserved benefits from Congress at the expense of those who were currently working. Instead of understanding that the U.S. government is in the pension business and is obliged to fulfill its contract with Social Security participants, these narrowly focused critics inappropriately evaluated Social Security as if it were just another investment, and concluded that

retirees were getting too big a "return" on what they "paid in."

Greene (1989) pointed out that Social Security is not merely a welfare transfer system in which wealth is transferred from people who presumably created it to people who presumably did not. It is an intergenerational transfer system that recognizes that the economic fruits we enjoy today come from a system created by the work of those who came before us. Indeed, the taxes that today's Social Security recipients paid constitute only a convenient index of the investment they made in the social system we all benefit from. As Greene pointed out:

Someone first had to withhold children from the labor market, and nurture and socialize them into reasonably healthy, socially constructive individuals. Someone then had to massively fund, both through tax payments and private purchase, the entire structure of public and private education through which human capital is created and maintained. In short, someone had to invest massively from their incomes and opportunities in order to create the stock of human capital from which most of our current wealth arises.

And who was it that made this massive investment? Who are these "human capitalists"? Who are they but our parents and our grandparents? And who are they but "the elderly"?

Are these the people we now want to argue are "on welfare" because they take more out of Social Security than their cash contributions?. . .

We must recognize that the claim of the elderly on the wealth produced by our economy is not a claim to welfare benefits, but the claim of major investors to a return on the human capital they sacrificed to create. . . . Any economic analysis that takes this fully into account will reveal the elderly to be taking from our society much *less* than a competitive return on what they have invested in it. On balance, they have probably taken better care of us than some of us now wish to take of them.

(1989,724)

Retirement Income Issues

In the 1980s, America's economic structure shifted in ways that widened the gap between the economically advantaged and the disadvantaged. Between 1980 and 1987, those who worked as managers, professionals, or technical workers in the global economy did well indeed, averaging real income gains of more than 20 percent (Reich 1991). At the same time, incomes of wage-earning factory and clerical workers suffered a real decline, and more than 2 million full-time employees sank into poverty. These changes are a reflection of the fact that the United States can no longer be thought of as having a single economy. America's professional, technical, and managerial classes share in the benefits of global research, production, and finance, whereas her routine production workers find themselves losing ground to low-paid clerical and production workers in Third World countries. Professional, technical, and managerial workers' pensions and health-care benefits remained basically unchanged in the 1980s, but the remainder of the middle and working classes suffered substantial losses. These

losses will result in an eventual drop in real retirement resources because retirement income is directly tied to earnings from employment.

The nature of private retirement benefit systems also changed in the 1980s. Most private pension programs initiated in the 1950s and 1960s were *defined benefit* programs that promised a specific amount of monthly retirement income, usually based on earnings and years of service. But in the 1980s, many employers began offering *defined contribution* programs under which a specified amount (usually a fixed percentage of earnings) is deposited into a tax-sheltered retirement account. Under a defined contribution program, the monthly retirement benefit is unpredictable because the value and growth of the retirement account rise and fall with financial markets. In 1990, about 35 percent of private retirement benefit systems were defined contribution plans, up from about 10 percent in 1970.

Many of the issues identified thus far stem from the fact that ours is not a single retirement income "system" but rather a large number of systems overlapping and supplementing one another for a minority of workers and ignoring a majority of workers. Proposals by the 1981 President's Commission on Pension Policy included the establishment of a minimum universal pension system covering all workers and funded by employer contributions. The commission felt such a system was necessary in order to give everyone access to pensions to supplement Social Security. However, the commission's recommendations were rejected. The conclusion we are left with is that the retirement income systems we now have continue and even increase the income gap between the haves and the have-nots. This gap is further widened by our tax system.

The Tax System

Government expenditures can be divided into two categories: direct expenditures and tax expenditures. Direct expenditures are easy to understand because they involve payments. Tax expenditures are more difficult to understand because they represent *potential* revenues intentionally *not* collected from a specific category of taxpayer. We refer to them colloquially as "tax breaks." In 1994, the following tax breaks were available to middle-aged and older people: partial exclusion of Social Security, Railroad Retirement, and veterans' benefits from taxable income; an additional tax exemption for each person 65 or older; and partial exclusion of capital gains on sales of homes of people 55 or older. In addition, the tax structure supported private pensions by allowing both employees and employers to deduct their pension contributions from their taxable income. As we have seen, private pensions are not available to most workers in the peripheral private sector. Table 18-2 shows the dollar breakdown of tax expenditures affecting older people in 1994.

Gary Nelson (1983) analyzed tax expenditures by income class and found that only 2 percent of 1982 tax expenditures went to older individuals with incomes of less than $5,000 per year, while 50 percent went to those with incomes over $20,000 per year. In other words, the federal government did without $98 in taxes from each older person with $5,000 in taxable income and did without $697 in taxes from those with incomes of over $20,000. Thus, not only do the systems that generate retirement income enlarge the gap between the well-off and the poor, but so does the tax system.

Direct taxes, particularly Social Security payroll taxes, sales taxes, and property taxes, increase the relative hardships low-income elders experience compared with middle- and upper-income elders. Because Social Security payroll taxes are paid on all earnings, no matter how small, Social Security beneficiaries who supplement low benefits with earnings must pay Social Security taxes even if they have incomes below the poverty level. Of course, low-income people of all ages experience this problem. However, there has not been serious discussion of eliminating Social Security taxes for low-income people.

Similarly, sales taxes usually impose a larger burden on low-income people than the general population. For example, Reschovsky (1989) found that in Massachusetts sales taxes amounted to 3.2 percent of income for elders with incomes of under $5,000 compared with less than 1 percent for elders with incomes of $15,000 or more.

Because property taxes are fixed by the value of land and the structures built upon it and because income tends to fall in later life as a result of retirement, property taxes as a proportion of income tend to increase with age. Reschovsky found that efforts to reduce the property tax burdens on low-income older people had been ineffective. For example, in Massachusetts, even with property tax credits for low-income elders, the property tax burden for older homeowners with incomes of under $5,000 amounted to 5 percent per year compared with only 2 percent for elders with incomes of $15,000 or more.

Table 18-2 Estimated U.S. tax expenditures affecting older adults: 1994.

Tax Expenditure	Amount (in $ billions)
Exclusion of contributions to retirement pension programs	70.5
Exclusion of Social Security retirement benefits	19.0
Exclusion of capital gains on home sales of people age 55 and over	6.4
Exclusion of Social Security survivors' benefits	3.7
Additional tax exemption for those 65 and over	1.9

Source: U.S. Bureau of the Census (1993).

Renters bore about 75 percent of the property tax burden on the property where they lived, and low-income elders paid a larger proportion of their income for rent than did middle- and upper-income older people.

A great deal of the perceived unfairness of our systems of taxation is related to our tendency to create complex tax laws that tax some kinds of income and not other kinds. For example, because incomes of Social Security beneficiaries were extremely low when the system was first introduced, Social Security benefits were excluded from federal income taxes, which meant that millions of recipients did not have to file tax returns and the Internal Revenue Service did not have to process them. But as retirement income systems improved, the proportion of Social Security recipients with middle and upper total incomes steadily increased. It is not obvious to most people why a retiree with an income of $30,000 per year should have $10,000 in Social Security benefits excluded from income taxes whereas an employee with earnings of $30,000 pays income tax on the entire $30,000. This perceived unfairness was used to justify taxation of half of the Social Security benefits of those with total taxable incomes above $25,000 ($32,000 for joint filers). The rationale for taxing only half was that taxes had already been paid on the employee's share of contributions into Social Security and were therefore due only on the employer contribution. However, like private pensions, Social Security returns the employee's total contributions with interest in less than four years, so why shouldn't all Social Security benefits be taxed after that, just as private pension benefits are? Also, Social Security income is the only type of income that comes under taxation only if people have total income from all sources above a threshold amount. For example, income from municipal bonds is not taxable, no matter what the income level, but income from these bonds is counted toward the threshold for determining when Social Security benefits are taxable.

Our tax laws are so complicated that thousands of lawyers and accountants make excellent incomes helping people negotiate the maze and avoid taxes. Under this system, creating approaches to taxation that would be seen as fair by a large majority of people is probably impossible. But currently our tax systems advantage upper-income elders and disadvantage low-income elders.

Retirement Income in the Future

The level of retirement (the proportion of the population retired) appears to be influenced primarily by two factors: the *minimum age of eligibility* for Social Security or other retirement benefits and the *financial adequacy* of retirement benefits. Opportunities for continued employment affect the level of retirement only slightly. Thus, as the minimum retirement age goes down and the adequacy of retirement benefits goes up, the level of retirement increases, which is what has been happening in the United States. Recent changes in mandatory retirement policies have had little effect on this picture. However, as the funds required to pay pensions for retired people increase, so do demands for contributions from the working population. For example, the recent increases in Social Security benefits required sharp increases in Social Security taxes over several years. The increased taxes lowered the disposable incomes of employed people unless benefits to other categories such as the handicapped or the unemployed were cut. For example, cutting out Social Security survivors' benefits to college students has partially offset the increased costs of retirement benefits.

As the disposable incomes of employed people go down (or at least lose ground to inflation), popular support for adequate retirement benefits goes down. If this lowered support for retirement benefits means these benefits become less adequate, then the level of retirement may go down as people delay retirement in order to boost their expected benefits.

Since 1965, American economic policy has consistently favored lowered retirement ages and more adequate retirement benefits. However, changes in the American economy may not permit this policy to continue. The costs may simply be too high. For example, Schulz (1992) found that when retirement age is dropped from age 65 to age 60, pension costs increase by about 50 percent. Aging of the population will increase the tax burden on the employed population by increasing the number of people collecting their Social Security IOUs. Any increases in longevity will increase the tax burden even further. At the same time, reduced fertility since 1967 means fewer employed people. Add to this the prospect that the decline in nonhuman energy resources may curtail the economy's capacity to produce an economic surplus as large as we are accustomed to. This combination of factors exerts pressure to increase the minimum retirement age and perhaps reduce the adequacy of retirement benefits as well. These forces were involved in 1982 when the Social Security Act was amended to gradually increase retirement age for full Social Security retirement benefits beginning at the turn of the century and to cut benefits immediately by delaying scheduled Social Security cost-of-living adjustments.

Because we may have to abandon the policy of encouraging "early" retirement, doing it sooner would be less traumatic economically. However, American business has a deeply ingrained preference for early retirement based on erroneous negative assumptions about the value and productivity of older workers. In addition, business has relied heavily on early retirement incentive plans to reduce the disruptions caused by the large numbers of companies that implemented reductions in the workforce in the late 1980s and early 1990s. It is difficult to envision what form the debate on this issue will take,

but it is sure to be spirited. Both employers and employees have come to view early retirement as economically beneficial, a view that is unlikely to change quickly. On the other hand, the realities of financing retirement may create an awareness that paying for early retirement may become prohibitively expensive.

Many state and municipal governments have pension systems funded completely or largely from current revenue, which has led to severe economic problems for many government systems. For example, some cities have bargained with municipal unions to keep down current salaries and wages by liberalizing retirement ages and benefits. But eventually these pension IOUs come due. And when they do, governments often have difficulty finding the necessary revenue. Part of this problem is no doubt related to the fact that the political regime creating the liability will probably not be the one that has to satisfy it.

Private pension systems are less likely to encounter these funding problems because their prepayment schedules are relatively sound. However, the ability of even the most well-designed pension program to meet its liabilities depends on a healthy economy. When economic times are hard, income generated by the assets of pension funds goes down, and with it go the reserves of the pension plan.

The merger mania of the 1980s left many American corporations saddled with substantial debt, and as a result, the Pension Benefit Guarantee Corporation (PBGC), which backs up private pensions, has seen a dramatic increase in pension plan failures. Most of these plans failed because the companies did not have the funds to continue to make the needed annual contributions. The expenses connected to early retirement incentive programs also played their part. As a result, the PBGC was forced to raise premiums employers were required to pay, which then reduced the number of employers offering private pensions. For example, in 1990, 16,000 defined benefit plans were terminated and less than 500 were initiated, which meant of course, that companies were taking money out of defined benefit plans, providing vested employees with an annuity, and either dropping pension coverage altogether or switching to less stringently controlled defined contribution plans, particularly profit sharing plans (Paine 1993).

In addition to the early retirement issue, policy makers may well have to deal with a large increase in retired population (even if the minimum retirement age goes up slightly), accompanied by a relative decrease in the number of employed persons available to help meet retirement pension obligations. However, if we estimate now that to fund retirement in the next century will require an increase in taxes of 10 percent and we begin now to phase in the increase, Schulz (1992) pointed out that we would only have to increase taxes 0.2 percent per year. This is not an unreasonable goal even if economic growth is limited to a very modest 2 to 3 percent per year.

Inflation can influence retirement income in several ways. To the extent that retirement income comes from assets that do not increase in value with inflation, such as savings, then the value of savings and savings income may be reduced. Likewise, income from fixed pensions and annuities is quickly eroded by inflation, and few private pensions are adjusted for inflation. Even if income sources such as Social Security or private pensions are adjusted for inflation, the adjustments usually lag behind, and thus real income is still reduced. Whenever income does not increase as fast as inflation, then real income (purchasing power) is reduced. Increases in earnings often lag behind inflation too.

Increases in Social Security benefits, the introduction of SSI and Medicare, and increases in in-kind benefits such as low-rent housing for older people have tended to improve the economic position of low-income older Americans over the past decade. However, these improvements will be only temporary if inflation continues to erode the value of supplements to

Social Security and we continue to cope with budget pressures by cutting benefits. Certainly, the explosion of benefits and programs for elders that occurred from 1965 to 1979 is very unlikely to occur again. Indeed, what is more likely is that these benefits, like Medicare, will be eroded by pressures on federal and state budgets.

Because retirement income can be generated from a large number of sources and those working for the core private sector and the federal government have access to several sources, including Social Security, private pensions, and government pensions, there is considerable unevenness in the delivery of retirement income. Economists estimate that retirement income equal to 70 percent of pre-retirement income will allow a retired household to maintain a comparable level of living in retirement (Schulz 1985). Under our present system, through multiple pensions in addition to Social Security, people can earn retirement benefits well in excess of 100 percent of preretirement earnings. At the same time, retired single individuals with only Social Security retirement benefits must survive on less than 40 percent of preretirement income on average.

I call this the *overdelivery/underdelivery problem*. That is, our system overdelivers retirement income to already economically advantaged workers and underdelivers retirement income to low-income workers. The delivery of maximum Social Security retirement benefits to high-income retirees has led to calls for means-testing of Social Security. However, such means-testing will not help those with inadequate Social Security unless the monies saved by means-testing are used to provide improved benefits to low-income workers. This is politically unlikely. Middle- and upper-income retirees feel that they have earned their Social Security benefits and see providing improved benefits to poor older people as the responsibility of the whole society, not just of middle- and upper-income older people.

Interestingly, retirement may also add to inflation. For example, the fact that people can sometimes work at a full-time job *and* draw full retirement benefits artificially increases the supply of money with no corresponding increase in production. In addition, large pension funds are among the nation's largest investors. They have an obligation to deliver a certain level of income; in periods of economic slowdown, these large investors are forced into rather conservative investment strategies that are not conducive to increased production. Very little systematic analysis has been made of the impact of retirement on inflation; this background work is needed to reevaluate retirement policies.

Is Aging Responsible for Our Economic Woes?

Much has been written in the past decade about the relationship between the aging of the population and the economic slowdown the United States has been experiencing. The federal budget deficit has taken much of the blame for the public's lack of confidence in the economy, but economists disagree sharply over how much of the economic slowdown is a result of the deficit and how much is the result of changing costs of energy and unfortunate corporate and government policies during the 1980s. For example, Barlett and Steele (1992) documented in great detail how tax policies, bankruptcy laws, deregulation of investment and banking markets, and corporate raiding combined in the 1980s to destroy the capacity of many American firms to take a long-range view of economic growth. To keep profits up in slow economic times, businesses cut their workforces, introduced more "labor saving" technology, and moved operations to areas of the world with cheaper labor costs. The result was an enormous reduction in the number of Americans with access to middle-class jobs, particularly in manufacturing, and the resulting loss of jobs and income created a monumental reduction in general government revenues.

Government expenditures were cut and/or needed increases were curtailed, but not as fast as the reduction in general tax revenue due to high unemployment and slumping wages and salaries for most segments of American society. This, coupled with a sharp increase in defense spending, is primarily what produced the sharp increase in the federal deficit in the 1980s.

Aging of the population had little to do with these difficulties. It is true that Social Security and Medicare represent a large portion of federal expenditures, and the sheer size of these programs makes them easy targets for political attack. But the fact is that the *Social Security retirement, disability, and survivors' programs have not added one dime to the federal budget deficit*. Social Security is not financed through general tax revenues; the retirement, disability, and survivors' trust funds take in more revenue each year than they pay out. And the tax money that goes into these programs cannot be used to finance discretionary budget expenditures. Lowering or eliminating Social Security taxes would not increase general tax revenues available for reducing the federal budget deficit.

The Medicare trust fund is partly supported by general tax revenue and has encountered difficulty staying within the funding provided by payroll taxes and premiums paid by elders covered by Medicare. As a result the general revenue support needed has been increasing. Medicare has thus contributed to the federal budget deficit, but this problem is not the result of the aging of the population or overutilization of health services by elders. Health-care inflation has been several times the size of general inflation for over a decade, and this has been the result of increased cost per unit of health-care service. Health-care inflation has created a crisis not only for Medicare but also for employer health insurance plans and individuals in the general population who find themselves without insurance or with decreased insurance benefits due to rapidly escalating costs of health insurance.

To single out pensions and health care for older people as the main contributors to our federal budget deficit is simply inaccurate, and as long as we have this misperception of the problem, solutions will be all the more difficult to find.

Another misperception is that our national productivity is lower than it should be because we have too many retired people. Those who espouse this position think the solution is to push older people back into the labor force. This proposal ignores the basic economic underpinnings of retirement in technologically advanced societies. There is no persuasive evidence that having a large population of retirees hampers the economy. In fact, the impact of retirees on local economies tends to point in just the opposite direction. The consumption of retirees can represent a substantial element of economic stability.

Between now and the year 2000 the greatest number of new jobs are likely to employ retail salespeople (4.6 million), general managers and top executives (3.5 million), janitors and cleaners (3.4 million), and general office clerks (3 million) (U.S. Bureau of the Census, 1991a,398). In terms of sheer proportional growth, the areas of employment that are expected to grow the fastest between now and 2000 are health, financial services, computers, human services, and corrections. What would be the advantage of forcing retirees back into the labor force to take these jobs? It would certainly take a substantial training investment to accomplish and, even with up-to-date training, would employers hire older workers for these jobs? The track record thus far suggests that they would not. More important, what effect would this policy have on the training and employment prospects of young people?

Even if a labor shortage results from the lower number of people coming into the labor force as a result of falling birth rates since 1966, hiring older workers may not be seen as the best solution. Employers can invest in more machines to replace human labor, lobby for liberalized immigration laws to bring in younger people with needed skills, or shift clerical production facilities to developing countries rather than hire older workers (Schulz 1992).

There is no persuasive evidence that aging of the population has a negative effect on economic vitality, that our economy cannot support the levels of retirement that we are likely to have over the next thirty years, that aging is responsible for the federal budget deficit, or that to survive economically we need to put all able-bodied elders back to work. As Schulz states:

Today, as in the past, the most important determinants of the future economic welfare of people of all ages are the longstanding factors influencing growth: labor-force incentives and participation levels, savings, investment in human and business capital, technological change, entrepreneurial initiatives, managerial skills, government provision of infrastructure and so forth. *Thus, the debate over how best to run an economic system is not primarily an aging discussion. In fact, the aging of populations may have little to do with the outcome.*

(1992,273) (Italics in the original.)

Private Enterprise and the Aging Population

Many elders' needs can be readily met in the marketplace. Most older people can purchase food, clothing, housing, and so on from local businesses, but a substantial minority are not served adequately by the private marketplace. Let us look at a few examples.

Many older people cannot find housing that is both suitable and affordable. There is a severe shortage of low-rent housing for elders, particularly in rural areas, and even middle-income elders have difficulty finding affordable apartments because private developers and lending institutions concentrate their resources where the most money can be made. There is much less money to be made in constructing or remodeling housing for low- or even middle-income people. This response of the housing industry is predictable, but is of little comfort to older people who cannot find a place to live.

When the government intervenes in this housing shortage, it is primarily through the construction of subsidized housing for low-income older people. For example, suppose that a private developer builds and manages a low-rent housing project for older people and that the builder has agreed to do so because the government guaranteed to subsidize the rents. Let us say that the residents are required to contribute 30 percent of their monthly incomes toward the rent, with the federal government paying the difference between that figure and the actual rental price. So far, so good. But let us also say that the rental rate structure of this privately built low-rent housing for older people has been set substantially higher than that of similar housing in the open market. Thus, for instance, Mrs. L. lives in a one-bedroom apartment for which she pays $130.20 per month—30 percent of her monthly SSI income. The rental rate used to compute the government subsidy is $400 per month—which means that the government is paying $269.80 per month for Mrs. L.'s apartment. But quite adequate one-bedroom apartments are available in Mrs. L.'s community for $300 per month. If private entrepreneurs can make a profit on housing at $300 per month, they must be doing well indeed on rents of $400 per month. This is a good example of government operating for the benefit of private business at public expense. But it is too easy to blame this situation on bad government. The fact is that it arose from a long-standing practice in the private sector of doing business with the government only at high rates of profit.

Whereas government programs may have led to overpriced housing subsidies, in the area of home-delivered personal care, government efforts at cost containment, especially in Medicare and Medicaid, have constricted the capacity of the private sector to provide services. In many states, Medicare and Medicaid personal care services are reimbursed at the minimum wage, transportation time or training costs are not reimbursed, and agencies often have to wait as long as 150 days to be paid (Kane 1989). Because public programs create an artificially low wage rate for home care, wages lag substantially behind other areas of unskilled employment, which results in high staff turnover and low staff morale. Kane looked at the financial performance of publicly held companies in the home-care business and found that the majority experienced financial losses on their home-care services and most were scaling back these operations and limiting their expansion plans to more lucrative medical equipment sales and high-technology medical services.

Of course, government policies are not the only factor that restricts the growth of home-care services. The population in need of home care is predominately made up of women living alone who are age 85 or older. As we saw in Chapter 8, this population is among the most economically deprived in our society. Thus, the need for home care is concentrated exactly in the population category least able to afford to pay for it themselves. The result is a sharply constrained private-pay market for home care. Yet at the same time, states are looking to home care as the main way to control public long-term care expenditures. If public payment policies do not support private-sector home care and only a small proportion of elders can afford private pay, how can we expect the private sector to play a significant role in the provision of home care?

These examples by no means exhaust the list of inadequate private services to elders. Many of the difficulties cited earlier, such as

Elders are a prime market for cruises.
Photograph by Ann Purcell/Words & Pictures

inadequate services to the disabled, lack of home health care, transportation problems, and the like, have arisen because the private sector has not responded—and cannot be expected to respond—to many needs of older people because there is not enough economic profit to be made. When we look at government programs in the next chapter, we need to keep these limits of the private sector in mind.

Aging People as Consumers

People in later maturity have traditionally been treated as an unimportant market by those in charge of developing and marketing goods and services. This view was based mainly on the notion that older people had relatively low incomes and relatively low consumption and expenditure patterns. However, most of the "facts" that supported this view came from an era when retirement income systems were in their infancy and many older Americans were indeed poor. As elders' in-

comes have improved, they have become a force to be reckoned with in the marketplace.

For example, in many nonmetropolitan areas of the United States, the pension and Social Security checks that in-migrating retirees bring are referred to as the "mailbox economy." In areas where the population has been depleted by the decline in agricultural employment and the consequent out-migration of young people, the migration of retirees into the area has literally saved the local economy. According to the Appalachian Regional Commission, in 1992 the average retirement migrant household's impact on the local economy is over $71,500 per year (Crispell and Frey 1993). In many nonmetropolitan counties, and not just in the Sunbelt, the mailbox economy is the largest source of disposable income in the area (Glasgow 1991).

Middle-aged and older consumers are important in mass markets too. In addition to food, clothing, and housing, older consumers are interested in products and services oriented around the home, health care, leisure, education, managing finances, and coping with the aging process. They are also potential consumers of retirement housing, new cars, auto repairs, fuel, restaurants, and entertainment. Middle-aged and older adults also tend to be mainstay contributors to local charities.

A great deal of the growth of the American consumer economy since World War II has been tied to a rapidly increasing population with a high proportion of young people. But since the 1960s, lower birthrates have meant fewer young people entering the population. The fastest-growing segment of the U.S. population is now older people, particularly the old-old. In the past decade, the consumer economy has begun to recognize the true consumer potential of mature markets. Allen (1981) reported that sporting goods companies were beginning to market products such as lightweight rifles and special golf clubs specifically designed to meet older consumers' needs. Travel agents and major carriers had begun aiming travel packages specifically at the older market. Allen also reported that households with heads 55 or over accounted for 80 percent of the deposits in savings and loan institutions and 28 percent of all discretionary money spent—money not spent on necessities—in the marketplace. This figure was nearly double the amount spent by households with heads 34 and under. The much-sought-after under-25 youth market accounted for only 1 percent.

Although it is commonly thought that older people are unwilling to spend money even if they have it, Allen found this notion inaccurate: the 55- to 64-year-old group was *the* most important consumer market in the country, leading other categories in sales for a wide variety of purchases ranging from vacation travel and restaurant meals to garden supplies and luxury automobiles. The 65-and-over population was also above the national averages for categories such as vacation travel, women's clothing, mobile homes, and magazine and newspaper subscriptions. High-income older households are the most conspicuous consumers of luxury goods and services such as furs and jewelry, expensive clothing, and cruises.

In the literature on aging and consumer behavior, "mature" consumers are usually defined as those age 50 or over. Given our discussions throughout this book of the diversity in the middle-aged and older population, it should be no surprise that the population age 50 and over consists of many potential consumer categories, each with its separate needs and consumer orientation. Marketing professionals generally use population data broken down by age and gender to estimate the size of potential markets within the middle-aged and older populations. However, Wolfe (1990) argued persuasively that age is not a particularly useful characteristic for defining mature markets. Pointing to a literature filled with warnings about the difficulties of marketing to mature adults, Wolfe con-

tended that adult development causes mature adults to become more individuated and to be less susceptible than teenagers and young adults to the herd instinct. Using developmental concepts borrowed from Erikson and Maslow, Wolfe posited that older adults tend to be less motivated by materialism and more influenced by the *experience value* of goods and services. However, age is not necessarily a good indicator of stage of adult development.*

Lesser and Kunkel (1991) found empirical evidence that, based on developmental changes, middle-aged and older shoppers have very different motivations from teenagers or young adults for shopping. Motivations that influence mature consumers more than the general adult population include personal comfort, especially in clothing and shoes; personal safety and emotional security; convenience; sociability, especially in the context of serving others; old-fashioned values such as independence; a sense of purpose, especially personal growth through education; and spirituality, though not necessarily through traditional religion (Schewe 1990).

Marketing efforts have tended to fall into several traps in trying to serve mature consumers. First, mature consumers respond to images of opportunity and positive experience, not to the negative realities of aging, which most know only too well. For example, one company markets an adult continence product by showing how it enables young-old people to lead active lives without worrying. Another markets a similar product by showing how rising from a sitting position pushes on the bladder and increases the chance of an "accident." Guess which company's product outsells the other by 5 to 1?

Advertising campaigns often employ stereotypes of mature consumers that insult

---

* For an excellent analysis of the relationship between adult development and marketing to middle-aged and older adults, see Wolfe (1990).

Older consumers are less susceptible to the herd instinct.
Photograph by Marianne Gontarz

the very people to whom they are attempting to appeal. Confused older people may be funny to young advertising professionals, but to most older people, there is nothing funny about mental confusion.

Wolfe (1990) pointed out that in order to market products and services to mature adults, marketers must have the capacity to empathize with the potential consumer. This is the key missing ingredient in most unsuccessful marketing campaigns aimed at mature markets. To remedy this problem is not easy, however, because the culture of marketing and advertising has been youth-oriented and ageist for a long time, and even seasoned professionals who are themselves retired may still buy into the common stereotypes about aging, just as the general older population does.

A major disadvantage of the discovery of "the gray market" in the 1980s may be its negative effect on social policy (Minkler 1989). There is evidence that the new stereotype of the older population as affluent consumers has undercut efforts to provide affordable housing to middle- and low-income older people. The drive to sell private long-term care insurance to upper-income elders may have had a dampening effect on the movement in Congress to provide public long-term care insurance.

In the coming years we can expect an expanding interest in developing and marketing new products for older people. Already, products are being developed that use available technology to help elders compensate for physical decrements. Home-delivered services such as meals, housekeeping, personal care, and health care aimed at older people able to pay will come when the concentration of people wanting such services increases to the payoff point. We can also expect to see personal care items that take into account age changes in the skin, such as makeup, hair-care products, soaps, and deodorants. And advertising campaigns for such products can be expected to make greater use of older models, who will in turn provide images of physical attractiveness appropriate to older people. The list of interesting possibilities seems endless.

Economic Exploitation of Elders

As consumers, older people are particularly vulnerable to fraud, deception, and medical quackery. To begin with, older people who depend on asset income are very susceptible to any "surefire" scheme to get a higher return on their savings. Also, loneliness and isolation sometimes make older people susceptible to deception by a friendly, outgoing person who takes an apparent interest in them. Finally, hopeless illness is more frequent among older people, and many unscrupulous people have exploited the desperation that such illness can evoke. Harris (1978) reported that more than 90 percent of confidence scheme victims in two large metropolitan areas were older people. A few examples illustrate the point.

The American Continental Corporation–Lincoln Savings and Loan Association scandal was a prominent case of fraud aimed at elders whose incomes depended on interest from savings (U.S. House Committee on Banking, Finance, and Urban Affairs 1989). Lincoln Savings and Loan, a federally insured thrift institution, was owned by a holding company, American Continental Corporation. In 1987, financially troubled American Continental began an aggressive campaign of selling high-risk "junk" bonds through sales agents located in branch offices of Lincoln Savings and Loan. Lincoln employees were told to place first priority on directing depositors to the bond salespeople. Depositors assumed that since they were in a federally insured institution, what they were sold there was federally insured. Purchasers could even have the interest from the bonds deposited directly into their Lincoln Savings passbook accounts, and the monthly statements they received showed the activity in both their Lincoln Savings accounts and the American Continental bond accounts. If asked, sales agents told bond purchasers that their investment was highly secure and federally insured and that they could redeem the bonds at any time. These statements were false. The bonds were highly speculative, they were not insured, and they could be redeemed only upon the death of the purchaser. At a time when interest rates on federally insured certificates of deposit were falling all over the country, people were being promised 10.5 percent interest on two-year bonds sold at their local branch of Lincoln Savings. This campaign was directed mainly at older clients. Indeed, 79 percent of the people who were sold these junk bonds and who subsequently lost their entire investment were age 50 or over. Over $200 million was lost by depositors in California alone.

Mrs. V.'s case is typical. Recently widowed, Mrs. V. had a modest Social Security survivors' benefit and about $30,000 in savings to provide the additional income she needed to live on. When she saw the ad promising 10.5 percent interest, she called her local branch of Lincoln Savings. She was connected with a sales agent, who she assumed was an employee of Lincoln Savings. He persuaded her to withdraw her $30,000 from another thrift institution and bring it to Lincoln Savings. In return for her $30,000, Mrs. V. was promised an income of $262.50 per month for two years. She opened a savings account at Lincoln Savings into which the interest would be deposited each month. She was assured by the FSLIC logo on the door that her investment would be protected. There was never any mention of loss or risk. Mrs. V. was told that should she need her money in an emergency she could bring the bonds back to Lincoln Savings and cash them in. In April 1989, Mrs. V. received a letter telling her that she would no longer be receiving the monthly check and that her

$30,000 investment was gone (U.S. House Committee on Banking, Finance, and Urban Affairs 1989). American Continental Corporation had gambled and lost Mrs. V.'s money, along with the investments of more than 14,000 other older people. Officers of the company were prosecuted and received jail sentences for fraud, but that is little consolation to Mrs. V.

The American Continental scheme appealed mainly to elders who needed income from a fixed amount of savings. They assumed that government regulations would prevent them from being exploited. However, the wheels of regulation grind very slowly at times, and in a deregulated financial services industry, con artists are often easily able to operate undetected until it is too late to protect their elderly victims. American Continental's lawyers and accountants stayed one step ahead of both the Federal Home Loan Bank Board and the Securities and Exchange Commission long enough to bilk thousands of elders of their life savings.

Another example of economic exploitation of older people is the "investment property" business. In these scams, promoters buy large tracts of cheap land—often swampy land in Florida or desert in Arizona—and carve the property up into the smallest parcels allowable. Beautiful brochures and advertisements in newspapers and magazines promote the land as an investment for resale later at a profit or as a place for a retirement home. Lots are usually sold on installment contracts, and no deed is recorded until the contract is paid off. Most contracts stipulate that the property reverts to the seller if the buyer misses one or two payments, and the seller is not required to notify the buyer that he or she is delinquent. Many of these lots are sold over and over again, year after year, as buyers stop their monthly payments for any number of reasons—they die, come upon hard times, see the land and realize that their "investment" is worthless, and so on. Many of the buyers are elders looking to buy land for retirement homes.

Many practices in the "preneed" funeral business are also fraudulent. People have been sold "complete burial service" for thousands of dollars, and all the survivors receive (if they are lucky) is a cheap casket. People have been sold crypts in nonexistent mausoleums and burial plots in cemeteries that exist only on paper. This type of deception thrives on the deep-seated desire of many older people to take care of burial arrangements ahead of time, to make things easier for their relatives.

It may be unethical for even reputable businesspeople to take money from older people for prearranged funerals since businesspeople fully realize the problems of prearranging funeral services in a society where people travel as much as they do in ours. If the person dies far away from the city in which the prearrangements were made, the relatives either lose the money spent for the prearranged funeral or must pay to have the remains shipped back for burial, which can be quite costly. By encouraging prepayment, the funeral director is merely relieving the older person of savings, often without paying interest on the money held.

In addition to fraudulent business practices, other confidence rackets prey on older people. One of the more frequent con games is the bank examiner gambit. A person posing as a bank examiner calls on the intended victim and explains that one of the bank employees is suspected of embezzlement but that unfortunately the bank officials have been unable to catch the employee in the act. The "examiner" asks the victim to go to the bank and withdraw all of his or her savings, to thereby force the suspect in the bank to alter account books and the bank would then "have the goods" on the employee. If the intended victim takes the bait and withdraws the savings, the so-called examiner telephones, says they have caught the suspect, thanks the victim, and then offers to send a "bank messenger" to pick up the money—"to save you the trouble of having to come all the way down here, since you have been so kind as to cooperate with us." The "messenger" then appears, takes the money, and gives the victim an official-looking deposit slip. That is usually the last the victim ever sees of the money. Unfortunately, a large number of older people fall for this ruse each year.

Home repair scams are also common. Elders are sold "driveway sealer" made of used motor oil, unneeded furnace repairs suggested by a "city inspector," carpet that disintegrates, and so on.

Today, many confidence rackets operate over the phone. Elders are told that they have won a resort vacation and are persuaded to use their credit card to guarantee reservations, "only as a formality." Later substantial unauthorized charges appear on their credit card bills. Elders are sold package tours that do not exist and products that never arrive.

Medical swindles and quackery represent a particularly vicious hazard to older people. Arthritis is a good case in point. Millions of elders suffer from arthritis. Many find themselves on mailing lists of organizations offering a wide range of useless treatments. And because they are desperate to escape the constant pain of arthritis, many elders are relieved of their money but not their suffering. Law enforcement officials have been trying to stamp out medical fraud and quackery for over one hundred years with little effect. By the time complaints are made, the "business" has moved on to another location and phone number.

The financial loss to quackery amounts to several billions of dollars per year, but the human cost is even higher. The chairman of the U.S. Senate Special Committee on Aging (1965) summed up the true cost of fraud and quackery:

There are losses that go far beyond the original purchase price for the phony treatment, the useless gadget, the inappropriate drug or pill. How can you measure the cost in terms of suffering, disappointment, and final despair?

One of the heaviest of these costs is surely the attitudes such practices create among older people: They become suspicious and reluctant to get involved with strangers.

Summary

Retirement is not only a life stage that must be financed, but also an important tool of employment policy. It helps society keep unemployment within acceptable bounds, helps reduce labor mobility in highly skilled jobs, gives workers incentive to endure undesirable or overly demanding jobs by limiting their duration, provides a graceful means of phasing out ineffective workers, and creates opportunities for younger workers. Accordingly, retirement benefits the entire society, not just currently retired individuals.

The need to provide retirement income has played an important role in capital formation since World War II. Pension funds now own substantial interests in the corporations of America. However, thus far these large pension funds have not been managed with the goal of maximizing pensions but of providing capital for low-growth corporations and government, freeing other investors to seek the high-yield, high-growth investments.

The economy is a very complex social institution, one that affects middle-aged and older people by governing their access to jobs, by providing the economic surplus and financial institutions needed to provide retirement income, and by its approach to aging people as consumers. The U.S. economy is influenced by a long-standing free-market ideology that seeks to increase the wealth of private individuals and sees the operation of this accumulated wealth in the private market as the most appropriate way to meet people's needs.

However, access to wealth is not equal in American society. A core private sector of the economy, composed of the largest business and financial corporations, has relatively plentiful opportunities to earn high incomes, amass substantial investments, and generate entitlement to high retirement incomes. Workers in this sector are predominantly white males; and at the upper echelons, where opportunities are greatest, they tend to be from the most well-to-do families. The peripheral private sector is made up of smaller, more competitive businesses that operate on lower profit margins and in which wages tend to be low and fringe benefits and private pension coverage inadequate or nonexistent. Most American workers are employed in the peripheral private sector. The government sector of the economy employs about 17 percent of the labor force and offers a broad range of wages, fringe benefits, and pension adequacy. Some government workers fare nearly as well as those in the core private sector, but most fall somewhere between the core and peripheral private sectors' benefits. Thus, access to fully adequate retirement income is not available to all Americans, only those in about the top 30 percent of jobs. The incomes that support older people are overwhelmingly *earned* income from wages, accumulated assets, and pension credits. Less than 4 percent of the incomes of elders comes from "unearned" sources such as Supplemental Security Income or Medicaid.

The income inequalities among people increase as they move into later maturity, a result of inequities in access to retirement pensions and assets and also of a tax structure that concentrates tax breaks for older people among the most well-to-do and concentrates property, sales, and Social Security taxes disproportionately on low-income people. There is a large minority of older people for whom the American economy does not deliver adequate retirement income.

Future trends will probably increase the difficulties of providing adequate retirement income. Unless we find an alternative to the payroll tax for meeting our Social Security and Medicare obligations, pressure to increase retirement ages and cut benefits will continue, particularly if inflation in medical care costs continues at its present very high rate. These difficulties can be expected to increase the problems of the low-income older population more than those of the middle class.

For many older people, the private economy is effective in delivering food, clothing, housing, transportation, recreation, and the like. But the private economy has not responded to many elders' needs, particularly housing, personal care, health care, and transportation, because meeting those needs was not profitable enough. Attempts at government intervention are sometimes ineffective and overly costly because government must work through private businesses committed to making high profits from work done for government. At the same time, government cost containment efforts may cause private enterprise to leave or never enter important areas of service to older people.

Older people have customarily been treated as a trivial market for consumer goods. However, with the declining numbers of young people, the slowing of population growth, and the improvement of retirement incomes for the middle class, business is beginning to show an interest in older markets. Indeed, the over 55-population is currently spending more discretionary money than any other age category, particularly on luxury goods and services. In the future, we can look for businesses to develop more products and services aimed directly at the older population.

Because older people are often trying to live on limited budgets, they are likely targets for investment and real estate frauds. They also are the targets of medical quackery. Problems such as ill health and isolation apparently make elders more vulnerable than the general population to such fraud and quackery.

19 ▮ Politics and Government

Power, the ability to realize goals even against opposition, is the central core of politics, and people engage in politics as a way of securing power. Practically all large-scale institutions have their power aspects and their political aspects, but we generally reserve the label *politics* for the relationships surrounding the struggle for power over the machinery of government.

Politics in a free society is based on three indispensable elements: the rule of law, democracy, and leadership (Dahrendorf 1988). The rule of law strives for equal treatment for all residents. Constitutional checks and balances and principles such as due process and judicial review protect each person's stake in maintaining the civil order that is essential in modern industrial societies. The rule of law also controls those in power and their administrations. Democracy provides systematic ways for popular views and interests to enter into the political process:

Democracy enables people to make their voices heard. It gives them the right, and at times of crisis the duty, to say what they want and what not. The positive side of democracy, what people want, is better suited for festive speeches about political culture, but the apparently negative side, control, criticism, and protest, may well be more important for freedom. Leadership keeps societies moving. It helps prevent them from getting stuck in the cage of bureaucracy; it also cuts through the perpetual discourse of democracy. By the interplay of democracy and leadership, civil societies remain open.

(Dahrendorf 1988,70–71)

Overview

The machinery of government is complex. Under the U.S. Constitution, the powers of the federal government are shared among three main branches: executive, consisting of the president and the president's administration; legislative, consisting of elected members of the House of Representatives and the Senate and their administrative staffs; and judicial, consisting of the U.S. Supreme Court and a network of subsidiary federal courts established by Congress. In very simple terms, each branch serves as a check on the others. No bill can become law unless both houses of Congress pass it. The president can veto laws passed by Congress, but Congress can override vetoes by a two-thirds majority. The president can recommend legislation to Congress, but only Congress can pass it. Congress alone can establish taxes and appropriate funds for government expenditure. The president has the responsibility to administer government, but only Congress can authorize the executive departments needed to do so. The Supreme Court reviews laws to ensure that they are in accord with the Constitution and has the power to interpret the law. Each of these major decision-making bodies of the national government is chosen by different constituencies, and because each body has different terms of office, a complete overturn of the government by one stroke is impossible. This often slow and cumbersome governmental system has stood well the test of time.

In addition, there is a federal bureaucracy staffed by nonelected, career employees of the federal Civil Service who manage various government agencies and carry out administrative details of running the government. In 1992, the federal government employed just over 3 million people throughout the country, nearly 1 million in the Department of Defense and 775,000 in the U.S. Postal Service alone. In addition, there were also 51 state governments, over 3,000 county governments, over 19,000 municipal governments, over 16,000 township governments, and nearly 15,000 local public school district governments. These governments employed nearly 16 million people (U.S. Bureau of the Census 1994). State and local governments are constructed in a variety of ways, but all are required to conform to the principles set forth in the U.S. Constitution.

Governments are responsible for a wide range of functions in modern societies. In the United States, for example, the federal government is responsible for, among other things, maintaining national defense; promoting international diplomacy; setting monetary policy; protecting public health and safety; safeguarding the environment; promoting economic growth; promoting commerce by establishing and maintaining a national network of airports and roads; protecting workers from exploitation by establishing and enforcing child labor laws, job safety regulations, and minimum wages consistent with our national level of living; and protecting our natural resources. State governments play a key role in higher education, public assistance, highways, and prisons. Local governments provide most of the resources for elementary and secondary education, hospi-

tals, courts, police and fire protection, sewage and sanitation, and government housing and community development. Politics involves resolving disputes over the relative priority to be given to these various functions as well as over the means to be used to achieve the functions.

As we saw in Chapter 18, the American economy is a capitalist one, in which a primary goal is the growth and accumulation of private wealth. Politics in a capitalist society is fundamentally influenced by the desire to have political decisions interfere as little as possible with the private economy.

As Dahrendorf (1988) skillfully pointed out, the need for material progress and the need for democratic access to opportunities go hand in hand. It does little good to have the most democratic society on earth if its economy does not produce enough food, housing, and clothing to meet the needs of the population. By the same token, societies that can produce every manner of consumer goods but whose mass of people do not earn enough to buy the goods are not satisfactory either. Thus, maintaining both economic growth and access by the people to the results of that growth are central issues of politics in free societies.

In the United States, some politicians and political interest groups emphasize maintaining economic growth through the application of private capital; others emphasize the need to be concerned with creating and maintaining democratic access to education and employment opportunities, particularly through support for civil rights. For example, Republicans generally emphasize economic growth and the accumulation of private capital; Democrats generally emphasize protecting freedom of access to education and employment opportunities and ensuring that there are mechanisms in place to protect the disadvantaged from poverty. These are not either/or positions; they are matters of emphasis. Members of both parties usually recognize that both access and economic growth are important. The parties differ mainly in the priority they attach to these goals. For both parties, preserving the existing economic system is the first priority (Estes 1991).

From 1960 to 1975, economic growth was so strong that it was taken as a given, and American politics was dominated by concerns surrounding the creation of more equal opportunities for minorities and women and the creation of social programs that would prevent the occurrence of poverty in an affluent society. It was during this period that many of today's programs benefiting older people came into being. However, in the late 1970s, dramatic increases in the cost of energy, shifts in international economic markets, recession, and high inflation sent a strong signal that for the United States, economic growth could not be taken for granted. Since then, maintaining economic growth has had the highest political priority, which has resulted in steady pressure to moderate or cut back programs such as Social Security, Medicare, Supplemental Security Income, Medicaid, and the Older Americans Act.

Scholars differ considerably in terms of the conceptual and theoretical frameworks they use to interpret politics and government. Three major perspectives have been used to explain the politics of age: interest group theory, political economy theory, and critical theory. As a theory of the political process, *interest group theory* presumes that politicians respond to pressures from competing interest groups.* Interest groups have paid staff and their operations can be financed by membership dues of voluntary associations or by patronage by wealthy sponsors. Power is dispersed both among elites and among the masses because they are divided on a wide variety of issues. Interest groups are the vehicles through which both elites and masses decide on their priorities and allegiances with

respect to specific, pragmatic goals. Thus, according to interest group theory, no single economic class or narrow elite dominates the political process. Interest groups are often issue-specific. For example, the pro-life (antiabortion) movement, the gun lobby, and the gay rights movement each has large numbers of supporters, but their support is focused on a specific and limited set of issues. The National Rifle Association cannot be presumed to have a well-defined position on the need for national health care reform, for example. We return to interest group theory when we discuss advocacy on behalf of aging and older people.

Political economy theory presumes that economic class conflict is at the root of political processes in capitalist democracies.* According to political economy theory, workers at the lower end of the socioeconomic spectrum are relatively powerless in the political process unless they organize into labor unions. Labor unions are necessary to oppose the exploitative posture employers are predisposed to take toward workers. Profit-driven private enterprise is assumed to have a built-in bias toward sharing as little of the profits of production with employees as possible. Workers can gain a fair share of the product of their labor only by making organized demands for fair treatment. Once organized, labor can also advance proposals in the political arena. Advocates of political economy theory contend that many of the gains American workers take for granted today, including the minimum wage, the forty-hour workweek, employer pensions, employer health insurance benefits, occupational health and safety standards, and even the Social Security system itself, would not exist had it not been for the political pressure exerted by organized labor. Political economy theory accepts the goal of economic progress; it simply asserts that class

---

* See Robert H Binstock (1972}.

* Estes (1991).

conflict results in a fairer distribution. We return to this perspective when we discuss governmental responses to issues concerning aging.

Critical theory presumes that the goal of the political process in a free society is to create and maintain a civil society that allows each individual a maximum amount of freedom. The early promise of industrial economic development was that it might free the mass of society from the drudgery of dehumanizing work and release time for the pursuit of higher human potentials. Yet this emancipation could not be completely individualistic because a certain amount of conformity is necessary to maintain the civil society. Critical theory engages the political process by looking at the claims of need asserted by political actors. In political dialogue, politicians claim that this or that point of view ought to carry the day based on statements about what is actually occurring in the society, what ought to occur, and what needs to be done to keep society on course toward what ought to occur.

For example, among other things the 1994 Republican "Contract with America" promised to reduce taxes, reduce the size of the federal government, reduce the amount of control exerted by national government, and balance the federal budget. The need for these actions was justified by assertions that federal taxes were too high, that the federal government was wasteful and oppressive, and that the federal budget deficit was *the* major source of America's economic problems. Based on these assertions, advocates of the "Contract with America" claimed the right to dominate the political process. And in the first year after the Republican Party gained control of both houses of Congress, this agenda did indeed dominate political decision making in Congress.

Critical theory suggests that we step back and examine such claims carefully, because every claim can be challenged in terms of its factual validity and each claim also has costs in terms of constraints it imposes and bene-

fits it provides in terms of emancipation. For example, the "Contract with America" might succeed in its objectives and still not improve the living situations of large numbers of Americans if the federal budget deficit turns out not to be the real source of the declining number of middle-class jobs in America. And lower federal taxes are not much benefit to the unemployed. Unlike interest group theory and political economy theory, the prime goal for critical theory is maximizing human freedom. Economic productivity is necessary for freedom, but *optimizing* economic growth may be preferable because it offers more promise of free time than a political-economic model that maximizes economic growth. We return to this perspective when we look at political trends.

In examining the relationship between aging and politics, we concentrate on three fundamental issues: the political *activity* of aging and older people, the political *influence* of older people and their advocates, and the political *response* to aging and older people as evidenced through governmental programs and policies.

Political Activity

Political activity takes many forms in the United States: expressing political opinions, voting, participating in voluntary associations revolving around politics, or holding political office.

Political Opinions

Forming political opinions requires the least involvement from the individual. Aging does not appear to cause political opinions to become more conservative or more liberal (Jacobs 1990). Interest in politics and the proportion holding political opinions both increase with age (Hudson and Strate 1985).

It is commonly thought that people become more politically conservative as they grow older. However, most available evi-

Elders are a visible part of election campaigns.
Photograph by Robert E. Daemmrich/Tony Stone Images, Inc.

dence does not support this notion. Campbell and Strate (1981) reviewed evidence from various studies spanning 1952 to 1980 and found that raw age differences in political orientation, where they existed at all, tended to be small. Although there is evidence to suggest that attitudes become somewhat less susceptible to change as people age, there is no evidence that their attitudes become more conservative (Jacobs 1990; Alwin, Cohen, and Newcomb 1991).

There is little evidence that older voters share any sort of age consciousness that causes them to vote as a bloc on issues other than those that affect older people of all types (Day 1990). Issues such as gender, race, religion, and socioeconomic status seem to divide older voters in much the same way they divide younger voters. Miller, Gurin, and Gurin (1980) found that even when age consciousness existed, it did not lead to political action, precisely because older people felt that, as older people, they lacked influence.

Using data from a longitudinal panel of women who were students at Bennington College in the late 1930s and who were resurveyed in 1961 and 1984, Alwin and his colleagues (1991) looked beyond opinions about specific issues or candidates to consider underlying political ideologies and symbolic dispositions and how they might change as a result of adult development. They found that parents and peers had a marked impact on the underlying political philosophies of these women and the resulting political orientations were remarkably consistent over the forty-five years spanned by the study. Thus, their results supported the notion that political orientation is malleable in young adulthood but then becomes increasingly stable with age. This is not to say that change did not occur. About 60 percent of the 1984 attitudes of the panel represented continuity of attitudes from the 1930s and 40 percent represented change. As we found earlier with respect to global ideas such as life satisfaction or self-esteem, global political orientations tended to remain stable whereas specific ideas that supported them contained change.

Opinion formation is no doubt related to interest and information seeking. Research results indicate that older people are interested in politics and make the effort to remain informed about political matters. In comparison with the young, older people give greater attention to political campaigns and are more likely to follow politics in the news (Comstock et al. 1978).

Party Affiliation

Political party affiliation affects individual political attitudes on a wide variety of issues ranging from domestic and foreign policy to interest and involvement in political campaigns (Hudson and Strate 1985). The proportion who identify strongly with a political party is higher in older age categories, but most of these age differences are due to declining party identification in more recent cohorts. For example, Alwin, Cohen, and Newcomb (1991) found that political party identification was strong and sustained over a period of forty-five years among the women who took part in the Bennington College studies of the 1930s. Thus, as younger, more independent cohorts age, we can expect a decline in party identification in the older population (N. Cutler 1981).

Among older people, there are about equal proportions of Republicans and Democrats. There is no evidence that people are more likely to become Republicans as they grow older (Hudson and Strate 1985).

Because power in party organizations tends to be related to the duration of active participation, older party faithful have greater influence in the nominating process. However, this is a function of length of service or seniority rather than age.

Voting

Voting shows a complex and interesting pattern of change with age. In general, the proportion of age cohorts voting is lowest at the age of 18; it builds to a plateau in the 50s and begins a gradual decline after the age of 75. People in their 80s are much more likely to vote than people in their early 20s. This pattern describes the voting histories of many cohorts. Men and women show about the same age curve of participation, and the overall proportion voting is lower only for women in the oldest cohorts (see Table 19-1).

From the time voting began to receive study, education has had a profound effect on the proportion voting. For example, Table 19-1 shows that voters' educational level had a significant impact on the age pattern of voting in the November 1992 election. Among men, age made much more difference in percentage voting for those with only elementary or some high school education than for those who graduated from college (read down the columns in Table 19-1). Among women, age made less difference, but the effect of education on the age pattern was similar. Among those with some college, the difference between men and women in terms of percentage voting was quite small, whereas it was greatest among older men and women who had not completed elementary school (see "Effect of Age" rows). Thus, *the higher the level of education, the less the relationship of both age and gender to voting.*

One major trend in American electoral politics since 1960 has been the decline in the proportion of those eligible to vote who actually register and vote. For example, the proportion of voting-age population voting in presidential elections dropped from 62.8 percent in 1960 to 55.2 percent in 1972 to 50.3 percent in 1988 (U.S. Bureau of the Census 1989b). This growing alienation from voting participation occurred mainly among people under 35. Among older people, the proportion voting increased.

However, the 1992 presidential election showed a dramatic increase in the proportion of adults of all ages who voted, but especially among adults age 21 to 24 (from 38 percent

Table 19-1 Voter participation (percent voting), by age, sex, and education: U.S. General Election, November 1992.

	Highest School Year Completed							Effect of More Education
	Elementary		High School		College			
Age	<5	5–8	9–11	12	1–3	4	5+	(% 5+ College – <5 Elementary)
Males								
22–24	—	6.4	14.0	35.3	54.7	69.1		—
25–44	4.4	11.0	25.0	46.7	64.6	75.5	77.6	+73.2
45–54	16.6	27.6	46.0	63.0	76.7	82.3	85.6	+69.0
55–64	30.5	51.1	58.6	71.5	81.7	86.1	87.4	+56.9
65–74	34.2	60.6	71.1	79.5	85.3	89.5	87.8	+53.6
75+	47.8	59.0	71.7	76.5	84.0	81.4	88.8	+41.0
Effect of Age (% 75+ – % 22–24)	—	+52.6	+57.7	+41.2	+29.3	+12.3	—	
Females								
22–24	—	2.9	18.8	33.4	59.5	74.0	75.1	—
25–44	7.6	14.3	29.0	52.7	68.9	80.1	82.8	+75.2
45–54	9.5	33.7	46.2	67.4	78.2	84.4	88.7	+79.2
55–64	21.2	41.4	59.8	73.8	83.4	86.4	88.4	+67.2
65–74	27.9	50.9	64.8	76.9	83.0	86.7	87.3	+59.4
75+	27.6	47.6	57.4	67.5	74.0	80.7	72.4	+44.8
Effect of Age (% 75+ – % 22–24)	—	+44.7	+38.6	+34.1	+14.5	+6.7	–2.7	

Source: U.S. Bureau of the Census (1993).

in 1988 to 45 percent in 1992). Whether this was an anomaly stemming from the strong independent candidacy of Ross Perot or the strong appeal of Bill Clinton to younger voters remains to be seen. The very small turnout for the 1994 congressional election (40 percent of eligible voters) suggests that political alienation is still a salient issue for the nation.

Holding Political Office

Older people hold various political offices. Leadership in all areas related to public affairs seems to be amply accessible to older political leaders. Presidents, cabinet ministers, and ambassadors usually acquire their positions in their late 50s and often retain them well beyond the age of 60. Supreme Court justices are also most likely to be appointed in their late 50s, and most continue to serve well beyond 65 since retirement usually depends on their personal desire. In the 103rd Congress (1993), 104 representatives (24 percent) and 34 senators (34 percent) were age 60 or over; 15 representatives and 12 senators were 70 or over (U.S. Bureau of the Census 1994). Proportions of older people in state and local offices (whether elected or appointed) are substantially higher than among the rank and file of other occupations. Since 1990, as pressures have increased to maintain government services without raising taxes, being a legislator has become an even more difficult job. There has been greater turnover among senior lawmakers, about evenly split between those who were not re-elected and those who decided not to seek re-election. This trend has resulted in a reduction in the average age of legislators at both the federal and state levels.

One reason why older members of political organizations are influential is the weight that tenure alone carries in politics. Of course, it is impossible to achieve substantial tenure without also aging. In politics, perhaps more than in any other institution, older people are still able to play the role of sage (in fact, the word *politic* means "wise" or "shrewd"), probably partly because political processes have felt the impact of industrialization and rationalization less than the economy or even the family has. The professionalization of politics is much less complete than that of education or the economy. As a result, political prowess is still felt to be largely learned from experience rather than out of a book or in a professional school, although a large proportion of legislators and judges are trained lawyers. There seems to be public appreciation of this factor. Turner and Kahn (1974) found age of a candidate an insignificant issue among voters of all ages. However, experience does seem to be a salient factor. Schlesinger and Schlesinger concluded that "American politics has not been especially cordial to the older political neophyte" (1981,235).

Thus, older people do not seem to be at any great disadvantage in terms of access to political roles. Yet what does this accessibility mean to older people? Unless they have paid their dues in the form of earlier participation in politics, elders are unlikely to gain access to positions within party organizations or government itself.

Political Influence

There are several competing perspectives concerning how the issues of aging gain or lose political influence. Some say that aging and older people serve primarily as a pressure group that attempts to coerce politicians into either enacting legislation beneficial to them or opposing legislation harmful to them. Others see the problems of aging as the central focus for the formation of a social movement involving not only older people but people of all ages in an attempt to solve these problems. Still others view older people as a category that, because of the discriminatory and categorical treatment it receives at the hands of society at large, is helped by coalitions whose support for the older population fits into their own agenda. All these approaches contain some truth; none is sufficient in itself to explain the political treatment of issues concerning aging.

Older People as an Interest Group

Political success has been more often a hope than a reality for political groups formed around older people's interests. As Carlie concluded about early old-age interest groups such as the Townsend Movement:*

We have indicated [previously] that the old age political organization left a great deal to be desired as pertains to such matters [as] organization (a failure to develop strong secondary leadership), pressure (if there were not too few members then they were regionally segregated), votes (the aspiring representatives of old age political programs failed to secure enough votes to place them in pivotal political positions), and money. . . .

It seems as though one of the necessary conditions for the formulation and maintenance of interest groups is a homogeneity of characteristics among the membership. An effective interest group, then, should have more in common than just age. Other important shared characteristics may be ethnicity, nativity, educational

---

* See Chapter 3 for more on the Townsend Movement.

background, occupational status, race, and rural-urban residency. Lacking similarity beyond age (and perhaps the state of retirement), the old age political movements were handicapped from their very inception.

(1969,259–263)

Effective political interest groups often purport to be able to deliver votes in a bloc. But there has never been an instance in the history of the United States in which the leadership of an old-age interest group could deliver what would even begin to approach "the old folks' vote." Thus, the view that older people constitute a unified interest group that can mobilize political pressure by bloc voting is an illusion and quite likely to remain so (Wallace et al. 1991).

Arnold Rose (1965) was perhaps the strongest proponent of the idea that patterns of aging in American society would produce a *subculture of the aged*. According to Rose, a subculture would develop when people in a given category interacted with each other more than they interacted with people in other categories. The subculture would grow as a result of the positive affinity that drew like people together and the discrimination that excluded these same people from interacting with other groups. Rose cited the growing numbers of older people with common problems—usually related to health—living out their retirement in self-segregated retirement communities where a relatively high standard of living made possible the development of a unique lifestyle. And he contended that older people were rapidly shifting from the status of a category to that of a group. He further stated that in addition to pure self-interest, there was an interactional basis for the development of an age-conscious group. This interaction resulted from our society's policy of phasing older people out of almost every other kind of group, thus forcing them to seek out each other. Rose felt that this trend cut across the subcultures based on occupation, sex, religion, and ethnic

identification that typify the middle-aged population. Old-age group identification, rather than any specific organization of older people, led Rose to conclude that "the elderly seem to be on their way to becoming a voting bloc with a leadership that acts as a political pressure group" (1965,14).

This type of analysis is intuitively attractive to many because it postulates the existence of the old-age pressure group without necessarily requiring that this group manifest itself in an overt form of organization. The "leaders" speak for a kind of silent group that recognizes itself and is capable of the bloc voting required to achieve political influence. But there are major problems with this approach. To begin with, most older people interact *across* generational lines as well as within them, simply because a frequent source of interaction is their adult children and grandchildren. Second, the stereotype that older people are flocking in masses to retirement communities is largely a myth. The data on migration clearly show that only a small minority of older people move. As a result, older people tend to be interspersed among the general population. Third, while there may be a rising number of older people who can afford a unique leisure lifestyle, lifestyles vary widely among older people. Thus, several of the key conditions on which Rose based his case for the subculture approach turn out to be questionable. In fact, Longino, McClelland, and Peterson (1980) found that although older people in retirement communities tended to develop a subculture that enhanced their self-esteem, the political attitudes of this subculture were *retreatist* and thus led to little likelihood of concerted political action.

Rosenbaum and Button (1989) looked at county and city politics in Florida, a main destination of elders moving to retirement communities. They found that although older people were politically active, they did not organize themselves to either promote their own interests or to oppose other interest groups. However, in a later study, Rosenbaum and Button (1993) found that both older and younger adults *perceived* older adults as a category that is more likely to promote their own selfish interests and to oppose issues that do not advantage them personally. Interestingly, the older respondents in their study were actually more likely to favor school tax increases than were younger people. Thus, the *actual* political behavior of older people tends to be much less age-centered than their public image gives them credit for. Rosenbaum and Button (1993) argued that this inaccurate stereotype of the self-centered older voter has the potential to fuel intergenerational animosity because it is based on *beliefs* about older adults' voting behavior, not their actual behavior.

Advocacy for Older People

The view that political action favoring older people has resulted from political involvement, not only of older people but of others as well, is the view that political scientists have most often expressed (Cottrell 1960a; Binstock 1974; Hudson and Binstock 1976; Hudson and Strate 1985; Day 1990). According to this approach, older people have a great deal of *potential* influence because in the more rural areas they constitute a larger proportion of the population and these rural areas are overrepresented in both the state legislatures and the U.S. Senate. The presumed reason that this potential has never been realized, however, is that the older population is divided among a great many interest groups, most of which do not make the welfare of older people their primary goal. But if this reasoning is valid, we may ask, how did legislation in behalf of older people manage to get passed?

The answer seems to be that old-age political organizations, while not effective as interest groups, have been effective in making needs of the older population politically visible. As a result, the "cause" of elders was picked up by groups not based on age, such

as unions, political parties, and politicians looking for votes. The readiness of various groups interested in the general welfare to commit themselves to programs for older people was caused largely by the existence, at almost every level of organization, of large numbers of middle-class adult children who recognized that they could not meet the income security and health-care needs of older family members without help from social insurance. For this reason, as well as the votes that elders themselves command, no major political party, business interest group, labor union, or other large-scale organization has thus far been successful in their attempts to create a major overhaul or to "privatize" Social Security, although some have tried and continue to do so.

This view of older people's power fits the facts of how legislation was initially created in the 1935 to 1975 period better than any other so far. It describes relatively accurately the political processes that brought passage of Social Security and Medicare, the two most sweeping legislative proposals affecting older people (Lubove 1968; Marmor 1970).

In addition to persistently keeping problems of the aged in the public eye, organizations such as the National Council of Senior Citizens (NCSC—5 million members) and the American Association of Retired Persons (AARP—33 million members in 1995, up from 6 million in 1974) have been reasonably effective in gaining a larger share for older people from existing federal programs. In a very lucid article, Pratt (1974) attributed the increased effectiveness of post-1960 voluntary associations for older people, compared with the old-age interest groups of the 1930s, to three factors: (1) sources of funds in addition to member dues, (2) reliance on bureaucratic performance rather than charisma as a criterion for selecting leaders, and (3) a sympathetic political climate for lobbying activities. The NCSC early-on derived nearly half its financial support from large labor unions, while AARP got most of its initial operating funds from membership dues and insurance programs and other services it offered its membership (Day 1990). Both organizations were headed by persons of proven administrative capability. Political sensitivity to the concerns of old-age interest groups was most readily apparent in the concessions extracted from the Nixon administration through a threatened boycott of the 1971 White House Conference on Aging, particularly the indexing of Social Security benefits to protect them from inflation.

In addition to the organizations already discussed, many others promote political attention to issues affecting the older population. For example, the American Society on Aging (10,000 members in 1995) and the National Council on the Aging (5,500 members) are professional organizations made up primarily of people who organize and provide services to older people. The Gerontological Society of America and the Association for Gerontology in Higher Education promote research and education about aging. There are also groups that do research and draw attention to issues unique to minority elders. Older African Americans, Latinos, Asian and Pacific Islanders, and Native Americans each have advocacy groups that concentrate on their interests. Organized by Maggie Kuhn, the Gray Panthers is a coalition of young and old people who promote and publicize alternative lifestyles for older people. The Older Woman's League promotes attention to issues of interest to older women, especially better Social Security retirement benefits to nonmarried elders, most of whom are women, and better private pension coverage for women.

By the late 1970s, more than two dozen organizations were advocating in Washington for issues involving older Americans or specific subgroups of them. During 1977 and 1978, the economic recession, high inflation, and a fast-growing federal deficit were all exerting pressures on Congress and the Carter administration to curtail growth in programs

for older people. In an effort to oppose what most saw as a premature halt to the development of federal funding support for older people, twenty-six organizations formed the Leadership Council of Aging Organizations. The prospect of a threat to the successes of the 1965 to 1975 period thus brought together a collection of organizations whose differing agendas, perspectives, constituencies, and organizational leadership styles had previously seemed impossibly incompatible (Pratt 1982). However, the Leadership Coalition of Aging Organizations was not able to sustain its cohesion and never developed an integrated approach, which of course supports interest group theory's tenet that power is dispersed by competing interests.

By the late 1980s, AARP was so big that it tended to go its own way rather than to rely on coalitions to the extent it had in the past. AARP became more bureaucratized, more isolated from other organizations in the field of aging, and less interested in multiple points of view. As Day (1990) pointed out, this strategy had the advantage of creating AARP's image as the "most powerful lobby in Washington." It also had several disadvantages. AARP was more likely to be seen as *the* enemy that had to be overcome by interest groups opposing Social Security and Medicare. The early 1990s saw several public attacks on AARP's integrity as a membership organization. These attacks were based on its revenue from insurance, pharmacy, investment, and travel services and its status as a nonprofit organization. And as AARP began to advocate for various public policies on its own, its membership became more conscious of its positions. It was no longer just one of several organizations in an advocacy coalition. Most members join AARP to receive its publications and to participate in its discount purchasing programs, not to take political action. AARP members tend to see a proper role for AARP as a watchdog group that can let members know about issues affecting them, but many are uncomfortable with

AARP taking positions on issues about which the membership is divided. And since the membership is divided on most issues except preservation of Social Security and Medicare, AARP has relatively little maneuvering room to operate as an interest group.

In fact, AARP is far from being a powerful lobbying force. Because it is a nonprofit association, AARP is barred by federal law from making contributions to political parties or political candidates, which means that it cannot create political IOUs through campaign contributions. Because it has a diverse membership that spans all political persuasions, AARP cannot deliver a voting bloc. Compared with other lobbying organizations such as the National Rifle Association, the American Medical Association, corporate political action committees, or the Christian Coalition—all of which focus major amounts of money and/or volunteer service on political parties and individual candidates—AARP is a weak political actor indeed.

During the era of Democratic administrations, AARP staff often were influential by providing leadership and information to Congressional committees and government departments. But AARP opposition seldom caused Congress or administration officials to change course on any issue after 1980. Part of this decline in influence with congressional and administrative staff was the result of changes in the political process (which we will discuss shortly). The image of AARP as the most powerful lobby in Washington was a myth created largely by opponents of Social Security, who did not want to admit that Social Security persisted because it had overwhelming public support among adults of all ages. If Social Security's resilience could be blamed on an "overwhelmingly powerful lobby" such as AARP, then the fight to dismantle Social Security could be portrayed as a noble effort to defeat "special interests."

Estes (1979) pointed out that the "aging enterprise"—a loosely connected array of personnel from government and nongovern-ment agencies serving older people—depends for survival on a strong federal commitment to programs for elders, especially those related to health and social services. These people can be expected to operate through various professional organizations to oppose cutbacks in programs for older people and to advocate expansion in many cases. However, Wallace et al. (1991) pointed out that the main effects of these policy efforts have been to preserve benefits and prevent poverty for middle-class elders. There has been relatively little progress in decreasing poverty or better serving minorities, women, and the oldest-old, who are the most disadvantaged within the older population.

Thus, while advocacy groups have not been, and are not likely to be, very effective in bringing about a drastic move toward equality for older people, they have so far been effective in improving results of, or preventing severe cutbacks in, current programs for older people.

Over the past several years, the track record of advocacy groups using this strategy has been mixed. From 1967 to 1990, cuts in federal programs benefiting older people were smaller than for other interest groups advocating for the disadvantaged. On the other hand, advocates for the interests of older people were unsuccessful in preventing decremental cuts in Social Security retirement benefits and Medicare; nor were they able to prevent widespread questioning of the intergenerational social contract that served for over fifty years as the moral underpinning of Social Security (Marmor, Mashaw, and Harvey 1990).

Political Legitimacy and Utility

We emerge from this survey of organizations of and for elders with the view that older people themselves have relatively little power *as older people*. Some powerful people are old, but they are not powerful *because* they are old. Most of the political power behind programs for older people is generated by others on behalf of older people rather than by older people themselves. Hudson (1978) identified two reasons why this has been so: the *legitimacy* of older people as a target for governmental efforts and the *utility* of older people as potential voters. Legislators can only legislate in areas that the voting public considers a legitimate concern of government. And no other constituency on the domestic scene has been seen as a more appropriate concern of government than elders. Hudson argued that this legitimacy has rested on widespread public acceptance of the belief that older people represented the most undeservedly disadvantaged category of people in American society in terms of income, health, and vulnerability. There are other factors too. Age discrimination in the economy means that older people cannot do anything about their situation through the job market. They do not participate in general increases in wages and salaries. No other category of adults is so systematically excluded from this means of coping with economic change. Support for the legitimacy of programs for older people also comes from families who fully realize that adequate financial assistance to retired family members and financing their health care is beyond the family's capability. Finally, the legitimacy of programs for elders comes also from the fact that everyone eventually gets old, so creating adequate social supports for older people is a form of self-protection.

The effectiveness of older people as a political force is a sometime thing. When the issues concern elders directly (Social Security, Medicare, and so on), older people and their advocates can indeed represent a sizable political force, *if* there is time to organize a response to legislative proposals. In other areas, such as aging research or direct services to older people, the political effectiveness of aging issues is less clear-cut. In addition, changes in the political process have reduced opportunities for interest groups to mobilize effective opposition to congressional action. The time frame between when the details of legislative proposals become known and when the proposals are voted on is simply too short to allow organized opposition or support to form.

Hudson cited several factors that may threaten older people's political legitimacy. Cost pressures, especially for Medicare and long-term care under Medicaid, have made it increasingly difficult to meet the needs of the older population with the resources Congress and state legislatures have been willing to appropriate for these programs. In addition, recent improvements in the economic circumstances of some older people have altered the public image of older people as a disadvantaged group, which could have especially dire consequences for older minorities and women, among whom large proportions are still quite disadvantaged. These pressures have led to a reexamination of our policy of providing social insurance coverage for all older Americans regardless of their personal resources.

Political Trends

In the early 1990s, several trends in the climate of American politics continued to affect the treatment of issues related to aging. These trends included growing dissatisfaction with the federal government; growing power of congressional factions dedicated to cutting the size and scope of government and to balancing the federal general revenue budget; declining power of congressional factions that place a high priority on access to economic opportunity and on problems of the disadvantaged; increasing decentralization of responsibility for governmental functions to state and local governments; increased use of Congress's annual budget resolutions as vehicles for bypassing public input concerning changes in federal government responsibilities, financing, and operations; and increasing use of Social Security and Medicare issues to divert attention in the media from central national issues such as educational, economic, and/or foreign policy.

Growing dissatisfaction with the federal government in the mid-1990s was related to a large variety of problems. News stories repeatedly focused on wasteful government spending, particularly in the Department of Defense. Hundreds of billions in government funding were required to correct the savings and loan disaster created primarily by the Reagan administration's misguided deregulation of finance and banking. Since 1980, Congress has persistently been unable to come to terms with the growing deficit in the general revenue budget and the explosive increase in both the national debt and interest on that debt. At a time when most Americans' first concern was the health of the economy, the Congress and the administration seemed preoccupied with seemingly endless hearings designed to discredit government officials and to showcase government failures. The agendas of both political parties seemed out of step with the issues that most concerned the public: the continuing elimination at all occupational levels of jobs with good pay and benefits and the small proportion of new jobs that carried good pay and benefits; runaway health-care inflation; stagnant real wages; increasing crime; and growing environmental concerns.

To be fair, Congress and the past several administrations have been caught between contradictory public expectations. On the one hand, the public expects government to continue offering benefits and services as usual. On the other hand, the public does not want to pay additional general taxes. These conflicting expectations provide very little leeway within which Congress or the administration can act. Contradictory expectations have most often led to inaction, and the public's reaction has been growing dissatisfaction and a "throw the bums out" mentality.

In our culture, we often resort to either/or thinking. Either the government is doing a good job or it isn't. Most of the time, however, the actual situation reflects much more complexity. For example, despite many flaws in the federal government and many failings of Congress, the Social Security retirement program continues to be a success story. It is the largest retirement program in the world, and it is quite efficient. Whereas well over 10 percent of the revenues of private pension programs are typically spent on administrative costs, Social Security spends less than 1 percent of its revenues on administration. Our desire to get rid of ineffective, inefficient, and unnecessary government programs is understandable, but we would be wise to remember that there are some functions that government does well and that neither private enterprise or state and local governments can be expected to perform. Our national social insurance programs are prime examples.

The political pendulum continued to swing toward the *politically conservative* side in the mid-1990s. This meant that the focus of attention at the national level was on reducing the size and authority of the federal government and reining in the federal general revenue budget deficit. These priorities favor increasing profits and decreasing government controls, and they presumably serve the public interest by increasing the strength of the economy. Certainly, a strong economy is needed to provide material necessities to the public. However, there is little evidence that the financial fruits of deregulated free enterprise trickle very far down the economic pyramid. The vast majority of the proceeds from economic growth since 1980 went to the top 5 percent of the occupational pyramid; whereas the real incomes of those in the middle and below remained stable or declined (Barlett and Steele 1992).

The continuing deficit in the federal general revenue budget is a troublesome issue. Most of this debt accumulated during the Reagan-Bush administrations as a result of large increases in military spending and the substantial costs of the savings and loan collapse, coupled with a substantial reduction in corporate and individual income taxes. In other words, the Congress and three successive administrations increased spending substantially and at the same time reduced tax revenues by cutting income taxes. The aging of the population played no significant role in the increase in the national debt.

However, the deficit in the federal general revenue budget was used as a ploy to attack Social Security, Medicare, SSI, and Medicaid. The logic goes something like this: We have not been able to reduce the federal budget deficit because we are spending too much money on "entitlements" (a code word that increasingly means Social Security and Medicare). Therefore, to bring the budget under control, we need to limit spending on Social Security and Medicare. Sounds logical, right? But the argument begins with a false premise. The federal budget deficit is not driven by Social Security and Medicare spending. These programs get all but a very small proportion of their revenues from payroll taxes and premiums paid by Medicare beneficiaries, not from general tax revenues. Social Security and Medicare revenues are not general revenues and therefore are not part of the general revenue budget. Only a small portion of Medicare, Part B funding comes from the general revenue budget. *If we eliminated the Social Security retirement, survivors and disability programs completely, we would not gain one tax dollar that could be put toward reducing the general revenue budget deficit or paying off the national debt.* *

The spending that actually drives the federal general revenue budget is earmarked for defense (29 percent); income security programs (19 percent), including government employee retirement pensions and disability insurance [6 percent], unemployment insurance [3 percent], AFDC, Food Stamps, SSI, and other income assistance [10 percent]; interest on the national debt (11 percent); and

education (7 percent). Health, veterans' benefits, and transportation get about 5 percent each; natural resources and environment, commerce, agriculture, community development, energy, and international affairs get 2 to 3 percent each. Defense expenditures are such an integral part of our economy that we cannot reduce them too quickly without risking economic upheaval. Even with the downsizing of the military and closing of military bases, strong pressures remain on Congress to continue defense expenditures. The remaining federal general revenue expenditures are spread over a large number of functions, none of which, even if completely eliminated, offers much relief for the chronic gap between general tax revenues and federal commitments to provide continued funding of programs the public wants maintained.

Interest group theory offers a plausible explanation for why we cannot bring ourselves to just cut programs across-the-board. A large array of well-financed interest groups work very hard to preserve their slice of the federal general revenue budget. That Social Security and Medicare, which are not even a part of the general revenue budget, continue to be the most likely to be substantially cut speaks volumes both about the public's confusion concerning the issues and the "gray lobby's" lack of actual political power.

Decentralization has shifted national responsibilities for education, health, mental health, social welfare and many other programs to state governments, which tend to pass them on to local governments. Thus, since 1985 state and local taxes have risen while federal taxes have remained at about the same rates (U.S. Bureau of the Census 1994). However, as Liebig (1994) pointed out, the federal government still sees itself as being held accountable by voters for the outcomes, so they seldom leave the states unmonitored. Even though the streams of federal funding have slowed markedly for a variety of programs, the federal government operates on a principle of state or local

---

* For a more detailed treatment of specific elements of the debate about Social Security, see Marmor, Mashaw, and Harvey (1990).

responsibility and shared federal, state, and local funding, but with a declining share of federal funding and continuing accountability to the federal government.

Decentralization has the advantage of allowing state and local governments to develop programs that fit their unique populations and economies. But, decentralized programs have an uneven record, particularly with respect to providing access to economic opportunity and attention to the problems of disadvantaged groups. Voters in a number of local areas have been willing to tax themselves to provide needed home-care services to low- and middle-income older people, but issues of aging have not fared as well in many state and local government health and human services bureaucracies (Spore and Atchley 1990).

Federal legislation is increasingly tied to budget bills. Congress must pass a budget authorization bill each year to appropriate the money needed to run the U.S. government and all of its programs. In recent years, the annual budget resolution legislation resembles a jerry-built spaceship. The large government departmental budgets and their annotations make up the main body of the spaceship, but all sorts of other legislation— new programs, program changes, new funding, and funding changes—are appended to this main body. Attaching legislation to budget resolution bills is attractive because budget resolutions are "fast-tracked" through Congress. Fast tracking is justified for budget bills because they must be passed to allow continuity of government operations. But fast-tracking results in a minimum of opportunity for debate or citizen input. By the time the public becomes aware of a proposed change, it is often already law.

Congressional leaders claim that this process is necessary to get anything accomplished, but the critical theory perspective would lead us to ask if these claims, which are used to shut off debate and are thereby oppressive, are valid or simply a way of disen-franchising some constituencies. Obviously, the fast-track legislative process favors the highly selected number of interest groups that are invited to offer input when legislative amendments to budget bills are being written. It disadvantages those interest groups, including the "gray lobby," that are only able to take a reactive stance. How and why some interest groups are invited to provide proactive input to budget megabills and others are not would be an excellent topic for research.

Increasingly, *Social Security and Medicare are portrayed in political rhetoric and in the media as a significant and unsustainable drain on our economy.* These arguments are often based on very unlikely demographic and economic projections and an intentional and inaccurate portrayal of Social Security and Medicare as "welfare for the rich." If we were not spending so much money providing benefits to people who do not need them, the arguments usually go, then we would be able to eliminate the budget deficit, help poor children, and reinvigorate the American economy all at the same time. As we saw in Chapter 18, "The Economy," the reality is not so simple. Social Security is a centerpiece of retirement policy, and without retirement, we would have substantially greater unemployment problems and costs. Lack of Medicare would devastate middle-income families, not the rich. In addition, Social Security and Medicare benefits circulate immediately within the local and national economy and provide a vital base of disposable income in most communities. Finally, if social insurance benefits to an aging population are an inevitable drain on the economy, then how do we explain the indisputable fact that industrial societies with populations much older than that of the United States have been able to sustain social insurance retirement and health-care programs quite well?

Finally, political *pressures are building to means-test Social Security and Medicare*—to make economic need the major criterion for eligibility, rather than lifelong participation in the social insurance program. This issue revolves around the fact that we have two very separate systems for providing income security and health care to the older population. Social Security and Medicare benefits are earned by participating for many years in the economic life of the country. They are social insurance programs, and in the case of Social Security retirement benefits, pensions are tied to the lifelong amount of individual payroll earnings that have been taxed to fund the programs. Supplemental Security Income and Medicaid, on the other hand, use general tax revenues to provide income security and health care to various categories of people who are poor, but primarily to older poor people. Benefits for these programs are means-tested, which means that both income and assets must fall to a very low level to activate eligibility for benefits. To understand why social insurance is not "welfare for the rich," we need to step back and look at the nature of social insurance.

Social Insurance

Insurance was created to make people feel secure in the face of potential losses. It is a financial arrangement that pools resources and spreads the costs of insurance benefits across a large group of insured persons. Insurance generally protects people against loss for financial sums per capita that are much smaller than the actual costs associated with the insured loss. For example, workers' compensation insurance pools small premiums paid by employers on behalf of each worker to provide relatively large amounts of compensation to the small proportion of workers who are unemployed as a result of job-related injury or illness. And large pension programs pool contributions from participants or their employers to provide more secure pensions than usually could be provided simply through individual savings.

Individuals and families use private insurance to protect against: *property losses* that can be caused by events such as automobile accidents, fires, theft, floods, or earthquakes; *loss of income* due to death or disability of family income earners; or the high potential financial liabilities associated with the *costs of health care*. Individual or family protection is based completely on willingness and ability to pay insurance premiums and the willingness of insurance companies to insure specific individuals. Employers can also use private insurance to protect their employees through group policies, usually in the form of group life insurance, disability insurance, health insurance, and/or dental insurance. Private insurance can be very effective for those who can afford it and have access to it, but insurance companies often will not insure those at highest risk of loss, and many workers do not have access to group insurance.

Social insurance is group insurance operated by government. Social insurance extends the functions of private insurance to circumstances where either private companies will not provide insurance or many people cannot afford to purchase adequate amounts of private insurance (Marmor, Mashaw, and Harvey 1990). In addition to providing protection to individuals and families, social insurance protects society against the social disruptions and instability that can come when large proportions of people in society are vulnerable to financial disaster related to unemployment, age discrimination in employment, ill health or injury, disability, or death of a family economic provider. Social insurance is mainly "income insurance." In modern societies, most people depend on jobs for income and social insurance provides modest income security for those who are separated from this primary income source.

Social insurance works best when it spreads the costs of benefits over the entire workforce by requiring all workers to be covered (Ball 1978; Schulz 1995). Instead of premiums deposited in individual accounts, social insurance uses payroll or other taxes to establish a pool of pay-as-you-go financing. Social insurance tax payments of today's workers are used to pay benefits to those who are eligible today. This pay-as-you-go method has important advantages: it is flexible and can be adjusted to reflect changes in demand for benefits; it does not involve government in the management of large pools of investment capital; and it does not require governments to keep millions of individual investment accounts. Social insurance programs are used in all industrial societies to provide basic unemployment, survivors', disability, health and retirement benefits.

Thus, Social Security provides retirement, survivors', and disability insurance. Medicare provides health insurance, but only to those age 65 or older. People age 65 or older who have not worked at a covered job long enough to qualify for a Social Security pension are not covered by Medicare, unless they pay a substantial premium to qualify for both Hospital and Supplementary Medical Insurance benefits.

Especially in the case of retirement and health benefits, there is an important intergenerational component of financial protection that comes through participation in Social Security and Medicare. Social Security retirement pensions reduce the potential economic distress younger generations in families might face should the older generations in the family experience a collapse of their private pension or savings income in retirement. This does not happen often, but when it does, Social Security retirement benefits constitute an important safety net, even for upper-income people. Likewise, Medicare dramatically limits the potential financial demands on younger generations when older family members face substantial hospital or doctor bills. By spreading the uneven economic demands of supporting older generations in the family across the entire working population, everyone pays a relatively modest part of the cost and relatively few families are wiped out by the acute health care expenses of their older members.

The contrast between acute care and long-term care is instructive here. Because social insurance covers acute care costs, very few families are required to spend substantial financial resources out-of-pocket to guarantee that their older family members get the care they need. But long-term care is not covered by social insurance, and once the assets of older family members have been exhausted, thousands of families must finance long-term care to older family members out of the younger generation's current income or savings.

The general public is very aware of these important intergenerational social insurance benefits of Social Security and Medicare. As a result, public support for these programs has remained *very* high. For example, Cook and Barrett (1992) found that a majority of adults favored increasing Medicare and Social Security retirement benefits, and 97 percent favored either maintaining or increasing benefits. This was in stark contrast to the 33 percent who favored increased Aid to Families with Dependent Children. Day (1993) reported that young adults (18 to 35) were as strong as any other age group in their support for Social Security and Medicare (97 percent favored maintaining or increasing benefits). Cook and Barrett found that support for social insurance programs was based primarily on four key perceptions: self-interest, social insurance as an acceptable solution, program recipients as deserving, and the program as effective. Day found that the small minority who favored reductions in Medicare and Social Security shared three main characteristics: Republican party identification, extreme political conservatism, and a belief that government should substantially reduce benefits rather than increase taxes at all.

Now that we have some background on what social insurance is and why it is supported by a large majority of the general public, we can examine two examples of the political dynamics of Social Security and Medicare: the movement to undermine confidence in Social Security retirement and the repeal of the Medicare Catastrophic

Coverage Act. These illustrations show how interest group politics can often lead to legislative or government action that is quite contrary to the desires of a large majority of the general public.

The Politics of Social Security

The late 1970s was a period of economic distress, and Social Security was affected by the same economic forces that were causing concern everywhere. Rising inflation and unemployment created short-term financing problems for Social Security's pay-as-you-go method of financing. Inflation increased government obligations because Social Security benefits were indexed to increase automatically with increases in the cost of living. Increases in unemployment reduced the number of people paying payroll taxes, in turn reducing Social Security's tax revenues. Long-term financing problems for Social Security were anticipated in connection with the decline in birthrates after 1966 (which meant smaller cohorts of payroll tax payers in the future) and the retirement of the large baby boom cohorts (which would drastically increase the amount of funding needed to pay benefits). Solving these problems required increasing Social Security taxes, particularly the amount of Social Security taxes paid by middle- and upper-income workers.

The projected shortfall in revenues needed to meet Social Security obligations opened a door of opportunity for business and financial interests that opposed the incremental but nevertheless dramatic changes in Social Security during the 1965 to 1975 period. Business interests had never been strong supporters of Social Security, but accepted it as long as economic growth was high and Social Security taxes were low. In 1972, Congress created an "indexing" system that required Social Security benefits to increase automatically whenever inflation exceeded a small amount. This provision meant that business was being locked into possible future increases in payroll taxes regardless of the

economic climate. In addition, the employer's share of the payroll tax, which began at only 1 percent on payroll in 1937, was scheduled to reach 7.65 percent of payroll by 1990.

Financial interests were also alarmed by the growth in Social Security taxes. In 1937, employees paid only 1 percent Social Security taxes, and then only on their first $1,000 in earnings. By 1990, not only was the tax rate to increase to 7.65 percent, but the rate would apply to the first $42,000 of earnings. This was of great concern to insurance companies, thrift institutions, and investment companies who saw these Social Security tax increases as cutting into discretionary incomes that otherwise could have been spent to purchase insurance or make investments.

Thus, the goal of rolling back or at least containing Social Security tax increases took its place alongside relaxing enforcement of environmental legislation, repealing consumer protection laws, "reforming" labor laws (to reduce union influence), and creating legislation favoring capital formation as the major objectives of a resurgent cooperative effort by American business and financial interests to influence the political process (Myles 1986). This effort involved the Business Roundtable and the U.S. Chamber of Commerce, which raised funds to establish organizations dedicated to research and media campaigns in support of a conservative economic and political agenda.

When Social Security's short-term financial problems became known, the business press depicted Social Security as being on the verge of bankruptcy. Social Security was assailed as unfair to younger workers because its future could not be assured. The "burden" of Social Security taxes was said to be too great for the nation to bear. For example, in May 1980, *Forbes,* an influential business magazine, described Social Security as "the monster that's eating our future." Various media news services picked up these themes. Politicians used the "crisis" in Social Security to get media exposure.

The conservative rhetoric shook the conventional image of Social Security, especially among the young. Faith in government was already at an all-time low in the period just following the Watergate disclosures. As a result, the organized effort to call into question the soundness of the Social Security system significantly altered younger workers' perceptions about whether they could count on receiving Social Security when they reached retirement age. By 1985, 66 percent of younger workers thought that it was somewhat or very likely that Social Security would no longer be there when they reached retirement age (Yankelovich and White, Inc. 1985).

Yet the fears the conservative rhetoric raised were not supported by a full view of the facts. Unfortunately, the complex facts about Social Security's plans for handling various actuarial concerns were much more difficult to present in the media than the image that Social Security was "going broke," and as a result most of the public was unaware of Social Security's plans.

The short-term financing problems of Social Security were solved by slightly advancing the effective date of tax increases that were already planned and delaying a cost-of-living adjustment for current retirees by six months. Long-term financing issues were handled by planning to build a *$12 trillion reserve* in the Social Security retirement trust fund by the year 2030. These reserves could then be used to supplement Social Security tax revenues and enable the system to fund retirement pensions for baby boom retirees without having to increase taxes on the smaller baby bust cohorts who would still be in the labor force. A slight increase in age of eligibility for full Social Security retirement benefits was also planned. These changes were enacted in 1983.

In the debate over Social Security, the claims of "unsoundness" were eventually discredited (Munnell 1977, 1982; Ball 1978; U.S. General Accounting Office 1986). Eventually, those opposing Social Security were forced to admit that as long as the public, through Congress, supported the concept of Social Security, there was no way that the Social Security system could become bankrupt or "go broke." Even with the doubts raised by the debate over Social Security, in 1991 public support for the program remained very high, with 97 percent of adults believing that Social Security was an important government program that should be maintained (Day 1993).

Political support for Social Security has remained strong. In a large national sample of adults age 18 and older, Cook and Barrett (1992) found that 97 percent would maintain or increase Social Security, Medicare, and Supplementary Security Income. More than 80 percent were not dissatisfied with the Social Security taxes they were paying, and 90 percent said they would oppose benefit cuts in Social Security. More than 70 percent said they would be willing to write a letter to their congressional representative and pay higher taxes in order to preserve Social Security.

This widespread support rests in large part on the fact that all socioeconomic classes of Americans benefit from Social Security. It is a systematic way for younger generations to pool resources to provide for the often uneven costs of meeting the health-care and income needs of the retired members of their families *while at the same time* generating entitlement to similar benefits for themselves in their own retirements. This dual role of Social Security is unique to social insurance, and its capacity to meet the needs of several generations at once throughout the entire population cannot be matched by private insurance, private pensions, or private investments because each of these private vehicles of protection deals only with the needs of individuals and cannot address *intergenerational* needs. This capacity of Social Security to address both the present and the future and to meet the needs of multiple generations for income protection creates a broad-based political coalition.

That the opposition's attacks on Social Security lacked substance is less important than what the attacks themselves symbolized. In the increasingly conservative political and economic climate since 1980, organizations felt free to attack the legitimacy of Social Security using rhetoric that would have been unthinkable in 1965. The process of chipping away at the political legitimacy of programs for older Americans was well under way.

The fact that over 90 percent of Americans have family members who rely on Social Security for income and health care would seem to create a huge constituency that would probably be difficult to move against directly. What happened instead was a continuation of attempts to gradually reduce the role of Social Security in small decrements. After all, this is how it was built up in the first place—gradually and incrementally. A continuation of this process of trying to erode Social Security's legitimacy will test the political power of organizations such as AARP, for which protecting Social Security is a major priority.

Another important development, one that has affected legislators' perceptions of Social Security and Medicare, was the passage and repeal of the Medicare Catastrophic Coverage Act.

Lessons from the Medicare Catastrophic Coverage Act

In 1988, Congress passed the Medicare Catastrophic Coverage Act, the most sweeping expansion of Medicare since its initial passage in 1965. This new legislation set a cap ($1,370 in 1990) on the amount of out-of-pocket costs elders would have to pay annually for hospital and physicians' services covered by Medicare, provided coverage for prescription drugs for the first time (subject to copayments and deductibles), required states to buy into Medicare for their Medicaid older population, protected assets for spouses of Medicaid-funded nursing home residents, and made very modest increases in

availability of Medicare coverage for long-term care. These new benefits were to be financed by a controversial new approach that would tax middle- and upper-income older people at higher rates in order to fund benefits for low-income elders.

Reaction to the new legislation within the older population was negative, swift, and loud. The new provisions were seen as helping hospitals and doctors more than older people. Catastrophic hospital and physicians' bills affected a very small proportion of elders, particularly compared with the vast number left unprotected by the fact that the new legislation did not cover long-term care in any significant way. The coverage of prescription drugs was seen as a valuable benefit. However, perhaps the most problematic aspect of the new law was its financing mechanism. Instead of using a combination of general tax revenues and premiums paid by beneficiaries, which was the case with Medicare, Part B, the Catastrophic Coverage Act was financed completely through premiums and taxes paid only by middle- and upper-income beneficiaries. Premiums were acceptable, but not the provisions that redistributed income from middle- and upper-income elders to low-income elders. The latter provisions redefined the intergenerational contract that had provided the moral underpinnings of Social Security for more than fifty years (Holstein and Minkler 1991). But how did Congress come to enact a law that bypassed high-priority needs (long-term care) in favor of less needed benefits (catastrophic hospital and physicians' costs) and that ignored grassroots opinion about how Social Security benefits (of which Medicare is a part) should be financed?

The Reagan administration was characterized by a disregard for data. Situations were defined as true or false based on ideological grounds, not on well-documented information. Thus, as we saw earlier, it was possible for the Reagan administration to define away poverty among elders, to define as real an intergenerational conflict that precluded intergenerational financing of needed benefits for elders, to define catastrophic medical insurance as an important benefit, and to define long-term care as too costly to be covered by Medicare. In fact, a large number of elders were living on incomes not far above the poverty line and could ill afford to pay for health benefits for themselves, much less pay for other elders in poverty. The intergenerational conflict was certainly on the minds of policy makers, thanks to a well-organized media campaign financed by business interests seeking to discredit Social Security. However, high levels of intergenerational support for increased Social Security taxes were consistently found in poll after poll (Marmor, Mashaw, and Harvey 1990). And long-term care insurance as part of Medicare might well be expensive, but no careful cost estimates were actually developed.

The Catastrophic Coverage Act was a prime example of what can happen when the administration, Congress, and advocates for programs for elders forget that the view from Washington needs to be supplemented by broad-based dialogue with the people. Even AARP was caught napping. No one bothered to ask older people what they thought of the new legislation either in terms of its benefits or its financing strategy. All parties were genuinely surprised at the negative reaction.

A large number of middle-class elders saw the Medicare Catastrophic Coverage Act as a law that would unfairly increase their health-care costs and provide little in the way of new benefits. At a time when tax cuts for everyone were being promised, only middle- and upper-income elders were being asked to pay new taxes. Older people who paid over $150 in federal income taxes in 1989 were to be assessed a "surcharge" (new tax) of $22.50 for each $150 of income tax up to a maximum payment of $800. The "surcharge" rate was to increase to $42 per $150 in federal tax and the maximum new tax was to increase to $1,050 by 1993 (U.S. Senate Special

Committee on Aging 1989b). This regressive tax would have hit hardest elders with incomes between $30,000 and $50,000. Holstein and Minkler (1991) reported that these new taxes on middle-income elders ($50,000 annual income) would have resulted in a marginal federal income tax rate of 42 percent, whereas for upper-income elders ($208,000 or higher annual income) the combination of limited new taxes (the Medicare surcharge) and significant tax cuts would result in a marginal tax rate of only 28 percent. This was one aspect of perceived unfairness.

A second aspect of perceived unfairness was asking mainly middle-income elders to finance benefits for low-income elders (Cox 1993). One could argue that the children and grandchildren of low-income beneficiaries would benefit greatly from the Medicare Catastrophic Coverage Act because they would be relieved of the responsibility for paying their parents' catastrophic medical bills, yet the younger generations were not being asked to pay. Instead, the burden of financing benefits to the poor was to be placed entirely outside the intergenerational context. By what moral principle could middle-income elders be expected to pay for their low-income age peers? The intergenerational moral principle is consistent with our concepts of moral justice, but an *intra*generational redistribution of resources has no comparable moral underpinnings (Holstein and Minkler 1991).

A third aspect of perceived unfairness concerned the role of Medicare gap-filling insurance. Most of the people who would be paying the surcharge already had private insurance to cover the copayments, deductibles, and costs beyond what Medicare covered. In this sense, the Medicare Catastrophic Coverage Act provided benefits they did not need. And because the new law still left gaps in coverage, middle-income Medicare beneficiaries would be required to pay higher Medicare premiums and a tax surcharge to fund benefits for low-income elders, and they would still need to purchase a Medicare gap filler. To make matters worse, insurance companies were telling people that the cost of Medicare gap fillers would be going up even though the insurance companies' liabilities were going down.

The concerns elders had about the benefits and fairness of the Medicare Catastrophic Coverage Act seem on the whole to have been legitimate. These concerns were communicated directly to Congress, and in 1989, most aspects of the new law were repealed. This could be taken as a victory for elders, and that is certainly true in terms of protecting middle-class elders from having to finance programs for low-income elders.

However, the reaction to repeal was not generally based on a reasoned assessment. Instead of understanding that the administration and Congress were being appropriately slapped on the wrist for enacting an inadequate piece of legislation in the absence of feedback from the electorate, the repeal was portrayed in the media as the rejection of a good program because elders were unwilling to pay for programs that benefit them. Absent from the media rhetoric was the issue of the unfairness of the funding mechanism and the marginality of the benefits in comparison with other needs of the older population, especially needs for long-term care. Once again, we see a set of complex issues that seem to be beyond the capacity of the mass media, especially television, to portray issues in enough depth that the public can come to an informed conclusion. The inaccurate image the public was left with was of an older population with an insatiable appetite for benefits at other people's expense, despite the fact that elders have never objected in large numbers to paying premiums for other Medicare benefits.

Government Response to Issues Concerning Aging

Government response to issues affecting aging and older people is based on, among other things, perceived needs of these population categories, priority placed on these needs, and definitions of government responsibility for meeting them. As we saw earlier, a large majority of the American public believes that income security and health care for the older population are legitimate and high-priority concerns of the federal government.

In 1980, Ronald Reagan was elected president of the United States mainly because he was seen as standing for policies that would restore economic growth. Reagan subscribed to a controversial supply-side economic philosophy, which held that increasing the capacity of upper-income people to invest was the way to restore economic growth. He also held a very restrictive view of the federal government's responsibility in areas such as environmental protection, education, and help to the disadvantaged.

The Reagan administration, which took office in 1981, moved very quickly to cut taxes to upper-income Americans and sharply increased spending for defense, resulting in the largest federal budget deficit in history. This huge deficit was then used to defend massive budget cuts, particularly in "social programs"—mainly defined as social insurance programs and federal assistance to disadvantaged people. All Americans were presumably being asked to make sacrifices in the name of economic recovery, but it was the disadvantaged who made the bulk of the actual sacrifices. For example, from fiscal years 1982 to 1985, expenditures for Social Security were reduced by $24.1 billion, the largest single reduction of any federal human services program. The third largest cut—$13.2 billion—was in Medicare. The Congressional Budget Office (1983) concluded that these changes mainly affected households with less than $10,000 annual income.

The Reagan administration defended its actions by defining assistance to the disadvantaged as inappropriate for federal attention, despite consistent polls reflecting that public opinion was to the contrary. Instead, state and local governments and private philanthropy were exhorted to assume the costs of providing such assistance. Families were also expected to "do their duty" to their disadvantaged members.

The assumptions of fact that officials of the Reagan administration made and used to defend its policies were seriously inaccurate in many ways. The administration's position was that the welfare of the older population was being provided for; that 25 percent of the federal budget was going to support them; and that families, communities, and private philanthropy could assume most of the costs of caring for those in need. However, in 1980 there were in fact 15 percent of older people still living in poverty and another 15 percent near that level. Of the 25 percent of total federal expenditures earmarked for programs benefiting older people, only 2.4 percent of the general revenue budget was allocated for public assistance to older people.

Although state, local, and philanthropic involvement and commitment are very important in an overall effort to meet the needs of older people in our society, the federal government's role remains the most important and cannot be abandoned without severe consequences for older people for two reasons. First, a national commitment has important symbolic value. It shows that we are willing to put our national money where a very large majority of our people want it to be put. It indicates through action our national value of caring about older people. Second, the federal government has vastly greater power and resources than state or local governments. For example, in 1992 the federal government had revenues of $1.3 trillion, all state governments combined collected $744 billion, and combined local government revenues were $648 billion. Private philanthropic organizations—United Way, churches, foundations, corporations, and so on—raised $124 billion, most of it earmarked for church programs. Only $5.1 billion of private philanthropic money was earmarked for public assistance programs (U.S. Bureau of the Census 1994). It should be clear that state and local governments and "the private sector" have relatively few resources compared with the national government. In addition, state and local governments must fund activities that the federal government plays little or no role in financing, such as primary and secondary education, transportation, fire and police protection, and local courts.

The assumption that families can do more than they are currently doing was also quite misinformed. More than 85 percent of the in-home care being provided to older people was already coming from family, and there was only so much more that family could be expected to do. In terms of financial aid, those older people whose families could afford to support them financially were the older people most unlikely to need such support, while those most in need were quite likely either to be without family or to have families who were also poor. Thus, many of the assumptions the Reagan administration used to justify program cutbacks were erroneous, and as a result the policies enacted were also misguided and brought about a substantial decline in the access of older Americans to adequate housing, health care, and income (Oriol 1983).

Fortunately, Congress heard the public hue and cry over cutbacks in Social Security, Medicare, and other programs for older people, and blunted the Reagan administration's attempt to lower the place of programs for older people in our national priority structure. However, these efforts were renewed in 1995, as a Republican-controlled Congress proposed severe cuts in Medicare ($270 billion) and Medicaid ($180 billion). Substantial operational changes were also proposed for both programs, particularly the use of "managed care" (mainly health maintenance organizations and preferred provider organizations) to control costs.

The manner in which support for these proposed changes in Medicare and Medicaid was maintained against severe opposition from the "gray lobby" and a host of other groups is an instructive lesson in how interest group theory works. First, the proposed changes to Medicare and Medicaid were attached to the annual appropriations bill, and only one day of hearings was scheduled by the House of Representatives to allow review and debate of the massive changes in the way services were to be delivered under Medicare and Medicaid and the potential effects of funding cuts on access to health care by older Americans and the poor. Second, because reimbursement to physicians and surgeons was slated to be cut more than other areas, the Republican House leadership made several extraordinary concessions to the American Medical Association to secure its support. The proposed legislation rescinded a federal law prohibiting doctors from referring Medicare or Medicaid patients to laboratories or therapy clinics in which they have a personal financial interest. The proposed bill also eliminated federal regulation of laboratories located in physicians' offices, thus allowing physicians to install and bill for expensive diagnostic equipment such as magnetic resonance imaging machines and expensive therapy devices such as kidney dialysis machines. It also placed a cap on the amount of awards in medical malpractice cases and eliminated a requirement that hospitals and doctors give discounts to public hospitals and community clinics that provided a large proportion of their services to low-income people. Thus, by giving the proposed bill to change Medicare its seal of approval, the American Medical Association was able to achieve in one fell swoop many legislative objectives it had been lobbying for many years to achieve.

Most health services researchers felt at the time that these concessions would provide an enormous opening for doctors to increase their Medicare billings and thus undercut the Medicare savings the 1995 legislation was designed to achieve. Instead of being legislation designed to control costs, the bill was more likely to simply shift more of the costs for ordinary doctors' services to consumers (mostly older people) and allow doctors to maintain high levels of income from the Medicare and Medicaid programs by providing expensive diagnostic and therapy services in their offices or associated facilities.

This episode illustrates very clearly the importance of politics in determining the life chances of older Americans. As the U.S. population continues to age, we can expect continuing discussion and debate over the position of various government programs in our national, state, and local political priorities.

Making Policy

It is well known that the U.S. government funds Social Security and Medicare primarily for older people. The goals of these programs are to provide income security and health-care financing for older people. Older people have other concerns, many of which governments attempt to address. Consider the following list:

Rehabilitation	Design (buildings, furniture, clothing, and so on)
Housing	
Transportation	Inflation
Taxes	Homemaker services
Recreation	
Mental health	Protective services
Consumer protection	Alcoholism
Referral services	Blindness and deafness
Independence	Education
Poverty	Dental care
Activity	Employment
Long-term care	Research and development
Nutrition	
Long-term care financing	Protection from age discrimination

*Representative Claude Pepper, age 82, holding a press conference on
Social Security.*
Photograph from UPI/Bettmann News Photos

How adequately can government respond to
these concerns? To satisfy them requires,
first, a commitment to spend government
money on older people and, second, some
consensus on priorities. The first requirement brings us to perhaps the most fundamental problem that politicians face:
choosing the things that are possible from
among the many things that people want,
given that government can almost never command enough resources to satisfy everybody.
As Cottrell pointed out:

They [the politicians] know that to gain anything
people must sacrifice something else they might
have had. They try to judge the worth of an objective in terms of what the voter is willing to sacrifice to achieve it. If they judge correctly, they can
continue to make policy; but if they err too
greatly, policy will be made by others who have
organized a more effective coalition in support of
[other] policies.

(1966,96)

To build the power necessary to design, develop, and implement governmental programs
for older people, politicians must find out the
areas in which older people's interests coincide enough with those of other groups to
form an effective coalition. Older people do
not by themselves have the necessary power.
Politicians must also stay out of areas, however strong the need, that most people consider none of the government's business.

There are some sizable obstacles to governmental programs intended only for older

people. The first obstacle is *conflict of interest*. Meeting the needs of older people often conflicts with meeting the needs of the young, such as choosing between aid to schools and aid to nursing homes. *Vested interests* also play their part. The role of the American Medical Association in the enacted changes in Medicare in 1995 is a good illustration.

The inertia of bureaucratic organizations also inhibits programs for older people. For instance, when first proposed, Social Security legislation was hampered by the lack of any recognized bureaucratic structure in the government around which those interested in supporting income programs for older people could rally. At the time, both the Townsend Movement and Eliminate Poverty in California (under Upton Sinclair) were going strong, but there was no Washington office through which their support could be funneled. At the time Medicare was being considered, there was an Office of Aging under the Department of Health, Education and Welfare (HEW) and the Special Committee on Aging in the U.S. Senate, which made gaining support for programs for older people easier because there was an office staff paid by the government to keep tabs on the interests of the aging.

The U.S. Senate Special Committee on Aging, created in 1961 as a temporary study group (as opposed to a committee with the authority to report legislation to the full Senate), has repeatedly survived its one-year mandate and become a de facto, twenty-one-member standing committee of the Senate. In an analysis of the committee's role, Vinyard (1972) concluded that it served several important functions: (1) as a watchdog, seeing that the interests of older Americans do not get lost or ignored in broad-ranging government programs; (2) as a legislative catalyst, supporting legislation and lobbying for older people within Congress; (3) as a rallying point for senators wishing to demonstrate to the public their interest and support; (4) as a frequent legislative ally of old-age interest groups; and (5) as a symbol of Congress' concern for the interests of older Americans. In addition, the committee compiled and summarized data on needs of the older population and statistics on the performance of programs serving elders.

Aside from the issue of how programs for older people can win political support, there is the equally important issue of how the programs should be organized and coordinated. The federal government can do a reasonable job of providing direct support to older people through agencies such as Social Security. However, solutions to a great many of the problems older people face must be based on personal considerations and the *local* situation. In cases where personalized service is necessary, only local agencies are capable of doing the job.

Cottrell outlined several important questions concerning the nature of governmental programs:

1. What should be the function of a central agency on aging and what are the relationships between it and other departments and agencies within the federal government?

2. What should be the functions of official state agencies on aging? Where should they be located, and what should be their relationships with other units of state government?

3 What type of agency is needed at the community level to serve as a focal point for broad action in aging? What should its functions be? From where should it derive its authority and financial support?

4. What is the most desirable method of integrating or interrelating the activities of overall agencies in aging at federal, state, and community levels?

5. How can government at each level best maintain working relationships with voluntary organizations and with the private sector of the economy?

6. What should be the division of responsibility among public and private agencies and organizations?

7. Should government take initiative in stimulating roles in aging on the part of organizations? If so, what types of organizations? (1971,1–2)

These questions are as relevant today as they were when they were written more than twenty-five years ago.

Another policy issue concerns the variety of different chronological ages used in the law to differentiate older people from the general public. Cain (1974) reported that legal definitions of "old" began as early as 45 and as late as 72. This sort of nonuniformity creates problems of confused eligibility for government services for older people.

The existing intergovernmental division of labor with respect to programs for older people has resulted from variations in the ability of various agencies to muster the political power needed to enact legislation. The record at the national level is familiar: Social Security, Medicare, and agencies on aging in various federal departments attest to a small but sometimes effective political power base. At the state and local levels, however, the picture has changed greatly since 1973. Prior to that time, most state programs for older people were very limited, and the average state agency for aging had a staff of three people. Various governors' conferences on aging, forums, and so on brought home to state politicians that the aging represent an interest group to be served. However, state involvement in programs for older people was hindered, primarily by the fact that the four major programs—Social Security, Medicare, Old-Age Assistance, and Federal Housing—all bypassed state government on their way to the people. At the local level, the only governmental agency serving older people was usually the county human services department. Given the negative connotations the public and older people themselves attached to welfare programs, this situation was highly unsatisfactory. Only in very large cities had the concentration of older people been great enough to produce a separate local office for their needs. Multipurpose senior centers were also sometimes an effective community agency translating public programs for older people into action.

Based on 1971 White House Conference on Aging recommendations, the 1973 Comprehensive Service Amendments to the Older Americans Act required each state to set up a network of Area Agencies on Aging in order to receive federal funds. The nutrition program funneled large amounts of federal funds through state agencies. In addition, a great deal of the responsibility for planning and administering training and demonstration programs was delegated to the states. This reorganization had sweeping effects. For the first time, State Units on Aging had administrative control over a sizable budget. Accordingly, state agencies have grown considerably in size and power since 1973. Also, the need for Area Agencies on Aging required that local areas designate a single agency to serve as local coordinator for both state and federal programs, thus improving service coordination at the local level. This reorganization created effective advocates for older people in many areas of the country where previously there had been none. In Chapter 15, we looked at some of the services that resulted from this system.

Summary

Politics is the route to political power, and political power means control over the machinery of government. We know that aging and older people are concerned with politics and government because they participate in politics through voting, working in political organizations, and holding office. Aging and older people are also the object of governmental programs.

Older people seek to remain informed about and interested in politics, particularly if they are well educated or involved in local affairs. Older people vote in about the same proportions as when they were middle-aged. Each age cohort apparently develops its own level of participation, which stays relatively stable throughout the life course.

Older people have stronger party identification than the young only if they have been associated with a party over many years. It is number of years of affiliation, not age itself, that produces a strong party identification. In politics, youth apparently still bows to experience, and experience rather than age is the crucial variable. Older people have equal opportunities in politics, but only if they have been lifelong participants; there is little room for the retired grocer who suddenly decides to get into politics.

In terms of political power, early interest groups of older people suffered from lack of leadership, regional segregation, heterogeneity of interests among older people, and lack of funds. These same factors have prevented older people from developing into a subculture or a genuine minority group. Action in behalf of older people has almost always depended on the political support of organizations not based on age, such as unions or political parties. The role of modern aging interest groups such as AARP is to make the need for action highly visible and to lobby for the interests of aging people within existing programs. In 1980, based on inaccurate assumptions about the actual situation of the older population, substantial cuts were made in Social Security, Medicare, and Medicaid. Since 1990, older Americans have seen these programs and the Older Americans Act be targeted for even deeper cuts.

Governmental programs for older people assign responsibility for meeting some of their needs to various levels of government. The federal government operates programs in income and health maintenance. But many other needs require involvement on the part of state and local governments. The 1973 reorganization of the Older Americans Act, which gave states and local areas authority, funds, and responsibility, was a large step toward creating local advocates for older Americans.

The biggest problem in creating governmental programs for older people is deciding how these programs should be organized and coordinated, particularly in terms of intergovernmental relations. Older people have benefited a great deal from the fact that they were the first target for large-scale social programs in the United States. It is unlikely that SSI, Medicare, or many of the other supports to older Americans would be there if such supports had to be provided to everyone. American politics is still very sensitive to "government intervention" on too large a scale.

In recent years, the legitimacy of federal programs for elders has been called into question. However, public support for programs such as Social Security and Medicare remains strong. In terms of organization, the best system for governmental programs seems to be the combination of a strong federal agency, capable of gathering support for national fund-raising legislation, and strong State Units on Aging to administer federally funded programs through local agencies.

However, as we have seen, interest group politics did not necessarily result in treatment for issues related to an aging population that is in line with the priorities of the public. Political economy theory pointed to a class component in the attacks on Social Security and Medicare, with a small segment of the affluent population financing propaganda campaigns designed to discredit Social Security and Medicare based on what that small minority saw as its own economic self-interest. Critical theory questioned the legitimacy of various methods being employed to stifle political debate on issues of aging and questioned the accuracy of the claims that were used to justify the need to cut programs that provide retirement income and health-care financing for older Americans.

20 ▮ Epilogue: Aging and the Future

Discussion about what may happen in the future must be approached with some humility, for events are bound to prove us wrong. But intelligent decision making demands that we have a sense of direction. To get our bearings, we need ideas about both the future we want and the future that is possible. We also need to make educated guesses about the future course of events. We begin this chapter with a look at some long-range directions social change may take because these macrochanges address fundamental shifts in the social environment within which aging and adaptation take place. We then look at some demographic clues to the future, for demographic projections are some of the best we have. Next, we take a speculative look at changes that may occur in the nature of individual aging. In doing this, we use the familiar framework of physical aging, psychological aging, and social aging. We then consider society's response to the aging and the aged in the future, including both general stances toward aging and older people and major policy issues such as retirement income and health care for the older population.

Directions of Social Change

The history of aging in America has been greatly affected by the visions we have had of our future. As we shifted our vision away from the traditional, agrarian society with its relatively rigid social structure toward a mercantile society with its fluid structure of entrepreneurs and later to the industrial society with its large-scale bureaucracies and concentration of wealth, our views of socialization, authority, and social process changed along with them. Thus, to comprehend our future, we need to have a sense of the central focus of economic activity and the scale of human organizations. In addition, we need a sense of how economic and political factors will influence one another. We must also have some vision of factors such as the environment that will both set boundaries and provide direction. These are all very general topics, and they cannot be addressed with a sure hand, but we must make some attempt to deal with them because they will have a profound effect on the resources of both aging individuals and the societies in which they live.

In the early 1990s, three important contending ideological forces were especially prominent: a preoccupation with economic vitality, concerns about the environment, and a politics of diversity. Each of these ideological forces has major implications for what work we do, the types of organizations within which we do it, the relation of economics and politics, our definitions of progress, and our view of the political processes through which we can pursue our goals together.

To a great extent, economic vitality ideology seeks to recapture America's greatness by returning to the economic ideas and forms that initially made the nation great. Environmentalism looks to the relation of humans to the planet to find directions that will lead us into a harmonious match of the needs of people and their organizations with the needs of the physical planet on which we live. It seeks to avoid planetary catastrophe and at the same time foster more humanistic living and working conditions for the world's populations. The politics of diversity seeks to accommodate a wide variety of sometimes conflicting ethnic value systems within the political process. Genuine respect for ethnic diversity requires that we develop better means of resolving intergroup conflict than we have so far been able to create. All these ideological approaches are contending with the same macrosocial trends: emergence of the postindustrial society, growing interdependence of nations in the world economy, greater centralization of political and economic power, increased emphasis on ethnic identity, increasing decentralization of economic production, growing political alienation within the general public, and the depleting of the fossil fuels that drive today's technology.

Economic vitality ideology argues that tried-and-true methods of the past will work in addressing these trends, whereas environmentalism argues that the methods of the past are based on exploitation of our planet's physical resources at a rate that cannot be sustained without destroying the world we live in. Environmentalism instead argues for the development of revolutionary new ways of thinking about how we deal with our planet and with each other (Capra 1982). Ethnic diversity advocates insist that solutions respect the multiplicity of values found among indigenous peoples and avoid oppression of them in the name of either economic progress or environmental restraint.

Bell (1973) described the emergence of the postindustrial society as a shift from an economy dominated by industrial manufacturing to one dominated by the performance of services. The postindustrial economy involves the use of knowledge to provide ser-

vice. Thus, in the postindustrial economy, human knowledge and skill are the commodities to be bought and sold, whereas goods are the major commodities of the industrial society. Technology serves production of goods in the industrial society, whereas technology serves development of knowledge and the provision of service in the postindustrial society. The services provided in the postindustrial society should not be viewed in a narrow sense. In the industrial society, service is a blue-collar concept, but in the postindustrial society, service increasingly involves the most talented and skilled members of the labor force (Reich 1991). For example, whereas the United States is not the premier producer of many types of consumer goods (television and stereos are good examples), it leads the world in the development and management of information systems. Computer technology *facilitates* this task, but the major medium of exchange is know-how.

Hage and Powers (1992) argued that the shift in emphasis from an economy centered around large, bureaucratic organizations involved in mass production to one centered around symbolic analysis and custom production performed by relatively small-scale work teams requires a new way of thinking about what the economy is trying to produce and how individuals fit into jobs. In the postindustrial economy, according to Hage and Powers, consumers value convenience, customization, variety, quality, and reasonable cost. These requirements are most easily met by small-scale, high-tech organizations.

In the industrial society, job descriptions tend to be standardized and impersonal. Many people do routine jobs designed to fit bureaucratic or assembly-line principles. In the postindustrial society, people tend to be employed in very complex roles as part of teams that last only as long as it takes to solve the specific problem the team was created to solve or produce the custom products that are in demand in the marketplace. When each cycle of problem solving or production is

This team developed Pictionary, a best selling board game.
Photograph by AP/Wide World Photos

over, the team members move on to new teams. In this new economic environment, work tasks are not defined in terms of conformity to preexisting and agreed-upon standards but in terms of effective information gathering, problem solving, creative idea generating, and flexible response to changing circumstances. Work roles are flexible and need to be periodically renegotiated, not just with one's supervisor but with coworkers as well. Work in teams also requires greater sensitivity to emotional dimensions of work life compared with the impersonal philosophy of bureaucratic organizations.

At the other extreme, Ritzer (1995) argued that most Americans are willing to sacrifice excellence in order to achieve predictability. His archetype is the fast-food restaurant. We eat at fast-food chains not because their food is excellent but because it is predictable, acceptable, moderately priced, and quick. To achieve this predictability, McDonald's and other fast-food chains have applied job standardization principles with a vengeance. Jobs are broken down and controlled in minute detail, the workers' tasks are mind-numbingly repetitive, and the pace of work is rapid. The alienation workers feel from the job is often extreme, but because of the high degree of job standardization, employee turnover is not seen as an important problem.

These two very different visions of trends in the job market are both accurate. The portrait painted by Hage and Powers (1992) fits well with what has happened to symbolic workers in what Reich (1991) called "the fortunate fifth," that portion of the workforce that participates in the global economy and that reaps most of the profits. At the same time, routine production work has been declining as a proportion of American jobs and has been replaced by low-paying service jobs rationalized along the lines described by Ritzer (1995). As Reich pointed out, a major strain in American society stems from a growing feeling on the part of the "fortunate fifth" that their life chances no longer depend on what happens to the rest of the population. Their loyalties are to other members of the symbolic analyst class of workers rather than to the nation as a whole. The danger in this myopic vision, of course, is that it promotes the creation of an increasingly large population with nothing to lose from the disintegration of the civil society. This is, of course, where the ethnic diversity agenda becomes relevant. To maintain the civil society requires that we find ways to integrate all ethnic groups and social classes into the political and economic life of the nation. Mostly, this means that access to education and employment must be at least adequate for minority, female, and working-class residents. Since 1960, efforts to bring this about have not been particularly successful. The challenge in the postindustrial era is to find effective ways to bring a larger proportion of our population into the world economy doing jobs that will economically support a middle-class lifestyle. This will be no mean feat.

Research organizations produce knowledge, which is the raw material of the "information society" (Naisbitt 1982). The growth of research enterprise in the United States has been dramatic, and its pace is still accelerating. Federal funding of research jumped from $15 billion in 1970 to over $68 billion in 1993. The "research university" has emerged as an organizational form dedicated as much to the generation of new knowledge as to the education of new generations. Private enterprise also makes major investments in research ($84 billion in 1993). Thus, a growing part of the service economy in the postindustrial society consists of the research efforts of the most highly skilled and highly educated people in our society.

The aging of society also influences the postindustrial economy by increasing the demand for a wide variety of services to people. For example, Relman (1980) contended that health care—of which older people are the major consumers—is the focus of an emerging "medical-industrial complex" involving for-profit hospitals, nursing homes, home care, medical equipment producers, and laboratory and diagnostic facilities. Long-term care is probably the fastest-growing field within the human services, and in the postindustrial society we can expect that a goodly share of both private and public economic resources will be used to support such services. But it remains to be seen if compensation of workers in that economic sector will rise to a real living wage. Such is not the case for many home-care workers at present.

Multinational corporations have been able to engender international cooperation and interdependence much more effectively than has the United Nations. A wide variety of economic functions ranging from investment banking and management to production and assembly are now being coordinated and performed in an international context. Whether this is good or not depends largely on the direction the global economy takes and on one's definition of progress. However, there can be no doubt that the global economy is a reality.

Although the industrial societies of the world have a head start on the others in dealing with aging populations, the entire world's population of older people is increasing rapidly. Part of the United States' contribution to the global economy of the future may well be made in areas of gerontology, such as compensation for physical losses connected with aging or the effective organization and provision of long-term care.

Big government and big business are getting bigger, and with each increment in resources these huge organizations command, their power and influence increase, bringing increased centralization of decision making. Several U.S. presidents have promised to control the growth of the federal government, but none has succeeded. The breakup of AT&T's virtual monopoly of the telecommunications industry stands in stark contrast to the fever pitch of corporate mergers and consolidation of financial power that occurred in the late 1980s. Adapting to major shifts in world economics may well require centralized decision making, but an unintended consequence may well be the political and economic alienation of a public largely excluded from the decision-making process (Reich 1991).

On a very different level, the use of fossil fuels, environmental pollution, and depletion of resources have created a set of issues whose solutions will greatly influence what life in the future will be like. On the one hand, many of the comforts and luxuries of modern society depend on the use of fossil fuels. Fossil fuels are used to produce both consumer goods and the electricity or fuel needed to enjoy them. On the other hand, fossil fuel consumption has caused widespread pollution of the environment. Further, they are a limited resource and cannot support today's urban lifestyle indefinitely. Thus, major issues for the future include finding ways to reduce and clean up environmental pollution and finding energy sources to replace fossil fuels.

The future of energy use will have a profound effect on the everyday life of many older people through the levels and types of technology our energy systems will support. For example, the degree to which technology can be developed to compensate for physical or psychological functional losses depends on the *existence* of energy to power that technology and the *cost* of that energy.

Beck (1992) pointed out that whereas industrial societies were preoccupied with the production of material progress with little thought about environmental risks, postindustrial societies are much more alert to environmental risks associated with industrial production and may be willing to forgo some amount of material comfort in order to achieve environmental safety. An appropriate global objective might thus be to identify and maintain a balance between economic vitality and environmental goals. This may come to pass at some point, but today's governments the world over seem to throw environmental caution to the wind whenever economic growth shows signs of decline.

Demographic Clues to the Future

Demographic indicators usually change slowly, both in magnitude and direction. For instance, birthrates seldom double overnight, and deathrates usually drop slowly. In addition, the population that will become old over the next fifty years has already been born. We know the current size of our population, and we can predict relatively well how many will survive to each age. Demographic projections can thus give us some of our best clues to the future.

As we saw in Chapter 2, population aging has only just begun in the United States. Over the next several decades, the size of the older population will increase dramatically. By the year 2050, there will probably be about 68.5 million Americans 65 and older, compared with 31.1 million in 1990. The proportion of older people in the population is expected to increase from 12.5 percent in 1990 to 22.9 percent in 2050. More than 50 percent of the population growth in the United States over the next fifty-five years is expected to occur in the older population as increasingly more people survive to fill the oldest age categories, and the proportional increases will be greatest in the 95-and-over age category.

Demographic analysis is useful in projecting elders' needs and resources in a number of areas, including health care, housing, and availability of kin. For example, the need for long-term care can be expected to increase dramatically. In 1990, there were about 500,000 people 85 and over living in long-term care facilities. At the 1990 rates of utilization—which may go up as a result of the aging of even the old-old population—we can expect nearly one million people 85 and over to need long-term care spaces by the year 2000, and by 2020 that number could be 1.5 million, and by 2050 it could be over 3 million! Thus, we could need about six times as many long-term care spaces for people 85 and over in 2050 as we had in 1990. Of course, these facilities may not be like the nursing homes of today. They may look more like independent apartments with service facilities attached and residents may have more independent decision making and receive a less standardized array of services compared with today's long-term care population.

In this case, the 85-and-over population is a better illustration than the 65-and-over population. Changes in the availability of in-home services could dramatically change the proportion of people under 85 needing long-term care space since they are much more likely to need only minimal assistance. At age 85 and over, however, care needs increase dramatically and are not as likely to be amenable to in-home services. Projections of health-care needs not only involve projecting population sizes but utilization rates and availability of alternative forms of health care.

Projecting elders' housing needs also is quite complex and begins with demographic projections. Soldo (1981) used current trends in housing for the older population to anticipate their future needs. She concluded that

the demand for independent types of living arrangements will increase in the near future. Even if the same age-specific patterns of living arrangements are maintained, by the turn of the century, there will be 8 million older persons . . . living alone—a net increase of 3 million since 1970.

(1981,505)

Soldo went on to say that increased aging of the older population might tend to reduce the proportion living alone but that this would probably be offset by improvements in the financial status of elders as a result of the mat-uration of private pension and individual retirement plans, especially among women. However, Soldo cautioned, this projected increase in housing units headed by older people living alone would occur only if there were a sufficient supply of moderately priced living units suitable for older people. Given the political climate of the mid-1990s, which saw the federal government backing away from all forms of government housing assistance, this is unlikely. This is another good example of how demographic information must be related to social trends in order to be useful in planning.

Since, as we have seen, caregiving by family members is currently a major source of in-home services for older people, the availability of kin in the future is an important issue. Hammel, Wachter, and McDaniel (1981) used population simulations to look at the future availability to elders of spouses, children, grandchildren, and siblings. They found that compared with 1950, the older people of the year 2000 will be more likely to have living kin: spouses, siblings, and larger numbers of children and grandchildren. But after the year 2000, the availability of kin can be expected to decline as a result of the lowered fertility that began in the 1960s. After the year 2000, siblings will be more prevalent among older people, while children and grandchildren will become less prevalent, which could have important ramifications for the capacity of families to care for their very old members. Note how several cohorts were used to make demographic projections of kin availability in this example.

Wolf (1994) pointed out that sheer numbers of kin are not the only concern in projecting potential kin care. Proximity of kin, especially coresidence, can have a strong influence on the capacity of family to provide care. Little is known about the influence of marital dissolutions in the parents' generation on coresidence and kin care later in life. In addition, Soldo and Freedman (1994) suggested that the future availability of kin care needs to be projected in a context that includes larger informal networks as well as the availability of formal service providers in the local area.

Demographic projections of a rapidly increasing older population often raise anxieties about whether the economics of caring for the older population will strain our economy beyond the breaking point. Easterlin (1991) pointed out that the aging of the population does not occur in a vacuum, it occurs alongside substantial reductions in the proportion of children in the population. As a result, the increase in elders using resources will be more than offset by the decline in number of children using resources. The general population will be expending significantly fewer private resources on the care and education of children, which should make available increased resources to meet the needs of elders. Thus, the issue is not likely to be whether the resources are there to provide for elders' needs but rather whether the population is willing to spend them for that purpose.

The Future of Physical Aging

The future of physical aging will be influenced by the impact of biomedical research, trends in morbidity and mortality, and the impact of the wellness movement.

The Impact of Biomedical Research

Butler (1995) identified a number of changes that are likely to result from biomedical research on aging. First, great strides are being made in identifying genetic markers of disease. Currently, it is possible to diagnose about two hundred diseases connected with single gene defects. This means that diagnosis, prevention, and treatment can begin well before obvious symptoms of disease appear. Butler expects that many illnesses can be avoided in the future by gene splicing, substituting normal genes to overcome the adverse

effects of defective ones. In addition, more complete mapping of the human genetic system will allow many more genetic markers to be identified by the year 2000.

Second, lower-risk and less expensive treatments will become more widely available. For example, instead of the enormously expensive coronary bypass surgery, which is a very common treatment for coronary artery disease now, much less expensive chemical "uncloggers" may be used in the future to clear arteries of fatty deposits. Future drugs may use more naturally occurring compounds rather than synthetics, which could reduce the incidence of undesirable side effects. Laser surgery may greatly improve our capacity to remove cancerous tumors completely.

Third, use of microprocessors in connection with prostheses will allow for better compensation for physical disabilities. For example, artificial hearing is nearing reality. Voice-activated "smart" limbs will provide better compensating prostheses for amputees. Use of microprocessors along with voice activation will also allow the development of "smart house" environments in which even severely disabled people can be more in control of meeting their own needs.

Fourth, we will probably make considerable progress toward the diagnosis, prevention, and treatment of Alzheimer's disease and other dementias. If this happens, it will revolutionize how people view their prospects in old age. It could cause a major reduction in the need for institutional long-term care.

Fifth, we will make progress toward stimulating the immune, endocrine, and neurotransmission systems. These changes would greatly improve the quality of later life. Better immune system response would decrease susceptibility to both acute and chronic diseases; improved endocrine response would improve physiological control mechanisms such as temperature control; and better neurotransmission would delay declines in psychological functions such as memory and speed of information processing.

In addition to the trends Butler identified, the growing amount of research on sleep disorders is likely to lead to identification of factors related to the decline with age in effectiveness of sleep. Interventions may evolve that will, through improving the quality of sleep, markedly improve the older body's capacity to recover physically during sleep.

Changing Patterns of Death and Illness

As we saw in Chapter 2, life expectancy at the age of 65 is expected to increase. Over the next two decades, shifts in the life expectancy at the older ages are more likely to result from delaying death from various chronic diseases such as heart disease or cancer than from eliminating these causes of death (Olshansky 1985). Delaying deaths from the leading causes could result in gains in life expectancy at the oldest ages on the order of 3 to 5 years, rather than the 1.2 years currently being projected.

There is as yet little evidence that we have reached a point where the mortality gradient will become steeper. The mortality gradient is not being compressed into the eighth decade as Fries (1980) predicted but instead is simply being steadily pushed higher in essentially its same form (Rothenberg, Lentzner, and Parker 1991). In 1980, the steepest part of the mortality gradient for women ranged from 77 to 93. By 1990, the range spanned about the same number of years, but began at 81 and ended at 97. We do not know how much higher the midpoint of the mortality gradient can go. We do know that the number of people surviving past the age of 95 is increasing very, very rapidly.

But if we postpone death from chronic diseases, are we not likely to have a drastic increase in disability at the oldest ages? On the one hand, there is evidence that even now the disabling effects of chronic diseases such as heart disease are much less severe than was

true just ten years ago (Myers and Manton 1985). For example, better treatment is allowing millions of people to resume normal lives. Earlier, heart disease would have made these people invalids. Manton and Stallard (1994) pointed out that active life expectancy at the older ages is on the increase. Spirduso (1995) drew attention to a growing number of physically elite elders who are able to continue athletic competition well into advanced old age. On the other hand, Verbrugge (1984) analyzed health data from 1958 to 1981 and found that the annual average number of days of restricted activity from chronic conditions had increased substantially since 1970. Verbrugge suggested that people are perhaps more likely to adopt the sick role than in the past. The more we study disability, the more we realize that it is a complex and gradual process, especially for older people and for women. The weight of the evidence currently points to increasing prevalence of nondisabled elders even among those of very advanced age (Manton and Stallard 1994).

Impact of the Wellness Movement

Since the early 1970s, there has been a growing social movement that encourages people to look at wellness as *optimal* physical, mental, and spiritual well-being. The wellness movement extends the concept of health beyond the old-fashioned notion of good health as merely the absence of disease (Spengler 1985). Key concepts used in the wellness movement include self-responsibility, physical fitness, nutritional awareness, and stress management. Self-responsibility encourages the person to seek information and actively decide how best to pursue optimal health. Individuals are encouraged to make their own plans for optimal health (in consultation with appropriate health practitioners) rather than simply delegate the responsibility to various health-care providers. Physical fitness includes exercise tailored to the individual's needs and capacities that promotes muscular and cardiovascular endurance, strength,

The spread of wellness activities could reduce the prevalence of disability in later life.
Photograph by A. Ramey/Woodfin Camp & Associates

flexibility, balance, and coordination. Nutritional awareness encourages individuals to become more aware of the effects of diet on physical and mental health. Stress management involves learning to identify physical and mental symptoms of stress and taking steps to minimize the potential negative effects of stress on physical and mental health.

Like many social movements, the wellness movement has tended to claim benefits of adhering to its tenets without adequately putting those tenets to the test. However, the research results are encouraging in a couple of areas. Buskirk (1985) carefully surveyed the evidence on the relation between regular exercise and health, and concluded that a wide variety of physiological functions are improved by regular exercise. Although the health benefits of exercise tend to be greatest at the younger ages, physical conditioning produces measurable improvements in performance in healthy people of any age, and the functional alterations exercise produces tend to blunt the downward physical trends that are usually associated with aging. In the area of stress management, relaxation techniques and meditation both have been shown to produce sustained reductions in bodily symptoms of stress in older adults (Garrison 1978; Carrington et al. 1980). However, Minkler and Pasik (1986) called for caution because the effectiveness of many specific approaches to the wellness movement's general goals have yet to be systematically evaluated. Thus, it would appear that growth of the wellness movement could be a substantial positive force in the physical health of future cohorts of older people, but just how substantial we cannot yet confidently estimate.

Much of the information needed to promote wellness is common-sense behavior that is not a mystery. The key ingredient is to maintain motivation for changing behavior patterns that undermine health and developing behavior patterns that promote health. Haber (1994) presented a general model for *health promotion* in an aging population that involved the following elements:

■ Collaboration between older clients and health professionals. Most elders have chronic health conditions that require articulation of treatment with efforts at health promotion.

- Health assessment to identify areas in greatest need of attention.

- Health education approaches that acknowledge the health information needs and obstacles confronting aging adults.

- Social support, especially peer support, to help clients sustain their motivation to promote health.

- Behavior and psychological management techniques such as self-monitoring, development of healthy behavior agreements or contracts, and stress management help clients bring about and maintain behavior change.

- Community-based health education programs that concentrate the educational efforts of local health professionals.

- Advocacy to provide constant pressure for more focus on health promotion. Health-care resources will continue to flow first to care of disease unless strenuous efforts are made to reserve a proportion of health-care dollars for the promotion of good health as opposed to the treatment of poor health. (Adapted by the author from Haber 1994, 20–22)

Haber cited the growth of health maintenance organizations, with their built-in incentives for health promotion, as a positive sign that momentum may be building to recognize the importance of health promotion as an essential strategy for controlling health-care costs in an aging population. Schmidt (1994,5) also argued that "healthy aging is a concept whose time has come." In Schmidt's view, primary prevention—health care designed to prevent the occurrence of disease or illness—is emerging as a keystone of health policy for an aging population. But it is important to recognize that the flurry of reorganizations that were taking place in the health-care industry in the mid-1990s were mostly about organizational survival and did not necessarily indicate a shift in the percep-

tions of health-care providers toward a more accepting attitude toward primary prevention. Thus, what makes sense as health policy must often contend with the perceived threat new policies represent to the status quo in the health-care industry, which is very politically active.

The Future of Psychological Aging

The future of psychological aging will be influenced by the aging of the older population, better understanding of cognitive aging and perhaps better interventions, changes in human development, and changing psychological needs. The aging of the older population will bring increased incidence of visual and auditory impairment. We may offset this somewhat by technological changes in our capacity to compensate for these impairments. As the older population ages, the incidence of dementia will increase, unless effective treatment is developed.

Research on cognitive aspects of aging is expanding at a breathtaking rate. As we increase our understanding of how people compensate for age changes in cognitive functioning, we will be able to design more effective training and retraining programs and strategies to prevent or delay decrements in functioning. We will learn much in this area from observing how older people compensate in everyday environments for psychological age changes.

The psychological relationship between the individual and the social structure of age categories is extremely complex. As the older population extends further and further into the upper reaches of the life span, more categories may develop. What this will imply for the individual's identification with a particular age category remains to be seen. In addition, as a greater proportion of the population survives to reach later life, the experience of reaching specific older ages—85, for example—may seem less an achievement than it does today. We can definitely expect the growing density of older people in the population to affect the subjective experience of aging. For example, Troll (1995) pointed out that most very old people today are relatively unique. They have outlived their spouses, siblings, and long-time friends and have survived as individuals. But in the future, more people will reach advanced age as members of a convoy of social support. Troll could already see signs of this in her respondents over age 85. This is an excellent topic for longitudinal research.

Our theories of human development have tended to assume that development results either from a predetermined inner program of some sort or from changing social circumstances. In recent years, however, the human potential movement has emerged alongside the wellness movement to advocate self-responsibility in the area of human development. Individuals are encouraged to take responsibility for their own development. The appeal of this idea can be seen in the size of the self-help section of any large book store. Alongside books on exercise, fitness, and diet are books devoted to expanding one's psychological potential.

Some key psychological concepts of the human potential movement are self-responsibility, the personal construction of reality, the relatively unlimited nature of human psychological capability, and appreciation for the uniqueness of each person and the contribution the person can make. The idea of self-responsibility here means that the individual is in charge of personal psychological reactions to events as well as plans for future psychological growth. Again, individuals are urged not to delegate their development to others. The concept of personal construction of reality means that although there may be an objective world, we respond to our *perceptions* of that world and that our perceptions can be under our own control to some extent. The relatively unlimited nature of human mental capability means that although there may be limits to the capabilities of the human mind, we do not know what these limits are. According to human potential advocates, we should assume that psychological growth is possible until convinced otherwise by substantial proof to the contrary. Finally, the human potential movement assumes that every person, at whatever age, can continue to evolve and to make a positive contribution to her or his world. The task for the individual thus becomes to explore the self thoroughly to identify that specific individual's best prospects for development. Research results on the effects of the human potential movement are conflicting, owing largely to the great difficulty of measuring psychological growth. However, if greater numbers of aging and older people come to *believe* that growth is possible regardless of age, then they are likely to look for more growth opportunities. The explosive growth in Elderhostel, cited in Chapter 10, could be interpreted as evidence for a growing appetite for psychological growth among healthy older people.

The future of psychological aging will also probably include the evolution of new developmental issues. For example, the people entering the older population in the year 2020 will probably have had many more job and career changes during their working lives compared with today's older population. To become comfortable with the idea of job impermanence as a way of life, adults may have to develop a much greater flexibility in what they expect to happen during their working years.

Another important developmental issue of later life, one that is becoming increasingly prevalent, is being able to balance one's desire for the freedom of retirement with one's sense of duty to provide care for older parents. In the future, achieving this balance should be facilitated by better availability of respite care programs that will allow adult offspring to enjoy periods of retirement freedom and at the same time provide much of the care to older parents.

Older people of the future will have psychological needs that are present today but that will be more prevalent tomorrow. For example, sex counseling is currently available to only a small proportion of the older population. Research has shown that most older people are uninformed about the nature of age changes in sexual drives and capacities and that sex counseling can greatly improve quality of life in this important area. Look for more demand in the future for sex counseling programs for older adults.

A second area of psychological needs concerns job counseling. If, as I expect, aging people of the future will have more flexible job opportunities, then there will be an increased demand for programs to help aging individuals systematically clarify their values about what they want from a job, inventory their skills and aptitudes, and identify areas where they may need or want additional training. By the year 2000, most large communities will probably have both proprietary and nonprofit organizations providing these services.

The Future of Social Aging

Social factors directly shaping the future of individual aging include changes in the life course, new roles for older people, changes in age norms, and a shift in the focus of socialization designed for aging and older people. As we saw in Chapter 6, the life course concept is society's road map for moving through the various life stages. Life course maps differ by gender, social class, ethnicity, and region of residence. How will the life course change? First, we can expect some shift in the timing of life course transitions and stages and in the extent to which various life course dimensions parallel one another with respect to timing. Second, life course transitions are occurring later for birth cohorts entering adulthood since 1970. Third, boundaries for retirement are becoming increasingly difficult to discern. Fourth, new life course stages will appear.

In the life course, there is an ebb and flow to the timing of entry and exit. Neugarten and Hagestad (1976) studied cohorts of women born between 1900 and 1939 and pointed to what they called the "quickening" of the stages of the family life course, with women in most recent cohorts marrying earlier and earlier and thus experiencing the empty nest earlier and earlier. Cohorts born since 1940 have shown a tendency to delay marriage; in 1982, women's age at first marriage (21) was similar to that in 1900. Childbearing today is much delayed compared with the 1950s. Today's young adults will probably experience grandparenthood at a significantly later chronological age than cohorts born in the 1920s.

Not only is the timing of life course events changing, but the proportion of people taking various routes is changing. For example, among women born between 1900 and 1910, about 20 percent never married. In the cohort born in the 1930s, only about 5 percent never married. Among the cohort born in the 1960s, it appears that the proportion never marrying has increased substantially. Compared with those marrying in the 1950s, many more couples in the 1980s planned to have no children. Thus, different eras have different ideas not only about timing of entry into various paths in the life course but also about the paths one can take. In the future, there will probably be greater dispersion around the mean ages of entry into various life course paths, and the distribution of people across various paths, particularly in the family life course, can be expected to shift from time to time.

The relationship between life course stages in various areas of life will also change. For example, delayed marriage is related to the prolongation of the period of preparatory education, as more middle-class young adults not only go to college but to graduate school as well. Delayed childbearing allows men and women to get established in careers before they must cope with the demands of parenthood. Most mothers continue their employment with only brief maternity leaves, which means that the availability of child care and its quality will continue to be important issues tied not just to individual decisions but to societywide shifts in ideas about the relation of jobs and parenthood.

At the same time, the concept of careers is changing. Whereas the prevailing concept of occupational careers in the 1960s involved climbing the career ladder of a large corporation or government bureaucracy, in the 1990s even the concept of occupation itself was becoming more difficult to define. With the transition to smaller units even within large corporations and temporary work groups based on negotiated division of labor depending of the combinations of talents each team member brought to the process, the old notion of occupations as a firmly fixed set of knowledge and skills gave way to a more fluid concept that implied constant upgrading of knowledge and skills throughout one's working life. Job security based on long job tenure became more difficult to achieve and was replaced by security based on capacity to learn new skills quickly and to be flexible.

Bolles (1981) likened the adult life course to three successive boxes. In the first box are physically mature men and women who have been socially defined as teenagers and young adults and for whom education is their "job," in the sense that we expect them to do their serious work in school. In box one, employment is expected to be secondary, and this is reflected in the fact that most teenage employment is temporary employment in the fast-food industry. People in box one see leisure as a way to gain status and pass time while they wait to become eligible to do work that can contribute to their career resume.

In box two are adults for whom employment is the major focus of their life. Education plays a very small part in their lives, and leisure is okay as long as it does not slight the job. Box two, if taken too seriously, can be hazardous to mental health because of its heavy responsibilities. In box three are the retired, people who have completed their employment years and who now focus their lives around leisure.

Bolles's framework drastically oversimplifies, but his major point is well taken. We have tended to concentrate education, employment, and leisure too heavily within specific life stages rather than spreading them more evenly across the life span. As we saw in Chapter 3, the concept of the life course that stresses education first, then job, then retirement, was based on a rapidly growing population and fast-growing economic prosperity. In the future we will have a stable population, and real economic growth will probably never return to the pace of the 1950s and 1960s.

There is currently much concern over the "birth dearth"—the much smaller birth cohorts following the baby boom—and the impact it will have on the economy. McDonald's is already mounting a major effort to lure retirees into the fast-food business.

As we saw in Chapter 10, as a social institution, public education has been tightly focused on the population below the age of 22. Adult education has become an important private business, particularly through self-help books and both audio and video tape cassettes. Short-term job training workshops abound. What seems to be missing most is liberal education for adults who are employed. Older adults can get into Elderhostel and thereby gain access to tremendous variety in liberal education. But there is little available for adults in the 25- to 55-year-old range. Look for this to be a growing field of opportunity in the 1990s. People recognize that the years of life experience they bring to education are significant and that reviewing the basics of politics, religion, relationships, the arts, and so on at 45 produces a different result than reviewing the same information at 20. The point is that people want access to education across the life span, not just at the beginning.

As we saw in Chapter 9, retirement is not all play. Indeed, for many, retirement means giving up employment of dubious personal value in order to devote oneself to more worthwhile work. Retirement pensions allow individuals the freedom to define worthwhile work for themselves. Nevertheless, the image of retirement is one of a life stage dominated by an overabundance of leisure. As with education, people would like to have leisure spread over their entire adult life, not necessarily concentrated in the last third of the life span. Fortunately, the short workweek and longer vacations that began to become commonplace in the 1950s have required people to become comfortable with managing lots of time off the job, and this in turn is excellent retirement preparation—at least for the leisure aspects of retirement. On the other hand, the growing need for middle-class Americans to work more hours and hold second jobs to make ends meet restricts the availability of leisure time. Age discrimination and job protectionism, however, deny older people opportunities to do productive unpaid work in most communities. There are substantial numbers of retired people in the United States who would welcome an opportunity to do work that they could see was important to the well-being of their communities. In the future, as the availability of young and middle-aged adults diminishes and public funds to pay people to do public work become scarcer, there may be more opportunities for older people to continue to gain satisfaction through part-time employment, either for pay or as volunteers (R. Morris 1989).

Thus, look for significant change to occur in "the three boxes of life." There may be more socially productive work among people of college age, perhaps coupled with concurrent education. There will continue to be a significant amount of leisure during the 20- to 60-year-old period, when full-time employment is most likely. Look for more demand for adult education to complement leisure. The need for employees in the economy may create more opportunities for partial retirement in the future, with an increase among retirees in part-time employment, both paid and unpaid. Along with this will be kinks to be worked out in the procedures, accounting, and human relations skills involved in the blending of paid and unpaid workers.

Structural lag theory (Riley and Riley 1994; Riley, Kahn, and Foner et al., 1994) posits a mismatch between the greatly increased active life expectancy we have achieved and will continue to achieve in the future and society's age-graded social structures, which continue to move people out of mainstream social positions in their 60s. The historical division of life into education, work, and retirement continues today, despite the fact that many aging people are returning to school, starting new careers, and continuing their productive involvement in their communities. Structural lag theory suggests that the behavior of the healthier and better educated cohorts of today as they move into advanced old age will challenge our social structures to become more adaptable and move away from age-differentiated social organizations toward age-integrated patterns.

There is no doubt that many elders will succeed in their attempts to retain productive attachments to the labor force, community organizations, informal social networks, and their families. However, as long as ageism and age discrimination continue to categorically exclude ordinary older people from opportunities in a wide variety of contexts, those who continue their involvement will probably be mostly from the socioeconomic elite within the older population.

New life stages will emerge in the life course. Let us look at just two: caring for older parents and home sharing. In Chapter 8, we discussed the fact that, in the future, caregivers of frail older people will themselves be older. Coupled with the unpredictable course that individual physical aging can take and the lower numbers of adult offspring for future cohorts of the oldest old, there may be more families that find it difficult to meet the care needs of their older members. In addition, as we mentioned earlier, the duty to care for older parents or family members may conflict with retired people's visions of what they want their lives in retirement to be like. It is quite possible that the need to provide care will be intermittent. The caregivers may be able to enjoy a period of relative freedom in retirement, followed by a period of caregiving, followed by another shorter period of retirement freedom, followed by a more intensive period of caregiving, followed by shorter respite periods of freedom. The unpredictability of the timing and length of caregiving episodes will undoubtedly present a challenge and increase people's sense of uncertainty in their planning for retirement. In the future, people may take potential caregiving into account in their retirement plans much more than they do now. This might mean that they would be less likely to make permanent moves at retirement and that they would set aside financial resources as a contingency reserve in case they need it for caring for an older family member. Although caregivers can technically be retired, the responsibilities of full-time caregiving tend to dominate one's lifestyle; as more older people find themselves in this situation, caregiving in later life may become a life stage all its own.

Right now, most older people live in independent households, either alone or with their spouse. Full-service retirement communities often offer independent living in a congregate context. Opportunities are currently limited for older people who would like to live with others, but on a small scale. Home sharing may be a much more common option in the future, with housing being built that is specifically designed to be shared by six to eight independent older adults. This will bring a new housing market, new jobs for

management of such housing, and new jobs for those who work there. If home sharing becomes widespread, a new stage in the residential life course may develop around small-scale home sharing that will be positioned between living in independent housing in the community and living in higher-density and more service-oriented facilities such as retirement communities, assisted living apartments, and nursing homes.

Adjustment to this new form of housing will bring new socialization needs. Coordinating schedules and establishing ground rules are just two of the functional needs of shared housing that are a major departure from how one thinks about time and space in one's own independent household.

New roles may emerge for older people in America, perhaps in response to employee shortages connected to the baby bust, and perhaps from the growing recognition that retired people are an important source of energy for community service. Most new roles probably will be built on the assumption that older people's retirement incomes allow them an amount of flexibility in the job market that other adults do not have. Kieffer (1983) identified a large number of areas where older people could be used effectively in both the economy and the community. He is right to point out that we will not really know what proportion of the retired population can be attracted to the labor force until we offer more opportunities. However, if the European experience is any indication, unless the paid jobs being offered retirees carry responsibility and lead to personal satisfaction, only economically desperate retirees will take them.

Just a sampling of the jobs that Kieffer saw as suitable for older people include quality control monitor, bank teller, fast-food worker, technician, maintenance worker, temporary worker (perhaps through a brokering agency), or tax collector. Kieffer also pointed out that many jobs in the military currently being done by military personnel could be effectively done, and at less cost, by older workers, which would relieve some of the pressure associated with maintaining all-volunteer military forces. In addition, there are many jobs that need to be done in education—working individually with students, helping with literacy and second languages, and doing clerical work in schools to take some of the burden off the teachers. Mediation services are needed to take some of the overload that currently clogs our court systems, and older people could play a very positive role in this area. Health-care extension services allow paraprofessionals to provide care and monitoring between visits by professional care providers, and older people might play a very vital role in these services. Well-trained older people also make excellent homemaker aides and home health aides, and programs using older people in home repair services have been very successful. Older people could serve as paralegal workers, helping other older people deal with forms, for example.

This incomplete sampling should show that there are many places where new roles could emerge for older people in American society. The major obstacles are ageism, age discrimination, and societal disengagement. These biases must be removed from our view of our possible futures; then older people could enjoy a variety of new and challenging roles.

Age norms are becoming more flexible, and this trend is likely to accelerate. The explosive growth of interest in "senior" professional tours in golf and tennis is just one indication that our concepts of how age affects even the most physically demanding of professions are changing.

There is widespread discussion about the appropriateness of retirement for people in their 60s or younger, and there may be a growing tendency to consider delaying retirement a few years. Those who wish to take this course will be helped by recent changes in the Age Discrimination in Employment Act, which made age-based mandatory retirement illegal.

These are just two examples of how our ideas are beginning to change about how age should influence our evaluations of "right and wrong" with respect to entry into, exit from, and appropriateness of a wide variety of social roles.

Many of the changes that we discussed in this section will require that older people be psychologically secure and independent, no longer the extremely other-directed and conforming "organization man" or "man in the gray flannel suit" that dominated our cultural images in the 1950s. Older people in the twenty-first century who succeed in adapting to aging are likely to be independent and autonomous managers of their own life goals, but flexible in that they will be willing to try their hand at a variety of tasks. The life course will offer an even fuzzier map of later life than it does now, which means that individual decisiveness will be required for effective adaptation to aging. Tomorrow's older people may be pursuing self-knowledge to a much greater extent than today, and they are likely to ask for some help along the way. This could lead to a new wave of peer counseling programs.

I have tried to give you a sense of the many directions that change in social aging may take. In any event, we will see numerous changes, and if we pay attention we could see quite a show.

Society's Future Response

The aging of the population can influence our definitions of aging and the aged, our feelings about age discrimination and societal disengagement, and our attitudes toward the aged. But perhaps more important is the effect the aging of society will have on major social policies such as work and retirement, retirement income, and health care.

As we have seen, ageism and societal disengagement are pervasive forces in American society now. Will they increase or decrease in the future? To the extent that there is an increase in the concentration of mentally and/or physically impaired older people living in the community, there may be an increase in both ageism and societal disengagement. On the other hand, an increase in the proportion of healthy, financially secure, and well-educated older people could lead to a decrease in ageism and societal disengagement.

Kiesler (1981) hypothesized that the increasing concentration of older people in the population, coupled with improvements in health, educational levels, and financial security among elders, would bring more Americans into contact with capable older people and would lead to improvements in attitudes toward older people. However, one could also argue that an increased number of older people with mental and physical disabilities will merely reinforce existing stereotypes, causing attitudes toward elders to become even more negative. But having some notion of which way attitudes might go, and why, is an important prerequisite for doing sound and relevant research in the future.

Neal Cutler (1981) looked at how social and demographic changes could be expected to alter the politics of aging in the future. He noted that there would be a larger proportion of older people and a drastic reduction in the educational differences between generations of voters. These trends would presumably give elders as a category relatively more political resources than they enjoy now. In addition, the trend toward lower levels of identification with the major political parties might lead to a focusing of political energies around other interests, such as age. But for age consciousness to become a more important focus for political action, people would have to identify with the category "older people" more than they do now. The existence of organizations such as AARP may lead to greater age identification, at least for purposes of political action. On the other hand, the fact that most people will probably still find their greatest satisfactions in their families may retard any tendency toward collective age-consciousness and political action. Again, we know something about what to look at in the future, but not precisely what we will find.

Neugarten (1975) was among the first to call for systematic differentiation by age within the older population. First, we spoke of the young-old (65 to 74) and the old-old (75 and over). Then we began to speak of the young-old (65 to 74), middle-old (75 to 84), and the old-old (85 and over). As the population over 85 increases by leaps and bounds over the next two decades, we may be hard put to find enough labels to apply to meaningful age categories within the older population. Personally, I hope to be part of the hyper-old. Although the shifting boundaries of chronological age categories is a source of amusement, it is at the same time a troublesome social issue not likely to disappear soon. As we saw in Chapter 1, all of the ways of measuring age have disadvantages, and for the time being it appears that chronological age will remain our most common operational definition of the older population and detailed age categories within the older population. However, as the population ages and the resources needed to provide services to older people expand, we will see a growing use of various measures of need to establish or modify entitlement to public benefits. For example, recent changes in the tax laws required high-income older people to pay income tax on half of their Social Security retirement pensions, and the taxes collected were used to help build up Social Security's Retirement Trust Fund. This amounts to a redistribution of Social Security retirement benefits from high-income retirees to low-income retirees. Retirees with middle and low incomes were not subject to this tax; this practice constituted a means-tested exemption, the first ever for Social Security.

Major Policy Issues for the Future

As the population ages, pressure to limit the commitment of public funds to provide benefits to the older population will mount. This is happening in all industrial nations, not just in the United States. Major public policy issues for the future include policies on employment and retirement, retirement income policies, and health-care policies. All the potential issues for the future in these areas of social policy cannot be dealt with here; instead, we look at a small sampling of issues within each area. All these policy areas can be expected to occupy much media space and time over the next two decades.

Employment and Retirement Declines in the birthrate since the late 1960s have resulted in a diminishing number of young adults entering the labor market. We have discussed some of the ways that older people might be encouraged to remain in the labor force and in what types of occupational areas this involvement might occur. However, we should not expect a flood of older people to stay in the labor market. The freedom and self-determination of retirement strike a chord that is centuries old in the American national character. Much of the rhetoric about the employment of older people in the future is based on the assumption that large numbers of retirees are eager to get back to employment and that only the opportunity for employment, any kind of useful employment, is required to mobilize this potential. Having studied retirement over thirty-five years, I believe that these assumptions drastically overestimate the proportion of retirees who wish to have ongoing job responsibilities. In my opinion, we are going to have to make working conditions very attractive indeed, especially in terms of satisfaction that comes from feeling one's work is important, and to minimize bureaucratic aggravation in order to attract more than a tiny proportion of the older population away from retirement.

Even the assumption that older people will be needed in the labor market is debatable. For example, changes in energy conversion such as a shift to solar power could increase automation to the point that continued em-

ployment of older people would not be wanted. At the heart of this, of course, is the recognition that ageism is deeply ingrained in our culture too.

Finally, much of the rhetoric about continued employment of older workers is based on the following logic: older people are physically capable of holding a job; therefore they should hold jobs. This same argument could be applied to people under 16. They are physically able to hold jobs, too. Should we go back to putting people on jobs at the ages of 8 or 10? The logic seems similar, does it not? The flaw in these arguments is that retirement policy is not mainly about whether one is physically or mentally capable of employment. The main issue is how to deal with a surplus of potential workers in a postindustrial society where there is high individual demand for retirement and robust retirement income systems that provide adequate retirement incomes.

As Schulz (1995) pointed out, employers have many options, such as increased use of technology, increased immigration, or increased work rates from existing workers, that they can use to cope with labor shortages without having to abandon their deeply ingrained preconceptions that old workers offer little in an increasingly technology driven and changing work environment. And remember, workers begin to experience age discrimination in their 40s and 50s, not their 70s or 80s.

As we saw in Chapter 3, retirement was seen first as a means for achieving compliance from workers by providing them something they would be afraid to lose: pensions. Then we enacted a federal retirement pension program to entice older workers out of the labor market and thereby lower unemployment. The prosperity of the 1950s and 1960s created the economic surplus necessary to put in place diverse retirement income systems in addition to Social Security, and people were gaining enough experience with leisure to begin to like the idea of being able to leave employment while still physically able to enjoy life. Once people have developed their taste for retirement, it may not be possible to take retirement away except in the case of the direst national need.

Retirement Income   In my opinion, the most pressing policy issues surrounding retirement income are these: (1) Our current retirement income mechanisms financed by the public, either directly by federal taxes or by tax expenditures to promote private retirement programs, deliver more than is needed to replace preretirement income for those at the top of the income distribution, and (2) these systems deliver much, much less retirement income than is needed to replace preretirement earnings at the bottom of the income distribution.

In our system, a person can draw all the public pensions he or she can manage to qualify for. Thus, the same older person may have a military pension, a federal civil service pension, a state civil service pension, and Social Security. There is no limit on the amount of public retirement income entitlement an individual may amass. Often there is no requirement that the individual wait until retirement age to draw her or his pension. For example, military "retirees" can draw retirement benefits as early as age 37. In the meantime, some individuals have a continuous record of full-time employment over a period of four decades, yet their Social Security retirement pensions are so low that they qualify for Supplemental Security Income, which in many states does not even provide enough income to bring such people up to the poverty level. These inequities need to be addressed because they undercut the credibility of what is in the main a very sound system for providing retirement income.

In the mid 1980s, Social Security was attacked on the grounds that, compared with what they could get from a private investment, younger workers covered by Social Security would never get back from Social Security a fair return on their investment. But as we saw in Chapter 19, Social Security is not designed simply as individual insurance. It is insurance for the society as a whole against the social upheaval that would result from having millions of unemployed and destitute older people. By casting the fairness argument only in individual terms, the "generational equity" advocates drew attention away from more fundamental issues.

What would families do if there were no Social Security or Medicare? This question points squarely at the intergenerational function of Social Security, a function that only Social Security is able to fulfill. Social Security is for families. Social Security is for all generations. It is no accident that *all* industrial societies, regardless of political or economic ideology, have developed systems similar to Social Security. Whether totalitarian or democratic, socialist or capitalist, industrial societies have created social insurance systems to finance retirement. In answering the questions "Who benefits?" and "Who pays?" we need to look past purely individual concerns to the concerns of intergenerational families and the society at large.

Financing Health Care for the Older Population   The aging of the population is applying the greatest pressure for change in the area of health-care financing. The struggle between the government and health-care providers to contain health-care costs is intensifying. For example, in an attempt to contain Medicaid nursing home costs, many states have limited severely the number of nursing home spaces that can be licensed. Government reimbursement rates for Medicaid nursing home residents are consistently lower than rates charged private-pay residents and in some cases do not cover documented actual costs of providing care. As a result, there are often more people trying to get into nursing homes than there are spaces (because of the artificially limited supply), and because nursing homes want to cover their costs, they try to limit the number of Medicaid residents in their facilities. Medicaid patients in hospitals awaiting nursing home placement wait 50 percent longer on the average than others, and many never get in. Chronic shortage of Medicaid nursing home spaces also contributes to difficulty in enforcing care standards. Inspectors know that if a Medicaid nursing home is closed down, there is nowhere for the residents to go.

Because Medicaid pays a large share of the long-term care bill for older people in nursing homes, states try to contain costs by not allowing nursing homes to be built. But even if home-care programs become widely available, the fast-growing older population and the aging of the older population will combine to increase the number of congregate long-term care spaces needed.

Although home care seems to have many advantages over institutional care, home care is not necessarily less costly, particularly if there are no family members to provide free labor. Home care sometimes maintains the social isolation of homebound older people, too. In addition, monitoring the quality of home-delivered care is more difficult than monitoring similar care in a retirement community, nursing home, or assisted living facility. Nevertheless, the public image of home care is that it is a cheaper and more acceptable alternative than nursing home care. Accordingly, look for substantial growth in the home-care industry.

Most analysts expect that health care for older people will be increasingly subjected to rationing systems. One state official told me in a private conversation in 1993:

There is no way that our state budget can handle the projected increase in need for long-term care. We can make modest changes, provide more home care, and so on, but the truth is that we are going to have to deny care to all but those in *extreme* need.

In other words, this state is not making plans to raise the tax revenue necessary to provide

nursing home care to all indigent older people with a documented need for such care. Even with better availability of home care, the state is planning to leave a large proportion of needs unmet because the leadership sees no acceptable political course of action that would allow the state to meet the need. There is no attempt to deny the need, only the capacity to meet it.

Binstock (1985) looked for two types of responses to the growing gap between the costs of care and the resources to pay for it. The first steps are incremental shifts of costs from government programs to older people and their families or to health insurance companies. Examples include cost caps for Medicare and Medicaid and steady increases in copayments, the share of cost the insured must pay. Because Medicare and Medicaid often do not reimburse hospitals for the full costs of care, hospitals build the unreimbursed portion of the bills for their Medicare and Medicaid patients into their "overall structure of charges"—the rates they charge private-pay patients or patients covered by private insurance. This shifting of cost to insurance companies provides fuel for legislative initiatives to contain hospital costs.

Binstock saw four possible ways to contain hospital costs: (1) reduce physicians' fees, (2) reduce costs of facilities and nonphysician personnel, (3) ration use of high-technology equipment and procedures, and (4) develop a two-tier health-care system in which those who are unable to pay are not given access to high-cost health care. Binstock saw the latter two possibilities as most likely. He expected that, by 2000, old age would take its place alongside ability to pay as the major criteria for the rationing of health care, because in making decisions framed in terms of whose life is worth what, the ageist assumption that an older person's life is worth less than a younger person's often carries the day. This issue comes up now in the rationing of kidney dialysis and scarce donor organs for organ transplant. In a two-tier system of health care, there would be a division of labor among hospitals, with the publicly owned hospitals taking care of the lion's share of the Medicaid patients.

This very incomplete survey of the issues of the future should convince you that one of the major benefits of having conceptual frameworks and information about trends is that they create the capability for planning. They help us identify potential needs of older people, potential problems, and areas needing research. The future of aging will also be influenced by the future of social gerontology.

The Future of Social Gerontology

Obviously, much work is needed in social gerontology. But where do we go from here? What are the implications of the facts, figures, and perspectives given in this book for the field of aging?

Research

There is no single area of social gerontology that does not need more answers to crucial questions. In fact, the past decade of work in social gerontology has enabled us only to begin asking the right questions. Yet there are some areas where the need for research is particularly pressing. We know relatively little about America's older minority group people. We still do not fully understand the retirement process. Very little is known about religion as it relates to older people. We still do not understand why some people are devastated by old age and others are not. We do not fully understand the dynamics of age differentiation. The vast amount of research on aging in the United States needs to be complemented by research in other areas of the world.

In addition to the many stones as yet unturned, there is a crying need for *replication.* Knowledge is built piece by piece, and it takes many repeated studies to establish a scientific

proposition. Social gerontology is loaded with conflicting research evidence, and only more high-quality research can give us the tools to sort it out. Thankfully, since the earlier editions of this book, the quality of research reports in the field of aging has risen steadily, as has the number of replications.

Because social gerontologists focus on only a portion of the total social reality, there are many opportunities for genuinely *interdisciplinary research*. The Kansas City Study of Adult Life and the Langley Porter Institute Studies in Aging were both noteworthy because their study designs brought together psychologists, social psychologists, sociologists, and social anthropologists to do simultaneous longitudinal studies on the same samples of older people. In fact, the interplay of various traditions in social gerontology could be viewed as a step in the direction of needed theoretical integration in the social sciences in general. The literature of social gerontology is full of cases where supposedly general social theories failed the test when applied to older people and where insights gained from theories of personality or developmental psychology have helped refine sociological theories that were otherwise found wanting and vice versa.

In particular, detailed *research on community systems* holds great promise for understanding how the various social institutions interact with aging. Often, institutions and organizations pick up each other's slack, and needs not being met by one will be met by another. This kind of give-and-take is most observable at the local level. It is also at the local level that the individual most often comes into direct contact with the economy, politics, religion, health and welfare institutions, and the individual's family, friends, and neighbors. The impact of any given institution or organization thus occurs in the context of a locally based *system* of institutions and organizations. We are just beginning to sort out the various types of community systems.

To date, there have been very few attempts to study the *interdependent situational context* in which the individual experiences later life. The work of Clark and Anderson (1967) in particular showed that both the individual's personal system and interaction with the social system can be studied successfully in a community context. More studies of this kind are needed, especially to offset the tunnel vision often found in large-scale survey research studies on specific topics, such as retirement, widowhood, or voting behavior, done by scholars within a single discipline. But large-scale studies are also necessary to get a view of a phenomenon, such as retirement or income, that is *representative* of an entire nation or set of nations and not just of a particular community. Earlier large-scale studies tended to suffer greatly from sampling problems, although since 1975 we have seen more well-designed national surveys.

The primary responsibilities for funding federally sponsored research on aging have been given to the Administration on Aging, the National Institute on Aging, and the Center on Aging of the National Institute of Mental Health. Congress has generally resisted pressure to cut budgets for aging research. I hope this indicates Congress' recognition of the vital role research must play in developing the information base teachers, program planners, and policy makers need. However, Lawton (1991) pointed out that one liability associated with the National Institutes of Health's being the primary funder of basic research in social gerontology has been a strong emphasis on illness-related social research on aging rather than a balance between research aimed at discovering links between social variables and illness, and social research documenting the relation between social variables and wellness or other positive outcomes.

Education and Training

Education about aging is important at a number of levels of training. First, there is a growing need for continuing education programs for people employed full-time in agencies that serve older Americans. A solid information base in gerontology is also needed by students being trained at vocational schools for service jobs in health care and other areas related to aging, by students enrolled in undergraduate programs in the helping professions, and by graduate students in a wide variety of areas.

Ideally, education in gerontology imparts several kinds of knowledge. *General knowledge* is useful for establishing a broad context within which specific problems or issues are embedded. For example, knowing that most older people do not find retirement difficult implies that we do not need to reform all of society in order to handle the problems of those who do find it difficult. *Theoretical perspectives* provide important diagnostic tools for figuring out what is going on—what is causing what. *Substantive knowledge* provides in-depth understanding of specific areas—cognition, widowhood, methodology, aging in literature, and the like. Knowledge of *issues* focuses on the boundaries and core of debate about *what to do* in specific areas of public policy regarding aging or the aged. Most educational programs in aging comprise all these varieties of knowledge, although emphasis may vary.

If there is to be an increased research effort in social gerontology, obviously people must be trained to do it. There has been an encouraging increase in the number of institutions offering research training in social gerontology at both the graduate and undergraduate levels, much of it financed through federal programs. Not only does the actual research depend on a continuing federal commitment to research in social gerontology, but so does the existence of trained researchers. In addition, more training in social gerontology is needed for professionals working in fields serving older people. In turn, there must be organizations to provide this training. Experience in recent years has shown that gerontology is attractive to students and that courses in aging can more than pull their weight in student demand. Such courses are essential for students in the human services professions.

Unfortunately, federal support for education in gerontology has all but disappeared. As a result, many financially troubled universities have found it difficult to sustain their relatively new gerontology programs when resources are scarce and the existence of even traditional disciplines seems threatened. In addition, Kerin, Estes, and Douglass (1989) found that the declining support for gerontology education that was available tended to be focused on biomedical education to the exclusion of other important types of training.

Jobs

Since 1973 there has been an explosion of employment in the field of aging. In the public sector, most of it related to the establishment of the Area Agencies on Aging, the strengthening of State Units on Aging, and dramatic increases in the number and size of local programs for the older population financed by Older Americans Act funds. Many of these jobs, if not most, went to people with no training in gerontology (Klegon 1977). As a result, AAAs across the country have mounted substantial efforts toward in-service training. Colleges and universities are offering night courses for people already employed in the field. In addition, gerontology training programs are turning out graduates well suited to entry-level jobs in the aging service networks, and the number of jobs in these networks is still on the increase, although at a slower rate.

As we saw in Chapter 18, business has discovered older people as a market, and one result of this discovery is business' growing need for employees who know something about aging and older people. Especially needed are people with knowledge about unique problems involved in aiming products or services to older adults, such as the changes in adver-

tising art that might be needed to appeal to an aging eye. Knowledge about the typical life circumstances of various socioeconomic categories of older people can help businesses target their products to the right people. Obviously, insight into people is central to the process of successfully developing and selling anything in a competitive economy. New entrepreneurs are aiming their efforts at older markets every day, and they often lack the background needed to understand older consumers. They need and want help.

The need for trained gerontologists in higher education is on the increase. Not only has there been a rapid growth in the number of courses and systematic curricula in aging offered on the nation's college campuses, but there has also been a proliferation of short-term training programs aimed at the aging network. Research in gerontology is continuing to increase, and much of this is done at colleges and universities as well. Departments of gerontology and gerontology centers have been established, and in the late 1980s, new doctoral programs began to enroll the first classes of students to receive Ph.D.s in social gerontology.

A final example of explosive growth in employment is in the area of long-term care. As we saw earlier, between 1990 and 2020 there is going to be an explosive growth in the number of people needing long-term care services, both at home and in long-term care facilities. As of 1995, the organizations, the facilities, and the trained labor force are far short of what are needed to do the job. Over the coming years, there will be a flurry of development. New corporations will emerge. New types of facilities may be developed. Training programs will proliferate. Staffing needs of all these new efforts will be enormous.

Summary

The aging of the population will take place in a changing society, and the continuing debates about the relative priority that should be given to national concerns such as economic vitality, environmental hazards, and cultural diversity will influence the resources and opportunities available to the older population of the future.

As the number of people in the oldest age categories increases dramatically, issues of long-term care, the availability of kin, and housing will increase in importance. At the same time, improvements in health promotion and treatment of chronic disease are likely to increase active life expectancy. Psychologically, being very old will shift from something people experience as unique individuals to a status that is experienced by an age cohort. Socially, the older population will become increasingly differentiated, so common social meanings of aging and old age may become even less relevant than they are today. Yet, there is no indication that ageism and age discrimination are diminishing, particularly in terms of assumed exchange resources for employment or meaningful volunteer work in the community. In addition, despite the continuation of a variety of social problems that are disproportionately found among segments of the older population, such as poverty, inadequate housing, or lack of access to health or personal care, the probability that national social policies will seriously address these issues seems to be growing dimmer. Those older people who cannot solve their own problems through earnings from employment or drawing on substantial savings are not likely to fare well in the social policy climate of the late 1990s.

Future growth of the older population is going to be explosive. With the older population growing so fast, there is little chance that interest in aging will lessen. In fact, during the coming decades, services to older people may well represent one of the fastest-growing areas of employment in the Western world. Yet, at the same time, unmet needs of older people may reemerge as a serious social problem in America. And the demand for knowledge in the field of social gerontology can be expected to grow accordingly. All this implies an active future for social gerontology.

There are plentiful career opportunities in social gerontology. We are just beginning to grapple with the problems in many areas of research and practice. The "establishment" in the field is relatively small, and interest in research results and innovative programs is high. Funding for gerontology research and demonstration projects is available, and gerontology centers have been established in many universities. This situation offers people ready to embark on a new career an opportunity to "get in on the ground floor." I hope this book motivates some of its readers to join me in this fascinating field.

One does not have to become a gerontologist to benefit from a general education in gerontology. Although many people in academia have considered gerontology a vocational field, our society is aging, and to the extent that liberal education is designed to provide perspectives and information that can help creatively solve our common human problems, gerontology is becoming recognized as an important part of liberal education.

Glossary

accommodation Altering one's behavior to more nearly conform to the demands of an external situation.

acculturation The process of learning a culture.

active life expectancy The number of years of functional well-being that people could expect given current age-specific rates of disability.

activity theory A theory that holds that older people have the same psychological and social needs as middle-aged people unless they are affected by poor health or disability.

acute condition An illness or injury expected to be temporary.

adaptation The process of adjusting oneself to fit a situation or environment.

adult development The gradual unfolding of personality and self throughout adulthood.

age changes For an individual, differences, from one age to another, that result from internal physical or psychological change.

age consciousness Individuals' perceptions of their own age. Includes both objective and subjective elements.

age differences Differences between people of different ages. May be due to age changes, cohort differences, period effects, or some combination of the three factors.

age discrimination The overt denial of opportunity on the basis of age.

age norms Norms tied to the life course that tell people of a given age what is allowed or not allowed for someone of that age.

age stratification The division of a population into age strata such as youth, adulthood, or old age and the culturally prescribed relations between people of differing age strata.

aged People 65 years of age or older.

aged dependency ratio See *dependency ratio, aged.*

ageism Prejudice based on age.

aging (adj.) In the process of becoming old. Does not imply that the person *is* old.

aging (n.) The physical, psychological, and social processes that over time cause changes in a person's functional capacities and influence social definitions.

aging, robust Refers to elders for whom aging is experienced as a continuation of good physical health, psychological well-being, cognitive competence, and productive activity.

Alzheimer's disease Chronic organic brain disease caused by deteriorative loss of brain cells (as opposed to loss through injury or arteriosclerosis). See also *dementia.*

anticipation Identifying in advance the rights, obligations, prerequisites, resources, and outlook of a position one will occupy in the future.

Area Agency on Aging (AAA) Local agencies charged with planning and coordinating services to older people and providing information and referral. Older Americans Act funds flow to local community agencies through the AAAs.

case management   Looking at the totality of a person's needs and resources and developing an integrated treatment plan. See also *long-term care*.

chronic condition   An illness or injury expected to be long-term or permanent.

chronological age   The number of years a person has lived.

cohort   An aggregation of people having a common characteristic, usually the time period in which they were born.

compensation   The processes whereby aging people or others around them offset detrimental effects of aging.

consolidation   Adapting to role or activity loss by redistributing time and energy to roles and activities that remain.

continuity, external   Living in familiar environments and interacting with familiar people.

continuity, internal   The persistence of a personal structure of ideas based on memory.

coping   Contending with or attempting to overcome everyday problems.

decision demand   A type of norm that requires people to choose a course of action *within* a specified time period and *from* an age-linked field of possibilities.

dementia   An organic brain disease characterized by mental confusion, poor memory, incoherent speech, and poor orientation to the environment.

dependency   The condition of receiving assistance from others for the necessities of life.

dependency ratio, aged   The number of people 65 and over divided by the number of people 15 to 64 and the result multiplied by 100.

dependency ratio, total   The number of people 0 to 14 plus the number of people 65 and over divided by the number of people 15 to 64 and the result multiplied by 100.

dependency ratio, youth   The number of people 0 to 14 divided by the number of people 15 to 64 and the result multiplied by 100.

disability   A condition that restricts a person's physical or mental capacity to engage in desired or expected roles or activities.

disengagement   Adapting to role or activity loss by withdrawal.

disengagement, differential   Adapting to loss by withdrawal in selected areas of life. See also *consolidation*.

disengagement, societal   A process whereby people who reach an arbitrary age are no longer encouraged to seek positions for which they are qualified or are not encouraged to remain in positions in which they are functioning well.

dying trajectory   The speed with which a person dies; the rate of decline in functioning.

elder   People age 65 or older.

elderly   People age 65 or older.

external continuity   See *continuity, external*.

functional age   The use of attributes such as appearance, mobility, strength, and mental capacity to assign people to broad age categories such as middle age and old age.

gerontologists   People who specialize in aging.

gerontology   The use of reason to study and understand aging.

gerontology, social   See *social gerontology*.

holistic model   A health-care model stressing self-responsibility, prevention, and wellness.

hospice   An organization that delivers services to dying persons and their families.

internal continuity   See *continuity, internal*.

later maturity   A life-course stage socially defined or typified by energy decline; awareness of sensory loss; onset of chronic health problems; loss of social contacts through retirement, widowhood, and movement of children; and freedom from responsibilities such as work and child rearing.

life course   An ideal sequence of events that people are expected to experience and positions that they are expected to occupy as they mature and move through life.

life expectancy   The *average* length of time a group of individuals of the same age will live, given current mortality rates. Life expectancy can be computed for any age; life expectancy at birth is the most common statistic.

life span The length of life that is biologically possible for a given species.

life stages Broad age categories loosely based on ideas about effects of aging. Examples include middle age, later maturity, and old age.

life structure The pattern or design of a person's life that results from the interaction between the person and the external world at a given time in the life course.

life table A demographic model used to project the average number of years of life remaining at various ages, given current age-sex mortality rates.

long-term care Long-term management of chronic illness or disability.

maturation The process of development. Can be physical, psychological, or social.

maturity A quality of being fully grown. Physical maturity usually occurs much earlier than psychological or social maturity. Indeed, many contend that psychological development is never complete in the sense that physical development is.

Medicaid A comprehensive health care program for the poor. Finances both acute and long-term care.

medical model A health-care model revolving around medical treatment, hospital-like environments, and care provided by physicians and nurses.

Medicare A program of national health insurance for persons who are 65 or older and covered by Social Security. Finances primarily acute care.

menopause The period of natural cessation of menstruation, usually occurring between the ages of 45 and 50.

middle age A stage of the life course socially defined or typified by energy decline, shifting from physical to mental activities, feelings of having reached a goal or plateau in one's career, awareness that life is finite, shrinking of family as children leave home, entry of women into the labor force, and employment troubles. See also *later maturity; old age.*

midlife Middle age.

modernization theory The theory that industrialization caused older people to lose status in society.

nursing home A facility that provides personal care plus health care such as administering medication.

old age A stage of the life course socially defined or typified by increasing frailty and disability, much introspection and concern over the meaning of life, distinct awareness of approaching death, and financial and physical dependency. See also *middle age; later maturity.*

Older Americans Act Legislation that created a national network of services and programs for older people. It also provides funds for senior centers, nutrition programs, and research, training, and demonstration projects.

older people People 65 or older.

organic mental disorders See *dementia.*

pension A periodic payment to a person or the person's family, given as a result of previous job service.

pension, employer Retirement pension available only through a specific position of employment and administered by a work organization, union, or private insurance company.

pension, retirement Income a retired person receives by virtue of having been employed at least a minimum number of years on a job covered by a pension system. See also *pension, employer.*

percentage of older people in the population The number of people 65 and older divided by the number in the total population and the result multiplied by 100.

period effects Differences resulting from measures having been taken at different time periods.

personality The unique pattern of attitudes, values, beliefs, habits, mannerisms, and preferences that allows one to think of oneself and others as individuals.

population pyramid A graph showing the distribution, in either numbers or proportions, of population by age and sex.

relative appreciation A favorable attitude toward oneself or one's situation based on a comparison with others whose characteristics or situations are less fortunate.

reliability The extent to which a measuring instrument produces consistent results.

retirement The period following a career of job holding, in which job responsibilities and often opportunities are minimized and economic support comes by virtue of having held a job for a minimum length of time in the past.

retirement community A community, most of whose residents are retired.

retirement community, de facto A retirement community that results from large-scale migration of retired households to small towns in certain regions of the country.

retirement community, full-service A planned retirement community that offers its residents a continuum of services and levels of care.

robust aging See *aging, robust.*

role adaptations A process of fitting role demands to an individual's capabilities.

role anticipation A process that involves learning the rights, obligations, resources, and outlook of a position one will occupy in the future.

role relationship The ground rules that define what the players of reciprocal social roles can expect from one another.

self A person's awareness of his or her own nature and characteristics. The person as an object of his or her own awareness.

sex ratio The number of males per 100 females in a population.

social gerontology A subfield of gerontology dealing with the developmental and group behavior of adults and the causes and consequences of having older people in the population.

socialization The processes through which a group re-creates in its members the distinctive way of life of the group.

social model A health-care model involving a wide range of types of caregivers in making decisions about health care and provision of health care.

social roles The expected or typical behavior associated with positions in the organization of a group.

Social Security Colloquially, the general public retirement pension administered by the federal government. Technically, Social Security also provides a number of other types of benefits to survivors and disabled people. It also administers Medicare.

social situation The total structure of social forces influencing an individual's behavior or experience at a particular time. Social situations vary over time and place.

stereotype A composite of beliefs about a category of people. May be either accurate or inaccurate.

Supplemental Security Income (SSI) A federal program of public assistance to indigent older people.

support network The total of a person's relationships that involve the receiving of assistance and that are viewed by both the giver and receiver as playing a significant part in maintaining the psychological, social, and physical integrity of the receiver.

theories Sets of interrelated principles and definitions used conceptually to organize observations, information, or communication about particular aspects of reality.

total dependency ratio See *dependency ratio, total.*

validity The correspondence between what a measuring instrument is supposed to measure and what it actually measures.

youth dependency ratio See *dependency ratio, youth.*

Bibliography

Abraham, Joseph D., and Robert O. Hansson: 1995 Successful aging at work: And compensation through impression management. *Journal of Gerontology: Psychological Sciences* 50B(2):P94–P103.

Achenbaum, W. Andrew: 1978 *Old Age in the New Land.* Baltimore, MD: Johns Hopkins University Press.

1983 *Shades of Gray: Old Age, American Values, and Federal Policies Since 1920.* Boston: Little, Brown.

1986 *Social Security: Visions and Revisions, A Twentieth Century Fund Study.* New York: Cambridge University Press.

1995 Age-based Jewish and Christian rituals. Pp. 201–217 in Melvin A. Kimble et al. (eds.), *Aging, Spirituality, and Religion: A Handbook.* Minneapolis: Fortress Press.

Achenbaum, W. Andrew, and Lucinda Orwoll: 1991 Becoming wise: A psycho-gerontological interpretation of the Book of Job. *International Journal of Aging and Human Development* 32:21–39.

Adams, Rebecca G.: 1985–1986 Friendship and aging. *Generations* 10(4):40–43.

1987 Patterns of network change: A longitudinal study of friendships of elderly women. *The Gerontologist* 27(2):222–227.

1989 Conceptual and methodological issues in studying friendships of older adults. Pp. 17–41 in Rebecca G. Adams and Rosemary Blieszner (eds.), *Older Adult Friendship.* Newbury Park, CA: Sage.

Ade-Ridder, Linda: 1990 Sexuality and marital quality among older married couples. Pp. 48–67 in T. H. Brubaker (ed.), *Family Relationships in Later Life,* 2d ed. Newbury Park, CA: Sage.

Agich, George J.: 1995 Chronic illness and freedom. Pp. 129–153 in S. K. Toombs, D. Bernard, and R. A. Carson (eds.), *Chronic Illness.* Bloomington, IN.: Indiana University Press.

Alexander, Charles N., and Ellen J. Langer (eds.): 1990 *Higher Stages of Consciousness.* New York: Oxford University Press.

Alexander, Francesca, and Robert W. Duff: 1988 Social interaction and alcohol use in retirement communities. *The Gerontologist* 28(5):632–636.

Alexander, Lloyd: 1964 *The Book of Three.* New York: Dell.

1965 *The Black Cauldron.* New York: Dell.

1966 *The Castle of Llyr.* New York: Dell.

1967 *Taran Wanderer.* New York: Dell.

1968 *The High King.* New York: Dell.

Allan, Graham A., and Rebecca G. Adams: 1989 Aging and the structure of friendship. Pp. 45–64 in Rebecca G. Adams and Rosemary Blieszner (eds.), *Older Adult Friendship.* Newbury Park, CA: Sage.

Allen, Carole B.: 1981 Measuring mature markets. *American Demographics* 3(3):13–17.

Allen, Karen R., and Robert S. Pickett: 1987 Lifelong family careers of single women. *Journal of Marriage and the Family* 49:517–26.

Alwin, Duane, R. L. Cohen, and T. M. Newcomb: 1991 *Aging, Personality, and Social Change: Attitude Persistence and Change Over the Life Span.* Madison: University of Wisconsin Press.

American Association of Retired Persons: 1986 *Work and Retirement: Employees Over 40 and Their Views.* Washington, D.C.: American Association of Retired Persons.

Ando, A., and F. Modigliani: 1963 The life-cycle hypothesis of saving. *American Economic Review* 53:55–84.

Ansello, Edward F.: 1977 Old age and literature: An overview. *Educational Gerontology* 2: 211–218.

Antonucci, Toni C.: 1990 Social supports and social relationships. Pp. 205–226 in Robert H. Binstock and Linda K. George (eds.), *Handbook of Aging and the Social Sciences*, 3d ed. New York: Academic Press.

1985–1986 Hierarchical mapping technique (social support networks). *Generations* 10(4): 10–12.

Applebaum, Robert: 1992 *Challenges to Regulating Adult Care Facilities.* Oxford, OH: Scripps Gerontology Center.

Applebaum, Robert A., Robert C. Atchley, and Carol H. Austin: 1987 *A Study of Ohio's PASSPORT Program.* Report to the Ohio Department of Aging, May 12.

Applebaum, Robert, and Lynn Ritchey: 1992 *Adult Care Homes in Ohio.* Oxford, OH: Scripps Gerontology Center.

Aquilino, William S.: 1990 The likelihood of parent–adult child coresidence: Effects of family structure and parental characteristics. *Journal of Marriage and the Family* 52:405–419.

Arenberg, David, and Elizabeth A. Robertson-Tchabo: 1980 Age differences and age changes in cognitive performance: New "old" perspectives. Pp. 139–157 in R. L. Sprott (ed.), *Age, Learning Ability, and Intelligence.* New York: Van Nostrand Reinhold.

Ariès, Philippe: 1981 *The Hour of Our Death.* New York: Knopf.

Armstrong, M. Jocelyn, and Karen S. Goldsteen: 1990 Friendship support of older American women. *Journal of Aging Studies* 4(4): 391–404.

Aronoff, Craig: 1974 Old age in prime time. *Journal of Communication* 24:86–87.

Ash, Phillip: 1966 Pre-retirement counseling. *The Gerontologist* 6:97–99, 127–128.

Atchley, Robert C.: 1967 Retired women: A study of self and role. Unpublished doctoral dissertation. Washington, D.C.: The American University.

1971a Retirement and leisure participation: Continuity or crisis? The *Gerontologist* 11(1, part 1):13–17.

1971b *Understanding American Society.* Belmont, CA: Wadsworth.

1972 *The Social Forces in Later Life.* Belmont, CA: Wadsworth.

1974 The meaning of retirement. *Journal of Communications* 24:97–101.

1975 Dimensions of widowhood in later life. *The Gerontologist* 15:176–178.

1976 *The Sociology of Retirement.* Cambridge, MA: Schenkman.

1979 Issues in retirement research. *The Gerontologist* 19:44–54.

1980a Aging and suicide: Reflection on the Quality of Life? Pp. 141–162 in S. G. Haynes and M. Feinleib (eds.), *The Epidemiology of Aging.* Bethesda, MD: National Institutes of Health.

1980b *Social Forces in Later Life,* 3d ed. Belmont, CA: Wadsworth.

1982a The process of retirement: Comparing women and men. Pp. 153–168 in M. Szinovacz (ed.), *Women's Retirement.* Beverly Hills, CA: Sage.

1982b Retirement: Leaving the world of work. *Annals of the American Academy of Political and Social Sciences* 464:120–131.

1982c Retirement as a social institution. *Annual Review of Sociology* 8:263–287.

1985 *Social Forces and Aging: An Introduction to Social Gerontology,* 4th ed. Belmont, CA: Wadsworth.

1989 A continuity theory of normal aging. *The Gerontologist* 29:183–190.

1992 Retirement and marital satisfaction. Pp. 145–158 in M. Szinovacz, D. J. Ekerdt, and B. H. Vinick (eds.), *Families and Retirement.* Newbury Park, CA: Sage.

1993a Critical perspectives on retirement. Pp. 3–19 in Thomas R. Cole et al. (eds.), *Voices and Visions: Toward a Critical Gerontology.* New York: Springer.

1993b Continuity theory and the evolution of activity in later life. Pp. 5–16 in John R. Kelly (ed.), *Activity and Aging.* Newbury Park, CA: Sage.

1994 Is there life between life course transitions? Paper presented at the Annual Meeting of the Gerontological Society of America. Atlanta, November.

1995a Continuity of the spiritual self. Pp. 68–73 in Melvin A. Kimble et al. (eds.), *Aging, Spirituality, and Religion: A Handbook.* Minneapolis: Fortress Press.

1995b Everyday mysticism: Spiritual development in later life. Paper presented at the 20th International Conference on the Unity of the Sciences. Seoul, Korea, August 24.

1995c *Antecedents and Consequences of Functional Limitation in Later Life.* Oxford, OH: Scripps Gerontology Center.

1995d Continuity theory. Pp. 227–230 in G. L. Maddox et al. (eds.), *Encyclopedia of Aging*, 2d ed. New York: Springer.

1995e Activity theory. Pp. 9–12 in G. L. Maddox et al. (eds.), *Encyclopedia of Aging*, 2d ed. New York: Springer.

Forthcoming Retirement. In James E. Birren (ed.), *Encyclopedia of Gerontology*. New York: Academic Press.

Atchley, Robert C., and Mark Dorfman: 1994 Gaining marketing insights from the Ohio Long–Term Care Insurance Survey. *Journal of the American Society of CLU and ChFC* 48(5):66–71.

Atchley, Robert C., Suzanne R. Kunkel, and Carl Adlon: 1978 *An Evaluation of Preretirement Programs: Results from an Experimental Study.* Oxford, OH: Scripps Gerontology Center.

Atchley, Robert C., and Sheila J. Miller: 1982–1983 Retirement and couples. *Generations* 7(2):28–29, 36.

1983 Types of elderly couples. Pp. 77–90 in T. H. Brubaker (ed.), *Family Relationships in Later Life*. Beverly Hills, CA: Sage.

Atchley, Robert C., Linda Pignatiello, and Ellen Shaw: 1979 Interactions with family and friends: Marital status and occupational differences among older women. *Aging Research* 1: 83–94.

Atchley, Robert C., and Judith L. Robinson: 1982 Attitudes toward retirement and distance from the event. *Research on Aging* 4:299–313.

Atchley, Sheila J.: 1986 Conceptualizing interpersonal relationships. *Generations* 10(4):6–9.

Babbie, Earl R.: 1995 *The Practice of Social Research,* 7th ed. Belmont, CA: Wadsworth.

Babchuck, Nicholas: 1978–1979 Aging and primary relationships. *International Journal of Aging and Human Development* 9:137–151.

Back, Kurt W.: 1971 Metaphors as a test of personal philosophy of aging. *Sociological Focus* 5:1–8.

Bahr, S. J.: 1973 Effects of power and division of labor on the family. Pp. 167–185 in Lois W. Hoffman and I. F. Nye (eds.), *Working Mothers*. San Francisco: Jossey-Bass.

Ball, Robert M.: 1978 *Social Security: Today and Tomorrow.* New York: Columbia University Press.

Ballweg, John A.: 1967 Resolution of conjugal role adjustment after retirement. *Journal of Marriage and the Family* 29:277–281.

Baltes, Margaret M., and Paul B. Baltes (eds.): 1986 *The Psychology of Control and Aging.* Hillsdale, NJ: Erlbaum.

Baltes, Paul B.: 1993 The aging mind: Potential and limits. *The Gerontologist* 33(5):580–594.

Baltes, Paul B., and Gisela V. Labouvie: 1973 Adult development of intellectual performance: Description, explanation, and modification. Pp. 157–219 in C. Eisdorfer and M. P. Lawton (eds.), *The Psychology of Adult Development and Aging*. Washington, D.C.: American Psychological Association.

Bankoff, Elizabeth A.: 1983 Aged parents and their widowed daughters: A support relationship. *The Gerontologist* 38:226–230.

Barbato, Carole A., and Jerry D. Feezel: 1987 The language of aging in different age groups. *The Gerontologist* 27(4):527–531.

Barber, Clifton E.: 1980 Gender differences in experiencing the transition to the empty nest: Reports of middle-aged and older men and women. *Family Perspective* 14(3):87–95.

Barlett, Donald L., and James B. Steele: 1992 *America: What Went Wrong?* Kansas City: Andrews and McMeel.

Bartoshuk, L. M., B. Rifkin, L. E. Marks, and P. Bars: 1986 Taste and aging. *Journal of Gerontology* 41:51–57.

Bassili, John N., and Jane Reil: 1981 On the dominance of the old-age stereotype. *Journal of Gerontology* 36:682–688.

Baumeister, Roy F., Todd F. Heatherton, and Dianne M. Tice: 1994 *Losing Control: How and Why People Fail at Self-Regulation.* New York: Academic Press.

Bausell, R. Barker: 1986 Health-seeking behavior among the elderly. *The Gerontologist* 26: 556–559.

Beck, Ulrich: 1992 *Risk Society: Towards a New Modernity.* Newbury Park, CA: Sage.

Becker, Gay: 1993 Continuity after a stroke: Implications of life-course disruption in old age. *The Gerontologist* 33(2):148–158.

1994 The oldest old: Autonomy in the face of frailty. *Journal of Aging Studies* 8(1):59–76.

Belgrave, Linda Liska: 1988 The effects of race differences in work history, work attitudes, economic resources, and health in women's retirement. *Research on Aging* 10(3):383–398.

Bell, A. P., and M. S. Weinberg: 1978 *Homosexualities: A Study of Diversity Among Men and Women.* New York: Simon & Schuster.

Bell, Daniel: 1973 *The Coming of Post-Industrial Society.* New York: Basic Books.

Bellah, Robert N., Richard Madsen, William L. Sullivan, Ann Swidler, and Stephen M. Tipton: 1985 *Habits of the Heart: Individualism and Commitment in American Life.* Berkeley: University of California Press.
1991 *The Good Society.* New York: Knopf.

Bengtson, Vern L.: 1985 Diversity and symbols in grandparent roles. Pp. 11–29 in Vern L. Bengtson and Joan F. Robertson (eds.), *Grandparenthood.* Beverly Hills, CA: Sage.
1989 The problem of generations: Age group contrasts, continuities, and social change. Pp. 25–54 in Vern L. Bengtson and K. Warner Schaie (eds.), *The Course of Later Life: Research and Reflections.* New York: Springer.

Bengtson, Vern L., Margaret N. Reedy, and Chad Gordon: 1985 Aging and self conceptions: Personality processes and social contexts. Pp. 544–593 in J. E. Birren and K. W. Schaie (eds.), *Handbook of the Psychology of Aging,* 2d ed. New York: Van Nostrand Reinhold.

Bengtson, Vern L., and J. F. Robertson (eds.): 1985 *Grandparenthood.* Beverly Hills, CA: Sage.

Bengtson, Vern L., Carolyn Rosenthal and Linda Burton: 1990 Families and aging: Diversity and heterogeneity. Pp. 263–287 in Robert H. Binstock and Linda K. George (eds.), *Handbook of Aging and the Social Sciences,* 3d ed. New York: Academic Press.

Benjamin, A. E., and David A. Lindeman: 1983 Health planning and long term care. Pp. 207–226 in Carroll L. Estes and R. J. Newcomer (eds.), *Fiscal Austerity and Aging: Shifting Government Responsibility for the Elderly.* Beverly Hills, CA: Sage.

Benjamin, A. E., and Robert J. Newcomer: 1986 Board and care housing: An analysis of state differences. *Research on Aging* 8:388–406.

Bennett, Ruth G.: 1980 *Aging, Isolation and Resocialization.* New York: Van Nostrand Reinhold.

Berardo, Felix M., Jeffrey Appel, and Donna H. Berardo: 1993 Age dissimilar marriages: Review and assessment. *Journal of Aging Studies* 7(1):93–106.

Berger, Raymond M.: 1982 *Gay and Gray: The Older Homosexual Man.* Chicago: University of Illinois Press.

Bernstein, Judith: 1982 Who leaves—who stays: Residency policy in housing for the elderly. *The Gerontologist* 22:305–313.

Berscheid, Ellen, and Letitia A. Peplau: 1983 The emerging science of relationships. Pp. 1–19 in Harold H. Kelley et al. (eds.), *Close Relationships.* New York: W. H. Freeman.

Bianchi, Eugene C.: 1992 *Aging as a Spiritual Journey.* New York: Crossroad.

Biggar, Jeanne C.: 1984 The graying of the Sunbelt: A look at the impact of U.S. elderly migration. *Population Trends and Public Policy,* No. 6. Washington, D.C.: Population Reference Bureau.

Billingsley, Andrew: 1968 *Black Families in White America.* Englewood Cliffs, NJ: Prentice-Hall.

Binstock, Robert H.: 1972 Interest-group liberalism and the politics of aging. *The Gerontologist* 12:265–280.
1974 Aging and the future of American politics. Pp. 199–212 in Frederick R. Eisele (ed.), *Political Consequences of Aging.* Philadelphia: American Academy of Political and Social Sciences.
1985 Health care of the aging. Pp. 3–15 in C. M. Gaitz et al. (eds.), *Aging 2000: Our Health Care Destiny,* Vol. II. New York: Springer.

Birren, James E.: 1989 My perspective on research on aging. Pp. 135–149 in V. L. Bengtson and K. W. Schaie (eds.), *The Course of Later Life: Research and Reflections.* New York: Springer.

Birren, James E., and K. Warner Schaie (eds.): 1985 *Handbook of the Psychology of Aging,* 2d ed. New York: Van Nostrand Reinhold.
1990 *Handbook of the Psychology of Aging,* 3d ed. San Diego, CA: Academic Press, Inc.

Birren, James E., R. Bruce Sloane, and Gene D. Cohen (eds.): 1992 *Handbook of Mental Health and Aging,* 2d ed. San Diego, CA: Academic Press.

Bishop, James M., and Daniel R. Krause: 1984 Depictions of aging and old age on Saturday morning television. *The Gerontologist* 24: 91–94.

Blau, Zena: 1961 Social constraints on friendship in old age. *American Sociological Review* 26: 429–439.

Blenkner, Margaret: 1965 Social work and family relationships in later life with some thoughts on filial maturity. Pp. 46–59 in Ethel Shanas and Gordon F. Streib (eds.), *Social Structure and the Family.* Englewood Cliffs, NJ: Prentice-Hall.

Blieszner, Rosemary: 1989 Developmental processes of friendship. Pp. 108–126 in Rebecca G. Adams and Rosemary Blieszner (eds.), *Older Adult Friendship.* Newbury Park, CA: Sage.

Bolles, Richard N.: 1981 *The Three Boxes of Life and How to Get Out of Them.* Berkeley, CA: Ten Speed Press.

Bornstein, P. E., P. J. Clayton, J. A. Halikas, W. L. Maurice, and E. Robbins: 1973 The depression of widowhood after thirteen months. *British Journal of Psychiatry* 122:561–566.

Borup, Jerry H.: 1981 Relocation: Attitudes, information network and problems encountered. *The Gerontologist* 21:501–511.

Bossé, Raymond, Carolyn M. Aldwin, Michael Levenson, Avron Spiro, III, and Daniel K. Mroczek: 1993 Change in social support after retirement: Longitudinal findings from the normative aging study. *Journal of Gerontology: Psychological Sciences* 48(4):P210–P217.

Bossé, Raymond, Carolyn M. Aldwin, Michael R. Levenson, and Kathryn Workman-Daniels: 1991 How stressful is retirement? Findings from the Normative Aging Study. *Journal of Gerontology* 46(1):P9–P14.

Bossé, Raymond, Carolyn M. Aldwin, Michael R. Levenson, Kathryn Workman-Daniels, and David J. Ekerdt: 1990 Differences in social support among retirees and workers: Findings from the Normative Aging Study. *Psychology and Aging* 5(1):41–47.

Bossé, Raymond, and David J. Ekerdt: 1981 Change in self-perception of leisure activities with retirement. *The Gerontologist* 21:650–654.

Botwinick, Jack: 1967 Cognitive Processes in Maturity and Old Age. New York: Springer.
1978 *Aging and Behavior,* 2d ed., New York: Springer.

Bound, John, Greg J. Duncan, Deborah S. Laren, and Lewis Oleinick: 1991 Poverty dynamics in widowhood. *Journal of Gerontology* 46(3): S115–124.

Brackbill, Yvonne, and Donna Kitch: 1991 Intergenerational relationships: A social exchange perspective on joint living arrangements among the elderly and their relatives. *Journal of Aging Studies* 5(1):77–97.

Bradburn, N. M., and D. Caplovitz: 1965 *Reports on Happiness.* Chicago: Aldine.

Bradshaw, J., M. Clifton, and J. Kennedy: 1978 *Found Dead: A Study of Old People Found Dead.* Mitcham Surrey, England: Age Concern England.

Brecher, Edward M.: 1984 *Love, Sex, and Aging.* Boston: Little, Brown.

Bressler, Dawn S.: 1985–1986 Widowed persons service. *Generations* 10(4):21–22.

Brody, Elaine M.: 1977 *Long-Term Care of Older People: A Practical Guide.* New York: Human Sciences Press.
1985 Parent care as a normative family stress. *The Gerontologist* 25:19–29.

Brubaker, Timothy H.: 1985 *Later Life Families.* Beverly Hills, CA: Sage.

Brubaker, Timothy H. (ed.): 1987 *Aging, Health, and Family: Long Term Care.* Newbury Park, CA: Sage Publications.

Brubaker, Timothy H., and Charles B. Hennon: 1982 Responsibility for household tasks: Comparing dual-earner and dual-retired marriages. Pp. 205–220 in M. Szinovacz (ed.), *Women's Retirement.* Beverly Hills, CA: Sage.

Buckholz, Michael, and J. E. Bynum: 1982 Newspaper presentation of America's aged: A content analysis of image and role. *The Gerontologist* 22:83–88.

Buckley, Walter: 1967 *Sociology and Modern Systems Theory.* Englewood Cliffs, NJ: Prentice-Hall.

Bulcroft, Kris, June Van Leynseele, and Edgar F. Borgatta: 1989 Filial responsibility laws. *Research on Aging* 11:374–393.

Bulcroft, Richard A., and Kris A. Bulcroft: 1991 The nature and functions of dating in later life. *Research on Aging* 13(2):244–260.

Burch, Thomas K.: 1990 Remarriage of older Canadians: Description and interpretation. *Research on Aging* 12(4):546–559.

Burkhauser, Richard V., Barbara A. Butriea, and Michael J. Wasylenko: 1995 Mobility patterns of older homeowners. *Research on Aging* 17(4): 363–384.

Burnside, Irene M.: 1981 *Nursing and the Aged,* 2d ed. New York: McGraw-Hill.

Burton, Linda M.: 1985 Early and on-time grandmotherhood in multigenerational black families. Unpublished doctoral dissertation. University of Southern California.

1992 Black grandparents rearing children of drug-addicted parents: Stressors, outcomes, and social service needs. *The Gerontologist* 32: 744–751.

Burton, Linda M., and Vern L. Bengtson: 1985 Black grandmothers: Issues of timing and continuity of roles. Pp. 61–77 in Vern L. Bengtson and Joan F. Robertson (eds.), *Grandparenthood.* Beverly Hills, CA: Sage.

Buskirk, Elsworth R.: 1985 Health maintenance and longevity: Exercise. Pp. 894–931 in Caleb E. Finch and Edward L. Schneider (eds.), *Handbook of the Biology of Aging,* 2d ed. New York: Van Nostrand Reinhold.

Butler, Robert N.: 1980 The alliance of advocacy with science. *The Gerontologist* 20:154–162. 1995 Future trends. Pp. 387–390 in George Maddox et al. (eds.), *Encyclopedia of Aging,* 2d ed. New York: Springer.

Butler, Robert N., and Myrna I. Lewis: 1982 *Aging and Mental Health,* 3d ed. St. Louis: Mosby.

Byrd, Mark, and Trudy Breuss: 1992 Perceptions of sociological and psychological age norms by young, middle-aged, and elderly New Zealanders. *International Journal of Aging and Human Development* 34(2):145–163.

Cain, Leonard D., Jr.: 1974 The growing importance of legal age in determining the status of the elderly. *The Gerontologist* 14:167–174.

Callahan, Daniel: 1987 *Setting Limits: Medical Goals in an Aging Society.* New York: Simon and Schuster.

Cameron, Paul: 1975 Mood as an indicant of happiness: Age, sex, social class, and situational differences. *Journal of Gerontology* 30:216–224.

Campbell, John Creighton, and John Strate: 1981 Are old people conservative? *The Gerontologist* 21:580–591.

Campbell, Miriam K., Trudy L. Bush, and William E. Hale: 1993 Medical conditions associated with driving cessation in community-dwelling, ambulatory elders. *Journal of Gerontology: Social Sciences* 48(4):S230–S234.

Campione, Wendy A.: 1988 Predicting participation in retirement preparation programs. *Journal of Gerontology: Social Sciences* 43(3): S91–S95.

Cantor, Marjorie H.: 1980 The informal support system: Its relevance in the lives of the elderly. Pp. 131–146 in N. G. McClusky and E. F.

Borgatta (eds.), *Aging and Society: Research and Policy Perspectives.* Beverly Hills, CA: Sage.

Capel, W. C., B. W. Goldsmith, K. J. Waddell, and G. T. Stewart: 1972 The aging narcotic addict: An increasing problem for the next decades. *Journal of Gerontology* 27:102–106.

Capel, W. C., and L. G. Peppers: 1978 The aging drug addict: A longitudinal study of known abusers. *Addictive Diseases* 3:389–404.

Capra, Fritjof: 1982 *The Turning Point.* New York: Simon & Schuster.

Carlie, Michael K.: 1969 The politics of age: Interest group or social movement? *The Gerontologist* 9(4, Part 1):259–263.

Carp, Frances M.: 1968 Differences among older workers, volunteers, and persons who are neither. *Journal of Gerontology* 23:497–501. 1975 Impact of improved housing on morale and life satisfaction. *The Gerontologist* 15: 511–515. 1978–1979 Effects of the living environment on activity and the use of time. *International Journal of Aging and Human Development* 9: 75–91.

Carrington, P., G. Collins, Jr., H. Benson, L. Wood, P. Lehrer, R. Woolfolk, and J. Cole: 1980 The use of meditative relaxation techniques for the management of stress in a working population. *Journal of Occupational Medicine* 22: 221–231.

Carstensen, Laura L.: 1991 Selectivity theory: Social activity in life-span context. Pp. 195–217 in K. W. Schaie and M. P. Lawton (eds.), *Annual Review of Gerontology and Geriatrics: 1991.* New York: Springer.

Cattanach, Lynn, and Jacob Kraemer Tebes: 1991 The nature of elder impairment and its impact on family caregivers' health and psychosocial functioning. *The Gerontologist* 31(2): 246–255.

Chambré, Susan Maizel: 1984 Is volunteering a substitute for role loss in old age? An empirical test of activity theory. *The Gerontologist* 24:292–298. 1993 Volunteerism by elders: Past trends and future prospects. *The Gerontologist* 33(2): 221–228.

Chappell, Neena L.: 1987 The interface among three systems of care: Self, informal, and formal. Pp. 159–179 in Russell A. Ward and Sheldon S. Tobin (eds.), *Health and Aging.* New York: Springer.

1991a In-group differences among elders living with friends and family other than spouses. *Journal of Aging Studies* 5(1):61–76.

Chappell, Neena L., and Mark Badger: 1989 Social isolation and well-being. *Journal of Gerontology* 44(5):S169–S176.

Charness, Neil, and Elizabeth A. Bosman: 1992 Human factors and aging. Pp. 495–552 in Fergus I. M. Craik and Timothy A. Salthouse (eds.), *The Handbook of Aging and Cognition.* Hillsdale, NJ: Lawrence Erlbaum.

Chatters, Linda M., and Robert Joseph Taylor: 1994 Religious involvement among older African-Americans. Pp. 196–230 in J. S. Levin (ed.), *Religion in Aging and Health.* Thousand Oaks, CA: Sage.

Cherlin, Andrew, and Frank F. Furstenberg, Jr.: 1985 Styles and strategies of grandparenthood. Pp. 97–116 in V. L. Bengtson and J. F. Robertson (eds.), *Grandparenthood.* Beverly Hills, CA: Sage.

1985–1986 Grandparents and family crisis. *Generations* 10(4):26–28.

Chevan, Albert: 1987 Homeownership in the older population: 1940–1980. *Research on Aging* 9(2):226–280.

Cicirelli, Victor G.: 1985 The role of siblings as family caregivers. Pp. 93–107 in William J. Sauer and Raymond T. Coward (eds.), *Social Support Networks and the Care of the Elderly.* New York: Springer.

1992 *Family Caregiving: Autonomous and Paternalistic Decision Making.* Newbury Park, CA: Sage.

Clark, Margaret, and Barbara Anderson: 1967 *Culture and Aging.* Springfield, IL: Thomas.

Clark, Robert L.: 1988 The future of work and retirement. *Research on Aging* 10(2):169–193.

Clark, W. A. V., and Suzanne Davies: 1990 Elderly mobility and mobility outcomes: Households in the later stages of the life course. *Research on Aging* 12(4):430–462.

Coe, Richard D.: 1988 A longitudinal examination of poverty in the elderly years. *The Gerontologist* 28(4):540–544.

Cohen, Carl I., and Jay Sokolovsky: 1980 Social engagement versus isolation: The case of the aged in SRO hotels. *The Gerontologist* 20:36–44.

Cohen, Carl I., Jeanne Teresi, Douglas Holmes, and Eric Roth: 1988 Survival strategies of older homeless men. *The Gerontologist* 28(1):58–65.

Cohen, Elias S.: 1988 The elderly mystique: Constraints on the autonomy of the elderly with disabilities. *The Gerontologist* 28 (Supplement):24–31.

Cohen, Gene D.: 1994 Journalistic elder abuse: It's time to get rid of fictions, get down to facts. *The Gerontologist* 34(3):399–401.

Cole, Thomas R.: 1992 *The Journey of Life: A Cultural History of Aging in America.* New York: Cambridge University Press.

Colsher, Patricia L., Lorraine T. Dorfman, and Robert B. Wallace: 1988 Specific health conditions and work-retirement status among the elderly. *Journal of Applied Gerontology* 7(4):485–503.

Comfort, Alex: 1976 Age prejudice in America. *Social Policy* 7(3):3–8.

Comstock, G., S. Chafee, N. Katzman, M. McCombs, and D. Roberts: 1978 *Television and Human Behavior.* New York: Columbia University Press.

Congressional Budget Office: 1983 Major Legislative Changes in Human Resources Programs since January, 1981: Staff Memorandum.

Consumer Reports: 1989 Beyond Medicare. *Consumer Reports*, June, pp. 375–391.

Cook, Fay Lomax, and Edith Barrett: 1992 *Support for the American Welfare State: The Views of Congress and the Public.* NY: Columbia University Press.

Cooney, Teresa M.: 1989 Co-residence with adult children: A comparison of divorced and widowed women. *The Gerontologist* 29(6):779–784.

Corso, John. F.: 1981 *Aging, Sensory Systems, and Perception.* New York: Praeger.

Costa, Paul T., Jr., and Robert R. McCrae: 1988 Personality in adulthood: A six-year longitudinal study. *Journal of Personality and Social Psychology* 54:853–863.

Cotman, Carl W., and Vicky R. Holets: 1985 Structural changes at synapses with age: Plasticity and regeneration. Pp. 617–644 in Caleb E. Finch and Edward L. Schneider (eds.), *Handbook of the Biology of Aging*, 2nd ed. New York: Van Nostrand Reinhold.

Cottrell, Fred: 1955 *Energy and Society.* New York: McGraw-Hill.

1960a Governmental functions and the politics of age. Pp. 624–665 in Clark Tibbitts (ed.), *Handbook of Social Gerontology.* Chicago: University of Chicago Press.

1960b   The technological and societal basis of aging. Pp. 92–119 in Clark Tibbitts (ed.), *Handbook of Social Gerontology*. Chicago: University of Chicago Press.

1966   Aging and the Political System. Pp. 77–113 in J. C. McKinney and F. T. de Vyver (eds.), *Aging and Social Policy*. New York: Appleton-Century-Crofts.

1971   Government and Non-Government Organization. Washington, D.C.: White House Conference on Aging.

Cottrell, Fred, and Robert C. Atchley: 1969 Women in Retirement: A Preliminary Report. Oxford, OH: Scripps Foundation.

Courtenay, B. C., L. W. Poon, P. Martin, G. M. Clayton, and M. K. Johnson: 1992 Religiosity and adaptation in the oldest-old. *International Journal of Aging and Human Development* 34:47–56.

Cousins, Norman: 1968 Art, adrenaline, and the enjoyment of living. *Saturday Review,* April 20, pp. 20–24.

Coward, Raymond T., and Stephen J. Cutler: 1991   The composition of multigenerational households that include elders. *Research on Aging* 13(1):55–73.

Coward, Raymond T., Stephen J. Cutler, and Frederick E. Schmidt: 1989 Differences in the household composition of elders by age, gender, and area of residence. *The Gerontologist* 29(6):814–821.

Coward, Raymond T., Claydell Horne, and Jeffrey W. Dwyer: 1992 Demographic perspectives on gender and family caregiving. Pp. 18–33 in J. W. Dwyer and R. T. Coward (eds.), *Gender, Families, and Elder Care*. Newbury Park, CA: Sage.

Cowgill, Donald O.: 1972 A theory of aging in cross-cultural perspective. Pp. 1–14 in Donald O. Cowgill and Lowell Holmes (eds.), *Aging and Modernization*. New York: Appleton-Century-Crofts.

1974   Aging and modernization: A revision on theory. Pp. 123–146 in Jaber F. Gubrium (ed.), *Late Life. Communities and Environmental Policy*. Springfield IL: Thomas.

1986   *Aging Around the World*. Belmont, CA: Wadsworth.

Cox, Donna M.: 1993 The influence of class on aging policy: Why catastrophic was repealed. *Journal of Aging Studies* 7(1):55–65.

Craik, Fergus I. M., and Timothy A. Salthouse: 1992 *The Handbook of Aging and Cognition*. Hillsdale, NJ: Lawrence Erlbaum.

Crimmins, Eileen M., and Dominique G. Ingegneri: 1990 Interaction and living arrangements of older parents and their children: Past trends, present determinants, future implications. *Research on Aging* 12(1):3–35.

Crispell, Diane, and William H. Frey: 1993 American maturity. *American Demographics,* March, pp. 31–42.

Crowley, Joan E.: 1985 Longitudinal effects of retirement on men's psychological and physical well-being. Pp. 147–173 in Herbert S. Parnes (ed.), *Retirement Among American Men*. Lexington, MA: D. C. Heath.

Crown, William H.: 1988 State economic implications of elderly interstate migration. *The Gerontologist* 28(4):533–539.

Crystal, Stephen: 1982 *America's Old Age Crisis: Public Policy and the Two Worlds of Aging*. New York: Basic Books.

Crystal, Stephen, and Dennis Shea: 1990 Cumulative advantage, cumulative disadvantage, and inequality among elderly people. *The Gerontologist* 30(4):437–443.

Cuba, Lee: 1991 Models of migration decision making reexamined: The destination search of older migrants to Cape Cod. *The Gerontologist* 31(2):204–209.

Cumming, Elaine, and William E. Henry: 1961 *Growing old: The Process of Disengagement*. New York: Basic Books.

Cutler, Neal E.: 1981 Political characteristics of elderly cohorts in the twenty-first century. Pp. 127–157 in Sara B. Kiesler et al. (eds.), *Aging: Social Change*. New York: Academic Press.

Cutler, Stephen J.: 1976 Age differences in voluntary association memberships. *Social Forces* 55(1):43–58.

1977   Aging and voluntary association participation. *Journal of Gerontology* 32(4):470–479.

Cutler, Stephen J., and Armin E. Grams: 1988 Correlates of self-reported everyday memory problems. *Journal of Gerontology: Social Sciences* 43(3):S82–S90.

Cutler, Stephen J., and Jon Hendricks: 1990 Leisure and time use across the life course. Pp.169–185 in Robert H. Binstock and Linda K. George (eds.), *Handbook of Aging and the Social Sciences,* 3d ed. New York: Academic Press.

Cutler, Stephen J., and J. J. Hughes: 1982 The age structure of voluntary association memberships: Some recent evidence. Paper presented at the Annual Meeting of the Gerontological Society of America, Boston, November 21.

Dahrendorf, Ralf: 1988 *The Modern Social Conflict: An Essay on the Politics of Liberty.* New York: Weidenfeld & Nicholson.

Dail, Paula W.: 1988 Prime-time television portrayals of older adults in the context of family life. *The Gerontologist* 28(5):700–706.

Danigelis, Nicholas L., and Alfred P. Fengler: 1990 Homesharing: How social exchange helps elders live at home. *The Gerontologist* 30(2):162–170.

Dannefer, Dale: 1988 Differential gerontology and the stratified life course: Conceptual and methodological issues. Pp. 3–36 in George L. Maddox and M. Powell Lawton (eds.), *Annual Review of Gerontology and Geriatrics,* Volume 8. New York: Springer.

Davidson, G. W.: 1979 Hospice care for the dying. Pp. 101–119 in H. Wass (ed.), *Dying: Facing the Facts.* New York: McGraw-Hill.

Davis, Karen, and Diane Rowland: 1991 Old and poor: Policy challenges in the 1990s. *Journal of Aging & Social Policy* 2(3/4):37–59.

Day, Alice T.: 1991 *Remarkable Survivors: Insights into Successful Aging Among Women.* Washington, D.C.: Urban Institute Press.

Day, Christine L.: 1990 *What Older Americans Think: Interest Groups and Aging Policy.* Princeton, NJ: Princeton University Press.

1993 Public opinion toward costs and benefits of Social Security and Medicare. *Research on Aging* 15(3):279–298.

Dean, Lois R.: 1962 Aging and the decline of affect. *Journal of Gerontology* 17:440–46.

Decharms, R., and G. H. Moeller: 1961 Values expressed in children's readers: 1800–1950. *Journal of Abnormal and Social Psychology* 64:136–142.

De Crow, Roger: 1975 *New Learning for Older Americans: An Overview of National Effort.* Washington, D.C.: Adult Education Association.

Dement, William, Gary Richardson, Patricia Prinz, Mary Carskadon, Daniel Kripke, and Charles Czeisler: 1985 Changes of sleep and wakefulness with age. Pp. 692–717 in Caleb E. Finch and E. L. Schneider (eds.), *Handbook of the Biology of Aging.* New York: Van Nostrand Reinhold.

Dennis, Helen: 1989 The current state of preretirement planning. *Generations* 13(2):38–41.

De Vries, Brian, Susan Bluck, and James E. Birren: 1993 The understanding of death & dying in a life span perspective. *The Gerontologist* 33(3):366–372.

Dill, Ann, Phil Brown, Desiree Ciambrone, and William Rakowski: 1995 The meaning and practice of self-care by older adults: A qualitative assessment. *Research on Aging* 17(1):P8–P41.

Dobson, Cynthia: 1983 Sex-role and marital-role expectations. Pp. 109–126 in Timothy H. Brubaker (ed.), *Family Relationships in Later Life.* Beverly Hills, CA: Sage.

Doering, Mildred, Susan R. Rhodes, and Michael Schuster: 1983 *The Aging Worker: Research and Recommendations.* Beverly Hills, CA: Sage.

Doone, Barbara R.: 1991 Drug Use Patterns, Knowledge, and Adherence Among Older Individuals Living in the Community. Unpublished MGS thesis. Oxford, OH: Miami University.

Douglass, Richard L.: 1982 *Heating or Eating? The Crisis of Home Heating, Energy Costs and Well-being of the Elderly in Michigan.* Ann Arbor, MI: University of Michigan Institute of Gerontology.

Dowd, James J.: 1975 Aging as exchange: A preface to theory. *Journal of Gerontology* 30(5): 584–594.

1980 Exchange rates and old people. *Journal of Gerontology* 35:596–602.

Drake, St. Clair, and Horace Cayton: 1962 *Black Metropolis.* New York: Harper & Row.

Drazga, Linda, Melinda Upp, and Virginia Reno: 1982 Low-income aged: Eligibility and participation in SSI. Social Security Bulletin 45(5):28–31.

Drucker, Peter F.: 1976 *The Unseen Revolution: How Pension Fund Socialism Came to America.* New York: Harper & Row.

Dubler, Nancy N. (ed.): 1994 Current ethical issues in aging. *Generations* 18(4):4–70.

Dubois, W. E. B.: 1915 *The Negro.* New York: Holt, Rinehart and Winston.

Dugan, Elizabeth, and Vira R. Kivett: 1994 The importance of emotional and social isolation to loneliness among very old rural adults. *The Gerontologist* 34(3):340–346.

Duncan, Greg J.: 1984 *Years of Poverty: Years of Plenty.* Ann Arbor, MI: Institute for Social Research.

Dunlop, Burton D.: 1979 *The Growth of Nursing Home Care.* Lexington, MA: D. C. Heath.

Easterlin, Richard A.: 1991 The economic impact of prospective population changes in advanced industrial countries: An historical perspective. *Journal of Gerontology* 46(6):S299–S309.

Eckert, J. Kevin: 1980 *The Unseen Elderly: A Study of Marginally Subsistent Hotel Dwellers.* San Diego: Campanile Press.

Eckert, J. Kevin, and Stephanie M. Lyon: 1991 Regulation of board-and-care homes: Research to guide policy. *Journal of Aging & Social Policy* 3(1/2):147–162.

Eggebean, David, and Peter Uhlenberg: 1985 Changes in the organization of men's lives: 1960–1980. *Family Relations* 34:251–257.

Eglit, Howard: 1989 Ageism in the work place: An elusive quarry. *Generations* 13(2):31–35.

Eisenberg, D. M., Ronald C. Kessler, Cindy Foster, Frances E. Norlock, David R. Calkins, and Thomas L. Delbanco: 1993 Unconventional medicine in the United States. *New England Journal of Medicine* 328: 246–252.

Ekerdt, David J.: 1986 The busy ethic: Moral continuity between work and retirement. *The Gerontologist* 26:239–244.
1987 Why the notion persists that retirement harms health. *The Gerontologist* 27(4):454–457.

Ekerdt, David J., Lynn Baden, Raymond Bossé, and Elaine Dibbs: 1983 The effect of retirement on physical health. *American Journal of Public Health* 73:779–783.

Ekerdt, David J., Raymond Bossé, and Robert J. Glynn: 1985 Period effects on planned age for retirement, 1975–1984: Findings from the Normative Age Study. *Research on Aging* 7:395–407.

Ekerdt, David J., Raymond Bossé, and Sue Levkoff: 1985 An empirical test for phases of retirement: Findings from the Normative Aging Study. *Journal of Gerontology* 40:95–101.

Ekerdt, David J., Raymond Bossé, and Joseph S. Locastro: 1983 Claims that retirement improves health. *Journal of Gerontology* 38: 231–236.

Ekerdt, David J., and Stanley De Viney: 1990 On defining persons as retired. *Journal of Aging Studies* 4(3):211–299.
1993 Evidence for a preretirement process among older male workers. *Journal of Gerontology: Social Sciences* 48(2):S35–S43.

Elder, Glen H., Jr.: 1991 Making the best of life: Perspectives on lives, times, and aging. *Generations* 15(1):12–18.

Elder, Glen H., Jr., and Eliza K. Pavalko: 1993 Work careers in men's later years: Transitions, trajectories, and historical change. *Journal of Gerontology: Social Sciences* 48(4):S180–S191.

Elliott, Joyce: 1984 The daytime television drama portrayal of older adults. *The Gerontologist* 24:628–633.

Ellison, Christopher G.: 1994 Religion, the life stress paradigm, and the study of depression. Pp. 78–121 in J. S. Levin (ed.), *Religion in Aging and Health.* Thousand Oaks, CA: Sage.

Emerson, R. M.: 1962 Power-dependence relations. *American Sociological Review* 27:31–41.
1972 Exchange theory. Pp. 38–87 in Jospeh Berger, M. Zelditch, and B. Anderson (eds.), *Sociological Theories in Progress.* Volume 2. Boston: Houghton Mifflin.

Emerson, Ralph Waldo: 1837 Progress of culture. Address to the Phi Beta Kappa Society, July 18.

Epstein, Seymour: 1980 The self-concept: A review and the proposal of an integrated theory of personality. Pp. 82–132 in E. Staub (ed.), *Personality: Basic Issues and Current Research.* Englewood Cliffs, NJ: Prentice-Hall.

Erikson, Erik H.: 1963 *Childhood and Society.* New York: MacMillan.

Erikson, Erik H., Joan M. Erikson, and Helen Q. Kivnick: 1986 *Vital Involvement in Old Age.* New York: Norton.

Erikson, R., and K. Eckert: 1977 The elderly poor in downtown San Diego hotels. *The Gerontologist* 17:440–446.

Estes, Carroll L.: 1979 *The Aging Enterprise.* San Francisco: Jossey-Bass.
1991 The new political economy of aging: Introduction and critique. Pp. 19–36 in Meredith Minkler and C. L. Estes (eds.), *Critical Perspectives on Aging.* Amityville, NY: Baywood.
1993 The aging enterprise revisited. *The Gerontologist* 33(3):292–298.

Eustis, Nancy N., and Lucy Rose Fischer: 1991 Relationships between home care clients and their workers: Implications for quality of care. *The Gerontologist* 31(4):447–456.

Eustis, Nancy, Jay Greenberg, and Sharon Patten: 1984 *Long-Term Care for Older*

Persons: A Policy Perspective. Belmont, CA: Wadsworth.

Evans, Leonard: 1988 Older driver involvement in fatal and severe traffic crashes. *Journal of Gerontology: Social Sciences* 43(6):S186–S193.

Evans, Linda, David J. Ekerdt, and Raymond Bossé: 1985 Proximity to retirement and anticipatory involvement: Findings from the Normative Aging Study. *Journal of Gerontology* 40: 368–374.

Farrell, Michael P., and S. D. Rosenberg: 1981 *Men at Midlife.* Boston: Auburn House.

Ferraro, Kenneth F.: 1984 Widowhood and social participation in later life: Isolation or compensation? *Research on Aging* 6:451–468.

Fethke, Carol C.: 1989 Life-cycle models of saving and the effect of the timing of divorce on retirement economic well-being. *Journal of Gerontology* 44(3):S121–S128.

Filinson, Rachael: 1993 The effect of age desegregation on environmental quality for elderly living in public/publicly subsidized housing. *Journal of Aging and Social Policy* 5(3):77–93.

Finch, Caleb E., and Edward L. Schneider (Eds.): 1985 *Handbook of the Biology of Aging,* 2d ed. New York: Van Nostrand Reinhold.

Finley, Nancy J., M. Diane Roberts, and Benjamin F. Banahan, III: 1988 Motivators and inhibitors of attitudes of filial obligation toward aging parents. *The Gerontologist* 28:73–83.

Fischer, David H.: 1978 *Growing Old in America,* expanded edition. New York: Oxford University Press.

Fischer, Lucy Rose, and Nancy N. Eustis: 1988 DRGs and family care for the elderly: A case study. *The Gerontologist* 28(3):383–390.

Fischer, Lucy Rose, Daniel P. Mueller, and Philip W. Cooper: 1991 Older volunteers: A discussion of the Minnesota Senior Study. *The Gerontologist* 31(2):183–194.

Fiske, Marjorie, and David A. Chiriboga: 1990 *Change and Continuity in Adult Life.* San Francisco: Jossey-Bass.

Fletcher, Wesla L., and Robert O. Hansson: 1991 Assessing the social components of retirement anxiety. *Psychology and Aging* 6(1):76–85.

Fowler, James W.: 1981 *Stages of Faith.* San Francisco: Harper & Row.

1991 *Weaving the New Creation: Stages of Faith and the Public Church.* San Francisco: Harper & Row.

Francher, J. Scott: 1973 It's the Pepsi generation . . . : Accelerated aging and the TV commercial. *The International Journal of Aging and Human Development* 4:245–255.

Francis, Doris: 1990 The significance of work friends in late life. *Journal of Aging Studies* 4: 405–424.

Frankel, B. Gail, and David Dewit: 1989 Geographic distance and intergenerational contact: An empirical examination of the relationship. *Journal of Aging Studies* 3(2):139–162.

Frankfather, Dwight: 1977 *The Aged in the Community.* New York: Praeger.

Fries, James F.: 1980 Aging, natural death, and the compression of morbidity. *New England Journal of Medicine* 300:130–135.

Gaberlavage, George: 1987 *Social Services to Older Persons—Under the Social Services Block Grant.* Washington, D.C.: American Association of Retired Persons.

Gadow, Sally: 1983 Frailty and strength: The dialectic in aging. *The Gerontologist* 23:144–147.

Galbraith, J. Kenneth: 1958 *The Affluent Society.* Boston: Houghton Mifflin.

Gallagher, Dolores E., James N. Breckenridge, Larry W. Thompson, and James A. Peterson: 1983 Effects of bereavement on indicators of mental health in elderly widows and widowers. *Journal of Gerontology* 38:565–571.

Gallagher, Dolores E., Jonathan Rose, Patricia Rivera, Steven Lovett, and Larry W. Thompson: 1989 Prevalence of depression in family caregivers. *The Gerontologist* 29(4): 449–456.

Garfein, Adam J., and Regula Herzog: 1995 Robust aging among the young-old, old-old, and oldest-old. *Journal of Gerontology: Social Sciences* 50B(2):577–587.

Garrison, J.: 1978 Stress management training for the elderly. *Journal of the American Geriatrics Society* 26:397–403.

Gee, Ellen M.: 1990 Preferred timing of women's life events: A Canadian study. *International Journal of Aging and Human Development* 31(4):279–294.

George, Linda K., and Elizabeth C. Klipp: 1991 Subjective components of aging well: Researchers need to reconsider. *Generations* 15:1:57–60.

George, Linda K., and Stephen J. Weiler: 1981 Sexuality in middle and late life. *Archives of General Psychiatry* 38:919–923.

Gersuny, Carl: 1987 Employment seniority and senior citizens. *The Gerontologist* 27(4):458–463.

Gibby, Douglas A., and C. Dorer: 1988 How much retirement income do employees need? *Compensation and Benefits Management* 4(Autumn):25–34.

Gibson, Rose C.: 1986 Older black Americans. *Generations* 10(4):36–39.
1987 Reconceptualizing retirement for black Americans. *The Gerontologist* 27(6):691–698.

Gilford, Rosalie, and V. Bengtson: 1979 Measuring marital satisfaction in three generations: Positive and negative dimensions. *Journal of Marriage and the Family* 41:387–398.

Glasgow, Nina: 1991 A place in the country. *American Demographics,* March, pp. 24–30.

Glass, Jennifer, Vern L. Bengtson, and Charlotte C. Dunham: 1986 Attitude similarity in families. *American Sociological Review* 51:685–698.

Glick, Ira O., Robert S. Weiss, and C. Murray Parkes: 1974 *The First Year of Bereavement.* New York: Wiley.

Glock, Charles Y., and Rodney Stark: 1965 *Religion and Society in Transition.* Chicago: Rand McNally.

Golant, Stephen M.: 1990 The metropolitanization and suburbanization of the U.S. elderly population: 1970–1988. *The Gerontologist* 30(1): 80–85.
1992 *Housing America's Elderly.* Newbury Park, CA.: Sage.

Gold, Deborah T.: 1989 Sibling relationships in old age: A typology. *International Journal of Aging and Human Development* 28:37–54.
1990 Late-life sibling relationships: Does race affect typological distribution? *The Gerontologist* 30(6):741–748.

Goldberg, Evelyn L., George W. Comstock, and Shoban D. Harlow: 1988 Emotional problems and widowhood. *Journal of Gerontology: Social Sciences* 43(6):S206–S208.

Goldberg-Alberts, Amy L.: 1986 Private long-term care insurance: Problems and possibilities. *Journal of Long-Term Care Administration* 14(4):11–15.

Goodman, Mariene: 1994 Social, psychological, and developmental factors in women's receptivity to cosmetic surgery. *Journal of Aging Studies* 8(4):375–396.

Gordon, Chad, Charles M. Gaitz, and Judith Scott: 1976 Leisure and lives: Personal expressivity across the life span. Pp. 310–341 in Robert H. Binstock and Ethel Shanas (eds.), *Handbook of Aging and the Social Sciences.* New York: Van Nostrand Reinhold.

Goudy, Willis J.: 1981 Changing work expectations: Findings from the Retirement History Study. *The Gerontologist* 21:644–649.

Grad, Susan: 1990a Income change at retirement. *Social Security Bulletin* 53(1):2–10.
1990b Earnings replacement rates of new retired workers. *Social Security Bulletin* 53(10): 2–19.

Graebner, William: 1980 *A History of Retirement: The Meaning and Function of an American Institution, 1885–1978.* New Haven, CT: Yale University Press.

Gratton, Brian: 1986 *Urban Elders: Family, Work, and Welfare Among Boston's Aged, 1890–1950.* Philadelphia: Temple University Press.

Gratton, Brian, and Carole Haber: 1993 In search of 'intimacy at a distance': Family history from the perspective of elderly women. *Journal of Aging Studies* 7(2):183–194.

Greenberg, Jan S., and Marion Becker: 1988 Aging parents as family resources. *The Gerontologist* 28(6):786–791.

Greenblum, Joseph, and Barry Bye: 1987 Work values of disabled beneficiaries. *Social Security Bulletin* 50(4):67–74.

Greene, Vernon L.: 1989 Human capitalism and intergenerational justice. *The Gerontologist* 29:723–724.

Greenwald, Anthony: 1980 The totalitarian ego: Fabrication and revision of personal history. *The American Psychologist* 35:603–618.

Groger, Lisa: 1994 Decision as process: A conceptual model of black elders' nursing home placement. *Journal of Aging Studies* 8:77–94.

Gubrium, Jaber F.: 1975 Being single in old age. *The International Journal of Aging and Human Development* 6:29–41.

Guilliland, Nancy, and Linda Havir: 1990 Public opinion and long-term care policy. Pp. 242–253 in D. E. Biegel and A. Blum (eds.), *Aging and Caregiving.* Newbury Park, CA: Sage.

Gurin, Gerald, Joseph Veroff, and Sheila Feld: 1960 *Americans View Their Mental Health: A National Interview Study.* New York: Basic Books.

**Haas, William H., III, and William J. Serow:
1993** Amenity retirement migration process: A
model and preliminary evidence. *The Gerontologist* 33(2):212–220.

Haber, Carole: 1978 Mandatory retirement in
nineteenth-century America: The conceptual
basis for a new work cycle. *Journal of Social
History* 12:77–96.
1983 *Beyond Sixty-Five.* Cambridge, MA:
Cambridge University Press.

Haber, Carole, and Brian Gratton: 1992 Aging in
America: The perspective of history. Pp.
352–370 in Thomas R. Cole, David D. Van
Tassel, and Robert Kastenbaum (eds.), *Hand-
book of the Humanities and Aging.* New York:
Springer.
1994 *Old Age and the Search for Security: An
American Social History.* Bloomington, IN.:
Indiana University Press.

Haber, David, 1994 *Health Promotion and Aging.*
New York: Springer.

Hage, Jerald, and Charles H. Powers: 1992
*Post-Industrial Lives: Roles and Relationships
in the 21st Century.* Newbury Park, CA: Sage.

Hagestad, Gunhild O.: 1984 The continuous bond:
A dynamic multigenerational perspective on
parent-child relations between adults. In M. Perlmutter (ed.), *Minnesota Symposium on Child
Psychology.* Hillsdale, NJ: Lawrence Erlbaum.
1985 Continuities and connectedness. Pp. 31–
48 in Vern L. Bengtson and Joan F. Robertson
(eds.), *Grandparenthood.* Beverly Hills, CA:
Sage.

**Hale, W. E., F. E. May, R. G. Marks, and R. B.
Stewart: 1987** Drug use in an ambulatory geri-
atric population. *Drug Intelligence and Clinical
Pharmacy* 21:530–535.

**Hammel, E. A., K. W. Wachter, and C. K.
Mcdaniel: 1981** The kin of the aged in A.D.
2000. Pp. 11–39 in Sara B. Keisler et al. (eds.),
Aging: Social Change. New York: Academic
Press.

**Hamon, Raeann R., and Rosemary Blieszner:
1990** Filial responsibility expectations among
adult child–older parent pairs. *Journal of
Gerontology* 45(3):P110–P112.

Hand, Jennifer: 1983 Shopping-bag women:
Aging deviants in the city. Pp. 155–177 in E. W.
Markson (ed.), *Older Women: Issues and
Prospects.* Lexington, MA: Lexington Books.

Hanlon, Mark, 1986 Age and commitment toward work. *Research on Aging* 8:289–316.

**Happel, Stephen K., Timothy D. Hogan, and
Elmer Pflanz: 1988** The economic impact of
elderly winter residents in the Phoenix area.
Research on Aging 10(1):119–133.

Harman, D.: 1956 Aging: A theory based on free
radical and radiation chemistry. *Journal of
Gerontology* 11:298–300.

**Harris, Adella J., and Jonathan F. Feinberg:
1977** Television and aging: Is what you see
what you get? *The Gerontologist* 17:464–466.

Harris, Charles S.: 1978 *Fact Book on Aging: A
Profile of America's Older Population.* Washington, D.C.: National Council on Aging.

Harris, Louis, and Associates: 1975 *The Myth
and Reality of Aging in America.* Washington,
D.C.: National Council on Aging.
1981 *Aging in the Eighties: America in Transition.* Washington, D.C.: National Council on
Aging.

Harris, Mary B.: 1994 Growing old gracefully: Age
concealment and gender. *Journal of Gerontology:
Psychological Sciences* 49(4):P149–P158.

Hartwigsen, Gail, and Roberta Null: 1989 Full-
timing: A housing alternative for older people.
*International Journal of Aging and Human
Development* 29:317–328.
1990 Full-timers: Who are these older people
who are living in their RVs? *Journal of Housing
for the Elderly.* 7:133–147.

Hausman, Perrie B., and Marc E. Weksler: 1985
Change in the immune response with age. Pp.
414–432 in C. E. Finch and E. L. Schneider
(eds.), *Handbook of the Biology of Aging.* New
York: Van Nostrand Reinhold.

Havighurst, Robert J.: 1963 Successful aging.
Pp. 299–230 in Richard H. Williams, Clark
Tibbitts, and Wilma Donohue (eds.), *Processes
of Aging: Social and Psychological Perspectives.*
New York: Atherton.

**Havighurst, Robert J., Bernice L Neugarten, and
Sheldon S. Tobin: 1963** Disengagement, per-
sonality, and life satisfaction. Pp. 319–324 in P.
Hansen (ed.), *Age with a Future.* Copenhagen:
Munksgaard.

Hayflick, Leonard: 1987 Biological theories of
aging. Pp. 64–68 in George L. Maddox (ed.), *The
Encyclopedia of Aging.* New York: Springer.

**Hayward, Mark D., Eileen M. Crimmins, and
Linda A. Wray: 1994** The relationship between retirement life cycle changes and older
men's labor force participation rates. *Journal of
Gerontology: Social Sciences* 49(5):S219–S230.

Hayward, Mark D., William R. Grady, and Steven D. McLaughlin: 1988 The retirement process among older women in the United States: Changes in the 1970s. *Research on Aging* 10(3): 358–382.

Heidbreder, Elizabeth M.: 1972 Factors in retirement adjustment: White-collar/blue-collar experience. *Industrial Gerontology* 12:69–79.

Heidrich, Susan M., and Carol D. Ryff: 1993 The role of social comparisons processes in the psychological adaptation of elderly adults. *Journal of Gerontology: Psychological Sciences* 48(3): P127–P136.

Heimstra, Roger: 1991 Self-directed learning for older adults. Pp. 101–112 in Robert Harootyan et al. (eds.), *Resourceful Aging: Today and Tomorrow.* Volume V. *Lifelong Education.* Washington, D.C.: American Association of Retired Persons.

Hellebrandt, Frances A.: 1978 The senile dement in our midst: A look at the other side of the coin. *The Gerontologist* 18:67–70.

Hendricks, Jon: 1982 Time and social science. Pp. 12–45 in E. H. Mizruchi, B. Glassner, and T. Pastorello (eds.), *Time and Aging.* New York: General Hall.

Henretta, John C.: 1986 Retirement and residential moves by elderly households. *Research on Aging* 8:23–37.

Herskovits, Elizabeth, and Linda Mitteness: 1994 Transgressions and sickness in old age. *Journal of Aging Studies* 8:327–340.

Herzog, A. Regula, James S. House, and James N. Morgan: 1991 Relation of work and retirement to health and well-being in older age. *Psychology and Aging* 6(2):202–211.

Herzog, A. Regula, Robert L. Kahn, James N. Morgan, James S. Jackson, and Toni C. Antonuccia: 1989 Age differences in productive activities. *Journal of Gerontology: Social Sciences* 44(4):S129–S138.

Heyman, Dorothy K., and Frances C. Jeffers: 1968 Wives and retirement: A pilot study. *Journal of Gerontology* 23:488–496.

Hickey, Tom, and Richard L. Douglass: 1981 Neglect and abuse of older family members: Professional's perspectives and case experiences. *The Gerontologist* 21:171–176.

Hill, Bette S., and Katherine A. Hinckley: 1991 The changing national policy system: Complexity, Medicare, and implications for aging groups. *Journal of Aging & Social Policy* 3(1/2):91–110.

Hill, Connie Dessonville, Larry W. Thompson, and Dolores Gallagher: 1988 The role of anticipatory bereavement in older women's adjustment to widowhood. *The Gerontologist* 28(6):792–796.

Himmelstein, David: 1990 The case for a national health care program. Speech to the Ohio Retirement Systems Health Care Seminar. March 9.

Hochschild, Arlie Russell: 1973 *The Unexpected Community.* Englewood Cliffs, NJ: Prentice-Hall.

Hodgson, Lynne Gershenson: 1992 Adult grandchildren and their grandparents: An enduring bond. *International Journal of Aging and Human Development* 34(3):209–225.

Hogan, Timothy D.: 1987 Determinants of the seasonal migration of the elderly to sunbelt states. *Research on Aging* 9(1):115–133.

Holden, Karen C.: 1989 Do retirement resources hold up? *Generations* 13(2):42–46.

Holmes, Ellen Rhoads, and Lowell D. Holmes: 1995 *Other Cultures, Elder Years.* 2d ed. Thousand Oaks, CA: Sage.

Holmes, Thomas H., and R. H. Rahe: 1967 The social readjustment rating scale. *Journal of Psychosomatic Research* 11:213–218.

Holstein, Martha, and Meredith Minkler: 1991 The short life and painful death of the Medicare Catastrophic Coverage Act. Pp. 189–206 in Meredith Minkler and C. L. Estes (eds.), *Critical Perspectives on Aging.* Amityville, NY: Baywood.

Homans, George F.: 1961 *Social Behavior: Its Elementary Forms.* New York: Harcourt Brace Jovanovich.

Hong, Lawrence, and Robert W. Duff: 1994 Widows in retirement communities: The social context of subjective well-being. *The Gerontologist* 34(3):347–352.

Hooker, Karen, and Deborah G. Ventis: 1984 Work ethic, daily activities and retirement satisfaction. *Journal of Gerontology* 39:478–484.

Hooper, Celia R.: 1994 Sensory and sensory integrative development. Pp. 93–106 in B. R. Bonder and M. B. Wagner (eds.), *Functional Performance in Older Adults.* Philadelphia: F. A. Davis.

Horn, Susan D., Phoebe D. Sharkey, Angela F. Chambers, and Roger A. Horn: 1985 Severity of illness within DRGS: Impact on prospective payment. *American Journal of Public Health* 75:1195–1199.

Hornbrook, Mark C., Victor J. Stephens, Darlene J. Wingfield, Jack F. Hollis, Merwyn R. Greenlick, and Marcia G. Ory: 1994 Preventing falls among community-dwelling older persons: Results from a randomized trial. *The Gerontologist* 34:16–23.

Horowitz, Amy: 1994 Vision impairment and functional disability among nursing home residents. *The Gerontologist* 34(3):316–323.

Howard, John H., Peter A. Rechnitzer, David A. Cunningham, and Allan P. Donner: 1986 Change in Type A behavior a year after retirement. *The Gerontologist* 26:643–649.

Hoyert, Donna L.: 1991 Financial and household exchanges between generations. *Research on Aging* 13:205–225.
1992 Factors related to the well-being and life activities of family caregivers. *Family Relations* 41:74–81.

Huber, Lynn W.: 1995 The church in the community. Pp. 285–305 in Melvin A. Kimble et al. (eds.), *Aging, Spirituality, and Religion: A Handbook*. Minneapolis: Fortress Press.

Hudson, Robert B.: 1978 The "graying" of the federal budget and its consequences for old-age policy. *The Gerontologist* 18:428–440.
1987 Policy analysis: Issues and practices. Pp. 526–528 in George L. Maddox (ed.), *The Encyclopedia of Aging*. New York: Springer.

Hudson, Robert B., and Robert H. Binstock: 1976 Political systems and aging. Pp. 369–400 in R. Binstock and E. Shanas (eds.), *Handbook of Aging and the Social Sciences*. New York: Van Nostrand Reinhold.

Hudson, Robert B., and John Strate: 1985 Aging and political systems. Pp. 554–585 in R. Binstock and E. Shanas (eds.), *Handbook of Aging and the Social Sciences*, 2d ed. New York: Van Nostrand Reinhold.

Hummert, Mary Lee, Teri A. Garstka, Jaye L. Shaner, and Sharon Strahm: 1994 Stereotypes of the elderly held by young, middle-aged, and elderly adults. *Journal of Gerontology: Psychological Sciences* 49(5):P240–P249.
1995 Judgments about stereotypes of the elderly: Attitudes, age associations, and typicality ratings of young, middle–aged, and elderly adults. *Research on Aging* 17(2):168–189.

Hushbeck, Judith C.: 1989 *Old and Obsolete: Age Discrimination and the American Worker, 1860–1920*. New York: Garland.

Huyck, Margaret Hellie: 1990 Gender differences and aging. Pp. 124–134 in James E. Birren and K. Warner Schaie (eds.), *Handbook of the Psychology of Aging*. New York: Academic Press.

Iams, Howard M.: 1986 Characteristics of the longest job for newly disabled workers: Findings from the New Beneficiary Survey. *Social Security Bulletin* 49(12):13–18.

Iams, Howard M., and John L. McCoy: 1991 Predictors of mortality among newly retired workers. *Social Security Bulletin* 54(3):2–10.

Jackson, James S., Toni C. Antonucci, and Rose C. Gibson: 1990 Cultural, racial, and ethnic minority influences on aging. Pp. 103–123 in James E. Birren and K. Warner Schaie (eds.), *Handbook of the Psychology of Aging*, 3d ed. New York: Academic Press.

Jacobs, Bruce: 1990 Aging and politics. Pp. 350–361 in R. H. Binstock and L. K. George (eds.), *Handbook of Aging and the Social Sciences*, 3d ed. New York: Academic Press.

James, William: 1890 *Principles of Psychology*. New York: Dover.

Jecker, Nancy S. (ed.): 1991 *Aging and Ethics*. Clifton, NJ: Humana Press.

Jeffreys-Fox, Bruce: 1977 *How Realistic Are Television's Portrayals of the Elderly?* University Park, PA: The Annenberg School of Communications.

Jendrek, Margaret Platt: 1992 Grandparents who provide care to grandchildren: Preliminary findings and policy issues. Paper presented at the Annual Meeting of Sociologists for Women in Society, Pittsburgh.

Johnson, Colleen Leahy: 1985 The impact of illness on late-life marriages. *Journal of Marriage and the Family* 47:165–172.

Johnson, Colleen Leahy, and Donald J. Catalano: 1981 Childless elderly and their family supports. *The Gerontologist* 21:610–618.

Johnson, Colleen Leahy, and Lillian E. Troll: 1994 Constraints and facilitators to friendships in late late life. *The Gerontologist* 34(1):79–87.

Johnson, Elizabeth S., and Barbara J. Bursk: 1977 Relationships between the elderly and their adult children. *The Gerontologist* 17: 90–96.

Johnson, Paul: 1991   *The Birth of the Modern: World Society 1815–1830.* New York: HarperCollins.

Johnson, Sheila K.: 1971   *Idle Haven: Community Building Among the Working-Class Retired.* Berkeley: University of California Press.

Kahn, Robert L., and Toni C. Antonucci: 1981 Convoys of social support: A life-course approach. Pp. 383–405 in Sara B. Kiesler et al. (eds.), *Aging: Social Change.* New York: Academic Press.

Kalish, Richard A.: 1976 Death and dying in a social context. Pp. 483–507 in R. H. Binstock and E. Shanas (eds.), *Handbook of Aging and the Social Sciences.* New York: Van Nostrand Reinhold.

1985 *Death, Grief, and Caring Relationships,* 2d ed. Monterey, CA: Brooks/Cole.

Kalish, Richard A., and D. K. Reynolds: 1981 *Death and Ethnicity: A Psychocultural Study.* Farmingdale, NY: Baywood.

Kane, Nancy M.: 1989 The home care crisis of the nineties. *The Gerontologist* 29(1):24–31.

Karp, David A.: 1987 Professionals beyond midlife: Some observations on work satisfaction in the fifty-to-sixty-year decade. *Journal of Aging Studies* 1(3):209–224.

1988 A decade of reminders: Changing age consciousness between fifty and sixty years old. *The Gerontologist* 28(6): 727–738.

1989 The social construction of retirement among professionals 50–60 years old. *The Gerontologist* 29(6):750–760.

Kart, Cary S., and Carol A. Engler: 1994 Predisposition to self-health care: Who does what for themselves and why? *Journal of Gerontology: Social Sciences* 49(6):S301–S308.

1995 Self-health care among the elderly: A test of the health-behavior model. *Research on Aging* 17(4):434–458.

Kastenbaum, Robert J.: 1969 Death and bereavement in later life. Pp. 28–54 in A. H. Kutscher (ed.), *Death and Bereavement.* Springfield, IL: Thomas.

1983 Time course and time perspective in later life. Pp. 80–101 in Carl Eisdorfer (ed.), *Annual Review of Gerontology and Geriatrics,* Volume 3. New York: Springer.

1987 Alcohol use. Pp. 24–25 in George L. Maddox (ed.), *Encyclopedia of Aging.* New York: Springer.

1992 *The Psychology of Death,* 2d ed. New York: Springer.

1993 *Encyclopedia of Adult Development.* Phoenix: Oryx Press.

Katzman, Robert: 1982–1983 The complex problems of diagnosis (dementia). *Generations* 7(1): 8–10.

Kaufman, Sharon R.: 1986 *The Ageless Self: Sources of Meaning in Late Life.* Madison, WI: University of Wisconsin Press.

1994 The social construction of frailty: An anthropological perspective. *Journal of Aging Studies* 8(1):45–58.

Kay, Edwin J., Barbara H. Jensen-Osinski, Peter G. Beidler, and Judith L. Aronson: 1983 The graying of the college classroom. *The Gerontologist* 23:196–199.

Kayne, Ronald C.: 1976 Drugs and the aged. Pp. 436–451 in Irene M. Burnside (ed.), *Nursing and the Aged.* New York: McGraw-Hill.

Kearl, Michael C.: 1982 An inquiry into the positive personal and social effects of old age stereotypes among the elderly. *The International Journal of Aging and Human Development* 14:277–290.

Keating, Norah C., and Priscilla Cole: 1980 What do I do with him 24 hours a day? Changes in housewife role after retirement. *The Gerontologist* 20:84–89.

Keith, Jennie, Christine L. Fry, Anthony P. Glascock, Charlotte Ikels, Jeanette Dickerson–Putnam, Henry C. Harpending, and Patricia Draper: 1994 *The Aging Experience: Diversity and Commonality Across Cultures.* Thousand Oaks, CA: Sage.

Keith, Pat M., Kathleen Hill, Willis J. Goudy, and Edward A. Powers: 1984 Confidants and well-being: A note on male friendship in old age. *The Gerontologist* 24:318–320.

Keith, Verna M.: 1993 Gender, financial strain, and psychological distress among older adults. *Research on Aging* 15(2):123–147.

Kelly, George A.: 1955 *The Psychology of Personal Constructs.* New York: Norton.

Kelly, John R.: 1983 *Leisure Identities and Interaction.* Boston: Allen & Unwin.

1987 *Peoria Winter: Styles and Resources in Later Life.* Lexington, MA: Lexington Books.

Kelly, John R. (ed.): 1993 *Activity and Aging.* Newbury Park, CA: Sage.

Kelly, John R., Marjorie W. Steinkamp, and Janice R. Kelly: 1986 Later life leisure: How they play in Peoria. *The Gerontologist* 26: 531–537.

Kenny, Kathleen, and Bronwyn Belling: 1987 Home equity conversion: A counseling model. *The Gerontologist* 27(1):9–12.

Kerckhoff, Alan C.: 1966 Norm-value clusters and the strain toward consistency among older married couples. Pp. 138–159 in Ida H. Simpson and John C. McKinney (eds.), *Social Aspects of Aging*. Durham, NC: Duke University Press.

Kerin, Pamela B., Carroll L. Estes, and Elizabeth B. Douglass: 1989 Federal funding for aging education and research: A decade analysis. *The Gerontologist* 29(5):606–614.

Kertzer, David I., and Peter Laslett: 1995 *Aging in the Past: Demography, Society, and Old Age*. Berkeley: University of California Press.

Kieffer, Jarold A.: 1983 *Gaining the Dividends of Longer Life: New Roles of Older Workers*. Boulder, CO: Westview Press.

Kiesler, Sara B.: 1981 The aging population, social trends, and changes in behavior and belief. Pp. 41–74 in Sara B. Kiesler et al. (eds.), *Aging: Social Change*. New York: Academic Press.

Kimble, Melvin A., Susan H. McFadden, James W. Ellor, and James J. Seeber (eds.): 1995 *Aging, Spirituality and Religion: A Handbook*. Minneapolis: Fortress Press.

Klapp, Orinn E.: 1973 *Models of Social Order*. Palo Alto, CA: National Press Books.

Klegon, Douglas A.: 1977 *Manpower in the Field of Aging*. Oxford, OH: Scripps Foundation Gerontology Center.

Kligman, Albert M., Gary L. Grove, and Arthur K. Balin: 1985 Aging of human skin. Pp. 820–841 in Caleb E. Finch and Edward L. Schneider (eds.), *Handbook of the Biology of Aging*, 2d ed. New York: Van Nostrand Reinhold.

Kline, Donald W., and Frank Schieber: 1985 Vision and aging. Pp. 296–331 in James E. Birren and K. Warner Schaie (eds.), *Handbook of the Psychology of Aging*, 2d ed. New York: Van Nostrand Reinhold.

Knutsen, Mary M.: 1995 A feminist theology of aging. Pp. 460–482 in Melvin A. Kimble et al. (eds.), *Aging, Spirituality, and Religion: A Handbook*. Minneapolis: Fortress Press.

Koenig, Harold G.: 1995a *Aging and God: Spiritual Pathways to Mental Health in Midlife and Later Years*. New York. Haworth Pastoral Press.

1995b *Research on Religion and Aging: An Annotated Bibliography*. Westport, CT: Greenwood Press.

1995c Religion and health in later life. Pp. 9–29 in Melvin A. Kimble et al. (eds.), *Aging, Spirituality, and Religion: A Handbook*. Minneapolis: Fortress Press.

Koenig, Harold G., Linda K. George, Dan G. Blazer, and J. Pritchett: 1993 The relationship between religion and anxiety in a sample of community-dwelling older adults. *Journal of Geriatric Psychiatry* 26:65–93.

Koenig, Harold G., J. N. Kvole, and C. Ferrel: 1988 Religion and well-being in later life. *The Gerontologist* 28:18–28.

Koff, Theodore H.: 1982 *Long Term Care: An Approach to Serving the Frail Elderly*. Boston: Little, Brown.

Kogan, Nathan: 1990 Personality and aging. Pp. 330–346 in James E. Birren and K. Warner Schaie (eds.), *Handbook of the Psychology of Aging*, 3d ed. New York: Academic Press.

Kohlberg, L.: 1973 Continuities in childhood and adult moral development revisited. Pp. 179–204 in Paul B. Baltes and K. Warner Schaie (eds.), *Life-Span Developmental Psychology: Personality and Socialization*. New York: Academic Press.

Krause, Neal: 1993 Early parental loss and personal control in later life. *Journal of Gerontology: Psychological Sciences* 48(3): P117–P126.

Krause, Neal, and Thanh Van Tram: 1989 Stress and religious involvement among older blacks. *Journal of Gerontology* 44(1):S4–S13.

Kronus, Sidney: 1971 *The Black Middle Class*. Columbus, OH: Merrill.

Krout, John A., Stephen J. Cutler, and Raymond T. Coward: 1990 Correlates of senior center participation: A national analysis. *The Gerontologist* 30(1):72–79.

Kübler-Ross, Elizabeth: 1969 *On Death and Dying*. New York: Macmillan.

Kunkel, Suzanne R.: 1979 Sex Differences in Adjustment to Widowhood. Unpublished M.A. thesis. Oxford, OH: Miami University.

1989 An extra eight hours a day. *Generations* 13(2):57–60.

Kunkel, Suzanne R., and Robert A. Applebaum: 1989 *Estimating the Prevalence of Long-Term Disability for an Aging Society*. Oxford, OH: Scripps Gerontology Center.

Lachman, Margie E.: 1986 Personal control in later life: Stability, change, and cognitive correlates. Pp. 207–236 in Margaret M. Baltes and Paul B. Baltes (eds.), *The Psychology of Control and Aging*. Hillsdale, NJ: Lawrence Erlbaum.

Lakin, Martin, and Carl Eisdorfer: 1962 A study of affective expression among the aged. Pp. 650–654 in Clark Tibbitts and Wilma Donahue (eds.), *Social and Psychological Aspects of Aging*. New York: Columbia University Press.

Lambing, Mary L. B.: 1972 Leisure-time pursuits among retired blacks by social status. *The Gerontologist* 12:363–367.

Larson, Martha Klein: 1995 A feminist perspective on aging. Pp. 242–252, in Melvin A. Kimble et al. (eds.), *Aging, Spirituality, and Religion: A Handbook*. Minneapolis: Fortress Press.

Larson, Reed, Jiri Zuzanek, and Roger Mannell: 1985 Being alone versus being with people: Disengagement in the daily experience of older adults. *Journal of Gerontology* 40:375–381.

Larue, Asenath, Connie Dessonville, and Lissy Jarvik: 1985 Aging and mental disorders. Pp. 664–702 in James E. Birren and K. Warner Schaie (eds.), *Handbook of the Psychology of Aging*, 2d ed. New York: Van Nostrand Reinhold.

Lauer, Robert H., Jeanette C. Lauer, and Sarah T. Kerr: 1990 The long-term marriage: Perceptions of stability and satisfaction. *International Journal of Aging and Human Development* 31(3):189–195.

Lawton, M. Powell: 1978 Leisure activities for the aged. *Annals of the American Academy of Political and Social Sciences* 438:71–80.
1983 Environment and other determinants of well-being in older people. *The Gerontologist* 23:349–357.
1991 Research on aging and human development: Hard times, 1991. *Journal of Aging & Social Policy* 3(4):1–4.

Lawton, M. Powell, Maurice Greenbaum, and Bernard Liebowitz: 1980 The lifespan of housing environments for the aged. *The Gerontologist* 20:56–64.

Lawton, M. Powell, and A. Regula Herzog (eds.): 1989 *Social Research Methods for Gerontology*. Amityville, NY: Baywood.

Lawton, M. Powell, Miriam Moss, and Miriam Grimes: 1985 The changing service needs of older tenants in planned housing. *The Gerontologist* 25:258–264.

Lebowitz, Barry D., Enid Light, and Frank Bailey: 1987 Mental health center services for the elderly: The impact of coordination with area agencies on aging. *The Gerontologist* 27(6): 699–702.

Lee, Annette T., and Anthony Cerami: 1990 Modifications of proteins and nucleic acids by reducing sugars: Possible role in aging. Pp. 116–130 in E. L. Schneider and J. W. Rowe (eds.), *Handbook of the Biology of Aging,* 3d ed. New York: Academic Press.

Lee, Gary R.: 1978 Marriage and morale in later life. *Journal of Marriage and the Family* 40:131–139.

Lee, Gary R., and M. Ishii-Kuntz: 1987 Social interaction, loneliness, and emotional well-being among the elderly. *Research on Aging* 9: 459–482.

Lefley, H. P.: 1987 Aging parents as caregivers of mentally ill adult children: An emerging social problem. *Hospital and Community Psychiatry* 38:1063–1070.

Lesser, Jack A., and Suzanne R. Kunkel: 1991 Exploratory and problem-solving consumer behavior across the life span. *Journal of Gerontology* 46(5):P259–P269.

Levin, Jack, and William C. Levin: 1981 Willingness to interact with an older person. *Research on Aging* 3:211–217.

Levin, Jeffrey S.: 1993 Age differences in mystical experience. *The Gerontologist* 33:507–513.

Levin, Jeffrey S. (ed.): 1994a *Religion in Aging and Health*. Thousand Oaks, CA: Sage.
1994b Investigating the epidemiologic effects of religious experience: Findings, explanations, and barriers. Pp. 3–17 in J. S. Levin (ed.), *Religion in Aging and Health*. Thousand Oaks, CA: Sage.

Levin, Jeffrey S., Linda M. Chatters, and Robert J. Taylor: 1995 Religious effects on health status and life satisfaction among black Americans. *Journal of Gerontology: Social Sciences* 50B(3): S154–S163.

Levin, Jeffrey S., and Robert Joseph Taylor: 1993 Gender and age differences in religiosity among black Americans. *The Gerontologist* 33:16–23.

Levin, Jeffrey S., Robert J. Taylor, and Linda M. Chatters: 1994 Race and gender differences in religiosity among older adults: Findings from four national surveys. *Journal of Gerontology: Social Sciences* 49(3):S137–S145.

Levin, William C.: 1988 Age stereotyping: College student evaluations. *Research on Aging* 10(1):134–148.

Levinson, Daniel J.: 1990 A theory of life structure development in adulthood. Pp. 35–53 in Charles N. Alexander and Ellen J. Langer (eds.), *Higher Stages of Human Development: Perspectives on Adult Growth*. New York: Oxford University Press.

Levinson, Daniel J., C. M. Darrow, E. B. Klein, M. H. Levinson, and B. McKee: 1978 *The Seasons of a Man's Life*. New York: Knopf.

Levy, Judith A.: 1989 The hospice in the context of an aging society. *Journal of Aging Studies* 3(4):385–399.

Lieberman, Morton A., and Sheldon S. Tobin: 1983 *The Experience of Old Age: Stress, Coping and Survival*. New York: Basic Books.

Liebig, Phoebe S.: 1994 Decentralization, aging policy, and the age of Clinton. *Journal of Aging and Social Policy* 6(1/2):9–26.

Liebow, Elliot: 1967 *Tally's Corner*. Boston: Little, Brown.

Lincoln, C. Eric, and L. Mamiya: 1990 *The Black Church in the African American Experience*. Durham, NC: Duke University Press.

Lipman, Aaron: 1985 1986 Homosexual relationships. *Generations* 10(4).51–54.

Litwak, Eugene: 1981 *The Modified Extended Family, Social Networks, and Research Continuities in Aging*. New York: Columbia University Center for Social Sciences.

1985 *Helping the Elderly: The Complementary Roles of Informal Networks and Formal Systems*. New York: Guilford.

1989 Forms of friendship among older people in an industrial society. Pp. 65–88 in Rebecca G. Adams and Rosemary Blieszner(eds.), *Older Adult Friendship*. Newbury Park, CA: Sage.

Litwak, Eugene, and Charles F. Longino, Jr.: 1987 Migration patterns among the elderly: A developmental perspective. *The Gerontologist* 27(3): 266–272.

Litwin, H., and A. Monk: 1987 Do nursing home patient ombudsmen make a difference? *Journal of Gerontological Social Work* 2:95–104.

Liu, Korbin, Maria Perozek, and Kenneth Manton: 1993 Catastrophic acute and long-term care costs: Risks faced by disabled elderly persons. *The Gerontologist* 33(3):299–307.

Livson, Florine B.: 1976 Patterns of personality development in middle aged women: A longitudinal study. *The International Journal of Aging and Human Development* 7:107–115.

Logue, Barbara J.: 1991 Women at risk: Predictors of financial stress for retired women workers. *The Gerontologist* 31(5):657–665.

Longino, Charles F., Jr.: 1981 Retirement communities. Pp. 391–418 in F. J. Berghorn and D. E. Schafer (eds.), *The Dynamics of Aging*. Boulder, CO: Westview Press.

1986 *The Oldest Americans: State Profiles for Data Base Planning*. Coral Gables, FL: University of Miami Center for Social Research on Aging.

Longino, Charles F., Jr., and William H. Crown: 1990 Retirement migration and interstate income transfers. *The Gerontologist* 30(6): 784–789.

Longino, Charles F., Jr., David J. Jackson, Rick S. Zimmerman, and Julia E. Bradsher: 1991 The second move: Health and geographic mobility. *Journal of Gerontology* 46(4):S218–S224.

Longino, Charles F., Jr., and G. C. Kitson: 1976 Parish clergy and the aged: Examining stereotypes. *Journal of Gerontology* 31:340–345.

Longino, Charles F., Jr., Kent A. McClelland, and Warren A. Peterson: 1980 The aged subculture hypothesis. Social integration, gerontophilia and self conception. *Journal of Gerontology* 35:758–767.

Lopata, Helena Z.: 1973 *Widowhood in an American City*. Cambridge, MA: Schenkman.

Loughman, Celeste: 1980 Eras and the elderly: A literary view. *The Gerontologist* 20:182–187.

Love, Douglas O., and William D. Torrence: 1989 The impact of worker age on unemployment and earnings after plant closings. *Journal of Gerontology* 44(5):S190–S195.

Lowenthal, Marjorie F., and Paul L. Berkman: 1967 *Aging and Mental Disorder in San Francisco*. San Francisco: Jossey-Bass.

Lowman, C., and C. Kirchener: 1979 Elderly blind and visually impaired persons: Projected numbers in the year 2000. *Journal of Visual Impairment and Blindness* 73:73–74.

Lowther, Malcolm A., Stephen J. Gill, and Larry C. Coppard: 1985 Age and the determinants of teacher job satisfaction. *The Gerontologist* 25:520–525.

Luborsky, Mark R.: 1994 The cultural adversity of physical disability: Erosion of full adult personhood. *Journal of Aging Studies* 8:239–254.

Lubove, Roy: 1968   *The Struggle for Social Security: 1900–1935.* Cambridge, MA: Harvard University Press.

Lugaresi, E., G. Coccagna, P. Farnetti, M. Mantosani, and F. Cirignotta: 1975   Snoring. *Electroencephalography and Clinical Neurophysiology* 39:59–64.

Luggen, Ann S.: 1985   The Pain Experience of Elderly Women. Unpublished doctoral dissertation. Cincinnati: University of Cincinnati.

Lund, Dale A., Michael S. Caserta, and Margaret F. Dimond: 1986   Gender differences through two years of bereavement among the elderly. *The Gerontologist* 26:314–20.

Maas, H. S., and J. A. Kuypers: 1974   *From Thirty to Seventy.* San Francisco: Jossey-Bass.

Machemer, Richard H., Jr.: 1992   The news in the biology of aging: The good, the bad and the confusing. Paper presented at the Annual Meeting of the Association for Gerontology in Higher Education, Baltimore.

Mac Rae, Hazel: 1992   Fictive kin as a component of the social networks of older people. *Research on Aging* 14:226–247.

Maddox, George L.: 1988   The future of gerontology in higher education: Continuing to open the American mind about aging. *The Gerontologist* 28(6):748–752.

1991   Aging with a difference. *Generations* 15:1:7–10.

Maddox, George L., and Richard T. Campbell: 1985   Scope, concepts, and methods in the study of aging. Pp. 3–31 in R. Binstock and E. Shanas (eds.), *Handbook of Aging in the Social Sciences,* 2d ed. New York: Van Nostrand Reinhold.

Maehr, Martin L., and Douglas A. Kleiber: 1981   The graying of achievement motivation. *American Psychologist* 36:787–793.

Malatesta, Carol Zander, and Michelle Kalnok: 1984   Emotional experience in younger and older adults. *Journal of Gerontology* 39: 301–308.

Maldonado, David: 1995   Religion and persons of color. Pp. 119–128 in Melvin A. Kimble et al. (eds.), *Aging, Spirituality, and Religion: A Handbook.* Minneapolis: Fortress Press.

Mancini, Jay A., and Rosemary Blieszner: 1989   Aging parents and adult children: Research themes in intergenerational relationships. *Journal of Marriage and the Family* 51: 275–290.

Manton, Kenneth G.: 1986   Past and future life expectancy increases at later ages: Their implications for the linkage of chronic morbidity, disability and mortality. *Journal of Gerontology* 41:672–81.

1988   A longitudinal study of functional change and mortality in the United States. *Journal of Gerontology: Social Sciences* 43(5): S153–S161.

Manton, Kenneth G., Larry S. Corder, and Eric Stallard: 1993   Estimates of change in chronic disability and institutional incidence and prevalence rates in the U.S. elderly population from the 1982, 1984, and 1989 National Long Term Care Survey. *Journal of Gerontology: Social Sciences* 48(4):S153–S166.

Manton, Kenneth G., and Beth J. Soldo: 1985   Dynamics of health changes in the oldest old, new perspectives and evidence. *Milbank Memorial Fund Quarterly* 63(2):206–285.

Manton, Kenneth G., and Eric Stallard: 1994   Medical demography: Interaction of disability dynamics and mortality. Pp. 217–278 in L. G. Martin and S. H. Preston (eds.), *Demography of Aging.* Washington, D.C.: National Academy Press.

Markides, Kyriakos S.: 1989   *Aging and Health: Perspectives on Gender, Race, Ethnicity and Class.* Newbury Park, CA: Sage.

Markides, Kyriakos S., and Neal Krause: 1985–1986   Older Mexican Americans (family relationships). *Generations* 10(4):32–35.

Markides, Kyriakos, and H. W. Martin: 1983   *Older Mexican Americans.* Austin, TX.: Center for Mexican American Studies.

Markus, Hazel R., and A. Regula Herzog: 1991   The role of self-concept in aging. *Annual Review of Gerontology and Geriatrics* 11:110–143.

Markus, Hazel R., and Paula Nurius: 1986   Possible selves. *American Psychologist* 41:954–969.

Marmor, Theodore R.: 1970   *The Politics of Medicare.* London: Routledge and Kegan Paul.

Marmor, Theodore R., Jerry L. Mashaw, and Philip L. Harvey: 1990   *America's Misunderstood Welfare State.* New York: Basic Books.

Marsh, Gail R.: 1980   Perceptual changes with aging. Pp. 147–168 in E. Busse and D. Blazer (eds.), *Handbook of Geriatric Psychiatry.* New York: Van Nostrand Reinhold.

Marshall, Victor W.: 1980   *Last Chapters: A Sociology of Aging and Dying.* Monterey, CA: Brooks/Cole.

Masters, William H., and Virginia Johnson: 1966 *Human Sexual Response*. Boston: Little, Brown.

Mathematica Policy Research: 1986 *The Evaluation of the National Long-Term Care Demonstration: Final Report Executive Summary*. Plainsboro, NJ: Mathematica Research, Inc.

Matthews, Sarah H.: 1986 *Friendships Through the Life Course*. Beverly Hills, CA: Sage.

Maves, Paul B.: 1960 Aging, religion, and the church. Pp. 678–749 in Clark Tibbitts (ed.), *Handbook of Social Gerontology*. Chicago: University of Chicago Press.
1986 *Faith for the Older Years: Making the Most Out of Life's Second Half*. Minneapolis: Augsberg.

Mcauley, W., M. Jacobs, and C. Carr: 1984 Older couples: Patterns of assistance and support. *Journal of Gerontological Social Work* 6:34–48.

McCarthy, B. W.: 1984 Strategies and techniques for the treatment of inhibited sexual desire. *Journal of Sex and Marital Therapy* 10:97–104.

McConnel, Charles E., and Firooz Deljavan: 1983 Consumption patterns of the related household. *Journal of Gerontology* 38:480–490.

McCoy, John L., and Kerry Weems: 1989 Disabled worker beneficiaries and disabled SSI recipients. *Social Security Bulletin* 52(5):16–28.

McCrae, Robert R., and Paul T. Costa, Jr.: 1982 Aging, the life course, and models of personality. Pp. 602–613 in T. M. Field et al. (eds.), *Review of Human Development*. New York: Wiley.

McEvoy, G. M., and W. F. Cascio: 1989 Cumulative evidence for the relationship between employee age and job performance. *Journal of Applied Psychology* 74:11–17.

McGrew, Kathryn B.: 1991 *Daughters' Decision Making about the Nature and Level of Their Participation in the Long-Term Care of Their Dependent Elderly Mothers: A Qualitative Study*. Oxford, OH: Scripps Gerontology Center.

McKain, Walter C., Jr.: 1969 *Retirement Marriage*. Storrs, CT: University of Connecticut Agriculture Experiment Station.

McKee, Patrick L., (ed.): 1982 *Philosophical Foundations of Gerontology*. New York: Human Sciences Press.

McNeely, R. L.: 1988 Age and job satisfaction in human service employment. *The Gerontologist* 28(2):163–168.

McTavish, Donald G.: 1971 Perceptions of old people: A review of research, methodologies and findings. *The Gerontologist* 11(4, Part 2): 90–101.

Mehdizadeh, Shahla A., and Robert C. Atchley: 1992 *The Economics of Long-Term Care in Ohio*. Oxford, OH: Scripps Gerontology Center.

Mehdizadeh, Shahla A., Suzanne R. Kunkel, Robert C. Atchley, and Robert A. Applebaum: 1990 *Projected Need for Long-Term Care: 1990–2010*. Oxford, OH: Scripps Gerontology Center.

Mellinger, Jeanne C.: 1989 Emergency housing for frail older adults. *The Gerontologist* 29(3): 401–404.

Menninger, Karl: 1938 *Man Against Himself*. New York: Harcourt Brace Jovanovich.

Mercier, Joyce McDonough, Lori Paulson, and Earl W. Morris: 1989 Proximity as a mediating influence on the perceived aging parent-adult child relationship. *The Gerontologist* 26(6):785–791.

Meyer, Judith W.: 1987 County characteristics and elderly net migration rates: A three-decade regional analysis. *Research on Aging* 9(3): 441–456.

Midanik, Lorraine, Krikor Sokhikian, Laura J. Ransom, and Irene S. Tekawa: 1995 The effect of retirement on mental health and health behaviors: The Kaiser Permanente Retirement Study. *Journal of Gerontology* 50B(1):S59–S61.

Middleton, Francisca: 1991 Computers for seniors. Pp. 75–78 in Robert Harootyan et al. (eds.), *Resourceful Aging: Today and Tomorrow*. Volume V. *Lifelong Education*. Washington, D.C.: American Association of Retired Persons.

Miller, Arthur H., Patricia Gurin, and Gerald Gurin: 1980 Age consciousness and political mobilization of older Americans. *The Gerontologist* 20:691–700.

Miller, Baila, and Andrew Montgomery: 1990 Family caregivers and limitations in social activities. *Research on Aging* 12(1):72–93.

Miller, Dulcy B., and Janet T. Barry: 1979 *Nursing Home Organization and Operation*. Boston: CBI Publishing.

Miller, Marv: 1978 Toward a profile of the older white male suicide. *The Gerontologist* 18: 80–82.

Miller, Richard A.: 1990 Aging and the immune system. Pp. 157–180 in E. L. Schneider and J. W. Rowe (eds.), *Handbook of the Biology of Aging*. New York: Academic Press.

Miller, Stephen J.: 1965 The social dilemma of the aging leisure participant. Pp. 77–92 in Arnold M. Rose and Warren A. Peterson (eds.), *Older People and Their Social World.* Philadelphia: Davis.

Mills, C. Wright: 1951 *White Collar: The American Middle Classes.* New York: Oxford University Press.

1959 *The Power Elite.* New York: Oxford University Press

1963 *Power, Politics & People: The Collected Essays of C. Wright Mills.* New York: Oxford University Press.

Minaker, Kenneth L., Graydon S. Meneilly, and John W. Rowe: 1985 Endocrine systems. Pp. 433–456 in C. E. Finch and E. L. Schneider (eds.), *Handbook of the Biology of Aging.* 2d ed. New York: Van Nostrand Reinhold.

Minkler, Meredith: 1989 Gold in gray: Reflections on business' discovery of the elderly market. *The Gerontologist* 29:17–23.

1990 Aging and disability: Behind and beyond the stereotypes. *Journal of Aging Studies* 4(3):245–260.

Minkler, Meredith, and Rena J. Pasik: 1986 Health promotion and the elderly: A critical perspective on the past and future. Pp. 39–54 in Ken Dychtwald (ed.), *Wellness and Health Promotion for the Elderly.* Rockville, MD: Aspen Systems.

Minkler, Meredith, Kathleen M. Roe, and Marilyn Price: 1992 The physical and emotional health of grandmothers raising grandchildren in the crack cocaine epidemic. *The Gerontologist* 32:752–761.

Mirotznik, Jerrold, and Teresa G. Lombardi: 1995 The impact of intrainstitutional relocation on morbidity in an acute care setting. *The Gerontologist* 35:217–224.

Mitchell, Jim, and Jasper C. Register: 1984 An exploration of family interaction with the elderly by race, socioeconomic status, and residence. *The Gerontologist* 24:48–54.

Molander, Bo, and Lars Bäckman: 1994 Attention and performance in miniature golf across the life span. *Journal of Gerontology: Psychological Sciences* 49(2):P35–P41.

Moody, Harry R.: 1992a Bioethics and aging. Pp. 395–425 in T. R. Cole, D. D. Van Tassel, and R. Kastenbaum (eds.), *Handbook of Humanities and Aging.* New York: Springer.

1992b *Ethics in an Aging Society.* Baltimore: The Johns Hopkins University Press.

1995 Mysticism. Pp. 87–101 in Melvin A. Kimble et al. (eds.), *Aging, Spirituality, and Religion: A Handbook.* Minneapolis: Fortress Press.

Moore, Loretta M., Christene R. Nielsen, and Charlotte M. Mistretta: 1982 Sucrose taste thresholds: Age related differences. *Journal of Gerontology* 37:64–69.

Mor, Vincent, and Susan Masterson-Allen: 1987 *Hospice Care Systems: Structure, Process, Costs, and Outcome.* New York: Springer Publishing Company.

Morgan, David L.: 1989 Adjusting to widowhood: Do social networks really make it easier? *The Gerontologist* 29(1):101–107.

Morgan, David L., Tonya L. Schuster, and Edgar W. Butler: 1991 Role reversals in the exchange of social support. *Journal of Gerontology* 46(5):S278–S287.

Morgan, Leslie A.: 1976 A re-examination of widowhood and morale. *Journal of Gerontology* 31(6):687–695.

1984 Changes in family interaction following widowhood. *Journal of Marriage and the Family* 46:323–331.

Morgan, William R., Herbert S. Parnes, and Lawrence J. Less: 1985 Leisure activities and social networks. Pp. 119–145 in Herbert S. Parnes (ed.), *Retirement Among American Men.* Lexington, MA: D. C. Heath.

Morris, Robert: 1989 Challenges of aging in tomorrow's world: Will gerontology grow, stagnate, or change? *The Gerontologist* 29(4):494–501.

Morris, William (ed.): 1982 *The American Heritage Dictionary of the English Language.* New York: Houghton Mifflin.

Moss, Miriam S., and M. Powell Lawton: 1982 Time budgets of older people: A window on four lifestyles. *Journal of Gerontology* 37:115–123.

Moss, Miriam S., Sidney Z. Moss, and Elizabeth L. Moles: 1985 The quality of relationships between elderly parents and their out-of-town children. *The Gerontologist* 25:134–140.

Mui, Ada C.: 1995 Caring for frail elderly parents: A comparison of adult sons and daughters. *The Gerontologist* 35(1):86–93.

Mulkay, M. J.: 1971 *Functionalism, Exchange, and Theoretical Strategy.* New York: Schocken.

**Mundorf, Norbert, and Winifred Brownell:
1990** Media preferences of older and younger
adults. *The Gerontologist* 30(5):685–691.

Munnell, Alicia H.: 1977 *The Future of Social
Security*. Washington, D.C.: Brookings.
1982 *The Economics of Private Pensions*.
Washington, D.C.: Brookings

**Myers, George C., and Kenneth G. Manton:
1985** Morbidity, disability, and mortality: The
aging connection. Pp. 25–39 in C. M. Gaitz, G.
Niederehe, and N. L. Wilson (eds.), *Aging
2000: Our Health Care Destiny*. Volume II:
Psychosocial and Policy Issues. New York:
Springer-Verlag.

Myles, John F.: 1986 *The politics of the retirement wage*. Ottawa, Canada: Carleton University. (photocopy)

Naisbitt, John: 1982 *Megatrends: Ten New
Directions Transforming Our Lives*. New York:
Warner Books.

National Center for Health Statistics: 1989a
Physical functioning of the aged: United States,
1984. *Vital and Health Statistics*, Series 10, No.
167.
1989b The National Nursing Home Survey:
1985 summary for the United States. *Vital and
Health Statistics*, Series 13, No. 97, Washington,
D.C.: U.S. Government Printing Office.
1991a Current estimates from the National
Health Interview Survey, 1990. *Vital and Health
Statistics*, Series 10, No. 181. Washington,
D.C.: U.S. Government Printing Office.
1991b *Vital Statistics of the United States,
1988. Volume II—Mortality, Part A*. Washington, D.C.: U.S. Government Printing Office.
1994a Current estimates from the National
Health Interview Survey. *Vital and Health
Statistics*, Series 10, No. 190. Washington,
D.C.: U.S. Government Printing Office.
1994b *Vital Statistics of the United States,
1990. Volume II—Mortality, Part A*. Washington, D.C.: U.S. Government Printing Office.
1995 *Vital Statistics of the United States, 1990.
Volume II—Mortality, Part A*. Washington,
D.C.: U.S. Government Printing Office.

Nelson, Gary M.: 1983 Tax expenditures for the
elderly. *The Gerontologist* 23:471–478.

**Nelson, H. Wayne, Ruth Huber, and Kathy L.
Walter: 1995** The relationship between volunteer long-term care ombudsmen and regulatory
nursing home actions. *The Gerontologist*
35(4):509–514.

Neugarten, Bernice L.: 1975 The future of the
young-old. *The Gerontologist* 15(1, Part 2):4–9.
1977 Personality and aging. Pp. 626–649 in
James E. Birren and K. Warner Schaie (eds.),
Handbook of the Psychology of Aging. New
York: Van Nostrand Reinhold.

Neugarten, Bernice L., and Nancy Datan: 1973
Sociological perspectives on the life cycle. Pp.
53–69 in Paul B. Baltes and K. Warner Schaie
(eds.), *Life-Span Developmental Psychology:
Personality and Socialization*. New York:
Academic Press.

**Neugarten, Bernice L., and Gunhild O.
Hagestad: 1976** Age and the life course. Pp.
35–55 in Robert H. Binstock and Ethel Shanas
(eds.), *Handbook of Aging and the Social
Sciences.* New York: Van Nostrand Reinhold.

**Neugarten, Bernice L., and Karol K. Weinstein:
1964** The changing American grandparent.
Journal of Marriage and the Family 26:
199–204.

**Newcomer, Robert J., and Charlene Harring-
ton: 1983** State Medicaid expenditures: Trends
and program policy changes. Pp. 157–
186 in Carroll L. Estes and R. J. Newcomer
(eds.), *Fiscal Austerity and Aging: Shifting
Government Responsibility for the Elderly*.
Beverly Hills, CA: Sage.

Noelker, Linda S., and Aloen Townsend: 1987
Perceived caregiving effectiveness. Pp. 58–79 in
T. H. Brubaker (ed.), *Aging, Health and
Family: Long-Term Care*. Newbury Park, CA:
Sage.

**Norburn, Jean E. Kincade, Shulamit L. Bernard,
Thomas R. Konrad, Alison Woomert, Gordon
H. DeFriese, William D. Kalsbeek, Gary G.
Koch, and Maria G Ory: 1995** Self-care and
assistance from others in coping with functional
status limitations among a national sample of
older adults. *Journal of Gerontology: Social
Sciences* 50B:S101–S109.

Nuessel, Frank H., Jr.: 1982 The language of
ageism. *The Gerontologist* 22:273–276.

O'Hare, William: 1985 Poverty in America:
Trends and New Patterns. *Population Bulletin*
40: No. 3.
1989 In the black: Affluent blacks are a rapidly
growing market. *American Demographics*
11(11):25–29.

Olshansky, S. Jay: 1985 Pursuing longevity:
Delay vs. elimination of degenerative diseases.
Paper presented at the Annual Meeting of the
American Sociological Association.

Olsho, Lynne W., Stephen W. Harkins, and Martin L. Lenhardt: 1985 Aging and the auditory system. Pp. 332–374 in James E. Birren and K. Warner Schaie (eds.), *Handbook of the Psychology of Aging,* 2d ed. New York: Von Nostrand Reinhold.

Olson, Laura Katz: 1982 *The Political Economy of Aging.* New York: Columbia University Press.

O'Rand, Angela, and John C. Henretta: 1982 Midlife work history and retirement income. Pp. 25–44 in M. Szinovacz (ed.), *Women's Retirement.* Beverly Hills, CA: Sage.

Oriol, William E.: 1982 *Aging in All Nations: A Special Report on the United Nations World Assembly on Aging.* Washington, D.C.: The National Council on the Aging.

1983 *Redefining the New Federalism.* Washington, D.C.: National Council on the Aging.

Orr, Alberta L. (ed.): 1992 *Vision and Aging: Crossroads for Service Delivery.* New York: American Foundation for the Blind.

Owens, W. A., Jr.: 1953 Age and mental abilities: A longitudinal study. *Genetic Psychology Monographs* 48:3–54.

Ozawa, Martha N., and Simon Wai-On Law: 1992 Reported reasons for retirement: A study of recently retired workers. *Journal of Aging and Social Policy* 4(3/4):35–51.

Paine, Thomas H.: 1993 The changing character of pensions: Where employers are headed. Pp. 33–40 in R. V. Burkhauser and D. L. Salisbury (eds.), *Pensions in a Changing Economy.* Washington, D.C.: Employee Benefit Research Institute.

Palmore, Erdman B.: 1978 When can age, period, and cohort be separated? *Social Forces* 57:282–295.

1980b Gender and life expectancy. Pp. 57–64 in S. G. Haynes and M. Feinlaub (eds.), *Epidemiology of Aging.* Bethesda, MD: National Institutes of Health.

Palmore, Erdman B., Bruce M. Burchett, Gerda G. Fillenbaum, Linda K. George, and Laurence M. Wallman: 1985 *Retirement: Causes and Consequences.* New York: Springer.

Pampel, Fred C., and Melissa Hardy: 1994 Status maintenance and change during old age. *Social Forces.* 73(1):289–314.

Pargament, Kenneth I., Kimberly S. Van Haitsma, and David S. Ensing: 1995 Religion and coping. Pp. 47–67 in Melvin A. Kimble et al. (eds.), *Aging, Spirituality, and Religion: A Handbook.* Minneapolis: Fortress Press.

Park, Denise C.: 1992 Applied cognitive aging research. Pp. 449–493 in Fergus I. M. Craik and Timothy A. Salthouse (eds.), *The Handbook of Aging and Cognition.* Hillsdale, NJ: Lawrence Erlbaum.

Parkes, C. Murray: 1975 Determinants of outcome following bereavement. *Omega* 6:303–323.

Parmalee, Patricia A., I. R. Katz, and M. Powell Lawton: 1989 Depression among the institutionalized aged: Assessment and prevalence estimate. *Journal of Gerontology* 44:M22–M29.

1991 The relation of pain to depression among institutionalized aged. *Journal of Gerontology* 46:P15–P21.

Parnes, Herbert S.: 1985 Conclusion. Pp. 209–224 in Herbert S. Parnes (ed.), *Retirement Among American Men.* Lexington, MA: D. C. Heath.

1989 Postretirement employment: How much is there? How much is wanted? *Generations* 13(2):23–26.

Parnes, Herbert S., and Randy King: 1977 Middle-aged job losers. *Industrial Gerontology* 4:77–96.

Parnes, Herbert S., and Lawrence Less: 1983 *From Work to Retirement: The Experience of a National Sample of Men.* Columbus, OH: Ohio State University Center for Human Resource Research.

1985 The volume and pattern of retirements, 1966–1981. Pp. 57–77 in Herbert S. Parnes (ed.), *Retirement Among American Men.* Lexington, MA: D. C. Heath.

Parnes, Herbert S., and Gilbert Nestel: 1981 The retirement experience. Pp. 132–154 in Herbert S. Parnes (ed.), *Work and Retirement: A Longitudinal Study of Men.* Cambridge, MA: MIT Press.

Parnes, Herbert S., and David G. Sommers: 1994 Shunning retirement: Work experience of men in their seventies and early eighties. *Journal of Gerontology: Social Sciences* 49(3):S117–S124.

Pascual-Leone, Juan: 1990 Reflections on lifespan intelligence, consciousness, and ego development. Pp. 258–285, in Charles N. Alexander and Ellen J. Langer (eds.), *Higher Stages of*

Human Development: New York: Oxford University Press.

Payne, Barbara Pittard, and Susan H. McFadden: 1994 From loneliness to solitude. Religious and spiritual journeys in later life. Pp. 13–27 in L. E. Thomas and S. A. Eisenhandler (eds.), *Aging and the Religious Dimension*. New York: Auburn House.

Peacock, James R.: 1990 Voices Which Dared Not Speak: An Older Cohort's Perception of Social Change and Individual Development Among Gay Males. Unpublished MGS thesis. Oxford, OH: Miami University.

Peacock, James R., and Margaret M. Paloma: 1991 Religiosity and life satisfaction across the life course. Paper presented at the Annual Meeting of the Society for the Scientific Study of Religion, Pittsburgh, November.

Pedrick-Cornell, Claire, and Richard J. Gelles: 1982 Elderly abuse: The status of current knowledge. *Family Relations* 31:457–465.

Perlmutter, Marion, Cynthia Adams, Jane Berry, Michael Kaplan, Denise Person, and Frederick Verdonik: 1987 Aging and memory. Pp. 57–92 in K. Warner Schaie (ed.), *Annual Review of Gerontology and Geriatrics,* Volume 7. New York: Springer.

Persson, Diane: 1993 The elderly driver: Deciding when to stop. *The Gerontologist* 33(1):88–91.

Petersen, David M.: 1988 Substance abuse, criminal behavior, & older people. *Generations* 12(4):63–67.

Petersen, M.: 1973 The visibility and image of old people on television. *Journal Quarterly* 50: 569–573.

Peterson, Candida C., and James L. Peterson: 1988 Older men's and women's relationships with adult kin: How equitable are they? *International Journal of Aging and Human Development* 27(3):221–231.

Peterson, David A., and Elizabeth B. Douglass: 1990 The impact of Administration on Aging funding on gerontology instruction. *Journal of Aging & Social Policy* 2(3/4):169–182.

Peterson, David A., and Donna Z. Eden: 1977 Teenagers and aging: Adolescent literature as an attitude source. *Educational Gerontology* 2:311–325.

Peterson, David A., and Elizabeth L. Karnes: 1976 Older people in adolescent literature. *The Gerontologist* 16:225–231.

Pfeiffer, Eric: 1977 Sexual behavior in old age. Pp. 130–141 in Ewald W. Busse and Eric Pfeiffer (eds.), *Behavior and Adaptation in Late Life,* 2d ed. Boston: Little, Brown.

Phillips, Paul D., Robert A. Applebaum, Sheila J. Atchley, and Rosalie McGinnis: 1989 Quality assurance strategies for home-delivered long-term care. *Quality Review Bulletin* 15: 156–162.

Pillemer, Karl: 1985 The dangers of dependency: New findings on domestic violence against the elderly. Paper presented at the Annual Meeting of the American Sociological Association, Washington, D.C., August.

Pillemer, Karl, and David Finkelhor: 1988 The prevalence of elder abuse: A random sample survey. *The Gerontologist* 28(1):51–57.

Pillemer, Karl, and J. Jill Suitor: 1991 "Will I ever escape my child's problems?" Effects of adult children's problems on elderly parents. *Journal of Marriage and the Family* 53:585–594.

Pitcher, Brian L., William F. Stinner, and Michael B. Toney: 1985 Patterns of migration propensity for black and white American men: Evidence from a cohort analysis. *Research on Aging* 7:94–120.

Population Reference Bureau: 1992 *World Population Data Sheet*. Washington, D.C.: Population Reference Bureau.

1994 *World Population Data Sheet*. Washington, D.C.: Population Reference Bureau.

Pratt, Henry J.: 1974 Old age associations in national politics. Pp. 106–119 in Frederick R. Eisele (ed.), *Political Consequences of Aging*. Philadelphia: American Academy of Political and Social Sciences.

1982 The "gray lobby" revisited. *National Forum* 62:31–33.

Quadagno, Jill: 1988 *The Transformation of Old Age Security: Class and Politics in the American Welfare State*. Chicago, IL: University of Chicago Press.

Quinn, Joan L.: 1987 Standards for health care programs and practice. Pp. 644–647 in George L. Maddox (ed.), *Encyclopedia of Aging*. New York: Springer.

Quinn, Joseph F.: 1981 The extent and correlates of partial retirement. *The Gerontologist* 21: 634–643.

Quinn, Joseph F., and Richard V. Burkhauser: 1990 Work and retirement. Pp. 303–327 in Robert H. Binstock and Linda K. George (eds.), *Handbook of Aging and the Social Sciences.* New York: Academic Press.
1994 Retirement and labor force behavior of the elderly. Pp. 50–101 in Linda G. Martin and Samuel Preston (eds.), *Demography of Aging.* Washington, D.C.: National Academy Press.

Quinn, Patrick Kaye, and Marvin Reznikoff: 1985 The relationship between death anxiety and the subjective experience of time in the elderly. *The International Journal of Aging and Human Development* 21:197–210.

Rabin, David C., and Patricia Stockton: 1987 *Long-Term Care for the Elderly: A Factbook.* New York: Oxford University Press.

Radner, Daniel B.: 1990 Assessing the economic status of the aged and managed using alternative income-wealth measures. *Social Security Bulletin* 53(3):2–14.
1991 Changes in the incomes of age groups, 1984–1989. *Social Security Bulletin* 54(12): 2–18.
1992 The economic status of the aged. *Social Security Bulletin* 55(3):3–23.
1993 Economic well-being of the old old. *Social Security Bulletin* 56(1):3–19.

Raskind, Murray A., and Elaine R. Peskind: 1992 Alzheimer's disease and other dementing disorders. Pp. 478–513 in James E. Birren, R. Bruce Sloane, and Gene D. Cohen (eds.), *Handbook of Mental Health and Aging,* 2d ed. New York: Academic Press.

Rathbone-McCuan, Eloise, and Raymond T. Coward: 1985 Respite and adult day-care services. Pp. 457–482 in A. Monk (ed.), *Handbook of Gerontological Services.* New York: Van Nostrand Reinhold.

Reff, Mitchell E.: 1985 RNA and protein metabolism. Pp. 225–254 in C. E. Finch and E. L. Schneider (eds.), *Handbook of the Biology of Aging,* 2d ed. New York: Van Nostrand Reinhold.

Regnier, Victor, and Jon Pynoos: 1992 Environmental interventions for cognitively impaired older persons. Pp. 764–792 in J. E. Birren, R. B. Sloane, and G. D. Cohen (eds.), *Handbook of Mental Health and Aging,* 2d ed. New York: Academic Press.

Reich, Robert B.: 1991 *The Work of Nations.* New York: Knopf.

Reichstein, Kenneth J., and Linda Bergofsky: 1983 Domiciliary care facilities for adults: An analysis of state regulations. *Research on Aging* 5: 25–44.

Relman, Arnold S.: 1980 The new medical-industrial complex. *The New England Journal of Medicine* 303:963–970.

Reschovsky, Andrew: 1989 State and local taxation of the elderly. *Journal of Aging & Social Policy* 1(3/4):143–170.

Reschovsky, James D., and Sandra J. Newman: 1991 Home upkeep and housing quality of older homeowners. *Journal of Gerontology* 46(5):S288–S297.

Retsinas, Joan, and Nicolas P. Retsinas: 1991 Accessory apartment conversion programs. *Journal of Aging & Social Policy* 3(1/2):73–89.

Riegel, Klaus F.: 1976 The dialectics of human development. *The American Psychologist* 31: 689–700.

Riesman, David: 1950 *The Lonely Crowd.* New Haven: Yale University Press.

Riley, Matilda W.: 1987 Age stratification. Pp. 20–22 in George L. Maddox (ed.), *Encyclopedia of Aging.* New York: Springer.

Riley, Matilda W., Marilyn Johnson, and Anne Foner: 1972 *Aging and Society.* Volume 3: *Sociology of Age Stratification.* New York: Russell Sage Foundation.

Riley, Matilda W., Robert L. Kahn, and Anne Foner (eds.): 1994 *Age and Structural Lag.* New York: Wiley Interscience.

Riley, Matilda W., and John W., Riley, Jr.: 1994 Age integration and the lives of older people. *The Gerontologist* 34(1):110–115.

Rindfuss, Ronald R.: 1991 The young adult years: Diversity, structural change, and fertility. *Demography* 28(4):493–512.

Rindfuss, Ronald R., C. Gray Swicegood, and Rachel A. Rosenfeld: 1987 Disorder in the life course: How common and does it matter? *American Sociological Review* 52:785–801.

Ritzer, George: 1995 *The McDonaldization of Society,* rev. ed. Thousand Oaks, CA: Pine Forge Press.

Rivlin, Alice M., and Joshua M. Weiner: 1988 *Caring for the Disabled Elderly: Who Will Pay?* Washington, D.C.: Brookings Institution.

Roberto, Karen A., and Johanna Stroes: 1992 Grandchildren and grandparents: Roles, influences, and relationships. *International Journal of Aging and Human Development* 34(3): 227–239.

Rodin, Judith, and Ellen Langer: 1977 Long-term effect of a control-relevant intervention. *Journal of Personality and Social Psychology* 36:12–29.

Rogers, Andrei: 1988 Age patterns of elderly migration: An international comparison. *Demography* 25:355–370.

Rogers, Andrei, and John F. Watkins: 1987 General versus elderly interstate migration and population redistribution in the United States. *Research on Aging* 9(4):483–529.

Rogers, Andrei, John F. Watkins, and Jennifer A. Woodward: 1990 Interregional elderly migration and population redistribution in four industrialized countries. *Research on Aging* 12(3):251–293.

Rogers, Richard G., Andrei Rogers, and Alain Belanger: 1989 Active life expectancy among the elderly in the United States: Multistate life-table estimates and population projections. *Milbank Quarterly* 67:370–411.

Rokeach, Milton: 1973 *The Nature of Human Values.* New York: The Free Press.
1978 *Understanding Human Values: Individual and Societal.* New York: The Free Press.

Rollins, Boyd C., and Harold Feldman: 1970 Marital satisfaction over the family life cycle. *Journal of Marriage and the Family* 32:20–28.

Roman, Paul, and Philip Taietz: 1967 Organizational structure and disengagement: The emeritus professor. *The Gerontologist* 7:147–152.

Rook, Karen S.: 1989 Strains in older adults' friendships. Pp. 166–194 in Rebecca G. Adams and Rosemary Blieszner (eds.), *Older Adult Friendship.* Newbury Park, CA: Sage.

Rooke, Constance: 1991 Old age in contemporary fiction: A new paradigm of hope. Pp 241–257 in T. R. Cole, D. D. Van Tassel, and R. Kastenbaum (eds.), *Handbook of the Humanities and Aging.* New York: Springer.

Rose, Arnold M.: 1965 The subculture of aging: A framework for research in social gerontology. Pp. 3–16 in Arnold M. Rose and Warren A. Peterson (eds.), *Older People and Their Social World.* Philadelphia: Davis.

Rosen, Benson, and Thomas H. Jerdee: 1976a The nature of job-related stereotypes. *Journal of Applied Psychology* 61:180–183.

1976b The influence of age stereotypes on managerial decisions. *Journal of Applied Psychology* 61:428–432.

Rosenbaum, Walter A., and James W. Button: 1989 Is there a grey peril?: Retirement politics in Florida. *The Gerontologist* 29(3):300–306.
1993 The unquiet future of intergenerational politics. *The Gerontologist* 33(4):481–490.

Rosenberg, George S.: 1970 *The Worker Grows Old.* San Francisco: Jossey-Bass.

Rosow, Irving: 1967 *Social Integration of the Aged.* New York: Free Press.
1974 *Socialization to Old Age.* Berkeley: University of California Press.

Ross, Ian K.: 1995 *Aging of Cells, Humans & Societies.* Boston: W. C. Brown.

Rothenberg, Richard, Harold R. Lentzner, and Robert A. Parker: 1991 Population aging patterns: The expansion of mortality. *Journal of Gerontology* 46(2):S66–S70.

Rothstein, Morton: 1987 Free radicals. Pp. 262–263 in George L. Maddox et al. (eds.), *Encyclopedia of Aging.* New York: Springer.

Rotter, J. B.: 1966 Generalized expectancies for internal versus external control of reinforcement. *Psychological Monographs* 80 (whole no. 609).

Rowe, John W., and Robert L. Kahn: 1987 Human aging: Usual and successful. *Science* 237:143–149.

Rubinstein, Robert L.: 1987 Never married elderly as a social type: Re-evaluating some images. *The Gerontologist* 27(1):108–113.

Ruhm, Christopher J.: 1989 Why older Americans stop working. *The Gerontologist* 29(3): 294–299.
1990 Career jobs, bridge employment, and retirement. Pp. 92–110 in Peter B. Doeringer (ed.), *Bridges to Retirement.* Ithaca, NY: ILR Press.

Sainer, J. S., and M. Zander: 1971 Guidelines for older person volunteers. *The Gerontologist* 11: 201–204.

Salthouse, Timothy A.: 1982 *Adult Cognition: An Experimental Psychology of Human Aging.* New York: Springer-Verlag.
1984 Effects of age and skill in typing. *Journal of Experimental Psychology*: General 113: 345–371.
1985 Speed of behavior and its implications for cognitions. Pp. 400–426, in James E. Birren and

K. Warner Schaie (eds.), *Handbook of the Psychology of Aging,* 2d ed. New York: Van Nostrand Reinhold.

1987 The role of experience in cognitive aging. Pp. 135–158 in K. Warner Schaie (ed.), *Annual Review of Gerontology and Geriatrics,* Volume 7. New York: Springer.

1990a Cognitive competence and expertise in aging. Pp. 311–319, in James E. Birren and K. Warner Schaie (eds.), *Handbook of the Psychology of Aging,* 3d ed. San Diego: Academic Press, Inc.

1990b Influence of experience on age differences in cognitive functioning. *Human Factors* 32(5):551–569.

1991a Cognitive facets of aging well. *Generations* 15(1):35–38.

1991b Mediation of adult age differences in cognition by reductions in working memory and speed of processing. *Psychological Science* 2(3): 179–183.

1991c *Theoretical Perspectives on Cognitive Aging.* Hillsdale, NJ: Lawrence Erlbaum.

Saunders, Cicely: 1976 St. Christopher's Hospice. Pp. 516–523 in E. S. Schneidman (ed.), *Death: Current Perspectives.* Palo Alto, CA: Mayfield.

Saxon, Sue V., and Mary Jean Etten: 1994 *Physical Change and Aging.* New York: Tiresias Press.

Schaie, K. Warner: 1980 Intelligence and problem solving. Pp. 262–284 in J. Birren and R. B. Sloane (eds.), *Handbook of Mental Health and Aging.* Englewood Cliffs, NJ: Prentice-Hall.

1987 Research methods in gerontology. Pp. 570–573 in George L. Maddox et al. (eds.), *Encyclopedia of Aging.* New York: Springer.

1989 The hazards of cognitive aging. *The Gerontologist* 29(4):484–493.

1990 Intellectual development in adulthood. Pp. 291–310 in James E. Birren and K. Warner Schaie (eds.), *Handbook of the Psychology of Aging, 3d ed.* San Diego: Academic Press, Inc.

Schaie, K. Warner (ed.).: 1987 *Annual Review of Gerontology and Geriatrics,* Volume 7. New York: Springer.

Schaie, K. Warner, and I. A. Parham: 1976 Stability of adult personality traits: Fact or fable? *Journal of Personality and Social Psychology* 34:146–158.

Schewe, Charles D.: 1990 Get in position for the older market. *American Demographics,* June, pp. 38–41; 61–63.

Schieber, Frank: 1992 Aging and the senses. Pp. 252–306 in James E. Birren, K. Warner Sloane, and Gene D. Cohen (eds.), *Handbook of Mental Health and Aging,* 2d ed. San Diego: Academic Press, Inc.

Schlesinger, Joseph A., and Mildred Schlesinger: 1981 Aging and opportunities for elective office. Pp. 205–239 in Sara B. Kiesler et al. (eds.), *Aging: Social Change.* New York: Academic Press.

Schmid, Hermann, Kim Manjee, and Tulshi Shah: 1994 On the distinction of suicide ideation versus attempt in elderly psychiatric inpatients. *The Gerontologist* 34(3):332–339.

Schmidt, Daniel F., and Susan M. Boland: 1986 Structure of perceptions of older adults: Evidence for multiple stereotypes. *Psychology and Aging* 1:255–260.

Schmidt, Robert M.: 1994 Healthy aging into the 21st century. *Contemporary Gerontology* 1(1): 3–5.

Schneider, Edward L., and John W. Rowe (eds.): 1990 *Handbook of the Biology of Aging,* 3d ed. New York: Academic Press.

Schonfield, David: 1982 Who is stereotyping whom and why? *The Gerontologist* 22: 267–272.

Schooler, Carmi: 1990 Psychosocial factors and effective cognitive functioning in adulthood. Pp. 347–358 in J. E. Birren and K. W. Schaie (eds.), *Handbook of the Psychology of Aging,* 3d ed. New York: Academic Press.

Schroots, Johannes J. F., and James E. Birren: 1990 Concepts of time and aging in science. Pp. 45–64 in James E. Birren and K. Warner Schaie (eds.), *Handbook of the Psychology of Aging.* New York: Academic Press.

Schuerman, Laurell E., Donna Z. Eden, and David A. Peterson: 1977 Older people in women's periodical fiction. *Educational Gerontology* 2: 327–351.

Schulz, James H.: 1985 To old folks with love: Aged income maintenance in America. *The Gerontologist* 25:464–471.

1992 *The Economics of Aging,* 5th ed. New York: Auburn House.

1995 *The Economics of Aging,* 6th ed. Westport, CT: Auburn House.

Scott, Jean Pearson, and Vira R. Kivett: 1985 Differences in the morale of older, rural widows and widowers. *The International Journal of Aging and Human Development* 21:121–36.

Seccombe, Karen: 1988 Financial assistance from elderly retirement-age sons to their aging parents. *Research on Aging* 10(1):102–118.

Seeber, James J.: 1995 Congregational models. Pp. 253–269 in Melvin A. Kimble et al. (eds.), *Aging, Spirituality, and Religion: A Handbook*. Minneapolis: Fortress Press.

Seefeldt, Carol, Richard K. Jantz, Alice Galper, and Kathy Serock: 1977 Using pictures to explore children's attitudes toward the elderly. *The Gerontologist* 17:506–512.

Seelbach, Wayne C.: 1977 Gender differences in expectations for filial responsibility. *The Gerontologist* 17:421–425.

Seltzer, Mildred M.: 1990 Role reversal: You don't go home again. *Journal of Gerontological Social Work* 15:5–14.

Seltzer, Mildred M., and Robert C. Atchley: 1971 The concept of old: Changing attitudes and stereotypes. *The Gerontologist* 11:226–230.

Seltzer, Mildred M., and Jane Karnes: 1988 An early retirement incentive program: A case study of Dracula and Pinocchio complexes. *Research on Aging* 10(3):342–357.

Serow, William J.: 1987 Determinants of interstate migration: Differences between elderly and nonelderly movers. *Journal of Gerontology* 42(1):95–100.

Settersten, Richard A.: 1992 Age Norms. Unpublished doctoral dissertation. Evanston, IL: Northwestern University.

Shapiro, Evelyn, and Robert Tate: 1988 Who is really at risk of institutionalization? *The Gerontologist* 28(2):237–245.

Sheehy, Gail: 1992 *The Silent Passage: Menopause*. New York: Random House.

Sherman, Sally R.: 1985 Reported reasons retired workers left their last job. *Social Security Bulletin* 48(3):22–30.

Sherrill, Kimberly A., and David B. Larson: 1994 The anti-tenure factor in religions research in clinical epidemiology and aging. Pp. 149–177 in J. S. Levin (ed.), *Religion in Aging and Health*. Thousand Oaks, CA: Sage.

Shock, Nathan W.: 1977 System integration. Pp. 639–665 in Caleb E. Finch and Leonard Hayflick (eds.), *Handbook of the Biology of Aging*. New York: Van Nostrand Reinhold.

Sill, John Stewart: 1980 Disengagement reconsidered: Awareness of finitude. *The Gerontologist* 20:457–462.

Silliman, Rebecca A., and Josef Sternberg: 1988 Family caregiving: Impact of patient functioning and underlying causes of dependency. *The Gerontologist* 28(3):377–382.

Silverstein, C.: 1981 *Man to Man: Gay Couples in America*. New York: William Morrow.

Silverstein, Merril, and Eugene Litwak: 1993 A task-specific typology of intergenerational family structure in later life. *The Gerontologist* 33(2): 258–264.

Simmons, Leo W.: 1945 *The Role of the Aged in Primitive Society*. New Haven, CT: Yale University Press.

Simon-Rusinowitz, Lori, and Brian F. Hofland: 1993 Adopting a disability approach to home care services for older adults. *The Gerontologist* 33(2):159–167.

Simonton, Dean K.: 1989 The swan song phenomenon: Last-works effects for 172 classical composers. *Psychology and Aging* 4:42–47.

1990a Creativity and wisdom in aging. Pp. 320–329 in James E. Birren and K. Warner Schaie (eds.), *Handbook of the Psychology of Aging*, 3d ed. San Diego: Academic Press, Inc.

1990b Creativity in later years: Optimistic prospects for achievement. *The Gerontologist* 30(5):020–031.

Simpson, Ida H., Kurt W. Back, and John C. McKinney: 1966 Continuity of work and retirement activities, and self-evaluation. Pp. 106–119 in Ida H. Simpson and John C. McKinney (eds.), *Social Aspects of Aging*. Durham, NC: Duke University Press.

Sinex, F. Marott: 1977 The molecular genetics of aging. Pp. 37–62 in C. E. Finch and L. Hayflick (eds.), *Handbook of the Biology of Aging*. New York: Van Nostrand Reinhold.

Singelakis, Andrew Thomas: 1990 Real estate market trends and the displacement of the aged: Examination of the linkages in Manhattan. *The Gerontologist* 30(5):658–666.

Sinnott, Jan D.: 1994 *Interdisiciplinary Handbook of Lifespan Learning*. Westport, CT: Greenwood Press.

Skinner, Ellen A., and James P. Connell: 1986 Control understanding: Suggestions for a developmental framework. Pp. 35–69 in Margaret M. Baltes and Paul B. Baltes (eds.), *The Psychology of Control and Aging*. Hillsdale, NJ: Lawrence Erlbaum.

Skolnick, Arlene: 1981 Married lives: Longitudinal perspectives on marriage. In D. Eichorn et al. (eds.), *Present and Past in Middle Age.* New York: Academic Press.

Smeeding, Timothy M.: 1990a Toward a knowledge base for long-term care finance. *The Gerontologist* 30(1):5–6.

1990b Economic status of the elderly. Pp. 362–381 in Robert H. Binstock and Linda K. George (eds.), *Handbook of Aging and the Social Sciences*, 3d ed. New York: Academic Press.

Smith, Adam: 1776 *The Wealth of Nations.* New York: Modern Library.

Smith, Carolyn H.: 1992 Images of aging in American poetry, 1925–1985. Pp. 215–240 in T. R. Cole, D. D. Van Tassel, and R. Kastenbaum (eds.), *Handbook of the Humanities and Aging.* New York: Springer.

Smith, David W. E.: 1989 One-story living. *American Demographics* 11(6):36–39.

Sohngen, Mary: 1977 The experience of old age as depicted in contemporary novels. *The Gerontologist* 17:70–78.

1981 The experience of old age as depicted in contemporary novels: A supplementary bibliography. *The Gerontologist* 21:303.

Sokolovsky, Jay: 1983 *Growing Old in Different Societies: Cross-Cultural Perspectives.* Belmont, CA: Wadsworth.

Soldo, Beth J.: 1981 The living arrangements of the elderly in the near future. Pp. 491–512 in Sara B. Kiesler et al. (eds.), *Aging: Social Change.* New York: Academic Press.

Soldo, Beth J., and Vicki A. Freeman: 1994 Care of the elderly: Division of labor among the family, market, and state. Pp. 195–216 in L. H. Martin and S. H. Preston (eds.), *Demography of Aging.* Washington, D.C.: National Academy Press.

Soldo, Beth J., and Martha S. Hill: 1993 Intergenerational transfers: Economic, demographic, and social perspectives. Pp. 187–216 in George. L. Maddox and M. P. Lawton (eds.), *Annual Review of Gerontology and Geriatrics*, Volume 13. New York: Springer.

Somers, Anne R., and Nancy L. Spears: 1992 *The Continuing Care Retirement Community.* New York: Springer.

Sommers, Tish, and Laurie Shields: 1979 Problems of the displaced homemaker. Pp. 86–106 in Ann F. Cahn (ed.), *Women in Midlife—Security and Fulfillment.* Washington, D.C.: House Select Committee on Aging.

Speare, Alden, Jr., and Judith W. Meyer: 1988 Types of elderly residential mobility and their determinants. *Journal of Gerontology: Social Sciences* 43(3):S74–S81.

Speas, Kathy, and Beth Obenshain: 1995 *Images of Aging in America: Final Report to AARP.* Chapel Hill, NC: FGI Integrated Marketing.

Spector, William D., and Peter Kemper: 1994 Disability and cognitive impairment criteria: Targeting those who need the most home care. *The Gerontologist* 34(5):640–651.

Spengler, Linda: 1985 Holistic Health, Wellness, and Aging: A Guide for Pre-Retirement Training. Unpublished Master of Gerontological Studies (MGS) thesis. Oxford, OH: Miami University.

Spirduso, Waneen W.: 1995 *Physical Dimensions of Aging.* Champaign, IL: Human Kinetics.

Spirduso, Waneen W., and Priscilla G. MacRae: 1990 Motor performance and aging. Pp. 184–200 in J. E. Birren and K. W. Schaie (eds.), *Handbook of the Psychology of Aging*, 3d ed. New York: Academic Press.

Spitze, Glenna, and John R. Logan: 1990 More evidence on women (and men) in the middle. *Research on Aging* 12(2):182–198.

Spore, Diana L., and Robert C. Atchley: 1990 Ohio community mental health center directors' perceptions of programming for older adults. *Journal of Applied Gerontology* 9:36–52.

Stanford, E. Percil, Catherine J. Happersett, Deborah J. Morton, Craig A. Molgaard, and K. Michael Peddecord: 1991 Early retirement and functional impairment from a multiethnic perspective. *Research on Aging* 13(1):5–38.

Starr, Bernard D.: 1985 Sexuality and aging. Pp. 97–126 in M. Powell Lawton and George L. Maddox (eds.), *Annual Review of Gerontology and Geriatrics*, Volume 5, 1985. New York: Springer.

Starr, Bernard D., and M. B. Weiner: 1981 *Sex and Sexuality in the Mature Years.* New York: McGraw-Hill.

Steinkamp, Marjorie W., and John R. Kelly: 1987 Social integration, leisure activity, and life satisfaction in older adults: Activity theory revisited. *International Journal of Aging and Human Development* 25(4):293–307.

Steinmetz, Suzanne K.: 1988 *Duty Bound: Elder Abuse and Family Care.* Newbury Park, CA: Sage.

Stenback, Aser: 1980 Depression and suicidal behavior in old age. Pp. 616–652 in James E. Birren and R. Bruce Sloane (eds.), *Handbook of Mental Health and Aging.* Englewood Cliffs, NJ: Prentice-Hall.

Sterns, Harvey L., and Ralph A. Alexander: 1987 Industrial gerontology: The aging individual and work. Pp. 243–264 in K. Warner Schaie (ed.), *Annual Review of Gerontology and Geriatrics,* Volume 7. New York: Springer.

Sterns, Harvey L., G. V. Barrett, R. A. Alexander, P. E. Panek, B. J. Avilio, and L. R. Forbinger: 1977 Training and Evaluation of Older Adult Skills Critical for Effective Driving Performance. Final Report. Akron, OH: University of Akron, Department of Psychology.

Sterns, Harvey L., and Michael A. McDaniel: 1994 Job performance and the older worker. Pp. 27–51 in Sara Rix (ed.), *Older Workers: How Do They Measure Up?* Washington, D.C.: AARP Public Policy Institute.

Stevens, D. P., and C. V. Truss: 1985 Stability and change in adult personality over 12 and 20 years. *Developmental Psychology* 21:568–584.

Stinnett, Nick, Linda M. Carter, and James E. Montgomery: 1972 Older persons' perceptions of their marriages. *Journal of Marriage and the Family* 34:665–670.

Stoller, Eleanor P.: 1993 Interpretations of symptoms by older people: A health diary study. *Journal of Aging and Health* 5:58–81.

Stoller, Eleanor P., and Michael A. Stoller: 1987 The propensity to save among the elderly. *The Gerontologist* 27(3):314–321.

Stone, Robyn, Gail Lee Cafferata, and Judith Sangl: 1987 Caregivers of the frail elderly: A national profile. *The Gerontologist* 27(5):616–626.

Storck, Patricia A., and Marion B. Cutler: 1977 Pictorial representation of adults as observed in children's literature. *Educational Gerontology* 2:293–300.

Strawbridge, William J., and Margaret I. Wallhagen: 1991 Impact of family conflict on adult child caregivers. *The Gerontologist* 31(6): 770–777.

Strehler, Bernard L.: 1977 *Time, Cells, and Aging,* 2d ed. New York: Academic Press.

Streib, Gordon, F., W. Edward Folts, and Anthony J. LaGreca: 1984b Autonomy, power, and decision making in thirty-six retirement communities. *The Gerontologist* 25:403–409.

Streib, Gordon F., and Clement J. Schneider: 1971 *Retirement in American Society.* Ithaca, NY: Cornell University Press.

Sudnow, David: 1967 *Passing on: The Social Organization of Dying.* Englewood Cliffs, NJ: Prentice-Hall.

Suitor, J. Jill: 1991 Marital quality and satisfaction with the division of household labor across the family life cycle. *Journal of Marriage and the Family* 53:221–230.

Sullivan, Deborah A., and Sylvia A. Stevens: 1982 Snowbirds: Seasonal migrants to the Sunbelt. *Research on Aging* 4:159–178.

Swenson, C., R. W. Eskew, and K. A. Kolhepp: 1981 Stage of the family life cycle, ego development, and the marriage relationship. *Journal of Marriage and the Family* 43:841–853.

Szinovacz, Maximiliane: 1982 Retirement plans and retirement adjustment. Pp. 139–150 in M. Szinovacz (ed.), *Women's Retirement.* Beverly Hills, CA: Sage.

Taeuber, Cynthia M.: 1992 Sixty-five Plus in the U.S.A. Unpublished report. Washington, D.C.: U.S. Bureau of the Census.

Taietz, Philip: 1975 Community facilities and social services. Pp. 145–156 in Robert C. Atchley (ed.), *Environments and the Rural Aged.* Washington, D.C.: Gerontological Society.

Talbott, Maria M.: 1990 The negative side of the relationship between older widows and their adult children: The mothers' perspective. *The Gerontologist* 30(5):595–603.

Taylor, Robert Joseph, and Linda M. Chatters: 1986a Church-based informal support among elderly blacks. *The Gerontologist* 26:637–642. 1986b Patterns of informal support to elderly black adults: Family, friends, and church members. *Social Work* 31:432–438.

Thibault, Jane M.: 1995 Congregation as a spiritual care community. Pp. 350–361 in Melvin A. Kimble et al. (eds.), *Aging, Spirituality, and Religion: A Handbook.* Minneapolis: Fortress Press.

Thomae, Hans: 1992 Emotion and personality. Pp. 356–375 in James E. Birren, R. B. Sloane, and G. D. Cohen (eds.), *Handbook of Mental Health and Aging,* 2d ed. New York: Academic Press.

Thomae, Hans (ed): 1975 *Patterns of Aging: Findings from the Bonn Longitudinal Study of Aging.* New York: S. Karger.

Thomas, L. Eugene: 1977 Motivations for mid-life career change. Paper presented at the Annual Meeting of the Gerontological Society, San Francisco.

1994a The way of the religious renouncer. Pp. 51–64 in L. E. Thomas and S. A. Eisenhandler (eds.), *Aging and the Religious Dimension*. Westport, CT.: Auburn House.

1994b Values, psychosocial development, and the religious dimension. Pp. 3–12 in L. E. Thomas and S. A. Eisenhandler (eds.), *Aging and the Religious Dimension*. Westport, CT: Auburn House.

Thomas, L. Eugene, and P. Cooper: 1978 Measurement and incidence of mystical experiences. *Journal for the Scientific Study of Religion* 17(4):433–437.

Thomas, L. Eugene, and Susan A. Eisenhandler (eds.): 1994 *Aging and the Religious Dimension*. Westport, CT.: Auburn House.

Thompson, Larry W., Dolores Gallagher-Thompson, Andrew Futterman, Michael J. Gilewski, and James Peterson: 1991 The effects of late-life spousal bereavement over a 30-month interval. *Psychology and Aging* 6(3): 434–441.

Thursby, Gene R.: 1992 Islamic, Hindu, and Buddhist conceptions of aging. Pp. 175–196 in T. R. Cole et al. (eds.), *Handbook of the Humanities and Aging*. New York: Springer.

Tibbitts, Clark (ed.): 1960 *Handbook of Social Gerontology*. Chicago: University of Chicago Press.

Tice, Raymond R., and Richard B. Setlow: 1985 DNA repair and replication in aging organisms and cells. Pp. 173–224 in C. E. Finch and E. L. Schneider (eds.), *Handbook of the Biology of Aging*. New York: Van Nostrand Reinhold.

Tobin, Sheldon S.: 1991 *Personhood in Old Age: Implications for Practice*. New York: Springer.

Tornstam, Lars: 1994 Gero–transcendence: A theoretical and empirical exploration. Pp. 203–229 in L. E. Thomas and S. A. Eisenhandler (eds.), New York: Auburn House.

Troll, Lillian E.: 1971 The family of later life: A decade review. *Journal of Marriage and the Family* 33:263–290.

1972 The salience of members of three-generation families for one another. Paper presented at the Annual Meeting of the American Psychological Association, Honolulu.

1983 Grandparents: The family watchdogs. Pp. 63–74 in T. H. Brubaker (ed.), *Family Relationships in Later Life*. Beverly Hills, CA: Sage.

1985 The contingencies of grandparenthood. Pp. 135–149 in Vern L. Bengtson and Joan F. Robertson (eds.), *Grandparenthood*. Beverly Hills, CA: Sage.

1986 (ed.) *Family Issues in Current Gerontology*. New York: Springer.

1995 Some psychological implications of an explosion of centenarians. Pp. 71–86 in M. M. Seltzer (ed.), *The Impact of Increased Life Expectancy: Beyond the Gray Horizon*. New York: Springer.

Tuckman, Jacob, and Irving Lorge: 1953 Attitudes toward old people. *Journal of Social Psychology* 37:249–260.

Turner, Barbara F., and Robert L. Kahn: 1974 Age as a political issue. *Journal of Gerontology* 29:572–580.

Uhlenberg, Peter I.: 1969 Study of cohort life cycles: Cohorts of native born Massachusetts women, 1830–1920. *Population Studies* 23: 407–420.

Uhlenberg, Peter I., and Teresa M. Cooney: 1990 Family size and mother-child relations in later life. *The Gerontologist* 30(5):618–625.

Uhlenberg, Peter I., Teresa M. Cooney, and Robert Boyd: 1990 Divorce for women after midlife. *Journal of Gerontology* 45(1):S3–S11.

Uhlenberg, Peter I., and Mary Anne P. Myers: 1981 Divorce and the elderly. *The Gerontologist* 21:276–282.

United Nations: 1990 *Demographic Yearbook: 1988*. New York: United Nations.

U.S. Bureau of the Census: 1954 *Statistical Abstract of the United States*, Washington, D.C.: U.S. Government Printing Office.

1975 *Historical Statistics of the United States: Colonial Times to 1970*. Washington, D.C.: U.S. Government Printing Office.

1977 *Current Population Reports*, Series P-25, No. 704. Washington, D.C.: U.S. Government Printing Office.

1978b *Statistical Abstract of the United States: 1978*. Washington, D.C.: U.S. Government Printing Office.

1981 Characteristics of households and persons receiving noncash benefits: 1979. *Current Population Reports*, Series P-23, Number 110. Washington, D.C.: U.S. Government Printing Office.

1989a *Current Population Reports*, Series P-25, No. 1018. Washington, D.C.: U.S. Government Printing Office.

1989b Voting and registration in the election of November 1988. *Current Population Reports*, Series P-20, No. 440. Washington, D.C.: U.S. Government Printing Office.

1989c *Statistical Abstract of the United States: 1989*. Washington, D.C.: U.S. Government Printing Office.

1990 Residents of farms and rural areas: 1989. *Current Population Reports*, Series P-20, No. 446. Washington, D.C.: U.S. Government Printing Office.

1991a *Statistical Abstract of the United States: 1991*. Washington, D.C.: U.S. Government Printing Office.

1991b Money income of households, families, and persons in the United States: 1990. *Current Population Reports*, Series P-60, No. 174. Washington, D.C.: U.S. Government Printing Office.

1991c Geographic mobility: March 1987 to March 1990. *Current Population Reports*, Series P-20, No. 456. Washington, D.C.: U.S. Government Printing Office.

1992a Age, sex, race and Hispanic origin information from the 1990 Census. Unpublished report 1990 CPH-L-74. Washington, D.C.: U.S. Government Printing Office.

1992b Educational attainment in the United States: March 1991 and 1990. *Current Population Reports*, Series P-20, No. 462. Washington, D.C.: U.S. Government Printing Office.

1992c Household and family characteristics: March 1991. *Current Population Reports*, Series P-20, No. 458. Washington, D.C.: U.S. Government Printing Office.

1992d Marital status and living arrangements: March 1991. *Current Population Reports*, Series P-20, No. 461. Washington, D.C.: U.S. Government Printing Office.

1992e Money income of households, families and persons in the United States: 1991. *Current Population Reports*, Series P-60, No. 180. Washington, D.C.: U.S. Government Printing Office.

1992f Poverty in the United States: 1991. *Current Population Reports*, Series P-60, No. 181. Washington, D.C.: U.S. Government Printing Office.

1992g *Current Population Reports*, Series P-25, No. 1092. Washington, D.C.: U.S. Government Printing Office.

1992h *Statistical Abstract of the United States: 1992*. Washington, D.C.: U.S. Government Printing Office.

1993 Money income of households, families, and persons in the United States: 1992. *Current Population Reports*, Series P60–184. Washington, D.C.: U.S. Government Printing Office.

1994 *Statistical Abstract of the United States, 1994*. Washington, D.C.: U.S. Government Printing Office.

U.S. Commission on Civil Rights: 1977 *The Age Discrimination Study*. Washington, D.C.: The Commission.

U.S. Department of Labor: 1982 Retired couples' budgets, final report, Autumn 1981. *Monthly Labor Review* 105(11):37–38.

U.S. General Accounting Office: 1982 *The Elderly Should Benefit from Expanded Home Health Care but Increasing These Services Will Not Insure Cost Reductions*. Washington, D.C.: U.S. General Accounting Office.

1986 Retirement *Before Age 65: Trends, Costs, and National Issues*. Washington, D.C.: U.S. General Accounting Office.

1987 *Stronger Enforcement of Nursing Home Requirements Needed*. Washington, D.C.: U.S. General Accounting Office.

U.S. House Committee on Banking, Finance, and Urban Affairs: 1989 *Investigation of Lincoln Savings and Loan Association*. Serial No. 101-59. Washington, D.C.: U.S. Government Printing Office.

U.S. Senate Special Committee on Aging: 1965 *Frauds and Deceptions Affecting the Elderly*. Washington, D.C.: U.S. Government Printing Office.

1986 *Developments in Aging: 1985*, Volume 3. Washington, D.C.: U.S. Government Printing Office.

1988 *Developments in Aging: 1987*, Volume 3. The Long-term Care Challenge. Washington, D.C.: U.S. Government Printing Office.

1989a *Developments in Aging: 1988*. Washington, D.C.: U.S. Government Printing Office.

1989b *Developments in Aging: 1988*. Volume 1. Washington, D.C.: U.S. Government Printing Office.

1990 *Aging America: Trends and Projections*. Washington, D.C.: U.S. Government Printing Office.

1991 *Developments in Aging: 1990.* Washington, D.C.: U.S. Government Printing Office.

1992 *Developments in Aging: 1991,* Volume I. Report No. 102–261. Washington, D.C.: U.S. Government Printing Office.

U.S. Social Security Administration: 1988 *Annual Statistical Bulletin.* Washington, D.C.: U.S. Government Printing Office.

1991a *Annual Statistical Supplement, 1991.* Washington, D.C.: U.S. Government Printing Office.

1991b *Social Security Bulletin. Annual Statistical Supplement, 1991.* Washington, D.C.: U.S. Government Printing Office.

1993 *Annual Statistical Supplement, 1993. Social Security Bulletin.* Washington, D.C.: U.S. Government Printing Office.

Van Den, Hoonaard, and Deborah Kestin: 1994 Paradise lost: Widowhood in a Florida retirement community. *Journal of Aging Studies* 8(2): 121–132.

Ventura-Merkel, Catherine: 1991 Community colleges in an aging society. Pp. 49–56 in Robert Harootyan et al. (eds.), *Resourceful Aging: Today and Tomorrow. Volume V: Lifelong Education.* Washington, D.C.: American Association of Retired Persons.

Vera, Hernan, Felix M. Berardo, and Joseph S. Vandiver: 1990 Age irrelevancy in society: The test of mate selection. *Journal of Aging Studies* 4(1):81–95.

Verbrugge, Lois M.: 1984 Longer life but worsening health? Trends in health and mortality of middle-aged and older persons. *Milbank Memorial Fund Quarterly* 62:475–519.

Veroff, Joseph, Daniel Reuman, and Sheila Feld: 1984 Motives of American men and women across the adult life span. *Developmental Psychology* 20:1142–1148.

Vinick, Barbara H.: 1979 Remarriage. Pp. 141–243 in R. H. Jacobs and Barbara H. Vinick, *Re-Engagement in Later Life.* Stamford, CT: Greylock.

Vinick, Barbara H., and David J. Ekerdt: 1989 Retirement and the family. *Generations* 13(2):53–56.

1991 Retirement: What happens to husband-wife relationships? *Journal of Geriatric Psychiatry* 24:16–23.

Vinyard, Dale: 1972 The Senate Special Committee on Aging. *The Gerontologist* 12:298–303.

Waldman, D. A., and B. J. Avolio: 1986 A meta-analysis of age differences in job performance. *Journal of Applied Psychology* 71:33–38.

Waldrop, Judith: 1992 Old money. *American Demographics* 14(4):24–32.

Wallace, Steven P.: 1990 The no-care zone: Availability, accessibility, and acceptability in community-based long-term care. *The Gerontologist* 30(2):254–261.

Wallace, Steven P., John B. Williamson, Rita G. Lung, and Lawrence A. Powell: 1991 A lamb in wolf's clothing? The reality of senior power and social policy. Pp. 95–114 in Meredith Minkler and C. L. Estes (eds.), *Critical Perspectives on Aging.* Amityville, NY: Baywood.

Wantz, Molly S., and John E. Gay: 1981 *The Aging Process: A Health Perspective.* Cambridge, MA: Winthrop.

Ward, Russell A.: 1984 The marginality and salience of being old: When is age relevant? *The Gerontologist* 24:227–232.

Weber, Nancy (ed.): 1991 *Vision and Aging: Issues in Social Work Practice.* The Hawthorne Press, Inc.

Weiffenbach, James M., Bruce J. Baum, and Rosemary Burghauser: 1982 Taste thresholds: Quality specific variation with human aging. *Journal of Gerontology* 37:372–377.

Weiland, Steven: 1995 Interpretive social science and spirituality. Pp. 589–611 in Melvin A. Kimble et al. (eds.), *Aging, Spirituality, and Religion: A Handbook.* Minneapolis: Fortress Press.

Weishaus, S., and D. Field: 1988 Half a century of marriage: Continuity or change? *Journal of Marriage and the Family* 50:763–774.

Weisman, Avery D.: 1972 *On Dying and Denying.* New York: Behavioral Publications.

Weiss, Lawrence J., and Robert A. Applebaum: 1994 *Assisted Living in Ohio: Policy Options and Program Recommendations.* Oxford, OH: Scripps Gerontology Center.

Wershow, Harold J.: 1977 Reality orientation for gerontologists: Some thoughts about senility. *The Gerontologist* 17:297–302.

Whitbourne, Susan K.: 1985 *The Aging Body: Physiological Changes and Psychological Consequences.* New York: Springer-Verlag.

1986 *The Me I Know: A Study of Adult Identity.* New York: Springer-Verlag.

Whitbourne, Susan K., and Comilda S. Weinstock: 1979 *Adult Development: The*

Differentiation of Experience. New York: Holt, Rinehart and Winston.

White, Lynn K., Alan Booth, and John N. Edwards: 1986 Children and marital happiness: Why the negative correlation? *Journal of Family Issues* 7:131–149.

Whittington, Frank J.: 1987 Drug abuse. Pp. 190–191 in George L. Maddox (ed.), *Encyclopedia of Aging*. New York: Springer.
1988 Making it better: Drinking and drugging in old age. *Generations* 12(4):5–8.

Whyte, William H., Jr.: 1956 *The Organization Man*. New York: Doubleday.

Wigdor, Blossom T.: 1980 Drives and motivations with aging. Pp. 245–261 in J. E. Birren and R. B. Sloane (eds.), *Handbook of Mental Health and Aging*. Englewood Cliffs, NJ: Prentice-Hall.

Williams, David R.: 1994 The measurement of religion in epidemiologic studies. Problems and prospects. Pp. 125–148 in J. S. Levin (ed.), *Religion in Aging and Health*. Thousand Oaks, CA: Sage.

Williams, Richard H., and Claudine Wirths: 1965 *Lives Through the Years*. New York: Atherton Press.

Windus, Victoria: 1987 Factors Influencing Administrators' Decisions about Early Retirement Incentive Plans in Ohio Public Universities and Community Colleges. Unpublished MGS thesis. Oxford, OH: Miami University.

Winkler, Mary G.: 1992 Walking to the stars. Pp. 258–284 in T. R. Cole, D. D. Van Tassel, and R. Kastenbaum (eds.), *Handbook of the Humanities and Aging*. New York: Springer.

Wolf, Douglas A.: 1994 The elderly and their kin: Patterns of availability and access. Pp. 146–194 in L. G. Martin and S. H. Preston (eds.), *Demography of Aging*. Washington, D.C.: National Academy Press.

Wolf, Rosalie S.: 1986 Major findings from three model projects on elder abuse. Pp. 218–238 in Karl A. Pillemer and Rosalie S. Wolf (eds.), *Elder Abuse: Conflict in the Family*. Dover, MA: Auburn House.

Wolf, Rosalie S., and Karl A. Pillemer: 1989 *Helping Elderly Victims: The Reality of Elder Abuse*. New York: Columbia University Press.
1994 What's new in elder abuse programming? Four bright ideas. *The Gerontologist* 34(1):126–129.

Wolfe, David B.: 1990 *Serving the Ageless Market: Strategies for Selling to the Fifty-Plus Market*. New York: McGraw-Hill.

Wolfsen, Connie R., Judith C. Barker, and Linda S. Mitteness: 1990 Personalization of formal social relationships by the elderly. *Research on Aging* 12(1):94–112.

Woodruff, Diana S.: 1985 Arousal, sleep and aging. Pp. 261–295 in James E. Birren and K. Warner Schaie (eds.), *Handbook of the Psychology of Aging*, 2d ed. New York: Van Nostrand Reinhold.

Woodruff-Pak, Diana S.: 1989 Aging and intelligence: Changing perspectives in the twentieth century. *Journal of Aging Studies* 3(2):91–118.

Woods, John R.: 1989 Pension coverage among private wage and salary workers. *Social Security Bulletin* 52(10):2–19.

Wright, Paul: 1989 Gender differences in adults' same- and cross-gender friendships. Pp. 197–221 in Rebecca S. Adams and Rosemary Blieszner (eds.), *Older Adult Friendship*. Newbury Park, CA: Sage.

Yankelovich, Skelly, and White, Inc.: 1985 *A Fifty-Year Report Card on the Social Security System: The Attitudes of the American Public*. Washington, D.C.: American Association of Retired Persons.

Ycas, Martynas A.: 1985 *Trends in Health and Retirement: New Findings from the Health Interview Survey*. Washington, D.C.: Social Security Administration.

Zuckerman, Harriet, and Robert K. Merton: 1972 Age, aging, and age structure in science. Pp. 292–456 in Matilda W. Riley, Marilyn Johnson, and Anne Foner (eds.), *Aging and Society, Volume 3: A Sociology of Age Stratification*. New York: Russell Sage Foundation.

Index